DEVELOPMENT OF THE NERVOUS SYSTEM

THIRD EDITION

DEVELOPMENT OF THE NERVOUS SYSTEM

THIRD EDITION

Dan H. Sanes

Thomas A. Reh

William A. Harris

AMSTERDAM • BOSTON • HEIDELBERG • LONDON
NEW YORK • OXFORD • PARIS • SAN DIEGO
SAN FRANCISCO • SINGAPORE • SYDNEY • TOKYO
Academic Press is an imprint of Elsevier

ELSEVIER

Academic Press is an imprint of Elsevier
30 Corporate Drive, Suite 400, Burlington, MA 01803, USA
The Boulevard, Langford Lane, Kidlington, Oxford, OX5 1GB, UK

Notices
Knowledge and best practice in this field are constantly changing. As new research and experience broaden our understanding, changes in research methods, professional practices, or medical treatment may become necessary.

Practitioners and researchers must always rely on their own experience and knowledge in evaluating and using any information, methods, compounds, or experiments described herein. In using such information or methods they should be mindful of their own safety and the safety of others, including parties for whom they have a professional responsibility.

To the fullest extent of the law, neither the Publisher nor the authors, contributors, or editors, assume any liability for any injury and/or damage to persons or property as a matter of products liability, negligence or otherwise, or from any use or operation of any methods, products, instructions, or ideas contained in the material herein.

Library of Congress Cataloging-in-Publication Data
Sanes, Dan Harvey.
    Development of the nervous system / by Dan Sanes, Thomas Reh, William Harris. -- 3rd ed.
        p. cm.
    Includes bibliographical references and index.
    ISBN 978-0-12-374539-2 (alk. paper)
    1. Developmental neurophysiology.   I. Reh, Thomas A.   II. Harris, William A.   III. Title.
QP356.25.S365 2012
612.8--dc22
2010041189

British Library Cataloguing-in-Publication Data
A catalogue record for this book is available from the British Library.

ISBN: 978-0-12-374539-2

For information on all Academic Press publications
visit our Web site at www.elsevierdirect.com

Printed in China

11 12 13 9 8 7 6 5 4 3 2

To our families

Contents

CONTENTS

Preface to the Third Edition

In a field of science where the tools of investigation continue to improve dramatically and the challenge is to understand the construction of what is, arguably, the most complex object in our known universe, it is not unexpected that this third edition of *Development of the Nervous System* required extensive revision. Moreover, it has become increasingly clear that, in many respects, the processes of neural development continue in the "mature" adult brain. Discoveries in adult neurogenesis and plasticity have profound implications for brain function throughout life. Moreover, abnormalities in developmental mechanisms lead to brain disorders that only become manifest in adulthood. Our understanding of these developmental processes holds the promise for emerging therapies, such as deriving neurons and glia from embryonic stem cells. In this way, the study of neural development has never been more relevant.

Experts in various subfields of neural development helped us by reviewing each chapter, telling us what they thought was missing, wrong, needed updating, or should be removed from the text. They also suggested where entire sections of the book should be approached afresh, emphasizing new conceptual angles. We took most of their excellent advice. However, we were mindful that many of the older studies in our field have stood the test of time, and continue to serve as the core knowledge of neural development. This core still forms the storyline of the textbook. We hope that those of you who were content with our second edition, particularly for teaching purposes, will be comfortable with the third edition. The book is built on the same foundation, yet we have embraced ideas that have gained in acceptance and included several new studies to convey the excitement that is part of a field where very recent discoveries continue to have enormous impact. We were cautious, however, about including too much of this new material for two reasons. First, we wanted to keep the size of the book the same. Second, experience has taught us that what is new and exciting will not always turn out to be as pivotal for the field as it now appears. The future will be the best judge of which studies become classics and which studies will form the core of future textbooks.

Therefore, though we are enormously grateful to many colleagues (listed below) who have contributed advice and material to this third edition, the choice of what to include and what not to include was ours alone. We accept responsibility for any deficits in concept or coverage. Our thanks go explicitly to the following people who helped us with the third edition: Michael Bate, John Bixby, Steve Burden, Martha Constantine-Paton, Ford Ebner, Gord Fischell, John Flanagan, Francois Guillemot, Christopher Henderson, Christine Holt, Chris Kintner, Lynne Kiorpes, Alex Kolodkin, Vibhakar Kotak, Matthias Landgraf, Jeff Lichtman, Tony Movshon, Alan Roberts, John Rubenstein, Peter Scheiffele, Josh Weiner, Ed Ziff, Lance Zirpel. And a special note of thanks to the editorial staff at Elsevier: Clare Caruana, Mica Haley, Johannes Menzel, and Melissa Turner.

Dan H. Sanes
Thomas A. Reh
William A. Harris
January 2011

Preface to the Second Edition

The human brain—perhaps the most complex object in our universe—is composed of billions of cells and trillions of connections. It is truly a wonder of enormous proportions. Although we are far from a thorough understanding of our brains, study of the way that the cellular constituents of the nervous system, the neurons and glia, work to produce sensations, behaviors, and higher order mental processes has been a most productive area of science. However, more and more, neuroscientists are realizing that we are studying a moving target-growth and that changes are integral to brain function, forming the very basis for learning, perception, and performance. To comprehend brain function, then, we must understand how the circuits arise and the ways in which they are modified during maturation. Santiago Ramón y Cajal, one of the founders of modern neuroscience, was able to make his remarkable progress in studies of the cellular makeup of the nervous system in large part because of his work with the embryonic brain, choosing to study "the young wood, in the nursery stage…rather than the…impenetrable…full grown forest."

The construction of the brain is an integrated series of developmental steps, starting with the decision of a few early embryonic cells to become neural progenitors and nearing completion with the emergence of behavior, which is the scope of this book. Interactions with the world continuously update and adapt synaptic connections within the brain, and the mechanisms by which these changes occur are fundamentally a continuation of the same processes that sculpted the emerging brain during embryogenesis.

Studies of development have also led to insights about the evolutionary relationships among organisms. The dogma of phylogeny and ontogeny of the last century has been superseded by our current deeper understanding of the ways in which evolutionary change can be effected through changes in development. The brain is no exception to these rules, and we can expect that much insight into the evolution of that which makes us most human will be gained from an appreciation of how developmental processes are modified over time.

The goal of this text is to provide a contemporary overview of neural development both for undergraduate students and for those who have some background in the field of biology. This intent is not compatible with a comprehensive review of the literature. In the first edition, we noted that there were about 54,000 papers published in this field between 1966 and 1999. Another 25,000 have appeared during the past 4 years (using the search string "neural or neuron or nervous" and "development or embryology or maturation" and 2000:2004). We charted a compromise between the need to update students and our strong inclination to hold their attention. The book does not contain exhaustive lists of molecular families, and the most current review articles must serve as an appendix to our text. Since the text does not encompass many exciting areas of research, students will find themselves quickly turning to specialized texts and reviews.

Among those who helped us through discussion and editorial comment are: Chiye Aoki, Michael Bate, Carla Shatz, Ford Ebner, Edward Gruberg, Christine Holt, Lynne Kiorpes, Vibhakar Kotak, Tony Movshon, Ron Oppenheim, Sarah Pallas, Sheryl Scott, Tim Tully, and Lance Zirpel.

Finally, we acknowledge our editor, Johannes Menzel, with particular gratitude, for his help, advice, and perseverance.

Dan H. Sanes
Thomas A. Reh
William A. Harris
July 2005

Preface to the First Edition

The human brain is said to be the most complex object in our known universe, and the billions of cells and trillions of connections are truly wonders of enormous proportions. The study of the way that the cellular elements of the nervous system work to produce sensations, behaviours, and higher order mental processes has become a most productive area of science. However, neuroscientists have come to realize that they are studying a moving target: growth and change are integral to brain function and form the very basis by which we can learn anything about it. As the behavioral embryologist George Coghill pointed out, "Man is, indeed, a mechanism, but he is a mechanism which, within his limitations of life, sensitivity and growth, is creating and operating himself." To understand the brain, then, we need to understand how this mechanism arises and the ways in which it can change throughout a lifetime.

The construction of the brain is an integrated series of developmental steps, beginning with the decision of a few early embryonic cells to become neural progenitors. As connections form between nerve cells and their electrical properties emerge, the brain begins to process information and mediate behaviors. Some of the underlying circuitry is built into the nervous system during embryogenesis. However, interactions with the world continuously update and adapt the brain's functional architecture. The mechanisms by which these changes occur appear to be a continua-tion of the processes that sculpt the brain during development. Since the text covers each of these developmental steps, it is relatively broad in scope.

An understanding of the development of the nervous system has importance for biologists in a larger context. Studies of development have led to insights into the evolutionary relationships among organisms. The dogma of phylogeny and ontogeny of the last century has been superseded by a deeper understanding of the ways in which evolutionary change can be effected through changes in development. The brain is no exception to these rules. We should expect that insight into the evolution of that which makes us most human will be gained from an appreciation of how developmental processes are modified over time.

The goal of this text is to provide a contemporary overview of neural development for undergraduate students or those who have some background in the filed of biology. This intent is not compatible with a comprehensive review of the literature. A recent MEDLINE search of publications in the field of neural development [(neural or neuron or nervous) and (development or embryology or maturation)] yielded 56,840 papers published between 1966 and 1999. We admit, up front, to having read only a fraction of these papers or of the thousands that were published before 1966. As a practical matter, we made use of authoritative books, contemporary review articles, hallway conversations, and e-mail consultations to select the experiments that are covered in our text. Even so, we expect that important contributions have been missed inadvertently. Therefore, advanced students will find themselves quickly turning to specialized texts and reviews. Another compromise that comes from writing an undergraduate biology book well after the onset of the revolution in molecular biology is that all subjects now have a rather broad cast of molecular characters. In addition, the most instructive experiments on a particular class of molecules have often been performed on non-neural tissue. Even if we chose to cover only the genes and proteins whose roles have been best characterized in the nervous system, most chapters would run the risk of sounding like a (long) list of acronyms. Therefore, we charted a compromise between the need to update students and our strong inclination to hold their attention. The book does not contain exhaustive lists of molecular families, and the most current articles must serve as an appendix to our text.

Among the many scientists who helped us through discussions, unpublished findings, or editorial comment are (in alphabetical order) Chiye Aoki, Michael Bate, Olivia Bermingham-McDonogh, John Bixby, Sarah Bottjer, Martin Chalfie, Hollis Cline, Martha Constantine-Paton, Ralph Greenspan, Voker Hartenstein, Mary Beth Hatten, Christine Holt, Darcy Kelley, Chris Kintner, Sue McConnell, Ilona Miko, Ronald Oppenheim, Thomas Parks, David Raible, Henk Roelink, Edwin Rubel, John Rubenstein, David Ryugo, Nancy Sculerati, Carla Shatz, and Tim Tully.

Neural induction

1

DEVELOPMENT AND EVOLUTION OF NEURONS

Almost as early as multicellular animals evolved, neurons have been part of their tissues. Metazoan nervous systems range in complexity from the simple nerve net of the jellyfish to the billions of specifically interconnected neuron assemblies of the human brain. Nevertheless, the neurons and nervous systems of all multicellular animals share many common features. Voltage-gated ion channels are responsible for action potentials in the neurons of Jellyfish as they are in people. Synaptic transmission between neurons in nerve nets is basically the same as that in the cerebral cortex in humans (**Figure 1.1**). This book describes the mechanisms responsible for the generation of these nervous systems, highlighting examples from a variety of organisms. Despite the great diversity in the nervous systems of various organisms, underlying principles of neural development have been maintained throughout evolution.

It is appropriate to begin a book concerned with the development of the nervous system with an evolutionary perspective. The subjects of embryology and evolution have long shared an interrelated intellectual history. One of the major currents of late-nineteenth-century biology was that a description of the stages of development would provide the key to the path of evolution of life. The phrase "ontogeny recapitulates phylogeny" was important at the start of experimental embryology (Gould, 1970). Although the careful study of embryos showed that they did not resemble the adult forms of their ancestors, it is clear that new forms are built upon the structures of biological predecessors. One aim of this book is to show how an understanding of the development of the nervous system will give us insight into its evolution. It is also wise to remember, as Dobzhansky pointed out, "nothing in biology makes sense except in the light of evolution."

EARLY EMBRYOLOGY OF METAZOANS

The development of multicellular organisms varies substantially across phyla; nevertheless, there are some common features. The cells of all metazoans are organized as layers. These layers give rise to the various organs and tissues, including the nervous system. These layers are generated from the egg cell through a series of cell divisions and their subsequent rearrangements. The eggs of animals are typically polarized, with differences in their cytoplasm from one "pole" to the other. Amphibians, for example, have an "animal pole" and a "vegetal pole" that is visible in the egg, since the vegetal pole contains the yolk, the stored nutrient material necessary for sustaining the embryo as it develops. Once fertilized by the sperm, the egg cell undergoes a series of rapid cell divisions, known as cleavages. There are many variations of cleavage patterns in embryos, but the end result is that a large collection of cells, the blastula, is generated over a relatively short period of time. In many organisms the cells of the blastula are arranged as a hollow ball, with an inner cavity known as a blastocoel. The rearrangement of this collection of cells into the primary (or germ) layers is called *gastrulation*. Gastrulation can occur via a variety of mechanisms, but all result in an inner, or endodermal, layer of cells, an outer layer of cells, the ectoderm, and a layer of cells between the two other layers, known as the mesoderm (Gilbert and Raunio, 1997). The middle layer can be derived from either the ectoderm (ectomesoderm) or the inner layer (endomesoderm). During the process of gastrulation, the cells of the mesoderm and endoderm move into the inside of the embryo, often at a single region, known as the blastopore. Once the endoderm and mesoderm are inside the ball, they usually obliterate the blastocoel and form a new cavity, the archenteron, or primitive gut. Animals can be divided in two on the basis of whether the mouth forms near the point of this blastopore (in protostomes) or at a distant site (in deuterostomes). Once these three primary germ layers are established, the development of the nervous system begins. A more detailed description of the development of the other organ systems is beyond the scope of this text. Nevertheless, one should keep in mind that the development of the nervous system does

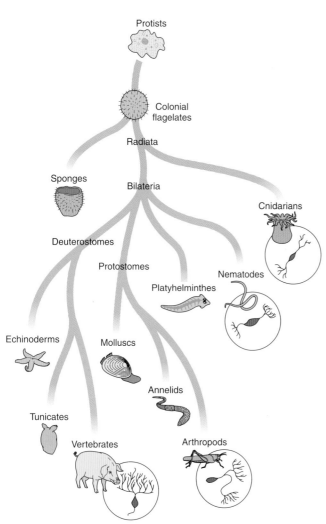

Fig. 1.1 Neurons throughout the evolution of multicellular organisms have had many features in common. All animals other than colonial flagellates and sponges have recognizable neurons that are electrically excitable and have long processes. The Cnidarians have nerve networks with electrical synapses, but synaptic transmission between neurons is also very ancient.

not take place in a vacuum, but is an integral and highly integrated part of the development of the animal as a whole.

The next three sections will deal with the embryology of several examples of metazoan development: nematode worms (*Caenorhabditis elegans*); insects (*Drosophila melanogaster*); and several vertebrates (frogs, fish, birds, and mammals). The development of these animals is described because they have been particularly well studied for historical and practical reasons. However, one should take these examples as representative, not as definitive. The necessity of studying many diverse species has become critical to the understanding of the development of any one species.

DERIVATION OF NEURAL TISSUE

The development of the nervous system begins with the segregation of neural and glial cells from other types of tissues. The many differences in gene expression between neurons and muscle tissue, for example, arise through the progressive

narrowing of the potential fates available to a blast cell during development. The divergence of neural and glial cells from other tissues can occur in many different ways and at many different points in the development of an organism. However, the cellular and molecular mechanisms that are responsible for the divergence of the neural and glial lineages from other tissues are remarkably conserved.

C. ELEGANS

The development of *C. elegans*, a nematode worm, highlights the shared lineage of the epidermal and neural cell fates. These animals have been studied primarily because of their simple structure (containing only about a thousand cells), their rapid generation time (allowing for rapid screening of new genetic mutants), and their transparency (enabling lineage relationships of the cells to be established). These nematodes have a rigid cuticle that is made of collagenous proteins secreted by the underlying cells of the hypodermis. The hypodermis is analogous to the epidermis of other animals, except that it is composed of a syncytium of nuclei rather than of individual cells. They have a simple nervous system, composed of only 302 neurons and 56 glial cells. These neurons are organized into nerve cords. The nerve cords are primarily in the dorsal and ventral sides of the animals, but there are some neurons that run along the lateral sides of the animal as well. The nematodes move by a series of longitudinal muscles, and they have a simple digestive system.

Nematodes have long been a subject for developmental biologists' attention. Theodore Boveri studied nematode embryology and first described the highly reproducible pattern of cell divisions in these animals in the late 1800s. Boveri's most famous student, Hans Spemann, whose work on amphibian neural induction will be described below, worked on nematodes for his Ph.D. research. The modern interest in nematodes, however, was motivated by Sydney Brenner, a molecular biologist who was searching for an animal that would allow the techniques of molecular genetics to be applied to the development of metazoans (Brenner, 1974).

Because of the stereotypy in the pattern of cell divisions, the lineage relationships of all the cells of *C. elegans* have been determined (Sulston et al., 1983). The first cleavage produces a large somatic cell, the AB blastomere, which gives rise to most of the hypodermis and the nervous system and the smaller germline P cell, which in addition to the gonads will also generate the gut and most of the muscles of the animal (**Figure 1.2**). Subsequent cleavages produce the germ cell precursor, P4, and the precursor's cells for the rest of the animal: the MS, E, C, and D blastomeres (Figure 1.2), and these cells all migrate into the interior of the embryo, while the AB-derived cells spread out over the outside of the embryo completing gastrulation (**Figure 1.3**). The next phase of development is characterized by many cell divisions and is known as the *proliferation phase*. Then an indentation forms at the ventral side of the animal marking the beginning of the morphogenesis stage, and as this indentation progresses, the worm begins to take shape (Figure 1.3). At this point, the worm has only 556 cells and will add the remaining cells (to the total of 959) over the four larval molts. The entire development of the animal takes about two days.

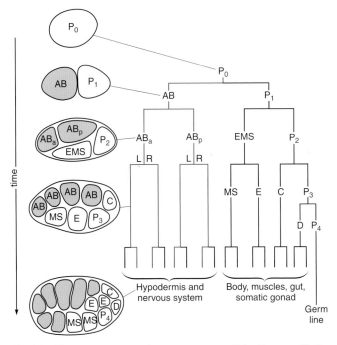

Fig. 1.2 The nervous system shares a common cellular lineage with the ectoderm. The cell divisions that generate the *C. elegans* nematode worm are highly reproducible from animal to animal. The first division produces the AB blastomere and the P1 blastomere. The germ line is segregated into the P4 blastomere within a few divisions after fertilization. The subsequent divisions of the AB blastomere go on to give rise to most of the neurons of the animal, as well as to the cells that produce the hypodermis—the epidermis of the animal.

The neurons of *C. elegans* arise primarily from the AB blastomere, in lineages shared with the ectodermally derived hypodermis. An example of one of these lineages is shown in Figure 1.3. The Abarpa blastomere can be readily identified in the 100-min embryo through its position and lineal history. This cell then goes on to give rise to 20 additional cells, including 9 neurons of the ring ganglion. The progeny of the Abarpa blastomere, like most of the progeny of the AB lineage, lie primarily on the surface of the embryo prior to 200 min of development. At this time, the cells on the ventral and lateral sides of the embryo move inside and become the nervous system, whereas the AB progeny that remain on the surface spread out to form the hypodermis, a syncytial covering of the animal. Most of the neurons arise in this way; of the 222 neurons in the newly hatched *C. elegans*, 214 arise from the AB lineage, whereas 6 are derived from the MS blastomere and 2 from the C blastomere.

DROSOPHILA

The development of the fruit fly, *Drosophila*, is characteristic of many arthropods. Unlike the embryos of the nematode, where cleavage of the cells occurs at the same time as nuclear divisions, the initial rounds of nuclear division in the *Drosophila* embryo are not accompanied by corresponding cell divisions. Instead, the nuclei remain in a syncytium up until just prior to gastrulation, three hours after fertilization. Prior to this time, the dividing nuclei lie in the interior of the egg, but they then move out toward the surface and a process

known as cellularization occurs, and the nuclei are surrounded by plasma membranes. At this point the embryo is known as a cellular blastoderm.

The major part of the nervous system of *Drosophila* arises from cells in the ventrolateral part of the cellular blastoderm (**Figure 1.4**, top). Soon after cellularization, the ventral furrow, which marks the beginning of gastrulation, begins to form (Figure 1.4, middle). At the ventral furrow, cells of the future mesoderm fold into the interior of the embryo. The process of invagination occurs over several hours, and the invaginating cells will continue to divide and eventually will give rise to the mesodermal tissues of the animal. As the mesodermal cells invaginate into the embryo, the neurogenic region moves from the ventrolateral position to the most ventral region of the animal (Figure 1.4). The closing of the ventral furrow creates the ventral midline, a future site of neurogenesis. On either side of the ventral midline is the neurogenic ectoderm, tissue that will give rise to the ventral nerve cord, otherwise known as the central nervous system (**Figure 1.5**). A continuation of the neurogenic region into the anterior of the embryo, is called the procephalic neurogenic region and gives rise to the cerebral ganglia or brain.

Drosophila neurogenesis then begins in the neurogenic region; some cells enlarge and begin to move from the outside layer into the inside of the embryo (Figure 1.5). At the beginning of neurogenesis, the neurogenic region is a single cell layer; the first morphological sign of neurogenesis is that a number of cells within the epithelium begin to increase in size. These larger cells then undergo a shape change and squeeze out of the epithelium. This process is called *delamination* and is shown in more detail in Figure 1.5. The cells that delaminate are called *neuroblasts* and are the progenitors that will generate the nervous system. In the next phase of neurogenesis, each neuroblast divides to generate many progeny, known as *ganglion mother cells* (GMCs). Each GMC then generates a pair of neurons or glia. In this way, the entire central nervous system of the larval *Drosophila* is generated. However, the *Drosophila* nervous system is not finished in the larva, but rather additional neurogenesis occurs during metamorphosis. Sensory organs, like the eyes, are generated from small collections of cells in the larva (called imaginal discs) that undergo a tremendous amount of proliferation during metamorphosis to generate most of what we recognize as an adult fly.

VERTEBRATES

All vertebrate embryos develop in a fundamentally similar way, though at first appearance they seem to be quite different. In this section we will review the development of several different vertebrates: amphibians, fish, birds, and mammals. In all of these animals, multiple cleavage divisions generate a large number of cells from the fertilized oocyte. However, while gastrulation in all of these animals is basically conserved, the details of the cellular movements during this phase can look quite different.

Amphibian eggs are like those of many animals in that the egg has a distinct polarity with a nutrient-rich yolk concentrated at the "vegetal" hemisphere and a relatively yolk-free "animal" hemisphere. After fertilization, a series of rapid cell divisions, known as cleavages, divides the fertilized egg into blastomeres. The cleavage divisions proceed

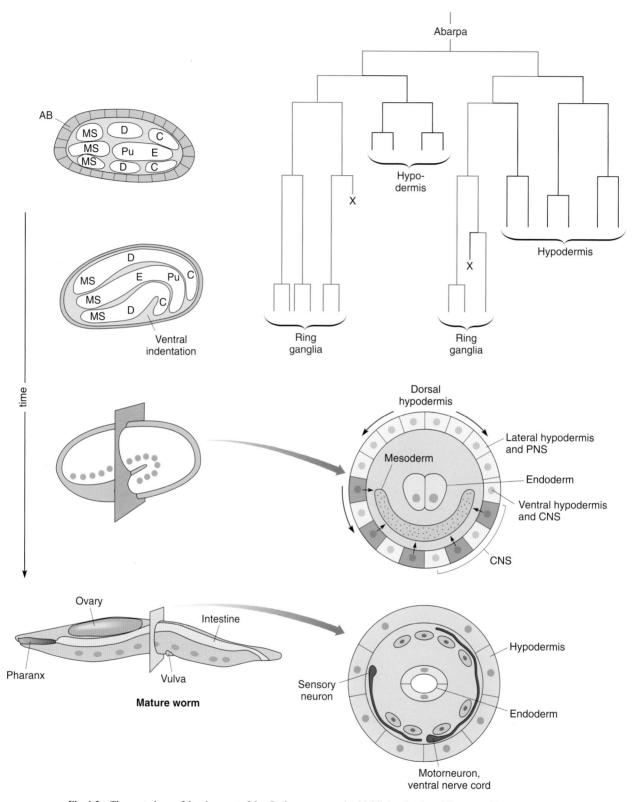

Fig. 1.3 The next phase of development of the *C. elegans* worm also highlights the shared lineages of hypodermis and neurons. During gastrulation, the MS, E, C, and D blastomeres all migrate into the interior of the embryo, while the progeny of the AB blastomeres spread out over the external surface. Once the embryo starts to take form, sections through the embryo show the relationships of the cells. The neurons are primarily derived from the ventrolateral surface, through the divisions of the AB progeny cells. As these cells are generated, they migrate into the interior and form the nerve rings. A typical lineage is also shown. The Abarpa blastomeres undergo five rounds of division, to generate 9 neurons and 10 hypodermal cells. Neural lineages are shown in red.

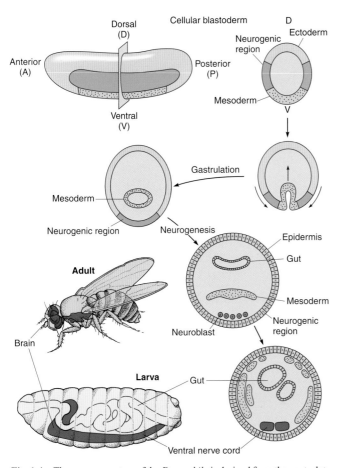

Fig. 1.4 The nervous system of the *Drosophila* is derived from the ventrolateral region of the ectoderm. The embryo is first (top) shown at the blastoderm stage, just prior to gastrulation. The region fated to give rise to the nervous system lies on the ventral-lateral surface of the embryo (red). The involution of the mesoderm at the ventral surface brings the neurogenic region closer to the midline. Scattered cells within this region of the ectoderm then enlarge, migrate into the interior of the embryo, and divide several more times to make neurons and glia. These neurons and glia then condense into the ganglia of the mature ventral nerve cord (or CNS) in the larva and the adult.

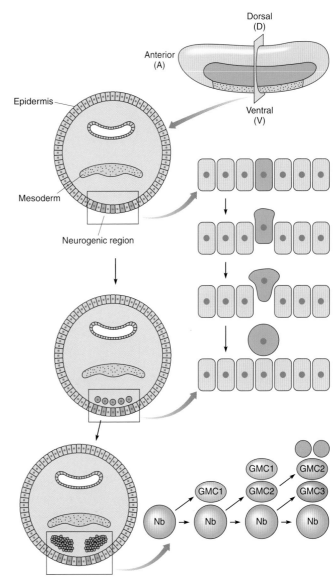

Fig. 1.5 The neuroblasts of the *Drosophila* separate from the ectoderm by a process known as delamination. The neuroblasts enlarge relative to the surrounding cells and squeeze out of the epithelium. The process occurs in several waves; after the first set of neuroblasts has delaminated from the ectoderm, a second set of cells in the ectoderm begins to enlarge and also delaminates. The delaminating neuroblasts then go on to generate several neurons through a stereotypic pattern of asymmetric cell divisions. The first cell division of the neuroblast produces a daughter cell known as the ganglion mother cell, or GMC. The first ganglion mother cell divides to form neurons, while the neuroblast is dividing again to make another GMC. In the figure, the same neuroblast is labeled through its successive stages as Nb, while the GMCs are numbered successively as they arise.

less rapidly through the vegetal hemisphere, and by the time the embryo reaches 128 cells, the cells in the animal half are much smaller than those of the vegetal half (**Figure 1.6**). The embryo is called a *blastula* at this stage. The process by which the relatively simple blastula is transformed into the more complex, three-layered organization shared by most animals is called *gastrulation.* During this phase of development, cells on the surface of the embryo move actively into the center of the blastula. The point of initiation of gastrulation is identified on the embryo as a small invagination of the otherwise smooth surface of the blastula, and this is called the *blastopore* (Figure 1.6). In amphibians the first cells to invaginate occur at the dorsal side of the blastopore (Figure 1.6), opposite to the point of sperm entry. The dorsal side of the blastopore has a special significance for the process of neural induction and much more will be said about this in this chapter.

The involuting cells lead a large number of cells that were originally on the surface of the embryo to the interior (Figure 1.6). The part of the blastula that will ultimately reside in the interior of the embryo is called the *involuting*

marginal zone (IMZ). Most of these cells will give rise to mesodermally derived tissues, like muscle and bone. The first cells to involute crawl the farthest and produce the mesoderm of the anterior part of the animal (i.e., the head). The later involuting IMZ cells produce the mesoderm of more posterior regions, including the tail of the tadpole. At this point in development, the neural plate of the vertebrate embryo still largely resembles the rest of the surface ectoderm. However, shortly after its formation, the neural plate begins to fold onto itself to form a tubelike structure, the

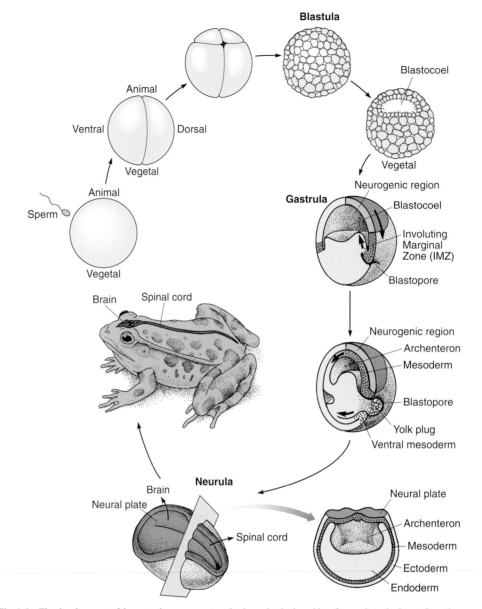

Fig. 1.6 The development of the central nervous system, brain, and spinal cord in a frog embryo is shown from the egg cell to the adult. After a series of cleavage divisions produce a blastula, a group of cells known as the involuting marginal zone, or IMZ, grow into the interior of the embryo at a point known as the blastopore. This process of gastrulation is shown in two cross sections. The involuting cells go on to form mesodermal tissues (blue) and induce the cells of the overlying ectoderm to develop into neural tissue, labeled as the neurogenic region (red). After the process of neural induction, the neurogenic region is known as the neural plate and is now restricted to giving rise to neural tissue. A cross section of the embryo at the neural plate stage shows the relationships between the tissues at this stage of development. The neural plate goes on to generate the neurons and glia in the adult brain and spinal cord.

neural tube (**Figure 1.7**). Much more will be said about the neural tube and its derivatives and shape changes in the next two chapters. For now, suffice it to say that this tube of cells gives rise to nearly all the neurons and glia of vertebrates. Another source of neurons and glia is the neural crest, a group of cells that arises at the junction between the tube and the ectoderm (Figure 1.7). The neural crest is the source of most of the neurons and glia of the peripheral nervous system, whose cell bodies lie outside the brain and spinal cord. This tissue is unique to vertebrates and has the capacity to generate many diverse cell types; we will have more to say about the neural crest in later chapters.

The complex tissue rearrangements that occur during gastrulation in the amphibian occur in other vertebrates in fundamentally the same way. However, the details of these movements can be quite different. Much of the difference in cell movements lies in differences in the amount of yolk in the egg. Fish and bird embryos have a substantial amount of yolk; since the cleavage divisions proceed more slowly through the yolk, these animals have many more cleavage divisions in the animal pole than in the vegetal pole. In zebrafish embryos, the blastomeres are situated at the top of the egg, and as development proceeds these cells divide and spread downward over the surface of the yolk cells; this downward spreading is called

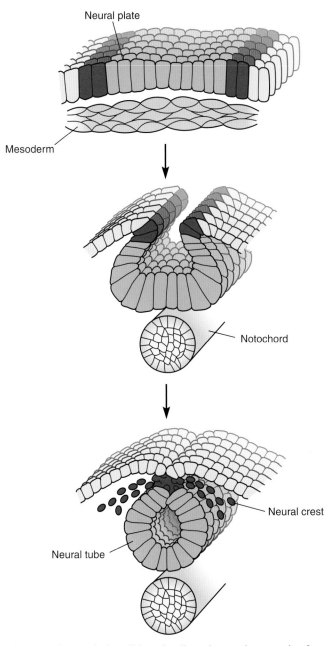

Neural plate

Mesoderm

Notochord

Neural crest

Neural tube

Fig. 1.7 The neural plate (light red) rolls up into a tube separating from the rest of the ectoderm. The involuting cells condense to form a rod-shaped structure—the notochord—just underneath the neural plate. At the same time, the neural plate begins to roll up and fuse at the dorsal margins. A group of cells known as the neural crest (bright red) arises at the point of fusion of the neural tube.

epiboly. When epiboly is about 50% complete (when the spread reaches the equator), there is a transient pause as the process of gastrulation begins at the future dorsal margin of the embryo, which is at this point called the *shield* (**Figure 1.8**). The shield begins to thicken as gastrulation commences. Prospective mesodermal cells delaminate, move inside the ectodermal layer, and begin migrating back toward the animal pole. The rest of the ectoderm then continues its migration to the vegetal pole until the yolk is completely enveloped at 100% epiboly. As the dorsal mesodermal cells migrate toward the animal pole on the shield side, the ectoderm above becomes committed to a neural fate and a definitive neural plate begins to form.

Avian embryos are an extreme example of a "yolky" egg. We all know how much yolk is in a chicken egg. The large amount of yolk makes it difficult for the cells to divide like they do in frogs, and so the cell cleavage divisions do not penetrate into this yolk but are restricted to the relatively yolk-free cytoplasm at the animal pole. These cleavages lead to a disc of cells, called the *blastodisc*, which is essentially floating on the yolk, not too different from what happens in the zebrafish. However, instead of these cells "surrounding" the yolk, like what happens in fish during epiboly, the developing chick embryo continues to sit on the surface of the yolk. After sufficient numbers of cells are generated in the blastodisc, the process of gastrulation occurs. The invagination of mesoderm occurs in this disc through a blastopore-like structure known as the "primitive streak." During this invagination, future mesoderm cells migrate into the interior of the embryo (**Figure 1.9**). The nervous system is induced to form over the involuting mesodermal cells, much like in frogs and fish. Another structure, called Hensen's node, is at the posterior end of the primitive streak and is analogous to the dorsal lip of the blastopore of amphibians.

What about mammalian embryos, which have essentially no yolk and derive all their nourishment from the placenta? The cleavage divisions of mammalian embryos are complete (**Figure 1.10**), and the resulting cells are equal in their potential; there is no obvious animal or vegetal pole. However, after a sufficient number of divisions, when the blastula forms, there are cells on the inside of the ball, called the *inner cell mass*, that produce the embryo, whereas the cells on the outside of the ball make the placenta and associated extra-embryonic membranes. Even though they lack yolk, mammalian embryos undergo a process of gastrulation that is similar to the avian embryo in that the developing mesodermal cells migrate through the primitive streak to reach the interior (Figure 1.10). The primitive streak runs along the anterior-posterior axis of the embryo, and the ectoderm laying above the ingressing mesodermal cells becomes the neural plate and subsequently the neural tube, much like that described above for the other vertebrate embryos. Although the processes of early neural development seem to vary dramatically from insects to people, we will see in this chapter and throughout the book that the underlying molecular mechanisms are highly conserved.

INTERACTIONS WITH NEIGHBORING TISSUES IN MAKING NEURAL TISSUE

As we have seen, the three basic layers of the embryo—the endoderm, mesoderm, and ectoderm—arise through the complex movements of gastrulation. These movements also create new tissue relations. For example, after gastrulation in the frog, presumptive mesoderm now underlies the dorsal ectoderm. A large number of experimental studies in the early part of the twentieth century revealed that these new tissue arrangements were of critical importance to the development of a normal animal. By culturing small pieces of embryos in isolation, it was possible to determine the time at which each part of the embryo acquired its character or fate. When the dorsal ectoderm of a frog embryo was cultured in isolation prior to gastrulation, the cells differentiated into epidermis. However, when roughly the same piece of tissue was isolated from gastrulating embryos, the piece of ectoderm

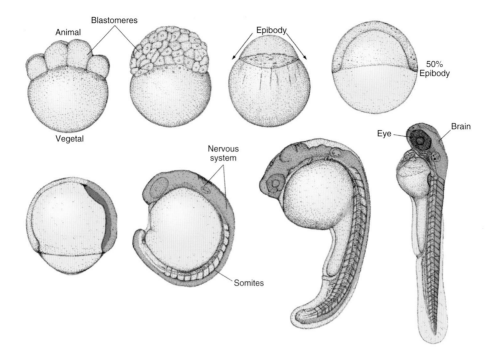

Fig. 1.8 Development of the zebrafish embryo. The zebrafish embryo develops primarily on top of a large ball of yolk. The cleavages are confined to the dorsal side. After multiple cleavage divisions, the cells migrate over the yolk in a process called epiboly. Once the cells have enclosed the yolk, the development of the nervous system proceeds much like that described for the frog. Neural tissue is colored red. (After Kimmel et al., 1995)

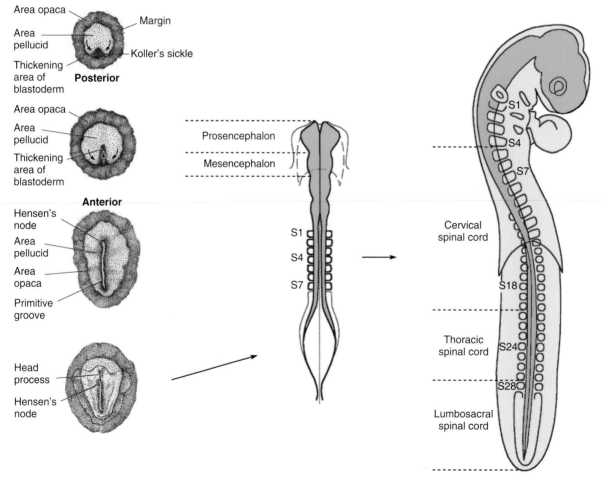

Fig. 1.9 Development of the chick embryo. The blastoderm (area opaca) sits on top of the large yolk and is the result of a large number of cleavage divisions. At the start of gastrulation, cells move posteriorly (arrows) and migrate under the area opaca. The embryo begins to elongate in the anterior-posterior axis, and the region where the cells migrate underneath the area opaca is now called the primitive groove and then primitive streak. The cells migrate into the blastocoel to form the mesoderm. At the anterior end of the primitive streak an enlargement of the streak is called *Hensen's node*. Later stages of chick embryo development show the expanding nervous system in red.

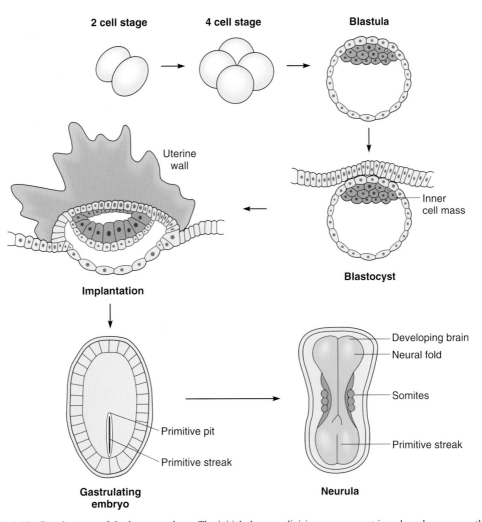

2 cell stage   **4 cell stage**   **Blastula**

Uterine wall

Inner cell mass

Blastocyst

Implantation

Primitive pit

Primitive streak

Gastrulating embryo

Developing brain

Neural fold

Somites

Primitive streak

Neurula

Fig. 1.10 Development of the human embryo. The initial cleavage divisions are symmetric and produce apparently identical blastomeres. There is not much yolk in the mammalian embryo, since most nutrients are derived from the placenta. After multiple cleavage divisions, the embryo is called a blastocyst and develops a distinct inner cell mass and an outer layer of cells. The inner cell mass will develop into the embryo, while the outer cells will contribute to the placenta. After implantation, the embryo begins to elongate, and develops a primitive streak much like that present in the chick embryo. The primitive streak is a line of cells migrating into the blastocoel that will form the mesoderm, and the neural tube (red) will form from the ectoderm overlying the involuting mesoderm. The tube rolls up and forms the brain and spinal cord in a process much like that described for the other vertebrates.

now differentiated into neural tissue, including recognizable parts of the brain, spinal cord, and even eyes (**Figure 1.11**). These results led Hans Spemann, a leading embryologist of the time, to speculate that the ectoderm became fated to generate neural tissue as a result of the tissue rearrangements that occur at gastrulation (Hamburger, 1969). One possible source of this "induction" of the neural tissue was the involuting mesoderm, known at the time as the archenteron roof. As noted above, the involuting tissue is led into the interior of the embryo by the dorsal lip of the blastopore. To test the idea that the involuting mesoderm induces the overlying ectoderm to become neural tissue, Spemann and Hilde Mangold carried out the following experiment. The dorsal lip of the blastopore was dissected from one embryo and transplanted to the interior of another embryo, and the latter embryo was allowed to develop into a tadpole. Spemann and Mangold found that an entire second body axis, including a brain, spinal cord, and

eyes developed from the place where they placed the new blastopore (**Figure 1.12**). To determine whether the new neural tissue that developed in these twinned embryos came from the transplanted tissue, they used a pigmented frog embryo as the donor and the embryo of a nonpigmented strain of frogs as the host. They found that the second body axis that resulted was made of mostly nonpigmented cells, indicating that it came largely from the host blastula, not the transplanted dorsal lip. Thus, the grafted blastopore cells have the capacity to *induce* neural tissues from a region of the ectoderm that would normally not give rise to a nervous system. In addition to the neural tissue in these embryos, they found that mesodermally derived structures also contributed to the twinned embryo. They concluded that the dorsal lip acts not only as a neural inducer but also as an "organizer" of the entire body axis. As a result of these experiments, this region of the embryo is known as the Spemann organizer.

9

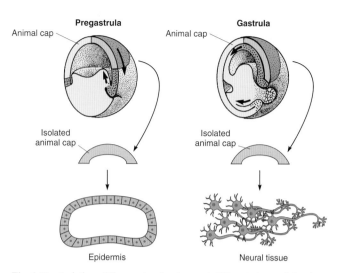

Fig. 1.11 Isolation of fragments of embryos at different stages of development demonstrates when tissue becomes committed to the neural lineage. If the animal cap is isolated from the rest of the embryo (left), the cells develop as epidermis, or skin. If the same region of the embryo is isolated a few hours later, during gastrulation (right), it will develop into neural tissue (shown in the figure as red neurons). Experiments like these led to the idea that the neural lineage arises during gastrulation.

THE MOLECULAR NATURE OF THE NEURAL INDUCER

There were many early efforts to define the chemical nature of the organizer activity. Bautzman, Holtfreter, Spemann, and Mangold showed that the organizer tissue retained its inductive activity even after the cells had been killed by heat, cold, or alcohol. Holtfreter subsequently reported that the neuralizing activity survived freezing, boiling, and acid treatment; however, the activity was lost at temperatures of 150°C. Several embryologists then set out to isolate the active principle(s) in the dorsal lip of the blastopore using the following three approaches: (1) extracting the active factor from the dorsal blastopore cells; (2) trying out candidate molecules to look for similar inductive activities; and (3) testing other tissues for inductive activities.

The initial attempts at direct isolation of the inducing activity from the blastopore lip cells were unsuccessful, largely because of the small amounts of tissue that could be obtained and the limited types of chemical analyses available at the time. From the initial report in 1932 to the late 1950s, over one hundred studies tried to characterize the neural inducing activity. The search for the neural inducer was one of the major preoccupations of developmental biologists in this period. Whereas one group reported that the active principle was lipid extractable, another would report that the residues were more active than the extracts. To obtain more tissue to work with, several investigators screened a variety of adult tissues for similar inducing activities. Although some found a certain degree of specificity, liver and kidney being the most potent neural inducers, others found that "fragments from practically every organ or tissue from various amphibians, reptiles, birds, and mammals, including man, were inductive" (Holtfreter and Hamburger, 1955). Perhaps most disconcerting to the investigators at the time were the results from the candidate molecule approach. Some of the factors found to

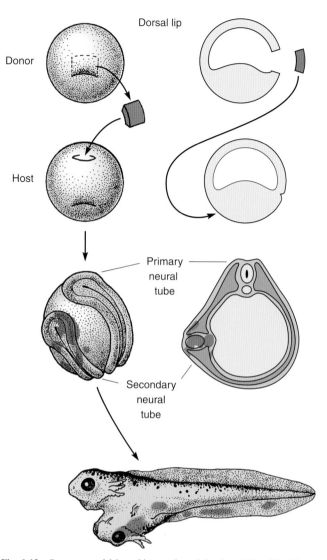

Fig. 1.12 Spemann and Mangold transplanted the dorsal lip of the blastopore from a pigmented embryo (shown as red) to a nonpigmented host embryo. A second axis, including the neural tube, was induced by the transplanted tissue. The transplanted dorsal blastopore lip cells gave rise to some of the tissue in the secondary axis, but some of the host cells also contributed to the new body axis. They concluded that the dorsal lip cells could "organize" the host cells to form a new body axis, and they named this special region of the embryo the organizer.

have neuralizing activity made some sense: polycyclic hydrocarbon steroids, for example. However, other putative inducers, such as methylene blue and thiocyanate, most likely had their effects through some toxicity or contamination.

In the early 1980s, a number of investigators began to apply molecular biological techniques to study embryonic inductions. The first of these studies attempted to test for factors that would trigger the process of mesoderm induction in the frog. As described above, the frog embryo is divided into an animal half and a vegetal half; the animal half will ultimately give rise to neural tissue and ectodermal tissue, while the vegetal half will give rise primarily to endoderm. The mesoderm, which will ultimately go on to make muscle and bone and blood, arises in-between these two tissues, from the cells around the embryo's equator (see **Figure 1.13**). It has been known for many years, based on the work of Peter Nieuwkoop, that the formation of

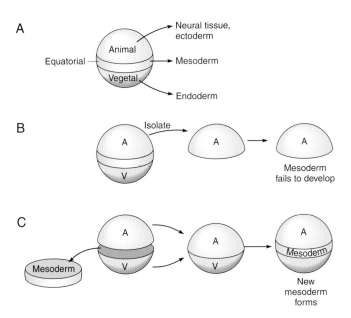

Fig. 1.13 Interactions between the animal and vegetal cells of the amphibian embryo are necessary for induction of the mesoderm. A. The regions of the amphibian embryo that give rise to these different tissue types are shown. The animal pole gives rise to epidermal cells and neural tissue, the vegetal pole gives rise to endodermal derivatives, like the gut, while the mesoderm (blue) arises from the equatorial zone. B. If the animal cap and vegetal hemispheres are isolated from one another, mesoderm does not develop. C. If the equatorial zone is removed from an embryo and the isolated animal and vegetal caps are recombined, a mesoderm forms at a new equatorial zone.

the mesodermal cells in the equatorial region requires some type of interaction between the animal and vegetal halves of the embryo. If this animal half or "cap" is isolated from the vegetal half of the embryo, no mesodermal cells develop. However, when Nieuwkoop (1973, 1985) recombined the animal cap with the vegetal half, mesodermal derivatives developed in the resulting embryos (Figure 1.13). He postulated that a signal from the vegetal half of the embryo induced the formation of mesoderm at the junction with the animal half of the embryo. The identification of the molecular basis for this induction has been the subject of intense investigation, and the

reader is referred to more general textbooks on developmental biology for the current model of this process.

At the same time these studies of mesodermal inducing factors were taking place, a number of investigators realized that the animal cap assay might also be a very good way to identify neural inducers. Not only do isolated animal caps fail to generate mesodermal cells, but they also fail to develop into neuronal tissue. Several factors added to animal caps caused the cells to develop into neural cells as well as mesodermal tissue. However, since the organizer at the dorsal lip of the blastopore is made from mesoderm, it was not clear whether the neural tissue that developed in animal caps was directly induced by the exogenous factor, or, alternatively, whether the factor first induced the organizer and subsequently induced neural tissue (**Figure 1.14**). Therefore, to refine the assay to look for direct neural induction, studies concentrated on identifying factors that would increase the expression of neural genes without the induction of mesoderm-specific gene expression.

The animal cap assay was used for the isolation of the first candidate neural inducer. Richard Harland and his colleagues (Lamb et al., 1993; Smith et al., 1993) used a clever expression cloning system to identify a neural inducing factor (**Figure 1.15**). The cloning was done by taking advantage of the fact that UV-irradiated frog embryos fail to develop a dorsal axis, including the nervous system, and instead develop only ventral structures. Nevertheless, the transplantation of a dorsal blastopore lip from a different embryo can restore a normal body axis to the UV-treated embryo, indicating that the UV embryo can still respond to the neural inducing factor(s). Furthermore, injection of mRNA from a hyperdorsalized embryo can also restore a normal body axis. Harland's group took advantage of this fact and used pools of cDNA isolated from the organizer region to rescue the UV-treated embryos. By dividing the pools into smaller and smaller collections, they isolated a cDNA that coded for a unique secreted protein, which they named Noggin. When Noggin was purified and supplied to animal caps, it was capable of specifically inducing neural genes, without inducing mesodermal genes. Noggin mRNA is expressed in gastrulating embryos, by the cells of the dorsal lip of the blastopore, precisely where

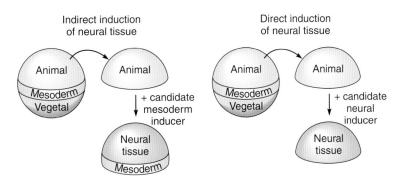

Fig. 1.14 Indirect neural induction versus direct neural induction. The organizer transplant experiments show that the involuting mesoderm has the capacity to induce neural tissue in the cells of the animal cap ectoderm. When assaying for the factor released from mesoderm that is responsible for this activity, it was important to distinguish between the direct and indirect induction of neural tissue when animal caps were treated with a candidate factor. In the first example (left), mesoderm (blue) is induced by the factor, and then neural tissue (red) is induced by the mesoderm. Thus, both mesoderm and neural genes are turned on in the animal caps. However, in the case of a direct neural inducer (right), neural genes are turned on (red), but mesoderm-specific genes are not expressed.

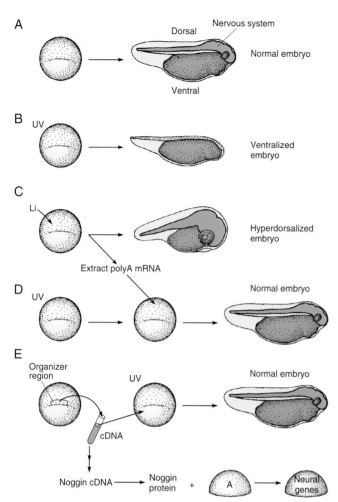

Fig. 1.15 The identification of Noggin as a neural inducer used an expression cloning strategy in *Xenopus* embryos. A. Normal development of a *Xenopus* embryo. B. UV light treatment of the early embryo inhibits the development of dorsal structures by disrupting the cytoskeleton rearrangements that pattern the dorsal inducing molecules prior to gastrulation (ventralized). C. Lithium treatment of the early embryo has the opposite effect; the embryo develops more than normal dorsal tissue (i.e., hyperdorsalized). D. If messenger RNA is extracted from the hyperdorsalized embryos and injected into a UV-treated embryo, the messages encoded in the mRNA can "rescue" the UV-treated embryo and it develops relatively normally. E. Similarly, cDNA from the organizer region of a normal embryo can rescue a UV-treated embryo. The Noggin gene was isolated as a cDNA from the organizer region that could rescue the UV-treated embryo when injected into the embryo, and subsequently, recombinant protein was made from this cDNA and shown to induce neural tissue from isolated animal caps, without any induction of mesodermal genes. Neural tissue is shown in red in all panels.

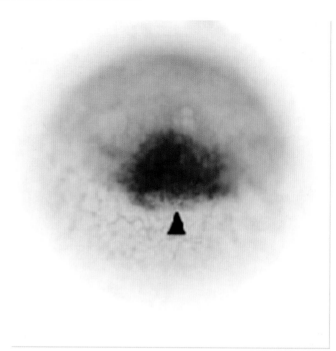

Fig. 1.16 Chordin expression in a frog embryo. The arrow points to the organizer region at the dorsal lip of the blastopore, and the blue label shows the cells of the pregastrula that express the gene Chordin. *(From Sasai et al., 1994)*

the organizer activity is known to reside. Injection of *Noggin* mRNA into UV-treated embryos at the four-cell stage can restore body axis and even hyperdorsalize the embryos to give bigger brains than normal.

At the same time that the Noggin studies were being done, other labs were using additional approaches to identify other neural-inducing molecules. DeRobertis was interested in identifying genes that were expressed in the dorsal blastopore lip organizer region. They isolated a molecule they named Chordin. Like Noggin, this is a secreted protein that is expressed in the organizer during the period when neural induction occurs (**Figure 1.16**). Over-expression of Chordin in the ventral part of the embryo causes a secondary axis to form, like Noggin over-expression, and the authors concluded that Chordin might be similar to Noggin as a second neural inducer.

When these first two neural inducers were identified, it was not immediately clear how they might work. However, it was not until the identification of a third candidate neural inducer, Follistatin, by Melton and his colleagues (Hemmati-Brivanlou et al., 1994) that a potential mechanism began to emerge. Prior to its connection with neural induction, Follistatin was known as a key regulator in the adult reproductive system, where it works as a regulatory factor by binding to and inhibiting activin, a member of the TGF-β family of proteins that controls FSH secretion from the pituitary gland. Follistatin was initially studied for a role in mesoderm induction (Maéno et al., 1994) and Melton was analyzing the mechanism of its action. To study the mechanism of activin action on mesoderm induction, he constructed a truncated Activin receptor that when misexpressed in embryos would interfere with normal endogenous Activin signaling (Hemmati-Brivanlou and Melton, 1994) (**Figure 1.17**). To the surprise of these investigators, interfering with Activin signaling not only disrupted normal mesoderm development, but it also induced the cells of the animal cap to develop as neurons without any additional neural-inducing molecule. They proposed that Activin—or something like it—normally inhibits neural tissue from differentiating in the ectoderm. They also suggested that perhaps neural induction occurred by inhibiting this neural inhibitor; in other words, that the Spemann organizer secretes factors that antagonize a neural inhibitor. These results led

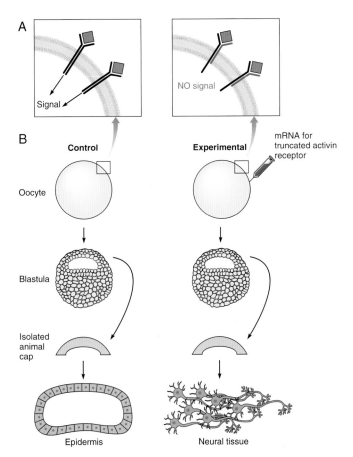

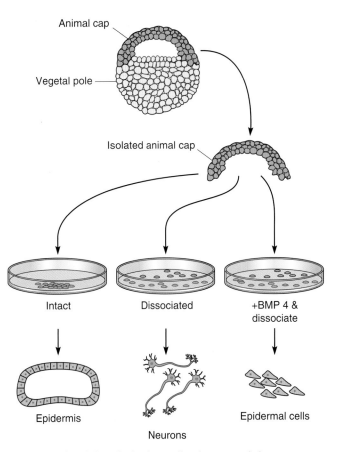

Fig. 1.17 Expression of a truncated activin receptor blocks normal signaling through the receptor and induces neural tissue. A. The normal activin receptor transmits a signal to the cell when activin binds the receptor and it forms a dimer. The truncated activin receptor still binds the activin (or related TGF-β), but now the normal receptor forms dimers with the truncated receptor. Lacking the intracellular domain to signal, the truncated receptor blocks normal signal transduction through this receptor. B. Oocytes injected with the truncated activin receptor develop to the blastula stage, and when the animal caps were dissected from these embryos, they developed into neural tissue without the addition of a neural inducer. This result indicated that inhibiting this signaling pathway might be how neural inducers function.

Fig. 1.18 Dissociation of animal cap cells prior to gastrulation causes most of them to differentiate into neurons in culture. Animal caps can be cultured intact (left) or dissociated into single cells by removing the Ca+2 ions from the medium (middle and right). If the intact caps are put into culture, they develop as epidermis (left). If the dissociated animal cap cells are cultured, they develop into neurons (red; middle). This result supports the hypothesis that the neural fate is actively suppressed by cellular associations in the ectoderm. If the cells are dissociated and then BMP is added to the culture dish, the cells do not become neurons, but instead act as if they are not dissociated and develop as epidermis (right).

to the idea that neural tissue is in some way the default state of the ectoderm and that it might be actively inhibited by Activin-like proteins of the TGF-β family. The idea that the ectoderm is actively inhibited from becoming neural tissue fit nicely with earlier observations that dissociation of the animal cap cells prior to neural induction resulted in most of the cells differentiating as neurons (Godsave and Slack, 1989; Grunz and Tacke, 1989) (**Figure 1.18**). Taking these lines of evidence together, it became clear that molecules that could inhibit activin signaling would make good neural inducers. Since Follistatin was already known to inhibit TGF-β signaling from the studies of these factors in the reproductive system, Melton and colleagues tested whether Follistatin could act as a neural inducer. They found that indeed Follistatin could cause a secondary axis when misexpressed, and recombinant Follistatin could induce neural tissue from animal caps. Follistatin is also expressed in the organizer region of the embryo at the time of neural induction, like Chordin and Noggin.

CONSERVATION OF NEURAL INDUCTION

Equally as fascinating as the fact that the three neural inducing factors may all act by a related mechanism was the discovery that this mechanism is conserved between vertebrates and invertebrates. Analysis of the amino acid sequence of Chordin revealed an interesting homology with a *Drosophila* gene called *short gastrulation* or *sog* (Sasai et al., 1995; DeRobertis and Sasai, 1996). *Sog* is expressed in the ventral side of the fly embryo, and mutations in this gene in *Drosophila* result in defective dorsal-ventral patterning of the embryo. In null mutants of *sog*, the epidermis expands and the neurogenic region is reduced; like Chordin, microinjection of *sog* into the nonneurogenic region of the embryo causes the formation of ectopic neural tissue. Thus, *sog* seems to be the functional homolog of Chordin. At this point the advantages of fly genetics were important. From analysis of other *Drosophila* mutants, it was possible to show that *sog* interacts with a gene called *decapentaplegic*, or *dpp*, a TGF-β-like protein related to the vertebrate genes known as bone-morphogenic proteins, BMPs. *Dpp* and *sog* directly antagonize one another in *Drosophila*. Mutations in *dpp* have the opposite phenotype as *sog* mutations; in *dpp* mutants the neurogenic region expands at the expense of the epidermis, and ectopic expression of *dpp* causes a reduction in neural tissue. These *Drosophila* studies motivated studies of the distribution of the BMPs at early stages of *Xenopus* development, and a similar pattern emerged. BMP4 is expressed throughout most of the gastrula, but at reduced levels in the organizer and neurogenic animal cap. As expected, recombinant BMP4 can suppress neural induction by Chordin, the vertebrate homolog of *sog*.

The studies of *sog/chordin* and *dpp/bmp4* lead to two conclusions. First, it appears that the dorsal-ventral axis of the developing embryo uses similar mechanisms in both the fly and the vertebrate. However, as discussed in the previous section, the neural tissue in the vertebrate is derived from the dorsal side of the animal, while the neurogenic region of the fly is on the ventral side (DeRobertis and Sasai, 1996; Holley et al., 1995). The idea that the vertebrate and arthropod body plans were inverted with respect to one another was first proposed by Geoffroy Saint-Hilaire from comparative anatomical studies, and this appears to be confirmed by these recent molecular studies (**Figure 1.19**). Second, the antagonistic mechanism between *sog* and *dpp* in the fly solidified the conclusion that the various neural inducers work through a common mechanism, the antagonism of BMP signaling. The following three key experiments all indicate that

this is indeed the case. First, BMP4 will inhibit neural differentiation of animal caps treated with Chordin, Noggin, or Follistatin. Second, BMP4 will also inhibit neural differentiation of dissociated animal cap cells. Third, antisense BMP4 RNA causes neural differentiation of animal caps without addition of any of the neural inducers (Hawley et al., 1995). The dominant-negative Activin receptor induction of neural tissue can also be understood in this context, since the Activin receptor is related to the BMP4 receptor, and additional experiments have shown that the expression of the truncated receptor also blocks endogenous BMP4 signaling (Wilson and Hemmati-Brivanlou, 1995).

Do all three of these neural inducers act equivalently to inhibit BMP4 signaling? Biochemical studies have demonstrated that Chordin blocks BMP4-receptor interactions by directly binding to the BMP4 with high affinity. Noggin also appears to bind BMP4 with an even greater affinity, while Follistatin can bind the related molecules BMP7 and Activin. Therefore, it is likely that at least these three neural inducers act by blocking the endogenous epidermalizing BMP4, thereby allowing neural differentiation of the neurogenic ectoderm (Piccolo et al., 1996) (**Figure 1.20**). This view of neural induction is now known as the "default model."

The studies described in *Xenopus* embryos have provided evidence that these factors are capable of inducing neural tissue, but it is more difficult technically to determine whether these factors are required for neural induction in *Xenopus*. To study the requirement for BMP inhibition in neural induction, several labs have examined animals that have mutations in one or more of the putative neural inducer genes. Zebrafish with mutations in the *Chordin* gene have reductions in both neural tissue and in

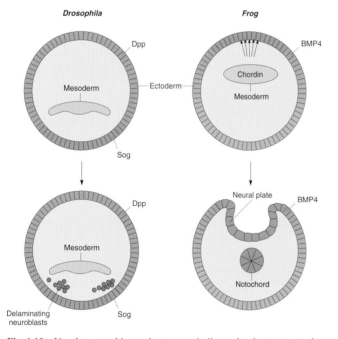

Fig. 1.19 Vertebrates and invertebrates use similar molecules to pattern the dorsal-ventral axis. The *Drosophila* embryo in cross section resembles an inverted *Xenopus* embryo. As described in **Fig. 1.6**, the neurogenic region is in the ventral-lateral *Drosophila* embryo, whereas in the vertebrate embryo, the neural plate arises from the dorsal side. In the *Drosophila*, a BMP-like molecule, dpp, inhibits neural differentiation in the ectoderm, and in the vertebrate embryo, the related molecules, BMP2 and BMP4, suppress neural development. In *Drosophila*, sog (short gastrula) promotes neural development by inhibiting the dpp signaling in the ectoderm in this region, while in the *Xenopus*, a related molecule, Chordin (chd), is one of the neural inducers released from the involuting mesodermal cells and in an analogous way inhibits BMP signaling, allowing neural development in these ectodermal cells.

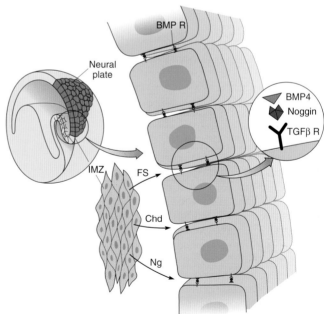

Fig. 1.20 The current model of neural induction in amphibian embryos. The involuting mesodermal cells of the IMZ release several molecules that interfere with the BMP signals between ectodermal cells. Chordin, Noggin, and Follistatin all interfere with the activation of the BMP receptor by the BMPs in the ectoderm and thereby block the antineuralizing effects of BMP4. In other words, they "induce" this region of the embryo to develop as neural tissue, ultimately generating the brain, spinal cord, and most of the peripheral nervous system.

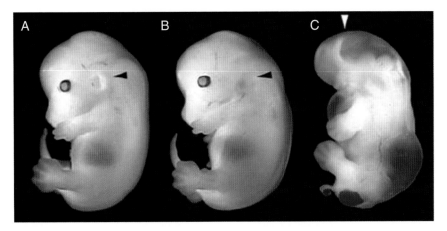

Fig. 1.21 Loss of *noggin* and *chordin* in developing mice causes severe defects in head development. A. Wild-type mouse embryo. B. Loss of *noggin* only. C. Loss of both *noggin* and *chordin*. Note that only mild defects are present in mice deficient in only *noggin*, but the head is nearly absent when both genes are knocked out. *(From Bachiller et al., 2000)*

other dorsal tissues (Schulte-Merker et al., 1997). In mice, targeted deletions have been made in the genes for *follistatin*, *noggin*, and *chordin*. Although deletion of any one of these genes has only minor effects on neural induction, elimination of both *noggin* and *chordin* genes has major effects on neural development. **Figure 1.21** shows the nearly headless phenotype of these animals. The cerebral hemispheres of the brain are almost completely absent. Nevertheless, some neural tissue forms in these animals. Thus, while antagonism of the BMP signal via secreted BMP antagonists is clearly required for the development of much of the nervous system, other factors are likely involved.

Are there additional neural inducers? Experiments with chick embryos and ascidians indicate that at least one of these additional factors is likely a member of the fibroblast growth factor (FGF) family of signaling molecules. In the chick embryo, Streit et al. (2000) found that neural induction actually occurs prior to gastrulation. Blocking FGF signaling in chick embryos with an FGF receptor inhibitor, called SU5402, prevented this early phase of neural induction. Evidence from the ascidian embryo further supports the role of FGF in neural induction (Bertrand et al., 2003). Ascidians are not vertebrates, but before becoming a sessile adult, they have a "tadpole" intermediate form that resembles a simple vertebrate-like larva, with a notochord and dorsal neural tube. In ascidians, BMP antagonists like Chordin and Noggin do not appear to be sufficient for neural induction. Instead FGF is the critical factor, as for the chick. These lines of evidence, and others, have led to the proposal that the "default model" does not completely explain the process of neural induction, and in recent years several studies have attempted to reconcile the relative roles of FGF and BMP antagonists (Delaune et al., 2005). Although all of the ways in which these two different signaling pathways interact with one another are not yet clear, there is evidence that Smad proteins, downstream pathway components of the BMP pathway (see "BMP and Wnt" sidebar, below) can be inhibited by phosphorylation by FGF signaling pathway components (Pera et al., 2003) (see sidebar). A recent study focusing on downstream transcription factors during neural induction has shed some additional light as to how the BMP and FGF pathways might converge (Marchal et al., 2009). Two transcription factors, Zic1 and Zic3, are expressed very early in the ectoderm that is going to become the neural plate. Both of these

transcription factors are needed for the ectodermal cells to become neural tissue (Mizuseki et al., 1998a). When embryos are injected with Noggin mRNA, Zic1 and Zic3 are expressed in the entire animal cap, but if FGF signaling is blocked (with SU5402), Zic3 expression is lost, while Zic1 expression is unaffected. Thus, BMP inhibition and FGF activation each control a different part of the process, but both are required for full neural induction, since both Zic1 and Zic3 are required for the neural plate to form (**Figure 1.22**).

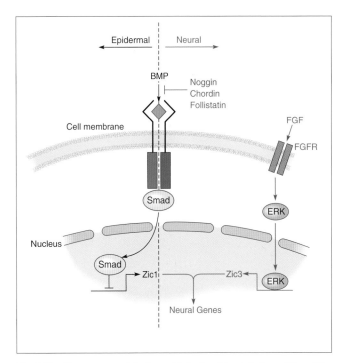

Fig. 1.22 The current model of neural induction. The default condition of the ectoderm is to make epidermis, through the activation of the BMP pathway. BMP in the ectoderm stimulates the receptor to activate a set of intracellular proteins (Smad; see "BMP and Wnt" sidebar, below) that repress Zic1 transcription. If the BMP signal is blocked by one of the many inhibitors in the organizer, the Smad pathway is inactivated, and Zic1 is no longer inhibited. When FGF receptors are activated, the ERK pathway activates transcription of Zic3. Together Zic1 and Zic3 activate downstream neural progenitor genes, like Sox2, directing them to a neural fate.

BMP and Wnt: An Introduction to Signaling in Development

Once an organism develops more than one cell, the cells need ways to communicate with one another. In multicellular organisms several key molecular signals have evolved. These signals are frequently proteins that are released from one cell and bind to receptors on adjacent cells. The binding of a factor to its receptor causes a series of intracellular changes in transducing molecules known as a signal transduction cascade. Changes in the molecules of a specific signal transduction cascade often have as their endpoint a change in gene expression of the cell. Although at first there may seem to be a bewildering number of signals and receptors described in this book, there are a few types of signaling systems that are repeatedly used during development. We assume that the reader has a basic understanding of these signaling systems, but some of the most critical ones will be highlighted in various chapters.

Bone-morphogenic proteins (BMPs) are members of a very large family of proteins, known as the TGF-beta family of factors, since the first protein of this group discovered was Transforming Growth Factor-beta. These proteins range in size from 10 kD to 30 kD and have a characteristic structure, known as the cystein-knot. They bind to a receptor composed of a type I receptor subunit and a type II receptor subunit. The type II receptor subunit is a *kinases*; that is, it can add a phosphate group (phosphorylation) onto specific serine or threonine amino acids on the adjacent type I receptor subunit. The phosphorylation of the serine or threonines on the type I receptor causes the further phosphorylation of another group of proteins, known as *R-smads*. The phosphorylated R-Smads then form complexes with closely related co-Smads. This complex moves to the cell's nucleus and binds to specific sequences in the cell's DNA and activates nearby genes. The specific DNA sequences are known as BMP *response elements*, and they occur in the promoter regions of genes that are expressed when this pathway is activated by the BMP.

Another signaling protein critical in the regulation of development is the Wnt pathway. Wnts are secreted molecules, but they are typically associated with the cell membrane and diffuse only limited distances from the cell that secretes them. The Wnt proteins bind to a receptor called Frizzled, an integral membrane protein, with seven transmembrane domains. Along with the Frizzled receptor, there is a second component to the receptor complex, the LRP protein. When Wnt is bound to its receptor complex, an intracellular protein, β-catenin, associates with several other proteins, including Axin, GSK3β, and APC. This complex is continually degraded, and so exists only transiently in the cell. However, when Wnt binds to Frizzled, another protein called Disheveled blocks the degradation of the complex and causes the β-catenin to accumulate. As the β-catenin accumulates, some of it moves to the nucleus, where it forms a complex with a different protein, TCF. The β-catenin/TCF complex can bind to DNA at specific sequences and activate target genes.

There are many similarities between the Wnt and BMP signaling pathways. Both rely on cell-surface receptors to cause a change in cytoplasmic components of the cell that eventually reach the nucleus to cause a change in gene transcription. In addition, there are several natural inhibitors of these pathways; for the BMP pathway, Follistatin, Noggin, and Chordin can interfere with the activation of the pathway by blocking BMP from binding to the receptor, and for the Wnt pathway, cerberus, FrzB, and Dkk prevent activation, most likely by blocking the Wnt from accessing the receptor. Throughout this book we will see that different signal-receptor systems are involved in nearly all developmental events, and while the types of proteins and details of the transduction cascades may vary, the fundamental features of all these pathways are similar.

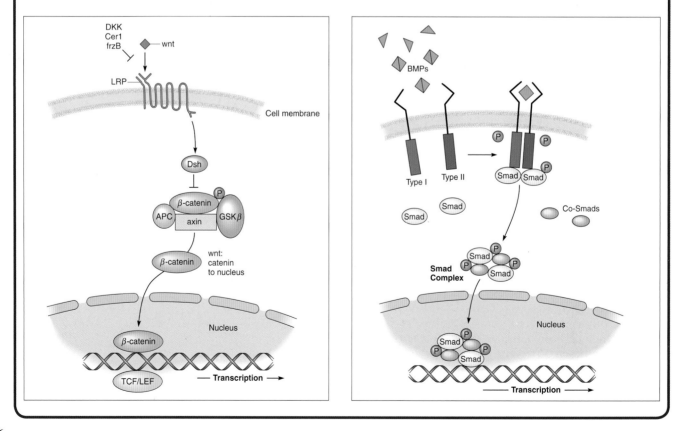

In addition to the Zic genes, Sox genes are other transcription factors that are also regulated by BMP signaling in the developing neural tissue of the embryo. Sasai and colleagues used differential screening for genes up-regulated in the animal caps after Chordin treatment. They found that a gene called Soxd (Sox15) is expressed prior to neural induction in the ectoderm, but it becomes restricted to the neural ectoderm at midgastrulation. Over-expression of Soxd causes neural differentiation in animal caps as well as in nonneural ectoderm in the embryo, similar to that observed from Neurod (Mizuseki et al., 1998b). Another Sox gene, Sox2, has a more restricted pattern of expression, and is expressed in the presumptive neural ectoderm from the late blastula onwards. Sasai and colleagues have found that overexpression of a dominant negative form of Sox2 blocks the effects of neural-inducing factors, and inhibits nervous system formation. The current model puts the Zic genes upstream of Sox2, though these connections are still not completely understood and await more definitive experiments to better define this key step in neural development.

INTERACTIONS AMONG THE ECTODERMAL CELLS IN CONTROLLING NEUROBLAST SEGREGATION

The generation of neurons from the neurogenic region of both vertebrates and invertebrates typically involves an intermediate step: a neural precursor cell is first produced, and this cell goes on to produce many neurons. The neural precursor is capable of mitotic divisions, whereas the neuron itself is usually a terminally postmitotic cell. In the previous section, we saw that in both *Drosophila* and *Xenopus* the antagonism of BMP/Dpp was critical in defining the neurogenic region of the embryo. This section describes some of the genes important in the next stage of nervous system development, the formation of neuroblasts in these neurogenic regions. In insects, this occurs in small regions called "proneural clusters" that are distributed across the surface of the neurogenic region (Figure 1.4). In vertebrates, we have already seen that the entire neurogenic region rolls up into a tube (Figure 1.7). Studies of mutations in genes involved in the formation of the proneural clusters in Drosophila led to the identification of many of the genes that are critical for neurogenesis and neuronal differentiation in all animals, and so this section will focus on this process in *Drosophila*.

In *Drosophila,* as described in the previous section, the neural precursors or neuroblasts form by a process that starts with their delamination: certain cells within proneural clusters enlarge and begin to move to the inside of the embryo. Next, each neuroblast divides to generate many progeny, known as *ganglion mother cells* (GMCs). Each GMC then generates a pair of neurons or glia (Figure 1.5). The neuroblasts form from the neurogenic ectoderm in a highly stereotyped array, and each neuroblast can be assigned a unique identity based on its position in the array, the expression of a particular pattern of genes, and the particular set of neurons and glia that it generates (Doe, 1992). The first neuroblasts to form are arranged in four rows along the anterior-posterior axis and in three columns along the mediolateral axis (**Figure 1.23**). The types of neurons and glia generated

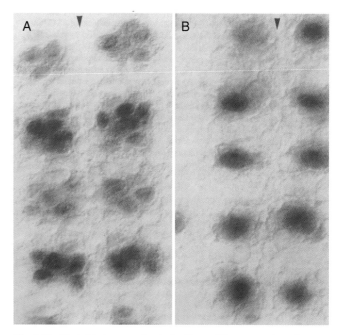

Fig. 1.23 Neuroblast segregation in the *Drosophila* neurogenic region proceeds in a highly patterned array. A. In this embryo stained with an antibody against *achaete-scute (as-c)* protein, clusters of proneural cells in the ectoderm express the gene prior to delamination. B. A single neuroblast develops from each cluster and continues to express the gene. The other proneural cells downregulate the *as-c* gene. (From Doe, 1992)

by a particular neuroblast depend on its position in the array, and so each neuroblast is said to have a unique identity. The next waves of neuroblasts to form are also organized in rows and columns, adjacent to the preceding waves of neuroblasts. The genes involved in controlling the identity of several of the neuroblasts have been defined and will be discussed in Chapter 4; however, at this point the mechanisms that control the segregation of the neuroblasts from the ectoderm will be described.

Among the molecules that are intimately involved in the segregation of the neuroblasts from the other epidermal cells are the members of the *achaete scute* gene complex. The *achaete scute* genes were identified for their effects on the development of the bristles, or chaete, on the fly, each of which contains a sensory neuron. These genes are organized in a complex of four (*achaete; scute; asense;* and *lethal of scute*) at a single locus in *Drosophila* (Alonso and Cabrera, 1988). Deletion of this locus results in the absence of most of the neuroblasts in the fly, in both the central and peripheral nervous systems (Cabrera et al., 1987), while animals with extra copies of these genes have ectopic neurons and sense organs (Brand and Campos-Ortega, 1988). Since these genes are required for the formation of neurons from the epidermal cells, they have been called the "proneural genes" (**Figure 1.24**). An additional *achaete scute*-related gene, *atonal*, is a proneural gene for the internal chordotonal sensory organs and the eye (Jarman et al., 1993).

How do the proneural genes function in neuroblast segregation? The proneural genes code for transcription factors of a particular class, known as the **b**asic-**h**elix-**l**oop-**h**elix, or bHLH, family (**Figure 1.25**). These proteins bind to specific short stretches of DNA, known as E-boxes, in the promoters

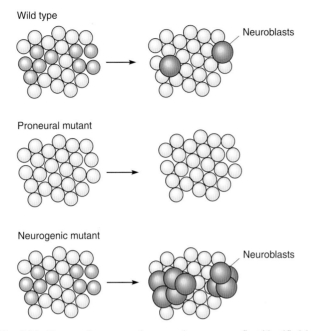

Wild type

Neuroblasts

Proneural mutant

Neurogenic mutant

Neuroblasts

Fig. 1.24 Neurogenic genes and proneural genes were first identified in the *Drosophila* due to their effects on neural development. In the wild-type embryo (top), only one neuroblast (red) delaminates from a given proneural cluster in the ectoderm. However, in flies mutant for proneural genes (middle), like *achaete scute*, no neuroblasts form. By contrast, in flies mutant for neurogenic genes (bottom), like *notch* and *delta*, many neuroblasts delaminate at the positions where only a single neuroblast develops in the wild-type animal. Thus, too many neurons delaminate—hence the name "neurogenic."

of target genes, and activate their transcription. Proneural bHLH transcription factors are members of a broader class of tissue-specific transcription factors (class B). Other tissues, like muscle, also have class B transcription factors, but instead of activating neural genes, the muscle-specific bHLH proteins activate muscle-specific genes. The first of these that was discovered was called *myod*, for myogenic determination factor. These tissue-specific class B transcription factors bind DNA as dimers; their dimer partners are similar, but more ubiquitously expressed, bHLH genes, known as class A. In *Drosophila* the class A gene is called *daughterless* (*da*), named for its role in the sex determination process (Caudy et al., 1988). The dimerization occurs through one of the helices (Figure 1.25), while the basic region is an extension of the other helix and interacts with the major groove of the DNA. The *achaete scute* transcription factors bind to E-boxes in the promoter regions of neuroblast-specific genes and activate their transcription, to make the cell into a neuroblast.

The process of neuroblast formation requires that a precise number of cells from the neurogenic region delaminate. Just prior to the delamination of the neuroblasts, *achaete* is expressed in a group of four to six epidermal cells, and this group of potential neuroblasts is called the proneural cluster (Skeath and Carroll, 1992; Figure 1.23). Under normal circumstances, only a single cell from each cluster will delaminate as a neuroblast. The cell that delaminates to form the neuroblast continues to express the *achaete scute* proneural genes, while all of the other cells in the proneural cluster downregulate their expression of *achaete scute* (Figure 1.23).

In experiments designed to determine whether interactions among the cells were necessary for singling out one of the cells of the proneural cluster for delamination as the neuroblast, Taghert et al. (1984) used a laser microbeam to destroy the developing neuroblast at various times during its delamination (**Figure 1.26**). They found that ablation of

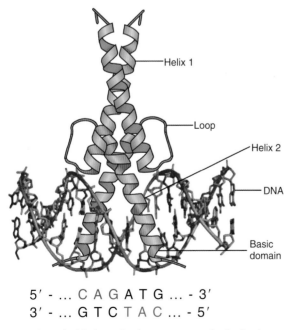

Helix 1

Loop

Helix 2

DNA

Basic domain

5′ - ... C A G A T G ... - 3′
3′ - ... G T C T A C ... - 5′

Fig. 1.25 Several critical proteins that are necessary for the development of specific cell and tissue types are members of the bHLH transcription factor family of molecules. bHLH denotes the "basic-helix-loop-helix" structure of these molecules. The bHLH transcription factors dimerize via their first helix and interact with DNA via their second helix and their basic region.

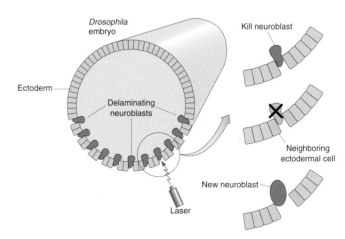

Drosophila embryo

Ectoderm

Delaminating neuroblasts

Laser

Kill neuroblast

Neighboring ectodermal cell

New neuroblast

Fig. 1.26 Ablation of the delaminating neuroblast with a laser microbeam directed to the ventral neurogenic region of the fly embryo causes a neighboring ectodermal cell to take its place. This experiment shows that the neuroblast inhibits neighboring cells from adopting the same fate via the mechanism of lateral inhibition.

the delaminating neuroblast with a laser microbeam causes one of the other cells in the cluster to take its place and delaminate. These results led to the idea that the expression of *achaete scute* genes, and hence neuroblast potential, is regulated by a system of lateral inhibition. The cell that begins to delaminate maintains its *achaete scute* expression and suppresses proneural gene activity function in the other cells of the cluster. The other cells then remain as epidermal cells, while the *achaete scute*-expressing cell delaminates as the neuroblast.

The mechanisms by which the cell that ultimately develops as the neuroblast is singled out from the original cluster have been the subject of intensive investigation. Studies of this process have uncovered a unique signaling pathway, which underlies lateral inhibitory processes in many regions of the embryo. Molecules that act prominently in this process are the Notch receptor and one of its ligands, Delta. Notch is a large transmembrane protein characterized by an extracellular portion with a large number of repetitive domains, known as EGF-repeats because of their similarity to the cysteine-bonded tertiary structure of the mitogen epidermal growth factor, EGF. However, despite this apparent structural similarity to an extended peptide mitogen, Notch has no apparent mitogenic activity, but rather acts to bind two ligands with somewhat similar structures, Delta and Serrate (Fehon et al., 1990). These proteins are expressed not only in the nervous system, but also in many other areas of the embryo where lateral inhibitory interactions define tissue boundaries. In fact, Notch and Delta, and the additional ligand, Serrate, were named for their effects on wing development, where lateral inhibition is also mediated by these molecules and is necessary for the proper development of wing morphology.

The Notch/Delta pathway is critical for singling out the neuroblast from the proneural cluster, and the fate of the cells in the neurogenic region depends on their level of Notch activity. Low Notch receptor activity in one of these cells causes it to become a neuroblast, while high activity results in the cell adopting an epidermal fate. In *notch* null mutants, nearly all of the cells in the neurogenic region become neuroblasts; as a result, the *notch* null embryos have defects in the epidermis (Figure 1.24). A similar phenotype occurs in *delta* null mutants. Because of the phenotypes the mutant animals show, the *notch* and *delta* genes have been termed *neurogenic*. What the Notch signal appears to do is to allow only a single cell in the proneural cluster to maintain *achaete-scute* expression.

How does Notch activation lead to the suppression of *achaete scute* in a neighboring cell? The mechanism of Notch signal transduction has been worked out over many years and while not completely understood, a highly simplified version of the current model is shown in **Figure 1.27** (Kopan and Ilagan, 2009). The activation of Notch by binding to Delta causes a proteolytic cleavage event that releases the intracellular domain of the protein into the cytoplasm. The Notch-intracellular domain has a nuclear-localization signal on it and so moves to the nucleus, where it forms a complex with a protein called Supressor of Hairless (SuH). The complex of SuH, the Notch-ICD, and another protein called Mastermind

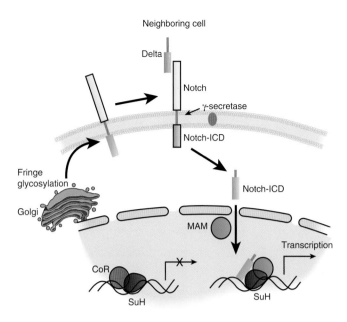

Fig. 1.27 The binding of Delta to Notch leads to a proteolytic cleavage of the molecule by a protease called gamma-secretase. This releases the intracellular part of the Notch molecule (called the Notch-ICD, for intracellular domain). The Notch-ICD interacts with another molecule, Suppressor of Hairless (SuH), and together they form a transcription activation complex to turn on the expression of downstream target genes, specifically *Enhancer of Split*. The E(spl) proteins are repressors of *Asc* gene transcription, and so they block further neural differentiation and reduce the levels of Delta expression.

acts as a transcriptional activator when it binds to the DNA sequence GTGGAA in the cis-regulatory regions of genes (Rebeiz et al., 2005).

Since Notch activation in the epidermal cells prevents them from delamination, we might expect that the genes that the SuH complex regulates are repressors of the proneural genes. This expectation has been confirmed by the demonstration that SuH directly regulates the transcription of another class of bHLH proteins, the *enhancer of split complex* (E(spl)) of genes. The *E(spl)* genes code for proteins that are similar to the proneural bHLH proteins of the *achaete scute* class, but instead of acting as transcriptional activators, they are strong repressors of transcription. There are seven *E(spl)* genes in *Drosophila,* and their expression patterns overlap considerably, so they are thought to be at least partly redundant. The proteins coded by these genes form heterodimers with the achaete scute proteins through their dimerization domains, but they can also form homodimers. The E(spl) proteins may directly interfere with achaete scute-mediated transcription, binding to DNA sequences called N-boxes, that, like E-boxes, are present in the cis-regulatory regions of the proneural genes. As might be expected by their function as transcriptional repressors, the E(spl) proteins are expressed in the cells surrounding the delaminating neuroblast, and they act to prevent cells from adopting the neural fate.

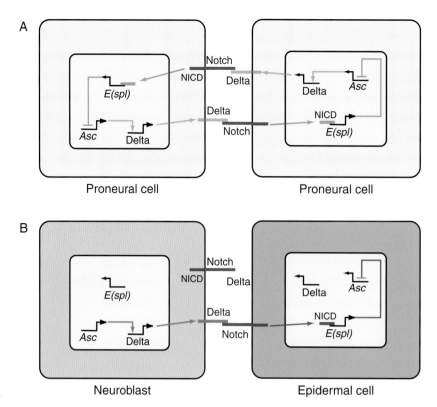

Fig. 1.28 The lateral inhibitory mechanism involves the neurogenic proteins, Notch and Delta. A. In the proneural cluster, all cells express achaete scute genes (Asc), and components of the Notch pathway. B. The Asc activates the expression of downstream neuroblast genes and the Notch ligand Delta. This pathway forms a positive feedback loop such that if a particular cell in the cluster expresses a higher level of Asc than its neighbors, the higher level of Asc will increase Delta expression in the cell, which will then activate the Notch pathway more strongly in the neighboring cells. The higher Notch activity in the neighboring cells causes them to increase expression of E(spl) which will inhibit their Asc expression and lower their Delta. This lateral inhibitory loop ensures that only a single cell in the proneural cluster becomes a neuroblast.

The model for how the Notch/Delta signaling pathway mediates the lateral inhibitory interactions among the cells of the proneural cluster is shown in **Figure 1.28**. Initially, all of the *achaete scute*-expressing cells express an equal amount of Notch and Delta (Hartley et al., 1987; Castro et al., 2005). If, the central cell in the proneural cluster expresses more *achaete scute* than the others, and *achaete scute* directly activates the Delta promoter, then this cell will also express a higher level of Delta than the other cells of the cluster. The high level of Delta in this cell activates Notch on the neighboring cells, causing more Notch-ICD translocation to the nucleus. The Notch-ICD complexes with the SuH protein, resulting in the expression of E(spl) in these cells. The E(spl) proteins bind to an N-box in the *achaete-scute* gene promoter and repress its transcription and this further downregulates Delta expression in these cells, and prevents them from differentiating as neuroblasts. In this way, only a single neuroblast develops from the proneural cluster at a particular location in the fly (**Figure 1.29**). But if that cell is experimentally deleted (or has difficulty forming naturally), a neighboring cell can take its place to ensure proper formation of the neuroblast.

As described at the beginning of this chapter, on the surface the process of neuroblast delamination from the ectoderm described for *Drosophila* is quite different from similar stages in vertebrate embryos. Nevertheless, the proneural and neurogenic pathways described in this section are common to all animals. Proneural genes have been identified in *C. elegans* and even Cnidarians, and so this seems to be a very ancient system for directing cells to a neural identity. Notch/Delta signaling pathway is also fundamental in the process by which neighboring cells become different from one another, not just in the nervous system, but throughout the embryo. The proneural and neurogenic have vertebrate homologs that play similar roles in the vertebrate. The first proneural gene to be discovered in vertebrates was named *mash1,* for mammalian **a**chaete **s**cute **h**omolog-1 (Johnson et al., 1990) and is now called *ascl1*. In fact, vertebrates have many more proneural genes than *Drosophila*. These genes are expressed in the developing nervous system in distinct subsets of neural progenitor cells. They have the same bHLH structure as the *Drosophila achaete scute* genes and can act as transcriptional activators as heterodimers with vertebrate Daughterless homologs, E12 and E47. Vertebrates

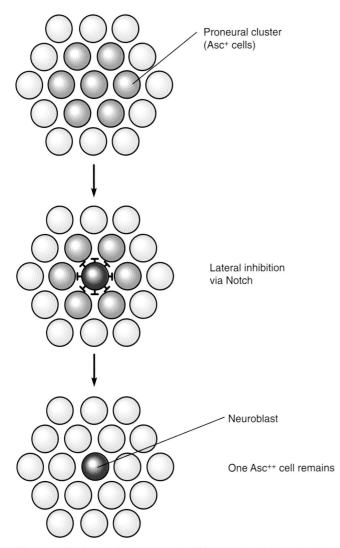

Proneural cluster (Asc⁺ cells)

Lateral inhibition via Notch

Neuroblast

One Asc⁺⁺ cell remains

Fig. 1.29 The lateral inhibitory network of Notch and Asc leads to the selection of a single cell from the proneural cluster to develop as the neuroblast, since Asc expression is restricted by Notch signaling to a single cell.

also have all the components of the Notch pathway. Instead of a single Notch receptor, vertebrates have upto four Notch genes, three of which are expressed in parts of the developing nervous system. In addition, vertebrates also have several ligands for Notch receptors, including two Delta ligands and two Serrate-like ligands, also known as *Jagged* in mammals (Weinmaster and Kintner, 2003). We will have more to say about both the proneural genes and the Notch pathway in Chapter 3, when the process of vertebrate neurogenesis is described.

SUMMARY

Classical embryology led to the conclusion that the nervous system arises during development through an inductive conversion of ectodermal cells to neural tissue via the Spemann organizer. In the past 25 years, a concerted effort has identified the molecular basis of this neural tissue induction as an inhibition of a tonic BMP-mediated repression. In both *Drosophila* and *Xenopus,* the antagonism of the TGF-beta-like factors BMP/Dpp are critical in defining the neurogenic region of the embryo, and a number of "neural inducers," BMP antagonists, are found in the organizer. The antagonism of BMP is both necessary and sufficient for nervous system induction; however, in *Drosophila,* the delamination of neuroblasts from the neurogenic region requires a second signaling system, the Notch/Delta system, to regulate the expression and function of a proneural bHLH class of transcription factors. The Notch/Delta/proneural system is conserved in vertebrates, though instead of being involved in the process of segregation of neuroblasts from the ectoderm, Notch signaling is critical for the production of neurons from the progenitors of the neural tube. We will have much more to say about this in Chapter 3. Nevertheless, the remarkable conservation of the basic processes of neural "induction" shows the ancient origin of these developmental mechanisms.

REFERENCES

Alonso, M. C., & Cabrera, C. V. (1988). The achaete-scute gene complex of Drosophila melanogaster comprises four homologous genes. *EMBO J, 7* (8), 2585–2591.

Bachiller, D., Klingensmith, J., Kemp, C., Belo, J. A., Anderson, R. M., May, S. R., et al. (2000). The organizer factors Chordin and Noggin are required for mouse forebrain development. *Nature, 403*(6770), 658–661.

Bertrand, V., Hudson, C., Caillol, D., Popovici, C., & Lemaire, P. (2003). Neural tissue in ascidian embryos is induced by FGF9/16/20, acting via a combination of maternal GATA and Ets transcription factors. *Cell, 115*(5), 615–627.

Brand, M., & Campos-Ortega, J. A. (1988). Two groups of interrelated genes regulate early neurogenesis in Drosophila melanogaster. *Roux's Arch Dev Biol, 197,* 457–470.

Brenner, S. (1974). The genetics of Caenorhabditis elegans. *Genetics, 77*(1), 71–94.

Cabrera, C. V., Martinez-Arias, A., & Bate, M. (1987). The expression of three members of the achaete-scute gene complex correlates with neuroblast segregation in Drosophila. *Cell, 50*(3), 425–433.

Castro, B., Barolo, S., Bailey, A. M., & Posakony, J. W. (2005). Lateral inhibition in proneural clusters: cis-regulatory logic and default repression by Suppressor of Hairless. *Development, 132*(15), 3333–3344.

Caudy, M., Vassin, H., Brand, M., Tuma, R., Jan, L. Y., & Jan, Y. N. (1988). Daughterless, a Drosophila gene essential for both neurogenesis and sex determination, has sequence similarities to myc and the achaete-scute complex. *Cell, 55*(6), 1061–1067.

Delaune, E., Lemaire, P., & Kodjabachian, L. (2005). Neural induction in Xenopus requires early

FGF signalling in addition to BMP inhibition. *Development, 132*(2), 299–310.

De Robertis, E. M., & Sasai, Y. (1996). A common plan for dorsoventral patterning in Bilateria. *Nature, 380*(6569), 37–40.

Doe, C. Q. (1992). Molecular markers for identified neuroblasts and ganglion mother cells in the Drosophila central nervous system. *Development, 116*(4), 855–863.

Fehon, R. G., Kooh, P. J., Rebay, I., Regan, C. L., Xu, T., Muskavitch, M. A., et al. (1990). Molecular interactions between the protein products of the neurogenic loci Notch and Delta, two EGF-homologous genes in Drosophila. *Cell, 61*(3), 523–534.

Gilbert, S. F., & Raunio, A. M. (1997). *Embryology: constructing the organism.* Sunderland, MA: Sinauer Associates, Inc. Publishers.

Godsave, S. F., & Slack, J. M. (1989). Clonal analysis of mesoderm induction in Xenopus laevis. *Developmental Biology*, *134*(2), 486–490.

Gould, S. J. (1970). *Ontogeny and phylogeny.* Cambridge, MA: Harvard University Press.

Grunz, H., & Tacke, L. (1989). Neural differentiation of Xenopus laevis ectoderm takes place after disaggregation and delayed reaggregation without inducer. *Cell Differentiation and Development*, *28*(3), 211–217.

Hamburger, V. (1969). Hans Spemann and the organizer concept. *Experientia*, Nov 15, *25*(11), 1121–1125.

Hartley, D. A., Xu, T. A., & Artavanis-Tsakonas, S. (1987). The embryonic expression of the Notch locus of Drosophila melanogaster and the implications of point mutations in the extracellular EGF-like domain of the predicted protein. *The EMBO Journal*, *6*(11), 3407–3417.

Hawley, S. H., Wünnenberg-Stapleton, K., Hashimoto, C., Laurent, M. N., Watabe, T., Blumberg, B. W., et al. (1995). Disruption of BMP signals in embryonic Xenopus leads to direct neural induction. *Genes & Development*, *9*(23), 2923–2935.

Hemmati-Brivanlou, A., Kelly, O. G., & Melton, D. A. (1994). Follistatin, an antagonist of activin, is expressed in the Spemann organizer and displays direct neuralizing activity. *Cell*, *77*(2), 283–295.

Hemmati-Brivanlou, A., & Melton, D. A. (1994). Inhibition of activin receptor signaling promotes neuralization in Xenopus. *Cell*, *77*(2), 273–281.

Holley, S. A., Jackson, P. D., Sasai, Y., Lu, B., De Robertis, E. M., Hoffmann, F. M., et al. (1995). A conserved system for dorsal-ventral patterning in insects and vertebrates involving sog and chordin. *Nature*, *376*(6537), 249–253.

Holtfreter, J., & Hamburger, V. (1955). In B. H. Willier, P. Weiss, & V. Hamburger, (Eds.), *Analysis of Development*, (pp. 230–296). Philadelphia, PA: Saunders.

Jarman, A. P., Grau, Y., Jan, L. Y., & Jan, Y. N. (1993). Atonal is a proneural gene that directs chordotonal organ formation in the Drosophila peripheral nervous system. *Cell*, *73*(7), 1307–1321.

Johnson, J. E., Birren, S. J., & Anderson, D. J. (1990). Two rat homologues of Drosophila achaete-scute specifically expressed in neuronal precursors. *Nature*, *346*(6287), 858–861.

Kimmel, C. B., Ballard, W. W., Kimmel, S. R., Ullmann, B., & Schilling, T. F. (1995). Stages of embryonic development of the zebrafish. *Developmental Dynamics*, *203*(3), 253–310.

Kopan, R., & Ilagan, M. X. (2009). The canonical Notch signaling pathway: unfolding the activation mechanism. *Cell*, *137*(2), 216–233.

Lamb, T. M., Knecht, A. K., Smith, W. C., Stachel, S. E., Economides, A. N., Stahl, N., et al. (1993). Neural induction by the secreted polypeptide noggin. *Science*, *262*(5134), 713–718.

Maéno, M., Ong, R. C., Suzuki, A., Ueno, N., & Kung, H. F. (1994). A truncated bone morphogenetic protein 4 receptor alters the fate of ventral mesoderm to dorsal mesoderm: roles of animal pole tissue in the development of ventral mesoderm. *Proceedings of the National Academy of Sciences of the United States of America*, *91*(22), 10260–10264.

Marchal, L., Luxardi, G., Thomé, V., & Kodjabachian, L. (2009). BMP inhibition initiates neural induction via FGF signaling and Zic genes. *Proceedings of the National Academy of Sciences of the United States of America*, *106*(41), 17437–17442.

Miázuseki, K., Kishi, M., Matsui, M., Nakanishi, S., & Sasai, Y. (1998a). Xenopus Zic-related-1 and Sox-2, two factors induced by chordin, have distinct activities in the initiation of neural induction. *Development*, *125*(4), 579–587.

Mizuseki, K., Kishi, M., Shiota, K., Nakanishi, S., & Sasai, Y. (1998b). SoxD: an essential mediator of induction of anterior neural tissues in Xenopus embryos. *Neuron*, *21*(1), 77–85.

Nieuwkoop, P. D. (1973). The organization center of the amphibian embryo: its origin, spatial organization, and morphogenetic action. *Advances in Morphogenesis*, *10*, 1–39.

Nieuwkoop, P. D. (1985). Inductive interactions in early amphibian development and their general nature. *Journal of Embryology and Experimental Morphology*, *89*(Suppl), 333–347.

Pera, E. M., Ikeda, A., Eivers, E., & De Robertis, E. M. (2003). Integration of IGF, FGF, and anti-BMP signals via Smad1 phosphorylation in neural induction. *Genes & Development*, *17*(24), 3023–3028.

Piccolo, S., Sasai, Y., Lu, B., & De Robertis, E. M. (1996). Dorsoventral patterning in Xenopus: inhibition of ventral signals by direct binding of chordin to BMP-4. *Cell*, *86*(4), 589–598.

Rebeiz, M., Stone, T., & Posakony, J. W. (2005). An ancient transcriptional regulatory linkage. *Developmental Biology*, *281*(2), 299–308.

Sasai, Y., Lu, B., Steinbeisser, H., & De Robertis, E. M. (1995). Regulation of neural induction by the Chd and Bmp-4 antagonistic patterning signals in Xenopus. *Nature*, *376*(6538), 333–336.

Sasai, Y., Lu, B., Steinbeisser, H., Geissert, D., Gont, L. K., & De Robertis, E. M. (1994). Xenopus chordin: a novel dorsalizing factor activated by organizer-specific homeobox genes. *Cell*, *79*(5), 779–790.

Schulte-Merker, S., Lee, K. J., McMahon, A. P., & Hammerschmidt, M. (1997). The zebrafish organizer requires chordino. *Nature*, *387*(6636), 862–863.

Skeath, J. B., & Carroll, S. B. (1992). Regulation of proneural gene expression and cell fate during neuroblast segregation in the Drosophila embryo. *Development*, *114*(4), 939–946.

Smith, W. C., Knecht, A. K., Wu, M., & Harland, R. M. (1993). Secreted noggin protein mimics the Spemann organizer in dorsalizing Xenopus mesoderm. *Nature*, *361*(6412), 547–549.

Streit, A., Berliner, A. J., Papanayotou, C., Sirulnik, A., & Stern, C. D. (2000). Initiation of neural induction by FGF signalling before gastrulation. *Nature*, *406*(6791), 74–78.

Sulston, J. E., Schierenberg, E., White, J. G., & Thomson, J. N. (1983). The embryonic cell lineage of the nematode Caenorhabditis elegans. *Developmental Biology*, *100*, 64–119.

Taghert, P. H., Doe, C. Q., & Goodman, C. S. (1984). Cell determination and regulation during development of neuroblasts and neurones in grasshopper embryo. *Nature*, *307*(5947), 163–165.

Weinmaster, G., & Kintner, C. (2003). Modulation of notch signaling during somitogenesis. *Annual Review of Cell and Developmental Biology*, *19*, 367–395.

Wilson, P. A., & Hemmati-Brivanlou,, A. (1995). Induction of epidermis and inhibition of neural fate by Bmp-4. *Nature*, *376*(6538), 331–333.

Polarity and segmentation

2

REGIONAL IDENTITY OF THE NERVOUS SYSTEM

Like the rest of the body in most metazoans, the nervous system is regionally specialized. The head looks different from the tail, and the brain looks different from the spinal cord. There are a number of basic body plans for animals with neurons, and in this section we will consider how the regional specialization of the nervous system arises during the development of some of these animals. At least some of the mechanisms that pattern the nervous system of animals are the same as those that pattern the rest of the animal's body. Similarly, many different types of tissues play key roles in regulating the development of the nervous system.

As was described in the previous chapter, the insect nervous system is made up of a series of connected ganglia known as the ventral nerve cord. In many insects, the ganglia fuse at the midline. The segmental ganglia of the ventral nerve cord each contain approximately the same number and types of neurons and glia, though there are some differences depending on where along the body axis the gang3lion is located. The neuroblasts that will generate the neurons and glia of each segment delaminate from the ventral neurogenic region in a very stereotypic pattern (**Figure 2.1**) that is repeated in each segment; about 30 neuroblasts delaminate in each thoracic and abdominal hemisegment. The neuroblast that arises at a specific position within a particular segment produces the same (or similar) neurons and glia in each segment (see

Chapter 4). In addition to the ventral nerve cord, the insect CNS also has a brain at the anterior end. The insect brain is composed of three regions, known as the protocerebrum, the deutocerebrum, and the tritocerebrum (Figure 2.1), although more recently these regions have been further subdivided as more details on the lineages of the neuroblasts are discovered. The brain contains about 100 neuroblasts, and although they still delaminate from the neurogenic region like those in the abdominal and thoracic segments, the pattern of delamination is not as regular. Between the thoracic segments and the tritocerebrum is a region of the CNS with neuroblasts organized more like the ganglia of the posterior segments (Urbach and Technau, 2004). These four segments are known as the gnathal segments and constitute a third region of the CNS, intermediate between the anterior brain regions and the thoracic and abdominal segments (Lichtneckert and Reichert, 2005).

In the vertebrate embryo, most of the neural tube will give rise to the spinal cord, while the rostral end enlarges to form the three primary brain vesicles: the prosencephalon (or forebrain), the mesencephalon (or midbrain), and the rhombencephalon (or hindbrain) (**Figure 2.2**). The prosencephalon will give rise to the large paired cerebral hemispheres, the mesencephalon will give rise to the midbrain, and the rhombencephalon will give rise to the more caudal regions of the brainstem. The three primary brain vesicles become further subdivided into five vesicles. The prosencephalon gives rise to both the telencephalon and the diencephalon. In addition to generating the thalamus and hypothalamus in the mature brain, important features of the diencephalon are the paired evaginations of the optic vesicles. These develop into the retina and the pigmented epithelial layers of the eyes. The mesencephalon remains as a single vesicle and does not expand to the same extent as the other regions of the brain. The rhombencephalon divides into the metencephalon and the myelencephalon. These two vesicles will form the cerebellum and the medulla of the adult brain, respectively.

The most caudal brain region is the rhombencephalon, the region that will develop into the hindbrain. At a particular time in the development of this part of the brain, the rhombencephalon becomes divided into segments, known as rhombomeres (see below). The rhombomeres are regularly spaced repeating units of hindbrain cells and are separated by distinct boundaries. Since this is one of the clearest areas of segmentation in the vertebrate brain, study of the genes that control segmentation in rhombomeres has received a lot of attention and will be discussed in detail in the next section as a model of how the anterior–posterior patterning of the nervous system takes place in vertebrates.

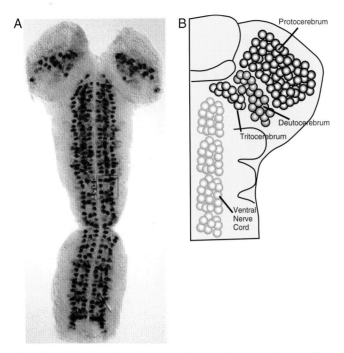

Fig. 2.1 The brain of the *Drosophila* develops from extensive neuroblast delamination in the head. There are about 100 brain neuroblasts total. A. Ventral view of Drosophila embryo showing the delaminating neuroblasts in the head and in the segmental ganglia along the rest of the body. Figure provided by James Skeath. B. Three basic divisions of the brain are known as the protocerebrum, the deutocerebrum, and the tritocerebrum. These divisions are similar to the segmental ganglia in that they are derived independently from delaminating neuroblasts in their respective head segments. *(Modified from Technau et al., 2006)*

THE ANTERIOR–POSTERIOR AXIS AND *HOX* GENES

In both vertebrates and invertebrates, the mechanisms that control the regional development of the nervous system are dependent on the mechanisms that initially set up the anterior–posterior axis of the embryo. These mechanisms were first discovered in the *Drosophila* embryo; however, many of the same genes are involved in the specification of the anterior–posterior axis in the vertebrate.

The anterior–posterior axis of the fly is primarily set up by the distribution of two molecules: a transcription factor known as *bicoid*, localized in the anterior pole of the embryo, and a gene that codes for an RNA-binding protein called *nanos*, localized primarily in the posterior pole of the embryo (Driever and Nusslein-Volhard, 1988). The mRNAs for these genes are localized in their distribution in the egg prior to fertilization by the nurse cells in the mother. Shortly after fertilization, these mRNAs are translated, resulting in opposing protein gradients of the two gene products (**Figure 2.3**). The levels of these two proteins determine whether a second set of genes, the gap genes, are expressed in a particular region of the embryo. The gap genes, in turn, control the striped pattern of a third set of genes, the pair rule genes. Finally, the pattern of expression of the pair rule genes controls the segment-specific expression of a fourth set of genes, the segment polarity genes. This developmental hierarchy progressively divides the embryo into smaller and smaller domains with unique identities (Small and Levine, 1991; Driever and Nusslein-Volhard, 1988). This chain of transcriptional activations produces the reproducible pattern of segmentation of the animal (Figure 2.3).

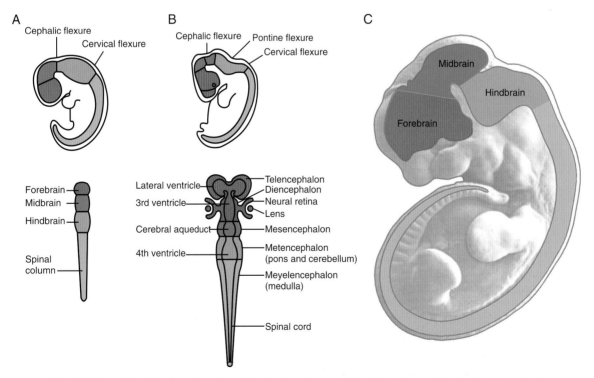

Fig. 2.2 The vertebrate brain and spinal cord develop from the neural tube. Shown here as lateral views (upper) and dorsal views (lower) of human embryos at successively older stages of embryonic development (A,B,C). A. The primary three divisions of the brain occur as three brain vesicles or swellings of the neural tube, known as the forebrain (prosencephalon), midbrain (mesencephalon), and hindbrain (rhombencephalon). B. The next stage of brain development results in further subdivisions, with the forebrain vesicle becoming subdivided into the paired telencephalic vesicles and the diencephalon, and the rhombencephalon becoming subdivided into the metencephalon and the myelencephalon. C. These basic brain divisions can be related to the overall anatomical organization of the embryo.

24

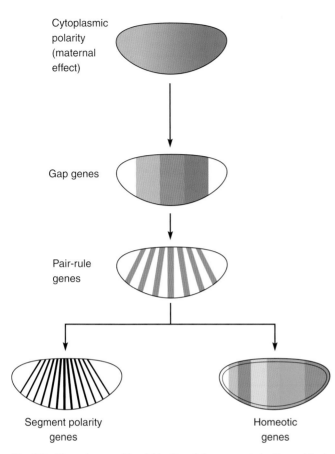

Cytoplasmic polarity (maternal effect)

Gap genes

Pair-rule genes

Segment polarity genes

Homeotic genes

Fig. 2.3 The unique positional identity of the segments in *Drosophila* is derived by a program of molecular steps, each of which progressively subdivides the embryo into smaller and smaller domains of expression. The oocyte has two opposing gradients of mRNA for the maternal effect genes; bicoid and hunchback are localized to the anterior half, while caudal and nanos messages are localized to the posterior regions. The maternal effect gene products regulate the expression of the gap genes, the next set of key transcriptional regulators, which are more spatially restricted in their expression. Orthodenticle (*otd*), for example, is a gap gene that is only expressed at the very high concentrations of bicoid present in the prospective head of the embryo. Specific combinations of the gap gene products in turn activate the transcription of the pair-rule genes, each of which is only expressed in a region of the embryo about two segments wide. The periodic pattern of the pair-rule gene expression is directly controlled by the *gap* genes, and along with a second set of periodically expressed genes, the segment polarity genes determine the specific expression pattern of the *homeotic* genes. In this way, each segment develops a unique identity.

At this point in the development of the fly, the anterior–posterior axis is clearly defined, and the embryo is parceled up into domains of gene expression that correspond to the different segments of the animal. The next step requires a set of genes that will uniquely specify each segment as different from one another. The genes that control the relative identity of the different parts of *Drosophila* were discovered by Edward Lewis (1978). He found mutants of the fly that had two pairs of wings instead of the usual single pair. In normal flies, wings form only on the second thoracic segment; however, in flies with a mutation in the *ultrabithorax* gene, another pair of wings forms on the third thoracic segment. These mutations transformed the third segment into another second segment. Mutations in another one of these homeotic genes—*antennapedia*—cause the transformation of a leg into another antenna. Elimination of all of the Hox genes in the beetle *Tribolium* results in an animal in which all parts of the animal look nearly identical (Stuart et al., 1993) (**Figure 2.4**). Analysis of many different types of mutations in this complex have led to the conclusion that, in insects, the homeotic genes are necessary for a given part of the animal to become morphologically different from another part.

The Homeobox genes in *Drosophila* are arranged in two clusters, both in linear arrays on the chromosomes in the order of their expression along the anterior–posterior axis of the animal (**Figure 2.5**). A total of eight genes are organized on the chromosome in the two complexes: the Antennapaedia (ANT-C) and Bithorax (BX-C) clusters (Duboule and Morata, 1994; Gehring, 1993). The Homeobox genes code for proteins of the homeodomain class of transcription factors and were the original members of this very large set of related molecules. All of the Homeobox proteins have a sequence of approximately 60 highly conserved amino acids. Like other types of transcription factors, the Homeobox proteins bind to a consensus sequence of DNA in the promoters of many other genes (Gehring, 1993; Biggin and McGinnis, 1997). A change in the morphology of a particular segment is likely to require the coordinated activation and suppression of numerous genes, and it has been estimated that between 85 and 170 genes are likely regulated by the *ubx* gene (Gerhart and Kirschner, 1997). Thus, it is likely that the Homeobox genes interact with other transcription factors to enhance their DNA-binding specificity (Affolter et al., 2008).

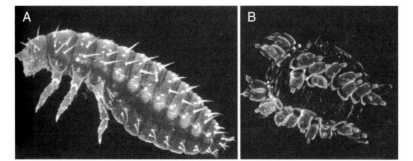

Fig. 2.4 Elimination of the *Hox* gene cluster in the *Tribolium* beetle results in all segments developing an identical morphology. A. shows the normal appearance of the beetle, and B. shows an animal without a *Hox* gene cluster. The normal number of segments develop, but all of the segments acquire the morphology of the antennal segment, showing the importance of the *Hox* genes in the development of positional identity in animals. *(Reproduced from Stuart et al., 1993, with permission)*

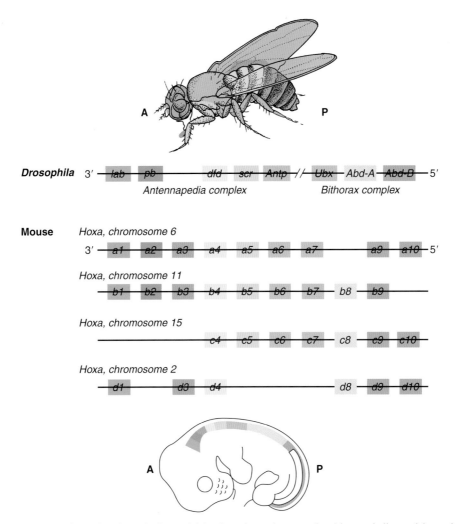

Fig. 2.5 Hox gene clusters in arthropods *(Drosophila)* and vertebrates (mouse embryo) have a similar spatial organization and similar order along the chromosomes. In *Drosophila*, the Hox gene cluster is aligned on the chromosome such that the anterior most expressed gene is 3′ and the posterior most gene is 5′. In the mouse, there are four separate Hox gene clusters on four different chromosomes, but the overall order is similar to that in arthropods: the anterior to posterior order of gene expression is ordered in a 3′ to 5′ order on the chromosomes.

Another striking feature of the Homeobox genes is their remarkable degree of conservation throughout the phyla. Homeobox clusters similar to those found in *Drosophila* have been identified in vertebrates, as well as nearly all the major classes of animals. Figure 2.5 shows the relationship between the *Drosophila* Hox genes and those of the mouse. Although there are many more Hox genes in the mouse than in *Drosophila*, in both flies and mice (and us) the position of a particular Hox gene on the chromosome is correlated with its expression along the anterior–posterior axis. The genes expressed by cells at the anterior end of the fly or mouse embryo (*labial* in the fly, or *hox a1, b1 or d1* in the mouse) are located at the most 3′ end of the cluster, whereas the genes expressed in the posterior end of the animal are located at the most 5′ end of the cluster (the *bithorax* genes in flies and the *hox a13,b13, c13 or d13* genes of the mouse). In mice and other vertebrates, *hox* genes that are in the same relative positions on each of the four chromosomes, and similar to one another in sequence, form paralogous groups. For example, *hoxa4, hoxb4, hoxc4,* and *hoxd4* make up the number 4 paralogous group. The striking organization of the genes in Hox clusters that reflects their expression has led to the idea that the

organization of these genes on the chromosome is part of the mechanism of anterior–posterior axis specification; however, not all animals have such well-organized Hox clusters and yet have perfectly good anterior–posterior axes (Duboule, 2007).

HOX GENE FUNCTION IN THE VERTEBRATE NERVOUS SYSTEM

The function of the Hox genes in controlling the regional identity of the vertebrate nervous system has been most clearly investigated in the hindbrain. The vertebrate hindbrain provides the innervation for the muscles of the head through a set of cranial nerves. Like the spinal nerves that innervate the rest of the body, some of the cranial nerves contain axons from motor neurons located in the hindbrain, as well as sensory axons from neurons in the dorsal root ganglia. However, we will primarily be concerned with the motor neurons for the time being. The cranial nerves of an embryo are shown in **Figure 2.6**. As noted above, during embryonic development, the hindbrain undergoes a pattern of "segment formation" that bears some resemblance to that which occurs in the fly embryo. In the developing

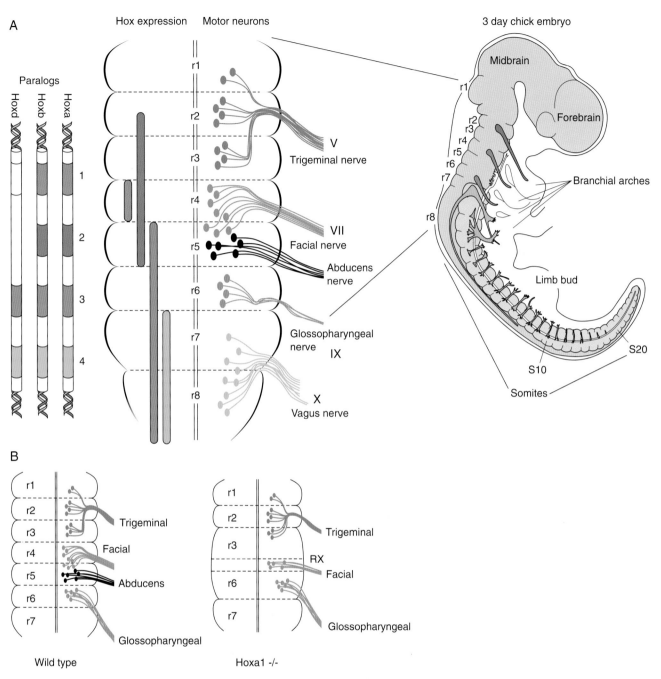

Fig. 2.6 Rhombomeres are repeated morphological subdivisions of the hindbrain. A. The rhombomeres are numbered from the anterior-most unit, *r1*, just posterior to the midbrain (mesencephalon), to the posterior most unit, *r7*, at the junction of the hindbrain with the spinal cord. The members of the *Hox* gene cluster are expressed in a 3′ to 5′ order in the rhombomeres. The segmentation in this region of the embryo is also observed in the cranial nerves, and the motoneurons send their axons through defined points at alternating rhombomeres. B. Rhombomere identity is determined by the *Hox* code. *Hox* gene knockouts in mice affect the development of specific rhombomeres. Wild-type animals have a stereotypic pattern of motoneurons in the hindbrain. The trigeminal (V) cranial nerve motoneurons are generated from *r2* and *r3*, while the facial nerve motoneurons are produced in *r4* and the abducens motorneurons are produced by *r5*. Deletion of the *Hoxa1* gene in mice causes the complete loss of rhombomere 5 and a reduction of rhombomere 4 (*rx*). The abducens motoneurons are lost in the knockout animals, and the number of the facial motoneurons is reduced.

hindbrain, the segments are called rhombomeres (Figure 2.6). The rhombomeres give rise to a segmentally repeated pattern of differentiation of neurons, some of which interconnect with one another within the hindbrain (the reticular neurons) and some of which project axons into the cranial nerves (Lumsden and Keynes, 1989). Each rhombomere gives rise to a unique set of motor neurons that control different muscles in the head. For

example, progenitor cells in rhombomeres 2 and 3 make the trigeminal motor neurons that innervate the jaw, while progenitor cells in rhombomeres 4 and 5 produce the motor neurons that control the muscles of facial expression (cranial nerve VII) and the neurons that control eye muscles (abducens nerve, VI). Rhombomeres 6 and 7 make the neurons of the glossopharyngeal nerve, which controls swallowing. Without differences in these segments, we would not have differential control of smiling, chewing, swallowing, or looking down. Clearly, correct rhombomere patterning is important.

How do these segments become different from one another? The pattern of expression of the paralogous groups of Hox genes coincides with the rhombomere boundaries (Figure 2.6), and the expression of these genes precedes the formation of obvious morphological rhombomeric boundaries. Members of paralogous groups 1–4 are expressed in the rhombomeres in a nested, partly overlapping pattern. Group 4 genes are expressed up to the anterior boundary of the seventh rhombomere, group 3 genes are expressed up to and including rhombomere 5, while group 2 genes are expressed in rhombomeres 2–5. These patterns are comparable in all vertebrates. As discussed below, loss of a single Hox gene in mice usually does not produce the sort of dramatic phenotypes seen in *Drosophila*. This is because of overlapping patterns of Hox gene expression from the members of the four paralogous groups. When two or more members of a paralogous group are deleted, say *Hoxa4* and *Hoxb4*, then the severity of the deficits increases. The deficits that are observed are consistent with the Hox genes acting much as they do in arthropods. That is, they control the relative identity of their region of expression.

Studies of Hox genes in neural development have concentrated on the hindbrain. Several studies have either deleted specific Hox genes or misexpressed them in other regions of the CNS and examined the effects on hindbrain development. Only a few examples will be given. Elimination of the *hoxa1* gene from mice results in animals with defects in the development of rhombomeres and the neurons they produce (Carpenter et al., 1993; Mark et al.,

1993). Specifically, the rhombomere 4 domain is dramatically reduced and does not form a clear boundary with rhombomere 3. Rhombomere 5 is completely lost, or fused with rhombomere 4, into a new region called "rx." The abducens motoneurons fail to develop in these animals, and the facial motor neurons are also defective. However, some of the neurons derived from this region of the hindbrain now begin to resemble the trigeminal motor neurons (Figure 2.6). Thus, when *hoxa1* is lost from the hindbrain, rhombomeres 4 and 5 are partly transformed to a rhombomere 2/3 identity. Thus, at least at some level, the Hox genes of mice confer regional anterior–posterior identity on a region of the nervous system in the same way as the homeotic genes of *Drosophila*.

Earlier in this section, we showed a picture of an arthropod that had no Hox genes; all segments were essentially identical. What would the hindbrain look like without Hox genes? Studies in both *Drosophila* and vertebrates have found that the specificity of the Hox genes for promoters on their downstream targets is significantly enhanced through their interactions with the *pbx* and *meis* homeodomain proteins. Moens and her colleagues (Waskiewisz et al., 2002) have taken advantage of this interaction to ask what the hindbrain would look like without any functional Hox function. By eliminating the *pbx* genes from the hindbrain of the zebrafish with a combination of genetic mutation and antisense oligonucleotide gene inactivation, they have found that the "ground state" or default condition of the hindbrain is rhombomere number 1. Embryos lacking both *pbx* genes that are normally expressed in the hindbrain during rhombomere formation lose rhombomeres 2–6, and instead these segments are transformed into one long rhombomere 1 (**Figure 2.7**).

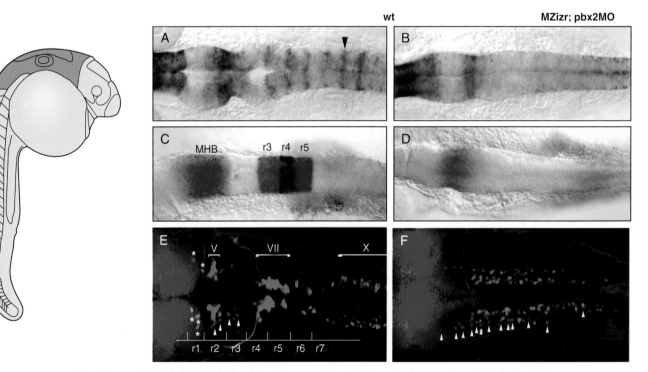

Fig. 2.7 What would the hindbrain look like without Hox genes? By eliminating the *pbx* genes from the hindbrain of the zebrafish with a combination of genetic mutation and antisense oligonucleotide gene inactivation, Moens and colleagues found that the "ground state" or default condition of the hindbrain is rhombomere number 1. To the left is a drawing of the fish for orientation, with the hindbrain highlighted in red. A, C, and E show the wild-type embryo, and panels B, D, and F show the mutant embryo hindbrain. In embryos lacking both *pbx* genes, all segments are transformed into one long rhombomere 1, and both the specific gene expression seen in rhombomeres 3, 4, and 5 (D) and the diversity of neurons that form in the hindbrain (E) are lost in the mutant. *(Modified from Waskiewicz et al., 2002)*

The remarkable conservation of Hox gene functioning in defining segmental identity in both *Drosophila* embryos and vertebrate hindbrain prompts the question whether similar mechanisms upstream of Hox gene expression are also conserved. As discussed above, a developmental cascade of genes—the gap genes, the pair-rule genes, and the segment polarity genes—parcel up the domains of the fly embryo into smaller and smaller regions, each of which has a unique Homeobox expression pattern. Does a similar mechanism act in the vertebrate brain to control the expression of the Hox genes? Although the final answers are not yet known, there are several key observations that indicate vertebrates may use somewhat different mechanisms to define the pattern of Hox expression.

One of the first signaling molecules to be implicated as a regulator of Hox expression was a derivative of vitamin A, retinoic acid (RA). This molecule is a powerful teratogen; that is, it causes birth defects. Retinoic acid is a common treatment for acne, and since its introduction in 1982, approximately one thousand malformed children have been born. The most significant defects involve craniofacial and brain abnormalities. The way in which RA works is as follows: RA crosses the cell membrane to bind a cytoplasmic receptor. The complex of RA and the retinoic acid receptor (RAR) moves into the nucleus, where it can regulate gene expression through interaction with a specific sequence in the promoters of target genes (the retinoic acid response element, or RARE). In the normal embryo, there is a gradient of RA concentration, with RA levels about 10 times higher in the posterior region of *Xenopus* embryos. When *Xenopus* embryos are treated with RA, they typically show defects in the anterior parts of the nervous system. When embryos are exposed to increasing concentrations of RA, they fail to develop head structures (**Figure 2.8 A, B**), and the expression of anterior Hox genes is inhibited (Durston et al., 1989).

Do the teratogenic effects of RA have anything to do with the control of regional identity in the CNS? In fact, it has been known for some time that retinoic acid can induce the expression of Hox genes when added to embryonic stem (ES) cells. With low concentrations of RA added to the ES cells, only those Hox genes normally expressed in the anterior embryo are expressed, while at progressively higher concentrations of RA, more posteriorly expressed Hox genes are expressed in the cells (Simeone et al., 1991). Targeted deletion of the RARs produces defects similar to those observed from pharmacological manipulation of this pathway (Chambon, 1995). Finally, both the *hoxa1* and the *hoxb1* promoters have RAREs, and these elements are both necessary and sufficient for the rhombomere-specific pattern of expression of these genes. These facts all point to the importance of RA signaling in hindbrain development, but where does the gradient of RA come from in normal embryos? Early models of gradient formation invoked a highly expressing source of the signal and a declining gradient from the source, possibly "sharpened" by an active degradation mechanism. Evidence from several labs now indicates that the source of the RA is the mesoderm that lies immediately adjacent to the neural tube. The paraxial mesoderm contains enzymes that synthesize the RA (**Figure 2.8 C,D**), and this then diffuses into the hindbrain neural tube to activate the correct pattern of Hox gene expression. The fact that the nonneural tissue outside the developing nervous system can have such a critical impact in its formation

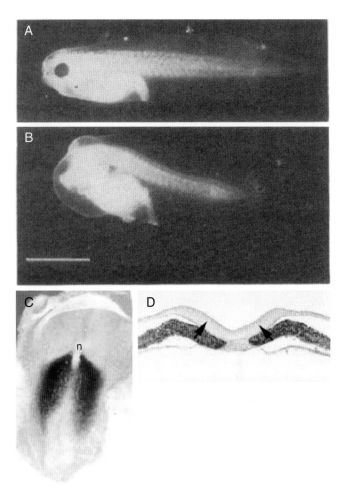

Fig. 2.8 Retinoic acid signaling is important for the anterior–posterior pattern of *Hox* gene expression. A, B. RA-treated embryos typically show defects in the anterior parts of the nervous system (B). When embryos are exposed to increasing concentrations of RA, they fail to develop head structures and the expression of anterior genes is inhibited. C, D. The mesodermal cells surrounding the neural tube pattern the neural tissue through RA. The cells that make the RA are marked in purple in C in a top-down view, and in a cross-sectional view in D. The neural tube is marked with an "n" in C and arrows in D. *(Modified from Maden, 2006)*

reminds us that the nervous system does not develop in a vacuum, but rather many important aspects of its development rely on interactions with adjacent nonneural tissues.

Overall, the similarity of body segmentation in *Drosophila* and hindbrain rhombomere development in vertebrates has led to a rapid understanding of both processes. However, the development of other regions of the vertebrate nervous system does not rely so heavily on the same mechanisms. Instead, other types of transcription factors control the development of the more anterior regions of the brain. In the next sections we will review how divisions in these other brain regions arise.

SIGNALING MOLECULES THAT PATTERN THE ANTERIOR–POSTERIOR AXIS IN VERTEBRATES: HEADS OR TAILS

The overall organization of the anterior–posterior axis of the nervous system in vertebrates is coupled with earlier events in axis determination and neural induction. As noted in Chapter 1, evidence from Spemann and others demonstrated that there may be separate "head" and "tail" organizers. This fact suggests that the very early inductive signals for neural development also influence the A-P axis. In a now classic experiment, Nieuwkoop (see Chapter 1) transplanted small pieces of ectodermal tissue from one embryo into a host at various positions along the anterior–posterior axis. In all cases, the transplanted cells developed anterior neural structures. However, when the cells were transplanted in the caudal neural plate, posterior structures, such as spinal cord, also developed. Therefore, he concluded that the initial signal provided by the organizer is to cause ectodermal cells to develop anterior characteristics, known as the "activator," while a second signal is required to transform a portion of this neural tissue into hindbrain and spinal cord, known as the "transformer."

Nieuwkoop's activator-transformer hypothesis is now widely accepted and supported by more recent studies. First, the neural inducers that have been identified (e.g., Noggin, Chordin, Follistatin) produce primarily anterior brain structures when added to frog animal caps (see Chapter 1). Second, as described in Chapter 1, targeted deletion of putative neural inducers, such as the *noggin/chordin* double knockout mouse, results in headless mice, but these animals usually still have posterior neural tissue. At the present time, three molecular pathways have been implicated as contributing to the "transformer" activity: retinoic acid, Wnts and FGFs. As described above, retinoic acid treatment can posteriorize embryos and is almost certainly responsible for the patterning of the hindbrain Hox gene expression. In the next section, the evidence for the other transformers is discussed. Other groups have found that there is an endogenous AP gradient of Wnt/beta-catenin activity in the embryo, with the highest levels in the posterior of the embryo (Kiecker and Niehrs, 2001).

The identification of Wnt signaling as a "transformer" came initially from the fact that manipulations in this pathway during development had profound effects on head development. Factors that inhibit the Wnt pathway were found to synergize with factors that inhibit the BMP pathway in promoting the development of head and brain structures in early frog embryos. In addition, several inhibitors of the Wnt pathway are expressed in the organizer region. One of the first factors specifically implicated in head induction was called *cerberus*, after the three-headed dog that guards the gates of Hades in Greek mythology. Injection of *cerberus* into *Xenopus* embryos causes ectopic head formation, without the formation of trunk neural tissue (Bouwmeester et al., 1996). A second Wnt inhibitor, known as FrzB, is a member of a family of proteins that are similar to the receptors for the Wnt proteins, known as Frizzleds. The FrzB proteins work by binding to the Wnt proteins and preventing them from binding to their signaling receptor. Injection of extra FrzB into *Xenopus* embryos also causes them to form heads larger than normal. A third Wnt inhibitor was isolated by a functional screen similar to that described for Noggin (Chapter 1); in this case, Niehrs and colleagues injected the truncated BMP receptor (tBMPR) along with pools from a cDNA library and looked for genes that would cause complete secondary axes, including heads, only when co-injected with the tBMPR. They identified a gene that was particularly effective in inducing head structures, *dickkopf,* for the German word meaning big-head or stubborn (**Figure 2.9**; Glinka et al., 1998). These three Wnt inhibitors are reminiscent of the BMP inhibitors described in Chapter 1

in that they are expressed in the organizer region during the time when the inductive interactions are taking place, and they all have head-inducing activity, particularly when combined with a BMP inhibitor (Figure 2.9). Other groups have found that there is an endogenous AP gradient of Wnt/beta-catenin activity in the embryo, with the highest levels in the posterior of the embryo (Kiecker and Niehrs, 2001).

The evidence that there are indeed several putative Wnt inhibitors in the organizer is good support for the model that a co-inhibition of Wnt and BMP signals leads to induction of the anterior neural structures, that is, the brain. In fact, the Cerberus protein can inhibit both the Wnt and BMP pathways. Additional support for the model was obtained from studies of mice in which the mouse homolog to *dickkopf, dkk1,* was deleted via homologous recombination. The mice lacking *dkk1* alone are similar to the double *Noggin/Chordin* knockout mice described above: they lack head and brain structures anterior to the hindbrain (Mukhopadhyay et al., 2001; **Figure 2.10**). Synergy between the BMP antagonist, Noggin, and the Wnt antagonist, Dkk1, can be seen by producing mice with a single allele of each of these genes. Although the loss of a single allele of either of these genes has no discernible effect on mice, the loss of a single allele of both of these genes causes severe head and brain defects, similar to those animals that have lost both alleles of the *dkk1* gene. Similarly, in the zebrafish mutant *masterblind*—where there is a loss of axin, a component of Wnt inhibitory signaling pathway—no forebrain develops. Taken together, these studies show that the Wnt and BMP antagonists work together to bring about the correct induction and pattern of the nervous system.

The third class of molecules that has been proposed as a "transformer" is FGF. FGFs are able to act as neural inducers in their own right (Chapter 1), but in addition are able to induce posterior gene expression in animal caps that have undergone experimental neural induction using a BMP antagonist (Slack and Tannahill, 1992). Moreover, FGF acts in a concentration-dependent manner, much like retinoic acid: In amphibian embryos, high concentrations of FGF induce more posterior neural genes (Kengaku and Okamoto, 1995) and in chick embryos, treatment with increasing concentrations of FGF induces more posterior members of the Hox cluster to be expressed (Liu et al., 2001). The specific FGF necessary for the endogenous transforming activity may vary depending on the species, but in chick embryos, Fgf8 is prominently expressed at the posterior of the developing neural tube. The three "transformer" activities are thought to be integrated during posterior neural development as shown in **Figure 2.11** (Maden, 2006).

In Chapter 1, we introduced the concept that signaling molecules act to create differences among cells by changing the pattern of gene expression via specific transcription factors. The signals that "anteriorize" or "posteriorize" the embryo are no exception to this rule and act to regulate the expression of transcription factors that control the identity of cells along this axis. We have already seen how the Hox genes are critical in the development of rhombomere identity in the hindbrain; however, an even more fundamental division of the CNS is mediated by two other homeodomain transcription factors—*otx2* and *gbx2* (Joyner et al., 2000). At late gastrula/early neural plate stages in the embryo, one can already see these genes expressed in domains adjacent to one another: *gbx2*-expressing cells extend from the posterior end of the embryo to the

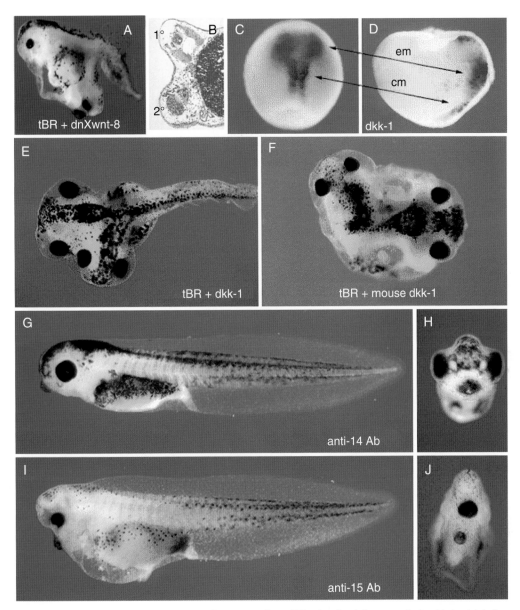

Fig. 2.9 Heads vs. tails: the role of Wnt signaling. Antagonism of Wnt signaling is important for head induction in frog embryos. A,B. Injection of four-cell embryo with both the truncated BMP receptor (tBR) and a dominant-negative form of Wnt8 *(dnXwnt8)* causes frog tadpoles to develop a second head. B shows a section through such a tadpole revealing both the primary and secondary brains. C,D. Expression of *dkk-1* in late gastrulae (stage 12) *Xenopus* embryos. *In situ* hybridization of embryo whole-mount (C) and section (D). Embryos are shown with animal side up, blastopore down. Arrows point to corresponding domains in C and D. The endomesoderm (em) is stained in a wing-shaped pattern, and most posterior expression is in two longitudinal stripes adjacent to the chordamesoderm (cm). E,F. Injection of either *Xenopus* or mouse *dkk-1* into the blastomeres of a four-cell-stage frog embryo causes an extra head to develop as long as the truncated BMP (tBR) receptor is co-injected. G–J. *Dkk-1* is required for head formation. Stage 9 embryos were injected with antibody (Ab) into the blastocoel and allowed to develop for three days. G,H. Embryos injected with a control (anti-14) antibody show no abnormalities. An anterior view is shown on the right. I,J. Embryos injected with anti-*dkk1* (anti-15) antibody show microcephaly and cyclopia. An anterior view is shown on the right. Note that trunk and tail are unaffected *(Modified from Glinka et al., 1997; Glinka et al., 1998)*

midbrain/hindbrain border, while *otx2* has the complementary pattern of expression, from the midbrain/hindbrain border to the anterior-most part of the brain (Hidalgo-Sánchez et al., 2005; Figure 2.11). Direct evidence that shows these genes are critical for this fundamental division of the CNS into anterior and posterior compartments come from mouse gene targeting experiments. Deletion of the *otx2* gene in mice results in animals without a brain anterior to rhombomere 3 (**Figure 2.12**;

Matsuo et al., 1995; Acampora et al., 1995). *otx2* is thus one of the earliest genes to be expressed in the developing forebrain and is both necessary and sufficient to activate genes that specify most of the brain. In mice without the *gbx2* gene, the converse result is observed: the mice lack the hindbrain region (Millet et al., 1999; Wassarman et al., 1997).

Another important gene family, the Iroquois genes, is also involved in the anterior–posterior axis specification in the central

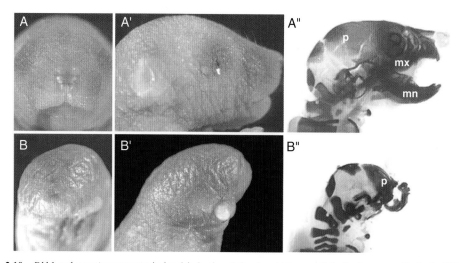

Fig. 2.10 *Dkk1* and *noggin* cooperate in head induction. Mice in which one allele for the genes for both *dkk1* and *nog* have been deleted have severe head defects. Frontal (*A*,*B*) and lateral (*A′*,*B′*) views of wild-type (*A*,*A′*) and mutant (*B*,*B′*), newborn animals. Lateral view of skeletal preparations from wild-type (*A″*) and severe mutant (*B″*) newborn heads reveal loss of maxillar (mx), mandibular (mn), and other bones anterior to the parietal bone (p). *(Modified from del Barco Barrantes et al., 2003)*

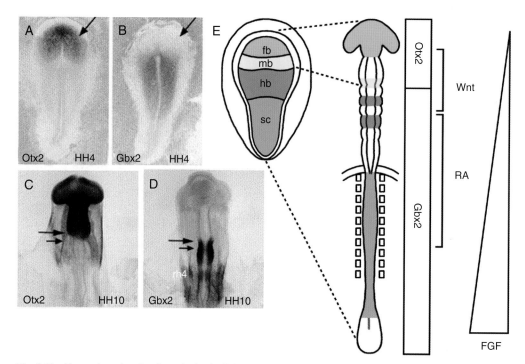

Fig. 2.11 Expression of *otx2* reflects the basic division between the rostral brain and the hindbrain and spinal cord. *Otx2* expression at various stages of embryonic development in the chick brain. *Otx2* is expressed in the anterior neural plate (A) and remains expressed in most of the brain throughout development (B–E). The arrowhead points to the midbrain–hindbrain boundary. E. The relationship between the three transformer signals and the expression of *otx2* and *gbx2*. *(From Millet et al., 1996)*

nervous system. Most vertebrates have six Iroquois genes, grouped as two paralog clusters of three genes (de la Calle-Mustienes et al., 2005), and these genes are expressed along the anterior-posterior axis of the nervous system (Rodriguez-Seguel et al., 2009). One of these genes, *irx1* (also called *xiro*), activates both *otx2* and *gbx2*, which then cross-repress one another to create a sharp border between them (Glavic et al., 2002). This type of cross repression of transcription factors is a widely used mechanism for the generation of distinct boundaries between expression domains in the embryo. As we shall see in the next section, the midbrain/hindbrain boundary becomes an important organizing center in its own right.

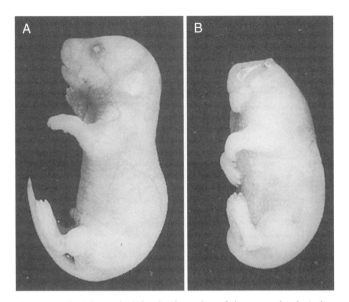

Fig. 2.12 *Otx2* is required for the formation of the mouse head. A dramatic illustration of the importance of the *otx2* gene is the development of the mouse forebrain and rostral head. If the gene is deleted using homologous recombination, embryos without either allele of the gene fail to develop brain regions rostral to rhombomere 3, a condition known as anencephaly. Since many of the bones and muscles of the head are derived from the neural crest, which also fails to form in these animals, the animals lack most of the head in addition to the loss of the brain. *(From Matsuo et al., 1995)*

ORGANIZING CENTERS IN THE DEVELOPING BRAIN

The division between the metencephalon and the mesencephalon is a fundamental one for the central nervous system. This boundary is a major neuroanatomical division of the mature brain as well; the metencephalon gives rise to the cerebellum, and the mesencephalon gives rise to the midbrain (superior and inferior colliculi) (**Figure 2.13**). But in addition to the important neural structures derived from this region, the midbrain/hindbrain border (or mesencephalon/met-encephalon border) has a special developmental function. Like the Spemann "organizer" of the gastrulating embryo, the midbrain/hindbrain border expresses signaling molecules that have an important organizing influence on the development of the adjacent regions of the neuroepithelium.

The idea that specific regions of the neural tube act as organizing centers for patterning adjacent regions was first put on a firm molecular basis through studies of the midbrain/hindbrain border. In a series of experiments designed to test the state of commitment of this part of the neural tube, Alvarado-Mallart and colleagues transplanted small pieces of the neuroepithelium from the midbrain/hindbrain border of chick embryos to similarly staged quail embryos (Alvarado-Mallart, 1993). Grafting between these two species allows the investigator to follow the fate of the transplanted cells. Although the chick and quail cells behave similarly and integrate well together in the tissues, molecular and histological markers can be used to tell them apart after histological processing (Figure 2.13). When the presumptive metencephalon region was transplanted from a quail to the metencephalon of a chick embryo, the transplanted cells developed as cerebellum (Figure 2.13C,D). When cells from the mesencephalon were transplanted to

a corresponding region of the chick embryo, the cells developed into midbrain structures, like the optic tectum (or superior colliculus). However, when cells from the metencephalon were transplanted to the forebrain, not only did cerebellum still develop from the transplants but, surprisingly, the transplanted tissue "induced" a new mesencephalon in the forebrain. In other words, the small piece of hindbrain neural tube was able to repattern the more anterior regions of the neural tube to adopt more posterior identities. This experiment is reminiscent of the organizer transplant of Spemann, in that a small region of specialized tissue is able to repattern the surrounding neuroepithelium when transplanted (see Chapter 1).

Several important signaling molecules have been localized to this region and are now known to play a key role in these patterning activities, including *wnt1, engrailed (en1),* and *fgf8.* A member of the Wnt gene family, *wnt1,* is expressed in this region (**Figure 2.14**), and when this gene is deleted in mice, the animals lose most of the midbrain and cerebellum (McMahon and Bradley, 1990). One of the earliest observed defects in these animals is the loss of expression of a transcription factor, *en1,* which is normally expressed in the region of the mesencephalon-metencephalon (mes-met) boundary. The expression of *en1* in this region has also been shown to be critical for normal development of midbrain and hindbrain structures. Mice homozygous for a targeted deletion in the *en1* gene are missing most of the cerebellum and the midbrain similar to the *wnt1*-deficient mice (Wurst et al., 1994). The homologs to *en1* and *wnt1* were first identified in *Drosophila* segmentation mutants; when either of these genes is defective in flies, the animals have defects in segmentation. Moreover, in *Drosophila* the homologous gene for *wnt1, wingless,* is required for maintaining the expression of the *Drosophila engrailed* gene at the segment boundaries. Thus, the midbrain–hindbrain boundary is another example where the same basic mechanisms as those used in segmentation in *Drosophila* create differences and boundaries in the brain. In addition to the *Wnt* and *Engrailed* patterning system, the midbrain–hindbrain junction also expresses another key signaling molecule, Fgf8, a receptor tyrosine kinase ligand (Figure 2.14). *Fgf8* is necessary for both setting up this boundary and maintaining it, since mice deficient in *fgf8* show defects in cerebellar and midbrain development similar to the *wnt1* and *en1* knockout animals (e.g., Meyers et al., 1998). *Fgf8, en1,* and *wnt1* are in an interconnected network; deleting any one of them affects the expression of the other two. Fgf8's role in patterning the tissue around the mes-met boundary was demonstrated in a remarkable experiment; Crossley et al. (1996) placed a bead coated with Fgf8 protein onto a more anterior region of the neural tube and found that this molecule was sufficient to induce the repatterning of these anterior tissues into midbrain and hindbrain structures (Figure 2.13C). Thus, the Fgf8 produced by midbrain/hindbrain acts like an "organizer" for the midbrain and hindbrain.

The current model of how the midbrain–hindbrain signaling center arises is as follows (**Figure 2.15**). Irx1 activates both Otx2 and Gbx2 in this region of the developing CNS. Gbx2 and Otx2 cross inhibit one another, and it is at this point of cross-inhibition that Fgf8 is expressed (Glavic et al., 2002). The interaction between Otx2 and Gbx2 maintains Fgf8 expression, and Fgf8 induces En1 in those cells that express both Irx1 and Otx2. Through these cross-regulatory loops between cells, the border is initially set up and maintained through development (Rhinn and Brand, 2001). The Fgf8

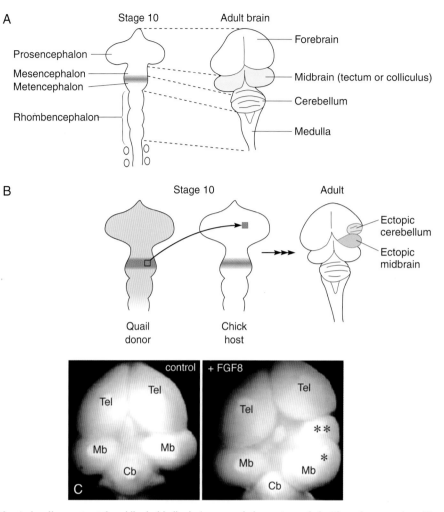

Fig. 2.13 A signaling center at the midbrain-hindbrain (mesencephalon-metencephalon) boundary organizes this region of the brain. A. During normal development, the region of the midbrain-hindbrain junction expresses the homeodomain transcription factor *engrailed* (red), and this region of the neural tube contains the progenitors of the midbrain (tectum) and the cerebellum. B. To determine whether these parts of the neural tube were restricted in their potential at this time in development, Alvarado-Mallart et al., 1990 transplanted a small piece of the quail metencephalon (red) to the forebrain of a similarly staged chick embryo. Cerebellum still developed from the metencephalon transplants, but in addition, the transplanted tissue had induced a new mesencephalon to develop from the adjacent forebrain neural tube cells. C. Fgf8 is a critical signal for the "organizer" activity of the mes-met boundary tissue. Crossley et al. placed a bead (shown in arrow) of Fgf8 onto the telencephalon of the chick embryo and found that this caused a new mes-met boundary to form with a mirror duplicated midbrain, similar to the transplant experiment of Alvarado-Mallart. *(Modified from Martinez et al., 1999)*

produced by this region then goes on to regulate growth of the progenitor cells in this region to ultimately produce the brain structures of the midbrain and hindbrain, including the cerebellum and the superior colliculus.

The unique signaling characteristics of the midbrain–hindbrain boundary suggest that such localized organizing centers may be a basic mechanism of brain patterning. Other key organizing regions exist between the dorsal and ventral thalamus and at the anterior pole of the neural tube. Moreover, as development proceeds and the brain expands, new organizers and signaling centers pattern the newly expanded regions. It may be that the appearance of new signaling centers coincides with the expansion of the neuroepithelium past the distance over which these molecules can signal. Once the number of cells exceeds the range over which the signal can act, a new signaling source arises. A second important signaling center occurs at the anterior pole of the brain, the anterior neural

ridge, where Fgf8 is again important in inducing forebrain regionalization (Shimamura and Rubenstein, 1997). This later role in the regionalization of the cerebral cortex will be further described later in this chapter.

FOREBRAIN DEVELOPMENT, PROSOMERES, AND *PAX* GENES

To this point, we have explored how Hox genes control the specification of anterior–posterior position in the nervous system. However, Hox gene expression stops at the anterior boundary of the metencephalon. Are there similar transcription factors that control positional identity in the rest of the brain? Many other types of homeodomain proteins are expressed in these more anterior regions of both vertebrate and invertebrate embryos, and they perform a role similar to

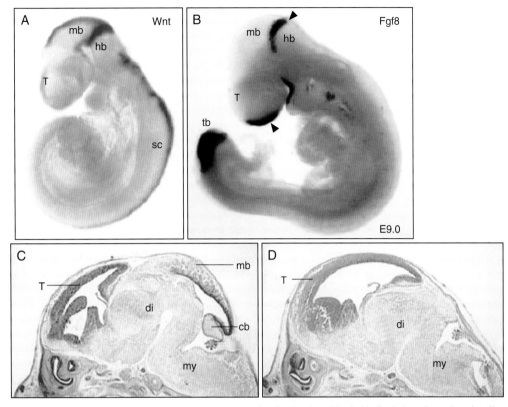

Fig. 2.14 Several important signaling molecules have been localized to the midbrain-hindbrain boundary, a key signaling center in the brain. *Wnt1* (A) *engrailed-1* and *fgf8* (B) form an interconnected network that specifies this boundary and is necessary for the growth of the midbrain and the cerebellum. Deletion of any of these molecules in mice results in a loss of the midbrain and reduction in cerebellar size. A section through the brain of a wild-type embryo is shown in C, while a *wnt1* knockout mouse brain is shown in D. Note the loss of the midbrain (mb) and cerebellum (cb) in the mutant brain. Other structures are normal (ch = cerebral cortex; cp = choroids plexus; di = diencephalon; my = myelencephalon). *(A, C, D modified from Danielian et al., 1997; B modified from Crossley and Martin, 1995)*

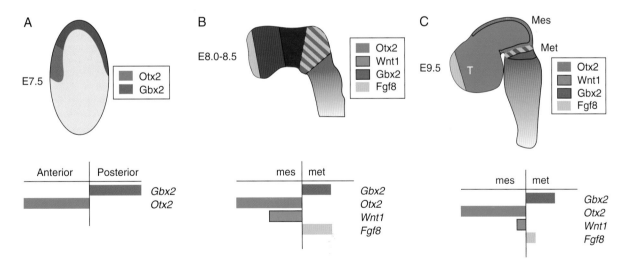

Fig. 2.15 The model of how the midbrain–hindbrain signaling center arises. A. The initial distinction between the anterior and posterior of the embryonic nervous system is reflected in expression of *otx2* and *gbx2*. B. At the boundary between these two factors, the mes-met boundary forms, and *wnt1, en1,* and *fgf8* are all expressed in this region and act in a regulatory network to maintain their expression and (C) refine this boundary. *(Modified from Joyner et al., 2000)*

that of Hox gene clusters in more caudal segments. Below, we explore the evidence that homeodomain proteins specify the structures that comprise the head and brain.

The most widely held view is that different parts of the brain are generated through the progressive subdivision of initially similar domains. The neural plate begins to show regional differences in the anterior–posterior direction at its formation. Embryologists at the beginning of the last century applied small amounts of dyes to specific parts of developing embryos and found that particular regions of the neural plate are already constrained to produce a particular part of the nervous system. Many embryologists have also used transplantation between species to define the contributions to the mature brain of particular regions of the neural tube. One particularly useful interspecific transplantation paradigm that was developed by Nicole Le Douarin (1982) is to transplant tissues between chick embryos and quail embryos, as described in the previous section. Since these species are similar enough at early stages of development, the transplanted cells integrate with the host and continue developing along with them (**Figure 2.16**). The chick and quail cells can be later distinguished since the quail cells contain a more prominent nucleolus, which can be identified following histological sectioning and processing of the chimeric tissue. Antibodies specific for quail cells have also been generated, and these are also useful for identifying the transplanted cells. The combination of vital dyes, cell injections, and chick-quail transplant studies have produced a description or "fate map" of the ultimate fates of the various cells of the embryo. **Figure 2.17** shows the fate maps for amphibian (Eagleson and Harris, 1990), avian (Couly and Le Douarin, 1987), and mammalian (Inoue et al., 2000) neural tubes, for the basic forebrain regions that have been derived from these fate-mapping studies. The basic pattern has been elaborated upon to generate the wide diversity of brains that are found in vertebrates.

Although fate-mapping studies provide information about the fate of the different neural tube regions, embryologists have also investigated whether the fate of the cells is fixed or can be changed. The goal of these experiments, in general, is to provide a timetable for understanding the moments in development when molecular mechanisms are actively directing a specific region of the neural tube to its specific fate (i.e., is "specified"). In order to determine at what point in development this "specification" occurs, pieces of the neural plate are transplanted to ectopic locations in the embryo. If the transplanted cells give rise to a particular brain region, we say that it has already been specified. For example, a piece of the anterior neural plate, near the eye, is transplanted to the presumptive flank of another embryo. After sufficient developmental time

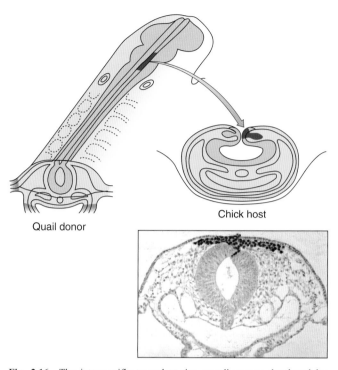

Fig. 2.16 The interspecific transplantation paradigm was developed by Nicole Le Douarin using chick embryos and quail embryos. Tissue is dissected from quail embryos and then placed into specific regions of live chick embryos. In this case, the dorsal ridge of the neural tube, the region that will give rise to neural crest, is transplanted to a similar region in the chick. The chick and quail are similar enough to allow the quail to contribute to the chick embryo, and the quail cells can be specifically identified with an antibody raised against quail cells (bottom). *(Modified from Le Douarin et al., 2004)*

has passed, the embryos are analyzed for the type of neural tissue that developed from the graft. In this case the finding is that, as early as late gastrula, a particular region of the neural plate will always give rise to the anterior brain, including the eye. This occurs regardless of where the tissue is placed in the host animal. A number of embryologists carried out these

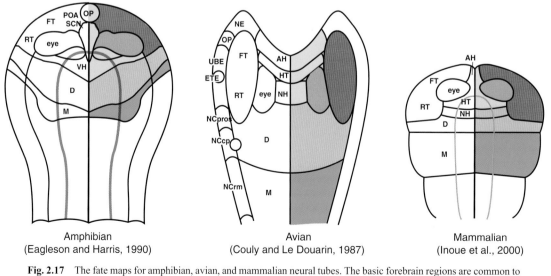

Amphibian
(Eagleson and Harris, 1990)

Avian
(Couly and Le Douarin, 1987)

Mammalian
(Inoue et al., 2000)

Fig. 2.17 The fate maps for amphibian, avian, and mammalian neural tubes. The basic forebrain regions are common to all vertebrates; however, the basic pattern has been elaborated upon to generate the wide diversity of brains that are found in vertebrates. The rhombomeric and prosomeric organization of the mouse brain can already be recognized at this early stage by the pattern of expression of certain genes.

types of experiments using various regions of the neural plate as the donor tissue, and the results consistently demonstrate that at some point in development, the cells of the neural plate take on a regional identity that cannot be changed by transplantation to some other place in the embryo. The fact that different regions of the neural plate are already committed to a particular fate has been extended in recent years by the observations that a number of genes are expressed in highly specific regions of the developing nervous system. In many cases the domain of expression of a particular transcription factor corresponds to that region of the neural tube that will ultimately give rise to one of the five brain vesicles, and the gene may continue to be expressed in that brain region throughout its development.

Embryologists have taken advantage of the patterns in gene expression in the forebrain to gain insight into the basis of its organization. In what has become known as the prosomeric model of forebrain development, there are longitudinal and transverse patterns of gene expression that subdivide the neural tube into a grid of different regional identities (Puelles and Rubenstein, 1993), somewhat analogous to the rhombomeres described above. The prosomere model is shown for the mouse embryo at two different stages of development (**Figure 2.18**). In many cases, the boundary of expression of a particular gene corresponds closely to the morphological distinctions between the prosomeres. For example, two genes of the Emx class are expressed in the telencephalon, one in the anterior half of the cerebral hemispheres (*emx1*) and the other in the posterior half of the hemispheres (*emx2*). Thus, the telencephalic lobes can be divided into anterior and posterior segments on the basis of the pattern of expression of these two genes. Analysis of the expression patterns of additional genes has led to the conclusion that the prosencephalon can be subdivided into prosomeres (Figure 2.18). They are numbered from caudal to rostral, and so prosomere (P1) is adjacent to the mesencephalon. P1, P2, and P3 subdivide caudal regions of the diencephalon (pretectum, thalamus, and prethalamus).

Prosencephalic subdivisions within the rostral diencephalon (hypothalamus) and the more dorsal telencephalon are more complex, but nonetheless exhibit distinct boundaries (see Puelles and Rubenstein, 2003).

While the studies of regional expression of transcription factors present a model of brain organization and evolution, the functional analyses of these factors have yielded remarkable evidence that these molecules are critically involved in defining the regional identity of the anterior brain. There are now many examples of regionally expressed transcription factors that have essential roles in brain development, but only a few will be mentioned. However, one principle that emerges is that several different classes of transcription factors are likely to be important in specifying the positional identity of cells in any particular region of the brain.

A key class of transcription factors that are critical for specifying regional differences in the nervous system are the Pax genes. These genes have a homeodomain region, and they also have a second conserved domain known as the paired box (named for its sequence homology with the *Drosophila* segmentation gene, *paired*). There are nine different Pax genes, and all but two, *pax1* and *pax9*, are expressed in the developing nervous system (Chalepakis et al., 1993). Several of these genes are also disrupted in naturally occurring mouse mutations and human congenital syndromes, and the defects observed in these conditions generally correspond to the areas of gene expression. *pax2*, for example, is expressed in the developing optic stalk and the optic vesicle of the embryo, and mutations in *pax2* in mice and humans cause optic nerve abnormalities, known as colobomas.

Perhaps the most striking example of Pax gene regulation of regional differentiation in the nervous system comes from the studies of *pax6*. This gene is expressed early in the development of the eye, when this region of the neural plate is specified to give rise to retinal tissue. Humans with mutations in one allele of this gene exhibit abnormalities in eye development, causing

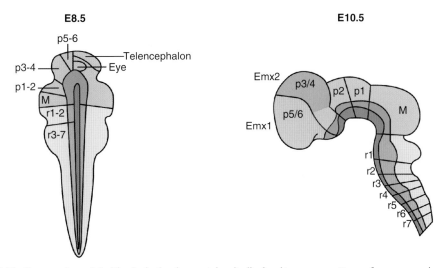

Fig. 2.18 Prosomeric model of forebrain development; longitudinal and transverse patterns of gene expression that subdivide the neural tube into a grid of regional identities shown at E8.5 and E10.5. Two genes of the *Emx* class are expressed in the telencephalon, one in the anterior half of the cerebral hemispheres *(Emx1)* and the other in the posterior half of the hemispheres *(Emx2)*. Analysis of the expression patterns of additional genes shows that the prosencephalon can be subdivided into six prosomeres. They are numbered from caudal to rostral, and so prosomere 1 is adjacent to the mesencephalon, *P2* and *P3* subdivide what is traditionally known as the diencephalon, and *P4*, *P5*, and *P6* subdivide the telencephalon. *(Modified from Puelles and Rubenstein, 2003)*

a condition known as aniridia (a lack of formation of the iris). However, when both alleles of *pax6* have mutations, the eyes fail to develop past the initial optic vesicle stage. A homologous gene, called *eyeless*, is found in *Drosophila* (as well as many other organisms), and loss-of-function mutations in this gene prevent eye formation in flies. Even more surprising, when the *eyeless* gene is experimentally expressed at inappropriate positions in the embryo, like the antennal imaginal disc, ectopic eyes are induced in the antenna (Halder et al., 1995; **Figure 2.19**). The ability of a single gene to direct the development of an entire sensory organ like the eye is striking, and when this was first discovered it was argued that Pax6 was at the top of a hierarchy, a master controller of eye development. However, it is now clear that Pax6 acts in a network of transcription factors, called the eye determination network. This includes Pax6/eyeless, Eya/eyes absent, Six3/sine oculis and possibly others. A similar network exists in vertebrates (**Figure 2.20**) and the onset of expression of these genes in the anterior region of the neural plate signals the determination of this region to the eye fate. Loss of any one gene in the eye determination network in vertebrates leads to defects in eye development, and in some cases, the eyes fail to develop entirely. Remarkably, experimental over-expression of some of these factors on their own can cause the formation of ectopic eyes in *Xenopus* frogs; over-expression "cocktails" of several of the factors together have much more potency in

inducing ectopic eyes (Zuber et al., 2003) (Figure 2.20). The eye determination network thus provides an excellent example of how regional patterning mechanisms in the nervous system have been highly conserved in evolution.

DORSAL–VENTRAL POLARITY IN THE NEURAL TUBE

The early neural tube looks fairly uniform, but we know that different types of neurons are located at highly specific positions in the mature brain and spinal cord. For example, in the spinal cord, motor neurons are collected together in the ventral horns, while sensory neurons are concentrated in the dorsal horns. How does this dorsal–ventral difference in spinal cord organization come about during development? At the neural plate stage, several mechanisms are set in motion that will define the overall organization of the neural tube. First, the most ventral part of the neural tube becomes flattened into a distinct "floorplate." Second, the most dorsal aspect of the neural tube develops into a tissue known as the roofplate. Third, a distinct fissure, the *sulcus limitans*, forms between the dorsal and ventral parts of the neural tube along most of its length (**Figure 2.21**). These structures are an early sign that the neural tube is differentiating along the dorsal–ventral axis. Later, the neural tube will become even more polarized along this axis; in the ventral part of the tube, motor neurons will begin to arise, while in the dorsal part, the sensory neurons form. The distinct polarity of the neural tube arises largely because of the interaction between the surrounding nonneural tissue and the neural tube. Experiments during the early part of the twentieth century by Holtfreter (1934) demonstrated that the basic dorsal–ventral polarity of the neural tube was dependent on an adjacent, nonneural structure, called the notochord. Isolation of the neural tube from the surrounding tissues resulted in an undifferentiated tube, without obvious motoneuronal differentiation in the ventral tube. However, when he transplanted a new notochord to a more dorsal location, this induced a second floorplate (**Figure 2.22**) and motoneuron differentiation in the dorsal neural tube. Thus, the notochord is both necessary and sufficient for the development of the dorsal–ventral axis of the spinal cord.

The studies that led to identification of the signals that control dorsal–ventral polarity in the developing spinal cord relied on the use of many molecular markers of cell identity that were obviously not around at the time Holtfreter was doing his experiments. These genes include the Pax class of transcription factors discussed in the previous section, as well as a variety of other genes that are restricted to particular populations of both differentiated and/or undifferentiated cells within the spinal cord. The expression of three of the critical genes that define particular domains along the dorsal–ventral axis of the spinal cord are shown in **Figure 2.23A**. Each of these genes, *pax7*, *olig2*, and *nkx2.2*, is expressed in a well-defined domain in the neural tube along the dorsal–ventral axis. To track down the polarity signal released by mesoderm, a cell culture system was devised in which the notochord and the neural tube were co-cultured in collagen gels (Figure 2.23B). A clue to the identity of the factor was uncovered in a rather roundabout manner. During a large screen for developmental mutants in *Drosophila* (Nusslein-Volhard and Wieschaus, 1980), a severely deformed mutant

Fig. 2.19 Ectopic eyes are formed when the *Drosophila Pax6* gene—*eyeless*—is misexpressed in other imaginal discs. Halder et al. (1995) misexpressed the *eyeless* gene in the leg disc in the developing fly and found that an ectopic eye was formed in the leg. This remarkable experiment argues for the concept that master control genes organize entire fields, or structures during embryogenesis, possibly by activating tissue-specific cascades of transcription factors. *(Courtesy of Walter Gehring, 1993)*

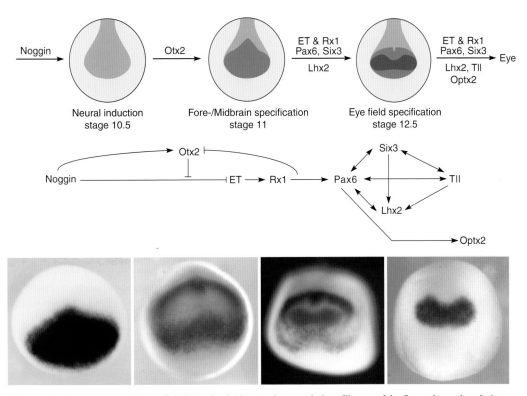

Fig. 2.20 Summary model of eye field induction in the anterior neural plate. The top of the figure shows dorsal views of the neural plate of *Xenopus* embryos at successively later stages of development from left to right. Light blue indicates the neural plate, blue shows the area of *otx2* expression, and dark blue represents the eye field. The diagram shows the complex relationships among the eye-determining transcription factors, including *Pax6, rx1, lhx2, six3, otx2,* and *Tll*. These genes act together to coordinate eye development in this specific region of the neural plate. The bottom panels show examples of *in situ* hybridizations for several eye transcription factors to show their specific patterns of expression in the presumptive eye-forming region of the embryo. *(Modified from Zuber et al., 2003)*

was found, named *hedgehog* for its unusual appearance. Subsequent cloning of the gene showed that this molecule was a secreted protein.

The link between hedgehog and the notochord-signaling molecule began with the identification of the mammalian homolog, called *sonic hedgehog* (*shh*). Jessell and colleagues found that *shh* is expressed in the notochord (Figure 2.22) at the time when the dorsal–ventral axis of the neural tube is being specified (Roelink et al., 1994) and soon after *shh* expression occurs in the floorplate as well. To determine whether Shh was indeed the inducer of dorso-ventral polarity in the spinal cord, a small aggregate of cells that were experimentally induced to express Shh was placed next to the neural tube. The Shh released from these cells was sufficient to induce a second floorplate, as well as other genes normally expressed in the ventral neural tube. In further experiments, simply adding recombinant Shh protein to explants of neural tube was sufficient to induce them to differentiate as ventral neural tissues, including floorplate and motor neurons (Figure 2.23B,C). These experiments thus show that Shh is sufficient to ventralize the neural tube during development. What was also very interesting about the activity of Shh in these experiments was that it works in a gradient: highest concentrations of the factor causes the neural tube to develop the most ventral fates (Nkx2.2+ cells), intermediate concentrations of Shh induces intermediate neural tube fates (Olig2+ cells) and treating the neural tube explants with little or no Shh allows the cells to develop dorsal fates (Pax7+). Since

Shh is produced by the notochord and floorplate cells, the highest concentrations are present in these cells, while there is a decreasing gradient in concentration in increasingly dorsal regions of the neural tube (Dessaud et al., 2008). Thus, Shh is acting as a morphogen in the neural tube, like retinoic acid was for the patterning of rhombomeres discussed in a previous section. Two additional results show that Shh is required during normal development to specify the dorsal–ventral axis of the neural tube. First, antibodies raised against Shh will block the differentiation of floorplate and motor neurons when added to neural tube explants. Second, targeted deletion of the *shh* gene in mice results in the failure of the development of the ventral cell types in the spinal cord (see Chapter 4).

In addition to its role in the ventralization of the neural tube, Shh is also expressed in the more anterior regions of the body axis immediately subjacent to the neural tube, in what is known as the prechordal mesoderm. Here the function of Shh is similar to that of the notochord and floorplate: it serves to induce ventral differentiation in the forebrain. In the forebrain, the growth of the different brain vesicles gives rise to complex anatomy, and so the induction of ventral forebrain is critical for a number of subsequent morphogenetic events. Consequently, the loss of Shh signaling in the prechordal mesoderm produces dramatic phenotypic changes in embryos and the resulting animals. One particularly striking phenotype that arises from the disruption of *Shh* in embryogenesis is cyclopia (Roessler et al., 1997). The eyes normally develop from paired evaginations of the ventral diencephalon (see above). However, in the neural plate, the eye

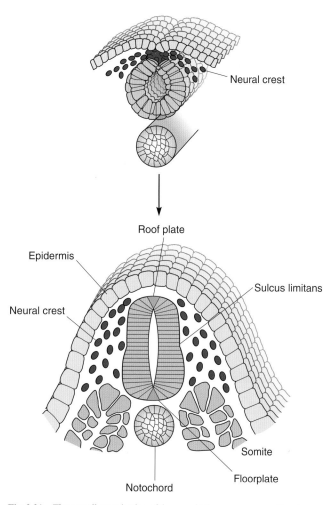

Neural crest

Roof plate

Epidermis

Neural crest

Sulcus limitans

Somite

Floorplate

Notochord

Fig. 2.21 The overall organization of the neural tube emerges soon after closure. The most ventral part of the neural tube becomes flattened into a distinct "floorplate." The most dorsal aspect of the neural tube develops into a tissue known as the roof plate. A distinct fissure, the *sulcus limitans*, forms between the dorsal and ventral parts of the neural tube along most of its length.

field is initially continuous across the midline and is split into two by the inhibition of eye-forming potential by Shh from the prechordal mesoderm. Shh represses Pax6 at the midline and induces Pax2. Pax6 and Pax2 cross repress, creating a sharp border between developing retinal fields (Pax6) and optic stalk region (Pax2) that separate the developing retinas. When this signal is interrupted, the eye field remains continuous and a single eye forms in the midline. The subsequent elaboration of the forebrain depends on correct midline development, and so deleting the Shh gene disrupts later stages of brain development as well, leading to a condition known as holoprosencephaly, where the normally paired cerebral hemispheres are fused into a single large structure.

DORSAL NEURAL TUBE AND NEURAL CREST

The experiments of Holtfreter (1934) and others showed that removal of the notochord resulted in a neural tube without much dorso-ventral polarity. This implies that the dorsal neural tube is in some way the default condition, whereas the ventral structures require an additional signal to develop their fates. However, the dorsal neural tube also requires signals for

its appropriate development. Before the neural tube closes, the future dorsal neural tube is continuous with the adjacent ectodermal cells (**Figure 2.21**). As the dorsal neural tube closes, a distinct group of cells, known as the neural crest, forms at the point of fusion of the neural tube margins. Thus, the neural crest is, in some sense, the most dorsal derivative of the neural tube, and has often been used as an indicator of dorsal differentiation. In addition, several genes specifically expressed in the dorsal neural tube at these early stages are important for neural crest development (e.g., *msx1*, *slug* and *snail*).

The neural crest is a remarkable collection of cells, which undergo extensive migration throughout the embryo and produce an impressive array of different tissues. In the trunk, the neural crest gives rise to the cells of the peripheral nervous system, including the neurons and glia of the sensory and autonomic ganglia, the Schwann cells surrounding all peripheral nerves, and the neurons of the gastric mucosal plexus. Several other cell types, including pigment cells, chromatophores, and smooth muscle cells, arise from the trunk neural crest. Neural crest also forms in the cranial regions, and here it contributes to most of the structures in the head. Most of the mesenchyme in the head, including that which forms the visceral skeleton and the bones of the skull, is derived from neural crest. The neurons and glia of several cranial ganglia, like the trigeminal sensory ganglia, the vestibulo-cochlear ganglia, and the autonomic ganglia in the head, are also derived largely from the progeny of the neural crest as well as from specialized regions of the head ectoderm called the cranial placodes.

Because of the extensive migration of the neural crest cells, and the great diversity of the tissues and cell types to which neural crest cells can contribute, the neural crest has been studied extensively as a model for these aspects of nervous system development. In the next sections we will review what is known about the origin of the neural crest and the factors that control the initial aspects of its differentiation. Chapters 3 and 4 will deal with the cellular determination of various crest derivatives and their migration throughout the body.

Classically, the neural crest has been thought to arise from the cells that form at the fusion of the neural folds when they become the neural tube. Vogt, using vital dyes to fate-map the different parts of the amphibian embryo, found that most of the neural crest forms from a narrow stripe of ectodermal cells at the junction between the neural plate and the epidermis. Subsequent studies using more sophisticated techniques have expanded this view. The chick-quail chimera system described above was used extensively by Le Douarin and her colleagues (2004) to track the fate of the neural crest that arises from the different regions along the neuraxis and to show the different types of tissues that are generated from different rostral-caudal regions (**Figure 2.24**). Bronner-Fraser and Fraser (1991) used single-cell injections to track the lineages of individual crest cells prior to their migration. The injected cells went on to divide, and they retained their lineage marker for several cell divisions. Many of the labeled cells went on to contribute to the tissues described above as the normal neural crest derivatives; however, some of the labeled cells that contributed to the neural crest also had progeny that populated the neural tube and the epidermis. Thus, although most of the cells in the neural crest field at the neural plate stage of development normally develop into neural crest, they are not restricted to this lineage. In addition, although in many embryos the neural crest develops at the fusion of the neural folds, there are regions of the neuraxis in

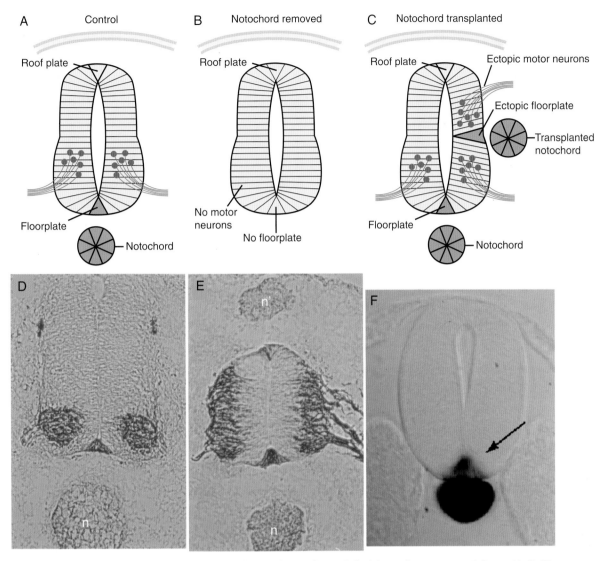

Fig. 2.22 Differentiation in the neural tube is dependent on factors derived from adjacent, nonneural tissues. (A, B, C) The diagrams at the top of the figure show that if the notochord, a mesodermally derived structure, is removed prior to neural tube closure, the neural tube fails to display characteristics of ventral differentiation, (B) such as the development of the floorplate (blue) and the spinal motoneurons (red). This shows that the notochord is necessary for the development of ventral neural tube fates. (C) If an additional notochord is transplanted to the lateral part of the neural tube at this same time in embryogenesis, a new floorplate is induced adjacent to the transplanted notochord. New motoneurons are also induced to form adjacent to the ectopic floorplate. Thus, the notochord is sufficient to specify ventral cell fates. (D, E) The experiment diagrammed at the top, the transplantation of an extra notochord, is shown next to a normal neural tube labeled with a marker for motor neurons. The extra notochord is labeled as *n'*. (F) the expression of *Sonic hedgehog* in the notochord and floorplate (arrow) of the neural tube is shown. *(D, E modified from Placzek et al., 1991; F courtesy of Henk Roelink)*

some species that do not form by the rolling of the neural plate. For example, in the fish, the neural tube forms first as a thickening of the neurectoderm, known as the neural keel, and tube formation occurs later by a process of cavitation, but the neural crest still forms from the lateral edges of the plate.

The first experimental studies to indicate that the induction of the neural crest may involve some of the same factors as those responsible for neural induction were those of Raven and Kloos (1945). They found that neural crest was induced from ectoderm by lateral pieces of the archenteron roof, whereas the neural tube was induced by medial pieces, such as the presumptive notochord. Similar results led Dalq (1941) to propose that a concentration gradient of a particular organizing substance originating in the midline tissue of

the archenteron roof could set up medial–lateral distinctions across the neural plate. Since the cells that will ultimately develop into the dorsal neural tube are initially immediately adjacent to the nonneural ectodermal cells, these could also provide a signal for dorsal differentiation similar to the notochord-derived Shh for ventralization of the neural tube.

Several lines of evidence now support the hypothesis that the ectoderm provides the molecular signals to promote dorsal differentiation in the lateral regions of the spinal cord. Moury and Jacobson (1990) tested whether interactions between the neural plate and the surrounding ectoderm were responsible for the induction of the neural crest by transplanting a small piece of the neural plate from a pigmented animal to the ventral surface of the embryo. When the embryo was

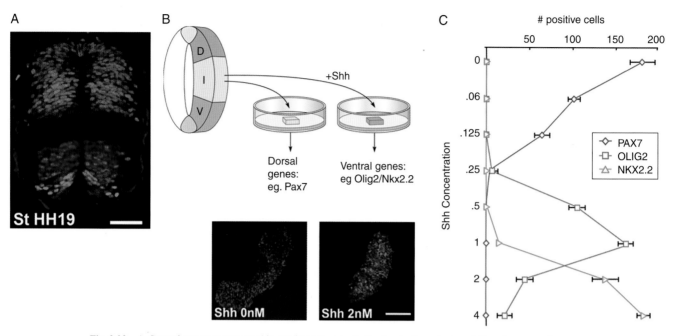

Fig. 2.23 A. Several genes are expressed in restricted domains in the developing spinal cord; these have served as useful markers for positional identity of cells in this region of the nervous system. *Pax7* (blue) is expressed in the intermediate and dorsal regions of the neural tube, while *Nkx2.2* (green) and *Olig2* (red) are expressed in the ventral neural tube. Markers like these and others allowed Jessell and colleagues to dissect the signals controlling the identity of the different types of neurons in the spinal cord (see also Chapter 4). B. A cell culture system in which the notochord and the neural tube were co-cultured in collagen gels was used to find the polarity signal released by the mesoderm. The signal was first shown to be diffusible since pieces of notochord could induce floorplate without touching the neural tube. Simply adding recombinant *Shh* protein to explants of neural tube was sufficient to induce them to differentiate as ventral neural tissues, including floorplate and motor neurons. C. Ventral markers (eg. Nkx2.2) increase with increasing concentrations of Shh. *(Modified from Dessaud et al., 2007)*

allowed to develop further, the transplant rolled into a small tube and at the margins gave rise to neural crest cells, as evidenced by the pigmented melanocytes that migrated from the ectopic neural tissue. These results were extended by similar experiments of Selleck and Bronner-Fraser (1995) in the chick embryo, and in addition, they used an explant culture system, in which the neural plate and epidermis were co-cultured and analyzed for proteins and genes normally expressed by the neural crest. They found that the neural crest was induced to form from the neural tube when placed adjacent to the epidermis. The initial steps toward identifying the crest inducer were made by Liem et al. (1995). BMPs, discussed in the previous chapter for their role in neural induction, also play important functions in specifying dorsal regional identity in the developing spinal cord. Liem et al. (1995) used a similar explant culture system as that used for the analysis of Shh effects on ventralization of the neural tube. The neural tube was dissected into a ventral piece, a dorsal piece, and an intermediate piece (**Figure 2.25**). They then analyzed the expression of genes normally restricted to either the dorsal neural tube or the ventral neural tube to determine whether these genes were specifically induced by co-culture with the ectoderm. They found that certain dorsally localized markers, such as Pax3 and Msx1, are initially expressed throughout the neural tube and are progressively restricted from the ventral neural tube by Shh from the notochord and floorplate. However, co-culture with the ectoderm was necessary to induce the expression of other, more definitive, dorsal markers, such as Hnk1 and slug. BMPs were found to effectively replace the ectodermally derived signal, since these could also activate

Hnk1 and slug, even from ventral explants. These experiments show that there is an antagonism between Shh from the ventral neural tube and BMPs from the dorsal neural tube; when BMP is added along with Shh to the explants, the Shh-induced motoneuron differentiation is suppressed.

In addition to the BMP signal that defines the border of the neural tube, there is evidence that both the Wnt signaling pathway and specific FGFs play a critical function in the specification of the neural crest fate (Deardorff et al., 2001; Garcia-Castro et al., 2002; Mayor et al., 1997). Treatment of neural plate explants with Wnt, like those described for BMPs, is also sufficient to induce neural crest markers in the cells (Garcia-Castro et al., 2002), while blocking Wnt signaling perturbs neural crest development. Several Wnt genes are expressed in the developing ectoderm, adjacent to the point of origin of the crest, including *wnt8* and *wnt6*. Using a transgenic zebrafish line with a heat-inducible inhibitor of Wnt signaling, Lewis et al. (2004) were able to precisely define the time in development when cells require the signal to become crest. They found a critical period when inhibiting Wnt signaling was able to prevent the neural crest development without affecting development of neurons in the spinal cord.

The model of dorsal–ventral polarity in the spinal cord that has emerged from these studies is as follows: BMPs and Wnts, expressed at the margin of the neural plate, induce the development of the neural crest at the boundary of the neural plate and the ectoderm (Sauka-Spengler and Bronner-Fraser, 2008; **Figure 2.26**). BMPs and Wnts are also important for the development of the dorsal fates within the neural tube. Shh,

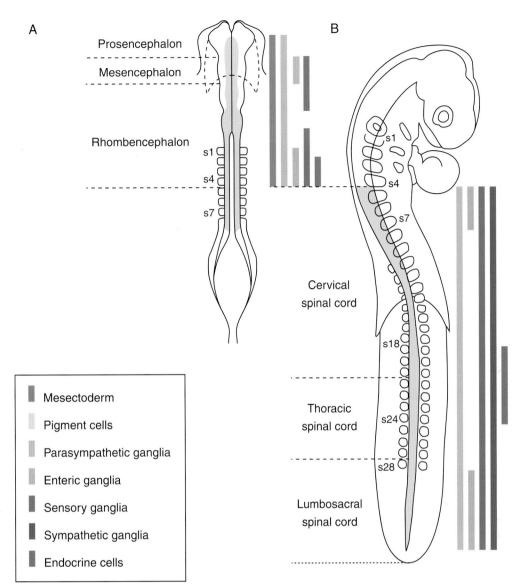

Fig. 2.24 The fate map of the neural crest in the chick embryo. Various types of tissues, including pigment cells, sensory ganglia, and endocrine cells, are derived from the neural crest. The cells migrating from the various positions along the neural tube give rise to different tissues; for example, the sympathetic ganglia arise from the neural crest of the trunk, but not from the head. Similarly, the parasympathetic ganglia arise from the neural crest of the head but not from the crest that migrates from most trunk regions. *(Modified from Le Douarin et al., 2004)*

expressed first in the notochord and later in the floorplate, induces ventral differentiation in the neural tube. The Shh and BMP/Wnt signals antagonize one another, and through this mutual antagonism they set up opposing gradients that control both the polarity of spinal cord differentiation and the amount of spinal cord tissue that differentiates into dorsal, ventral, and intermediate cell fates.

On a final note, the analysis of the dorsal–ventral patterning of the *Drosophila* nervous system has led to the uncovering of a fascinating evolutionary conservation (Cornell and Ohlen, 2000). As noted in Chapter 1, the insect nervous system arises from delaminating neuroblasts from the ventral neurogenic region. These neuroblasts form distinct rows, and the progeny of the neuroblasts from each row develop into specific subtypes of neurons. Each neuroblast of the *Drosophila* nervous system is thus precisely patterned along a Cartesian grid. Three

key transcription factors that uniquely define position along the dorsal–ventral axis are *msh*, *ind*, and *vnd*. These genes are necessary to specify the identity of the neuroblasts, and if one of them is missing, say *vnd*, then the ventral-most neuroblasts from each segment fail to develop. As described in the previous chapter, a gradient of the morphogen, Dpp, suppresses neural differentiation in more dorsal regions in the fly embryo, and it is this same morphogen that determines which of the three different genes, *vnd*, *ind*, or *msh*, is expressed in a specific row of neuroblasts. *Vnd* is expressed when there is little or no Dpp, while *msh* is expressed at higher levels of Dpp. The dorsal–ventral axis is thus patterned by the conversion of graded levels of Dpp into distinct differences in expression of specific transcription factors, very similar to what was just described for dorsal–ventral patterning in the vertebrate embryo. What is even more striking is that the genes *vnd*, *ind*, and *msh* are

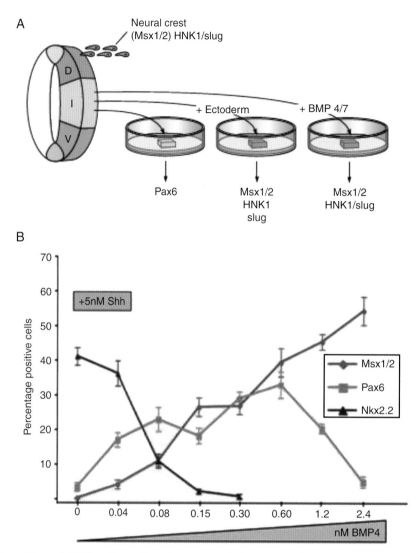

Fig. 2.25 A. Liem et al. (1995) used an explant culture system to define the signals that specify dorsal cell fates. The neural tube was dissected into a ventral piece, a dorsal piece, and an intermediate piece, and the expression of genes normally restricted to either the dorsal neural tube or the ventral neural tube was used to determine whether these genes were specifically induced by co-culture with the ectoderm. They found that certain dorsally localized genes are progressively restricted from the ventral neural tube by *Shh* from the notochord and floorplate; however, co-culture with the ectoderm was necessary to induce the expression of other, more definitive, dorsal markers, like *HNK1* and *slug*. BMPs were found to effectively replace the ectodermally derived signal, since these could also activate *HNK1* and *slug*, even from ventral explants. B. Shh and BMP interact to pattern the expression of dorsal-ventral genes in a dose dependent manner. *(Modified from Mizutani et al., 2006)*

homologs to the same genes that are used to define the different dorsal–ventral domains in the vertebrate neural tube, *nkx2.2*, *gsh*, and *msx*, respectively (**Figure 2.27**), and as you might recall from Chapter 1, the homolog of Dpp in vertebrates is BMP. The remarkable conservation in both the signaling molecules and the downstream transcription factors activated by graded levels of the signals provides strong evidence that the common ancestor of both vertebrates and flies likely had a nervous system that was patterned in a similar manner.

PATTERNING THE CEREBRAL CORTEX

The most obvious part of the mammalian brain is the cerebral cortex. In humans, this part of the brain constitutes most of what people think of when they see a brain and our sophisticated mental functions, like foresight and planning, are possible because of the enormous expansion of this region. The cerebral cortex is derived from the dorsocaudal telencephalic vesicle, and this has greatly expanded in mammals, when compared with other vertebrates (Rakic, 1988; 2006). The mammalian cerebral cortex is unique among animals, in that it has six layers of neurons and is therefore called the neocortex, to distinguish it from laminated forebrain structures that are found in all vertebrates, like the hippocampus. Although all neocortical areas have six layers of neurons, the cerebral cortex is not a homogeneous structure, but rather has many distinct regions, each of which has a dedicated function. It has been known for over 100 years that there are significant variations in the cellular structure (cytoarchitecture) of the cortex from region to region. The different regions of the cerebral cortex were exhaustively classified into approximately 50 distinct areas by Brodmann (1909). The relative number of cells in each layer and the size of the cells are quite variable and

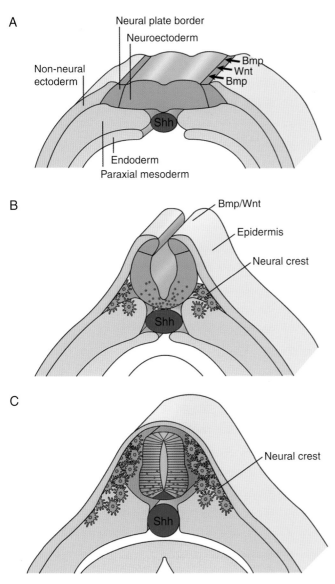

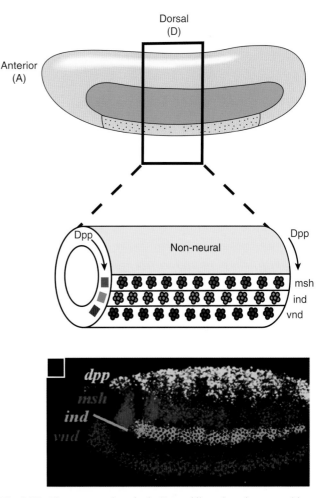

Fig. 2.26 *Shh* is expressed first in the notochord and later in the floorplate and induces ventral differentiation in the neural tube. BMPs are expressed in the ectoderm overlying the neural tube and then in the dorsal neural tube cells later in development. These two signals antagonize one another, and through this mutual antagonism they set up opposing gradients that control both the polarity of spinal cord differentiation and the amount of spinal cord tissue that differentiates into dorsal, ventral, and intermediate cell fates. *(Modified from Sauka-Spengler and Bronner-Fraser, 2008)*

Fig. 2.27 The neuroectoderm in the Drosophila embryo is patterned by a gradient in Dpp, which represses neuroectoderm at high concentrations (light yellow) induces msh expression at moderate concentrations and allows ind and vnd expression at progressively lower concentrations. The vertebrate homologs of these Drosophila genes are similarly patterned by the Dpp homolog BMP. *(lower panel from Mizutani et al., 2006)*

specialized to the specific function of that area. For example, the visual cortex, a primary sensory area, has many cells in layer IV, an input layer, whereas the motor cortex has very large neurons in layer V, an output layer.

Although for many years it has been thought that these different specializations occur later in development, as a consequence of the specific connections with other brain regions, more recent data indicate that the different areas have distinct identities much earlier in development, and these identities are not altered by changes in innervation (see Grove and Fukuchi-Shimogori, 2003). Like the other brain regions we have been discussing, the cerebral cortex arises from a layer of progenitors that comprises the early neural tube. In the specific case of the cerebral cortex, the anterior-most part of the

neural tube, the telencephalon, is the source of these progenitors (Figure 2.1). The regional identities of the cortical areas can be monitored through the analysis of transcription factor expression. Two transcription factors that appear to have a role in the specification of regional identities in the cortex are Pax6 (which we have already encountered for its role in eye development) and Emx2 (Bishop et al., 2000; Mallamaci et al., 2000; Muzio et al., 2002). These two genes are expressed in opposing gradients across the cortical surface (**Figure 2.28**). Emx2 is expressed most highly in the caudo-medial pole, while Pax6 is expressed highest at the rostral–lateral pole. Mutations in *pax6* cause an expansion of Emx2's domain of expression and ultimately an expansion of the areas normally derived from the caudal medial cortex, such as the visual cortex. Mutations in *emx2*, by contrast, cause the Pax6-expressing domain to expand, and ultimately result in an expansion of the frontal and motor cortical regions.

The graded patterns of expression of Emx2 and Pax6, along with the many examples of signaling centers we have already encountered in other regions of the developing nervous system, have led many investigators to postulate that similar signaling centers adjacent to the cortex regulate the

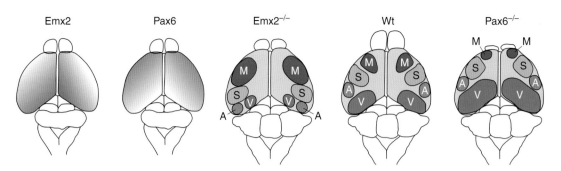

Fig. 2.28 Two transcription factors critical for the specification of regional identities in the cortex are *pax6* and *emx2*. *Emx2* is expressed primarily in the posterior cerebral cortex and then gradually diminishes in expression toward the rostral cortical pole; *pax6* has the complementary pattern of expression. Loss of either the *pax6* gene or the *emx2* gene affects the cerebral cortical pattern of development. In the wild-type (wt) animal, the motor cortex (M) is primarily located in the rostral cortex, and the other sensory areas for somatosensation (S), auditory sensation (A), and visual perception (V) are located in the middle and posterior cortex, respectively. In the *emx2*-deficient mice, the pattern is shifted caudally, and a greater area is occupied by the motor cortex; by contrast, in the *pax6*-deficient mice, the visual cortex is expanded and the motor cortex is severely reduced. *(Modified from Muzio and Mallamaci, 2003)*

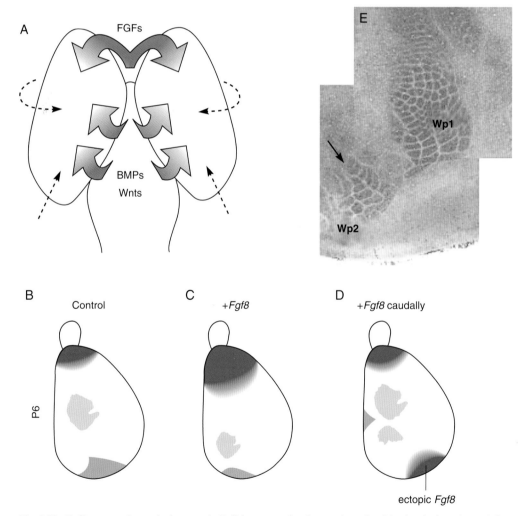

Fig. 2.29 Fgf8 patterns the cerebral cortex. A. Fgf8 is expressed at the anterior pole of the developing telencephalon, while BMPs and *wnt* genes are expressed in the posterior pole. Grove and her colleagues have misexpressed Fgf8 in different positions within the developing cortex. B. In the normal mouse, the barrel fields of the somatosensory map (yellow) are located near the middle of the cerebral cortex while Fgf8 (red) is expressed anteriorly and BMP (blue) is expressed posteriorly. C. Increasing the amount of Fgf8 (red) in the anterior pole causes a caudal shift in the cortical regions, including the somatosensory map. D. Placing a bead of Fgf8 in the caudal cortex causes the formation of a duplicated, mirror image of the somatosensory map. E. Micrograph of duplicated somatosory maps after the addition of an ectopic Fgf8 bead. Wp1 is the original map and Wp2 is the new map. *(Modified from Grove and Fukuchi-Shimogori, 2003)*

regional expression of these transcription factors. We have already encountered the two most well-studied cortical patterning signals, FGF and retinoic acid. The most dramatic results have come from the studies of Fgf8. Fgf8, along with related FGFs, Fgf17, and Fgf18, are all expressed at the anterior pole of the developing telencephalon. To analyze the role of the FGFs in specifying cortical areal identity, Grove and her colleagues have misexpressed Fgf8 in different positions within the developing cortex (Grove and Fukuchi-Shimogori, 2003). These studies have monitored the identity of cortical regions both using the expression of region-specific transcription factors, like Pax6 and Emx2, as well as analyzing later-developed properties of a region, like the barrel fields of the somatosensory map. Increasing the amount of Fgf8 in the anterior pole causes a downregulation of Emx2 and a caudal shift in the cortical regions, with an expansion of the rostral regions (**Figure 2.29**). Blocking the endogenous Fgf8 signal, by expressing a nonfunctional FGF receptor to bind up all the available Fgf8, causes the opposite result, a rostral-wards shift in the cortical regional identities. Most dramatically, placing a source of Fgf8 in the caudal cortex causes the formation of a duplicated, mirror image of cortical regions.

The graded pattern of expression of Emx2 and Pax6, in part regulated by Fgf8 and other *FGFs* from the anterior pole, appears to represent an early stage in the process by which areas of the cerebral cortex become specialized for different functions. For example, Fgf17, also expressed at the anterior pole of the telencephalic vesicle, is necessary for the development of the prefrontal cortex, the cortical region in humans that is critical for decision-making and planning (Cholfin and Rubenstein, 2007). Thus, as we saw for the segmentation of the fly embryo at the beginning of the chapter, patterning is often accomplished by an initial gradient of expression that becomes further subdivided into finer and finer regions over time. The drive toward increased diversification and specialization seems to be fundamental to biology at all levels, from cells, tissues, organisms, and biological communities, and the cerebral cortex, arguably the basis for human success, is no exception.

SUMMARY

The understanding of how the basic pattern of the nervous system is established has been put on a solid molecular ground in the past decade. One of the basic principles that has emerged from this work is that graded concentrations of antagonizing diffusible molecules are critically involved in setting up these patterns. These diffusible signaling molecules act to restrict the expression of specific transcription factors, which go on to regulate the expression of downstream target genes specific for the regional identity of part of the nervous system. One particularly well-conserved class of transcription factors, the Hox genes, is important in establishing and maintaining the regional identity of cells and tissues along the anterior–posterior axis of vertebrates throughout the hindbrain and likely the spinal cord. This conceptual framework holds true for vertebrates and invertebrates, and indeed, many of the molecular systems for generating specific parts of the nervous system have been highly conserved over the millions of years of evolution and considerable morphological diversity of animals.

REFERENCES

Acampora, D., Mazan, S., Lallemand, Y., Avantaggiato, V., Maury, M., Simeone, A., et al. (1995). Forebrain and midbrain regions are deleted in Otx2-/- mutants due to a defective anterior neuroectoderm specification during gastrulation. *Development, 121*(10), 3279–3290.

Affolter, M., Slattery, M., & Mann, R. S. (2008). A lexicon for homeodomain-DNA recognition. *Cell, 133*(7), 1133–1135.

Alvarado-Mallart, R. M., Martinez, S., & Lance-Jones, C. C. (1990). Pluripotentiality of the 2-day-old avian germinative neuroepithelium. *Dev Biol, 139*(1), 75–88.

Alvarado-Mallart, R. M. (1993). Fate and potentialities of the avian mesencephalic/metencephalic neuroepithelium. *Journal of Neurobiology, 24*(10), 1341–1355.

Biggin, M. D., & McGinnis, W. (1997). Regulation of segmentation and segmental identity by Drosophila homeoproteins: the role of DNA binding in functional activity and specificity. *Development, 124*(22), 4425–4433.

Bishop, K. M., Goudreau, G., & O'Leary, D. D. (2000). Regulation of area identity in mammalian neocortex by Emx2 and Pax6. *Science, 288*, 344–349.

Bouwmeester, T., Kim, S., Sasai, Y., Lu, B., & De Robertis, E. M. (1996). Cerberus is a head-inducing secreted factor expressed in the anterior endoderm of Spemann's organizer. *Nature, 382*(6592), 595–601.

Brodmann, K. (1909). *Vergleichende Lokalisationslehre der Großhirnrinde in ihren Prinzipien dargestellt auf Grund des Zellenbaues*. Leipzig: Barth.

Bronner-Fraser, M., & Fraser, S. E. (1991). Cell lineage analysis of the avian neural crest. *Development*, (Suppl. 2), 17–22.

Carpenter, E. M., Goddard, J. M., Chisaka, O., Manley, N. R., & Capecchi, M. R. (1993). Loss of Hox-A1 (Hox-1.6) function results in the reorganization of the murine hindbrain. *Development, 118*(4), 1063–1075.

Chalepakis, G., Stoykova, A., Wijnholds, J., Tremblay, P., & Gruss, P. (1993). Pax: gene regulators in the developing nervous system. *Journal of Neurobiology, 24*(10), 1367–1384.

Chambon, P. (1995). The molecular and genetic dissection of the retinoid signaling pathway. *Recent Progress in Hormone Research, 50*, 317–332.

Cholfin, J. A., & Rubenstein, J. L. (2007). Patterning of frontal cortex subdivisions by Fgf17. *Proceedings of the National Academy of Sciences of the United States of America, 104*(18), 7652–7657.

Cobos, I., Shimamura, K., Rubenstein, J. L., Martinez, S., & Puelles, L. (2001). Fate map of the avian anterior forebrain at the four-somite stage, based on the analysis of quail-chick chimeras. *Developmental Biology, 239*(1), 46–67.

Cornell, R. A., & Ohlen, T. V. (2000). Vnd/nkx, ind/gsh, and msh/msx: conserved regulators of dorsoventral neural patterning? *Current Opinion in Neurobiology, 10*, 63–71.

Couly, G. F., & Le Douarin, N. M. (1987). Mapping of the early neural primordium in quail-chick chimeras. I. Developmental relationships between placodes, facial ectoderm, and prosencephalon. *Dev Biol, 110*(1985), 422–439.

Crossley, P. H., & Martin, G. R. (1995). The mouse Fgf8 gene encodes a family of polypeptides and is expressed in regions that direct outgrowth and patterning in the developing embryo. *Development, 121*(2), 439–451.

Crossley, P. H., Martinez, S., & Martin, G. R. (1996). Midbrain development induced by FGF8 in the chick embryo. *Nature, 380*(6569), 66–68.

Dalq, A. (1941). Contributions a l'etude du potentiel morpogenetique chez les Anoures. III. *Arch. De Biol, 53*, 2–124.

Danielian, P. S., Echelard, Y., Vassileva, G., & McMahon, A. P. (1997). A 5.5-kb enhancer is both necessary and sufficient for regulation of Wnt-1 transcription in vivo. *Dev Biol, 192*(2), 300–309.

Deardorff, M. A., Tan, C., Saint-Jeannet, J. P., & Klein, P. S. (2001). A role for frizzled 3 in neural crest development. *Development, 128*(19), 3655–3663.

de la Calle-Mustienes, E., Feijóo, C. G., Manzanares, M., Tena, J. J., Rodríguez-Seguel, E., Letizia, A., et al. (2005). A functional survey of the enhancer activity of conserved non-coding sequences from vertebrate Iroquois cluster gene deserts. *Genome Research, 15*(8), 1061–1072.

del Barco Barrantes, I., Davidson, G., Grone, H. J., Westphal, H., & Niehrs, C. (2003). Dkk1 and noggin cooperate in mammalian head induction. *Genes & Development, 17*(18), 2239–2244.

Dessaud, E., Yang, L. L., Hill, K., Cox, B., Ulloa, F., Ribeiro, A., et al. (2007). Interpretation of the sonic hedgehog morphogen gradient by a temporal adaptation mechanism. *Nature, 450*(7170), 717–720.

Dessaud, E., McMahon, A. P., & Briscoe, J. (2008). Pattern formation in the vertebrate neural tube: a sonic hedgehog morphogen-regulated transcriptional network. *Development, 135*(15), 2489–2503.

Driever, W., & Nusslein-Volhard, C. (1988). The bicoid protein determines position in the Drosophila embryo in a concentration-dependent manner. *Cell, 54*(1), 95–104.

Duboule, D. (2007). The rise and fall of Hox gene clusters. *Development, 134*(14), 2549–2560.

Duboule, D., & Morata, G. (1994). Colinearity and functional hierarchy among genes of the homeotic complexes. *Trends in Genetics, 10*(10), 358–364.

Durston, A. J., Timmermans, J. P., Hage, W. J., Hendriks, H. F., de Vries, N. J., Heideveld, M., et al. (1989). Retinoic acid causes an anteroposterior transformation in the developing central nervous system. *Nature*, *340*(6229), 140–144.

Eagleson, G. W., & Harris, W. A. (1990). Mapping of the presumptive brain regions in the neural plate of Xenopus laevis. *Journal of Neurobiology*, *21*(3), 427–440.

Garcia-Castro, M. I., Marcelle, C., & Bronner-Fraser, M. (2002). Ectodermal Wnt function as a neural crest inducer. *Science*, *297*(5582), 848–851.

Gehring, W. J. (1993). Exploring the homeobox. *Gene*, *135*(1–2), 215–221.

Gerhart, J., & Kirschner, M. (1997). Cells, embryos, and evolution. Blackwell Science.

Glavic, A., Gomez-Skarmeta, J. L., & Mayor, R. (2002). The homeoprotein Xiro1 is required for midbrain-hindbrain boundary formation. *Development*, *129*(7), 1609–1621.

Glinka, A., Wu, W., Delius, H., Monaghan, A. P., Blumenstock, C., & Niehrs, C. (1998). Dickkopf-1 is a member of a new family of secreted proteins and functions in head induction. *Nature*, *391*(6665), 357–362.

Glinka, A., Wu, W., Onichtchouk, D., Blumenstock, C., & Niehrs, C. (1997). Head induction by simultaneous repression of Bmp and Wnt signalling in Xenopus. *Nature*, *389*(6650), 517–519.

Grove, E. A., & Fukuchi-Shimogori, T. (2003). Generating the cerebral cortical area map. *Annual Review of Neuroscience*, *26*, 355–380.

Halder, G., Callaerts, P., & Gehring, W. J. (1995). Induction of ectopic eyes by targeted expression of the eyeless gene in Drosophila. *Science*, *267*(5205), 1788–1792.

Hidalgo-Sánchez, M., Millet, S., Bloch-Gallego, E., & Alvarado-Mallart, R. M. (2005). Specification of the meso-isthmo-cerebellar region: the Otx2/Gbx2 boundary. *Brain Research Reviews*, *49*(2), 134–149.

Holtfreter, J. (1934). Formative Reize in der Embryonalentwicklung der Amphibien dargestellt an Explantationsversuchen. *Arch exp. Zellforsch. 15*, 281–301.

Joyner, A. L., Liu, A., & Millet, S. (2000). Otx2, Gbx2 and Fgf8 interact to position and maintain a mid-hindbrain organizer. *Current Opinion in Cell Biology*, *12*(6), 736–741.

Kengaku, M., & Okamoto, H. (1995). bFGF as a possible morphogen for the anteroposterior axis of the central nervous system in Xenopus. *Development*, *121*(9), 3121–3130.

Kiecker, C., & Niehrs, C. (2001). A morphogen gradient of Wnt/beta-catenin signalling regulates anteroposterior neural patterning in Xenopus. *Development*, *128*(21), 4189–4201.

Le Douarin, N. (1982). *The Neural Crest*. New York: Cambridge University Press.

Le Douarin, N. M., Creuzet, S., Couly, G., & Dupin, E. (2004). Neural crest cell plasticity and its limits. *Development*, *131*(19), 4637–4650.

Lewis, E. B. (1978). A gene complex controlling segmentation in Drosophila. *Nature*, *276*(5688), 565–570.

Lewis, J. L., Bonner, J., Modrell, M., Ragland, J. W., Moon, R. T., Dorsky, R. I., et al. (2004). Reiterated Wnt signaling during zebrafish neural crest development. *Development*, *131*(6), 1299–1308.

Lichtnecker, R., Reichert, H. (2005). Insights into the urbilaterian brain: conserved genetic patterning mechanisms in insect and vertebrate brain development. *Heredity*, *94*(5), 465–477.

Liem, K. F., Jr., Tremml, G., Roelink, H., & Jessell, T. M. (1995). Dorsal differentiation of neural plate cells induced by BMP-mediated signals from epidermal ectoderm. *Cell*, *82*(6), 969–979.

Liu, D. X., & Greene, L. A. (2001). Regulation of neuronal survival and death by E2F-dependent gene repression and derepression. *Neuron*, *32*, 425–438.

Lumsden, A., & Keynes, R. (1989). Segmental patterns of neuronal development in the chick hindbrain. *Nature*, *337*(6206), 424–428.

Maden, M. (2006). Retinoids and spinal cord development. *Journal of Neurobiology*, *66*(7), 726–738.

Mallamaci, A., Muzio, L., Chan, C. H., Parnavelas, J., & Boncinelli, E. (2000). Area identity shifts in the early cerebral cortex of Emx2−/− mutant mice. *Nature Neuroscience*, *3*(7), 679–686.

Mark, M., Lufkin, T., Vonesch, J. L., Ruberte, E., Olivo, J. C., Dollé, P., et al. (1993). Two rhombomeres are altered in Hoxa-1 mutant mice. *Development*, *119*(2), 319–338.

Martinez, S., Crossley, P. H., Cobos, I., Rubenstein, J. L., & Martin, G. R. (1999). FGF8 induces formation of an ectopic isthmic organizer and isthmocerebellar development via a repressive effect on Otx2 expression. *Development*, *126*(6), 1189–1200.

Matsuo, I., Kuratani, S., Kimura, C., Takeda, N., & Aizawa, S. (1995). Mouse Otx2 functions in the formation and patterning of rostral head. *Genes & Development*, *9*(21), 2646–2658.

Mayor, R., Guerrero, N., & Martínez, C. (1997). Role of FGF and noggin in neural crest induction. *Developmental Biology*, *189*(1), 1–12.

McMahon, A. P., & Bradley, A. (1990). The Wnt-1 (int-1) proto-oncogene is required for development of a large region of the mouse brain. *Cell*, *62*(6), 1073–1085.

Meyers, E. N., Lewandoski, M., & Martin, G. R. (1998). An Fgf8 mutant allelic series generated by Cre- and Flp-mediated recombination. *Nature Genetics*, *18*(2), 136–141.

Millet, S., Bloch-Gallego, E., Simeone, A., & Alvarado-Mallart, R. M. (1996). The caudal limit of Otx2 gene expression as a marker of the midbrain/hindbrain boundary: a study using in situ hybridisation and chick/quail homotopic grafts. *Development*, *122*(12), 3785–3797.

Millet, S., Campbell, K., Epstein, D. J., Losos, K., Harris, E., & Joyner, A. L. (1999). A role for Gbx2 in repression of Otx2 and positioning the mid/hindbrain organizer. *Nature*, *401*(6749), 161–164.

Moury, J. D., & Jacobson, A. G. (1990). The origins of neural crest cells in the axolotl. *Developmental Biology*, *141*(2), 243–253.

Mukhopadhyay, M., Shtrom, S., Rodriguez-Esteban, C., Chen, L., Tsukui, T., Gomer, L., et al. (2001). Dickkopf1 is required for embryonic head induction and limb morphogenesis in the mouse. *Developmental Cell*, *1*(3), 423–434.

Muzio, L., DiBenedetto, B., Stoykova, A., Boncinelli, E., Gruss, P., & Mallamaci, A. (2002). Emx2 and Pax6 control regionalization of the pre-neuronogenic cortical primordium. *Cerebral Cortex*, *12*(2), 129–139.

Muzio, L., & Mallamaci, A. (2003). Emx1, emx2 and pax6 in specification, regionalization and arealization of the cerebral cortex. *Cerebral Cortex*, *13*(6), 641–647.

Nusslein-Volhard, C., & Wieschaus, E. (1980). Mutations affecting segment number and polarity in Drosophila. *Nature*, *287*(5785), 795–801.

Placzek, M., Yamada, T., Tessier-Lavigne, M., Jessell, T., & Dodd, J. (1991). Control of dorsoventral pattern in vertebrate neural development: induction and polarizing properties of the floor plate. *Development*, Suppl. 2, 105–122.

Puelles, L., & Rubenstein, J. L. (1993). Expression patterns of homeobox and other putative regulatory genes in the embryonic mouse forebrain suggest a neuromeric organization. *Trends in Neurosciences*, *16*(11), 472–479.

Puelles, L., & Rubenstein, J. L. (2003). Forebrain gene expression domains and the evolving prosomeric model. *Trends Neurosci*, *26*(9), 469–476.

Rakic, P. (1988). Specification of cerebral cortical areas. *Science*, *241*(4862), 170–176.

Rakic, P. (2006). A century of progress in corticoneurogenesis: from silver impregnation to genetic engineering. *Cerebral Cortex*, *16*(Suppl. 1), i3–i17.

Raven, C. P., & Kloos, J. (1945). Induction by medial and lateral pieces of the archenteron roof with special reference to the determination of the neural crest. *Acta Neerl Morphol*, *5*, 348–362.

Rhinn, M., & Brand, M. (2001). The midbrain-hindbrain boundary organizer. *Current Opinion in Neurobiology*, *11*(1), 34–42.

Rodríguez-Seguel, E., Alarcón, P., & Gómez-Skarmeta, J. L. (2009). The Xenopus Irx genes are essential for neural patterning and define the border between prethalamus and thalamus through mutual antagonism with the anterior repressors Fezf and Arx. *Developmental Biology*, *329*(2), 258–268.

Roelink, H., Augsburger, A., Heemskerk, J., Korzh, V., Norlin, S., Ruiz i Altaba, A., et al. (1994). Floor plate and motor neuron induction by vhh-1, a vertebrate homolog of hedgehog expressed by the notochord. *Cell*, *76*(4), 761–775.

Roessler, E., Belloni, E., Gaudenz, K., Vargas, F., Scherer, S. W., Tsui, L. C., et al. (1997). Mutations in the C-terminal domain of Sonic hedgehog cause holoprosencephaly. *Human Molecular Genetics*, *6*(11), 1847–1853.

Sauka-Spengler, T., & Bronner-Fraser, M. (2008). Evolution of the neural crest viewed from a regulatory perspective. *Genesis*, *46*(11), 673–682.

Selleck, M. A., & Bronner-Fraser, M. (1995). Origins of the avian neural crest: the role of neural plate-epidermal interactions. *Development*, *121*(2), 525–538.

Shimamura, K., & Rubenstein, J. L. (1997). Inductive interactions direct early regionalization of the mouse forebrain. *Development*, *124*(14), 2709–2718.

Simeone, A., Acampora, D., Nigro, V., Faiella, A., D'Esposito, M., Stornaiuolo, A., et al. (1991). Differential regulation by retinoic acid of the homeobox genes of the four HOX loci in human embryonal carcinoma cells. *Mechanisms of Development*, *33*(3), 215–227.

Slack, J. M., & Tannahill, D. (1992). Mechanism of anteroposterior axis specification in vertebrates. Lessons from the amphibians. *Development*, *114*(2), 285–302.

Small, S., & Levine, M. (1991). The initiation of pair-rule stripes in the Drosophila blastoderm. *Current opinion in Genetics & Development*, *1*(2), 255–260.

Stuart, J. J., Brown, S. J., Beeman, R. W., & Denell, R. E. (1993). The Tribolium homeotic gene Abdominal is homologous to abdominal-A of the Drosophila bithorax complex. *Development*, *117*(1), 233–243.

Technau, G. M., Berger, C., & Urbach, R. (2006). Generation of cell diversity and segmental pattern in the embryonic central nervous system of *Drosophila*. *Developmental Dynamics*, *235*, 861–869.

Urbach, R., & Technau, G. M. (2004). Neuroblast formation and patterning during early brain development in Drosophila. *Bioessays*, *26*(7), 739–751.

Waskiewicz, A. J., Rikhof, H. A., & Moens, C. B. (2002). Eliminating zebrafish pbx proteins reveals a hindbrain ground state. *Developmental Cell*, *3*(5), 723–733.

Wassarman, K. M., Lewandoski, M., Campbell, K., Joyner, A. L., Rubenstein, J. L., Martinez, S., et al. (1997). Specification of the anterior hindbrain and establishment of a normal mid/hindbrain organizer is dependent on Gbx2 gene function. *Development*, *124*(15), 2923–2934.

Wurst, W., Auerbach, A. B., & Joyner, A. L. (1994). Multiple developmental defects in Engrailed-1 mutant mice: an early mid-hindbrain deletion and patterning defects in forelimbs and sternum. *Development*, *120*(7), 2065–2075.

Zuber, M. E., Gestri, G., Viczian, A. S., Barsacchi, G., & Harris, W. A. (2003). Specification of the vertebrate eye by a network of eye field transcription factors. *Development*, *130*(21), 5155–5167.

Genesis and migration 3

The human brain is made up of approximately 100 billion neurons and even more glial cells. The sources of all these neurons and glia are the cells of the neural tube, described in the previous chapters. Neurogenesis and gliogenesis, the generation of neurons and glia during development, is collectively also called histogenesis. Once the neurons and glia are generated by the progenitors during development, they almost always migrate over some distance from their point of origin to their final position. This chapter describes the cellular and molecular principles by which the appropriate numbers of neurons and glia are generated from the neural precursors, and gives an overview of some of the complex cellular migration processes involved in the construction of the brain. The number of cells generated in the developing nervous system is likely regulated at several levels. In some cases, the production of neurons or glia may be regulated by an intrinsic limit in the number of progenitor cell divisions. The level of proliferation and ultimately the number of cells generated can also be controlled by extracellular signals, acting as mitogens, promoting progenitor cells to reenter the cell cycle or alternatively as mitotic inhibitors that induce progenitor cells to exit from the cell cycle. However, as we will see in Chapter 7, the number of neurons and glia in the mature nervous system is a function not only of cell proliferation, but also of cell death.

As we saw in Chapter 1, the nervous system of *C. elegans* (as well as the rest of the animal) is derived from a highly stereotyped pattern of cell divisions. Therefore, in these animals, the lineages of the cells directly predict their numbers. The regulation of these cell divisions appears to depend less on interactions with surrounding cells than is the case in vertebrates. The lineages of the *C. elegans* progenitor cells also predict the particular types of neurons that are generated from a particular precursor, and it appears that the information to define a given type of cell resides largely in factors derived directly from the precursors. The same is true for the neuroblasts that produce the *Drosophila* central nervous system: the production of neurons from the neuroblasts is highly stereotypic. The neuroblasts of the insect CNS delaminate from the ventral–lateral ectoderm neurogenic region in successive waves (see Chapter 1). In *Drosophila*, about 25 neuroblasts delaminate in each segment, and they are organized in four columns and six rows (Doe and Smouse, 1990). Once the neuroblast segregates from the ectoderm, it undergoes several asymmetric divisions, giving rise to approximately five smaller ganglion mother cells. Each ganglion mother cell then divides to generate a pair of neurons. These neurons make up the segmental ganglia of the ventral nerve cord and have stereotypic numbers and types of neurons.

In the vertebrate, the situation gets considerably more complex. The neural tube of most vertebrates is initially a single layer thick. As neurogenesis proceeds, the progenitor cells undergo a large number of cell divisions to produce a much thicker tube. A section through the developing spinal cord is shown in **Figure 3.1A**, and an example of a progenitor cell is shown as a schematic in **Figure 3.1B** and in the actual neural tube in **Figure 3.1C**, labeled with a fluorescent protein to visualize the cell as it progresses through a cell division. At this stage of development, almost all the cells in the neural tube resemble those shown in Figure 3.1B,C, with a simple bipolar shape. They extend one process to the central canal of the neural tube (named the ventricular surface because it is continuous with the ventricles of the brain) and they extend their other process to the outer surface of the neural tube. If one were just to look at the nuclei of the neural tube at this stage, there would appear to be many cell layers, and at first, the early neurohistologists thought this was the case. However, in the early 1900s it was recognized that the cells of the neural tube move their nuclei from the inside of the neural tube to the outside during each cell cycle. This movement can be directly observed using time lapse recording of cells labeled with fluorescent proteins (Figure 3.1C). This constant nuclear movement is termed interkinetic nuclear migration. In this process, the nuclei move to the inner, ventricular surface moment just before mitosis, and divide into two daughter cells; then the nuclei of these daughter cells move away from this surface during S-phase; but wherever they are just before the next mitosis, they rapidly move back to the ventricular surface to complete division (Norden et al., 2009). In the developing chick spinal cord, this process can take as little as six hours (Figure 3.1C). Although the function of interkinetic nuclear migration is unknown, it may be necessary for the

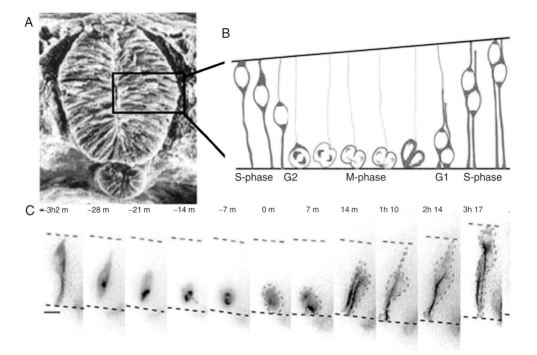

Fig. 3.1 The neural tube contains the progenitors of all the neurons and glia in the mature brain and spinal cord. A. These progenitors have a simple bipolar morphology. B. The nuclei of the progenitor cells undergo an "in and out" movement (shown here as an "up and down" movement) as they progress through the cell cycle. The cells move to the inside of the neural tube in the G2 phase of the cycle, go through the M-phase at the inner surface (also called the ventricular surface) and then move out again during S-phase. C. An actual time lapse recording showing a fluorescent labeled progenitor in the embryonic chick spinal cord undergo a mitotic division in a little over six hours. The M-phase lasts only about 20 minutes, while the S-phase can last much longer. *(A, Courtesy of Kathryn Tosney; B,C, Modified from Wilcock et al., 2007)*

progenitor cells to receive specific signals at different times in the cell cycle. On the other hand, Norden et al. (2009) propose that the migration to the ventricular surface prior to the mitotic division may be what is critical, and that the nuclei move away during S-phase to allow other nuclei access to that surface for their divisions. They compare this to people at a crowded pub, jostling around the room most of the time, but quickly returning to the bar for a refill.

The cells of the ventricular zone are the precursors of the differentiated neurons and glia of the central nervous system. In most other areas of the developing neural tube, once neurons and glia are generated by the progenitor cells, the neurons and glia migrate away from the ventricular surface to continue their differentiation elsewhere, and much more will be said of this process. Several methods have been developed to track the daughter cells of the progenitors once they leave the ventricular zone. One of the best ways to track the progeny of a progenitor is using a retrovirus that permanently marks all the daughter cells (**Figure 3.2**). The retroviral labeling technique takes advantage of the fact that retroviruses will only successfully infect and integrate their genes into cells that are going through the cell cycle. The genome of these viruses can be modified to contain genes that code for proteins not normally present in the nervous system but can be detected easily, such as green fluorescent protein (GFP). Once the virus infects a cell, and the viral genome is integrated into the cell's DNA, the viral genes are inherited in all the daughter cells of the originally infected cell. Another important feature of this technique is that the virus is typically modified so that it is incapable of making more virus in the infected cells and spreading the infection to other cells. This means that only the daughter cells of the *originally infected* progenitors will express the viral genes. If a retrovirus that contains DNA coding for GFP is then injected into the developing brain and infects some of the proliferating progenitor cells, the progeny of the infected cells will still express the GFP gene even in adult animals. This technique was used in the developing neural tube, and the distribution of clones has been analyzed shortly after injections as well as in the more mature spinal cord (Figure 3.2). When the neural tube was analyzed one day after the retroviral infection, a few progenitors expressed the reporter gene, and their progeny were clustered together, since they had not yet had time to migrate or differentiate; however, if the embryos were allowed to develop several more days to allow the spinal cord to mature, then the cells that express the reporter gene were typically more dispersed.v These cells could be shown to have developed from a single progenitor, and are therefore considered a "clone" and the technique is sometimes called "clonal analysis." In the case of the chick spinal cord, many of the clones contain both motoneurons and glial cells in the white matter. In the example shown in Figure 3.2, both astrocytes and oligodendrocytes are derived from the same progenitor cell that gave rise to the motoneurons (Leber et al., 1990). The clonal analysis method therefore not only allows one to track the cells as they migrate to other regions of the nervous system, but it also shows that the progenitors can make neurons and both types of macroglia found in the brain, the astrocytes and the oligodendrocytes. As a result, these progenitors are sometimes referred to as "multipotent progenitors," or at times "neural stem cells." (We prefer the former term, because the

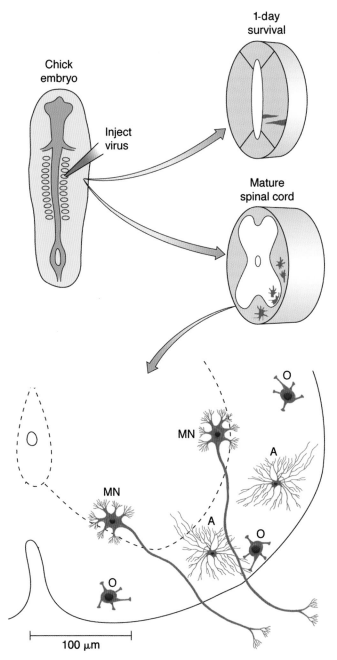

Fig. 3.2 Clonal analysis of progenitor cells in the chick neural tube. Injections of a retrovirus with a reporter gene are made into the chick embryo neural tube. After either short or long postinjection survival periods, the spinal cord is sectioned and analyzed for the labeled progeny of the few infected progenitor cells. In the case shown, a single progenitor cell has been infected at this level of the spinal cord, and it has gone through a single-cell division to give rise to two daughter cells after one day. If the embryo is allowed to survive to a point where the spinal cord is relatively mature, and neurons and glial cells can be identified, the labeled cells can be assigned to specific cell classes. In the case shown, the progeny of the infected cell include motoneurons (MN), astrocytes (A), and oligodendrocytes (O). *(Modified from Leber et al., 1990)*

term "stem cell" implies a self-renewal property that is difficult to assess. We will revisit this issue at the end of the chapter in the discussion of adult neurogenesis.)

In addition to the retroviral method for tracking the progeny of progenitors, a more classic method uses detectable analogs to the nucleotide, thymidine, to follow all cells generated at a particular

stage of development. This method is called "³H-thymidine birthdating." This technique, pioneered by Richard Sidman (1961), works as follows: While the progenitor cells are actively dividing, they are synthesizing DNA and incorporate thymidine into the new strand. If an animal is injected with an isotopically labeled form of thymidine (eg. ³H-thymidine), this labeled form of thymidine gets incorporated into the DNA of the mitotic cells. ³H-thymidine is incorporated into DNA during replication like ordinary thymidine, but because it is radioactive, it can be traced with a technique called autoradiography. Typically, a single injection of the ³H-thymidine is given and therefore it is available for only a few hours. The progenitor cells that were in the S-phase of the cell cycle at the time the injection was made will be labeled with the ³H-thymidine; however, if they continue to proliferate over many days, the labeled DNA will be diluted over time. By contrast, those cells that withdraw from the cycle and become postmitotic soon after the ³H-thymidine was administered will remain heavily labeled with the radioactive nucleotide. Thus, the postmitotic neurons generated, or "born," within a day after the ³H-thymidine injection will have heavily labeled nuclei, and neurons generated later in development will be more lightly labeled. Unlabeled cells are those that withdrew from the cell cycle before the ³H-thymidine injection. More recently, ³H-thymidine labeling has been replaced by bromo-deoxyuridine (BrdU), since this thymidine analog is also incorporated by S-phase cells and can be detected using an antibody and immunofluorescence, rather than the more complicated autoradiography technique.

The ³H-thymidine birthdating has been used extensively to track the migration and birthdates of the different neuronal and glial populations in the nervous system. These studies revealed that the process of neurogenesis is remarkably well-ordered. In many areas of the developing brain there are spatial and temporal gradients of neuron production. In general, there are well-conserved and orderly sequences of generation of different types of neurons and glia. For example, in the cerebral cortex, the neurons are arranged in layers or lamina. If mice are labeled with ³H-thymidine at successively later stages in their development (**Figure 3.3A**), the labeled cohort of newborn neurons forms a layer more superficial than the previous one. This "inside-out" pattern of neurogenesis is found in the cerebral cortex of all mammals, from mice to monkeys (**Figure 3.3B**), and presumably people as well.

Several additional generalizations can also be derived from the large number of thymidine birthdating studies that have been carried out in the different regions of the vertebrate CNS. As noted above, in many areas of the developing CNS, distinct types of neurons originate in a fairly invariant timetable. Often, the entire population of one type of neuron, like the spinal motoneurons, becomes postmitotic within a relatively short period of development. In general, large neurons are generated before small neurons in the same region. For example, pyramidal cells become postmitotic before granule cells in the hippocampus, cerebral cortex, and olfactory bulb; and in the cerebellum, Purkinje cells are generated prior to granule cells (Jacobson, 1977). Interestingly, it appears that the patterns of neuronal generation are also consistent with the hypothesis that phylogenetically older parts of the brain develop before the more recently evolved structures.

The picture of neurogenesis that has emerged from the thymidine birthdating studies and the retroviral lineage studies has led to many questions about the process: What controls the number of neurons and glia produced by the progenitors? How does a progenitor "decide" whether to make a neuron or a glial cell?

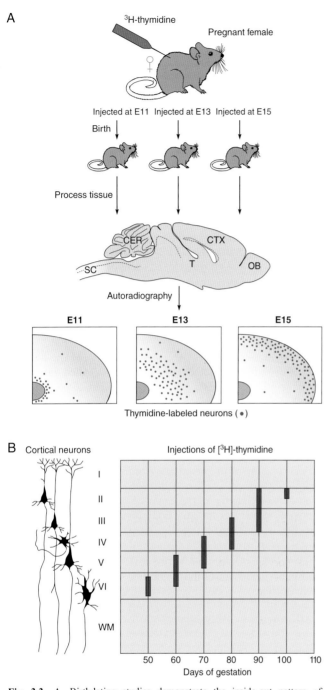

Thymidine-labeled neurons (•)

Fig. 3.3 A. Birthdating studies demonstrate the inside-out pattern of cerebral cortical histogenesis. Pregnant female rats are given injections of [3H]-thymidine at progressively later stages of gestation. When the pups are born, they are allowed to survive to maturity, and then their brains are processed to reveal the labeled cells. Neurons that have become postmitotic on embryonic day 11 are found primarily in the subplate (now in the subcortical white matter), while neurons "born" on day E13 are found in deep cortical layers, that is, V and VI, and neurons generated on E15 are found in more superficial cortical layers, that is, IV, III, and II. The most superficial layer, layer I, contains only the remnants of the preplate neurons (not shown). *(Modified from Angevine and Sidman, 1961)* B. Similar thymidine birthdating results in the monkey show this pattern more clearly than in the mouse due to the longer period of gestation.

What controls the migration of the cells from the ventricular zone to their ultimate location in the brain? The next sections will highlight what is known about these questions.

WHAT DETERMINES THE NUMBER OF CELLS PRODUCED BY THE PROGENITORS?

The thymidine birthdating studies and the retroviral lineage tracing studies described above provided a wealth of information about the migrations and cell types generated by the progenitors; however, they also provided information about how cell numbers are regulated during development. For example, with the thymidine method, it is possible to determine the length of the cell cycle, and it was shown that overall length of the cell cycle increases progressively during embryogenesis. Progenitor cells from the chick brain, for example, have an overall cell cycle time of 8 hours on embryonic day 3, but this increases to 15 hours by embryonic day 6. A similar increase in cell-cycle period occurs in the mammal, as rat cortical progenitor cells increase their cell-cycle time from 11 hours on embryonic day 12 to 19 hours at embryonic day 18. The second generality that can be made is that the increase in the cell-cycle period is largely due to increases in the G1 phase. As shown in **Figure 3.4**, the M and G2 phases of the cell cycle change little from embryonic

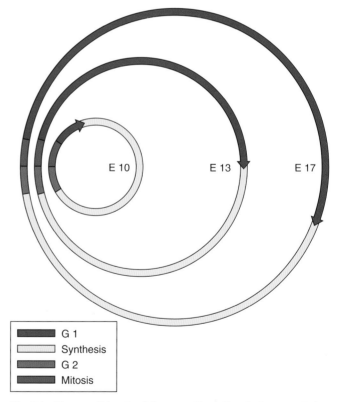

	G 1
	Synthesis
	G 2
	Mitosis

Fig. 3.4 The overall length of the progenitor cell cycle increases during embryogenesis. The cell cycles of progenitor cells from the mouse cerebral cortex are plotted as circles of increasing size from E10 to E17. The increase in the cell-cycle period is largely due to an increase in the G1 phase, which nearly triples in length (shown in red).

day 10 to embryonic day 19 in mouse cerebral cortex progenitor cells; however, the G1 phase nearly triples in length. The lengthening of the G1 period likely reflects some regulatory process that restricts or slows reentry of the progenitor cells into the S-phase from G1, consistent with the idea that a limiting supply of growth factor controls this step (see next section).

Labeling individual progenitor cells with retroviruses at different stages of brain development has shown directly that the number of progeny generated by a ventricular zone cell declines over the period of neurogenesis. For example, retroviral infections of the progenitor cells in the early embryonic brain result in very large clones of labeled cells, but retroviral infections of progenitor cells in the brains of late staged embryos gives much smaller clones. Simply counting the rate of overall expansion of the nervous system over time has also led to insight into the process of neurogenesis. In the early embryonic cerebral cortex, for example, the number of cells doubles each day. Since it takes approximately 12 hours for a progenitor cell to generate two daughters, more than half of the progeny must continue to divide; that is, many of the cell divisions must produce two mitotically active daughters. During this early "expansion phase" of the progenitor cells, most of the cell divisions are symmetric, generating two additional progenitor cells (**Figure 3.5**). As development proceeds, and the cell-cycle time becomes progressively longer, the number of new cells generated per day declines. Fewer cell divisions are symmetric and result in two progenitor cells at later stages of development, compared to the early stages of embryogenesis. Instead, in the later stages of neurogenesis, a greater proportion of the progenitor cells differentiate into neurons and glia. By the end of neurogenesis, nearly all of the cells leave the cell cycle, and very few remain to generate new neurons. From these results then it seems that the answer to the question of how cell numbers are regulated

during development of the brain might be divisible into two subquestions: (1) What factors account for the gradual lengthening of the cell cycle during development? (2) What factors control the shift from symmetric "expansion phase" cell divisions of the progenitors to their neurogenic, asymmetric divisions? As we will see below, regulation of these aspects of the process can have profound effects on the total number of cells produced by the progenitors.

The answers to these questions came in part from the identification of the molecular machinery that powers the mitotic cell cycle. Many of the same molecular mechanisms that control the proliferation of progenitors in the nervous system are also important for the control of cell division in other tissues (**Figure 3.6**). Through the analysis of mutations in yeast cells that disrupt normal cell cycle, a number of the components of the molecular machinery controlling cell cycle have been identified. An intricate sequence of protein interactions controls and coordinates the progress of a cell through the stages of cell replication. This molecular mechanism has been conserved over the millions of years of evolution from the simplest eukaryotic cells, like yeast, to more complex animals and plants. Key components of the cell cycle control process are called cyclins, a group of proteins that show dramatic changes in their expression levels that correlate with specific stages of the cell cycle. The association of cyclins with another class of proteins, called the cyclin-dependent kinases (Cdk), causes the activation of these kinases and the subsequent phosphorylation of substrate proteins necessary for progression to the next phase of the cell cycle. Different cyclin/Cdk pairs are required at different stages of the cell cycle. For example, the binding of cyclinB to Cdc2 forms an active complex that causes a cell to progress through the M-phase of the cycle, while the association of cyclinD and Cdk4 or Cdk6 regulates a critical step in the progression from G1- to S-phase.

One of the critical steps in the control of the cell cycle is at the transition from the G1 stage to the S-phase, and as noted above, cyclinD is an important regulator of this step (Figure 3.6). The cyclinD/Cdk4 complex causes cells to enter S-phase by phosphorylating a protein called retinoblastoma or Rb. This phosphorylation causes the Rb protein to release another transcription factor, E2F, and allows the E2F protein to activate many genes that push the cell into S-phase. The Rb protein received its name from a childhood tumor of the retina, retinoblastoma, since defects in this gene cause uncontrolled retinal progenitor proliferation. In fact, the Rb gene was the first of a class of genes called tumor suppressors to be identified. Children who inherit a mutant copy of the *Rb* gene develop retinoblastoma when the second allele of this gene is mutated in a progenitor cell in the retina. E2F is then free to activate the genes that cause the progenitor to progress through the cell cycle, and there is no active Rb around to stop the process. Thus, the regulation of progenitor proliferation is critical both for making a normal retina and for preventing the uncontrolled cell proliferation that leads to cancer. There are also proteins that inhibit the cell cycle. Two of these, p27^{kip} and p21, are also expressed in the nervous system, and they are expressed in the final mitotic cycle of a progenitor, causing it to exit the cell cycle and differentiate into neurons or glia. The p27^{kip} and p21 gene products are therefore called CdkIs (for Cdk inhibitors).

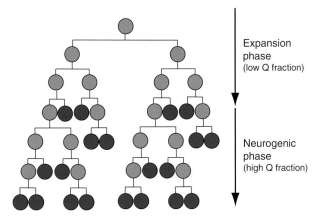

Fig. 3.5 Stages of neurogenesis. Early in development, the progenitor cells (green) divide symmetrically to produce two more progenitor cells. During this expansion phase of histogenesis, the progenitor population expands rapidly. In the middle phase of histogenesis, the progenitor cells divide asymmetrically to produce another progenitor and a postmitotic neuron (red) (sometimes call the Q or quit fraction because they do not reenter the mitotic cycle). In this neurogenic phase, the progenitor pool is stable, but not expanding. However, neurons are being produced and so the total cell number is increasing. At the end of histogenesis, the progenitors produce two postmitotic progeny (either neurons or glia) and the progenitor pool is depleted.

Expansion phase
(low Q fraction)

Neurogenic phase
(high Q fraction)

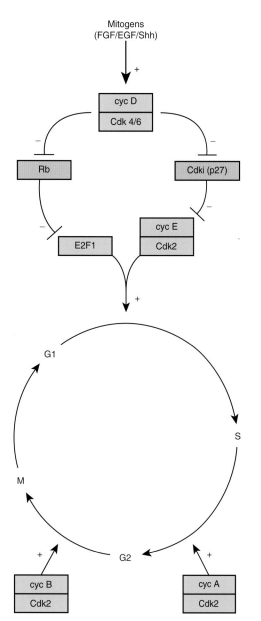

Caviness and his colleagues (Caviness et al., 2003) have developed quantitative models to investigate the role that cell cycle regulators can have in controlling cell number during neurogenesis in cerebral cortex. They have found that p27^{kip} plays a key role in this process. Total cell output from the mitotic divisions of the cortical progenitors can be expressed as the P (or progenitor fraction) + Q (or the quit fraction). The Q fraction is composed of postmitotic neurons, and so the daughters of a progenitor division that choose a neuronal fate no longer contribute to the production of additional neurons. In the early embryonic cerebral cortex, the percentage of the total in the Q fraction is relatively small, and those cells in the P fraction continue to divide and produce more cells. However, as development proceeds, the percentage of cells in the Q fraction increases and the overall growth rate of the cortex declines. If p27^{kip} is experimentally reduced, by knocking out the gene in mice, a smaller percentage of cells enter the Q fraction, and the resulting cortex is noticeably thicker (**Figure 3.7**). On the other hand, the converse experiment of

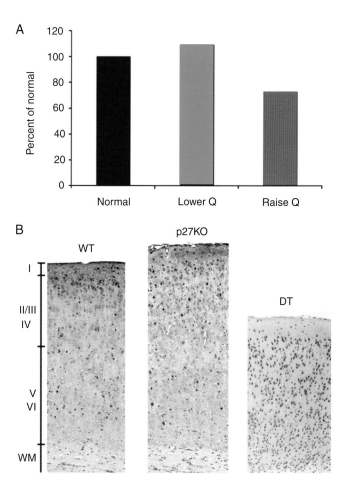

Fig. 3.6 Basic molecular mechanisms of the mitotic cell cycle. Molecules that promote cell proliferation are shown in green and those that inhibit the cell cycle are shown in red. The entry of a cell into the S-phase is one of the key check points on mitosis. E2F1 and cyclinE/Cdk2 complexes cause cells to enter S-phase. However, there are "brakes" on S-phase entry, the Rb protein and cyclin dependent kinase inhibitors, like p27^{kip}. Mitogens that stimulate cells to enter the cell cycle, like EGF and FGF, stimulate cyclinD expression or stabilization, which then inhibits the "brakes" on the system and promotes S-phase.

Fig. 3.7 Cerebral cortex growth and cell cycle regulators. A. The total number of neurons produced in the cerebral cortex is a function of the number of cell divisions of the progenitors that produce more progenitors (P) and the divisions that produce neurons (the "quit" or Q fraction). Mathematical modeling of this process shows that a slightly lower Q fraction produces an increase in the number of neurons in the cortex (green), while a higher Q fraction depletes the progenitor pool earlier in development and results in a smaller total cell number (red). B. Actual cortical thickness of mice in which p27^{kip} was genetically deleted (KO), resulting in a lower Q fraction and a larger brain. By contrast, if the Q fraction is experimentally raised in mice, the cortex is markedly thinner. *(Modified from Caviness et al., 2003)*

over-expressing p27^{kip} leads to a greater percentage of cells in the Q fraction, and a markedly thinner cortex. Therefore, the level of the CdkI p27^{kip} modulates the probability that a cell will enter the Q fraction. Since different cortical layers are generated at different times during development (see below), the level of p27^{kip} expression also affects the relative numbers of cortical cells in the various layers.

The studies of cyclins and their regulators have indeed revealed part of the answer to the question that began this section; however, we have really just pushed the question back a step and you may be wondering: "what regulates the cyclins?" Once again, the progenitors use many of the same regulatory factors as other tissues in the body. In many tissues in the body, secreted signaling factors have been identified that stimulate or inhibit the progress of mitotically active cells through the cell cycle. The signals that stimulate the proliferation of the mitotic cells are called growth factors or mitogens and were named for the tissue or cell type where they were first found to have mitogenic effects. For example, fibroblast growth factor (FGF) was first found to promote the proliferation of fibroblasts in cell cultures, whereas epidermal growth factor (EGF) was discovered as a mitogen for epidermal cells in vitro. These growth factors most commonly act to control the progression from G1- to S-phase of the cell cycle, in part by controlling the level of expression of cell cycle regulator proteins like cyclinD1. One potential explanation for the gradual lengthening of the G1 phase of the cell cycle in the progenitor cells at later stages of development (above) is an increasing dependence on these mitogenic growth factors for progression through the cell cycle as development proceeds. The factors that have been shown to act as mitogens for the the progenitor cells of the vertebrate CNS are primarily those peptides that act on receptor tyrosine kinases, including FGFs, TGF-alpha, EGF, and insulin-like growth factors. However, there are many other types of signaling molecules that act on progenitor cells in the nervous system and also play a role in their proliferation. Sonic hedgehog and members of the Wnt protein family are examples of molecules that are involved in patterning the nervous system (reviewed in Chapter 2), but are also critical for the regulation of progenitor proliferation at later stages of brain development. Progenitor cells express receptors for the various mitogenic factors, and depending on their location and stage of development, they are more responsive to one mitogen or another. Mitogenic factors like EGF and FGF stimulate cell division by the upregulation of the S-phase cyclins (Figure 3.7), such as cyclinD. On the other hand, there are also signaling molecules that act as "stop signals" for proliferation, like TGF-beta. These work through surface receptors to upregulate expression of cell cycle inhibitors, like p27^{kip}. The progenitors must integrate these signals to determine whether they progress to the next S-phase and in this way the extracellular signals are connected with the intrinsic cell cycle regulation machinery to allow for the correct cell numbers in each region of the brain.

In this section we have seen that the regulation of the numbers of neurons and glia in the developing brain is influenced by factors that cause the gradual lengthening of the cell cycle during development and factors that control the shift from symmetric "expansion phase" cell divisions of the progenitors to their neurogenic, asymmetric divisions. In fact, there is some evidence that the two processes might be intimately connected; Calegari et al. (2005) proposed that, as the cell cycle length gets progressively longer over the period of neurogenesis, this *causes* the progenitors to switch from generating additional progenitors, to the generation of neurons, though the mechanisms for this connection are not yet known. In the next section, we will discuss the mechanisms that control the developmental decision of the progenitor to produce neurons, glia or both. This process also appears to be tied to the developmental stage of the cell.

THE GENERATION OF NEURONS AND GLIA

Retroviral lineage studies have shown that, for many regions of the nervous system, neurons, astrocytes, and oligodendrocytes can arise from a single infected progenitor cell. In addition, these studies also showed that the ratio of the different types of cells produced by a progenitor is quite variable. One multipotent progenitor might produce only a few neurons, but many astrocytes, while another might generate mostly neurons. Therefore, the lineages of multipotent progenitors are considered to be "indeterminant" in the vertebrate CNS. Early in development, most of the progenitors are multipotent, but in some regions of the brain, there are committed progenitors that produce only neurons or only glia. What controls the relative number of these different types of cells made from the multipotent progenitors? What distinguishes the multipotent progenitors from the committed progenitors? Tracking the potency of single progenitor cells in vitro has shed some light on these questions. **Figure 3.8** shows the lineages of four different progenitor cells that were isolated from the developing brain, and then maintained in tissue culture over several days; the cell divisions of each progenitor cell were followed by direct observation, and the cell types they generated were confirmed at the end of the culture period by labeling each cell type with an antibody that uniquely recognizes each type of cell (Figure 3.8A). The lineage diagram (Figure 3.8 B) shows that each of the progenitor cells produces a different number of neurons and glia, and so the lineages are indeterminant in vitro as they are in vivo. It can also be seen from the figure that at the beginning of the observation period, two of the four progenitors that were followed over time produced neurons, astrocytes, and oligodendrocytes, while one generated only oligodendrocytes and one produced neurons and oligodendrocytes. Looking more closely, sometimes two different cell types were produced by the last cell division, but more commonly, the multipotent progenitor cells eventually produce bipotent and then unipotent progenitor cells. For example, cell number 2 is tri-potent at the start, but after one generation produces two bipotent progenitor cells, one of which makes neurons and oligodendrocytes and the other makes neurons and astrocytes. These data suggest that the potential of progenitor cells becomes progressively restricted over time and that unipotent progenitor cells are derived from multipotent progenitors.

What controls the progressive restriction in potential of the progenitor cells? Cell culture studies indicate that both extracellular signaling factors, like those that control cell proliferation of the progenitor and intrinsic processes within the cells, play important roles in regulating the potential of the progenitor cells to either a neuronal or glial lineage. In cell cultures, one can add defined factors and assay the effects on the production of either neurons or glia from the progenitor cells. These kinds of studies have led to some general

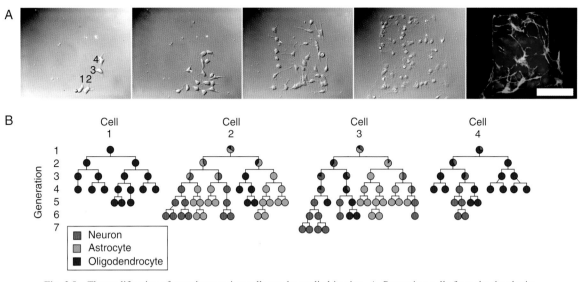

Fig. 3.8 The proliferation of neural progenitor cells can be studied in vitro. A. Progenitor cells from the developing CNS can be studied in cell culture by dissociating them into single cells, diluting them to only a few cells in each well of a tissue culture dish, and then examining them daily for increases in their numbers. B. These micrographs, taken daily, document the proliferation of progenitors. Labeling the culture with antibodies against cell-specific protein shows that several of the new cells have developed into neurons, while others express antigenic markers of either oligodendrocytes or astrocytes. (*A, B, Modified from Ravin et al., 2008*)

principles. In general, the addition of fibroblast growth factors (FGFs) to culture of neural progenitors causes them to increase their differentiation into neurons. By contrast, when cultures of progenitors are cultured in the presence of epidermal growth factor (EGF), ciliary neuronotrophic factor (CNTF), or bone morphogenic proteins (BMPs), the cells are more likely to develop as astrocytes. Still other factors, like PDGF (platelet-derived growth factor), promote oligodendroglial development when added to similar cultures (Raff et al., 1988). These pathways are summarized in **Figure 3.9**. While there are many exceptions to these generalizations, the effects of CNTF on astrocyte development have been particularly well worked out (Bonni et al., 1997; Rajan and McKay, 1998). The activation of the CNTF receptor leads to phosphorylation of a downstream signaling molecule, STAT3. The active STAT3 goes directly into the nucleus, binds to the promoter and activates the glial specific genes GFAP and S100. Thus, this provides a direct transcriptional connection between the signaling molecule and a glial-specific gene. BMP synergizes with the CNTF to give an even more robust response. The early progenitors are relatively unresponsive to CNTF signaling, and hence few glia are produced early in development. However, as embryogenesis continues, glia begin to be generated. What accounts for the increased response of these late progenitors to the gliogenic signal CNTF? The responsiveness of the cell to the CNTF signal is an intrinsic property of the progenitor that changes over development. The DNA in the promoter of GFAP is methylated in the early progenitors, so that the STAT3 cannot bind there and activate GFAP expression (Takizawa et al., 2001; Fan et al., 2005). A similar block in access is present in the promoters of other glial genes, and hence, early progenitors are blocked from producing astrocytes. This interplay between signaling factors in the local environment of the progenitors, along with intrinsic properties of the cells, allows for the developmental program to respond to the surrounding cells.

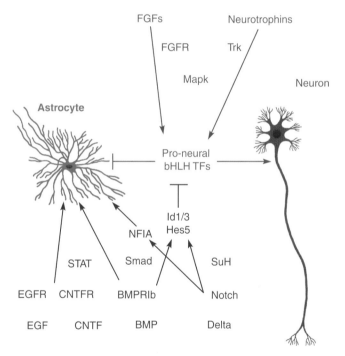

Fig. 3.9 Various mitogenic factors control proliferation of the different types of progenitors in the nervous system. Neurogenesis and gliogenesis are regulated by many growth factors, and these are summarized in the figure. *FGF2* and *Neurotrophin3* promote progenitor cells isolated from the brain to develop primarily as neurons, likely through the increase in expression of proneural *bHLH* genes, such as *NeuroD1*, EGF, and CNTF, which cause the progenitor cells to develop as astrocytes, and at least for CNTF this is known to work through the activation of the STAT transcription factor, which binds to the promoter of the glial-specific gene, GFAP. BMPs can synergize with CNTF to promote glial development, partly through the STAT pathway and partly through a direct inhibition of the proneural genes via the *Hes* pathway. Notch activation drives expression of the astrocyte-specific transcription factor, NFIA, and at the same time also activates the *Hes* pathway to promote gliogenesis and inhibit neurogenesis.

Another important pathway that regulates the production of neurons and glia is the Notch pathway. As we saw in the previous chapters, the Notch signaling pathway and the proneural transcription factors are important in the early stages of nervous system formation. These genes also play critical roles in the process of neurogenesis (Bertrand et al., 2002). The components of the Notch pathway and the proneural transcription factors are expressed in the progenitors and the differentiating neurons. The progenitor cells express several proneural transcription factors, including Mash1, Neurogenin, and Olig1/2. These proneural transcription factors are important for maintaining the progenitors by activating the expression of the Notch ligands, Dll1/3 (Kageyama et al., 2008). The Notch ligands, in turn, activate the Notch receptors on the progenitor cells, and activate the expression of the Hes genes, Hes1/5, and related factors. The Hes genes, and the Notch receptor itself, are necessary for the maintenance of the progenitor state in the cells. Blocking the Notch receptor, either genetically or with specific inhibitors, leads to the premature differentiation of these cells into neurons (Nelson et al., 2007). Over-expression of activated Notch causes the opposite: the progenitor cells fail to differentiate into neurons, and either remain progenitors, or become glia (see below). If all the progenitors have approximately equal levels of Neurogenin, Dll, and Hes1, the progenitor pool is maintained; however, if one of the daughter cells from a mitotic division expresses a higher level of Neurogenin than the other, it will also express

more Dll; this will activate Notch in the sister progenitor cell at a higher level, and lower its level of Dll, leaving the cell with more Neurogenin free to differentiate as a neuron. Creating even a small bias in the two daughter cells in their expression of Neurogenin, its repressor, Hes1, or the activity of the Notch receptor, would lead to the amplification of the difference because of this feedback between the two cells. In this way, the decision of one cell to become a neuron, while the surrounding cells remain as progenitors is similar to the developmental decision in the *Drosophila* of one cell from the proneural cluster to become the neuroblast while the surrounding cells remain as epidermal cells (**Figure 3.10**).

In addition to the basic feedback loop between cells created by the Notch pathway, Kageyama's group found that Hes1 protein represses its own transcription, and this leads to a simple feedback loop that causes the protein levels to oscillate within each progenitor cell every 2–3 hours (Shimojo et al., 2008). This Hes1 oscillation within each progenitor cell causes a counter-oscillation of the Neurogenin and Dll in each cell. Now when two progenitors, each with their own Hes1/Neurogenin cycle are brought in contact, they should cycle in opposite phase to one another. A second oscillation in the progenitor cell expression of Notch pathway activity also occurs with the mitotic cell cycle. Hes1 levels are higher when the cell is in the S-phase and lower as the cells enter the M-phase and G1 phase of the cell cycle (near the ventricular surface). The inhibition of Notch signaling at the M-phase appears to

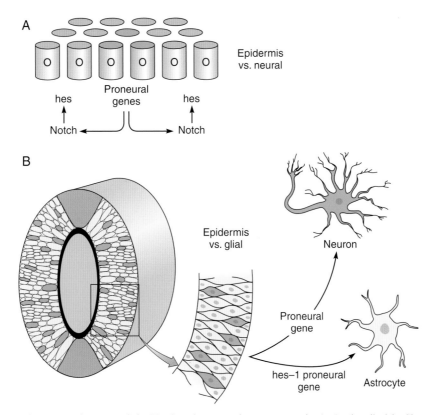

Fig. 3.10 The proneural genes and the Notch pathway regulate neurogenesis. A. As described in Chapter 1, the proneural genes are important in the initial segregation of neural tissue from the epidermis in both *Drosophila* and vertebrates and function through an inter-cellular feedback loop to amplify small biases between cells. Proneural transcription factors induce expression of Notch ligands (DL) which then activate the Notch receptor in adjacent cells (N) to drive *Hes1* expression, which inhibits the proneural genes. B. Within the progenitor population in the ventricular zone, all the cells express these genes at some level. Those cells that express higher levels differentiate as neurons (red).

be due to the release of the Notch pathway inhibitor ACBD3 from the Golgi as the cell divides (Zhou et al., 2007). How are the oscillations in Hes1 and Notch signaling related to neurogenesis? Although the molecular mechanisms underlying the oscillations in Notch signaling are known, it is not clear whether these are critical for the process of neurogenesis.

The Notch pathway is also critical for regulating gliogenesis. Over-expression of activators of this pathway in progenitors (e.g., Hes1, Hes5 or activated forms of the Notch receptor; Vetter and Moore, 2001) can either maintain the cells in the progenitor state or, in some cases, cause the cells to become glia, primarily astrocytes. Since astrocytic glia are frequently produced later than neurons in most areas of the nervous system, it could be that the activation of the Notch pathway simply prevents progenitors from differentiating until other signals that induce glial differentiation are produced. As we have seen above, signaling molecules like BMP and CNTF promote gliogenesis, through their activation of glial-specific genes. However, there is other evidence that over-activating Notch plays an instructive role in gliogenesis as well as this more permissive role. As noted above, the early progenitors are blocked from making astrocytes because the DNA in the promoters of glial-specific genes is methylated, inhibiting access of STAT3. Activated Notch induces the demethylation of the STAT3 binding site of the GFAP promoter, so that CNTF can better activate this pathway (Namihira et al., 2009). Along with the demethylation of the GFAP promoter, Notch activation also induces the expression of a glial promoting transcription factor: NFIA. NFIA is expressed by the late staged progenitors during the time when astrocytes are being generated (Deneen et al., 2006). NFIA is both necessary and sufficient for induction of astrocyte genes: knocking this gene out in mice leads to a reduction in astrogliogenesis, while over-expressing this gene leads to an increase in astrocyte production by the progenitors. At the same time the NFIA activates the gliogenesis program (Figure 3.9), it is also important in the repression of neurogenesis from the progenitors. The ability of NFIA to repress neurogenesis is mediated at least in part by Hes5, which as we saw in the previous section, is downstream of Notch signaling. This nicely ties together the findings that Notch and Hes5 can promote the glial fate.

In addition to astrocytes, the other type of macroglia in the central nervous system is the oligodendrocytes. What are the mechanisms that control their formation during neural development? As noted above, the lineage studies in vitro and in vivo showed that progenitor clones could contain neurons, astrocytes, and oligodendrocytes. This means that at least early in development, there are tripotent progenitors. It was observed several years ago, though, that the oligodendrocytes only arise from the ventral part of the spinal cord, and further studies found they only arise from the progenitors in a relatively small part of the ventral ventricular zone, called the pMN, since it is also the region that produces the motor neurons (MN). As we saw in the previous chapter, the signaling molecule Shh is critical for the determination of ventral fates in the spinal cord, and so is also necessary for the specification of this zone and the production of motor neurons (see Chapter 4). A search for the genes that are necessary for oligodendrocyte development led to the discovery of two transcription factors called Olig1 and Olig2. These transcription factors, are specifically expressed in the pMN

zone, and when they are knocked out in mice, the oligodendrocytes fail to develop. In addition, over-expression of Olig1 or Olig2 can induce additional oligodendrocytes, and so it can be considered part of the transcriptional network that controls development of this cell type. However, the Olig1/2 knockout mice have an additional problem: the motor neurons don't develop either. These results would indicate that the progenitors in this region are the ones that are competent to make both neurons and oligodendrocytes. Birthdating studies and retroviral lineage studies have both shown that the motoneurons are produced prior to the oligodendrocytes. Taking these data together, it would appear that the same progenitors make both cell types, but they initially generate motoneurons and then switch to oligodendrocytes production. What accounts for the switch in cell type production by these cells? At the time the cells make motoneurons, they express the proneural gene, Neurog2 (Kessaris et al., 2001; Zhou et al., 2001). Neurog2 and Olig1/2 combine their activity to generate motoneurons. Later in development, however, the cells turn off Neurog2 and produce a very different type of transcription factor, Nkx2.2; Nkx2.2 is a repressor of motoneuron genes (see Chapter 4), but it does not repress Olig1/2 expression in this region. Now the progenitors that express both Olig1/2 and Nkx2.2 start making oligodendrocytes (**Figure 3.11**). This molecular switch enables progenitor cells to produce different types of progeny at different times in development. A progenitor cell from this region of the neural tube would initially make neurons, but after one or two generations, it would progress to a state where it was generating oligodendrocytes.

The foregoing two sections have shown one of the common themes in the process of neurogenesis: intrinsic changes in the progenitor cells over developmental time determine the cell's responsiveness to signaling factors produced by neighboring cells. The developmental dance between cell-intrinsic and cell-extrinsic regulation of cell division and neuron or glial production allows for great flexibility in the numbers and relative proportions of neurons and glia in the different regions of the brain and in different species. This theme continues in Chapter 4 as we consider the question of neuronal diversity.

CEREBRAL CORTEX HISTOGENESIS

The cerebral cortex has been particularly instructive in elucidating the principles of histogenesis in the developing brain. Histogenesis is the process by which architecturally organized regions of the brain, such as the six-layered cortex, can be understood in terms of the timing of neurogenesis. The human neocortex has been called the "most complicated object in the universe," and there is no doubt that this structure endows us with remarkable cognitive abilities. The six-layered neocortex (**Figure 3.12**) is a uniquely mammalian structure, and it reaches its most extensive elaboration in humans. There has been a dramatic increase in size during evolution; comparing the surface area of the mouse, macaque monkey and human cerebral cortex gives ratios of 1 to 100 to 1000, and this has been accompanied by an increase in the number of distinct, identifiable regions, based on the relative numbers of neurons in each layer. The layers are numbered from the most superficial, layer I, to the deepest, layer VI. While all regions of the neocortex have these six layers, there are variations in the relative numbers of neurons in the different cortical regions,

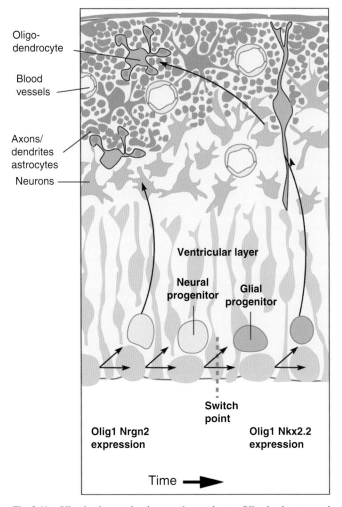

Oligo-
dendrocyte

Blood
vessels

Axons/
dendrites
astrocytes

Neurons

Ventricular layer

**Neural
progenitor**   **Glial
progenitor**

**Switch
point**

**Olig1 Nrgn2
expression**   **Olig1 Nkx2.2
expression**

Time ➡

Fig. 3.11 Oligodendrocyte development in vertebrates. Oligodendrocytes and neurons are derived from the same pool of stem cells that divide in the ventricular layer of the developing neural tube (pMN). At early stages, when the progenitors express both Olig1/2 and Neurog2, they generate motor neurons. Later in development, the same population of progenitors begin to express Nkx2.2, and down-regulate their expression of Neurog2. This acts as a molecular switch to cause the cells to start making oligodendrocytes instead of motor neurons.

Fig. 3.12 Drawing of a Golgi stained section showing the neurons in the human cerebral cortex from Ramón y Cajal, 1952. The layering and complex dendritic processes are critical for the processing units of the cerebral cortical columns.

depending on their function. For example, regions devoted to processing sensory information, like the visual cortex, have relatively large numbers of layer IV cells, which form the input layer, while regions important in the "output" of information from the cortex, such as the primary motor cortex, have large layer V pyramidal neurons, and relatively few layer IV cells. As the size of the cortex increased in evolution, there was an increase in specialization of the various regions of cortex, such that in people there are as many as 50 different regions that can be identified on the basis of their distinct "cytoarchitecture" (i.e., differences in the relative numbers of neurons in each of the six layers; see Chapter 2). So what are the mechanisms that have enabled us to develop the amazing neural tissue responsible for the accomplishments of Shakespeare and Einstein?

As noted in the previous chapter, the cerebral hemispheres develop from the wall of the telencephalic vesicle. The neocortex begins as a relatively simple neuroepithelium, similar to that which we have already encountered in the posterior regions of the neural tube—the spinal cord. The early embryonic neocortex is made up of morphologically homogeneous cells that span the width of the epithelium and have a simple bipolar shape and undergo extensive rounds of mitosis as the cerebral vesicle expands (**Figure 3.13**). Within a few days (in the mouse) additional cell types can be identified in the developing cerebral cortex, including postmitotic, migrating neurons and additional types of proliferating cells. The two most well-characterized types of proliferating cells are the apical progenitors (also called the radial glia, for reasons described below) and the intermediate precursor cells (IPCs or basal progenitors). The apical progenitors act as the primary "stem cell" of the cerebral cortex, with the capacity to generate all types of neurons and glia (see below). They divide asymmetrically in the region of the cortex adjacent to the lateral ventricle (a.k.a., the ventricular zone) and one of the cells from the mitotic division is a postmitotic neuron, which migrates along the basal process of the apical progenitor to its final position in one of the layers of the cortex (Figure 3.13).

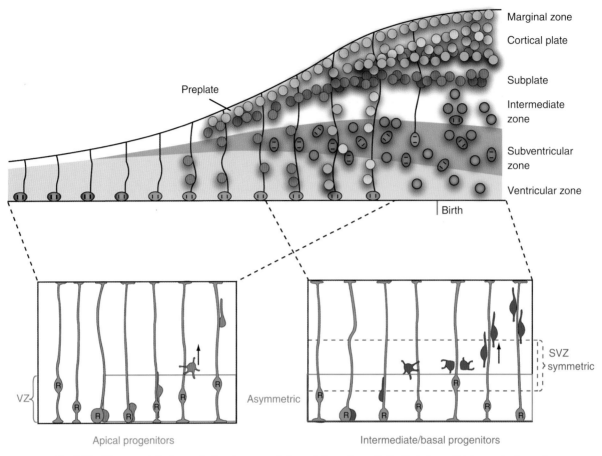

Fig. 3.13 Histogenesis in the cerebral cortex proceeds through three stages. In the first stage of histogenesis, the wall of the cerebral cortex is made up of the progenitor cells, which occupy the ventricular zone (VZ). In the next stage of development, the first neurons exit the cell cycle (red) and accumulate in the preplate, adjacent to the pial surface. The neurons of the preplate can be divided into the more superficial Cajal-Retzius cells and the subplate cells. In the next stage of cortical histogenesis, newly generated neurons (red) migrate along radial glial fibers to form a layer between the Cajal-Retzius cells and the subplate. This layer is called the cortical plate, and the majority of the neurons in the cerebral cortex accumulate in this layer. *(Modified from Noctor et al., 2004; 2008)*

The first neurons that are generated from the ventricular zone migrate a short distance to form a distinct layer known as the preplate, just beneath the outer surface of the cortex (Figure 3.13). The preplate consists of two distinct cell types: a more superficial marginal zone, containing a group of large, stellate-shaped cells, known as Cajal-Retzius cells, and a deeper zone of cells, called the subplate cells. The next stage of cortical development is characterized by a large accumulation of newly postmitotic neurons within the preplate (Marin-Padilla, 1998). These new neurons are called the cortical plate. The cortical plate divides the preplate into the superficial marginal zone (Cajal-Retzius cells) and the intermediate zone (subplate cells and increasing numbers of incoming axons). The developing cortex is thus described as having four layers: the ventricular zone, the intermediate zone, the cortical plate, and the marginal zone (Figure 3.13).

Although many of the mitotic divisions of the apical progenitors in the ventricular zone generate postmitotic cortical neurons directly, others produce progeny that can continue to undergo additional mitotic divisions after they leave the ventricular zone. As noted above, these cells are called intermediate progenitor cells (IPCs) and after leaving the ventricular zone, they migrate a short distance to a specialized zone

between the ventricular zone and the neurons of the cortex called the subventricular zone (or SVZ; Figure 3.13). Once an IPC has migrated to the SVZ, it divides symmetrically and usually generates two neurons, but may divide up to three times, making as many as six neurons.

The next phase of cortical histogenesis is characterized by the gradual appearance of defined layers within the cortical plate. As increasing numbers of newly generated neurons migrate from the ventricular zone into the cortical plate, they settle in progressively more peripheral zones. Meanwhile, the earlier-generated neurons are differentiating. Thus, later-generated neurons migrate past those generated earlier. As noted earlier in the chapter, this results in an inside-out development of cortical layers (Figure 3.3).

Pasko Rakic (1988) hypothesized that the progeny of each apical progenitor (ie. radial glial cell) form a column of differentiated neurons. As mammals evolved, they acquired more and more of these radial columns to expand the cortical processing power. This rather simple amplification strategy could explain the relative ease with which this region of the brain can undergo remarkable expansions during evolution. For example, the difference between monkey and human cortex can be viewed as a tenfold increase in the number of these radial units. This could come about as a result of symmetric mitotic divisions of the apical progenitor cells to expand their number and the ventricular surface they occupy. As the number of these units increases, the resulting increase in cerebral cortical surface area requires that sulci and gyri form and the cortex becomes folded. The radial unit hypothesis has recently been modified to include the IPCs (Pontious et al., 2008). Changes in the IPC amplification factor (1–3 cycles) at different times of development in different regions of cortex might explain the differences in cell numbers in the different areas of cortex, and might also provide a mechanism for the variation in relative cell numbers in the various regions of cerebral cortex. The way this could work for motor cortex, a region with many layer V projection neurons, versus visual cortex, a region with relatively more layer IV input neurons, is as follows: 1. during the early stage of neurogenesis, when layer V cells are being made, the IPCs in the motor cortex divide more times and therefore generate extra layer V neurons; 2. in the sensory cortex, however, the IPCs do not divide as many times early in cortical development, but instead have most of their divisions during the middle stage of cortical development, when layer

IV neurons are generated. In the last part of neurogenesis, both regions have the same number of IPC divisions, and so have the same number of upper layer (2–3) neurons. The final cytoarchitecture of the two regions reflects the difference in timing of maximum IPC proliferation, with the sensory area having more layer IV input neurons and the motor cortex having more layer V output neurons. A great deal more will be said about the molecular mechanisms that specify the various types of neurons in the cerebral cortex in Chapter 4.

The neurons produced in the ventricular zone then have to migrate to their final destinations in the different layers of the cerebral cortex. Since the processes of the apical progenitor cells span the entire thickness of the cortex, the cortical neurons that are generated use the predominantly radial orientation of their processes to guide their migration. Serial section electron microscopic studies by Rakic first clearly demonstrated the close association of migrating neurons with the processes of these cells, though at the time they were thought to be glial cells and so were given the name, "radial glia" (**Figure 3.14**). The migrating neurons wrap around their processes like you would if you were climbing up a pole. It has been possible to

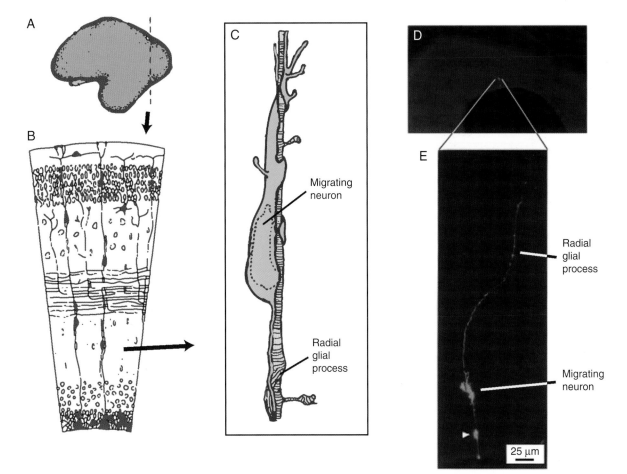

Fig. 3.14 A.B. Migration of neurons along radial glia. The radial glial fibers extend from the ventricular zone to the pial surface of the cerebral cortex. A section through the cerebral cortex at an intermediate stage of histogenesis shows the relationship of the radial glia and the migrating neurons. C. The postmitotic neurons wrap around the radial glia on their migration from the ventricular zone to their settling point in the cortical plate. *(From Rakic, 1972)* D,E. Live imaging of GFP labeled radial glia shows that radial glia are the same cells as the progenitors. Noctor et al. (2002) used a retrovirus to label small numbers of cortical progenitor cells in slice cultures of the cerebral cortex of mice. In this example, they found that the radial glia (arrowhead) has undergone several cell divisions, and the progeny are migrating immature neurons. The neurons migrate along the radial glia that generated them. *(From Noctor et al., 2001)*

directly observe the process of neuronal migration in vitro in slices of cerebral cortex kept alive in culture for several days. In these studies, cells in the ventricular zone were labeled using a GFP-expressing retrovirus to mark a subpopulation of the newly generated neuroblasts. As these cells left the ventricular zone, their leading processes were visible. Time-lapse imaging of neuroblasts shows clearly that many of the neuroblasts migrate just as predicted from the EM reconstructions of Rakic (Noctor et al., 2002).

While confirming the EM studies of Rakic, the direct visualization of neuronal migration gave rise to a surprise. As noted above, for many years it was thought that the radial glia and the progenitor cells were two separate populations. The radial glia were thought to have been generated early in development, and then as postmitotic cells, providing a scaffold to guide the newly generated neurons to the correct laminar position. However, the time-lapse imaging of the GFP-labeled radial glia revealed a surprising result: Noctor et al. (2002) found that the radial glia themselves *were* the neuronal progenitors. Figure 3.14 shows an example of one of the clones they found. When the slice was viewed on the first day, the labeled cell was a single radial glial cell, with a process extending the entire width of the cerebral cortex; however, as they continued to analyze the clone on subsequent days, they found that the radial glia underwent several cell divisions, and the progeny were not

additional radial glia, but migrating immature neurons. These neurons migrated along the radial glia that generated them. In addition to having the morphology of neurons, these migrating neurons labeled for neuron-specific markers, while the radial glial cell that generated them expressed proteins typical of radial glia. This finding, and critical findings from the labs of Magdelena Götz (Malatesta et al., 2000), Nat Heintz and Gord Fischell (Anthony et al., 2004) led to the model that was presented earlier in this section, that the radial glia and the ventricular zone apical progenitors are one and the same, and most neurons in the cortex are derived from them.

In addition to the predominantly radial migration of the newly generated neurons, however, there are also some cells that migrate tangential to the cortical surface, in the intermediate zone. Some of these cells arise in the ventricular zone from the cortical progenitors, but most of them are not derived from the cortical ventricular zone at all, but instead migrate all the way from the ventricular zone of a subcortical forebrain region (**Figure 3.15**). These tangentially migrating neurons are a special subpopulation of the neurons in the cortex. Although most of the neurons of the cerebral cortex are pyramidal in shape and use the neurotransmitter glutamate, there are other populations of neurons in the cerebral cortex that are stellate-shaped and use GABA as their transmitter. These GABA+ cells are not derived from the cortical ventricular

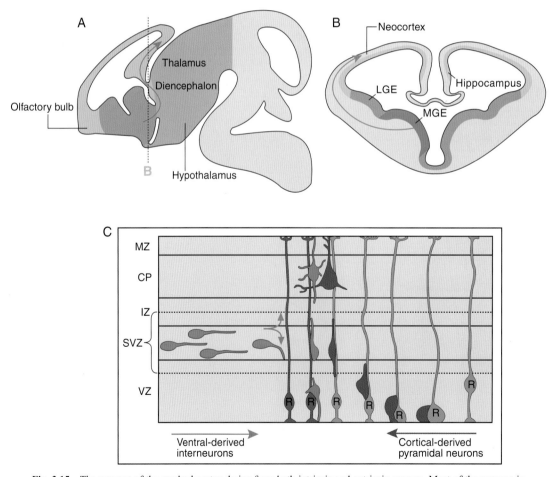

Fig. 3.15 The neurons of the cerebral cortex derive from both intrinsic and extrinsic sources. Most of the neurons in the cortex are derived from the ventricular zone cells immediately below their adult location. A-C. This figure shows the paths taken by the neurons from the ganglionic eminence in red at low (A,B) and high magnification (C) and the neurons generated within the cortex in green. *(Modified from Kriegstein and Noctor, 2004)*

zone, but instead they are the tangentially migrating cells that were produced by the progenitors in the subcortical zone known as the medial ganglionic eminence (MGE). Although the primary role of the MGE during development is to produce the neurons and glia of the basal ganglia (deep forebrain nuclei), they also produce these special neurons for the cortex. This was directly shown by the following experiment: when cortical slices were cultured without the MGE attached, the number of GABA neurons in the cortex was greatly reduced as compared to cultures that contained the MGE. The migration of these cells has also been directly visualized by labeling the premigratory population in the MGE, and tracking their migration to the cerebral cortex. Thus, it is now accepted that the precursors of most GABA-containing interneurons in the cerebral cortex migrate all the way from the subcortical progenitor zones (Corbin et al., 2001; Nakajima, 2007).

CEREBELLAR CORTEX HISTOGENESIS

As noted above, the cerebellum is a large, highly convoluted part of the brain that is critical for control of our movements, particularly our balance. Cerebellar function is particularly susceptible to ethanol; the weaving motion of alcoholics is likely due to alcohol's effects on cerebellar function. The mature cerebellum is made up of several distinct cell types, each repeated in an almost crystalline array (**Figure 3.16**). The two most distinctive of these cell types are the giant Purkinje cells and the very small granule cells. Purkinje cells are the principal neurons of the cerebellar cortex, sending axons out of the cortex to the deep cerebellar nuclei. The cerebellar granule neurons are much more numerous than the Purkinje cells. In fact, the cerebellar granule cells are the most numerous type of neuron in the brain. In the mature cerebellum, they form a layer deep to the Purkinje cells, and their axons extend past the Purkinje cell layer into the molecular layer. The axons of the granule cells bifurcate in the molecular layer, into a T-shape, and these axons extend in the molecular layer for a considerable distance, synapsing on the Purkinje cell dendrites. One can think of the Purkinje cells as telephone poles and the granule cell axons as the telephone wires.

The generation of the intricate cerebellar architecture is a complex process. The large Purkinje neurons are generated from a ventricular zone near the fourth ventricle of the brainstem, in a manner similar to the way in which the neurons of the cerebral cortex are produced. Once they have finished their final mitotic division, the Purkinje cells migrate a short distance radially to accumulate as an irregular layer, known as the cerebellar plate. As the cerebellum expands, these cells become aligned to form a single, regularly spaced layer. The Purkinje cells then grow their elaborate dendrites. In addition to the Purkinje cells, the ventricular zone generates several other cerebellar interneurons, such as the stellate and basket cells. In contrast to the somewhat standard pattern of neurogenesis of the Purkinje cells and the stellate and basket cells, the granule cells arise from a completely separate progenitor zone, known as the rhombic lip (**Figure 3.17**). The granule cell precursors are initially generated near the rim of the fourth ventricle but then migrate away from the ventricular zone, over the top of the developing Purkinje cells to form a secondary zone of neurogenesis, called the *external granule layer*. The cells in this layer continue to actively proliferate, generating an enormous number of granule cell progeny, thus increasing the thickness of the external granule layer considerably. The external granular layer persists for a considerable

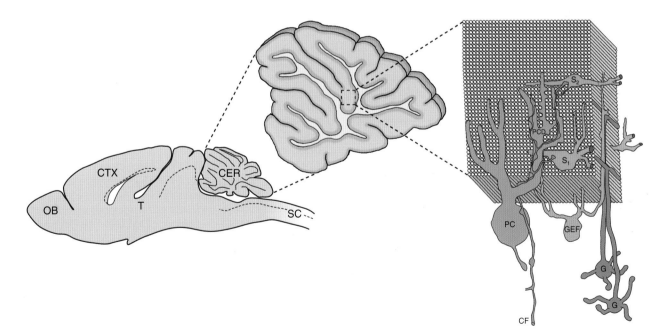

Fig. 3.16 The neurons of the cerebellar cortex are arranged in a highly ordered fashion. In the mature cerebellum, the very large Purkinje cells (PC) lie in a single layer (P) and have an extensive dendritic elaboration that lies in a single plane. The granule cells (red) lie below the Purkinje cells in the granule cell layer (purple) (G) and have a T-shaped axon that runs orthogonal to the plane of the Purkinje cell dendrites, like phone wires strung on the Purkinje cell dendritic "poles" in the molecular layer (M). In addition to these distinctive cell types, the cerebellar cortex also contains other cell classes, the stellate cells (S) and the Golgi epithelial cells (GECs). *(Modified from Rakic, 1971b)*

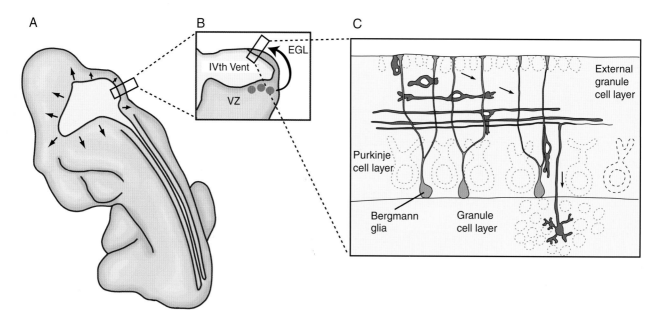

Fig. 3.17 The precursors of the cerebellar granule cells come from a region of the rhombencephalon known as the rhombic lip. The rhombic lip is a region of the hindbrain that lies adjacent to the fourth ventricle. Cells from this region migrate over the surface of the cerebellum to accumulate in a multicellular layer—the external granule cell layer. A.B. This dorsal view of the developing brain shows the migratory path of the granule cell precursors from the rhombic lip of the rhombencephalon to the surface of the cerebellum (arrows). Granule cell production in the external granule cell layer is followed by the migration of these cells to ultimately lie deep to the Purkinje cell layer. C. Arrows show the migratory path a single neuron would take from its birth to the granule cell layer. The Bergmann glial cells are shown in blue and function as guides for the migrating neurons. The migration of a granule cell is thought to take place along a single glia fiber, but in the diagram the migrating neuron is shown to be associated with several glial cells for clarity. (*Modified from Ramón y Cajal, 1952*)

time after birth in most mammals and continues to generate new granule neurons. There are still granule neurons migrating from the external granule layer as late as two years after birth in humans (Jacobson, 1978).

Although the granule neurons are generated superficially in the cerebellar cortex, they come to lie deep to the Purkinje cells in the mature cerebellum. The developing granule neurons must therefore migrate past the Purkinje cells. Figure 3.17 shows what this process looks like, as originally described by Ramón y Cajal (1952). Soon after their generation, after their final mitotic division, the granule cells change from a very round cell to take on a more horizontal-oriented shape as they begin to extend axons tangential to the cortical surface. Next, the cell body extends a large process at right angles to the axon. As this descending process grows deep into the cerebellum, the cell body and nucleus follow, leaving a thin connection to the axon. Meanwhile, the axons have been extending tangentially, and so the cell assumes a T-shape. The cell body eventually migrates past the Purkinje cell layer and then begins to sprout dendrites in the granule cell layer. The migration of the granule cells is another example of the importance of radial glia in CNS histogenesis. As they migrate, a specialized type of radial glia, known as the Bergmann glia, guides the granule cells. EM studies, similar to those described for the cerebral cortex, first demonstrated the relationship between the migrating granule cells and the Bergmann glia (Rakic, 1971b). Throughout the migration of the granule cells, they are closely apposed to the Bergmann glial processes. Hatten and her colleagues (1985 and 1990) have been able to demonstrate directly the migration of granule cells on Bergmann glia

using a dissociated culture system. When the external granule cell layer is removed from the cerebellum and the cells are cultured along with cerebellar glia, the granule glial cells migrate along the extended glial fibers in vitro. Time-lapse video recordings have even captured the granule cell migration in action.

A factor first encountered in the context of patterning the nervous system (Chapter 2), Sonic hedgehog, is also a key mitogen for nervous system progenitors, and this has been best demonstrated in the cerebellum. The way in which Shh acts in neurogenesis demonstrates the way in which differentiated neurons can feed back on the progenitors to maintain their proliferation and ensure that the correct number of neurons is generated during development (Wechsler-Reya and Scott, 1999). The Purkinje cells produce the mitogen, Shh, while the granule cell progenitors express the Shh receptors, patched and smoothened (named for *Drosophila* mutants defective in the homologous genes). The Shh released from the Purkinje cells stimulates the granule cell progenitors to make more granule cells. If the Shh pathway is experimentally blocked, fewer granule cells are produced. If the Shh pathway is activated, granule cell production is increased. In this way, Shh is used by the developing nervous system to mediate the cell interactions between the differentiated Purkinje neurons and the neural progenitors. This pathway also provides another example of how a childhood tumor can result from a misregulation of neurogenesis. Children with mutations in the Shh receptor, *patched*, that mediates Shh signaling will develop a tumor called medulloblastoma, in which granule cell production is fatally uncontrolled (Goodrich et al., 1997).

MOLECULAR MECHANISMS OF NEURONAL MIGRATION

As the previous two sections have shown, the processes of neurogenesis and cell migration are frequently closely coupled in the developing brain. In both the cerebral cortex and the cerebellum, the process of glial-guided migration has been the subject of very intense investigation over more than 20 years. In this section, we will describe some of what is known about the molecular mechanisms that underlie the correct positioning of neurons in laminated structures like the cerebral and cerebellar cortices.

Some of the greatest advances in our understanding of the molecular mechanisms of cell migration have come about by analysis of naturally occurring mouse mutations that disrupt the normal migration of neurons. One important function of the cerebellum is to maintain an animal's balance. Lesions to the cerebellum in humans frequently produce a syndrome that includes unsteady walking, known as ataxia. Genetic disruptions of the cerebellum in mice produce a similar syndrome, and therefore they can be identified and studied. By screening large numbers of mice for motor abnormalities, several naturally occurring mutations have been identified that disrupt cerebellar development (Caviness and Rakic, 1978). Because of the nature of the symptoms, these mutant mouse strains have names like *reeler*, *weaver*, and *staggerer*. The mutant genes that underlie these phenotypes have been identified, and one of these mutants, *reeler*, has been particularly informative in understanding neuronal migration. The *reeler* mutant mouse has ataxia and a tremor. Histological examination of individually labeled neurons in *reeler* mutant cerebral and cerebellar cortex revealed gross malpositioning of the cells. In the cerebellar cortex, the Purkinje cells are reduced in number and, instead of forming a single layer, appear to form aggregates instead. There are fewer granule cells, and most of them fail to migrate from the external granule cell layer to their normal mature position below the Purkinje cells.

The effects of the *reeler* mutation have been particularly well-studied in the cerebral cortex, where instead of the normal inside-out pattern of neurons that was described in the previous section, in the *reeler* mutant the later generated neurons fail to migrate past those generated earlier and so the mice have an outside-in organization of their cerebral cortex (**Figure 3.18**). The defective molecule underlying the *reeler* phenotype was identified several years ago. It is a large glycoprotein, named Reelin, containing over 3000 amino acids, and it bears similarities to some extracellular matrix proteins (D'Arcangelo et al., 1995). The Reelin protein is expressed by the most superficial neurons of the cortex, the Cajal-Retzius cells. Insight into the molecular mechanisms by which Reelin controls migration have come from the identification of additional components in its unique signal transduction pathway. Mutations in the genes coding for a tyrosine kinase called disabled or Dab1, the VLDLR (very low density lipoprotein receptor), and ApoER2 (apolipoprotein E, receptor 2) all cause defects in cerebral cortical neuroblast migration similar to those found in *reeler* mice (Jossin et al., 2003), where the newly generated neurons fail to move past the previously generated ones. The VLDLR, along with ApoER2, form a receptor complex that phosphorylates the Dab1 protein upon Reelin binding. Once phosphorylated, disabled can recruit other second messengers in the tyrosine kinase pathway and activate a host of cellular

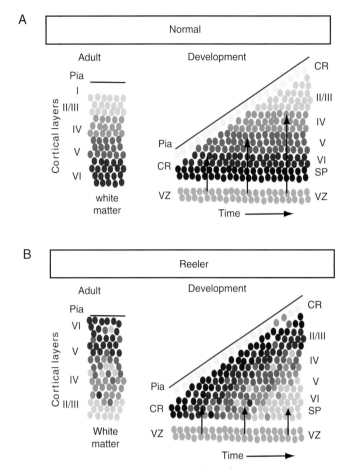

Fig. 3.18 The function of Reelin in the cerebral cortex. A. In normal cortical development the early generated neurons bypass the previously generated neurons to produce the inside-out developmental ordering to the lamination. B. In *reeler* mice, this orderly lamination process is disrupted, and the newly generated neurons cannot pass the previous ones and the lamination is reversed. *(Modified from Cooper, 2008)*

responses. The VLDLR and ApoER2 receptors are expressed in the migrating neuroblasts and in the radial glia themselves, while Reelin is made by the Cajal-Retzius cells at the cortical surface. The observed cellular expression pattern of Reelin and its receptors has led to two basic classes of hypotheses for its function during cortical development: (1) Reelin might be a chemoattractant in the cerebral cortex, causing the migrating neuroblasts to move toward the source of Reelin in the Cajal Retzius cells in the superficial layers of the cortex; (2) alternatively, Reelin could act as a stop signal at the cortical surface, telling the neuroblasts to "get off the track" and form a new cortical layer. Many experiments have been carried out to test these ideas, both in vivo and in vitro, but until recently, no clear answer emerged, and there was support for and against both hypotheses. For example, in support of the second hypothesis, Dulabon et al. (2000) found that adding Reelin to migrating neuroblasts in cell culture causes them to stop their migration. However, this result is not inconsistent with Reelin playing a role as a chemoattractant, since adding Reelin to cell culture surrounds the migrating neuroblasts with the potential attractant and thus causes them to be attracted equally in all directions and to stop moving. To distinguish between these two possibilities, Curran's group generated a transgenic mouse that has Reelin expressed under the control of the Nestin promoter, and therefore expressed in the radial glia themselves (Magdeleno et al., 2002). These mice express Reelin all along the path of migration of the neurons, including the Cajal-Retzius cells. If Reelin is a stop signal or an attractant, the migrating neuroblasts should never leave the ventricular zone. However, they found that the Nestin-Reelin mice were essentially normal. The migrating neuroblasts still left the ventricular zone on schedule and in fact made basically normal layers. They also mated the Nestin-Reelin mice with *reeler* mice. These mice only have the Reelin in the radial glia and no longer express any Reelin in the Cajal-Retzius cells. Although the cortical lamination was not perfect, it was improved over that of the *reeler* mouse. Therefore, it looks as if it is less important where the Reelin is localized in the developing cortex, as long as there is some Reelin around.

More recently, studies on the components of the Reelin signal transduction pathway have shed some light on this puzzle. Most signaling pathways have a built-in negative feedback mechanism to limit the time that the signal is active in a cell. This is particularly important in development, where timing can be critical. For the Reelin pathway, once the Dab1 has been activated, it is targeted to the proteosome by a protein called Cullin5 (Cul5), and then degraded. The Cooper lab has knocked down the level of Cul5 in the migrating neurons (Feng et al., 2007), which caused the activated Dab1 to persist in the neurons longer than it normally would. In these mice, the migrating neurons fail to stop once they have passed the previously generated one, and overshoot their target layer to end up with the Cajal-Retzius cells. Since loss of Dab1 entirely leads

to a failure of the newly generated neurons to migrate past the previous layer, while over-activating the system causes them to migrate too far, the simplest interpretation is that Reelin initially causes the activation of Dab1 to promote neuronal migration, but then causes Dab1 degradation, which causes the cells to stop. In this way, Reelin can be both a "go" signal and a "stop" signal (**Figure 3.19**).

In addition to Reelin, a large number of molecules have been implicated in neuroblast migration in the cerebral and cerebellar cortices, including astrotactin, integrins, and neuregulins. Integrins are cell adhesion molecules that allow many different types of cells to attach to the proteins in the extracellular matrix. Since these adhesion receptors are necessary for the pial extracellular matrix formation, it is not surprising that they are required for the appropriate formation of the glial scaffold and hence the migration of the neuroblasts and correct positioning of the cerebral cortical neurons. It is as if you were trying to climb a ladder without a wall to lean it against. Another class of molecules, the neuregulins and their receptors, likely has a very different role. Neuregulin, or glial growth factor, activates receptor tyrosine kinases called ErbBs on the glial cell surfaces and promotes the appropriate differentiation and/or survival of the glial cells. Without the glial cells adopting their elongate morphology, the neuroblast migration is abnormal. Again comparing this with a ladder, it is as if you were trying to climb a ladder made of rubber.

In sum, many cellular and molecular interactions are necessary for proper arrangement of the neurons in the complex neuronal structures that make up the mature brain. Nearly all the neurons in the brain end up some distance from where they were generated in the ventricular zone, and the mature neuronal

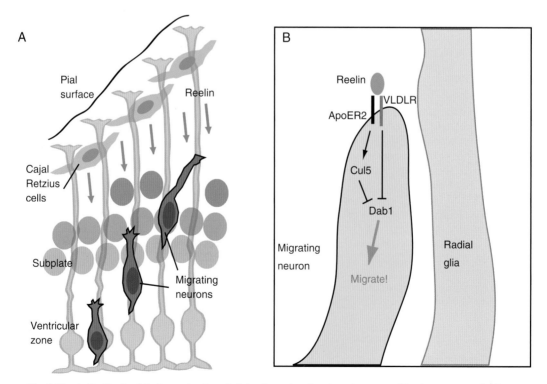

Fig. 3.19   A. Reelin, the defective molecule underlying the *reeler* phenotype, is expressed by the most superficial neurons of the cortex, the Cajal-Retzius cells (green). B. Higher magnification view of the leading process of a migrating neuron to show the signaling complex. Dab1 promotes cell migration. Binding of Reelin to ApoER2 and VLDLR results in the phosphorylation and ubiquitinlyation of Dab1, which leads to its degradation and inhibition of migration.

circuitry depends on cells getting to the right place at the right time. Mice with mutations in genes critical for neuronal migration have motor deficits, but it is likely that more subtle deficits are caused by less dramatic changes in neuronal migration. Several inherited mental retardation syndromes in humans are now known to be caused by defective migration of cortical neuroblasts. The beautiful choreography of neuronal migration is clearly an essential part of building a working nervous system (see Box 3.1).

POSTEMBRYONIC AND ADULT NEUROGENESIS

The process of neurogenesis ceases in most regions of the nervous system in most animals. Neurons themselves are terminally differentiated cells, and there are no well-documented examples of functional neurons reentering the mitotic cycle. However, it has long been appreciated that in most species some new neurons are generated throughout life. There is considerable remodeling of the nervous system of insects during metamorphosis. Much of this remodeling occurs through cell death, but new neurons are also produced.

Many amphibians also go through a larval stage. Frogs and toads have tadpole stages where a considerable amount of body growth takes place prior to metamorphosis into the adult form. During larval stages, many regions of the frog nervous system continue to undergo neurogenesis similar to that in embryonic stages. One of the most well-studied examples of larval frog neurogenesis is in the retinotectal system. The eye of the tadpole, like that of the fish, increases dramatically in size after embryonic development is complete. During this period, the animal uses its visual system to catch prey and avoid predators. The growth of the retina, however, does not occur throughout its full extent, but rather is confined to the periphery (**Figure 3.20**). Similar to the way in which a tree grows, the retina adds new rings of cells at the preexisting edge of the retina. This provides a way for new cell addition to go on at the same time the central retina functions normally. As the new retinal cells are added, they are integrated into the circuitry of the previously differentiated retina, into a seamless structure. The zone of cells responsible for adding the new neurons at the edge of the frog retina is called the ciliary marginal zone or CMZ. The cells of this region act as true "retinal stem cells," in the sense that they can generate all the different types of retinal neurons, and they seem to be inexhaustible: most of the retina of the mature frog or fish is actually generated by these cells, not by the embryonic progenitor cells of the retina. The CMZ is organized in a gradient with the most primitive stem cells located most peripherally, next come the multipotent progenitors, and then most centrally are the differentiating neurons, immediately adjacent to the mature retina. At the same time that new cells are added to the peripheral retina, the optic tectum also adds new neurons. The coordination between neurogenesis in these two regions likely involves their interaction via the retinal ganglion cell projection of the tectum. The growth of the optic tectum, the brain center to which the retina sends its axons, occurs at its caudal margin, so the axons of the ganglion cell must shift caudally during this time.

One of the most well-studied examples of neurogenesis in mature animals comes from studies of song birds. In 1980,

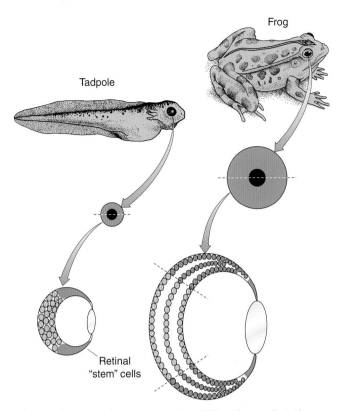

Fig. 3.20 The eyes of frogs grow by the addition of new cells to the margin. The neural retina of the frog tadpole is derived from the neural tube, as described in a previous chapter. The initial retinal neurons are generated during embryogenesis. However, as the eye grows, the neural retina grows by means of a specialized ring of retinal stem cells at the peripheral margin of the eye (red). The retinal stem cells generate all the different types of retinal neurons to produce new retina that is indistinguishable from the retina generated in the embryo, and thoroughly integrated with it. In the newly post-metamorphic *Rana pipiens* frog, nearly 90% of the retina has been generated during the larval stages; all this time the retina has been fully functional. This process continues even after metamorphosis but much more slowly.

Fernando Nottebohm reported that there was a seasonal change in the size of one of the brain nuclei important for song production in adult male canaries. In songbirds, specific nuclei in the telencephalon of the brain are critical for the production of the song. The HVC nucleus is of particular importance for both song learning and song production (see Chapter 10). The HVC is almost twice as large in the spring, when male canaries are generating normal adult song, than in the Fall, when they no longer sing. Nottebohm initially proposed that this change in size might be due to seasonal changes in the numbers of synapses. In further studies of the HVC in male and female canaries, Nottebohm also noticed that it was larger in males, which learn complex songs, than in female birds, which do not sing. Moreover, if adult females were given testosterone injections, the HVC nucleus grew by 90% and the female birds acquired male song (Nottebohm, 1985; 2002); later studies demonstrated that ^{3}H-thymidine neurons were in fact generated in the mature bird (Paton and Nottebohm, 1984).

To determine whether new neurons were added to the nucleus in response to the testosterone, female birds were injected with ^{3}H-thymidine as well as testosterone, and the animals were sacrificed for analysis five months later.

The researchers found that in both the testosterone-treated and control birds there were many thymidine-labeled cells, and many of these had morphological characteristics of neurons. They also analyzed birds immediately after the injections and found that the new neurons were not produced in the HVC itself, but rather were generated in the ventricular zone of the telencephalon and migrated to the nucleus, analogous to the way in which the nucleus is initially generated during embryogenesis. Subsequent studies have shown that the newly produced neurons migrate along radially arranged glial processes from the ventricular zone to the HVC (Garcia-Verdugo et al., 1998). Thus, ventricular zone neurogenesis is a normally occurring phenomenon in adult canaries. The progeny of the cells produced in the SVZ migrate to the HVC soon after their generation. There they differentiate into neurons, about half of which differentiate into local interneurons and half into projection neurons, which send axons out of the nucleus to connect with other neurons in the brain and form part of the functional circuit for song learning.

Thus, there appears to be a seasonally regulated turnover of neurons in the HVC and in other song control nuclei in the brain of the adult songbird. The turnover of neurons may correlate with periods of plasticity in song learning. Canaries modify their songs each year; each spring breeding season, they incorporate new syllables into the basic pattern, and then in the late summer and fall they sing much less frequently. Combining ^{3}H-thymidine injections with measures of cell death and overall neuronal number in the HVC over a year, one can see two distinct periods of cell death, and each one is followed by a burst in the number of new neurons in the nucleus. Both of these periods of high neuronal turnover correlate with peaks in the production of new syllables added to the song. The neurogenesis is balanced by cell death, and during periods of new song learning the nucleus adds cells, while during periods when no song is generated, the song-related nuclei undergo regression. Is the rate of neurogenesis in the ventricular zone controlled by the seasonal changes in testosterone in the male birds? When the number of labeled cells in the ventricular zone is compared in testosterone-treated and untreated female birds, there are no differences—indicating that the rate of neurogenesis does not change in response to the hormone. However, it appears instead that the survival of the neurons in the HVC is seasonally regulated—neurons generated in the spring have a much shorter average lifespan than those generated in the fall. Thus, the seasonal changes in neuron number in the songbird HVC are not dependent on changes in the number of newly added cells, but rather relate to seasonally and hormonally regulated differences in the survival of the newly produced neurons.

Neurogenesis also occurs in the mature mammalian brain. Although for many years this view was regarded as somewhat heretical, it has become well-accepted in recent years. The ^{3}H-thymidine birthdating studies of Altman, described at the beginning of this chapter, thoroughly documented the time and place of origin of neurons and glia of many regions of the rodent brain. It was found that many brain neurons are generated after birth in rodents. He next extended the labeling period to the second and third postnatal weeks and found that in one particular region, the olfactory bulb, thymidine-labeled cells were still found up to four weeks postnatally. These cells were generated in the subventricular zone (SVZ) in the forebrain and then migrated to the olfactory bulb (Altman, 1962; Altman and Das, 1965; **Figure 3.21**). These early studies of Altman were discounted, in part, because they could not prove that the cells that were generated in the adult formed functional neurons. More recent studies (Lois and Alvarez-Buylla, 1993; Luskin, 1993; Reynolds and Weiss, 1992) (see below) have now used better methods to confirm Altman's findings that neurogenesis occurs in specialized regions of the adult mammalian brain.

Most of the cells generated in the SVZ during the neonatal period and in mature rodents migrate to the olfactory bulb, in what is known as the rostral migratory stream (RMS) (Figure 3.21; Lois et al., 1996). The new neurons that migrate to the olfactory bulb in the RMS are generated at the lateral ventricles of the telencephalon, by cells with astrocytic properties, sometimes called B cells. These B cells have a single cilium that extends into the ventricle, and another process that contacts nearby blood vessels. Both the blood vessel contact and the cilium in the ventricle are thought to be important for the property of these cells that allow them to persist as "neural stem cells" throughout the animal's lifetime, like those of the frog CMZ (above). The B cells slowly self-renew, but at the same time generate C cells, a transit amplifying population, and these then make the neuroblasts, which migrate through the RMS. When the neuroblasts reach the olfactory bulb, they differentiate as granule cells and periglomerular cells, two types of GABAergic neurons. Although for many years it was thought that only inhibitory interneurons were generated in the adult mouse, recent studies have shown that a small number of excitatory glutamergic neurons are produced in this system as well.

How do these cells manage to migrate all the way from the lateral wall of the telencephalic ventricle to the olfactory bulb? The rostral migratory stream has a fascinating structure. The neuroblasts migrate in chains, along extended astrocyte networks. These networks are complex but in general have rostral–caudal orientation. One might imagine that the association of migrating SVZ cells is analogous to the migration of cortical neurons along radial glia; however, the SVZ cells do not appear to require the glia. The migration of SVZ cells has been termed *chain migration* and is distinct from the migration of neurons along radial glia. The SVZ cells form a chain *in vitro*, even in cultures devoid of glial cells, and migrate by sliding along one another.

In addition to the SVZ, one other region of the adult mammalian brain, the dentate gyrus of the hippocampus, also generates new neurons throughout life. Fred Gage and his colleagues have shown that the new neurons generated in the hippocampus are functionally integrated into the circuitry (Van Praag et al., 2002). To assay the function of the newly generated neurons, they used retroviral labeling in adult rats, similar to that which was described in the beginning of the chapter for labeling progenitors in the developing brain. Since a retrovirus will only infect and integrate into mitotically active cells, they were able to label the mitotically active hippocampal precursors with a retrovirus expressing the green fluorescent protein. When the authors examined the GFP-labeled cells after only 48 hours, the cells had a very immature morphology and resembled progenitors, like those found in the developing brain. However, when the animals were allowed to survive for four weeks, many of the GFP-labeled cells now expressed markers of differentiated neurons. Over the next three months these neurons continue to mature. To what

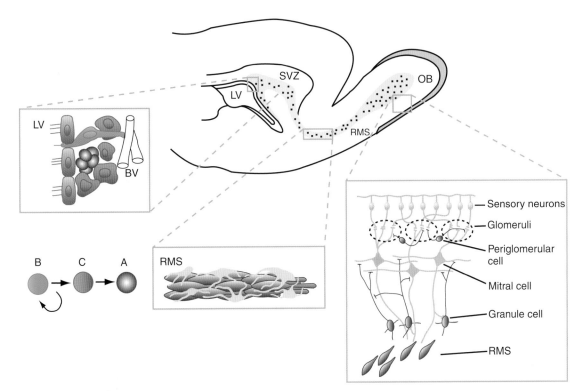

Fig. 3.21 Current model of adult neurogenesis in mice. New neurons are generated from stem cells lining the lateral ventricle (LV) in the subventricular zone (SVZ). The stem cells, which have a single cilium extending into the ventricle, and express GFAP (green) also have a process that contacts a blood vessel (BV). The stem cells, also called B cells, give rise to transit amplifying C cells (red) which then produce the A cells (purple) that migrate to the olfactory bulb via the rostral migratory stream (RMS). When these cells reach the olfactory bulb (OB), they differentiate into either periglomerular cells or granule cells, two types of interneurons in the bulb. *(Modified from Saghatelyan)*

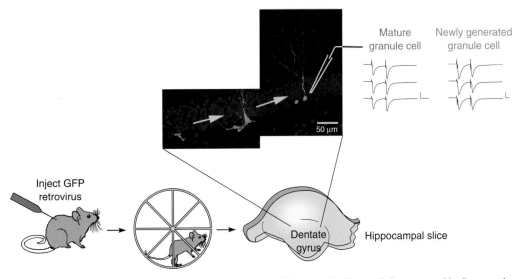

Fig. 3.22 Adult-generated hippocampal neurons are functionally integrated with preexisting neurons. Van Praag et al. labeled proliferating cells in the hippocampus with a GFP-expressing retrovirus, let the animals run on wheels to increase their production of new neurons, and then recorded from the GFP-labeled cells in hippocampal slices. The adult-generated neurons integrated into the hippocampal circuit and showed electrophysiological responses similar to their mature granule cell neighbors. *(Modified from Reh, 2002)*

extent are the new GFP cells functionally integrated into the hippocampal circuitry? The hippocampus can be sliced into thin sections while still functionally active, and the electrophysiological activity of the neurons monitored with microelectrodes (**Figure 3.22**). The newly generated granule cells had electrophysiological properties similar to those found in mature granule neurons, and they receive inputs from the major afferent pathway. Thus, newly generated neurons in the adult hippocampus integrate into the existing circuitry and function like those neurons generated during embryogenesis.

69

How many of the neurons in the adult brain are generated after we are born? As we have seen above, the most active regions of neurogenesis in the mammalian brain are the hippocampus and the subventricular zone, which produces neurons for the olfactory bulb. To determine what percentage of the neurons in these structures are generated in adulthood, Götz and her colleagues (Ninkovic et al., 2007) used a drug (tamoxifen) inducible form of Cre-recombinase that was specifically targeted to the progenitor cells (and astrocytes). These animals were mated with other mice that had been engineered with a reporter gene (beta-galactosidase) that could be activated by the Cre-recombinase. When the drug tamoxifen was given to the animals at three months of age, the Cre-recombinase was directed to the nucleus where it could induce a recombination of the DNA leading to the expression of the beta-galactosidase, permanently marking all the progeny of the adult progenitor cells. They found that up to one third of the neurons in the glomerular layer of the olfactory bulb were made by the adult stem cells, whereas only 14% of the granule neurons of the hippocampus were generated in adult mice.

Since new neurons continue to be produced throughout life by these adult stem/progenitor cells, there must be substantial death of many of the cells in order to maintain a stable ratio of new:old neurons in these brain regions.

What about the rest of our brains? Are new neurons generated during our lives outside of the hippocampus and olfactory bulb? A fascinating "natural" experiment has given insight into this question. During the Cold War, the U.S.A. and the Soviet Union routinely tested nuclear weapons above ground. This led to a global increase in the levels of ^{14}C in the atmosphere until 1963. Frisen's group in Sweden measured the level of ^{14}C in the DNA of neurons in the brains of individuals born during this period (Spalding et al., 2005). On the basis of their analysis of postmortem brains from individuals born before or during the period of nuclear testing, they concluded that virtually all the neurons in the adult (60 year-old) human occipital cortex are generated during fetal development and last a person's lifetime.

Why do mammals generate new neurons only in the hippocampus and olfactory bulb? Frogs and fish have eyes that

BOX 3.1 Neural Crest Cells: The Great Explorers

Neural crest cells are the great explorers of the vertebrate body. The migration of these cells was first demonstrated by Detwiler (1937) by labeling the premigratory cells with vital dye and seeing dye-stained descendents moving throughout the body. Their neuronal and nonneuronal descendents can be found almost everywhere. No other cell type undergoes such extensive migration during development. As crest cells migrate, they become exposed to a variety of extrinsic factors that influence both their journey and their fate. But crest cells are intrinsically specified to become multipotent explorers. It is, as usual, the balance between these intrinsic tendencies and the environmental influences that determines where and what any particular descendent of the neural crest will become. This box gives a brief view of what is known about the mechanisms of neural crest migration.

To begin to migrate, crest cells must first leave their home port: the neuroepilium in which they arise. As the neural fold closes, crest progenitors, which were at the lateral borders of the neural plate, become situated in the dorsalmost part of the neural tube (Chapter 2). At this point they begin to behave differently from the rest of the cells in the neuroepithelium. They go through what is called an epithelial to mesenchymal transition (EMT) (**Figure 3.23**A) (Kuriyama et al., 2008). Mesenchymal cells are loosely packed nonspecialized cells, usually of mesodermal origin, that move around the body associated with connective tissue and extracellular matrix. Migrating neural crest cells share many of these mesenchymal properties. To leave the neuroepithelium, neural crest cells first lose their apical tight junctions to each other and to their neighbors. A major component of tight junctions is the protein Occludin. Occludin is dramatically downregulated in premigratory neural crest cells and as a result these begin to lose their apicobasal polarity and detach from their neighbors on either side. But they are not yet ready to let go of these neighbors. To do this they must also downregulate cell adhesion molecules. Neural Cadherin (N-Cad) is a homophilic cell adhesion molecule expressed on the membranes of all cells of the neural tube including premigratory crest cells. It works as a kind of tissue-specific glue. Interference with N-Cad function through expression of a dominant negative version of the protein causes these cells to lose contact with each other (Kintner, 1992). As we learned in Chapter 2, early crest cells are specified by the expression of

the transcription factor genes *slug* and *snail*. The premigratory neural crest cells also express another transcription factor called *twist*. These transcription factors directly downregulate N-Cad and tight junctional proteins such as Occludin and so are important for the EMT of neural crest cells.

Once their tight junctions and homophilic adhesions with their neighbors have been lost, the premigratory crest cells face one more barrier before they can leave the confines of the neural tube. They must break through the heavy basal lamina composed of extracellular matrix proteins that completely surrounds the neural tube. To do this, they secrete special proteases, called matrix metalloproteases (MMPs). These proteases digest the extracellular matrix of the basal lamina, creating a hole for the crest cells to escape through so that they can begin their explorations of the rest of the body (Duong and Erickson, 2004).

The EMT of neural crest cells is in many ways similar to the first steps of metastasis in invasive cancers. Nonmetastatic tissues like the cells of the neuroepithelium have an apicobasal polarity with tight junctions and cadherins linking cells to their neighbors, keeping them in place. In early stages of metastatic cancer, tight junction proteins like Occludin and cadherins are downregulated. These cells may also then express MMPs and digest the basal lamina that is holding them in. Slug and snail expression are often upregulated in metastatic cancers. When metastatic cancer cells have made a full EMT, they are able to spread throughout the body, often via the blood stream, and invade other tissues. Of course, one of the key differences between normal neural crest cells and metastatic cancer cells is that crest cells are still restricted in terms of their proliferative potential, whereas cancer cells have somehow lost their growth control regulation.

Once the neural crest cells have left the neural tube, they begin to migrate. Large extracellular matrix molecules, such as collagen, laminin, and fibronectin, are known to support the migration of neural crest cells, for when these cells are dissociated from the embryo and plated onto tissue culture dishes coated with extracellular matrix, they migrate freely. The receptors for these extracellular matrix molecules, heterodimers of integrin proteins, are expressed by the migrating neural crest cells, and perturbation of these receptors inhibits neural

BOX 3.1 Neural Crest Cells: The Great Explorers—Cont'd

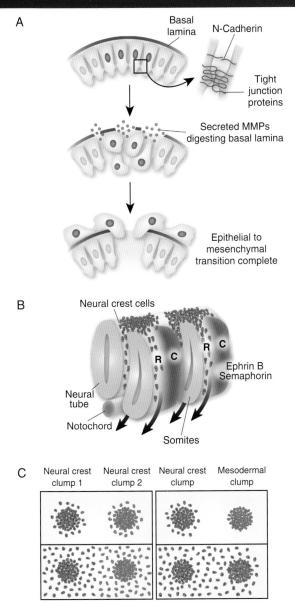

A
Basal lamina

N-Cadherin

Tight junction proteins

Secreted MMPs digesting basal lamina

Epithelial to mesenchymal transition complete

B
Neural crest cells

R C R C

Ephrin B Semaphorin

Neural tube

Notochord

Somites

C
Neural crest clump 1 Neural crest clump 2 Neural crest clump Mesodermal clump

Fig. 3.23 Migration of the neural crest. A. The epithelial to mesenchymal transition (EMT). Crest cells (red nuclei) in top panel are situated in the dorsal neural tube and attached to each other and to their neighbors by tight junctions and N-cadeherin. In the middle panel, under the influence of the transcription factors Slug, Snail and Twist, crest cells downregulate tight junctional proteins such as ocludin, cell adhesion proteins such as N-cadherin, and begin to secrete MMPs to digest the overlying basal lamina of the neural tube. In the bottom panel, the crest cells complete the EMT by escaping through the hole digested in the basal lamina. B. Crest cells migrate in streams supported by extracellular matrix material, but expressing the receptors EphB and Neuropilin, they avoid the caudal half of each somite which expresses the repulsive ligands Ephrin B and Semaphorin. C. Crest cells exhibit contact inhibition. Left. When two clumps of crest cells are cultured near each other, the cells migrating away from the centers of each clump do not mix when they encounter each other. Right. When a crest cell clump is cultured next to another tissue type such as mesoderm, the migrating crest cells do not avoid this tissue but rather migrate into and over it.

crest migration. If either β1-integrin, or its heterodimeric partner, α4-integrin, are blocked with specific antibodies, neural crest migration is blocked (Lallier et al., 1994; Kil et al., 1998).

In the cranial region, the cells migrate along the mandibular, hyoid, and branchial arches to their various regions in the head and neck. The neural crest from the trunk takes two basic routes from the neural tube: the ventral route, along which the cells that will form the sensory, enteric, and autonomic ganglia follow, and the dorsal or lateral route, in which the cells that will form the pigment cells in the epidermis predominate (Weston, 1963). A characteristic feature of emerging neural crest cells is that they migrate in interspersed streams. This is particularly clear in the trunk region where neural crest cells migrate through the rostral half of each somite but avoid the caudal half, which remains crest free (Figure 3.23B). What is responsible for these channeled migratory routes? The Eph receptors and Ephrin ligands were first identified for their roles in repulsive guidance of axonal growth (see Chapter 5). EphrinB is expressed by the caudal halves of the somites, while EphB the receptor for this repulsive ligand, is expressed by the migrating neural crest cells (Krull et al., 1997; Wang and Anderson, 1997). Semaphorins are also repulsive guidance molecules for axon growth (discussed in more detail in Chapter 5), and expressed in the caudal half of each somite. The Semaphorin receptor, Neuropilin, is expressed in the migrating crest cells. The same molecules are used to create crest-free zones in the cephalic region creating the streams of crest that flow between the branchial arches. These inhibitory guidance factors thus form the molecular riverbanks that break the flow of migrating crest cells into streams. *In vitro*, cells at the edges of neural crest clusters have active filopodia and migrate rapidly away from the center of the cluster, while the cells in the middle of the cluster have few filopodia and do not show such direction in their migration. This is because neural crest cells display contact inhibition of locomotion as demonstrated by the fact that when two migrating crest cells meet, they collapse their protrusions, transiently stop, and then migrate away from each other (Carmona-Fontaine et al., 2008). When a neural crest cell meets another cell type, such as a mesodermal cell, however, it does not show contact inhibition. Similarly, explants of neural crest cells do not invade each other, whereas cells migrating from a neural crest explant infiltrate explants of other cell types (Figure 3.23C). This homotypic contact inhibition of locomotion means that cells at the front of migration streams tend to move forward, and the lack of heterotypic contact inhibition means that when leading crest cells encounter a different type of tissue, they invade it. This behavior is again reminiscent of that shown by metastatic cancer cells.

The promotion of migration by extracellular matrix proteins, the guidance of these cells into streams by repellent molecules such as ephrins and semaphorins, and the contact inhibition that causes these cells to move away from each other and invade different kinds of tissues, should be sufficient to account for the far and wide migration patterns of neural crest cells. But on top of all this, there may also be positive, chemoattractive cues released from targets of neural crest migration, although there is as yet no convincing evidence for chemotaxis of crest cells up a gradient of such factors.

The route that crest cells take is to some degree determined by the environment in which they find themselves. For example, the neural crest from the most anterior part of the developing spinal cord migrates into the gut to form the enteric nervous system, while the neural crest from somewhat more caudal levels of the spinal cord never migrates in to the gut, but instead collects near the aorta and

Continued

BOX 3.1   Neural Crest Cells: The Great Explorers—Cont'd

forms the sympathetic ganglion chain. Transplantation of neural crest cells from anterior (enteric ganglion forming) levels of the embryo to more posterior regions results in the anterior crest cells following the posterior pathways and making sympathetic neurons instead of enteric neurons (Le Douarin et al., 1975; Le Douarin, 2004). As noted in Chapter 2, the neural tube has a considerable amount of pattern controlled in part by the regional expression of Hox transcription factors. The neural crest that migrates from the cranial regions of the neural tube also has positional identity, and this is also dependent on the Hox code. The figure shows the migration of the neural crest from the rhombomeres. The cranial crest contributes many cells to three branchial arches. The neural crest that migrates into these arches will give rise to most of the skeletal and cartilage of the skull and face. The unique contribution of the different regions of cranial neural crest has provided an opportunity to test for the specification of these cells and their migratory patterns. The crest cells from rhombomeres *r1* and *r2* migrate into the first (mandibular) arch, the crest from *r4* into the second (hyoid) arch, and the crest from *r6* and *r7* into the third

(branchial) arch (Kontges and Lumsden, 1996). The crest in each of these arches differentiates into specific skeletal elements of the face or jaw (**Figure 3.24**). The neural crest from each rhombomere continues to express the same pattern of *Hox* genes as it migrates from the neural tube, and thus has a unique identity. This unique identity can be demonstrated by transplantation experiments where the crest from one rhombomere is transplanted to the region of another, and its migration and further development are monitored (Noden, 1975). Crest cells that would normally populate the third arch were excised and replaced with first arch crest cells. The transplanted crest cells migrated into the third arch, but instead of making neck cartilage, they formed beaklike projections from the neck and a complete, duplicate first arch skeletal system in their new location. Thus, it appears that the patterning of branchial arch skeletal and connective tissues is an intrinsic property of the cells of the neural crest prior to their emigration from the neural tube. Although they can use the same cues to migrate through the branchial arches, they will differentiate in accord with the Hox code specific for their position or origin.

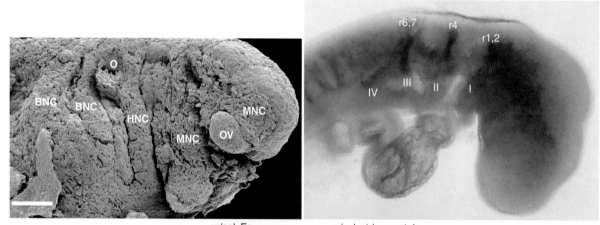

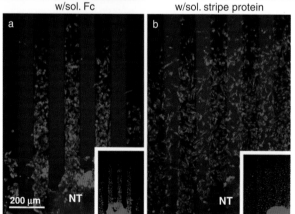

Fig. 3.24   Streams of neural crest. A. A scanning electron microscopic view of neural crest cells migrating in the first/mandibular (MNC), the second/hyoid (HNC), and the third and fourth branchial (BNC) arches in a salamander embryo. O = otic vesicle, OV = optic vesicle. B. A similar pattern is seen in a chick embryo stained with the neural crest specific marker HNK-1. Rhombomeres 1 and 2 (r1,2) contribute to the first arch, r4 contributes to the second arch and r6,7 contribute to the third and fourth arches. Left of that the most anterior of the segmental spinal streams can be seen. C. Crest cells in culture avoid stripes coated with Ephrin B (left), but if soulable EphrinB is added to the medium, this binds the EphB receptors on the crest cells and makes them insensitive to the coated stripes which they then no longer avoid. *(A from Falck et al., 2002; B and C from Mellott and Burke, 2008.)*

grow, birds learn a new song. What is the advantage to the mammal? Since both the olfactory bulb and the hippocampus are involved in the formation of olfactory memories, the neuronal turnover in these regions could be important in a seasonal change in nests or mates. Altman observed that the neurogenesis of the brain proceeds in two basic phases. The large projection neurons (or macroneurons) are generated early in embryonic development, while the smaller neurons (or microneurons) are generated later in development, through the postnatal period and even into adulthood. These later generated microneurons are then integrated into the framework provided by the macroneurons. Altman pictured this second stage of neurogenesis as a way for environmental influences to regulate the neurogenesis and produce a brain ideally suited to its environment. In one of the last reviews of his work, Altman (Altman and Das, 1965) summed up his hypothesis: "We postulate that this hierarchic construction process endows the brain with stability and rigidity as well as plasticity and flexibility." Although it has been difficult to prove that persistent neurogenesis in a particular region of the brain is necessary for behavioral plasticity in that brain region, recent studies are consistent with Altman's hypothesis. Selective deletion of the newly generated neurons in the SVZ does lead to some functional deficits, particularly with olfactory memory tests (Imayoshi et al., 2008). Imayoshi et al. used a tamoxifen inducible Cre-recombinase strategy similar to that described above, except in addition to labeling the progeny of the adult neural progenitor cells, they mated the inducible Cre mice to mice engineered with a Cre-activated toxin. This effectively killed all the progeny of the adult neural progenitor cells, and led to a decline in the total number of granule cells in the olfactory bulb over time. In the hippocampus, the addition of new neurons was also blocked in these mice. Although the mice could no longer make new neurons, they did retain the ability to make new *olfactory* memories; however, the mice showed a significant impairment in their ability to make new hippocampal-dependent *spatial* memories. This elegant study demonstrates the importance of adult neurogenesis, at least in the hippocampus, and suggests that when you learn directions to a new place, you have neurogenesis to thank.

SUMMARY

The enormous numbers of neurons and glia in the brain are generated by progenitor cells of the neural tube and brain vesicles. The progenitor cells from the early embryonic nervous system undergo many symmetric cell divisions to make more progenitor cells, while the progenitor cells in the late embryo are more likely to undergo an asymmetric division to generate neurons and glia. The production of both neurons and glia from the progenitor cells is under tight molecular control, and this allows the proper numbers of both neurons and glia to be produced for the proper functioning of the brain. Interactions between the neurons and the progenitor cells regulate their proliferation in both positive and inhibitory ways. Overall, a remarkable coordination takes place to regulate proliferation in the nervous system during development, and mutations in specific genetic pathways involved in neurogenesis can lead to childhood tumors and gliomas in adults. Once the developmental period of neurogenesis is complete, most areas of the brain do not generate new neurons, even after damage. This has led to the concept that you are born with all the neurons that you are going to ever have. However, in recent years, it has become clear that certain regions of the brain, the hippocampus and the olfactory bulb, continue to add new neurons throughout life. This continual addition of neurons in these regions may allow for greater plasticity in these specific brain circuits.

REFERENCES

Altman, J. (1962). Are new neurons formed in the brains of adult mammals? *Science, 30*(135), 1127–1128.

Altman, J., & Das, G. D. (1965). Post-natal origin of microneurones in the rat brain. *Nature, 207,* 953–956.

Angevine, J. B., Jr., & Sidman, R. L. (1961). Autoradiographic study of cell migration during histogenesis of cerebral cortex in the mouse. *Nature, 192,* 766–768.

Anthony, T. E., Klein, C., Fishell, G., & Heintz, N. (2004). Radial glia serve as neuronal progenitors in all regions of the central nervous system. *Neuron, 41*(6), 881–890.

Bertrand, N., Castro, D. S., & Guillemot, F. (2002). Proneural genes and the specification of neural cell types. *Nature Reviews. Neuroscience, 3*(7), 517–530.

Bonni, A., Sun, Y., Nadal-Vicens, M., Bhatt, A., Frank, D. A., Rozovsky, I., et al. (1997). Regulation of gliogenesis in the central nervous system by the JAK-STAT signaling pathway. *Science, 278*(5337), 477–483.

Calegari, F., Haubensak, W., Haffner, C., & Huttner, W. B. (2005). Selective lengthening of the cell cycle in the neurogenic subpopulation of neural progenitor cells during mouse brain development. *The Journal of Neuroscience, 25*(28), 6533–6538.

Carmona-Fontaine, C., Matthews, H. K., Kuriyama, S., Moreno, M., Dunn, G. A., Parsons, M., et al. (2008). Contact inhibition of locomotion in vivo controls neural crest directional migration. *Nature, 456*(7224), 957–961.

Caviness, V. S., Jr., Goto, T., Tarui, T., Takahashi, T., Bhide, P. G., & Nowakowski, R. S. (2003). Cell output, cell cycle duration and neuronal specification: a model of integrated mechanisms of the neocortical proliferative process. *Cerebral Cortex, 13*(6), 592–598.

Caviness, V. S., Jr., & Rakic, P. (1978). Mechanisms of cortical development: a view from mutations in mice. *Annual Review of Neuroscience, 1,* 297–326.

Cooper, J. A. (2008). A mechanism for inside-out lamination in the neocortex. *Trends in Neurosciences, 31*(3), 113–119.

Corbin, J. G., Nery, S., & Fishell, G. (2001). Telencephalic cells take a tangent: non-radial migration in the mammalian forebrain. *Nature Neuroscience, 4*(Suppl.), 1177–1182.

D'Arcangelo, G., Miao, G. G., Chen, S. C., Scares, H. D., Morgan, J. I., & Curran, T. (1995). A protein related to extracellular matrix proteins deleted in the mouse mutant reeler. *Nature, 374*(6524), 719–723.

Deneen, B., Ho, R., Lukaszewicz, A., Hochstim, C. J., Gronostajski, R. M., Anderson, D. J., et al. (2006). The transcription factor NFIA controls the onset of gliogenesis in the developing spinal cord. *Neuron, 52*(6), 953–968.

Detwiler, S. R. (1937). Observations upon the migration of neural crest cells and upon the development of the spinal ganglia and vertebral arches in Amblystoma. *Am. J. Anat., 6*(1), 63–94.

Doe, C. Q., & Smouse, D. T. (1990). The origins of cell diversity in the insect central nervous system. *Seminars in Cell Biology, 1*(3), 211–218 Review.

Dulabon, L., Olson, E. C., Taglienti, M. G., Eisenhuth, S., McGrath, B., Walsh, C. A., et al. (2000). Reelin binds alpha3beta1 integrin and inhibits neuronal migration. *Neuron, 27*(1), 33–44.

Duong, T. D., & Erickson, C. A. (2004). MMP-2 plays an essential role in producing epithelial-mesenchymal transformations in the avian embryo. *Developmental Dynamics, 229*(1), 42–53.

Fan, G., Martinowich, K., Chin, M. H., He, F., Fouse, S. D., Hutnick, L., et al. (2005). DNA methylation controls the timing of astrogliogenesis through regulation of JAK-STAT signaling. *Development, 132*(15), 3345–3356.

Falck, P., Hanken, J., & Olsson, L. (2002). Cranial neural crest emergence and migration in the Mexican axolotl (Ambystoma mexicanum). *Zoology (Jena), 105,* 195–202.

Feng, L., Allen, N. S., Simo, S., & Cooper, J. A. (2007). Cullin 5 regulates Dab1 protein levels and neuron positioning during cortical development. *Genes & Development, 21*(21), 2717–2730.

Garcia-Verdugo, J. M., Doetsch, F., Wichterle, H., Lim, D. A., & Alvarez-Buylla, A. (1998). Architecture and cell types of the adult subventricular zone: in search of the stem cells. *Journal of Neurobiology, 36*(2), 234–248.

Goodrich, L. V., Milenkovic, L., Higgins, K. M., & Scott, M. P. (1997). Altered neural cell fates and medulloblastoma in mouse patched mutants. *Science, 277*(5329), 1109–1113.

Hatten, M. E. (1985). Neuronal regulation of astroglial morphology and proliferation in vitro. *J Cell Biol, 100,* 384–396.

Hatten, M. E. (1990). Riding the glial monorail: a common mechanism for glial-guided neuronal migration in different regions of the developing mammalian brain. *Trends Neurosci, 13*(5), 179–184.

Imayoshi, I., Sakamoto, M., Ohtsuka, T., Takao, K., Miyakawa, T., Yamaguchi, M., et al. (2008). Roles of continuous neurogenesis in the structural and functional integrity of the adult forebrain. *Nature Neuroscience, 11*(10), 1153–1161.

Jacobson, M. (1977). Mapping the developing retinotectal projection in frog tadpoles by a double label autoradiographic technique. *Brain Res, 127*(1), 55–67.

Jacobson, M. (1978). Mapping the developing retinotectal projection in frog tadpoles by a double label autoradiographic technique. *Brain Research, 127*(1), 55–67.

Jossin, Y., Bar, I., Ignatova, N., Tissir, F., De Rouvroit, C. L., & Goffinet, A. M. (2003). The reelin signaling pathway: some recent developments. *Cerebral Cortex, 13*(6), 627–633.

Kageyama, R., Ohtsuka, T., Shimojo, H., & Imayoshi, I. (2008). Dynamic Notch signaling in neural progenitor cells and a revised view of lateral inhibition. *Nature Neuroscience, 11*(11), 1247–1251.

Kessaris, N., Pringle, N., & Richardson, W. D. (2001). Ventral neurogenesis and the neuron-glial switch. *Neuron, 31,* 677–680.

Kil, S. H., Krull, C. E., Cann, G., Clegg, D., & Bronner-Fraser, M. (1998). The alpha4 subunit of integrin is important for neural crest cell migration. *Developmental Biology, 202*(1), 29–42.

Kintner, C. (1992). Regulation of embryonic cell adhesion by the cadherin cytoplasmic domain. *Cell, 69*(2), 225–236.

Kontges, G., & Lumsden, A. (1996). Rhombencephalic neural crest segmentation is preserved throughout craniofacial ontogeny. *Development, 122,* 3229–3242.

Kriegstein, A. R., & Noctor, S. C. (2004). Patterns of neuronal migration in the embryonic cortex. *Trends in Neurosciences, 27*(7), 392–399 Review.

Krull, C. E., Lansford, R., Gale, N. W., Collazo, A., Marcelle, C., Yancopoulos, G. D., et al. (1997). Interactions of Eph–related receptors and ligands confer rostrocaudal pattern to trunk neural crest migration. *Current Biology, 7*(8), 571–580.

Kuriyama, S., & Mayor, R. (2008). Molecular analysis of neural crest migration. *Philosophical Transactions of the Royal Society of London. Series B, Biological Sciences, 363*(1495), 1349–1362.

Lallier, T., Deutzmann, R., Ferris, R., & Bronner-Fraser, M. (1994). Neural crest cell interactions with laminin: structural requirements and localization of the binding site for alpha 1 beta 1 integrin. *Developmental Biology, 162*(2), 451–464.

Le Douarin, N. M. (2004). The avian embryo as a model to study the development of the neural crest: a long and still ongoing story. *Mechanisms of Development, 121*(9), 1089–1102.

Le Douarin, N. M., Renaud, D., Teillet, M. A., & Le Douarin, G. H. (1975). Cholinergic differentiation of presumptive adrenergic neuroblasts in interspecific chimeras after heterotopic transplantations.

Proceedings of the National Academy of Sciences of the United States of America, 72, 728–732.

Leber, S. M., Breedlove, S. M., & Sanes, J. R. (1990). Lineage, arrangement, and death of clonally related motoneurons in chick spinal cord. *The Journal of Neuroscience, 10*(7), 2451–2462.

Lois, C., & Alvarez-Buylla, A. (1993). Proliferating subventricular zone cells in the adult mammalian forebrain can differentiate into neurons and glia. *Proceedings of the National Academy of Sciences of the United States of America, 90*(5), 2074–2077.

Lois, C., Garcia-Verdugo, J. M., & Alvarez-Buylla, A. (1996). Chain migration of neuronal precursors. *Science, 271*(5251), 978–981.

Luskin, M. B. (1993). Restricted proliferation and migration of postnatally generated neurons derived from the forebrain subventricular zone. *Neuron, 11*(1), 173–189.

Magdaleno, S., Keshvara, L., & Curran, T. (2002). Rescue of ataxia and preplate splitting by ectopic expression of Reelin in reeler mice. *Neuron, 33*(4), 573–586.

Malatesta, P., Hartfuss, E., & Götz, M. (2000). Isolation of radial glial cells by fluorescent-activated cell sorting reveals a neuronal lineage. *Development, 127*(24), 5253–5263.

Marin-Padilla, M. (1998). Cajal-Retzius cells and the development of the neocortex. *Trends in Neuroscience, 21*(2), 64–71.

Mellott, D.O., & Burke, R.D. (2008). Divergent roles for Eph and ephrin in avian cranial neural crest. *BMC Dev Biol, 8,* 56.

Nakajima, K. (2007). Control of tangential/non-radial migration of neurons in the developing cerebral cortex. *Neurochemistry International, 51*(2–4), 121–131.

Namihira, M., Kohyama, J., Semi, K., Sanosaka, T., Deneen, B., Taga, T., et al. (2009). Committed neuronal precursors confer astrocytic potential on residual neural precursor cells. *Developmental Cell, 16*(2), 245–255.

Nelson, B. R., Hartman, B. H., Georgi, S. A., Lan, M. S., & Reh, T. A. (2007). Transient inactivation of Notch signaling synchronizes differentiation of neural progenitor cells. *Developmental Biology, 304*(2), 479–498.

Ninkovic, J., Mori, T., & Götz, M. (2007). Distinct modes of neuron addition in adult mouse neurogenesis. *The Journal of Neuroscience, 27*(40), 10906–10911.

Noctor, S. C., Flint, A. C., Weissman, T. A., Dammerman, R. S., & Kriegstein, A. R. (2001). Neurons derived from radial glial cells establish radial units in neocortex. *Nature, 409*(6821), 714–720.

Noctor, S. C., Flint, A. C., Weissman, T. A., Wong, W. S., Clinton, B. K., & Kriegstein, A. R. (2002). Dividing precursor cells of the embryonic cortical ventricular zone have morphological and molecular characteristics of radial glia. *The Journal of Neuroscience, 22*(8), 3161–3173.

Noctor, S. C., Martenez-Cerdeno, V., Ivic, L., & Kriegstein, A. R. (2004). Cortical neurons arise in symmetric and asymmetric division zones and migrate through specific phases. *Nat Neurosci, 7*(2), 136–144.

Noctor, S. C., Martenez-Cerdeno, V., & Kriegstein, A. R. (2008). Distinct behaviors of neural stem and progenitor cells underlie cortical neurogenesis. *J Comp Neurol, 508*(1), 28–44.

Noden, D. M. (1975). An analysis of migratory behavior of avian cephalic neural crest cells. *Developmental Biology, 42,* 106–130.

Noden, D. M. (1983). The embryonic origins of avian cephalic and cervical muscles and associated connective tissues. *The American Journal of Anatomy, 168*(3), 257–276.

Norden, C., Young, S., Link, B. A., & Harris, W. A. (2009). Actomyosin is the main driver of interkinetic nuclear migration in the retina. *Cell, 138*(6), 1195–1208.

Nottebohm, F. (1985). Neuronal replacement in adulthood. *Annals of the New York Academy of Sciences, 457,* 143–161.

Nottebohm, F. (2002). Why are some neurons replaced in adult brain? *The Journal of Neuroscience, 22*(3), 624–628.

Paton, J. A., & Nottebohm, F. N. (1984). Neurons generated in the adult brain are recruited into functional circuits. *Science, 225*(4666), 1046–1048.

Pontious, A., Kowalczyk, T., Englund, C., & Hevner, R. F. (2008). Role of intermediate progenitor cells in cerebral cortex development. *Developmental Neuroscience, 30*(1–3), 24–32.

Raff, M. C., Lillien, L. E., Richardson, W. D., Burne, J. F., & Noble, M. D. (1988). Platelet-derived growth factor from astrocytes drives the clock that times oligodendrocyte development in culture. *Nature, 333*(6173), 562–565.

Rajan, P., & McKay, R. D. (1998). Multiple routes to astrocytic differentiation in the CNS. *The Journal of Neuroscience, 18*(10), 3620–3629.

Rakic, P. (1988). Specification of cerebral cortical areas. *Science, 241*(4862), 170–176.

Rakic, P. (1971a). Guidance of neurons migrating to the fetal monkey neocortex. *Brain Research, 33*(2), 471–476.

Rakic, P. (1971b). Neuron-glia relationship during granule cell migration in developing cerebellar cortex. A Golgi and electronmicroscopic study in Macacus Rhesus. *The Journal of Comparative Neurology, 141*(3), 283–312.

Rakic, P. (1972). Mode of cell migration to the superficial layers of fetal monkey neocortex. *J Comp Neurol, 145,* 61–84.

Ramón y Cajal, S. (1952). Histologie du systeme nerveux de l'homme et des vertebres. In *Consejo Sup Invest Cient.* (p. 589). Madrid.

Ravin, R., Hoeppner, D. J., Munno, D. M., Carmel, L., Sullivan, J, Levitt, D. L., et al. (2008). Potency and fate specification in CNS stem cell populations in vitro. *Cell Stem Cell, 3*(6), 670–680.

Reh, T. A. (2002). Neural stem cells: form and function. *Nature Neuroscience, 5*(5), 392–394.

Reynolds, B. A., & Weiss, S. (1992). Generation of neurons and astrocytes from isolated cells of the adult mammalian central nervous system. *Science, 255*(5052), 1707–1710.

Shimojo, H., Ohtsuka, T., & Kageyama, R. (2008). Oscillations in notch signaling regulate maintenance of neural progenitors. *Neuron, 58*(1), 52–64.

Sidman, R. L. (1961). Histogenesis of the mouse retina studied with tritiated thymidine. In G. K. Smelser (Ed.), *The structure of the eye.* New York: Academic Press.

Spalding, K. L., Bhardwaj, R. D., Buchholz, B. A., Druid, H., & Frisén, J. (2005). Retrospective birth dating of cells in humans. *Cell, 122*(1), 133–143.

Takizawa, T., Yanagisawa, M., Ochiai, W., Yasukawa, K., Ishiguro, T., Nakashima, K., & Taga, T. (2001). Directly linked soluble IL-6 receptor-IL-6 fusion protein induces astrocyte differentiation from neuroepithelial cells via activation of STAT3. *Cytokine, 13*(5), 272–279.

van Praag, H., Schinder, A. F., Christie, B. R., Toni, N., Palmer, T. D., & Gage, F. H. (2002). Functional neurogenesis in the adult hippocampus. *Nature, 415*(6875), 1030–1034.

Vetter, M. L., & Moore, K. B. (2001). Becoming glial in the neural retina. *Developmental Dynamics, 221*(2), 146–153.

Wang, H. U., & Anderson, D. J. (1997). Eph family transmembrane ligands can mediate repulsive guidance of trunk neural crest migration and motor axon outgrowth. *Neuron, 18,* 383–396.

Wechsler-Reya, R. J., & Scott, M. P. (1999). Control of neuronal precursor proliferation in the cerebellum by Sonic hedgehog. *Neuron, 22*(1), 103–114.

Weston, J. A. (1963). A radioautographic analysis of the migration and localization of trunk neural crest in the chick. *Dev Biol, 6,* 279–310.

Wilcock, A. C., Swedlow, J. R., & Storey, K. G. (2007). Mitotic spindle orientation distinguishes stem cell and terminal modes of neuron production in the early spinal cord. *Development, 134*(10), 1943–1954.

Zhou, Q., Choi, G., & Anderson, D. J. (2001). The bHLH transcription factor Olig2 promotes oligodendrocyte differentiation in collaboration with Nkx2.2. *Neuron, 31*(5), 791–807.

Zhou, Y., Atkins, J. B., Rompani, S. B., Bancescu, D. L., Petersen, P. H., Tang, H., et al. (2007). The mammalian Golgi regulates numb signaling in asymmetric cell division by releasing ACBD3 during mitosis. *Cell, 129*(1), 163–178.

Determination and differentiation

4

The nervous system is a coral reef of the body where evolution and development have collaborated to produce an extraordinary diversity of cell types. Neurons show enormous variety in anatomy, function, chemistry, and connectivity. The numerous granule cells of the cerebellum are tiny, they have simple dendritic arbors, bifurcated axons, and release the excitatory transmitter glutamate, whereas cerebellar Purkinje cells nearby in the cerebellum are gigantic by comparison, have impressively complex dendritic trees, single axons, and release the inhibitory neurotransmitter GABA. These differences could hardly be greater. Yet of course, the differences between other neurons can be quite subtle. All the motor neurons of the spinal cord, for example, share a common morphology, and they all innervate skeletal muscle cells where they release the neurotransmitter acetylcholine, yet they are specified into hundreds of molecularly distinct motor pools so that they connect with particular presynaptic partners and specific muscles.

The fates of some neurons are largely the products of particular lineages. The fates of others appear to depend more on the local environment, but the determination of cell type for almost all neurons is a product of both factors. The adoption of a particular fate is a multistep sequential process that involves a sequence of intrinsic and extrinsic influences. A progenitor cell may be externally influenced to take a step along a particular fate pathway, and so the unborn daughter of that cell has also, in a sense, taken the same step. A signal from the environment may act upon this daughter cell to refine its fate further, and the response of the daughter cell to the signal is to express an intrinsic factor consistent with its limited fates. So as an ancestor cell and its successive line of descendents go through their rounds of division to produce a lineage tree, the environment often influences the way the tree branches, and thus the lineage history.

To find whether particular aspects of cell fate are determined more by lineage or environment, scientists often transplant progenitor cells from a donor animal to a genetically different host animal, or into a different cellular environment, or even into tissue culture (**Figure 4.1**). If the phenotype of the transplanted cell is unaltered by putting it into these new environments, then the cell phenotype is "intrinisically determined" at the time of transplantation. If, however, the cell adopts a new fate, for example consistent with the position into which it is transplanted, then the fate at the time of transplantation is still plastic or flexible and can be influenced by external factors in the environment. In culture, one can add or take away factors, and such a strategy has been very valuable in discovering factors that influence cell fate.

The environment in which neural progenitors divide and give rise to neurons is extremely rich with diffusible molecules, cell surface proteins, and extracellular matrix factors, which by themselves or together in particular combinations may affect neural progenitor fates. These extrinsic signals are usually transduced by transmembrane receptors, which then signal to the nucleus to influence the genes these progenitors express. These genes then act like intrinsic factors. Transcription factors are often the targets of such signaling systems. Transcription factors direct the expression of other transcription factors, either by activation or repression, in the progenitors resulting in a transcriptional cascade that may restrict fate choices in their descendents. The intrinsic

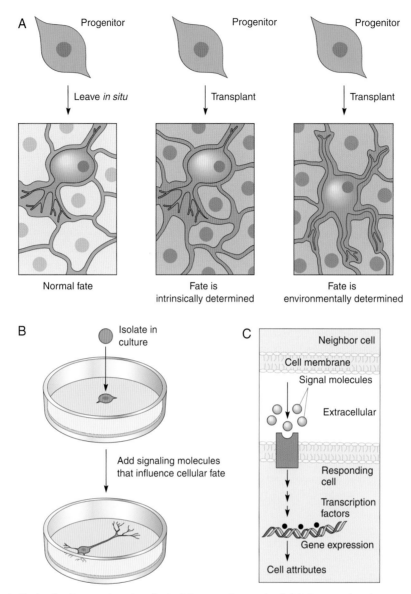

Fig. 4.1 A. Testing fate by transplantation. On the left, a neural progenitor left in its normal environment turns into a particular type of neuron. In the middle, an intrinsically determined progenitor's fate is unchanged by transplantation to a different environment. On the right is an example of a progenitor whose fate is determined extrinsically and so is changed by transplantation to a different environment. B. An undifferentiated progenitor is placed into a culture dish, and signaling molecules are tested, which may influence the fate that the cell takes as it differentiates into a neuron. C. An extracellular signal originating from one cell can influence the fate of nearby cells by causing the responding cell to change its gene expression pattern.

factors inherited by a neuron at the moment of its birth (when it permanently leaves the cell cycle) regulates the program of differentiation into a particular cell morphology, chemistry, and initial circuitry.

The number of genes used to carry out this task of neuronal specification throughout the nervous system is impressive. It has been estimated that half of an organism's genes are expressed exclusively in the nervous system, and most of these are involved in various aspects of neuronal determination and differentiation. To identify the genes that influence determination, mutational and transgenic analyses are the key tools. Mutations in particular genes can alter the fate of certain types of neurons. With a genetic approach it is possible

to show where and when a gene is expressed and needed for a particular fate decision, and importantly it is also possible to identify the gene product in question. Forward genetics uses random mutagenesis to define new genes that have effects on neural differentiation, while reverse genetics uses molecular engineering to knock out or overexpress particular genes (see Chapter 2) that are candidates for roles in neuronal fate determination or differentiation.

This chapter examines the several facets of cell fate determination investigated with such techniques in several systems, so that one can begin to appreciate the full range of molecular and cellular mechanisms that lead relatively simple-looking progenitor cells to their varied neuronal fates.

TRANSCRIPTIONAL HIERARCHIES IN INVARIANT LINEAGES: *C. ELEGANS* NEURONS

Time-lapse studies of the development of the nervous system of the nematode *C. elegans* show that all neurons arise from a relatively invariant set of lineages (**Figure 4.2**). In this system, the progenitors are all uniquely identifiable by their position and characteristic patterns of division. Ablation of one of these progenitors usually leads to the loss of all the neurons in the adult animal that arise from that progenitor, indicating that neighboring progenitors cannot fill in the missing fates. This is called mosaic development, and suggests that the nervous system of *C. elegans* is built using highly cell intrinsic mechanisms, suggestive of phenotypic programming by transcriptional cascades.

To discover the transcription factors that are involved, a genetic approach can be very powerful. A forward genetic screen for mutants that interfere with the development of particular neurons can not only define the factors but can also provide the biological material to help dissect the mechanisms involved. One of the best examples of such an analysis is that of the specialized mechanosensory cells in nematodes studied by Martin Chalfie and his colleagues (Chalfie and Sulston, 1981; Chalfie and Au, 1989; Chalfie, 1993; Ernstrom and Chalfie, 2002). Most nematodes wiggle forward when touched lightly on the rear and backward when touched on the front. By prodding mutagenized nematodes with an eyelash hair attached to the end of a stick, Chalfie and colleagues were able to find mutants that had lost the ability to respond to touch. Many touch insensitive worms have mutations in a group of genes involved in the specification of the mechanosensory cells. Mutations in the gene *unc-86* result in the failure of the mechanosensory neurons to form. Unc-86 is a transcription factor that is expressed transiently in many neural precursors and particularly in the lineage produced by a cell called Q. In wild-type animals, Q divides into two Q1 daughter cells, Ql.a and Ql.p (**Figure 4.3**A). Both of these cells divide once more, but only Ql.p produces a touch cell. The *unc-86* gene is turned on only in the Ql.p. and a mutation in *unc-86* (Figure 4.3B) results in the "transformation" of Ql.p into a cell that behaves like its mother, Q, and is therefore called Q'. This transformed cell continues to divide, producing Q1.a' and Q1.p', but in the continued absence of *unc-86* function, the Q1.p' transforms into Q", which continues to behave like its mother Q' and its grandmother Q. Thus, mutations in *unc-86* affect the lineage of touch cells such that mechanosensory neurons are never born.

Another gene uncovered in Chalfie's screen of touch mutants is named *mec-3*. In these mutants (Figure 4.3C), the cells that would be touch sensitive are born, but they do not differentiate into mechanosensory neurons. Instead, they turn into

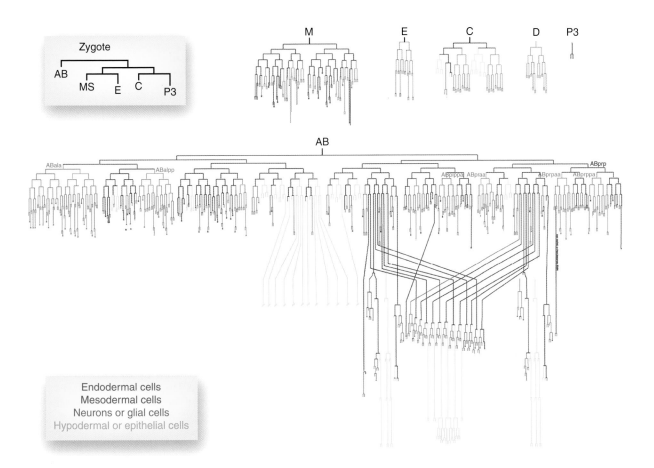

Fig. 4.2 The location of neurons in the lineage of *C. elegans*. (Based on Sulston et al., 1983; Hobert, 2005). Upper left panels indicate embryonic blastomeres. All other lineages are descendants of these blastomeres. Strike-through indicates that the lineage has not been drawn out to completion.

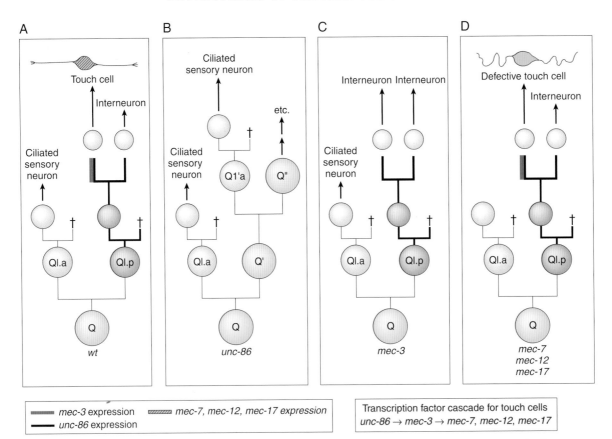

Fig. 4.3 Intrinsic determinants at different steps on a neural lineage of touch cells in *C. elegans*. The transcription factor cascade (Unc-86 -> Mec-3 -> Mec-7/-12/-17 outlined at bottom) that underlies the cell lineage is deduced from the mutant phenotypes. A. Normal lineage. B. If Q1.p cannot express *unc-86*, it becomes Q', a copy of its mother, Q. The result is a repeat of the previous division, which results only in Q progenitors and ciliated sensory neurons. C. Mec-3 needs to be expressed in the touch cell. If it is not, as in *mec-3* mutants, this cell turns into an interneuron like its sister. D. In *mec-7, mec-12,* and *mec-17* mutants, the touch neuron is correctly specified but differentiates missing critical components of its morphology or function.

interneurons. Thus, the *mec-3* mutation affects neural subtype determination. The *mec-3* gene codes for a transcription factor that is a member of the LIM-homeodomain family. Illustrating the principle of a transcriptional hierarchy, *mec-3* is directly regulated by the Unc-86 transcription factor. Cells fated to become touch cells all express Unc-86 at first, and this factor binds to the regulatory sequence of DNA that controls the transcription of the *mec-3* gene. Thus, *unc-86* mutants do not express Mec-3. In normal animals, however, the protein Unc-86 leads to the expression of Mec-3, and when these two proteins are expressed in the same cell (the cell that will become a mechanosensory neuron), they physically interact to make a heterodimeric transcription factor with new specificity that activates genes that neither Mec-3 nor Unc-86 can activate on their own. Several of these target genes are defined by mutations that also cause touch insensitivity, such as the *mec-7, mec-12,* and *mec-17* genes. These three genes encode proteins that are not transcription factors, but are building blocks used in the differentiation of the specialized touch cell cytoskeleton (Figure 4.3D). This is an example of a simple hierarchical cascade of transcription factors, one regulating and interacting with the next, the end result of which is to turn on genes that the cell uses to realize its fate. Chalfie and colleagues (Zhang et al., 2002) have found many more genes involved in the touch cell pathway by looking for differences in profiles of all expressed genes in normal animals versus *mec-3* mutants.

This approach identified approximately 50 new genes in the pathway, all downstream of *mec-3*, that are likely to be important for the function of the mechanosensory neurons.

Another set of *C. elegans* neurons that have been studied in detail are the egg-laying neurons (Desai et al., 1988; Desai and Horvitz, 1989). Mutants in genes involved in the determination or function of these neurons are unable to lay eggs. The result is that the self-fertilized eggs hatch inside the mother and begin to feed within their mother. The larvae proceed to devour their mother from the inside. Eventually, with only her epidermis intact, she becomes a bag of wriggling larval worms that in their hunger eventually eat through her cuticle into the world. Mutant lines in these genes thus seem to give their offspring a protected start in life characteristic of viviparous species but at what appears to be a mother's ultimate sacrifice. The "bag of worms" mutants have a phenotype that is easy to detect, and so a large collection of such mutants has been identified. Twenty or so genes have been found to define hierarchical transcriptional cascades affecting egg-laying neuron development. One of these genes is also necessary for the proper development of the touch cells. This is our friend *unc-86*. The role of *unc-86* in the determination of the egg-laying neurons, however, is quite different. Instead of controlling lineage as it does in mechanosensory cells, it regulates neurotransmitter expression and axon outgrowth in the egg-laying neurons.

That *unc-86* is involved in specification of both mechanosensory neurons and egg-laying neurons shows that particular transcription factors may be involved in specification and differentiation of multiple very different cell types. This important concept will be reiterated using other transcription factors in this chapter. What will become clearer as these stories unfold is that the activity of these transcription factors is heavily determined by the cellular context, for example with which other transcription factors they interact. In mechanosensory progenitors Unc-86 interacts with Mec-3 to regulate a specific set of target genes, yet in egg-laying neurons, *mec-3* is not expressed and Unc-86 regulates other targets. Except for *unc-86*, there is, in fact, surprisingly little overlap in the genes that are involved in these two systems, suggesting that the molecular cascades of neuronal determination must be complex and highly individualized.

There are also, however, similarities that are worth emphasizing. In the case of both the egg-laying and the mechanosensory neurons, there is a hierarchical pathway, rich in transcription factors that operate through the specific lineages. These factors regulate other intrinsic transcription factors in a molecular cascade whereby the lineage, the specification, the differentiation, and finally the physiological properties of the neurons are

established through a series of successive stages. This molecular strategy, we will see, is also used in the determination of neurons in most other species. Transcriptional mechanisms, whether in simple worms or mammals, have a central role in cell fate decisions. Complex transcriptional hierarchies, such as those discussed above, often make fate decisions progressive, incremental, and irreversible, so that once a progenitor turns on a particular set of intrinsic factors, it can no longer revert to a more primitive state. We can also note at this point that a single progenitor may express a variety of transcription factors at the same time, which act as a transcription factor code to restrict the developmental potential of progenitors.

Another remarkable and instructional example of transcriptional control comes from two morphologically and physiologically very similar neurons, a bilateral pair of gustatory neurons, called ASE neurons (**Figure 4.4**) studied by Hobert and colleagues (Hobert et al., 2002; Johnston and Hobert, 2003; Johnston et al., 2005). These two neurons are mirror images of each other, yet the one on the left (ASEL) is responsible for the worm's ability to taste sodium and the one on the right (ASER) allows the worm to taste chloride. This asymmetry begins with a signal from the blastomere P2 to the ABp rather than ABa blastomere at the 4-cell stage of

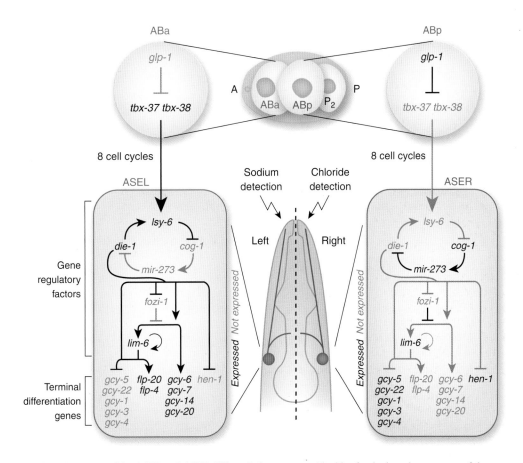

Fig. 4.4 Summary of the ASEL and ASER differentiation program. The bias for the lateral asymmetry of these two mirror image gustatory neurons can be traced back to the four-cell stage in which ABa and ABp are instructed by a signal, emanating from the P2 cell, to adopt distinct fates. Eight cell cycles later the precursors of ASEL and ASER arise. These cells express transcription factor genes *die-1* and *cog-1* and the microRNAs *lys-6* and *mir-273* that together form a bistable regulatory feedback loop which assures that ASEL only expresses Die-1, which then sits on top of a regulatory hierarchy that activates the expression of a whole set of genes involved in tasting sodium and represses the set of genes that are necessary for tasting chloride. This bistable regulatory network, composed of a double negative feedback loop, assures that ASEL does not express Cog-1 and that ASER does not express Die-1. *(From Sarin et al., 2007)*

development. Eight cell cycles later, the two ASE cells arise and are initially equivalent in gene expression, but as a result of this early asymmetric signal, ASEL the distant descendant of ABa begins to preferentially express the transcription factor Die-1, and ASER the similarly distant descendent begins to preferentially express the transcription factor Cog-1. Die-1 is responsible for the expression of the proteins involved in transducing the sodium taste and Cog-1 controls the expression of the proteins involved in chloride detection. Die-1 and Cog-1 both have other very interesting targets, which are the promoters of microRNAs called *lys-6* and *mir-273* respectively. MicroRNAs are short sequences of RNA that target the RNAs of particular genes for repression or destruction (Bartel, 2009). *lys-6* microRNA in ASEL targets and represses Cog-1 mRNA, while *mir-273* targets and represses the mRNA for Die-1. Thus cells that express Cog-1 cannot express Die-1 and vice-versa. This example demonstrates a double negative feedback loop (Figure 4.4), which makes it impossible for a cell to have the tastant characteristics of both ASEL and ASER, even though these two cells are otherwise identical twins. Simple and complex feedback loops involving transcription factors, in addition to linear activation hierarchies, are commonly used in specifying neural fates.

SPATIAL AND TEMPORAL COORDINATES OF DETERMINATION: *DROSOPHILA* CNS NEUROBLASTS

The CNS of the fly and all other arthropods develops from a set of individually identifiable neuroblasts that enlarge within the epithelium of the neurogenic region of the blastoderm and then delaminate to the inside, forming a neuroblast layer. All these neuroblasts, we learned in Chapter 1, express proneural transcription factors of the *Achaete-Scute* family that give them a common "neural" specification. However, each neuroblast is an individual and through successive divisions reproducibly gives rise to a unique set of neurons (**Figure 4.5**). How do these neuroblasts get their specific

identities? The first answer is a spatial one. They are arranged in reproducible columns and rows and so can be identified by their position. In Chapter 2, we discussed how the *Drosophila* embryo is finely subdivided in the anterior to posterior axis into stripes of expression of particular combinations of gap genes, pair-rule genes, *Hox* genes, and segment polarity genes. These genes provide neuroblasts with intrinsic positional information that reflects their location along the antero-posterior axis (**Figure 4.6**). *Hox* genes are expressed in the middle and posterior portions of the neural primordium, and the "head gap" genes are expressed more anteriorly in nested domains and provide positional information to the neuroblasts that give rise to particular brain regions or segments. Segment polarity genes control positional information within each individual segment (Bhat, 1999). These anterior-posterior (AP) positional identity genes

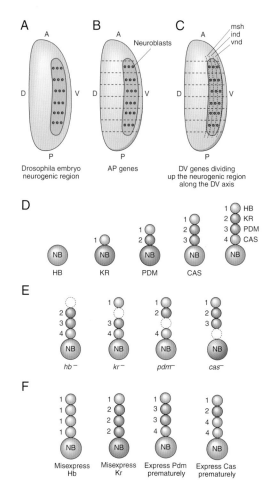

Fig. 4.6 Positional and temporal coordinates of neuroblasts in *Drosophila*. A. The neurogenic region of a *Drosophila* embryo showing the rows and columns of neuroblasts. B. The embryo and the neurogenic region is divided segmentally into AP domains as described in Chapter 2. C. The neurogenic region is further divided into DV domains by the expression of transcription factors such as Msh, Ind, and Vnd, creating a grid whereby each neuroblast has its own specific spatial coordinates. D. Neuroblasts transiently express temporally coordinated transcription factors Hb, then Kr, then Pdm, then Cas, but their progeny the GMCs maintain the transcription factor profile that was present at their birth. E. In *hb, kr, pdm* and *cas* mutants, certain GMCs are missing. F. In animals that are made to misexpress Hb or Kr or express Pdm or Cas prematurely, GMC fate is changed accordingly.

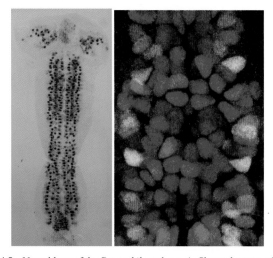

Fig. 4.5 Neuroblasts of the *Drosophila* embryo. A. Shows the rows of neuroblasts labeled with an antibody to a late neuroblast-specific protein called Snail. B. Shows neuroblasts labeled with three different antibodies to the different neuroblast-specific proteins Hunchback, Eagle, and Castor. *(Photos courtesy of Skeath and Doe)*

play important roles role in determining the identity of the neuroblasts as illustrated by the loss and/or duplications of particular sets of neurons in mutants for these genes. For example, Wingless *(wnt)* and Hedgehog proteins activate the expression of a gene called *huckebein* in some neuroblasts, and the transcription factors *Engrailed* and *Gooseberry* repress *huckebein* expression in other neuroblasts, thus establishing the precise pattern of Huckebein protein in specific neuroblast lineages (McDonald and Doe, 1997).

Another set of spatially expressed transcription factors divides the embryo and the nervous system along the dorsoventral axis. Three homeobox genes, *vnd*, *ind*, and *msh*, are expressed in longitudinal stripes within the neural ectoderm (Cornell and Ohlen, 2000). *vnd* is expressed in neuroblasts closest to the ventral midline, *msh* is expressed in the most dorsolateral stripe of the neurectoderm, and *ind* is expressed in an intermediate stripe between these two (Figure 4.6). As is the case for the AP genes, mutations in these genes lead to loss of the neuroblast fates that normally express the mutated gene. The mechanism responsible for setting up these stripes involves responses of the promoters of these genes to threshold levels of the morphogen Dpp, which forms a gradient of expression from dorsal (high) to ventral (low). Once set up, the boundaries between the stripes of *vnd, ind,* and *msh* are sharpened by mutual repression.

A neuroblast in any position can thus be uniquely identified by expression of these spatial coordinate markers of "latitude" and "longitude" (Figure 4.6). The genes specifying position information along these two Cartesian axes collaborate to specify a unique positional identity for each central neuroblast in the developing organism. Once expressed in a neuroblast, the spatial coordinate genes are inherited by the progeny of these cells, and act as intrinsic determinants of neuronal fate.

Each spatially identified neuroblast divides asymmetrically to produce a copy of itself and a smaller ganglion mother cell (GMC). The neuroblast divides several times sequentially giving rise in ordered succession to a set of GMCs (see Chapter 1). Each GMC can thus be identified not only by the position of the neuroblast from which it arises, but also importantly from the relative order of its generation (e.g., whether it is the first, second, or third GMC to arise from a particular neuroblast). The first GMCs of a neuroblast lineage tend to lie deeper in the CNS and generate neurons with long axons, whereas the later arising GMCs stay closer to the edge of the CNS and produce neurons with short axons. In generating sequential GMCs, neuroblasts go through a temporally conserved program of transcription factor expression: Hunchback (Hb) expression is followed by Krueppel (Kr), which is followed by Pdm, which is followed by Castor (Cas) (Figure 4.6 D and E). In the stage when neuroblasts express Hb, the GMC generated at this time inherits this Hb. Later, when the same neuroblast turns off Hb expression and replaces it with Krueppel (Kr), the GMC generated at this stage inherits Kr, but not Hb, expression. If Hb expression is eliminated from the neuroblasts when they are making GMCs, the neuroblasts generate GMCs that cannot make the earliest fates but only later neuron fates. Similarly, if Kr is eliminated, then the neurons that arise from the GMC that normally expresses

Kr are lost. If instead, Hb is experimentally maintained in the neuroblasts at the stages when they would normally start expressing Kr, the neuroblasts keep making the earliest GMCs, which make early neuron types (Figure 4.6) (Isshiki et al., 2001). Such loss and gain of function studies in the lineage of NB7-1 show that Hb and Kr are responsible for making the early motor neurons, and similar studies show that Pdm and Cas are required for development of the late-born motor neurons that arise from NB7-1 (Figure 4.6) (Grosskortenhaus et al., 2006). The expression of the successive transcription factors is linked to cell cycling, and driven by an interesting transcriptional network where each transcription factor activates the next factor in the pathway while repressing the previous one. This ratchets the pathway forward in such a way that the competence to make each neuron type only lasts during a limited time window.

The expression of both spatial and temporal coordinate genes in neuroblasts is preserved in their progeny, the GMCs, and forms part of the increasingly rich transcriptional and intrinsic inheritance of each developing neural progenitor. Thus, the ontogenetic roots of any CNS neural progenitor in an arthropod such as a fly can be read in the combination of transcription factors it expresses, and these factors in turn influence the cell's eventual phenotype.

ASYMMETRIC CELL DIVISIONS AND ASYMMETRIC FATE

The GMC daughter of a *Drosophila* neuroblast inherits the spatial and temporal coordinates expressed by the parental neuroblast at the time of birth. But why does the parent neuroblast divide to produce two kinds of daughters, one a large neuroblast that will go on to divide several more times and the other a small GMC that will divide just one more time? And what is it that makes the fates of these two daughters of the same mother cell so different? The answer to this question has attracted a great deal of research, which has revealed a remarkable dance of molecular components that orient the cell division and differentially specify the two daughters.

Neuroblasts in *Drosophila* arise in a polarized epithelium with a basal and and apical side. The same is true for neural progenitor cells in all species, including vertebrates. At the apical pole of all epithelial cells is a complex of proteins known as the Par complex. The asymmetric localization of cell fate determinants begins here. The Par complex comprises three proteins, Bazooka (Baz, a.k.a. Par-3 in other species), Par-6, and atypical protein kinase C (aPKC). As the neuroblasts prepare to enter mitosis, the Par complex remains localized to the apical cortex and an adaptor protein called Inscuteable (Insc) starts to be expressed and it binds to the Par complex. Inscutable then recruits its partner, called Partner of Inscutable (Pins), which through a G-protein coupled pathway attracts one of the spindle pole bodies. The result of this is that the axis of the division plane is apicobasal; it is orthogonal to the plane of the epithelium.

Once the orientation of cell division is set up, a kinase, called Aurora-A, phosphorylates Baz, causing it to leave the Par complex. Meanwhile aPKC in the Par complex phosphorylates a cytoskeletal protein called Lethal giant larvae

(Lgl), which replaces Baz. The remodeled complex is then able to phosphorylate Numb, an inhibitor of Notch signaling (Chapter 1). In the meantime, two other proteins, Miranda and Partner of Numb (Pon) begin to accumulate on the basal cortex of the dividing neuroblast. Pon traps the phosphorylated Numb, and Miranda traps the homeobox transcription factor Prospero (Kaltschmidt and Brand, 2002; Wirtz-Peitz et al., 2008). Thus when the neuroblast finally divides, both Numb and Prospero are inherited only by the GMC and not its parental Neuroblast (**Figure 4.7**). In the absence of Notch signaling, the basal GMC is free to move down the determination pathway. The more apical cell remains a neuroblast capable, upon production of more Miranda, Prospero, and Numb, to split off another GMC at the next division (**Figure 4.8**).

An interesting example of a situation in which daughter cells adopt different fates due to asymmetric inheritance of Numb are the small sensory organs called sensilla, scattered over the body surface in *Drosophila*. The cells that compose each sensillum are usually clonal descendants of a single sensory organ precursor, SOP. The SOP cells are like the CNS neuroblasts in that they delaminate from the ectoderm during development in much the same way, dependent on proneural genes and Notch and Delta interactions, as described in Chapter 1. Each specified SOP undergoes an invariant pattern of cell divisions. This division pattern has been investigated in detail for the external mechanosensilla (Guo et al., 1996; Schweisguth et al., 1996). The primary SOP (spI) for each macrochaete divides into an anterior daughter called spIIb and a posterior daughter called spIIa (**Figure 4.9**). SpIIa produces the outer two accessory cells: the socket cell and the shaft cell. The anterior daughter SpIIb divides into a neuron and a support cell, after first giving rise to a glial cell. These different fates arise through the reuse of the Notch signaling system during each of the cell divisions. When the SOP divides into spIIa and spIIb, these two cells interact with each other via Notch. The SpIIb fate is dominant, which is shown by the fact that if spIIb is ablated, spIIa will transform into spIIb. In *Notch* mutants, both cells become spIIb, and the result is no bristles or sockets, and when Notch is experimentally activated in both cells, they both turn into spIIa and there are no neurons or glia. Several intrinsic determinants,

including the asymmetrically inherited Numb (see above), control the fate of spIIa versus spIIb (Figure 4.9). In this case, Numb is distributed to spIIb upon cell division. In the absence of Numb, the Notch pathway is active in spIIb, and it is transformed into spIIa; neither neurons nor support cells appear, but the sensilla form instead with double sockets and shafts. *Numb* mutants are thus insensitive to touch because the sensory bristles are uninnervated, hence the origin of the name.

In the vertebrate CNS, neuroepithelial cells first tend to divide symmetrically, each giving rise to two dividing progenitor cells, but as neurogenesis proceeds, some neuroepithelial cells, like

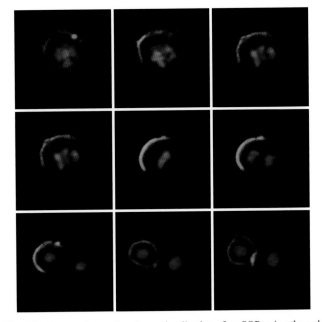

Fig. 4.8 Frames from a time-lapse visualization of an SOP going through an asymmetrical division. The green label follows Numb, and the red label follows the chromosomes. In this sequence (*courtesy of F. Schweisguth*), it is easy to see Numb localized to one pole of the SOP and then inherited by a single daughter, spIIb.

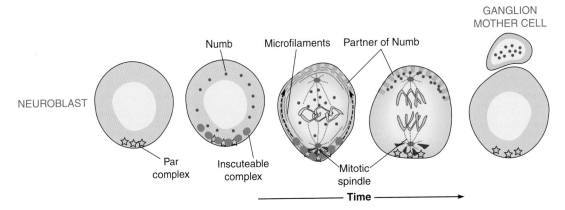

Fig. 4.7 Control of asymmetrical cell division in *Drosophila*. The Par complex is localized apically. The Par complex then recruits the Inscuteable complex to the apical pole of the neuroblast where it orients the mitotic spindle. Numb is phosphorylated (not shown) by the Par complex and then binds Partner of Numb (PON) which has been transported to the basal crescent, causing in turn the basal localization of Numb, which blocks the Notch pathway in the GMC that inherits it.

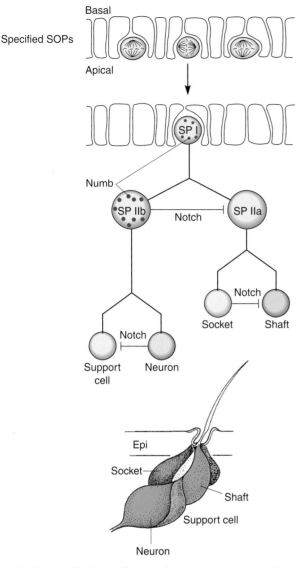

Fig. 4.9 Lineage of a *Drosophila* external mechanosensory organ. From top to bottom: a sensory organ precursor (SOP) enlarges and delaminates from the epithelium. It divides asymmetrically into spIIb, which inherits Numb, and spIIa, which does not. SpIIa divides again asymmetrically, as does spIIb slightly later. Notch signaling between daughters is involved in all these asymmetric divisions so that four daughter cells of the SOP have four different fates: support cell, sensory neuron, socket cell, and shaft cell.

Drosophila neuroblasts, begin to divide asymmetrically, giving rise to one neuroepithelial progenitor and one differentiated neuron. Time-lapse observations of neuroepithelial cells in organotypic slices of the developing mammalian cortex suggest that divisions along the horizontal plane tend to produce cells that stay in contact with the ventricular (apical) surface, whereas more apicobasal divisions produce cells that migrate away from the ventricular surface (Chenn and McConnell, 1995). More recent work has shown that cortical stem cells do indeed tend to divide mostly along the horizontal plane of the ventricular surface and produce at least one daughter, which remains as a stem cell. Intermediate progenitors, however, which divide in the subventricular zone tend to divide apicobasally and produce two immature neurons (Noctor et al., 2008). Mammalian

homologues of many of the *Drosophila* genes, discussed above, work similarly in vertebrate neurogenesis. Thus, interfering with the function of mInsc (the mammalian homolog of Insc) or Asg3 (the mammalian homolog of Pins), leads to the loss of apico-basal spindle orientation in the mammalian nervous system, and results in too many cells dividing symmetrically. This causes the production of large clones, which can be explained if these symmetrically dividing cells produce two progenitors rather than one. The mammalian homolog of Numb, mNumb, also plays a key role, as knockout mice suggest that it is essential for cortical neuroepithelial cells to remain in a symmetrically dividing, early progenitor, state (Cayouette and Raff, 2003; Huttner and Kosodo, 2005; Zigman et al., 2005; Konno et al., 2008).

Interestingly, the environment may play a role in the orientation of cell division and thus whether a cell divides symmetrically or not, and recent studies in both insects and vertebrates have shown that extrinsic signals, perhaps coming from recently postmitotic cells, can influence the plane of cleavage of a neighboring progenitor. This is one mechanism by which the specific types of neuron may work through a feedback control mechanism to control the appropriate numbers of neurons generated (Poggi et al., 2005; Siegrist and Doe, 2006).

GENERATING COMPLEXITY THROUGH CELLULAR INTERACTIONS: THE *DROSOPHILA* RETINA

The compound eye of an insect is composed of a large number of identical unit eyes, called facets or ommatidia, each with its own lens and array of cell types. This has proved to be an excellent experimental system in which to investigate how specific cells get their fates through communicating with neighboring cells. Each of the 800 ommatidia in a *Drosophila* eye possesses 8 photoreceptors and 12 accessory cells. The 8 photoreceptors (R1–R8) are specialized sensory neurons (**Figure 4.10**). Among the accessory cells, there are cone cells that form the lens of each ommatidium and pigment cells that surround the photoreceptors, optically shielding one ommatidium from light that enters neighboring ommatidia. Each type of cell within a facet expresses a unique set of intrinsic determinants. For example, *rough* (*ro*) is expressed in R2, R5, R3 and R4; *Bar* appears in R1 and R6; *Seven-up* (*svp*) in R1, R6, R3, R4; *Prospero* (*pros*) in R7 and cone cells; *lozenge* (*lz*) in R1, R6, R7, and cone cells; and *Pax2* in cone cells only (Figure 4.10). Each of these factors is linked to the normal differentiation of the respective cells in which it is expressed, as revealed by the fact that a particular cell type fails to develop in an eye disc that lacks the corresponding gene.

In this system, lineage is *not* the main determining feature of the specification of the different cell types. This was first shown by the fact that when all the clonal descendants of any single projenitor are genetically labelled, one can find no common lineages. Any cell of any cell type can have a sister who is also of any cell type (Ready et al., 1976). In this system, signaling pathways operating between the postmitotic cells are primarily responsible for specifying cell fate, and inherited determinants have little role (Banerjee and Zipursky, 1990). Even after a cell has become postmitotic, it remains temporarily uncommitted to any particular differentiated fate. The mechanism controlling retinal cell fate diversification depends on the fact that the cells do not

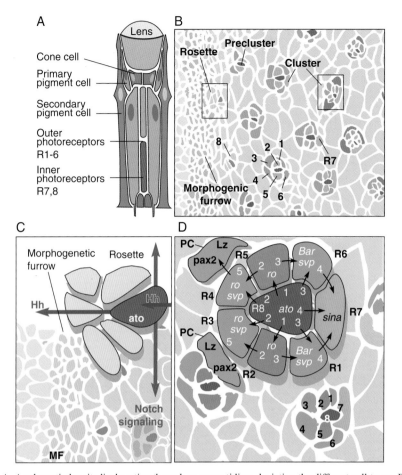

Fig. 4.10 A. A schematic longitudinal section through an ommatidium depicting the different cell types. B. Diagram showing a surface view of part of the eye disc at a stage when photoreceptor clusters become assembled. C. A precluster in which the R8 precursor expresses *ato* in response to a previously generated Hedgehog (Hh) signal. The R8 cell produces its own Hh and by this relay starts the next R8. R8s spread themselves out through the Notch lateral inhibition pathway. D. Cascade of photoreceptor determination in a developing ommatidial cluster. R8, the first cell to be determined, expresses *atonal* (*ato*) and signals neighboring cells to become R2 and R5, which then express *Rough* (*ro*). R2 and R5 in combination with R8 then signal the next set of neighboring cells to join the cluster of five becoming R3 and R4, which express *Seven-up* (*svp*). A cluster of seven cells is formed when R1 and R6 are induced to express *Bar*, *svp*, and *Lozenge* (*Lz*) by R2, R5, and R8. Finally R8, in combination with R1 and R6, induce the final photoreceptor R7 expressing *sevenless in absentia* (*sina*) and *prospero* (*pros*) to join. After the photoreceptors have joined, pigment cells expressing *lz* and *pax-2* are induced to join the cluster.

differentiate all at once, but follow a precise and reproducible temporal sequence. During late larval life, a wave of differentiation slowly passes over the eye disc in a posterior to anterior direction (Figure 4.10). The front of this wave is morphologically visible as a narrow groove formed by apical constriction of the eye disc cells. This moving groove is called the morphogenetic furrow (MF). It is at the MF where ommatidial differentiation first begins. As the MF advances, cells in its wake aggregate into "rosettes" that foreshadow the regular ommatidial pattern. One cell in each rosette is singled out by a lateral inhibition mechanism that involves Notch signaling. This cell becomes the photoreceptor known as R8, which expresses the bHLH proneural transcription factor Atonal.

The current molecular explanation for the advancement of the furrow involves a molecular signaling loop in which Atonal turns on the signaling protein Hedgehog (Hh) a homolog of vertebrate Shh (discussed in Chapter 2) and Hh signals to turn on Atonal in nearby cells. As the R8 cells begin to differentiate just posterior to the furrow, they emit Hh. Hh diffuses across to the anterior side of the MF to induce the next set of Atonal expressing cells, and so on. This very simple positive feedback loop propels the morphogenetic furrow from posterior to anterior across the entire eye disc (Kumar and Moses, 2000)

After R8 begins to differentiate, it also sends a signal to its immediate neighbors, two of which then begin to differentiate as R2 and R5. These cells in combination with R8 then help recruit the next cells that join the cluster, R3 and R4, and then R1 and R6, which aid in the recruitment of R7 (Ready, 1989). This has many aspects of the formation of a chemical crystalline lattice and so can be thought of as a kind of cellular crystallization process in which ommatidial clusters incorporate neighboring cells into an organized structure in a manner analogous to how growing crystals incorporate molecules into their structure. As new cells encounter the growing crystal front (the morphogenetic furrow) they become appropriately assimilated. In this way, the determination of a specific fate moves as a wave of crystallization across the developing eye.

Shortly following its own determination, R8 puts out signals that activate two different signaling pathways, the Notch pathway and the Ras pathway (Freeman, 1997; Brennan and Moses, 2000). The Notch pathway, as described in Chapter 1, is activated by Delta. The Ras pathway is a highly conserved biochemical cascade of cytoplasmic kinases (Ras, Raf, MPK), which in this case are activated by the epidermal growth factor receptor (EGFR). R8 emits an EGF-like molecule, Spitz (Spi), that activates this receptor. Activation of these signaling cascades spreads concentrically from R8 to R2 and R5, R3, R4, R1, R6, and R7, and then the remainder of the ommatidial cells. The precise, temporally regulated activation of the EGFR and Notch signaling pathways is critical for assigning distinct phenotypes to the cells that join the ommatidial clusters.

The last cell to join the inner photoreceptor cluster, R7, uses an additional signaling pathway, the discovery of which took advantage of the fact that only R7 is sensitive to UV light. Thus, mutagenizing flies and screening for offspring that are blind to UV light yielded a mutant that lacked the R7 cell in every ommatidium and was therefore called *sevenless* (*sev*) (Harris et al., 1976), (**Figure 4.11**). Sev is a receptor, the lack of which causes cells that, due to the time and place they joined the cluster, should have become R7s to develop into lens secreting cells that normally join the cluster later. In several follow-up screens, a large number of signaling molecules activated by the Sev receptor were identified (Rubin, 1991; Hafen et al., 1994) (**Figure 4.12**). Among them were the *Drosophila* homologs of Ras, Raf, MEK, MAPK, and Gap1, all parts of the Ras signaling pathway, as well as several new components. This pathway regulates the activity of transcription factors Yan and Pointed that control the differentiation of R7. One mutant with a *sevenless* phenotype was called *bride of sevenless* (*boss*). Boss is a transmembrane protein that binds and activates the Sev receptor. Boss is

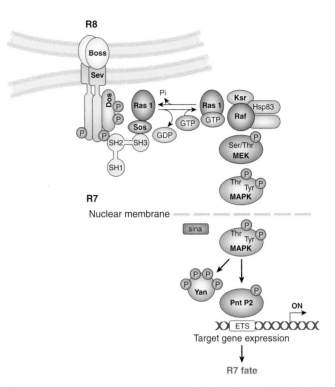

Fig. 4.12 Components of the sevenless transduction pathway, including the seven transmembrane signaling molecule bride of sevenless (Boss) which is expressed by R8, the Sevenless receptor molecule (Sev), and a number of downstream components of the signaling cascade such as daughter of sevenless (Dos) and Son of sevenless (Sos) and various members of the Ras Raf pathway ending in the activation of the *yan* and *pointed* genes. *(Courtesy of E. Hafen)*

expressed specifically in R8 cells, so when a cell expressing Sev touches R8, the Ras signaling pathway fires in this cell and, *presto*, it becomes R7 (Reinke and Zipursky, 1988).

SPECIFICATION AND DIFFERENTIATION THROUGH CELLULAR INTERACTIONS AND INTERACTIONS WITH THE LOCAL ENVIRONMENT: THE VERTEBRATE NEURAL CREST

We have seen how transcriptional cascades and cell–cell interactions work to give neural cells specific fates. Now, using the example of the vertebrate neural crest, we add another component: the local environment. The vertebrate neural crest is a transient stem cell population that arises along the lateral edges of the neural plate induced by the convergence of secreted signals (notably Wnts, BMPs and FGFs), at the juxtaposition of neural plate, lateral epidermis, and subjacent paraxial mesoderm (see Chapters 1 and 2 for details of crest induction). As described in Chapter 3, these cells migrate from their site of origin at the dorsal-most part of the neural tube, along stereotypic pathways through the embryo. The neural crest progenitors continue to divide as they migrate until they coalesce at their destinations. Crest cells generate a variety of cell types. Not only does the crest produce the entire peripheral nervous system, including the autonomic and sensory ganglia, and the

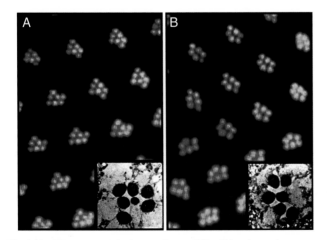

Fig. 4.11 Photoreceptors in the eye of normal flies (A) and *sevenless* mutants (B). The red images show light that is piped up through the clusters of receptors in each facet. The inserts are electron micrographs through a single facet and show cross sections of the photoreceptor array. Notice that the seventh central photoreceptor is missing in each facet of the mutant eye.

peripheral glia (Schwann cells), but it also produces endocrine chromaffin cells of the adrenal medulla, smooth muscle cells of the aorta, melanocytes, cranial cartilage, and teeth, and a variety of other nonneural components. Because of its variety of descendants, crest has been an important model for investigating mechanisms that generate cell diversity (Le Douarin, 1982).

To test if crest cells are already committed to particular fates before they begin to migrate, Le Douarin and colleagues transplanted the crest between different anterior-posterior positions (**Figure 4.13**) in classic experiments that took advantage of the chick–quail chimeric system described in Chapter 2. The results show that crest cells acquire their characteristic fates during their migrations, as well as when they arrive at their final destinations. Crest cells from the trunk normally give

rise to adrenergic cells of the sympathetic nervous system, whereas the more anterior crest cells normally give rise to cholinergic parasympathetic neurons that innervate the gut. When anterior crest cells from quail embryos were transplanted into the trunk region of chicken embryos, the transplanted crest migrated along the trunk pathways and differentiated into adrenergic neurons in sympathetic ganglia. Similarly, trunk crest cells that were transplanted to the anterior region gave rise to cholinergic neurons of the gut. Similar experiments have been done to test the competence of crest cells to form a variety of different cell types, and the general conclusion is that crest cells display great flexibility in responding to local environmental cues.

An alternative possibility, however, has to be considered. This is that each region of the crest contains a full complement of specified progenitors, but that only some of these survive to proliferate in each location. This boils down to the question of whether the local environment provides instructive or selective mechanisms. One of the factors that influences neural crest fates is the signaling molecule Wnt, which promotes sensory neuron fate. In the absence of Wnt signaling, experimentally achieved by selective knock-out of the intracellular β-catenin effector of the canonical Wnt pathway (see Chapter 1), neural crest cells migrate, proliferate, and differentiate normally, except that they do not make very many sensory neurons. When, however, Wnt signaling is prolonged, too many neural crest cells develop as sensory neurons and there is a deficit of other crest derivatives. To test if this is due to an instructional or selective mechanism, single premigratory crest cells were individually cultured with or without Wnt in the medium. Without Wnt, the individual cells proliferated and gave rise to clones of cells that contained a mixture of crest derived cell types. In the Wnt treated cultures, cells generated clones composed of only sensory neurons (Lee et al., 2004a). The minimal cell death seen in these experiments in either scenario excludes a selective mechanism and favors instead an instructional one.

Migrating crest cells, it seems, become exposed to a sequence of such instructive environments, each with a unique set of factors, and the migrating cells respond to these factors and each other in a way that sequentially limits their potential as they make the sometimes long journey to their targets. Initially, neural crest cells are multipotent, and labeling of single progenitors at the earliest stages of migration shows that these cells can give rise to a wide variety of derivatives (Bronner-Fraser and Fraser, 1988) (**Figure 4.14**). But as the cells migrate along particular routes, they segregate into several classes of more specialized progenitors. Thus, as development proceeds, they become more restricted. In the trunk region, an early decision influenced by Wnt causes some to become the sensory progenitors, which remain in the somitic mesodermal region and express the proneural *bHLH* transcription factor Neurogenin (Nrgn2), whereas autonomic progenitors that are not exposed to Wnt do not express Nrgn2 (Lo et al., 2002). Once they have made this decision, it is irreversible, and transplantation studies with these two types of progenitors show they can no longer make the full array of cell types. In addition to the instructional Wnt signal, sensory cells of the dorsal root ganglia (DRG) inhibit other crest cells from assuming this fate. If the cells that normally make the DRGs are ablated, then later migrating crest cells will differentiate into sensory neurons (Zirlinger et al., 2002).

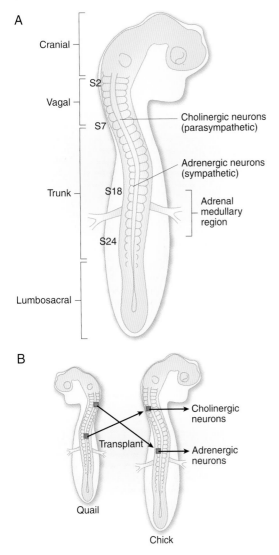

Fig. 4.13 The environment influences the fate of neural crest cells. A. Crest cells from the trunk normally give rise to adrenergic cells of the sympathetic nervous system, whereas the more anterior crest cells give rise to cholinergic parasympathetic neurons that innervate the gut. B. When anterior crest cells from quail embryos are transplanted into the trunk region of chicken embryos, they differentiate into adrenergic neurons. Similarly, trunk crest cells that are transplanted anteriorly give rise to cholinergic neurons. *(After Le Douarin et al., 1975)*

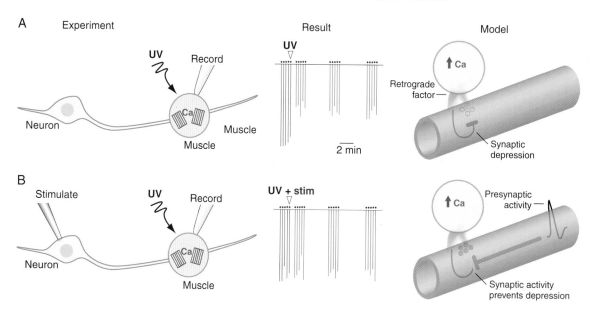

Fig. 9.29 Synapse depression depends on postsynaptic calcium. A. A whole-cell recording is made from an innervated muscle cell that is filled with "caged" calcium. Intracellular calcium is elevated when UV light releases the calcium from its "cage" (left). Under these conditions, nerve-evoked synaptic currents become depressed (middle). Baseline nerve-evoked synaptic currents (black downward deflections beneath each dot which represent the stimuli) are first recorded from the muscle. When intracellular calcium is elevated by exposure to UV light, the nerve-evoked synaptic currents (red) become depressed within seconds. This suggests a model in which elevated calcium triggers the release of a retrograde factor that leads to reduced presynaptic transmitter release (right). B. When the presynaptic nerve is stimulated during the UV-evoked rise in calcium (left), synaptic depression is prevented (middle). In this case, the model suggests that presynaptic activity can interfere with the retrograde signal (right). *(Adapted from Cash et al., 1996)*

(Rabacchi et al., 1992). In frogs, the innervation pattern of retinal afferents can be disrupted by NMDAR blockers. For example, when an extra eye is transplanted into a tadpole, the retinal axons project to the midbrain and form a "striped" innervation pattern, similar to the ocular dominance columns in mammalian visual cortex. The segregation of afferents into stripes was blocked completely in the presence of an NMDAR antagonist (Figure 9.30B). This effect is reversible: when the NMDAR blocker is removed, the fibers from each eye became segregated. Interestingly, exposing the midbrain to NMDA leads to a sharpening of the borders between eye-specific stripes (Cline et al., 1987; Cline and Constantine-Paton, 1990).

One can also observe the influence of NMDA receptor activity on the development of individual neuron coding properties. When the hamster superior colliculus is treated with an NMDAR blocker during development, the single neuron receptive field sizes (i.e., the portion of visual space that activates the neuron) are more than 60% larger than normal (Huang and Pallas, 2001). Therefore, NMDAR blockade interferes with synapse elimination in the central nervous system, leading to less specific afferent projections. It is also significant that NMDARs play an important role in one neural analog of learning, called *long-term potentiation* (see Box 9.1, "Remaining Flexible").

CALCIUM-ACTIVATED SECOND MESSENGER SYSTEMS

Since calcium plays an important role in developmental plasticity, it is reasonable to ask what calcium interacts with in the cytoplasm. One calcium-binding protein, called *calmodulin*, is a major constituent of the postsynaptic density. Together with calcium, it serves to activate a cytoplasmic kinase called *Ca²⁺/calmodulin-dependent protein kinase II (CaMKII)*. Once activated, the CaMKII becomes autophosporylated, and its activity then becomes independent of Ca^{2+} and calmodulin binding.

To test the role of CaMKII in synapse elimination, transgenic fruit flies can be engineered to constitutively express an inhibitor of this enzyme. The motor nerve terminals of these transgenic animals have numerous sprouts and a greater number of presynaptic sites, as compared to wild-type flies (Wang et al., 1994). The opposite effect is produced in the frog optic tectum by causing activated CaMKII (i.e., catalytically active in the absence of calcium) to be expressed constitutively in postsynaptic neurons (Zou and Cline, 1996). Retinal axons make simpler arborizations when CaMKII is active, suggesting that connections are being eliminated more rapidly than normal. Finally, the number of connections per neuron can be decreased in vitro by transfecting cells with a constitutively active form of CaMKII. The transfected neurons display a net reduction in the number of presynaptic partners (Pratt et al., 2003). Thus, CaMKII participates in decisions about the number of connections to be made, making it a good candidate for the synapse elimination mechanism.

The activation of CaMKII can exert an influence on synapse stability, in part, by regulating cell adhesion molecules at the synapse. As we learned in Chapter 8, *discs large* (DLG) is a clustering protein at the *Drosophila* nerve–muscle junction. One of the synaptic proteins that DLG regulates is the cell adhesion molecule, FasII. To test whether CaMKII signaling can influence the clustering of DLG and FasII, mutant

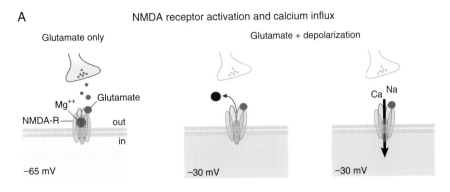

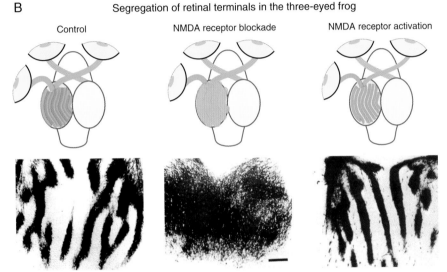

Fig. 9.30 NMDA receptors are involved in synaptic plasticity. A. The NMDA receptor is activated by a combination of glutamate binding and membrane depolarization. A positively charged magnesium ion blocks the NMDA receptor channel at rest. Depolarization causes the magnesium ion to be ejected, and this permits sodium and calcium ions to flow in. B. When a third eye is implanted into a frog embryo, the tectum becomes co-innervated by two sets of afferents (blue and red). The afferents from each eye segregate into stripes, similar to primary visual cortex (left). When three-eyed frogs are treated with an NMDA receptor antagonist (either APV or MK801), the afferents do not segregate, and stripe formation is prevented (middle). When three-eyed frogs are treated with NMDA, stripe formation is enhanced (right), and the borders between stripes is sharper. *(Adapted from Cline et al., 1987; Cline and Constantine-Paton, 1990)*

flies expressing a constitutively active form of the kinase were examined (Koh et al., 1999). In these flies, DLG was apparently displaced from the synaptic region. Since DLG can be phosphorylated by CaMKII, these results suggest a model in which activity mediates synapse stability by modifying the level of adhesion (**Figure 9.31**A). An influx of calcium may activate CaMKII, leading to phosphorylation of DLG and declustering of FasII. Taken together with the results from flies in which CaMKII is inhibited, it appears that the activity of this enzyme is one factor that controls synaptic sprouting and stabilization.

Many other neurotransmitters and second messenger systems could play a similar role during synaptogenesis. The developing neuromuscular synapse releases neurotrophins and calcitonin gene-related peptide (CGRP), and each has been implicated in the competition for postsynaptic space (Nelson et al., 2003). CGRP elevates postsynaptic cAMP, leading to activation of PKA and phosphorylation of AChRs (Miles et al., 1989; Lu et al., 1993). A role for the cAMP-PKA signaling pathway is suggested by mutant flies in which cAMP

levels are persistently increased. This leads to a decrease in FasII expression and a 70% increase in boutonal endings on muscles 6 and 7 (Figure 9.31B). When *dunce* flies are engineered to carry a transgene that maintains high levels of FasII expression, the sprouting of motor terminals does not occur (Schuster et al., 1996). Thus, synaptic activity may also influence the adhesion of pre- and postsynaptic cell through cAMP signaling. In the CNS, glutamate also activates metabotropic glutamate receptors (mGluRs) which lead to phospholipid metabolism or cAMP production. Mice deficient for a particular mGluR subunit exhibit significantly less synapse elimination at cerebellar Purkinje cells compared to controls (Kano et al., 1997; Ichise et al., 2000). Here, the phospholipase (PLC) signaling pathway has been implicated. Mice deficient for PLCβ4 also display reduced climbing fiber elimination (Kano et al., 1998).

Finally, neuromodulatory transmitters have long been known to affect synaptic plasticity during development. The effects of monocular deprivation are markedly reduced when either cholinergic, noradrenergic, or serotonergic terminals are

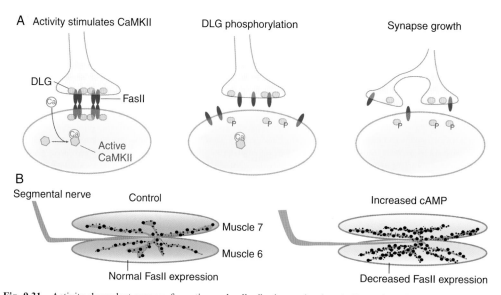

Fig. 9.31 Activity-dependent synapse formation and cell adhesion molecules. A. Synaptic activity leads to calcium influx and activation of the kinase, CaMKII (left). The active CaMKII phosphorylates a protein that mediates clustering of synaptic proteins (DLG), including the cell adhesion molecule, FasII (middle). The phosphorylated DLG is not as restricted to the synaptic complex, and FasII may no longer be located at the synapses, leading to less adhesion and sprouting of new boutons. B. In wild-type flies, muscles 6 and 7 receive about 180 boutons, and the expression of FasII is relatively high (left). In a mutant strain of flies (*dunce*) with increased cAMP production, FasII levels are lower and about 70% more boutonal endings are formed (right). *(Adapted from Budnik et al., 1996; Zhong et al., 1992; Schuster et al., 1996; Koh et al., 1999)*

eliminated from the developing nervous system even though these transmitters do not mediate visually evoked activity (Kasamatsu and Pettigrew, 1976; Bear and Singer, 1986; Gu and Singer, 1995). In fact, the projections from the relevant brainstem nuclei (e.g., the raphe, the locus coeruleus, and the nucleus basalis) arborize widely in the brain, and can modify synaptic transmission in adult and developing animals.

GAIN CONTROL

To this point, our discussion has focused on the concept that synapses can be weakened and eliminated by the activity of neighboring connections. However, developing synapses can also undergo long-lasting alterations in strength following a period of stimulation. For example, synaptic activity can lead to a long-lasting decrease in the response amplitude, called *long-term depression* (LTD). In many areas of the brain, LTD is more prominent during early development (**Figure 9.32**A). For example, LTD is present in Layer IV of the visual cortex in juvenile cats and guinea pigs. However, it is virtually absent in adult animals (Dudek and Friedlander, 1996). To test whether this form of plasticity plays a role in synapse elimination, the contacts between pre- and postsynaptic elements can be monitored before and after induction of LTD (Figure 9.32B). In the developing hippocampus, LTD is correlated significantly with a reduction in the synaptic contact area (Bastrikova et al., 2008).

An activity-dependent increase in synaptic potentials, called *long-term potentiation* (LTP) has also been found at many developing synapses. One possibility is that LTP is an important mechanism in the adult nervous system and simply appears during development without playing any particular role in synaptogenesis. However, LTP appears transiently in some areas of the brain, and is thought to regulate synaptic stabilization (Crair and Malenka, 1995; Boettiger and Doupe, 2001; Ding and Perkel, 2004). An extreme case of synaptic potentiation occurs when the physical contacts between nerve cells display absolutely no transmission ("silent synapses"), yet can be turned on by using them (Dubinsky and Fischbach, 1990; Lewis et al., 1990; Maggi et al., 2003).

It is possible that many glutamatergic synapses are silent during development because activation of the NMDAR requires membrane depolarization (Figure 9.30A). Intracellular recordings from the neonatal hippocampus show that most synapses have only NMDARs, and they do not respond when membrane potential is held at −60 mV. However, neural activity can enhance synaptic transmission by rapidly recruiting new functional AMPA-type glutamate receptors (Durand et al., 1996). In fact, eye opening is associated with an increase in the percentage of "silent" NMDAR-containing synapses in the rat superior colliculus. Furthermore, the AMPAR:NMDAR ratio increases during the next 24 hours of vision, and there is a commensurate reduction in the number of retinal inputs per postsynaptic neuron (Lu and Constantine-Paton, 2004).

How can LTP and LTD play a role in the selective stabilization of inputs as development progresses? To address this question, whole-cell recordings were obtained from *Xenopus* tectal neurons in vivo, and stimulating electrodes were placed on two different RGCs that converged on the recorded neuron (Zhang et al., 1998). When the RGCs are stimulated synchronously for 20s, the EPSCs evoked by stimulation at each site display LTP. The situation is a bit more complicated when the two retinal sites are stimulated asynchronously. When stimulation of one retinal site produces a postsynaptic action potential, and the other site produces a subthreshold excitatory synaptic event, the precise timing of the two inputs is

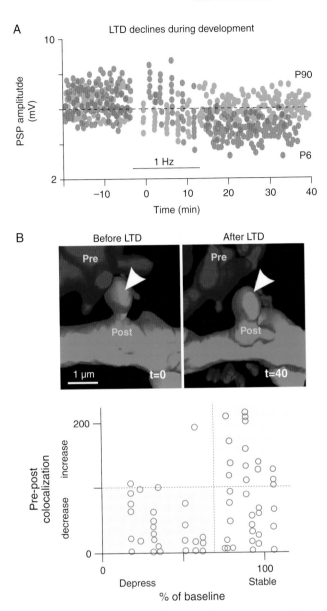

Fig. 9.32 Transient expression of long-term depression may mediate synapse elimination A. Long-term depression (LTD) declines with age in Layer IV of visual cortex. At postnatal day 6, low-frequency stimulation of the thalamic afferents leads to a long-lasting depression of excitatory synaptic currents (red dots). By P90, the same treatment has no effect (yellow dots). The probability of detecting LTD is nearly 90% in tissue from juvenile animals but is negligible in young adults. B. Contacts between presynaptic terminals (red) and postsynaptic spines (green) are monitored before and after the induction of LTD in a hippocampal slice (top). The presynaptic contact withdraws following LTD. When this experiment is repeated many times, it is found that those synapses displaying the greatest depression are associated with a decrease in colocalization of pre- and postsynaptic elements (light red box). In contrast, those synapses displaying the greatest stability of response are equally likely to increase their colocalization (light green box). *(Adapted from Dudek and Friedlander, 1996; Bastrikova et al., 2008)*

of their activity with respect other inputs, and this is called *spike timing dependent plasticity (STDP)*.

STDP can be demonstrated in vivo using natural stimuli. Whole-cell recordings are again obtained from tectal neurons in developing tadpoles. The retina is then stimulated with a visual pattern either just before or immediately after a spike has been generated through the patch electrode. This paradigm can induce significant changes in a tectal neuron's coding properties, and suggests how LTP and LTD contribute to the refinement of neural connections (Mu and Poo, 2006; Vislay-Meltzer et al., 2006). As visual stimuli (or spontaneous activity waves) sweep across the retina, a postsynaptic tectal neuron will be sequentially activated by a series of retinal ganglion cells. When the most effective retinal ganglion cell is activated, it will cause the tectal neuron to fire an action potential. Thus, any retinal ganglion cell that produces a subthreshold event within the next 20 ms or so may become depressed and subject to elimination.

HOMEOSTATIC PLASTICITY: THE MORE THINGS CHANGE, THE MORE THEY STAY THE SAME

Synapses and ion channels can each be regulated by the average level of postsynaptic activity, referred to as *homeostatic plasticity* (Marder and Prinz, 2002; Burrone and Murthy, 2003; Pozo and Goda, 2010). Its purpose is to keep the average postsynaptic discharge rate at about the same level (Royer and Pare, 2003). Thus, when postsynaptic activity is increased or decreased, voltage- and ligand-gated channels are adjusted to resist the manipulation. For example, when spiking activity is blocked in cortical cultures, membrane currents that support excitation increase (e.g., sodium, glutamate-evoked) and those that suppress excitation decline (e.g., potassium, GABA-evoked) (Rao and Craig, 1997; Desai et al., 1999; Murthy et al., 2001; Kilman et al., 2002; Burrone et al., 2002). Conversely, when activity is increased, the glutamate-evoked excitation declines (O'Brien et al., 1998). Presynaptic mechanisms are also modified homeostatically: activity blockade can increase the size and efficacy of terminals (Murthy et al., 2001). Even ionic equilibrium potentials are effected: inhibitory strength can decline due to down-regulation of the potassium chloride cotransporter, KCC2 (Vale et al., 2003).

Although homeostatic plasticity occurs throughout life, mechanistic differences exist between the developing and adult nervous system (Desai et al., 2002; Goel and Lee, 2007). This can be demonstrated by blocking action potentials at two different stages of development, and then determining whether the same changes occur. When in vivo blockade of activity is initiated in juvenile rats (P15), excitatory synaptic currents decline in amplitude. However, this change is not observed when the same manipulation is performed at P30 (Echegoyen et al., 2007). Similarly, when activity is blocked in hippocampal cultures during the period of synaptogenesis, postsynaptic glutamate receptors are increased. When the same manipulation is performed after synaptogenesis is complete, there is no such increase; rather the neurons respond by increasing presynaptic glutamate release (Han and Stevens, 2009).

The intracellular signaling cascades that initiate and sustain homeostatic plasticity are poorly understood. The process is

crucial to the outcome. When the subthreshold event *leads* the retinally evoked action potential by <20 ms, then it displays LTP. When the subthreshold event *follows* the retinally evoked action potential by <20 ms, then it displays LTD. Thus, inputs that evoke subthreshold synaptic potentials can either be strengthened or weakened, depending on the timing

controlled with great precision: when neural activity is altered, large clusters of synaptic proteins are up- or downregulated in synchrony. In addition to the expression of new proteins, the orderly turnover and degradation of existing proteins is a primary mechanism for activity-dependent remodeling of the synapse (Ehlers, 2003). Another interesting possibility is that changes in neural activity may modulate neuroligin-neurexin signaling which has been implicated in synaptogenesis (see Chapter 8). Chronic blockade of excitatory synaptic activity in hippocampal cultures leads to a suppression of neuroligin-2 activity which is normally associated with the addition of inhibitory synapses (Chubykin et al., 2007).

Most of these homeostatic changes can be demonstrated throughout the CNS following in vivo manipulations that decrease spontaneous or driven activity, including blindness and deafness. Direct measures of EPSC and IPSC amplitudes made in a brain slice preparation after surgically induced hearing loss reveal that compensatory changes occur in both the midbrain and cortex (Vale and Sanes, 2000, 2002; Kotak et al., 2005; Takesian et al., 2010). Excitatory synapses become stronger and inhibitory synapses become much weaker (**Figure 9.33**). Similar observations have been made in the visual system (Desai

et al., 2002; Morales et al., 2002). Alterations of excitatory and inhibitory synaptic strength measured directly in a brain slice are closely correlated with single neuron sensory coding properties recorded in vivo. For example, when facial whiskers are trimmed during development, the receptive field of barrel cortex neurons is subsequently found to have much larger excitatory and weaker inhibitory receptive fields (Shoykhet et al., 2005).

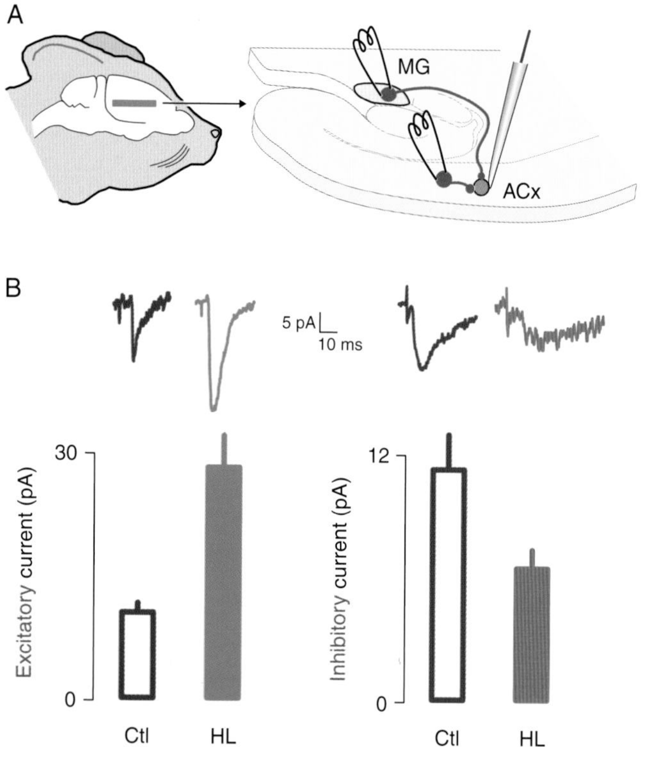

PLASTICITY OF INHIBITORY CONNECTIONS

Inhibitory synapses contribute to neural processing in approximately equal weight to synaptic excitation. Despite this, we are just now beginning to learn about their developmental plasticity. Is inhibitory transmission required for normal maturation and refinement of these synapses? In fact, synaptic transmission at inhibitory terminals can directly influence maturation, just as it does at excitatory connections (Chattopadhyaya et al., 2007). When GABA synthesis is decreased by eliminating the synthetic enzyme GAD67, cortical inhibitory neurons display much smaller axonal arbors and a 50% reduction in the number of synaptic contacts that they form on cortical pyramidal cells (**Figure 9.34**). Inhibitory synapse maturation is also impaired by a postsynaptic manipulation that abolishes transmission. When GABA receptor function is eliminated by genetic deletion of the α1 subunit, there is a 75% reduction of GABAergic synapses between cerebellar stellate cells and Purkinje cell dendrites (Fritschy et al., 2006). The remaining

Fig. 9.33 Homeostatic plasticity of excitatory and inhibitory synapse function in the developing auditory cortex. A. The schematic shows the orientation of a brain slice containing the projection from thalamus (MG) to auditory cortex (ACx). Stimuli delivered to the MG activate excitatory afferents (green) and excitatory synaptic currents can be recorded in the ACx. Stimuli delivered within the ACx activate inhibitory afferents (red) and inhibitory synaptic currents can be recorded. B. A comparison is made between evoked synaptic currents in control animals (Ctl) and those raised with bilateral hearing loss (HL). Electrical stimuli are delivered just above threshold to activate a single excitatory or inhibitory afferent. The traces and bar graph illustrate that developmental HL induces an increase in the amplitude of evoked excitatory synaptic currents (green), yet causes a decrease in the amplitude of inhibitory currents (red). *(Adapted from Kotak et al., 2005; Kotak et al., 2008)*

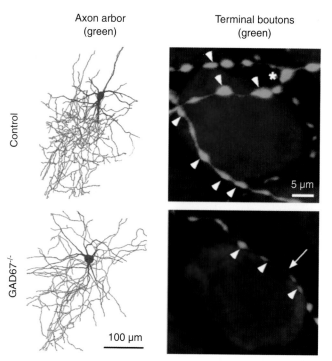

Fig. 9.34 GABA transmission is required for normal inhibitory synapse formation. The axonal projections of cortical inhibitory neurons were analyzed in control mice, and compared to animals in which the GABA synthesizing enzyme, GAD67, was inactivated. The axonal projections (green) of control inhibitory cells (top left) are much denser as compared to those in GAD67-/- mice (bottom left). On average, the control axons target about twice as many postsynaptic cell bodies. The number of inhibitory terminal boutons per postsynaptic pyramidal cell also declines by 50% in GAD-/- mice. *(Adapted from Chattopadhyaya et al., 2007)*

GABA terminals synapse at glutamate receptor-containing sites on the dendritic spine, a location reserved for excitatory terminals in wild-type animals. Thus, inhibitory transmission is not required absolutely for synapse formation, but appears to be essential for key maturational events.

The molecular mechanism by which neuronal activity regulates inhibitory synapse development includes BDNF, a neurotrophin whose expression is, itself, dependent on activity (Huang et al., 1999; Marty et al., 2000; Lu, 2003). This can be demonstrated by genetically eliminating a BDNF gene promotor that drives expression when bound by the activity-regulated factors, CREB or Npas4. In wild-type mice that are reared in the dark, there is a dramatic increase in BDNF expression when they are exposed to light for a few hours. This response declines by 50% when the activity-dependent BDNF promotor is missing. Furthermore, the number of inhibitory synapses in cortex is significantly reduced, despite a similar number of inhibitory neurons (Hong et al., 2008; Lin et al., 2008). These studies suggest that inhibitory synapse development is controlled largely by depolarizing activity, yet there may be a direct role for GABA signaling too. First, GABA-evoked potentials are depolarizing during very early brain development due to a reversed chloride potential (see Chapter 8). Second, GABA has been found to increase BDNF expression directly, through activation of $GABA_B$ receptors. Furthermore, when this signaling pathway is disrupted genetically, the number of inhibitory synapses is decreased in the developing hippocampus (Fiorentino et al., 2009).

Does synaptic activity at inhibitory contacts mediate their elimination or stabilization? This issue has been addressed for an inhibitory projection in the auditory brainstem. The medial nucleus of the trapezoid body (MNTB) contains inhibitory neurons that are activated by one ear, and which project to two different auditory nuclei that encode sound location. In one postsynaptic nucleus that encodes the intensity difference between two ears (LSO), the inhibitory terminals go through a two-stage process of refinement that results in a precise tonotopic map (**Figure 9.35**A and B). From birth to the onset of hearing, there is a striking refinement of functional contacts between MNTB and LSO neurons (Kim and Kandler, 2003). Recordings made from individual LSO neurons while focally

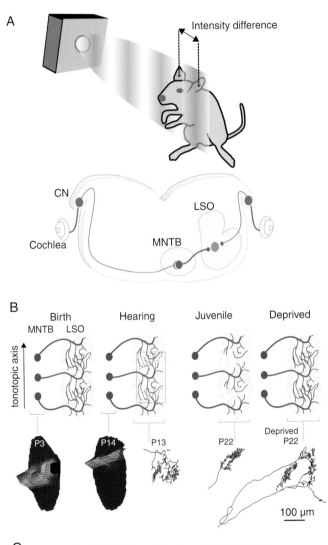

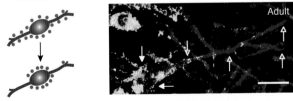

Fig. 9.35 Activity-dependent elimination of inhibitory connections during development. A. Animals can locate a sound source based on the intensity difference between the two ears (top). A nucleus in the auditory brainstem, the lateral superior olive (LSO), encodes intensity differences by integrating excitatory input driven by the ipsilateral ear with inhibition driven by the contralateral ear (bottom). This inhibition is mediated by MNTB neurons which map tonotopically onto the LSO. B. The development of MNTB arbors is shown schematically (top). During the period prior to hearing, the number of functional connections between MNTB and LSO decreases dramatically (dashed lines). This is demonstrated by the reduced MNTB region that elicits an inhibitory current in LSO (bottom left). During the period after hearing onset, MNTB axonal arbors are refined anatomically, as assessed by the distribution of their terminal boutons (red dots, bottom middle). When MNTB neurons are deprived of activity by removal of the contralateral cochlea, the axonal refinement does not occur (right). C. MNTB terminals are also eliminated in a second postsynaptic target, the MSO. At birth, inhibitory terminals are located on the soma and dendrites of MSO neurons. However, most of the dendritic synapses are eliminated during postnatal development (top left). The micrograph (top right) shows stained MSO neurons and glycine receptors from an adult animal. The glycine receptors (yellow) are largely restricted to the soma, and very few remain on the dendrites (blue). When animals are deafened unilaterally during development, the elimination of inhibitory synapses fails to occur (bottom left). The micrograph shows that significant glycine receptor staining (yellow) is now found on the dendrites (bottom right). Scale bars are 20 μm. *(Adapted from Sanes and Siverls, 1991; Sanes and Takács, 1993; Kapfer et al., 2002; Kim and Kandler, 2003)*

activating small areas within the MNTB demonstrate that 75% of the functional contacts are eliminated (Figure 9.35B, left). After the onset of hearing, individual axonal arborizations are pruned back by about 30% and come to occupy a narrow portion of the tonotopic axis (Figure 9.35B, middle). When MNTB neurons are deprived of excitatory input during development by removing the contralateral cochlea, then axonal refinement fails to occur (Figure 9.35B, right) (Sanes and Siverls, 1991; Sanes and Takács, 1993).

A complementary phenomenon occurs at a second target nucleus of the MNTB, the MSO. Inhibitory terminals withdraw from MSO dendrites during postnatal development and become restricted to the cell body in the adult (Figure 9.35C). The staining pattern of glycine-containing boutons and glycine receptor clusters demonstrate that both pre- and postsynaptic elements of the inhibitory synapse are eliminated during normal development, and this fails to occur when MNTB neurons are deprived of input (Kapfer et al., 2002). Finally, some inhibitory projections segregate into a banded pattern in the auditory midbrain, similar to visual cortex "stripes" (Figure 9.4). These inhibitory bands emerge from a diffuse projection pattern during the first two postnatal weeks, and this depends on an intact cochlea, suggesting once again that spontaneous activity plays a role (Gabriele et al., 2000; Henkel et al., 2007; Franklin et al., 2008). Thus, inhibitory synapses undergo a period of refinement, much as excitatory systems do, during which they attain a precise pattern of innervation. A growing number of studies suggest that spontaneous or sensory-evoked inhibitory transmission can influence this process.

Is the inhibitory synapse elimination associated with a weakening in their strength? In fact, MNTB inhibitory terminals release glutamate at first, and this excitatory signal is required for the early phase of synapse elimination. In mice with a genetic deletion of the glutamate transporter, VGLUT3, MNTB terminals do not release glutamate, and consequently fail to undergo the normal refinement of functional connectivity with LSO (Noh et al., 2010). Recordings made from LSO neurons at hearing onset reveal that they are innervated by twice as many MNTB neurons when glutamatergic transmission is missing. Furthermore, the inhibitory synapses fail to undergo a normal increase in strength. Somewhat later in development, MNTB synapses also display an activity-dependent form of long-term depression that is most prominent during the period of axonal refinement (Kotak and Sanes, 2000). This inhibitory LTD is mediated by GABA release from MNTB and can be blocked with a GABA$_B$ receptor antagonist (Kotak et al., 1998; Chang et al., 2003).

SYNAPTIC INFLUENCE ON NEURON MORPHOLOGY

Synaptic activity plays an extremely important role in regulating postsynaptic neuron morphology. During early development, even if denervation does not result in cell death (see Chapter 7), then it certainly leads to shrinkage of cell body size, atrophy of dendritic processes, or loss of dendritic spines in most areas of the central nervous system (Globus and Scheibel, 1966; Valverde, 1968; Rakic, 1972; Harris and Woolsey, 1981; Vaughn et al., 1988). Furthermore, changes in the amount of sensory experience given to an animal during development (which presumably affects neural activity) also

lead to measurable alterations in nerve cell morphology. Thus, young rats that are reared in an enriched, social environment have more dendritic branching than rats reared alone in an impoverished environment (Fiala et al., 1978). More precise manipulations of the sensory environment, such as sound attenuation or vertical stripe rearing, have also been associated with specific changes in auditory or visual neuron morphology, respectively.

Although it is common to perform a manipulation and then wait days or weeks to look for a change in the central nervous system, the effects of denervation occur at a surprising rate. In the chick *nucleus laminaris* (NL), it is possible to denervate the ventral dendrites while leaving the dorsal dendrites untouched (**Figure 9.36**). The entire NL dendritic arborization is then visualized with a Golgi stain, beginning one hour after

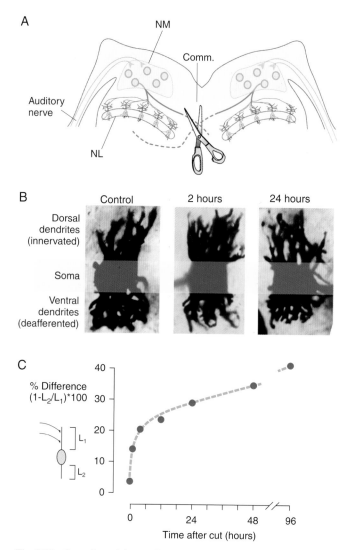

Fig. 9.36 Synaptic activity regulates dendrite length in the chick auditory brainstem. A. Neurons of the *nucleus laminaris* (NL) receive afferents from ipsilateral NM to their dorsal dendrites and from the contralateral NM to their ventral dendrites. The ventral dendrites of NL neurons were denervated by cutting NM afferent axons at the midline. The dorsal dendrites remained fully innervated. B. When NL neurons are stained with the Golgi technique, the ventral dendrites are found to be significantly shorter than those on the dorsal side within one hour of the manipulation. C. The ventral dendrites shrink by almost 40% by 96 hours after the lesion. *(Adapted from Deitch and Rubel, 1984)*

the afferents to the ventral dendrites are cut. Even at the earliest time point, there is a 14% decrease in the ventral dendrites compared to the dorsal set (Deitch and Rubel, 1984). Since this is well before terminals are actually removed from the postsynaptic neuron, the dendritic atrophy probably results from the sudden cessation of synaptic activity. Electron microscopy revealed that this change in morphology was correlated with a dramatic change in cytoskeletal structures, such as microtubules.

Dendrite morphology can be studied in vivo by imaging visual cortical neurons in transgenic mice that express a fluorescent protein. Images obtained over a two-hour period show that dendrites extend and retract filopodial processes and spines while they are developing (**Figure 9.37**A and B). However, the number of filopodia declines dramatically, and the spines become quite stable over the first two postnatal months (Grutzendler et al., 2002). When animals are visually

deprived during the critical period, spine motility increases (Figure 9.37C). However, deprivation in the somatosensory system, induced by trimming the whiskers during the first two weeks of development, decreases spine elimination in barrel cortex. Together, these results suggest that dendrites are maximally sensitive to synaptic activity during this period and that they participate in the activity-dependent formation and elimination of synaptic connections (Majewska and Sur, 2003; Zuo et al., 2005).

As with cell death, the effect of synaptic transmission can be tested directly by manipulations that block glutamate receptors. When chick embryos are treated with an NMDA receptor antagonist, cerebellar Purkinje cell dendrites do not develop as many branches and occupy a smaller cross-sectional area (Vogel and Prittie, 1995). A similar effect has been observed in frog optic tectal neurons and in spinal motor neurons. In hippocampal cultures, the number of dendritic spines, but not the number of branches, is dependent on glutamatergic synaptic activity (Kossel et al., 1997). The effects of excitatory transmission can change during development as new signaling systems are added to the cytoplasm. For example, the decrease in the growth of optic tectum dendrites correlates with increased expression of CaMKII, and the rate of growth is experimentally increased when animals are treated with a CaMKII inhibitor (Wu and Cline, 1998). It is also likely that other growth-promoting factors are co-released with neurotransmitter. In organotypic cultures of visual cortex, function-blocking antibodies against the TrkB receptor decrease the amount of basal dendritic growth, indicating that endogenously released BDNF promotes postsynaptic growth (McAllister et al., 1997).

Spine morphology is controlled by the cytoskeletal component, actin, which is highly dynamic in developing systems (Star et al., 2002). Glutamatergic synaptic activity can influence actin polymerization and stabilize spines by raising intracellular calcium (Fischer et al., 2000). In fact, dendrite stability is very sensitive to local calcium levels. In the embryonic chick retina, dendritic retractions can be prevented by locally raising calcium levels within the process (Lohmann et al., 2002). A local rise in calcium may serve to stabilize actin filaments, perhaps by activating gelsolin, the calcium-dependent actin-binding protein. In general, GTP-binding proteins (Rho GTPases) have been implicated in regulating neuron morphology by affecting the stability and assembly of actin filaments (Luo, 2002).

Given that the effects of excitatory denervation are so dramatic, it would be surprising if inhibitory synapses did not have a trophic influence on postsynaptic maturation. When dendrite growth first begins, inhibitory synapses elicit a depolarizing response because the chloride battery is not yet mature (see Chapter 8). To test whether depolarizing GABA-evoked depolarization is critical for dendrite development, neurons were transfected with the chloride transporter (KCC2) prior to migration. Although migration itself is unaffected, these neurons display a tremendous reduction in dendrite branching and length (Cancedda et al., 2007). In contrast, synaptic inhibition appears to have the opposite effect once the potentials becomes hyperpolarizing. This is demonstrated by depriving neurons of functional inhibition after their dendrites have formed. Dendrite branching increases significantly in an auditory brainstem nucleus, LSO (see Figure 9.35A), when inhibition is reduced in vivo by removing the contralateral

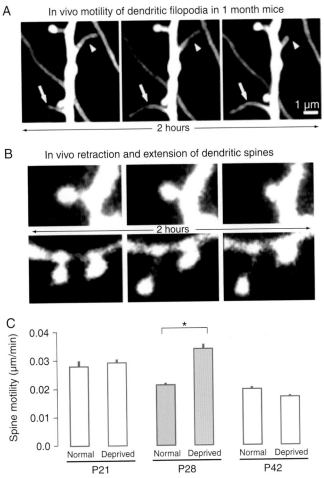

A. In vivo motility of dendritic filopodia in 1 month mice

1 µm

— 2 hours —

B. In vivo retraction and extension of dendritic spines

— 2 hours —

C

Spine motility (µm/min)

Normal Deprived Normal Deprived Normal Deprived
P21   P28   P42

Age and treatment

Fig. 9.37 Two-photon imaging of dendritic spine and filopodia motility in vivo. A. A set of time-lapse images (one hour intervals) in a 1-month-old animal shows that filopodia undergo rapid extension (arrows) and retraction (arrowheads). B. and C. Dendritic spine motility is more prominent at P21. Two time-lapse sequences are shown (acquired over two hours). In the first, a spine is retracted (top), and in the second a spine elongates (bottom). Binocular deprivation from P14 significantly increases spine motility at P28 by 60%. However, there is no change in motility when deprivation begins at P21. There is a slight reduction in motility in mice deprived at P42. *(From Grutzendler et al., 2002; Majewska and Sur, 2003)*

cochlea. Furthermore, when organotypic cultures of the LSO are grown in the presence of the inhibitory antagonist, strychnine, dendrites are twice as long as those grown in normal media (Sanes et al., 1992; Sanes and Hafidi, 1996). Thus, synaptic activity regulates postsynaptic morphology by either depolarizing or hyperpolarizing the growing dendrites.

SUMMARY

Synapses can form in the absence of synaptic transmission (Cohen, 1972; Duxson, 1982). Presynaptic cholinergic terminals are even able to differentiate in a zebrafish mutant lacking functional AChR clusters (Westerfield et al., 1990). In contrast, sensory deprivation studies suggest that the disuse of a synapse leads to its weakening or elimination, especially if competing synapses are active. Therefore, the presence of synaptic transmission is not an absolute requirement for the initial formation of a synapse but influences its subsequent development and stability.

At one level, the purpose of synapse elimination seems perfectly obvious: to create the optimal connections between neurons based upon their use. Perhaps the nervous system cannot take full advantage of the plasticity mechanisms unless it generates extra cell bodies, a surplus of dendritic branches, and a profusion of presynaptic arbors. In reality, the purpose of synapse elimination will remain enigmatic until we can produce a specific alteration in a single set of adult connections, let's say two climbing fibers per Purkinje cell, and determine the exact behavioral outcome. Therefore, our understanding of developmental plasticity is intimately tied to our insight into how the CNS encodes sensory information and controls behavior. What is the optimal pattern of connectivity for running, singing, perfect pitch, speed-reading, learning, and so forth? Is it even possible to have a nervous system that is optimized for diverse motor, sensory, and cognitive tasks? Unfortunately, we have yet to devise an experiment that tests whether extra cell bodies or small arborization errors actually affect animal behavior.

At present, we believe that synapses can be weakened or lost if they are not activated correctly during development. This might be particularly important for animals, such as humans, that inhabit a wide range of environments. For example, it is plausible that our central auditory system is shaped by the spoken language(s) to which we are exposed as infants (see Chapter 10). However, the experimental manipulations used to demonstrate an influence of environment or neural activity remain rather flagrant. For example, it is unlikely that any animal sees only vertical stripes or hears only one sound frequency during development. Nonetheless, we know that developing humans do experience many "extreme" rearing environments, such as blindness, deafness, malnutrition, and many others that result from genetic or epigenetic causes. Therefore, the clinical importance of understanding developmental plasticity is enormous.

REFERENCES

Adelsberger, H., Garaschuk, O., & Konnerth, A. (2005). Cortical calcium waves in resting newborn mice. *Nature Neuroscience, 8*, 988–990.

Antonini, A., Stryker, M. P. (1993). Development of individual geniculocortical arbors in cat striate cortex and effects of binocular impulse blockade. *The Journal of Neuroscience, 13*, 3549–3573.

Bagust, J., Lewis, D. M., & Westerman, R. A. (1973). Polyneuronal innervation of kitten skeletal muscle. *The Journal of Physiology, 229*, 241–255.

Balice-Gordon, R. J., & Lichtman, J. W. (1993). In vivo observations of pre- and postsynaptic changes during the transition from multiple to single innervation at developing neuromuscular junctions. *The Journal of Neuroscience, 13*, 834–855.

Balice-Gordon, R. J., & Lichtman, J. W. (1994). Long-term synapse loss induced by focal blockade of postsynaptic receptors. *Nature, 372*, 519–524.

Bastrikova, N., Gardner, G. A., Reece, J. M., Jeromin, A., & Dudek, S. M. (2008). Synapse elimination accompanies functional plasticity in hippocampal neurons. *Proceedings of the National Academy of Sciences of the United States of America, 105*, 3123–3127.

Bear, M. F., & Singer, W. (1986). Modulation of visual cortical plasticity by acetylcholine and noradrenaline. *Nature, 320*, 172–176.

Bennett, M. R., Pettigrew, A. G. (1974). The formation of synapses in reinnervated and cross-reinnervated striated muscle during development. *J Physiol (Lond), 241*, 547–573.

Berardi, N., Pizzorusso, T., & Maffei, L. (2000). Critical periods during sensory development. *Current Opinion in Neurobiology, 10*, 138–145.

Blakemore, C., & Van Sluyters, R. C. (1975). Innate and environmental factors in the development of the kitten's visual cortex. *The Journal of Physiology, 248*, 663–716.

Blankenship, A. G., & Feller, M. B. (2010). Mechanisms underlying spontaneous patterned activity in developing neural circuits. *Nature Reviews. Neuroscience, 11*, 18–29.

Blasdel, G., Obermayer, K., & Kiorpes, L. (1995). Organization of ocular dominance and orientation columns in the striate cortex of neonatal macaque monkeys. *Visual Neuroscience, 12*, 589–603.

Bliss, T. V. P., & Lomo, T. (1973). Long-lasting potentiation of synaptic transmission in the dentate area of the anesthtized rabbit following stimulation of the perforant path. *The Journal of Physiology, 232*, 331–356.

Boettiger, C. A., & Doupe, A. J. (2001). Developmentally restricted synaptic plasticity in a songbird nucleus required for song learning. *Neuron, 31*, 809–818.

Brainard, M. S., & Knudsen, E. I. (1993). Experience-dependent plasticity in the inferior colliculus: A site for visual calibration of the neural representation of auditory space in the barn owl. *The Journal of Neuroscience, 13*, 4589–4608.

Budnik, V., Koh, Y. H., Guan, B., Hartmann, B., Hough, C., Woods, D., et al. (1996). Regulation of synapse structure and function by the Drosophila tumor suppressor gene dlg. *Neuron, 17*, 627–640.

Burrone, J., O'Byrne, M., & Murthy, V. N. (2002). Multiple forms of synaptic plasticity triggered by selective suppression of activity in individual neurons. *Nature, 420*, 414–418.

Busetto, G., Buffelli, M., Tognana, E., Bellico, F., & Cangiano, A. (2000). Hebbian mechanisms revealed by electrical stimulation at developing rat neuromuscular junctions. *The Journal of Neuroscience, 20*, 685–695.

Campbell, G., & Shatz, C. J. (1992). Synapses formed by identified retinogeniculate axons during segregation of eye input. *The Journal of Neuroscience, 12*, 1847–1858.

Cancedda, L., Fiumelli, H., Chen, K., & Poo, M. M. (2007). Excitatory GABA action is essential for morphological maturation of cortical neurons in vivo. *The Journal of Neuroscience, 27*, 5224–5235.

Cash, S., Zucker, R. S., & Poo, M.-m. (1996). Spread of synaptic depression mediated by presynaptic cytoplasmic signaling. *Science, 272*, 998–1001.

Casticas, S., Thanos, S., & Clarke, P. G. H. (1987). Major role for neuronal death during brain development: refinement of topographic connections. *Proceedings of the National Academy of Sciences of the United States of America, 84*, 8165–8168.

Chang, E. H., Kotak, V. C., & Sanes, D. H. (2003). Long-term depression of synaptic inhibition is expressed postsynaptically in the developing auditory system. *Journal of Neurophysiology, 90*, 1479–1488.

Chattopadhyaya, B., Di Cristo, G., Wu, C. Z., Knott, G., Kuhlman, S., Fu, Y., et al. (2007). GAD67-mediated GABA synthesis and signaling regulate inhibitory synaptic innervation in the visual cortex. *Neuron, 54*, 889–903.

Chiba, A., Shepherd, D., & Murphey, R. K. (1988). Synaptic rearrangement during postembryonic development in the cricket. *Science, 240*, 901–905.

Chubykin, A. A., Atasoy, D., Etherton, M. R., Brose, N., Kavalali, E. T., Gibson, J. R., et al. (2007). Activity-dependent validation of excitatory versus inhibitory synapses by neuroligin-1 versus neuroligin-2. *Neuron, 54*, 919–931.

Cline, H. T., & Constantine-Paton, M. (1990). NMDA receptor agonist and antagonists alter retinal ganglion cell arbor structure in the developing frog retinotectal projection. *The Journal of Neuroscience, 10*, 1197–1216.

Cline, H. T., Debski, E. A., & Constantine-Paton, M. (1987). N-methyl-D-aspartate receptor antagonist desegregates eye-specific stripes. *Proceedings of the National Academy of Sciences of the United States of America, 84*, 4342–4345.

Cohen, M. W. (1972). The development of neuromuscular connexions in the presence of D-tubocurarine. *Brain Res, 41*, 457–463.

Connold, A. L., Evers, J. V., & Vrbova, G. (1986). Effect of low calcium and protease inhibitors on synapse elimination during postnatal development in the rat soleus muscle. *Developmental Brain Research, 28*, 99–107.

Conradi, S., & Ronnevi, L. O. (1975). Spontaneous elimination of synapses on cat spinal motoneurons after birth: Do half of the synapses on the cell bodies disappear? *Brain Research, 92*, 505–510.

Crair, M. C., Gillespie, D. C., & Stryker, M. P. (1998). The role of visual experience in the development of columns in cat visual cortex. *Science, 279*, 566–570.

Crair, M. C., Horton, J. C., Antonini, A., & Stryker, M. P. (2001). Emergence of ocular dominance columns in cat visual cortex by 2 weeks of age. *The Journal of Comparative Neurology, 430*, 235–249.

Crair, M. C., & Malenka, R. C. (1995). A critical period for long-term potentiation at thalamocortical synapses. *Nature, 375*, 325–328.

Crowley, J. C., & Katz, L. C. (1999). Development of ocular dominance columns in the absence of retinal input. *Nature Neuroscience, 2*, 1125–1130.

Crowley, J. C., & Katz, L. C. (2000). Early development of ocular dominance columns. *Science, 290*, 1321–1324.

Dan, Y., & Poo, M.-m. (1992). Hebbian depression of isolated neuromuscular synapses in vitro. *Science, 256*, 1570–1573.

DeBello, W. M., Feldman, D. E., & Knudsen, E. I. (2001). Adaptive axonal remodeling in the midbrain auditory space map. *The Journal of Neuroscience, 21*, 3161–3174.

Deitch, J. S., & Rubel, E. W. (1984). Afferent influences on brain stem auditory nuclei of the chicken: Time course and specificity of dendritic atrophy following deafferentation. *The Journal of Comparative Neurology, 229*, 66–79.

Demas, J., Eglen, S. J., & Wong, R. O. (2003). Developmental loss of synchronous spontaneous activity in the mouse retina is independent of visual experience. *The Journal of Neuroscience, 23*, 2851–2860.

Desai, N. S., Cudmore, R. H., Nelson, S. B., & Turrigiano, G. G. (2002). Critical periods for experience-dependent synaptic scaling in visual cortex. *Nature Neuroscience, 5*, 783–789.

Desai, N. S., Rutherford, L. C., & Turrigiano, G. G. (1999). Plasticity in the intrinsic excitability of cortical pyramidal neurons. *Nature Neuroscience, 2*, 515–520.

Ding, L., & Perkel, D. J. (2004). Long-term potentiation in an avian basal ganglia nucleus essential for vocal learning. *The Journal of Neuroscience, 24*, 488–494.

Dubin, M. W., Stark, L. A., & Archer, S. M. (1986). A role for action potential activity in the developement of neuronal connections in the kitten retinogeniculate pathway. *The Journal of Neuroscience, 6*, 1021–1036.

Dubinsky, J. M., & Fischbach, G. D. (1990). A role for cAMP in the development of functional neuromuscular transmission. *Journal of Neurobiology, 21*, 414–426.

Dudek, S. M., & Friedlander, M. J. (1996). Developmental down-regulation of LTD in cortical layer IV and its independence of modulation by inhibition. *Neuron, 16*, 1097–1106.

Durand, G. M., Kovalchuk, Y., & Konnerth, A. (1996). Long-term potentiation and functional synapse induction in developing hippocampus. *Nature, 381*, 71–75.

Duxson, M. J. (1982). The effect of post synaptic block on the development of the neuromuscular junction in postnatal rats. *Journal of Neurocytology, 11*, 395–408.

Echegoyen, J., Neu, A., Graber, K. D., & Soltesz, I. (2007). Homeostatic plasticity studied using in vivo hippocampal activity-blockade: synaptic scaling, intrinsic plasticity and age-dependence. *PLoS ONE, 2*(1), e700.

Ehlers, M. D. (2003). Activity level controls postsynaptic composition and signaling via the ubiquitin-proteasome system. *Nature Neuroscience, 6*, 231–242.

Eyre, J. A. (2007). Corticospinal tract development and its plasticity after perinatal injury. *Neuroscience and Biobehavioral Reviews, 31*, 1136–1149.

Fagiolini, M., Fritschy, J. M., Low, K., Mohler, H., Rudolph, U., & Hensch, T. K. (2004). Specific GABAA circuits for visual cortical plasticity. *Science, 303*, 1681–1683.

Fagiolini, M., & Hensch, T. K. (2000). Inhibitory threshold for critical-period activation in primary visual cortex. *Nature, 404*, 183–186.

Fiala, B. A., Joyce, J. N., & Greenough, W. T. (1978). Environmental complexity modulates growth of granule cell dendrites in developing but not adult hippocampus of rats. *Experimental Neurology, 59*, 372–383.

Fiorentino, H., Kuczewski, N., Diabira, D., Ferrand, N., Pangalos, M. N., Porcher, C., et al. (2009). GABA(B) receptor activation triggers BDNF release and promotes the maturation of GABAergic synapses. *The Journal of Neuroscience, 29*, 11650–11661.

Fischer, M., Kaech, S., Wagner, U., Brinkhaus, H., & Matus, A. (2000). Glutamate receptors regulate actin-based plasticity in dendritic spines. *Nature Neuroscience, 3*, 887–894.

Florence, S. L., & Casagrande, V. A. (1990). Development of geniculocortical axon arbors in a primate. *Visual Neuroscience, 5*, 291–309.

Forehand, C. J., & Purves, D. (1984). Regional innervation of rabbit ciliary ganglion cells by the terminals of preganglionic axons. *The Journal of Neuroscience, 4*, 1–12.

Franklin, S. R., Brunso-Bechtold, J. K., & Henkel, C. K. (2008). Bilateral cochlear ablation in postnatal rat disrupts development of banded pattern of projections from the dorsal nucleus of the lateral lemniscus to the inferior colliculus. *Neuroscience, 154*, 346–354.

Frey, U., & Morris, R. G. M. (1997). Synaptic tagging and long-term potentiation. *Nature, 385*, 533–536.

Fritschy, J. M., Panzanelli, P., Kralic, J. E., Vogt, K. E., & Sassoe-Pognetto, M. (2006). Differential dependence of axo-dendritic and axo-somatic GABAergic synapses on GABAA receptors containing the alpha1 subunit in Purkinje cells. *The Journal of Neuroscience, 26*, 3245–3255.

Gabriele, M. L., Brunso-Bechtold, J. K., & Henkel, C. K. (2000). Development of afferent patterns in the inferior colliculus of the rat: projection from the dorsal nucleus of the lateral lemniscus. *The Journal of Comparative Neurology, 416*, 368–382.

Galli, L., & Maffei, L. (1988). Spontaneous impulse activity of rat retinal ganglion cells in prenatal life. *Science, 242*, 90–91.

Garaschuk, O., Linn, J., Eilers, J., & Konnerth, A. (2000). Large-scale oscillatory calcium waves in the immature cortex. *Nature Neuroscience, 3*, 452–459.

Glanzman, D. L. (2010). Common mechanisms of synaptic plasticity in vertebrates and invertebrates. *Current Biology: CB, 20*, R31–R36.

Glanzman, D. L., Kandel, E. R., & Schacher, S. (1991). Target-dependent morphological segregation of Aplysia sensory outgrowth in vitro. *Neuron, 7*, 903–913.

Globus, A., & Scheibel, A. B. (1966). Loss of dendritic spines as an index of presynaptic terminal patterns. *Nature, 212*, 463–465.

Goel, A., & Lee, H. K. (2007). Persistence of experience-induced homeostatic synaptic plasticity through adulthood in superficial layers of mouse visual cortex. *The Journal of Neuroscience, 27*, 6692–6700.

Gordon, T., Perry, R., Tuffery, A. R., & Vrbova, G. (1974). Possible mechanisms determining synapses formation in developing skeletal muscles of the chick. *Cell and Tissue Research, 155*, 13–25.

Grutzendler, J., Kasthuri, N., & Gan, W. B. (2002). Long-term dendritic spine stability in the adult cortex. *Nature, 420*, 812–816.

Gu, Q., & Singer, W. (1995). Involvement of serotonin in developmental plasticity of kitten visual cortex. *European Journal of Neuroscience, 7*, 1146–1153.

Guo, Y., & Udin, S. B. (2000). The development of abnormal axon trajectories after rotation of one eye in Xenopus. *The Journal of Neuroscience, 20*, 4189–4197.

Han, E. B., & Stevens, C. F. (2009). Development regulates a switch between post- and presynaptic strengthening in response to activity deprivation. *Proceedings of the National Academy of Sciences of the United States of America, 106*, 10817–10822.

Harris, R. M., & Woolsey, T. A. (1981). Dendritic plasticity in mouse barrel cortex following postnatal vibrissa follicle damage. *The Journal of Comparative Neurology, 196*, 357–376.

Henkel, C. K., Keiger, C. J., Franklin, S. R., & Brunso-Bechtold, J. K. (2007). Development of banded afferent compartments in the inferior colliculus before onset of hearing in ferrets. *Neuroscience, 146*, 225–235.

Hensch, T. K. (2004). Critical period regulation. *Annual Review of Neuroscience, 27*, 549–579.

Hensch, T. K., & Stryker, M. P. (2004). Columnar architecture sculpted by GABA circuits in developing cat visual cortex. *Science, 303*, 1678–1681.

Holland, R. L., & Brown, M. C. (1980). Postsynaptic transmission block can cause terminal sprouting of a motor nerve. *Science, 207*, 649–651.

Hong, E. J., McCord, A. E., & Greenberg, M. E. (2008). A biological function for the neuronal activity-dependent component of Bdnf transcription in the development of cortical inhibition. *Neuron, 60*, 610–624.

Huang, L., & Pallas, S. L. (2001). NMDA antagonists in the superior colliculus prevent developmental plasticity but not visual transmission or map compression. *Journal of Neurophysiology, 86*, 1179–1194.

Huang, Z. J., Kirkwood, A., Pizzorusso, T., Porciatti, V., Morales, B., Bear, M. F., et al. (1999). BDNF regulates the maturation of inhibition and the critical period of plasticity in mouse visual cortex. *Cell, 98*, 739–755.

Hubel, D. H., & Wiesel, T. N. (1962). Receptive fields, binocular interaction and functional architecture in the cat's visual cortex. *The Journal of Physiology, 160*, 106–154.

Hubel, D. H., & Wiesel, T. N. (1965). Binocular interaction in striate cortex of kittens reared with artificial squint. *Journal of Neurophysiology, 28*, 1041–1059.

Hubel, D. H., & Wiesel, T. N. (1970). The period of susceptibility to the physiological effects of unilateral eye closure in kittens. *The Journal of Physiology, 206*, 419–436.

Huberman, A. D., Speer, C. M., & Chapman, B. (2006). Spontaneous retinal activity mediates development of ocular dominance columns and binocular receptive fields in v1. *Neuron, 52*, 247–254.

Huttenlocher, P. R., & de Courten, C. (1987). The development of synapses in striate cortex of man. *Human Neurobiology, 6*, 1–9.

Ichise, T., Kano, M., Hashimoto, K., Yanagihara, D., Nakao, K., Shigemoto, R., et al. (2000). mGluR1 in cerebellar Purkinje cells essential for long-term depression, synapse elimination, and motor coordination. *Science, 288*, 1832–1835.

Innocenti, G. M., Fiore, L., & Caminiti, R. (1977). Exuberant projection into the corpus callosum from the visual cortex of newborn cats. *Neuroscience Letters, 4*, 237–242.

Insanally, M. N., Kover, H., Kim, H., & Bao, S. (2009). Feature-dependent sensitive periods in the development of complex sound representation. *The Journal of Neuroscience, 29*, 5456–5462.

Iwai, Y., Fagiolini, M., Obata, K., & Hensch, T. K. (2003). Rapid critical period induction by tonic inhibition in visual cortex. *The Journal of Neuroscience, 23*, 6695–6702.

Jackson, H., & Parks, T. N. (1982). Functional synapse elimination in the developing avian cochlear nucleus with simultaneous reduction in cochlear nerve axon branching. *The Journal of Neuroscience, 2*, 1736–1743.

Jansen, J. K. S., Lømo, T., Nicolaysen, K., & Westgaard, R. H. (1973). Hyperinnervation of skeletal muscle fibers: dependence on muscle activity. *Science, 181*, 559–561.

Jones, T. A., Jones, S. M., & Paggett, K. C. (2001). Primordial rhythmic bursting in embryonic cochlear ganglion cells. *The Journal of Neuroscience, 21*, 8129–8135.

Jones, T. A., Leake, P. A., Snyder, R. L., Stakhovskaya, O., & Bonham, B. (2007). Spontaneous discharge patterns in cochlear spiral ganglion cells before the onset of hearing in cats. *Journal of Neurophysiology, 98*, 1898–1908.

Kaang, B. K., Kandel, E. R., & Grant, S. G. N. (1993). Activation of cAMP-responsive genes by stimuli that produce long-term facilitation in Aplysia sensory neurons. *Neuron, 10*, 427–435.

Kandel, E. R., & Tauc, L. (1965b). Mechanism of heterosynaptic facilitation in the giant cell of the abdominal ganglion of Aplysia depilans. *The Journal of Physiology, 181*, 28–47.

Kano, M., Hashimoto, K., Chen, C., Abeliovich, A., Aiba, A., Kurihara, H., et al. (1995). Impaired synapse elimination during cerebellar development in PKC mutant mice. *Cell, 83*, 1223–1231.

Kano, M., Hashimoto, K., Kurihara, H., Watanabe, M., Inoue, Y., Aiba, A., et al. (1997). Persistent multiple climbing fiber innervation of cerebellar Purkinje cells in mice lacking mGluR1. *Neuron, 18*, 71–79.

Kano, M., Hashimoto, K., Watanabe, M., Kurihara, H., Offermanns, S., Jiang, H., et al. (1998). Phospholipase cbeta4 is specifically involved in climbing fiber synapse elimination in the developing cerebellum. *Proceedings of the National Academy of Sciences of the United States of America, 95*, 15724–15729.

Kapfer, C., Seidl, A. H., Schweizer, H., & Grothe, B. (2002). Experience-dependent refinement of inhibitory inputs to auditory coincidence-detector neurons. *Nature Neuroscience, 5*, 247–253.

Kasamatsu, T., & Pettigrew, J. D. (1976). Depletion of brain chatecholamines: failure of ocular dominance shift after monocular occlusion in kittens. *Science, 194*, 206–209.

Keating, M. J., & Feldman, J. (1975). Visual deprivation and intertectal neuronal connections in Xenopus laevis. *Proceedings of the Royal Society of London. Series B, 191*, 467–474.

Kennedy, C., Des Rosiers, M. H., Jehle, J. W., Reivich, M., Sharpe, F., & Sokoloff, L. (1975). Mapping of functional neural pathways by autoradiographic survey of local metabolic rate with (14C)deoxyglucose. *Science, 7*, 850–853.

Kilman, V., van Rossum, M. C., & Turrigiano, G. G. (2002). Activity deprivation reduces miniature IPSC amplitude by decreasing the number of postsynaptic GABA(A) receptors clustered at neocortical synapses. *The Journal of Neuroscience, 22*, 1328–1337.

Kim, G., & Kandler, K. (2003). Elimination and strengthening of glycinergic/GABAergic connections during tonotopic map formation. *Nature Neuroscience, 6*, 282–290.

Kleinschmidt, A., Bear, M. F., & Singer, W. (1987). Blockade of NMDA receptors disrupts experience-dependent plasticity of kitten striate cortex. *Science, 238*, 355–358.

Koh, Y. H., Popova, E., Thomas, U., Griffith, L. C., & Budnik, V. (1999). Regulation of DLG localization at synapses by CaMKII-dependent phosphorylation. *Cell, 98*, 353–363.

Kossel, A. H., Williams, C. V., Schweizer, M., & Kater, S. B. (1997). Afferent innervation influences the development of dendritic branches and spines via both activity-dependent and non-activity-dependent mechanisms. *The Journal of Neuroscience, 17*, 6314–6324.

Kotak, V. C., Fujisawa, S., Lee, F. A., Karthikeyan, O., Aoki, C., & Sanes, D. H. (2005). Hearing loss raises excitability in the auditory cortex. *The Journal of Neuroscience, 25*, 3908–3918.

Kotak, V. C., Korada, S., Schwartz, I. R., & Sanes, D. H. (1998). A developmental shift from GABAergic to glycinergic transmission in the central auditory system. *The Journal of Neuroscience, 18*, 4646–4655.

Kotak, V. C., Sadahiro, M., & Fall, C. P. (2007). Developmental expression of endogenous oscillations and waves in the auditory cortex involves calcium, gap junctions, and GABA. *Neuroscience, 146*, 1629–1639.

Kotak, V. C., & Sanes, D. H. (1995). Synaptically-evoked prolonged depolarizations in the developing auditory system. *Journal of Neurophysiology, 74*, 1611–1620.

Kotak, V. C., Takesian, A. E., & Sanes, D. H. (2008). Hearing loss prevents the maturation of GABAergic transmission in the auditory cortex. *Cerebral Cortex, 18*, 2098–2108.

Kotak, V. K., & Sanes, D. H. (2000). Long-lasting inhibitory synaptic depression is age- and calcium-dependent. *The Journal of Neuroscience, 20*, 5820–5826.

Kuffler, D., Thompson, W., & Jansen, J. K. S. (1977). The elimination of synapses in multiply-innervated skeletal muslce fibres of the rat: dependence on distance between end-plates. *Brain Research, 138*, 353–358.

Larrabee, M. G., & Bronk, D. W. (1947). Prolonged facilitation of synaptic excitation in sympathetic ganglia. *Journal of Neurophysiology, 10*, 139–154.

Leake, P. A., Snyder, R. L., Rebscher, S. J., Moore, C. M., & Vollmer, M. (2000). Plasticity in central representations in the inferior colliculus induced by single- vs. two-channel electrical stimulation by a cochlear implant after neonatal deafness. *Hearing Research, 147*, 221–241.

Lendvai, B., Stern, E. A., Chen, B., & Svoboda, K. (2000). Experience-dependent plasticity of dendritic spines in the developing rat barrel cortex in vivo. *Nature, 404*, 876–881.

LeVay, S., Stryker, M. P., & Shatz, C. J. (1978). Ocular dominance columns and their development in layer IV of the cat's visual cortex. *The Journal of Comparative Neurology, 179*, 223–244.

Lewis, C. A., Ahmed, Z., & Faber, D. S. (1990). Developmental changes in the regulation of glycine-activated Cl-channels of cultured rat medullary neurons. *Developmental Brain Research, 51*, 287–290.

Li, M. X., Jia, M., Jiang, H., Dunlap, V., & Nelson, P. G. (2001). Opposing actions of protein kinase A and C mediate Hebbian synaptic plasticity. *Nature Neuroscience, 4*, 871–872.

Li, Y., Fitzpatrick, D., & White, L. E. (2006). The development of direction selectivity in ferret visual cortex requires early visual experience. *Nature Neuroscience, 9*, 676–681.

Li, Y., Van Hooser, S. D., Mazurek, M., White, L. E., & Fitzpatrick, D. (2008). Experience with moving visual stimuli drives the early development of cortical direction selectivity. *Nature, 456*, 952–956.

Lichtman, J. W. (1977). The reorganization of synaptic connexions in the rat submandibular ganglion during post-natal development. *The Journal of Physiology, 273*, 155–177.

Lichtman, J. W., & Purves, D. (1980). The elimination of redundant preganglionic innervation to hamster sympathetic ganglion cells in early post-natal life. *The Journal of Physiology, 301*, 213–228.

Lin, Y., Bloodgood, B. L., Hauser, J. L., Lapan, A. D., Koon, A. C., Kim, T. K., et al. (2008). Activity-dependent regulation of inhibitory synapse development by Npas4. *Nature, 455*, 1198–1204.

Lippe, W. R. (1994). Rhythmic spontaneous activity in the developing avian auditory system. *The Journal of Neuroscience, 14*, 1486–1495.

Lischalk, J. W., Easton, C. R., & Moody, W. J. (2009). Bilaterally propagating waves of spontaneous activity arising from discrete pacemakers in the neonatal mouse cerebral cortex. *Developmental Neurobiology, 69*, 407–414.

Lloyd, D. P. C. (1949). Post-tetanic potentiation of response in monosynaptic reflex pathways of the spinal cord. *The Journal of General Physiology, 33*, 147–170.

Lo, Y. J., Lin, Y. C., Sanes, D. H., & Poo, M. M. (1994). Depression of developing neuromuscular synapses induced by repetitive postsynaptic depolarizations. *The Journal of Neuroscience, 14*, 4694–4704.

Lo, Y.-j., & Poo, M.-m. (1991). Activity-dependent synaptic competition in vitro: Heterosynaptic supression of developing synapses. *Science, 254*, 1019–1022.

Lo, Y.-j., Poo, M.-m. (1994). Heterosynaptic supression of developing neruomuscular synapses in culture. *J Neurosci, 14*, 4684–4693.

Lohmann, C., Myhr, K. L., & Wong, R. O. (2002). Transmitter-evoked local calcium release stabilizes developing dendrites. *Nature, 418*, 177–181.

Löwel, S., & Singer, W. (1992). Selection of intrinsic horizontal connections in the visual cortex by correlated neuronal activity. *Science, 255*, 209–212.

Lu, B. (2003). BDNF and activity-dependent synaptic modulation. *Learning & Memory, 10*, 86–98.

Lu, W., & Constantine-Paton, M. (2004). Eye opening rapidly induces synaptic potentiation and refinement. *Neuron, 43*, 237–249.

Luo, L. (2002). Actin cytoskeleton regulation in neuronal morphogenesis and structural plasticity. *Annual Review of Cell and Developmental Biology, 18*, 601–635.

Maggi, L., Le Magueresse, C., Changeux, J. P., & Cherubini, E. (2003). Nicotine activates immature "silent" connections in the developing hippocampus. *Proceedings of the National Academy of Sciences of the United States of America, 100*, 2059–2064.

Majewska, A., & Sur, M. (2003). Motility of dendritic spines in visual cortex in vivo: changes during the critical period and effects of visual deprivation. *Proceedings of the National Academy of Sciences of the United States of America, 100*, 16024–16029.

Marder, E., & Prinz, A. A. (2002). Modeling stability in neuron and network function: the role of activity

in homeostasis. *BioEssays: News and Reviews in Molecular, Cellular and Developmental Biology, 24,* 1145–1154.

Mariani, J., & Changeux, J. P. (1981). Ontogenesis of olivocerebellar relationships. I. Studies by intracellular recordings of the multiple innervation of Pukinje cells by climbing fibers in the developing rat cerebellum. *The Journal of Neuroscience, 1,* 696–702.

Marty, S., Wehrle, R., & Sotelo, C. (2000). Neuronal activity and brain-derived neurotrophic factor regulate the density of inhibitory synapses in organotypic slice cultures of postnatal hippocampus. *The Journal of Neuroscience, 20,* 8087–8095.

McAllister, A. K., Katz, L. C., & Lo, D. C. (1997). Opposing roles for endogenous BDNF and NT-3 in regulating cortical dendritic growth. *Neuron, 18,* 767–778.

McCasland, J. S., Bernardo, K. L., Probst, K. L., & Woolsey, T. A. (1992). Cortical local circuit axons do not mature after early deafferentation. *Proceedings of the National Academy of Sciences of the United States of America, 89,* 1832–1836.

McLaughlin, T., Torborg, C. L., Feller, M. B., & O'Leary, D. D. (2003). Retinotopic map refinement requires spontaneous retinal waves during a brief critical period of development. *Neuron, 40,* 1147–1160.

Mears, S. C., & Frank, E. (1997). Formation of specific monosynaptic connections between muscle spindle afferents and motoneurons in the mouse. *The Journal of Neuroscience, 17,* 3128–3135.

Meister, M., Wong, R. O. L., Baylor, D. A., & Shatz, C. J. (1991). Synchronous bursts of action potentials in ganglion cells of the developing mammalian retina. *Science, 252,* 939–943.

Miles K, Greengard, P., Huganir, R. L. (1989) Calcitonin gene-related peptide regulates phosphorylation of the nicotinic acetylcholine receptor in rat myotubes. *Neuron, 2,* 1517–1524.

Morales, B., Choi, S. Y., & Kirkwood, A. (2002). Dark rearing alters the development of GABAergic transmission in visual cortex. *The Journal of Neuroscience, 22,* 8084–8090.

Mu, Y., & Poo, M. M. (2006). Spike timing-dependent LTP/LTD mediates visual experience-dependent plasticity in a developing retinotectal system. *Neuron, 50,* 115–125.

Murphy, G. G., & Glanzman, D. L. (1999). Cellular analog of differential classical conditioning in Aplysia: disruption by the NMDA receptor antagonist DL-2-amino-5-phosphonovalerate. *The Journal of Neuroscience, 19,* 10595–10602.

Murthy, V. N., Schikorski, T., Stevens, C. F., & Zhu, Y. (2001). Inactivity produces increases in neurotransmitter release and synapse size. *Neuron, 32,* 673–682.

Nelson, P. G., Lanuza, M. A., Jia, M., Li, M. X., & Tomas, J. (2003). Phosphorylation reactions in activity-dependent synapse modification at the neuromuscular junction during development. *Journal of Neurocytology, 32,* 803–816.

Noh, J., Seal, R. P., Garver, J. A., Edwards, R. H., & Kandler, K. (2010). Glutamate co-release at GABA/glycinergic synapses is crucial for the refinement of an inhibitory map. *Nature Neuroscience, 13,* 232–238.

O'Brien, R. A., Ostberg, A. J., & Vrbová, G. (1978). Observations on the elimination of polyneuronal innervation in developing mammalian skeletal muscle. *The Journal of Physiology, 282,* 571–582.

O'Brien, R. J., Kamboz, S., Ehlers, M. D., Rosen, K. R., Fischbach, G. D., & Huganir, R. L. (1998). Activity-dependent modulation of synaptic AMPA receptor accumulation. *Neuron, 21,* 1067–1078.

O'Leary, D. D. M., Stanfield, B. B., & Cowan, W. M. (1981). Evidence that the early postnatal restriction of the cells or origin of the callosal projection is due to

the elimination of axonal collaterals rather than to the death of the neurons. *Brain Research. Developmental Brain Research, 1,* 607–617.

Pasternak, T., Schumer, R. A., Gizzi, M. S., & Movshon, J. A. (1985). Abolition of visual cortical direction selectivity affects visual behavior in cats. *Experimental Brain Research. Experimentelle Hirnforschung. Experimentation Cerebrale, 61,* 214–217.

Penfield, W., & Rasmussen, T. (1950). *The cerebral cortex of man: a clinical study of localization of function.* New York: Macmillan.

Personius, K. E., Chang, Q., Mentis, G. Z., O'Donovan, M. J., & Balice-Gordon, R. J. (2007). Reduced gap junctional coupling leads to uncorrelated motor neuron firing and precocious neuromuscular synapse elimination. *Proceedings of the National Academy of Sciences of the United States of America, 104,* 11808–11813.

Personius, K. E., Karnes, J. L., & Parker, S. D. (2008). NMDA receptor blockade maintains correlated motor neuron firing and delays synapse competition at developing neuromuscular junctions. *The Journal of Neuroscience, 28,* 8983–8992.

Pozo, K., & Goda, Y. (2010). Unraveling mechanisms of homeostatic synaptic plasticity. *Neuron, 66,* 337–351.

Pratt, K. G., Watt, A. J., Griffith, L. C., Nelson, S. B., & Turrigiano, G. G. (2003). Activity-dependent remodeling of presynaptic inputs by postsynaptic expression of activated CaMKII. *Neuron, 39,* 269–281.

Rabacchi, S., Bailly, Y., Delhaye-Bouchaud, N., & Mariani, J. (1992). Involvement of the N-methyl D-aspartate receptor in synapse elimination during cerebellar development. *Science, 256,* 1823–1825.

Rakic, P. (1972). Mode of cell migration to the superficial layers of fetal monkey neocortex. *The Journal of Comparative Neurology, 145,* 61–84.

Rao, A., & Craig, A. M. (1997). Activity regulates the synaptic localization of the NMDA receptor in hippocampal neurons. *Neuron, 19,* 801–812.

Razak, K. A., Richardson, M. D., & Fuzessery, Z. M. (2008). Experience is required for the maintenance and refinement of FM sweep selectivity in the developing auditory cortex. *Proceedings of the National Academy of Sciences of the United States of America, 105,* 4465–4470.

Redfern, P. A. (1970). Neuromuscular transmission in new-born rats. *The Journal of Physiology, 209,* 701–709.

Redondo, R. L., Okuno, H., Spooner, P. A., Frenguelli, B. G., Bito, H., & Morris, R. G. (2010). Synaptic tagging and capture: differential role of distinct calcium/calmodulin kinases in protein synthesis-dependent long-term potentiation. *The Journal of Neuroscience, 30,* 4981–4989.

Reh, T. A., & Constantine-Paton, M. (1984). Retinal ganglion cell terminals change their projection sites during larval development of Rana pipiens. *The Journal of Neuroscience, 4,* 442–457.

Ridge, R.M.A.P., & Betz, W. J. (1984). The effect of selective, chronic stimulation on motor unit size in developing rat muscle. *The Journal of Neuroscience, 4,* 2614–2620.

Riquimaroux, H., Gaioni, S. J., & Suga, N. (1991). Cortical computational maps control auditory perception. *Science, 1,* 565–568.

Role, L. W., Roufa, D. G., & Fischbach, G. D. (1987). The distribution of acetylcholine receptor clusters and sites of transmitter release along chick ciliary ganglion neurite-myotube contacts in culture. *The Journal of Cell Biology, 104,* 371–379.

Royer, S., & Pare, D. (2003). Conservation of total synaptic weight through balanced synaptic depression and potentiation. *Nature, 422,* 518–522.

Ruthazer, E. S., Akerman, C. J., & Cline, H. T. (2003). Control of axon branch dynamics by correlated activity in vivo. *Science, 301,* 66–70.

Salmelin, R., Service, E., Kiesila, P., Uutela, K., & Salonen, O. (1996). Impaired visual word processing in dyslexia revealed with magnetoencephalography. *Annals of Neurology, 40,* 157–162.

Sanes, D. H. (1993). The development of synaptic function and integration in the central auditory system. *The Journal of Neuroscience, 13,* 2627–2637.

Sanes, D. H., & Constantine-Paton, M. (1985). The sharpening of frequency tuning curves requires patterned activity during development in the mouse, Mus musculus. *The Journal of Neuroscience, 5,* 1152–1166.

Sanes, D. H., & Hafidi, A. (1996). Glycinergic transmission regulates dendrite size in organotypic culture. *Journal of Neurobiology, 4,* 503–511.

Sanes, D. H., Markowitz, S., Bernstein, J., & Wardlow, J. (1992). The influence of inhibitory afferents on the development of postsynaptic dendritic arbors. *The Journal of Comparative Neurology, 321,* 637–644.

Sanes, D. H., & Siverls, V. (1991). Development and specificity of inhibitory terminal arborizations in the central nervous system. *Journal of Neurobiology, 22,* 837–854.

Sanes, D. H., & Takacs, C. (1993). Activity-dependent refinement of inhibitory connections. *The European Journal of Neuroscience, 5,* 570–574.

Schuster, C. M., Davis, G. W., Fetter, R. D., & Goodman, C. S. (1996). Genetic dissection of structural and functional components of synaptic plasticity. II. Fasciclin II controls presynaptic structural plasticity. *Neuron, 17,* 655–667.

Seebach, B. S., & Ziskind-Conhaim, L. (1994). Formation of transient inappropriate sensorimotor synapses in developing rat spinal cords. *The Journal of Neuroscience, 14,* 4520–4528.

Sherman, S. M., & Spear, P. D. (1982). Organization of visual pathways in normal and visually deprived cats. *Physiological Reviews, 62,* 738–855.

Shoykhet, M., Land, P. W., & Simons, D. J. (2005). Whisker trimming begun at birth or on postnatal day 12 affects excitatory and inhibitory receptive fields of layer IV barrel neurons. *Journal of Neurophysiology, 94,* 3987–3995.

Simon, D. K., & O'Leary, D. D. (1992). Development of topographic order in the mammalian retinocollicular projection. *The Journal of Neuroscience, 12,* 1212–1232.

Snyder, R. L., Rebscher, S. J., Cao, K., Leake, P. A., & Kelly, K. (1990). Chonic intracochlear electrical stimulation in the neonatally deafened cat. I: Expansion of central representation. *Hearing Research, 50,* 7–34.

Southwell, D. G., Froemke, R. C., Alvarez-Buylla, A., Stryker, M. P., & Gandhi, S. P. (2010). Cortical plasticity induced by inhibitory neuron transplantation. *Science, 327,* 1145–1148.

Sretavan, D. W., & Shatz, C. J. (1986). Prenatal development of retinal ganglion cell axons: segregation into eye-specific layers within the cat's lateral geniculate nucleus. *The Journal of Neuroscience, 6,* 234–251.

Stanfield, B. B. (1992). The development of the corticospinal projection. *Progress in Neurobiology, 38,* 169–202.

Star, E. N., Kwiatkowski, D. J., & Murthy, V. N. (2002). Rapid turnover of actin in dendritic spines and its regulation by activity. *Nature Neuroscience, 5,* 239–246.

Stellwagen, D., & Shatz, C. J. (2002). An instructive role for retinal waves in the development of retinogeniculate connectivity. *Neuron, 33,* 357–367.

Stern, E. A., Maravall, M., & Svoboda, K. (2001). Rapid development and plasticity of layer 2/3 maps in rat barrel cortex in vivo. *Neuron, 31,* 305–315.

Stryker, M. P., & Harris, W. A. (1986). Binocular impulse blockade prevents the formation of ocular

dominance columns in cat visual cortex. *The Journal of Neuroscience, 6,* 2117–2133.

Sur, M., Weller, R. E., & Sherman, S. M. (1984). Development of X- and Y-cell retinogeniculate terminations in kittens. *Nature, 310,* 246–249.

Takesian, A. E., Kotak V. C., Sanes, D. H. (2010). Presynaptic GABA(B) receptors regulate experience-dependent development of inhibitory short-term plasticity. *J Neurosci, 30,* 2716-2727.

Thompson, W., Kuffler, D. P., & Jansen, J. K. S. (1979). The effect of prolonged, reversible block of nerve impulses on the elimination of polyneuronal innervation of new-born rat skeletal muscle fibers. *Neuroscience, 4,* 271–281.

Tian, N., & Copenhagen, D. R. (2003). Visual stimulation is required for refinement of ON and OFF pathways in postnatal retina. *Neuron, 39,* 85–96.

Trachtenberg, J. T., & Stryker, M. P. (2001). Rapid anatomical plasticity of horizontal connections in the developing visual cortex. *The Journal of Neuroscience, 21,* 3476–3482.

Tritsch, N. X., Yi, E., Gale, J. E., Glowatzki, E., & Bergles, D. E. (2007). The origin of spontaneous activity in the developing auditory system. *Nature, 450,* 50–55.

Udin, S. B., & Keating, M. J. (1981). Plasticity in a central nervous pathway in Xenopus: Anatomical changes in the isthmotectal projection after larval eye rotation. *The Journal of Comparative Neurology, 203,* 575–594.

Vale, C., & Sanes, D. H. (2000). Afferent regulation of inhibitory synaptic transmission in the developing auditory midbrain. *The Journal of Neuroscience, 20,* 1912–1921.

Vale, C., & Sanes, D. H. (2002). The effect of bilateral deafness on excitatory synaptic strength in the auditory midbrain. *European Journal of Neuroscience, 16,* 2394–2404.

Vale, C., Schoorlemmer, J., & Sanes, D. H. (2003). Deafness disrupts chloride transport and inhibitory synaptic transmission. *The Journal of Neuroscience, 23,* 7516–7524.

Valverde, F. (1968). Structural changes in the area striata of the mouse after enucleation. *Experimental Brain Research. Experimentelle Hirnforschung. Experimentation Cerebrale, 5,* 274–292.

Van der Loos, H., & Woolsey, T. A. (1973). Somatosensory cortex: alterations following early insury to sense organs. *Science, 179,* 395–398.

Vaughn, J. E., Barber, R. P., & Sims, T. J. (1988). Dendritic development and preferential growth into synaptogenic fields: a quantitative study of golgi-impregnated spinal motor neurons. *Synapse, 2,* 69–78.

Vislay-Meltzer, R. L., Kampff, A. R., & Engert, F. (2006). Spatiotemporal specificity of neuronal activity directs the modification of receptive fields in the developing retinotectal system. *Neuron, 50,* 101–114.

Vogel, M. W., & Prittie, J. (1995). Purkinje cell dendritic arbors in chick embryos following chronic treatment with an N-methyl-D-aspartate receptor antagonist. *Journal of Neurobiology, 26,* 537–552.

Walsh, M. K., & Lichtman, J. W. (2003). In vivo time-lapse imaging of synaptic takeover associated with naturally occurring synapse elimination. *Neuron, 37,* 67–73.

Wang, J., Renger, J. J., Griffith, L. C., Greenspan, R. J., W. C. (1994). Concomitant alterations of physiological and developmental plasticity in Drosophila CaM kinase II-inhibited synapses. *Neuron 13,* 1373–1384.

Watt, A. J., Cuntz, H., Mori, M., Nusser, Z., Sjöström, P. J., & Häusser, M. (2009). Traveling waves in developing cerebellar cortex mediated by asymmetrical Purkinje cell connectivity. *Nature Neuroscience, 12,* 463–473.

Westerfield, M., Liu, D. W., Kimmel, C. B., & Walker, C. (1990). Pathfinding and synapse formation in a zebrafish mutant lacking functional acetylcholine receptors. *Neuron, 4,* 867–874.

Wiesel, T. N., & Hubel, D. H. (1963a). Single-cell responses in striate cortex of kittens deprived of vision in one eye. *Journal of Neurophysiology, 26,* 1003–1017.

Wiesel, T. N., & Hubel, D. H. (1963b). Effects of visual deprivation on morphology and physiology of cells in the cat's lateral geniculate body. *Journal of Neurophysiology, 26,* 978–993.

Wiesel, T. N., & Hubel, D. H. (1965). Comparison of the effects of unilateral and bilateral eye closure on cortical unit responses in kittens. *Journal of Neurophysiology, 28,* 1029–1040.

Wong, R. O., Meister, M., & Shatz, C. J. (1993). Transient period of correlated bursting activity during development of the mammalian retina. *Neuron, 11,* 923–938.

Wu, G. Y., & Cline, H. T. (1998). Stabilization of dendritic arbor structure in vivo by CaMKII. *Science, 279,* 222–226.

Xia, S., Miyashita, T., Fu, T.-F., Lin, W.-Y., Wu, C.-l., Grady, L., et al. (2005). NMDA receptors mediate olfactory learning and memory in Drosophila. *Current Biology: CB, 15,* 603–615.

Yang, F., Je, H. S., Ji, Y., Nagappan, G., Hempstead, B., & Lu, B. (2009). Pro-BDNF-induced synaptic depression and retraction at developing neuromuscular synapses. *The Journal of Cell Biology, 185,* 727–741.

Yang, X., Hyder, F., & Shulman, R. G. (1996). Activation of single whisker barrel in rat brain localized by functional magnetic resonance imaging. *Proceedings of the National Academy of Sciences of the United States of America, 93,* 475–478.

Yin, J. C., Del Vecchio, M., Zhou, H., & Tully, T. (1995). CREB as a memory modulator: induced expression of a dCREB2 activator isoform enhances long-term memory in Drosophila. *Cell, 7,* 107–115.

Yin, J. C., Wallach, J. S., Del Vecchio, M., Wilder, E. L., Zhou, H., Quinn, W. G., et al. (1994). Induction of a dominant negative CREB transgene specifically blocks long-term memory in Drosophila. *Cell, 7,* 49–58.

Young, S. R., & Rubel, E. W. (1986). Embryogenesis of arborization pattern and topography of individual axons in n. laminaris of the chicken brain stem. *The Journal of Comparative Neurology, 254,* 425–459.

Zhang, L. I., Tao, H. W., Holt, C. E., Harris, W. A., & Poo, M.-m. (1998). A critical window for cooperation and competition among developing retinotectal synapses. *Nature, 395,* 37–44.

Zhang, L. I., Bao, S., & Merzenich, M. M. (2002a). Disruption of primary auditory cortex by synchronous auditory inputs during a critical period. *Proceedings of the National Academy of Sciences of the United States of America, 99,* 2309–2314.

Zhang, Y., Ma, C., Delohery, T., Nasipak, B., Foat, B. C., Bounoutas, A., et al. (2002b). Identification of genes expressed in C. elegans touch receptor neurons. *Nature, 418,* 331–335.

Zhong, Y., Budnik, V., & Wu, C. F. (1992). Synaptic plasticity in Drosophila memory and hyperexcitable mutants: role of cAMP cascade. *The Journal of Neuroscience, 12,* 644–651.

Zou, D. J., & Cline, H. T. (1996). Expression of constituitively active CaMKII in target tissue modifies presynaptic axon arbor growth. *Neuron, 16,* 529–539.

Zou, D. J., Feinstein, P., Rivers, A. L., Mathews, G. A., Kim, A., Greer, C. A., et al. (2004). Postnatal refinement of peripheral olfactory projections. *Science, 304,* 1976–1979.

Zuo, Y., Lin, A., Chang, P., & Gan, W. B. (2005). Development of long-term dendritic spine stability in diverse regions of cerebral cortex. *Neuron, 46,* 181–189.

Behavioral development 10

BEHAVIORAL ONTOGENY

We often think of behavior in terms of postnatal or post-hatching animals because that is what we normally observe, but behavior in a mammal like you begins well before birth. As a fetus, you were making all sorts of movements. You were kicking, swallowing, putting your thumb in your mouth, and gyrating to songs that you heard through the wall of the womb.

With ultrasound, we can now see the emergence of behaviors in humans, and even diagnose behavioral deficits before birth. A bird embryo in its shell also moves and peeps. Even fly embryos wiggle in their egg cases before they hatch. These first movements that an animal makes seem simpler, and often less coordinated, than the sophisticated movements of an adult. How do these early behaviors first emerge and what do they reveal about the developing nervous system that controls them? Do these embryonic behaviors serve any strategic purpose in the further development of the nervous system, or are they merely the consequences of a nervous system that is wiring up and becoming electrically active? Which behaviors arise first in the embryo, how does the repertoire of behavior grow, is there any logic in the sequence that reflects a logic of neural development? These are very intriguing questions for a developmental neurobiologist.

Early behaviors have been classified as anticipatory, adaptive, or substrative (Oppenheim, 1981). Clearly, some behaviors develop in an *anticipatory* manner with reference to actions that will be of value in later life (Carmichael, 1954). For instance, prematurely born humans can respond to light although they would normally not be exposed to light in the womb. Since the nervous system is constructed over a period of time, behaviors cannot be slapped together exactly at the moment they are first needed, but arise with the developing circuitry. The case of sensing light is an excellent example of this, as the last cell types to mature and join the basic circuitry of the visual system are the photoreceptors; the rest of the circuit from eye to brain and brain to motor output is ready and waiting for this input. Some early behaviors are anticipatory in the sense that they reflect ancestral nervous systems and modes of behavior evolving toward later forms. There is evidence that mammalian embryonic spinal cords have rhythmic activity similar to swimming patterns of their fish-like ancestors. At early stages even human embryos display sinusoidal swimming movements. The grasp reflex is another interesting example of this type of early behavior (**Figure 10.1**). Touching the palm of a baby causes inward curling of the fingers and a forceful hold on any object. This reflex, present at birth in a human infant and disappearing at about 3 months of age, is thought to be a relic of a more primitive primate nervous system in which clinging to branches or mother was important for survival. These early behaviors that appear to be vestiges of ancestral behavioral traits may reflect important steps in the development of the neural control of our more derived behavioral repertoire.

Some embryonic and juvenile behaviors are *adaptive*, and serve specific functions at particular stages of development.

Fig. 10.1 The grasp reflex demonstrated in a newborn baby.

Hatching behavior is a good example (Oppenheim, 1982). Quail embryos make clicking noises in the shell, which helps to synchronize hatching (Vince and Salter, 1967, Vince 1979). More frequent clicks accelerate hatching, and less frequent clicks retard it. In reptiles, birds, and insects, hatching is composed of repeated stereotypical movements designed to break out of the egg case. These behaviors are transient, specific to that stage of life and clearly adaptive. In human babies, the rooting reflex, turning the face toward a touch on the cheek, is a transient reflex important for breast-feeding. It is important to recognize that embryonic and adult animals often live in very different environments. Each animal tends to exhibit stage-specific morphological, molecular, and behavioral adaptations to its specific environment. This is particularly obvious in the case of animals that go through metamorphosis, such as moths and frogs. Here, the larval and adult forms have a radically distinct appearance and behavior that are adapted to the very different ecological niches in which they live. The nervous systems, the substrates of these behaviors, undergo substantial modifications in response to metamorphic hormones (ecdysone for insects and thyroxine for amphibians), including death of larval neurons and genesis of new adult neurons.

What about *substrative* behaviors? Early behaviors can be considered substrative if they form a basis upon which more complex behaviors can be built. Such behaviors are crucial for the continued maturation of the nervous system itself (Carpenter, 1874). Through function and feedback, neural circuits can become finely tuned. The learning that we perform as adults can be considered as a continuation of the mechanisms used to adjust the embryonic nervous system. The behavioral patterns that we see in embryonic and juvenile animals are the integrated beginnings of more complex patterns that continue to develop. We must crawl before we can walk. If this view is correct, disruption of these early behaviors should have a significant impact on the development of later behaviors. A good example of this is the fine motor skills of the forearm. Experience is thought to be important for the development of these motor skills. Preventing normal forelimb use during early development causes defects in fine forelimb movements that last throughout life. Is there any obvious neural basis of such substrative behavioral development? Yes. In this last example, motor deprivation produces defective

development of the axonal terminals of the corticospinal neurons that control these movements (Martin et al., 2004).

In this chapter, we will examine the way behavior develops, especially as it reflects on the maturation of neural circuits. We will look in detail at the development of various specific motor, sensory, and social behaviors, and try to understand from a neural, molecular, and genetic perspective how and why they develop in the order that they do. We will also examine the mechanisms by which behaviors become more precise and skillful, and the role of sensory and motor experience in the neural circuitry that controls them.

THE FIRST MOVEMENTS ARE SPONTANEOUS

The first simple twitch that an animal makes tells us something about the beginning of its functional motor system. Part of the motor system is the musculature itself, and in some species, the first twitches that the embryo makes are myogenic, that is, they originate in the spontaneous activity of muscle cells themselves. In *Drosophila*, for example, the first movements occur within just 14 hours after egg laying and before any neurons begin to fire action potentials (**Figure 10.2**). The initial movements appear on schedule even when all synaptic input onto the muscles is experimentally blocked by the misexpression of tetanus toxin (Crisp et al., 2008).

In many other species, the earliest movements are neurogenic, i.e., caused by spontaneous activity of motor neurons. The muscles are ready for action but need input and if motor neurons have already made a functional synapse onto a muscle, the result is a movement when the motor neuron first fires. Movements elicited by sensory stimulation also emerge early in development. All that is needed in addition to the above is a synapse between the sensory neuron and the motor neuron to complete a reflex arc.

A question that has intrigued those who study the origins of behavior is whether the very first neurogenic movements of an animal are *spontaneous* and involve no sensory input, or whether they are *reflexive* movements in response to sensory stimulation. Sometimes it is difficult to tell when a movement is truly spontaneous. For example, leech embryos show movements in their egg cases, and the frequency of these unprovoked movements increases during the latter half of embryonic leech development (Reynolds et al., 1998). This increase corresponds to the time of eye formation, suggesting that the embryos may actually be responding to light. And indeed, if the embryos are observed in red light, rather than white light, this increase in "spontaneous" behavior is not observed. So is it really spontaneous if it is evoked by light?

By shining light through chick eggs, "candling" them as it is called, it is possible to see the embryo moving inside of its shell. If one observes the later stages of development, chick embryos are very active, moving their wings and legs within their shell (Preyer, 1885). Careful studies of chick embryos raised in glass dishes, rather than shells, reveal that totally undisturbed animals exhibit a variety of behaviors (**Figure 10.3**). Neuroanatomical studies done at the same stages suggest that the earliest of these behaviors occurs prior to the functional maturation of spinal sensory cells. And indeed, for several days after their first occurrence, sensory stimulation does not change the frequency of these behaviors, nor is sensory stimulation able to

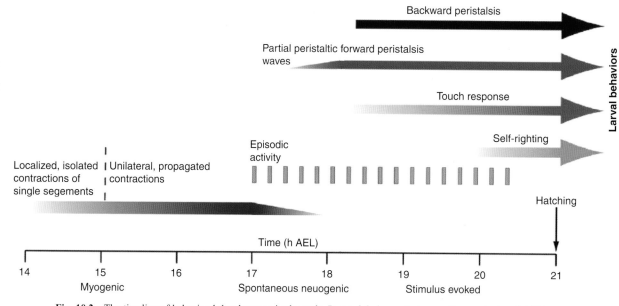

Fig. 10.2 The timeline of behavioral development in the early *Drosophila* larvae in hours after egg laying (hAEL). Myogenic movements start at about 14–15 hAEL, the spontaneous neurogenic episodes start at 17 hAEL and quickly transform into episodes of forward and then backward peristaltic crawling. Finally, between 18–19 hAEL, reflexes such as touch responses are seen. *(From Crisp et al., 2008)*

evoke a motor response at all. Thus sensory input seems to be unnecessary for and ineffective at influencing emergent motor behaviors. To resolve this issue thoroughly, Victor Hamburger surgically deafferented chick embryos by removing the neural crest cells that give rise to sensory DRG cells (Hamburger etal., 1966). In such embryos, which were sensory-deprived *ab initio*, he saw movements that began at the same stage as in unoperated controls. Moreover, these movements were indistinguishable in frequency or quality for several days (Figure 10.3). These experimental findings are supported by electrophysiological studies of the spinal cord that reveal no detectable synaptic input onto motor neurons when sensory neurons are stimulated during this prereflexogenic period of behavior.

In *Drosophila*, after the period of initial myogenic movements, at about 17 hours after egg laying, there begin episodes of muscle contractions that depend on neural activity. These episodes are poorly organized at first, but after just an hour they become coordinated crawling sequences. This is not a reflex to sensory input, because it occurs on schedule when sensory input is experimentally blocked by the misexpression of tetanus toxin in all sensory cells (Crisp et al., 2008). The movement even becomes fairly, if not perfectly, coordinated over the next hours, in the complete absence of sensory neuron activity (Suster and Bate, 2002). Thus, the motor circuit is assembled and begins to drive behavior in the absence of external stimuli or sensory feedback.

THE MECHANISM OF SPONTANEOUS MOVEMENTS

If sensory stimulation is not involved in early motor behavior, then how are these spontaneous and sometimes coordinated movements generated? Provine (1972) was the first to appreciate that the beginnings of electrical activity in the spinal cord of a chick were connected with the development of spontaneous behavior. Again, it is a similar story in *Drosophila*, where the beginnings of electrical activity in the CNS are first apparent just an hour before the motor program that drives episodic contractions begins (Baines and Bate, 1998). To get insight into the early central circuitry that drives these spontaneous behaviors, the chick embryonic spinal cords were isolated and the activity patterns of particular motor outputs recorded (O'Donovan et al., 1998). The motor nerve roots of such isolated cords produce activity patterns that closely mimic their output in an intact animal that is moving its wings or legs (**Figure 10.4**). In both intact animals and isolated embryonic chick spinal cords, one observes alternating bursts of action potentials from extensor and flexor motor neurons, which would be consistent with rhythmic movements of the limbs. Electrophysiological records from the neurons themselves reveal that there are depolarizing synaptic inputs onto both flexor and extensor motor neurons. Both neuronal populations begin to depolarize toward threshold at the same time and begin to fire synchronously. However, at the peak of their depolarization phase, the flexor motor neurons receive synaptic input that stop them from firing just when the extensor motor neurons are firing most rapidly.

As expected from Hamburger's experiments, removing the dorsal half of the chick cord, where sensory axons travel and many interneurons reside, does not affect the rhythmic motor episodes. Local interneurons in the ventral cord called R-interneurons, or Renshaw cells, receive cholinergic input from motor axon collaterals and feedback GABAergic input onto flexor motor neurons (Wenner and O'Donovan, 2001). In the adult, this is a negative feedback loop that prevents motor neurons from firing too much. But in the embryo, GABA is excitatory (Chapter 8), and so, acting more like a positive than a negative feedback, firing of motor neurons leads to massive excitatory activity among all the connected neurons,

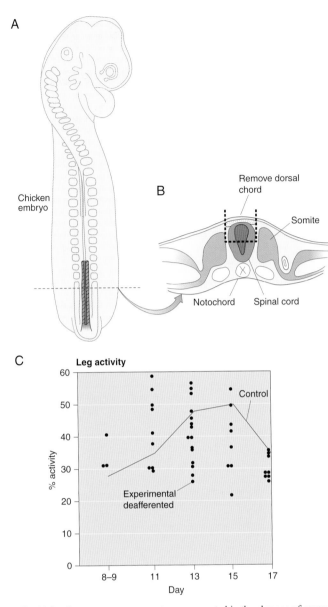

Fig. 10.3 Spontaneous movements are generated in the absence of sensory input. A. The cross-hatched region of the lumbar spinal cord of this premotile chicken embryo is shown in cross section in (B), which shows the operation done *in ovo*, which removes the dorsal spinal cord and neural crest containing all the sensory neurons in this region. C. Activity monitor of average normal leg movements (line) and leg movements generated by legs without any sensory input in operated animals (circles). *(Adapted from Hamburger et al., 1966)*

which gradually tapers off. Blocking both the GABAergic and glutamatergic synapses in an isolated cord preparation stops the rhythmic activity (Chub and O'Donovan, 1998).

The spontaneous initiation of movements starts early in the mammalian (mouse) fetus too. In isolated mouse spinal cord–limb preparations, spontaneous activity begins half-way through gestation at E11 with motor neurons themselves playing a critical role. As in chicks, cholinergic transmission from motor neurons onto R-like interneurons, as well as excitatory GABAergic input, is required for these early bursts of motor activity (Hanson and Landmesser, 2003).

MORE COMPLEX BEHAVIOR IS ASSEMBLED FROM THE INTEGRATION OF SIMPLE CIRCUITS

When an embryonic leech, still in its egg case, is prodded on one side, it leads to an excitation of all the muscles in the prodded segments of the animal, which causes a weird behavior, called circumferential indentation (**Figure 10.5**). This embryonic behavior is only transient for, in a short while, it is replaced by a more reasonable-looking adult behavior called local bending. In local bending, only the muscles on the prodded side contract, leading to the body segments prodded efficiently moving away from the stimulus. Circumferential indentation serves no purpose, so why does this behavior develop? The leech is a superb system to study this question because all the neurons involved in the circumferential indentation and local bending circuits are known and can be recorded from using electrophysiological and optical recording techniques (Marin-Burgin et al., 2008). Such studies show that all connections between the neurons in the circuit begin as excitatory electrical synapses so that activity in any sensory neuron will end up exciting all the motor neurons—resulting in circumferential indentation. Some of these electrical synapses, however, then become replaced with inhibitory chemical synapses and their appearance causes the embryo to switch from circumferential indentation to local bending. If the GABA blocker, bicuculline, is applied when local bending begins, the behavior reverts to circumferential indentation (Marin-Burgin et al., 2005).

The transition from circumferential indentation to local bending in the leech as well as the initial spontaneous movements of the chicks and mice illustrate two important features of behavioral development at the circuit level. The first is that circuits are often first established with electrical synapses or positive feedback that cause all the components of the circuit to burst together. The later development of inhibitory signaling changes the phase relationships between elements of the circuit so that the behavioral output is molded into a more useful pattern for the developing embryo. The second is that the circuit is built in stages. This is easy to appreciate in the leech, where the circuit subserving circumferential indentation prefigures the functional connectivity of the mature circuit, so that only a few changes need to be made in order for the more mature local bending circuit to emerge.

In the 1920s, Coghill, one of the very early pioneers in the investigation of how neural circuits developed to generate behavior, favored the notion that behavior develops in an integrated fashion as circuits mature and become more complex (Coghill, 1929). He began by studying the first movements of the salamander embryo because, like all amphibian embryos, they grow from shell-less eggs in water and are accessible for observation from the earliest stages of development. Coghill also chose salamanders because, in the 1920s, extensive neuroanatomical information became available on these embryos. Of particularly interest to him were studies of the embryonic nervous system anatomy, as revealed by Golgi-type silver staining. Putting the anatomy together with the behavior, Coghill thought he could see the maturation of the circuits that were responsible for integrating the new components of the behaviors. Coghill found that slow bending of the head to one side was the very first movement executed by the salamander embryo (**Figure 10.6**). The movement involves

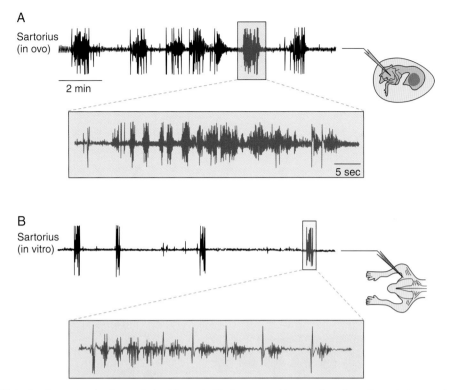

Fig. 10.4 In vitro motor development physiology. A. Electromyographic recordings from the sartorius muscle of a chick embryo in ovo. B. A piece of the embryo kept in a culture dish. Very similar spontaneous movements begin to occur in these two situations, although the bursts of activity in the in vitro preparation are shorter and less frequent. *(Adapted from O'Donovan et al., 1998)*

the trunk muscles situated immediately behind the head. As development proceeds, muscles further and further down the body become involved so that the "bend," which starts as a slow movement at the neck region, becomes a coiling of the entire body. Examination of coiling behavior revealed that the movements start at the neck region and then proceed down the body, such that the sequence of each movement recapitulates the developmental progression of the movement as well as the progression of neuronal maturation (e.g., first in the hindbrain and then down the spinal cord). Both bending and coiling can be stimulated by a light touch of the skin on the side opposite the contraction.

Coghill's anatomical explanation for this chronology is that sensory and motor neurons, which innervate the skin and muscles, are present in prereflexogenic stages. However, at the time that bending away from a light touch emerges, a set of interneurons appear to form the first commissural pathways from sensory neurons on one side to motor neurons on the other. Longitudinal ipsilateral tracts extend down the motor columns and up the sensory columns, but the commissural interneurons appear only in the rostral cord and hindbrain. Thus, a neural signal from a touch anywhere on the skin travels ipsilaterally up the sensory path, crosses over in the neck region, and stimulates motor neurons in that region of the other side, causing a bending of the neck away. The signal then proceeds down the motor pathway on that side of the spinal cord involving successively more posterior segments. After the animal is capable of coiling, the next component of new behavior is the "S" phase. This arises as the coiling movements become faster and alternate from side to side as new

interneurons are added that communicate between the left and right sides. As sustained swimming system develops, more spinal neurons are added to the circuit. If a right-hand coiling movement proceeds only halfway down the body before a left-hand coiling movement starts at the neck region, the result is an "S"-shaped wave proceeding caudally, and the propagation of the animal forward: the beginning of swimming. This was about as far as Coghill was able to surmise at the time and, unfortunately, without experimental tools, he could only guess about which neurons were doing what. Nevertheless, Coghill's work inspired later biologists to study the system with more definitive electrophysiological methods.

In *Xenopus* embryos, single cell physiology combined with labeling has provided a much more accurate picture of the major neuronal types in the early spinal cord and the wiring among them that generates the early behaviors in the tadpole embryo (**Figure 10.7**A–D) (Roberts, 2000). Pairwise recordings from all of these neurons in the embryonic spinal cord show that synapses can be found between all types of neurons, but the probability of contact could be predicted accurately simply by the anatomical overlap of their axons and dendrites (Figure 10.7E). This suggests that the initial synaptic connectivity in this circuit is set up by the paths that axons of these neurons take in the spinal cord (Li et al., 2007), that is, those molecules discussed in Chapter 5 that are responsible for correct axon guidance.

The zebrafish is an excellent system for studying the emergence of motility, not only because the embryo is transparent, but also because genetic knowledge and techniques are making it more possible to see, to record from, and to interfere

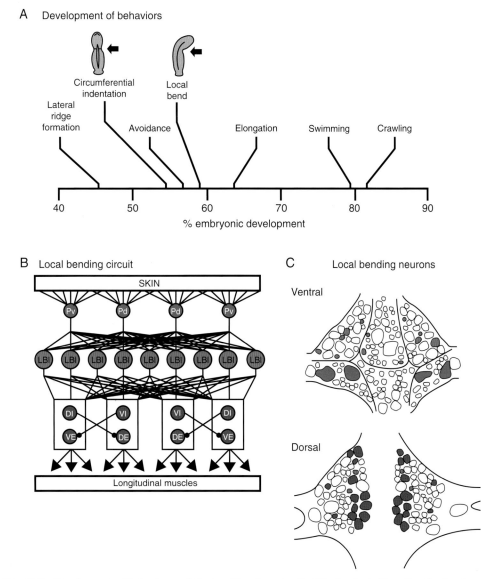

Fig. 10.5 Development of behaviors and the local bending circuit of the leech. A. Time line for development of behavior in leech embryos. Developmental progress is expressed as a percentage of total embryonic development (% ED); at room temperature development from egg deposition to hatching takes 30 days, so each day corresponds to roughly 3% ED. Drawings illustrate the transition from circumferential indentation (CI) to local bending (LB); arrow represents moderate touch in the midbody of the animal, which generates CI early and LB later. B. Diagram of the neuronal circuit in a single midbody segment that produces the longitudinal component of local bending in that segment. The cross-section of the body wall is represented at the top and bottom of the diagram as though the segment has been cut longitudinally along the ventral midline and spread out. Four P mechanosensory neurons (P cells, in green) innervate the ventral (Pv) and dorsal (Pd) regions on each side of the body wall and make excitatory chemical synapses onto 17 identified local bending interneurons (LBI; only 9 LBIs are shown, and they are in red). LBIs make excitatory chemical synapses onto excitatory (dorsal excitatory, DE; and ventral excitatory, VE) and inhibitory (DI and VI) motor neurons (in blue), all of which innervate longitudinal muscle fibers in the segment. The only known inhibitory synaptic connections in the circuit (represented by black circles at the end of the lines connecting the neurons) are made between motor neurons: inhibitory motor neurons synapse onto muscles in the periphery, but they also make inhibitory chemical synaptic connections onto excitatory motor neurons that project to the same muscle. Excitation of a particular site of the skin excites one or two particular P cells. That P cell excites some of the LBI that will cause activation of excitatory motor neurons innervating that side. In addition, those LBI also activate inhibitory motor neurons on the contralateral side. As a result, the longitudinal muscles ipsilateral to the touched skin contract and the ones contralateral relax, producing the LB. C. Standard maps of the ventral and dorsal surfaces of a midbody ganglion of the leech showing where the somata of the neurons that compose the local bending circuit are located. Colors indicate the same classes of neurons in panels B. and C. *(From Marin-Burgin et al. 2008)*

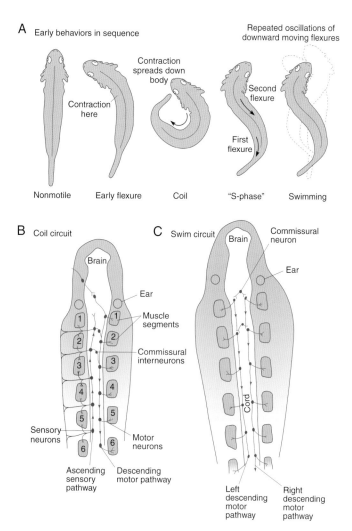

Fig. 10.6 Coghill's sequence of early amphibian behavior and its proposed neural basis. A. An axolotl embryo at five stages of behavioral development. B. The neural circuit for the coil response. A stimulus anywhere on one side of the body is transmitted to the contralateral spinal motor pathway by commissural cells in the anterior cord or hindbrain where the neural signal descends, stimulating primary motor neurons of the cord. C. The early swim circuit (the sensory mechanism) is omitted but is the same as in B. Motor excitation travels down one side of the cord, but by this stage in development, some reciprocally exciting commissural neurons that cross the floorplate in the hindbrain have developed so that excitation on one side at the neck region can cross over after a delay to excite the contralateral motor pathway, leading to coiling on one side being quickly followed by coiling on the other side. *(Adapted from Coghill, 1929)*

with the activity of the various classes of neurons as they develop (Higashijima et al., 2004) (**Figure 10.8**). Similar to what Coghill described for the salamander, zebrafish embryonic behaviors appear sequentially and include an early period of spontaneous tail coilings, followed by very brief swim-like events in response to touch, and finally sustained swimming. Electrical recording from specific neurons *in vivo* reveal that an electrically coupled network generates spontaneous tail coiling, whereas a glutamatergic and glycinergic synaptic drive and input from the hindbrain are needed for touch responses which elicit the brief swimming events. Thus, the transition

from coiling to brief swims is, like the transition from circumferential indentation to local bending in the leech, reflected in the emergence of chemical circuitry (Brustein et al., 2003; Fetcho et al., 2008; McLean and Fetcho, 2008).

Mauthner neurons of the hindbrain (Chapter 5) receive input ipsilaterally, cross the midline, and project posteriorly onto contralateral motor neurons. In the embryonic zebrafish, as early as 40 hours after fertilization, the Mauthner cells and homologous neurons in other hindbrain segments can initiate a directional escape response away from a stimulus followed by a series of strong tail flexures. The transparency of larval zebrafish has enabled physiologists to use calcium imaging in the intact fish to observe the activity of the Mauthner cell during behavior (**Figure 10.9**). Such work shows that, during an escape, these cells are indeed activated in patterns that are exactly predicted by behavioral studies (O'Malley et al., 1996). Swimming becomes sustained in larvae once serotonergic neuromodulatory effects are integrated into the circuit.

Serotonin, released by neurons in the *Xenopus* hindbrain, increases the duration of these rhythmic swimming bursts. This input is tuned, remarkably, by the earlier spontaneous activity of these neurons, for the calcium that accompanies these spikes lowers the expression of Lmx1b, a transcription factor controlling the synthesis of serotonin in these neurons (Demarque and Spitzer, 2010). When a tadpole bashes its head against a solid surface, it stops swimming. This is accomplished by sensory neurons in the trigeminal ganglion that excite descending inhibitory GABAergic neurons of the hindbrain. These descending neurons synapse directly onto motor neurons and reliably stop swimming when they fire (Perrins et al., 2002).

In the chick embryo, Hamburger (1963) describes the early movements as "uncoordinated twisting of the trunk, jerky flexions, extensions and kicking of the legs, gaping and later clapping of the beak, eye and eyelid movements and occasional wing-flapping…performed in unpredictable combinations." The integration of movements between limbs, such that the left leg alternates with the right during walking, or the right and left wings beating synchronously during flight, do not emerge until later in development. Thus, the random thrashings and reflexes of individual parts of the chick embryo, as well as the mammalian embryo, are brought under control as more circuitry develops. After the chick hatches, circuits across the midline synchronize the right and left wings so that they beat together, and if one wing is weighted down so that it moves more slowly, the contralateral wing will follow the slower pattern (Provine, 1982).

Even in humans, the precursors to mature locomotion can be seen long before infants take their first steps. One can see left-right motor coordination in the crawling movements that babies make even at pre-crawling stages, when they are put on a gentle downhill slope. The spinal circuits that control repetitive left-right movements in walking or swimming are conserved in vertebrates, because the mechanisms that generate the variety of spinal neurons and their wiring up is also conserved. So it is likely that, at the earliest stages, there are great similarities in the way movements are built up and integrated. In humans, movements can become very complex and demanding, as in the fine finger movements of a piano player.

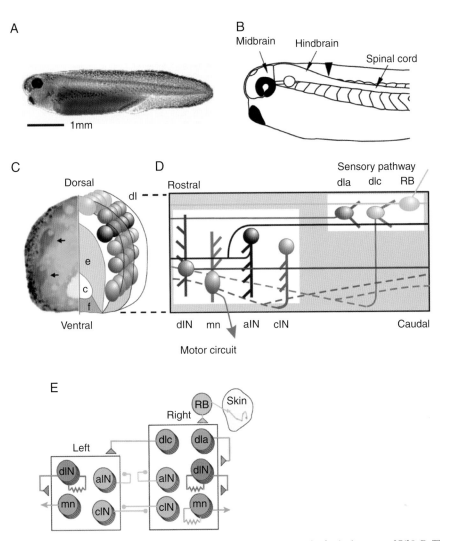

Fig. 10.7 Hatchling *Xenopus* tadpole, nervous system and neurons. A. Photograph of tadpole at stage 37/38. B. The main parts of the CNS with arrowhead at hindbrain/spinal cord border. C. Transverse section of the spinal cord with the left side stained to show glycine immunoreactive cell bodies (arrows) and axons (in the marginal zone). Diagrammatic right side shows the main regions: neural canal (c) bounded by ventral floorplate (f) and ependymal cell layer (e); lateral marginal zone of axons (mauve), layer of differentiated neuron cell bodies arranged in longitudinal columns (colored circles) lying inside the marginal zone except in dorsolateral (dl) and dorsal positions. D. Diagrammatic view of the spinal cord seen from the left side, showing characteristic position and features of seven different neuron types. Each has a soma (solid ellipse), dendrites (thick lines) and axon(s) (thin lines). Commissural axons projecting on the opposite right side are dashed. E. Model networks with probabilistic connectivity. The network has a single sensory RB neuron exciting neurons in the right half-centre, which also has sensory pathway dlc and dla interneurons. There are ten of each neuron type in each half center. The broad pattern of connections is shown by the axons from groups of neurons onto the half-centers (triangles are excitatory and circles are inhibitory synapses). The actual synaptic connections are determined probabilistically for each neuron. (Li et al. 2007)

Descending pathways from the cortex bring movements under voluntary control and it is not until the corticospinal tract develops fully that macaque monkeys are able to make fine finger movements and exhibit mature manual dexterity (Armand et al., 1994).

THE ROLE OF ACTIVITY IN THE EMERGENCE OF COORDINATED BEHAVIOR

Do the seemingly spastic first movements and motor patterns of embryos have a role in the development of coordinated behavior? More than a century ago, Harrison (1904) raised some salamander embryos in an anesthetic solution throughout the period of bending, coiling, "S" movements, and early and sustained swimming. He then transferred the embryos to anesthetic-free solution. He found that the long-term anesthetized embryos were able to begin to swim as soon as the anesthetic was washed out. After a short while, he reported, the previously drugged embryos were behaviorally indistinguishable from the controls. More recently, similar experiments with anesthetic and with alpha-Bungarotoxin (which specifically blocks the nerve-to-muscle synapse) have been done in combination with more quantitative analysis of behavioral,

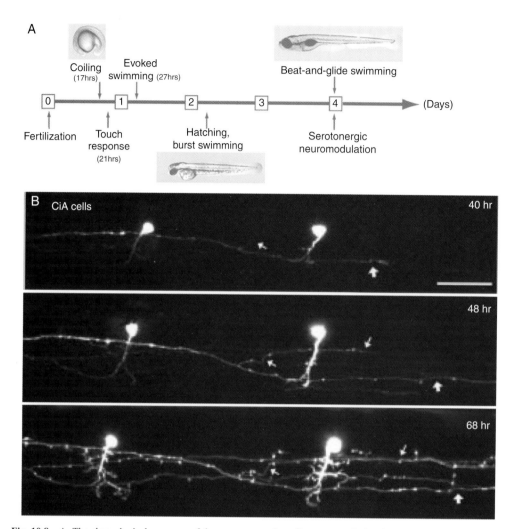

Fig. 10.8 A. The chronological sequence of the appearance of motility patterns during development of the zebrafish. Images emphasize key stages of zebrafish embryo development before and after hatching (52 h). B. Tracking the development of identified classes of spinal interneuron in vivo. Confocal time-lapse imaging of two inhibitory circumferential ascending (CiA) cells labeled with GFP using En1 as a promoter. CiAs are so named due to their primary ascending axon, however, they also have a secondary descending one. Successive images at different time points (hours postfertilization noted in respective images) illustrate the growth of the descending axon (at thin and thick white arrows), as well as the elaboration of the dendrites. Scale bar is 50 lm. *(from Higashijima et al., 2004)*

electrophysiological and physiological recovery, (Haverkamp and Oppenheim, 1986; Haverkamp 1986) (**Figure 10.10**). These results show that after a few hours, the treated embryos have fully recovered, but that for up to two hours immediately after drug withdrawal the embryos moved and swam more slowly and fitfully, indicating a possible delay in motor development. Note that none of the drugs used in these studies completely blocked activity in the CNS, and so the importance of central activity was not tested by these experiments. Perhaps, as in the early development of the visual system discussed in Chapter 9, the basic early connections are determined by the molecular signals that direct the placement of axons and dendrites, and it is the fine-tuning of the motor program that requires activity.

In many organisms, it is critical that certain motor programs are ready to go when the animal is born or hatches. In herd mammals, for example, a newborn has just a few minutes to get to its feet and follow its mother, and a mouse has to be able to suckle effectively within hours of its birth or it

will die. Even *Drosophila*—an organism that is clearly dedicated to extremely rapid development—spends more than 15% of embryogenesis moving energetically in the eggshell. *Drosophila* has about 24 hours to find food after hatching or else it starves to death. So, it seems that many motor programs are revved up in the embryo or at particular stages of development so that when they are called into action, they are executed well. The song of the cricket is a good example. Elements of the song circuitry are put together and function during the molts of a cricket's life; at each molt, the song pattern, read in the firing of the central neurons that generate the pattern, becomes more adult-like, but this is all in the cricket's CNS—it is fictive singing, like humming to himself. The cricket doesn't actually produce sound until after its final molt into full adulthood, and when it sings for the first time, the song has already been perfected (Bentley and Hoy, 1972).

To assess whether such practice, be it internal without actual movement or real but uncoordinated, is significant, it is necessary for the experimenter to silence or alter these

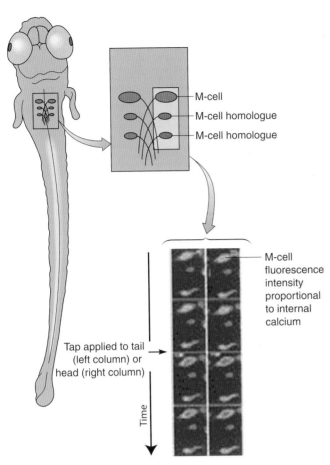

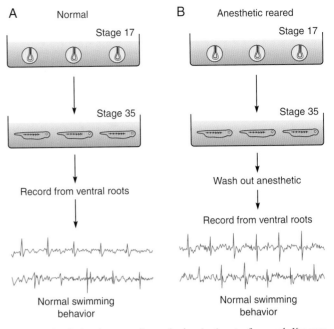

Fig. 10.9 Visualizing the neural basis of behavior in zebrafish. In a newly post-hatched zebrafish larva, the descending cells including the giant Mauthner neuron (M-cell) and its two homologues in the hindbrain have been filled with a fluorescent calcium indicator. The panels below show trials consisting of a sequence of images during which an escape was elicited by an ipsilateral touch to the tail or the head. The color scale represents fluorescence intensity (blue, lowest; red, highest). Simultaneous imaging of both the Mauthner cell (top of each panel) and one of its segmental homologues (bottom of each panel). The left column shows that with a tail tap only the M-cell responds. The right column shows the result from a head stimulus. In this case, both the M-cell and its homologue respond. *(Adapted from O'Malley et al., 1996)*

Fig. 10.10 Swimming out of anesthesia. A. A set of normal *Xenopus* embryos put in a culture dish at the late neural plate stage and raised till the swimming stage in normal pond water. They swim normally, and electrophysiological records from ventral roots on opposite sides of the spinal cord show the expected alternating pattern of activity. B. A set of sister embryos raised in an anesthetic solution until stage 35, at which point some of the embryos were immediately put into a recording chamber. The pattern of activity and the swimming behavior look essentially normal. *(Adapted from Haverkamp, 1986)*

patterns in the developing embryo and then look carefully for delays and perturbations in the development of motor patterns. Above, we discussed how motor neurons that control limb movements in the chick begin to burst and the role of cholinergic transmission in these first semirhythmic patterns: the motor neurons use acetylcholine to activate Renshaw cells which feed back positive excitation through the early excitatory action of GABA. In mice lacking choline acetyltransferase (ChAT), the enzyme that makes acetylcholine, such bursting does not occur during these early phases, and this has specific effects on the fine-tuning of the locomotor circuit that forms in these mutants: the duration of each cycle period is elongated, and right-left and flexor-extensor coordination are abnormal. In contrast, blocking acetylcholine receptors after the locomotor network is wired does not affect right-left or flexor-extensor coordination. This suggests that cholinergic driven activity is important during a transient period of development to mediate proper assembly of spinal locomotor circuits (Myers et al., 2005). Similar experiments in Drosophila show that if the early episodes of neurally induced muscle contractions are blocked during just this period by a temperature sensitive protein, which stops all synaptic activity in the CNS, there is a delay in the development of the coordinated motor program of approximately the same amount of time as the activity was blocked, suggesting that this early period of activity, though useless in terms of locomotion, was nevertheless important practice time and has a role in fine-tuning the motor program.

By what mechanisms does this fine-tuning work? If one reduces the spontaneous network activity by injecting a little lidocaine into the egg of a chick at the early bursting period, there is an initial silencing, but then the embryonic circuitry increases in the efficiency of its excitatory synaptic connections, thus compensating for the lack of activity by becoming more excitable—and this leads to the relative maintenance of the appropriate activity levels (Gonzalez-Islas and Wenner, 2006). Such a homeostatic mechanism depends on feedback, testing out the level of excitability through activity, so that it can be adjusted appropriately.

STAGE-SPECIFIC BEHAVIORS

Some early motor behaviors are neither merely anticipatory nor substrative, serving no other purposes than building and fine-tuning the motor program for later use; some are also adaptive, having stage-specific functions. One such early behavior that serves a clear purpose is the hatching

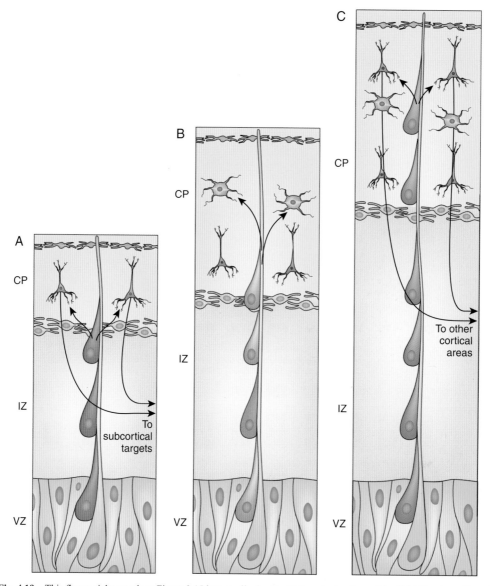

Fig. 4.19 This figure picks up where Figure 3.15 leaves off, after the birth and migration of the Cajal-Retzius cells and the subplate cells. A. The next neurons to be generated in the cortex are the pyramidal neurons of the deep layers, V and VI (red), whose axons project to subcortical targets. B. The next neurons to be born are the local interneurons in Layer IV of the cortex (blue). C. Finally, the pyramidal cells of the upper layers, II and III (green), are generated. They send axons to other cortical areas. VZ = ventricular zone, IZ = intermediate zone, CP = cortical plate.

bipolars, amacrines, and Müllers in the same general birth order, size, and composition as they would have done had they been left *in vivo* (Cayouette et al., 2003).

The nature of the intrinsic program in these progenitors appears to be established by temporally regulated transcription factors. Retinal progenitors may share with *Drosophila* neuroblasts the transient expression of genes such as the early temporal coordinate gene *hb*, (discussed above). Indeed, Ikaros, a mouse ortholog of Hb, is expressed in all early retinal progenitors, but is then downregulated in late progenitors. *ikaros* mutants show a reduction in early-born cell types, while misexpression of Ikaros generates more such neurons (Elliott et al., 2008). The transcription factor Vsx2 is also initially expressed in all retinal progenitors, but is then downregulated in all progenitors except those giving rise to a certain class of bipolar cell and all Müller cells. Vsx2 is a repressor (Clark

et al., 2008), whose targets include transcription factors such as the bHLH transcription factor Ath5, which restricts the fate of progenitors to RGCs, horizontal cells, amacrine cells and photoreceptor fates. Vsx2 also represses Foxn4, which is expressed in the progenitors of amacrine and horizontal cells. Thus, when Vsx2 is downregulated, Vsx2-negative progenitors escape Vsx2 repression and become able to express factors that take them down other specific lineages (Vitorino et al., 2009). This suggests that there is not a single global lineage through which all retinal progenitors pass, but rather a set of parallel lineages that are travelling through developmental time side by side, with cells in different lineages having different temporally regulated competences.

Intrinsic transcriptional factors that come on at or near the terminal division of retinal progenitors are very influential in retinal fate choice. For example, Ath5, which turns on after the

93

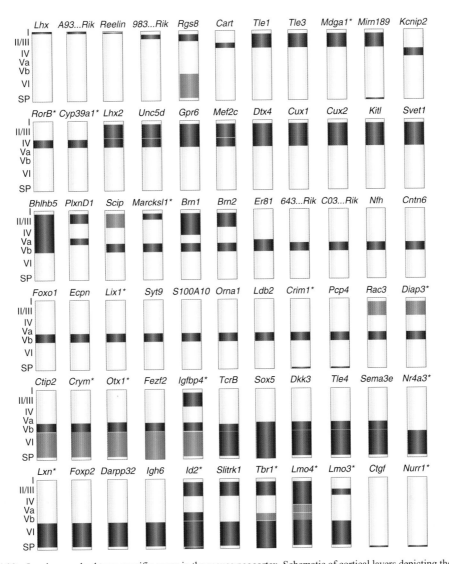

Fig. 4.20 Laminar- and subtype-specific genes in the mouse neocortex. Schematic of cortical layers depicting the laminar-specific expression of 66 genes within the neocortex. Dark blue and light blue indicate higher and lower relative levels of expression, respectively. Genes for which laminar or subtype expression varies by area within the neocortex are indicated by an asterisk. Some expression patterns are dynamic. For example, Ctip2 is initially expressed at much higher levels in layer V than layer VI, but is equivalent in the two layers later in development. SP, subplate. *(Molyneaux et al., 2007)*

final S-phase in retinal progenitors, is critical for RGC fate. The targets of Ath5 in the retina are a range of molecules including components of neurotransmission, cell cycle inhibitors and transcription factors, which in turn regulate more aspects of the differentiated phenotype (Liu et al., 2001; Logan et al., 2005; Matter-Sadzinski et al., 2005; Mu et al., 2005). This is not simply a linear sequence, as Ath5 and its downstream transcription factors activate common targets or each other. Such positive feedback loops ensure that once the cell starts differentiating, the flip to the differentiated fate is all-or-nothing. But Ath5 is at the top of this transcriptional program, so in animals where Ath5 is not expressed, the RGCs simply do not arise and the cells that would have become RGCs become other cell types instead. Other transcription factors that get turned on in particular lineages tend to specify particular fates in similar ways. Ptf1a is turned on in some retinal precursors immediately following their final mitosis. Ptf1a is necessary and sufficient to

switch cells that would have otherwise become one of the main excitatory cell types in the retina (photoreceptors, bipolars, RGCs) to one of the two inhibitory cell types in the retina (horizontal or amacrine). Similarly the postmitotically expressed transcription factor Nrl is expressed in rod photoreceptor precursors only. In the absence of Nrl, these precursors become cones, but if Nrl is expressed in all photoreceptor precursors, they all become rods (Oh et al., 2007).

Synergies between transcription factors, mostly between the homeodomain and bHLH families, suggest there is combinatorial coding of the specification of neuronal subtype identities (Hatakeyama and Kageyama, 2004; Wang and Harris, 2005). A good example of this concerns Mash1, Math3 and Vsx2, all of which are expressed by differentiating bipolar cells. There is a reduction in the number of bipolar cells in Mash1 or Math3 mutant embryos and a mutation in Vsx2 also results in a loss of these neurons. Misexpression of Mash1 or Math3 in retinal

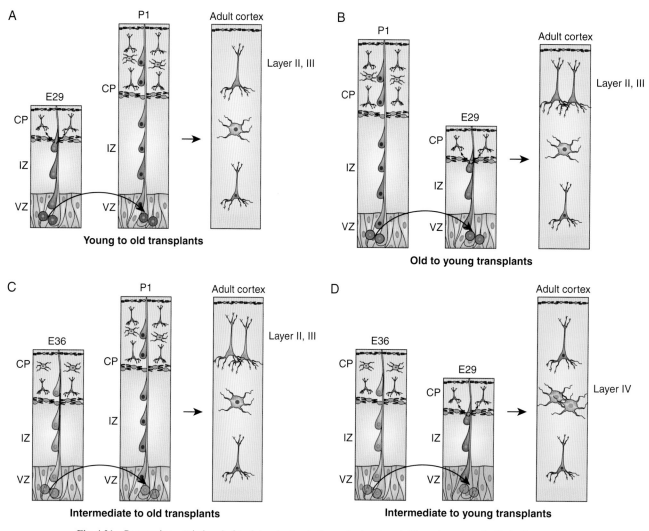

Fig. 4.21 Progressive restrictions in fate determination in the cerebral cortex. A. Transplantation of cells (yellow nuclei) from the VZ of an E29 ferret to a P1 ferret leads these cells to change from a deep layer (early) to a superficial layer (late) fate. B. But when P1 cells are transplanted to E29 hosts, they retain their superficial layer fates. C. Intermediate E36 generated cells normally destined for Layer IV can assume later fates when transplanted into an older host. D. But they cannot assume younger fates when transplanted into younger hosts. VZ = ventricular zone, IZ = intermediate zone, CP = cortical plate.

progenitors, however, results in the production of photoreceptors rather than bipolars, suggesting that these two bHLH proteins are not sufficient to specify bipolar cells by themselves (Hatakeyama et al., 2001). Misexpression of Vsx2 produces some extra bipolar cells, but also increases Müller cells and undifferentiated cells, indicating again that Vsx2 expression by itself is also not sufficient to promote the bipolar cell identity efficiently. Importantly, misexpression of Vsx2 with either Mash1 or Math3 strongly promotes the generation of bipolar cells. Such loss and gain of function studies in mouse and frogs with other pairs and trios of other transcription factors suggest that a combinatorial transcription factor code is involved in the specification of the different classes of retinal cells.

Ath5 is first expressed in progenitors that divide just once more to generate one RGC and one other post-mitotic retinal neuron by an asymmetric cell division (Poggi et al., 2005) (**Figure 4.24**A). Seeing transcriptionally specified lineage patterns, however, does not mean that retinal cell determination is immune from extrinsic influence. Indeed, when retinal cells are

removed at the stage when RGCs are being born, mixed into aggregates with an excess of retinal cells several days older, and cultured *in vitro*, they often become photoreceptors instead of RGCs (Watanabe and Raff, 1990) (**Figure 4.25**). This work shows that individual cells have the capacity to differentiate into different cell types, and the fate they choose depends on the environment in which they are born. This makes good sense, for as retinal development proceeds, the environment changes as a result of various retinal cell types being generated, much as in the *Drosophila* retina discussed above. Indeed, there is strong evidence for feedback in influencing the fate of RGCs. Young embryonic chick progenitors are inhibited in their ability to produce RGCs when cultured adjacent to older retinas in which there are lots of RGCs, and depletion of the RGCs from these older retinal cell populations abolishes this inhibition (Waid and McLoon, 1998). Similarly, while Ath5-positive cells in normal embryos generate two cell clones, only one of which tends to be an RGC daughter (see above), these same cells when transplanted into mutant retinas that lack RGCs often generate

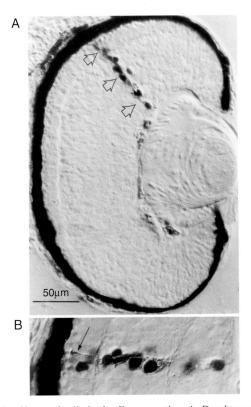

Fig. 4.22 Clones of cells in the *Xenopus* retina. A. Daughters of a single retina progenitor in *Xenopus* injected with horseradish peroxidase are seen to form a column that spans the retinal layers and contributes many distinct cell types. B. p, photoreceptor; b, bipolar cell; m, Müller cell; a, amacrine cell; and g, ganglion cell.

two RGC daughters, providing evidence that such lineages are sensitive to negative feedback signals that regulate the number of RGCs (Figure 4.24B). Similar negative feedback systems have been discovered that regulate the expression of Ptf1a and thus the relative number of inhibitory to excitatory cells in the retina, as well as feedback signals that regulate the number of specific subtypes of amacrine cell, suggesting that there are amazing homeostatic mechanisms at work, adjusting the relative amounts of various transcription factors that are made in retinal precursors so that just the right number of each cell type is made.

What could these feedback signals be? RGCs express Shh, and the Shh knockout mouse contains increased numbers of RGCs, suggesting that Shh provides a feedback inhibition signal in the developing mouse retina, echoing what happens in the *Drosophila* retina where R8 cells express Hh, which regulates the expression of *ato* in the progression of the morphogenetic furrow (discussed above). The growth differentiation factor 11 (Gdf11) is another factor secreted by RGCs, and it acts by limiting the temporal window during which progenitors are competent to express Ath5 and thus produce RGCs. The Notch signaling pathway is also involved. When Notch signaling is compromised in early retinal progenitors, too many cells differentiate as RGCs, whereas later interference with Notch signaling leads to an increase in later born cell types such as rod cells (Perron and Harris, 2000; Kim et al., 2005; Wallace, 2008). When cells are released from

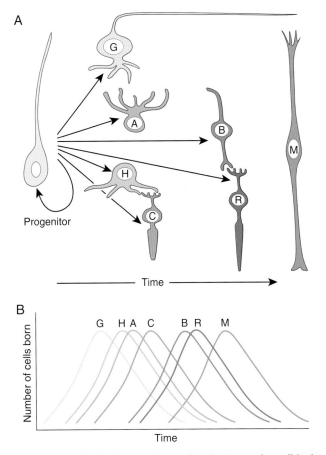

Fig. 4.23 Determination in a vertebrate retina. A. A progenitor cell in the neuroepithelium divides several times and gives rise to clones of cells that contain all the major cell types of the retina, including ganglion cells G, amacrine cells A, horizontal cells H, bipolar cells B, rods R, cones C, and Müller cells M. B. These cells tend to be born at different developmental times, indicating a rough histogenetic order B.

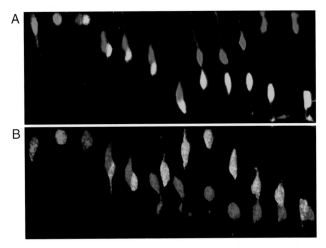

Fig. 4.24 A. Time-lapse images of an RGC progenitor in zebrafish dividing at the apical surface of the retina to produce two neurons, one of which (yellow) becomes an RGC. The other cell (red) is becoming a photoreceptor. B. A similar RGC progenitor in an *ath5* mutant retina in which there are no host RGCs. This cell divided to produce two daughters, both of which become RGCs (arrow points to emerging axon).

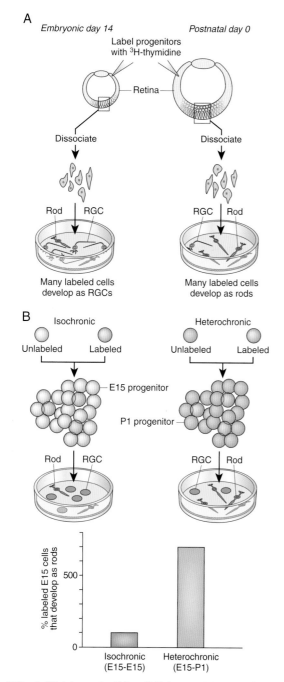

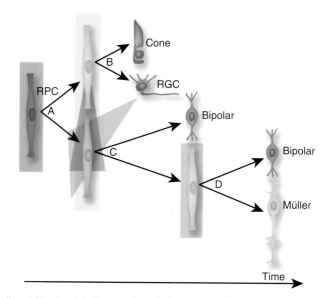

Fig. 4.26 Possible lineage of a retinal progenitor, illustrating aspects of its life: The proliferative potential, illustrated by the intensity of the cells' red color, generally decreases with time, and an early progenitor is more likely to give rise to proliferative divisions (A), whereas later the proportion of terminal divisions rises (D) but terminal divisions can also take place early (B). The change in the propensity to differentiate may be due to cell intrinsic changes or changes in a cell's environment, as seen by the background color. The two siblings from division (A) find themselves generally in the same environment (yellow background) but may be different due to asymmetric division or because of receiving different signals, for example through lateral inhibition, and therefore produce different progeny (B vs C). Differentiated cells also negatively signal to progenitors (green triangle). With time, progenitors produce late neuronal types (D), and eventually some Müller cells. *(From Agathocleous and Harris, 2009)*

Fig. 4.25 A. Birthdate and cell fate. Cells born on E14, even if dissociated into tissue culture, tend to differentiate into retinal ganglion cells (RGCs), while cells born on P0, when dissociated, tend to differentiate into rods. B. Early cell fate in the vertebrate retina is flexible and influenced by extrinsic factors. E15 cells labeled with thymadine and mixed with other E15 cells (isochronic) will not tend to differentiate as rods, while the same cells, if mixed with P0 cells (heterochronic) will. *(After Reh and Kljavin, 1989 and Watanabe and Raff, 1990)*

the inhibition mediated by the Notch pathway, proneural basic helix-loop-helix (bHLH) transcription factors, such as Ath5, bias cells toward particular fates, such as RGCs. The Notch signaling pathway, by allowing only a certain number of cells to differentiate at any one time, is thus important in creating cellular diversity in the vertebrate retina (Dorsky et al., 1997).

General models of retinal cell fate determination, such as that shown in **Figure 4.26**, incorporate many features of the findings discussed above. It is worthwhile pointing out that, as cells become determined to become specific cell types, they also must exit the cell cycle. A variety of external growth factors influence cell-cycle progression by affecting the expression of cell-cycle components (see Chapter 3). The mechanisms of coordination between cell cycle exit and cell fate specification are not well understood, but there are some intriguing examples to consider. In the retina, as in many other parts of the CNS, glia are the last cells to pull out of the cell cycle. In the *Xenopus* retina, this is by virtue of an accumulation of a cell-cycle inhibitor called p27Xic1. p27Xic1 is, surprisingly, a bifunctional molecule. It has a cyclin kinase inhibitor domain to take the cells out of the cell cycle, and it has a separate Müller glial determination domain. In the developing retina, when p27Xic1 builds up to high enough levels in cells, it forces them to leave the cell cycle and to become Müller cells (Ohnuma et al., 1999). Similarly Prox1, the mammalian homolog Prospero (mentioned above in the context of asymmetric cell division), also has separate cell cycle and cell determination functions, repressing genes involved in proliferation and activating genes involved in horizontal cell differentiation. Proneural transcription

factors, like Neurogenin, have been shown to help committed neural progenitors out of the cell cycle by stabilizing cyclin kinase inhibitors. Extrinsic cues may also have a role in this coordination as trophic factors may become less available or progenitor cells may become insensitive to the proliferative effects of these factors. These are some illustrative examples and ideas of how progenitors move from a dividing to a differentiated state. There is much more to learn about how this important transition is achieved (Agathocleous and Harris, 2009).

INTERPRETING GRADIENTS AND THE SPATIAL ORGANIZATION OF CELL TYPES: SPINAL MOTOR NEURONS

The vertebrate spinal cord is composed of a variety of cell types, including motor neurons, local interneurons, and projection neurons. The embryonic spinal cord even contains a set of sensory neurons, called Rohon–Beard cells. These cells die by adult stages, and sensory input to the spinal cord is then supplied by dorsal root ganglion neurons. The spinal cord primordium begins as a rectangular sheet of neural plate epithelium centered above the notochord. Lineage tracing experiments show that cells in the lateral edges of the plate tend to give rise to Rohon–Beard cells and dorsal interneurons, while cells in the medial plate tend to give rise to motor neurons and ventral interneurons (Hartenstein, 1989) (**Figure 4.27**). Occasional clones are composed of different cell types, so it is thought that local position, rather

than lineage, is involved in the generation of neuronal diversity in the spinal cord.

As we saw in Chapter 2, tissues outside the nervous system often provide critical signals that influence development within the CNS. In that chapter, we also reviewed the evidence that the notochord and the floorplate played a key role in establishing the dorsal–ventral axis of the neural tube by providing a gradient of Shh (Jessell et al., 1989). The ventrolateral region of the tube, which gives rise to motoneurons, is exposed to high levels of Shh, while further dorsally, where interneurons develop, the dose of Shh is lower (**Figure 4.28**A,B). The most ventral neurons require the highest doses of Shh, and successively more dorsal ones require correspondingly less. How do cells at different dorso-ventral levels interpret their exposure to different levels of Shh? Particular threshold levels of Shh turn on homeodomain genes of the Class II type and turn off Class I homeodomain genes (Jessell, 2000) (Figure 4.28C). Thus, the read-out of the Shh level is first registered in neural tube neurons by expression of specific homeodomain proteins, which are either turned on or off at particular Shh thresholds. In this way, the ventral boundaries of Class I expression and the dorsal boundaries of Class II expression set up unique domains in which certain homeodomain transcription factors are expressed. The boundaries between these domains are sharpened through cross-repression of the two classes of genes. For example, the ventral border of the Class I *Pax-6* gene initially overlaps the dorsal border of the Class II *Nkx2.2* gene, but the cross-repression between these two transcription factors means that only one of these two genes is eventually expressed in any particular cell, so the border between them becomes sharp. By repeating this process in other pairs of transcription factors, specific domains uniquely express particular combinations of Class I and Class II homeodomain transcription factors. By similar mechanisms, an opposing dorsal to ventral gradient of BMP helps establish and extend these domains dorsally, such that the entire neural tube is patterned into transcriptionally unique domains from which the various classes of spinal neurons that will make the early circuitry for locomotion arise (see **Figure 4.29** and Chapter 10).

Motor neurons come from the pMN domain that uniquely expresses Class I Nkx6.1 but not Irx3 or Dbx2, and Class II Pax6 but not Nkx2.2 or Nkx2.9. Throughout the neural tube, there is a general activation of the genes that determine neuronal identities, but Nkx6.1 and Pax6 repress the determinants of other neuronal types in the pMN domain. Similarly, Nkx2.2, Irx3, and Dbx2 repress motor neuron determinants in the non-pMN domains (Muhr et al., 2001; Lee et al., 2004b). This derepression network results in the expression of Olig2, a bHLH transcription factor that is critically required for motor neuron differentiation in just the pMN domain. Olig2 in turn activates the expression of motor neuron specific transcription factors such as Mnr2, Hb9, Lim3, and Isl1/2. Once expressed, Mnr2 can regulate its own expression and is sufficient to drive spinal progenitor cells down a motor neuron pathway. This is shown by experiments in which motor neurons arise dorsally when Mnr2 is expressed ectopically in dorsal progenitors. This system, beautifully worked out by Tom Jessell and colleagues, is a classic example of combinatorial derepression in a hierarchical transcription factor cascade.

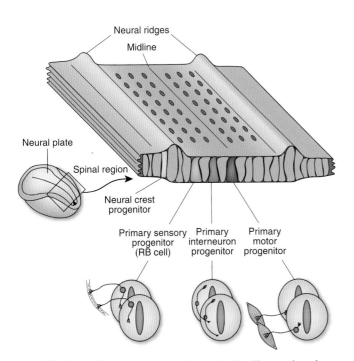

Fig. 4.27 Rows of primary neurons in the neural plate. The most lateral row gives rise to sensory neurons (Rohon-Beard cells), the middle row gives rise to interneurons, and the most medial row gives rise to primary motor neurons. *(After Hartenstein, 1989)*

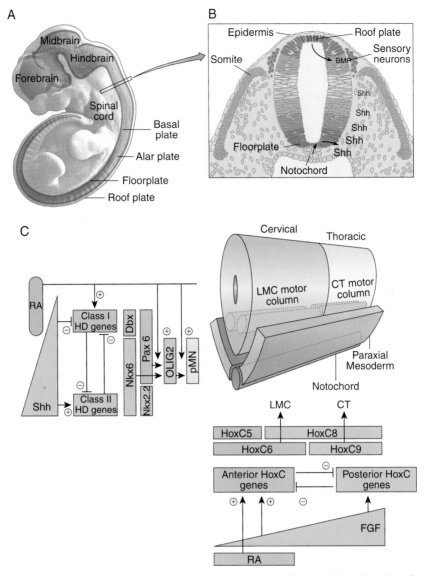

Fig. 4.28 Specification of motor neurons in the vertebrate spinal cord. A. The neural tube, shown here for a mouse, is subdivided into four longitudinal domains: the floorplate, basal plate, alar plate, and roof plate. Motor neurons are derived from the basal plate. B. Schematic cross section of the neural tube. The notochord, which is located underneath the floorplate, releases Sonic hedgehog (Shh) which induces the floorplate to release its own Shh. This forms a gradient in the neural tube with high concentrations ventrally and low concentrations dorsally. BMP molecules released from the dorsal epidermis and roof plate form an opposing gradient. C. (Left) A gradient of Shh emanates from the notochord and floorplate; threshold levels of Shh turn on Class II HD genes. Retinoic acid (RA) expressed by the paraxial mesoderm induces the expression of Class I HD genes that are turned off more ventrally by threshold levels of Shh. Class I and Class II HD transcription factors cross-repress each other, creating sharp definitive boundaries at different dorso-ventral levels in the cord. Thus the boundary between Dbx and Nkx6 is more dorsal than the boundary between Pax6 and Nkx2.2. Between these boundaries, the Olig2 *bHLH* transcription factor necessary for motor neuron specification is turned on by the concerted action of RA, Nkx6, and Pax6. Olig2 and RA are necessary for the expression of motor neuron differentiation genes of the pMN family. (Below) A gradient of FGF8 emanates from the mesoderm. High levels of FGF8 turn on more caudal *hoxC* genes, whereas RA and low levels of FGF8 turn on rostral *hoxC* genes. Rostral and caudal *hoxC* transcription factors cross-repress each other, creating sharp definitive boundaries at different rostrocaudal levels in the cord. The boundary between HoxC6 and HoxC9 establishes the boundary between the LMC of the cervical cord and the CT of the thoracic cord.

In Olig2 knockout mice, neither motor neurons nor oligodendrocytes develop, so Olig2 is also essential for oligodendrocyte development (Chapter 3). How do the same progenitors make both motoneurons and oligodendrocytes? The cells that express Olig2 initially also express the proneural gene Neurogenin2 (Nrgn2) (Kessaris et al., 2001; Zhou et al., 2001).

Nrng2 and Olig combine their activity to generate motoneurons. Later in development, however, Nkx2.2 moves dorsally into the pMN domain and, as we have seen above, Nkx2.2 is a repressor of motoneuron determinants such as Nrng2, but it does not repress Olig1 expression in this region. So now the progenitors that express both Olig1 and Nkx2.2 start making oligodendrocytes.

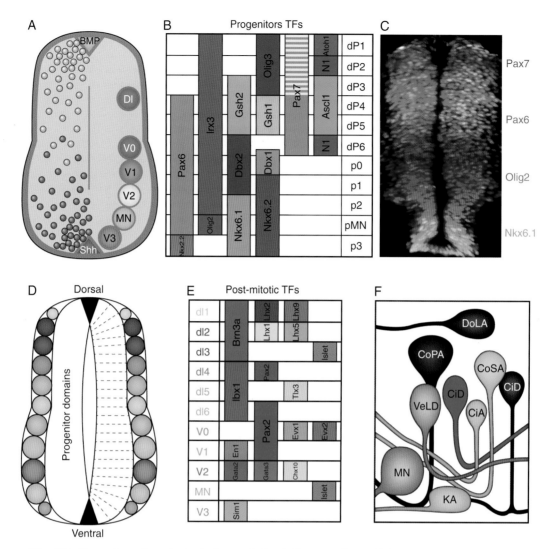

Fig. 4.29 Neuronal type determination in the neural tube. A. Countergradients of BMP and Shh at various threshold concentrations turn on and off sets of transcription factors that occupy domains in the dorso-ventral axis of the neural tube (B.) such that progenitor cells have dorso-ventral regional identity. C. The pattern of gene expression (shown for four transcription factors by *in situ* hydridization) determines the type of progenitor (shown in the right side of A.). D. As cells in the spinal cord begin to differentiate along the dorso-ventral axis, they begin to express post-mitotic transcript factors (E.) as a result of the particular region their progenitors occupied. F. Interneurons and motor neurons are specified by the transcriptional code. In this case the neurons are color coded (KA interneurons in pink come from V3; MN=motor neurons in grey come from MN, etc.) The interneurons in black have not yet been clearly associated with specific regions, but based on their position in the spinal cord, we can take an educated guess about the transcriptional code that specifies them. *(With thanks to James Briscoe and Kate Lewis.)*

One of the features of such transcriptional cascades is that the acquisition of specific neuronal fates is coordinated with the acquisition of generic neuronal features. This strategy was seen in the layer 5 neurons of the cortex, discussed above, but is even more clearly established in molecular detail in the motor neurons of the spinal cord. Here, proneural bHLH genes not only promote generic aspects of neuronal fate such as cell cycle arrest and the expression of pan neural markers, but they also interact directly by interacting with other cell fate determinants such as homeobox proteins to promote specific fates. A good example of such an interaction is the synergy between the bHLH transcription factor NeuroM and the LIM-homeodomain proteins Isl1 and Lhx3,

which is mediated by the ubiquitous adapter protein NLI. This interaction is necessary for induction of Hb9 expression, which is specific to motor neurons (Lee and Pfaff, 2003) (**Figure 4.30**).

The motor neurons in the spinal cord are organized into functional columns that project to different muscle groups in the mature animal along the anterior to posterior axis of the body. The signaling factors that were necessary for A-P patterning at earlier stages of development (Chapter 2) continue to play a similar role as motoneuron fates are being determined. Thus, the Lateral Motor Column (LMC) in the cervical region innervates forelimb muscles, and the Column of Terni (CT) in the mid-thoracic region innervates

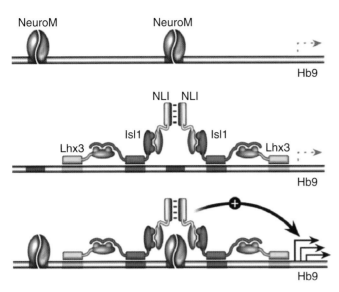

Fig. 4.30 Combinatorial specification by the cooperation of transcription factors. Motor neuron progenitor cells express unique combinations of bHLH neurogenic proteins and homeodomain factors. The bHLH transcription factors contribute to motor neuron differentiation in two ways: they regulate neurogenesis, a process whereby cells acquire generic neuronal properties, and they cooperate with LIM homeodomain factors to specify motor neuron subtype identity. The functional interaction between specific bHLH transcription factors such as NeuroM:E47 and higher order LIM-HD complexes comprised of NLI, Isl1, and Lhx3 is mediated by the dimerization of NLI. bHLH and LIM-HD factors bind independently to DNA but interact synergistically to activate the promoter of Hb9 and so promote motor neuron differentiation. *(From Lee and Pfaff, 2003)*

the sympathetic chain. The anterior to posterior patterning of motor neurons into motor columns is accomplished in response to a gradient of FGF8 (high FGF8 posteriorly to low FGF8 anteriorly) secreted by the paraxial mesoderm. This gradient establishes domains of different *Hox* genes (Dasen et al., 2003). In parwise combinations, anterior *Hox* genes inhibit the expression of the posterior *Hox* genes, and vice versa, so that sharp borders are established. The boundaries between these domains become the boundaries of the different motor columns. We can appreciate that this logic is strikingly similar to that governing the positioning of the motor neurons in the ventral region of the spinal cord. Thus, exposure to gradients in both axes leads to the differential expression of *homeobox* genes that through cross-repression establish sharp borders and different motor columns in the ventral spinal cord (Figure 4.28C).

Motor columns are subdivided into classes that innervate different groups of muscles. Transcription factors of the LIM homeodomain (LIM-HD) family are involved in this further specification, as motor neurons expressing different combinations of LIM-HD factors such as Islet-1, Islet-2, and Lim-3 are expressed in different motor neuron classes that innervate different groups of muscles. This is particularly well-illustrated in the primary motor neurons (Rop, Mip, and Cap) of zebrafish, which can be easily distinguished by their relative rostral to caudal position in each spinal segment as well as their axon trajectories to distinct muscle regions (**Figure 4.31**). These three neurons express different combinations of LIM transcription factors. The identity of these neurons,

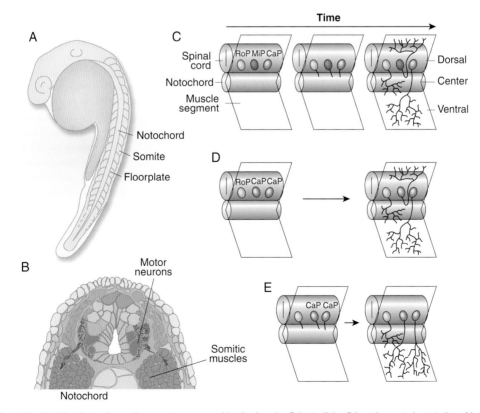

Fig. 4.31 Position determines primary motor neuron identity in zebrafish. A. Zebrafish embryo at about 1 day old. B. Schematic cross section showing the location of the primary motor neuron and the somites that give rise to the axial musculature. C. The rostral (RoP), middle (MiP), and caudal (CaP) primary motor neurons of a single segment develop over a course of about 24 hours. D. If CaP is transplanted to the MiP position before axonogenesis begins, it develops a MiP axonal projection. E. However, if the transplant is done several hours later after the axons have begun to grow, the axonal fates are fixed.

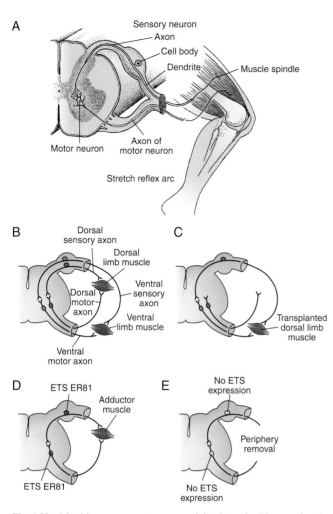

A
Sensory neuron
Axon
Cell body
Dendrite
Muscle spindle
Motor neuron
Axon of
motor neuron
Stretch reflex arc

B
Dorsal
sensory axon
Dorsal
limb muscle
Ventral
sensory
axon
Dorsal
motor
axon
Ventral
limb muscle
Ventral
motor axon

C
Transplanted
dorsal limb
muscle

D
ETS ER81
Adductor
muscle
ETS ER81

E
No ETS
expression
Periphery
removal
No ETS
expression

Fig. 4.32 Matching sensory motor connectivity determined by muscles. A. and B. Spindle afferents terminate on homonymous motor neurons. C. If the sensory fibers that normally innervate ventral muscles are forced to innervate dorsal muscles, they switch synaptic partners to the motor neurons that normally innervate dorsal muscles. D. Different muscles seem to induce or maintain ETS molecules on both the motor and sensory neurons that innervate it. Thus, the motor and sensory neurons that innervate the adductor muscle express the ETS ER81 molecule. E. If the peripheral muscles are removed, the motor and sensory neurons no longer express ETS molecules. *(After Lin et al., 1998)*

as judged by their projection to different muscle regions, is influenced by the position in which they develop. These projection patterns can be changed by transplantation in which the position of these neurons is changed prior to axonongenesis. Transplantations that lead to such switches in projection correlate with induced changes in LIM-HD code, suggesting a causal link.

Motor neurons that innervate specific muscles are grouped into motor pools, which are distinguished by the expression of distinct members of the ETS family of transcription factors. For example, the ETS gene *ER81* is expressed in the motor neurons that innervate the limb adductor muscle in chicks, while the iliotrochanter motor neurons express the ETS gene,

PEA3 (**Figure 4.32**). The initial expression of these ETS genes depends on peripherally derived signals such as GDNF and coincides with the arrival of motor neuron terminals in the vicinity of their specific targets. In PEA3 knockout mice, the motoneurons that are normally PEA3 positive fail to develop normally, so muscles that are normally targets of these neurons are hypoinnervated and become severely atrophic. Not only do these motor neuron pools fail to branch normally within their target muscles, the cell bodies and dendrites of these motor neurons are also mispositioned within the spinal cord which is critical for the control of their reflexive and descending innervation (Figure 4.32). Recent work suggests that these ETS genes regulate the expression of cell adhesion and axon guidance factors that help motor neurons recognize their target. Interestingly, the sensory afferents that innervate the stretch receptors in particular muscles express the same ETS gene as the motor neurons that innervate that muscle, and this helps the sensory neuron axons find the dendrites of these motor neurons, completing the monosynaptic stretch reflex. We will discuss the formation of functional neural circuits in more detail in Chapter 10.

SUMMARY

In this chapter we have looked at neuronal determination in several different systems and have seen common themes such as successive restrictions in potency and potential as progenitor cells develop and divide. There is immense variation in the role of lineage versus environment in neuronal determination, with the general rule that invertebrates are more dominated by lineage mechanisms and less by diffusible signals, while vertebrates are more dominated by diffusible signals and cellular interactions though they also use transcriptional cascades. Each determination pathway brings its own mix of lineage-dependent and lineage-independent mechanisms. Of the transcription factors, bHLH factors of the proneural class help tell cells to become neurons and are antagonized by the Notch pathway, which favors late differentiation of glia. Homeobox and paired domain transcription factors are often used to restrict neurons to certain broad classes linked to their position or coordinates of origin. Finally, POU, LIM, and ETS domain transcription factors may restrict cellular phenotypes even further. These transcription factors often work in interactive circuits, such as those that reinforce decisions by mutual cross-activation, or those that sharpen boundaries by cross-repression, or those that specify fate through combinatorial coding, or those that drive forward progression by a mixture of activation of the next step and repressing the previous one. Of the signaling molecules, we find a particularly important role for the Notch pathway, but also roles for BMPs, FGFs, and Hedgehog proteins. The last phases of determination involve interactions with synaptic targets, which may provide the final differentiative signals for maturing neurons. At the end of this process, the neuron becomes an individual cell with its own biochemical and morphological properties and its unique set of synaptic inputs and outputs.

REFERENCES

Adler, R., & Hatlee, M. (1989). Plasticity and differentiation of embryonic retinal cells after terminal mitosis. *Science, 243*, 391–393.

Agathocleous, M., & Harris, W. A. (2009). From progenitors to differentiated cells in the vertebrate retina. *Annual Review of Cell and Developmental Biology, 25*, 45–69.

Anderson, D. J. (1993). Molecular control of cell fate in the neural crest: the sympathoadrenal lineage. *Annual Review of Neuroscience, 16*, 129–158.

Anderson, D. J., Groves, A., Lo, L., Ma, Q., Rao, M., Shah, N. M., et al. (1997). Cell lineage determination and the control of neuronal identity in the neural crest. *Cold Spring Harbor Symposia on Quantitative Biology, 62*, 493–504.

Arlotta, P., Molyneaux, B. J., Chen, J., Inoue, J., Kominami, R., & Macklis, J. D. (2005). Neuronal subtype-specific genes that control corticospinal motor neuron development in vivo. *Neuron, 45*, 207–221.

Banerjee, U., & Zipursky, S. L. (1990). The role of cell-cell interaction in the development of the Drosophila visual system. *Neuron, 4*, 177–187.

Bartel, D. P. (2009). MicroRNAs: target recognition and regulatory functions. *Cell, 136*, 215–233.

Bhat, K. M. (1999). Segment polarity genes in neuroblast formation and identity specification during Drosophila neurogenesis. *BioEssays, 21*, 472–485.

Brennan, C. A., & Moses, K. (2000). Determination of Drosophila photoreceptors: timing is everything. *Cellular and Molecular Life Sciences, 57*, 195–214.

Britsch, S., Goerich, D. E., Riethmacher, D., Peirano, R. I., Rossner, M., Nave, K. A., et al. (2001). The transcription factor Sox10 is a key regulator of peripheral glial development. *Genes & Development, 15*, 66–78.

Bronner-Fraser, M., & Fraser, S. E. (1988). Cell lineage analysis reveals multipotency of some avian neural crest cells. *Nature, 335*, 161–164.

Bronner-Fraser, M., & Fraser, S. E. (1991). Cell lineage analysis of the avian neural crest. *Development*, (Suppl. 2), 17–22.

Cayouette, M., Barres, B. A., & Raff, M. (2003). Importance of intrinsic mechanisms in cell fate decisions in the developing rat retina. *Neuron, 40*, 897–904.

Cayouette, M., & Raff, M. (2003). The orientation of cell division influences cell-fate choice in the developing mammalian retina. *Development, 130*, 2329–2339.

Cepko, C. L., Austin, C. P., Yang, X., Alexiades, M., & Ezzeddine, D. (1996). Cell fate determination in the vertebrate retina. *Proceedings of the National Academy of Sciences of the United States of America, 93*, 589–595.

Chalfie, M. (1993). Touch receptor development and function in Caenorhabditis elegans. *Journal of Neurobiology, 24*, 1433–1441.

Chalfie, M., & Au, M. (1989). Genetic control of differentiation of the Caenorhabditis elegans touch receptor neurons. *Science, 243*, 1027–1033.

Chalfie, M., & Sulston, J. (1981). Developmental genetics of the mechanosensory neurons of Caenorhabditis elegans. *Developmental Biology, 82*, 358–370.

Chenn, A., & McConnell, S. K. (1995). Cleavage orientation and the asymmetric inheritance of Notch1 immunoreactivity in mammalian neurogenesis. *Cell, 82*, 631–641.

Clark, A. M., Yun, S., Veien, E. S., Wu, Y. Y., Chow, R. L., Dorsky, R. I., et al. (2008). Negative regulation of Vsx1 by its paralog Chx10/Vsx2 is conserved in the vertebrate retina. *Brain Research, 1192*, 99–113.

Cornell, R. A., & Ohlen, T. V. (2000). Vnd/nkx, ind/gsh, and msh/msx: conserved regulators of dorsoventral neural patterning? *Curr Opin Neurobiol, 10*, 63–71.

Dasen, J. S., Liu, J. P., & Jessell, T. M. (2003). Motor neuron columnar fate imposed by sequential phases of Hox-c activity. *Nature, 425*, 926–933.

Desai, A. R., & McConnell, S. K. (2000). Progressive restriction in fate potential by neural progenitors during cerebral cortical development. *Development, 127*, 2863–2872.

Desai, C., Garriga, G., McIntire, S. L., & Horvitz, H. R. (1988). A genetic pathway for the development of the Caenorhabditis elegans HSN motor neurons. *Nature, 336*, 638–646.

Desai, C., & Horvitz, H. R. (1989). Caenorhabditis elegans mutants defective in the functioning of the motor neurons responsible for egg laying. *Genetics, 121*, 703–721.

Dorsky, R. I., Chang, W. S., Rapaport, D. H., & Harris, W. A. (1997). Regulation of neuronal diversity in the Xenopus retina by Delta signalling. *Nature, 385*, 67–70.

Dorsky, R. I., Moon, R. T., & Raible, D. W. (2000). Environmental signals and cell fate specification in premigratory neural crest. *BioEssays, 22*, 708–716.

Edlund, T., & Jessell, T. M. (1999). Progression from extrinsic to intrinsic signaling in cell fate specification: a view from the nervous system. *Cell, 96*, 211–224.

Elliott, J., Jolicoeur, C., Ramamurthy, V., & Cayouette, M. (2008). Ikaros confers early temporal competence to mouse retinal progenitor cells. *Neuron, 60*, 26–39.

Ernstrom, G. G., & Chalfie, M. (2002). Genetics of sensory mechanotransduction. *Annual Review of Genetics, 36*, 411–453.

Francis, N. J., & Landis, S. C. (1999). Cellular and molecular determinants of sympathetic neuron development. *Annual Review of Neuroscience, 22*, 541–566.

Freeman, M. (1997). Cell determination strategies in the Drosophila eye. *Development, 124*, 261–270.

Goridis, C., & Rohrer, H. (2002). Specification of catecholaminergic and serotonergic neurons. *Nature Reviews. Neuroscience, 3*, 531–541.

Grosskortenhaus, R., Robinson, K. J., & Doe, C. Q. (2006). Pdm and Castor specify late-born motor neuron identity in the NB7–1 lineage. *Genes & Development, 20*, 2618–2627.

Groves, A. K., & Anderson, D. J. (1996). Role of environmental signals and transcriptional regulators in neural crest development. *Developmental Genetics, 18*, 64–72.

Guo, M., Jan, L. Y., & Jan, Y. N. (1996). Control of daughter cell fates during asymmetric division: interaction of Numb and Notch. *Neuron, 17*, 27–41.

Hafen, E., Dickson, B., Brunner, D., & Raabe, T. (1994). Genetic dissection of signal transduction mediated by the sevenless receptor tyrosine kinase in Drosophila. *Progress in Neurobiology, 42*, 287–292.

Harris, W., Stark, W., & Walker, J. (1976). Genetic dissection of the photoreceptor system in the compound eye of Drosophila melanogaster. *The Journal of Physiology, 256*, 415–439.

Hartenstein, V. (1989). Early neurogenesis in Xenopus: the spatio-temporal pattern of proliferation and cell lineages in the embryonic spinal cord. *Neuron, 3*, 399–411.

Hatakeyama, J., & Kageyama, R. (2004). Retinal cell fate determination and bHLH factors. *Semin Cell Developmental Biology, 15*, 83–89.

Hatakeyama, J., Tomita, K., Inoue, T., & Kageyama, R. (2001). Roles of homeobox and bHLH genes in specification of a retinal cell type. *Development, 128*, 1313–1322.

Hobert, O. (2005). *The C. elegans Research Community*. WormBook. doi:10.1895/wormbook.1.12.1 WormBook *www.wormbook.org*.

Hobert, O., Johnston, R. J., Jr., & Chang, S. (2002). Left-right asymmetry in the nervous system: the Caenorhabditis elegans model. *Nature Reviews. Neuroscience, 3*, 629–640.

Holt, C. E., Bertsch, T. W., Ellis, H. M., & Harris, W. A. (1988). Cellular determination in the Xenopus retina is independent of lineage and birth date. *Neuron, 1*, 15–26.

Huttner, W. B., & Kosodo, Y. (2005). Symmetric versus asymmetric cell division during neurogenesis in the developing vertebrate central nervous system. *Current Opinion in Cell Biology, 17*, 648–657.

Isshiki, T., Pearson, B., Holbrook, S., & Doe, C. Q. (2001). Drosophila neuroblasts sequentially express transcription factors which specify the temporal identity of their neuronal progeny. *Cell, 106*, 511–521.

Jessell, T. M. (2000). Neuronal specification in the spinal cord: inductive signals and transcriptional codes. *Nature Reviews. Genetics, 1*, 20–29.

Jessell, T. M., Bovolenta, P., Placzek, M., Tessier-Lavigne, M., & Dodd, J. (1989). Polarity and patterning in the neural tube: the origin and function of the floor plate. *Ciba Foundation Symposium, 144*, 255–276; discussion 276–280, 290–255.

Johnston, R. J., & Hobert, O. (2003). A microRNA controlling left/right neuronal asymmetry in Caenorhabditis elegans. *Nature, 426*, 845–849.

Johnston, R. J., Jr., Chang, S., Etchberger, J. F., Ortiz, C. O., & Hobert, O. (2005). MicroRNAs acting in a double-negative feedback loop to control a neuronal cell fate decision. *Proceedings of the National Academy of Sciences of the United States of America, 102*, 12449–12454.

Kaltschmidt, J. A., & Brand, A. H. (2002). Asymmetric cell division: microtubule dynamics and spindle asymmetry. *Journal of Cell Science, 115*, 2257–2264.

Kessaris, N., Pringle, N., & Richardson, W. D. (2001). Ventral neurogenesis and the neuron-glial switch. *Neuron, 31*, 677–680.

Kim, J., Wu, H. H., Lander, A. D., Lyons, K. M., Matzuk, M. M., & Calof, A. L. (2005). GDF11 controls the timing of progenitor cell competence in developing retina. *Science, 308*, 1927–1930.

Konno, D., Shioi, G., Shitamukai, A., Mori, A., Kiyonari, H., Miyata, T., et al. (2008). Neuroepithelial progenitors undergo LGN-dependent planar divisions to maintain self-renewability during mammalian neurogenesis. *Nature Cell Biology, 10*, 93–101.

Kumar, J. P., & Moses, K. (2000). Cell fate specification in the Drosophila retina. *Results and Problems in Cell Differentiation, 31*, 93–114.

Landis, S. C. (1992). Cellular and molecular mechanisms determining neurotransmitter phenotypes in sympathetic neurons. In M. Shanland, E. Macagno (Eds), Determinants of Neural Identity (pp. 497–523). San Diego: Academic Press.

Le Douarin, N. (1982). *The Neural Crest*. New York: Cambridge University Press.

Le Douarin, N. M., Renaud, D., Teillet, M. A., & Le Douarin, G. H. (1975). Cholinergic differentiation of presumptive adrenergic neuroblasts in interspecific chimeras after heterotopic transplantations. *Proceedings of the National Academy of Sciences of the United States of America, 72*, 728–732.

Lee, H. Y., Kleber, M., Hari, L., Brault, V., Suter, U., Taketo, M. M., et al. (2004a). Instructive role of Wnt/beta-catenin in sensory fate specification in neural crest stem cells. *Science, 303*, 1020–1023.

Lee, S. K., Jurata, L. W., Funahashi, J., Ruiz, E. C., & Pfaff, S. L. (2004b). Analysis of embryonic motoneuron gene regulation: derepression of general

103

activators function in concert with enhancer factors. *Development, 131,* 3295–3306.

Lee, S. K., & Pfaff, S. L. (2003). Synchronization of neurogenesis and motor neuron specification by direct coupling of bHLH and homeodomain transcription factors. *Neuron, 38,* 731–745.

Leimeroth, R., Lobsiger, C., Lussi, A., Taylor, V., Suter, U., & Sommer, L. (2002). Membrane-bound neuregulin1 type III actively promotes Schwann cell differentiation of multipotent Progenitor cells. *Developmental Biology, 246,* 245–258.

Lin, J. H., Saito, T., Anderson, D. J., Lance-Jones, C., Jessell, T. M., & Arber, S. (1998). Functionally related motor neuron pool and muscle sensory afferent subtypes defined by coordinate ETS gene expression. *Cell, 95,* 393–407.

Liu, W., Mo, Z., & Xiang, M. (2001). The Ath5 proneural genes function upstream of Brn3 POU domain transcription factor genes to promote retinal ganglion cell development. *Proceedings of the National Academy of Sciences of the United States of America, 98,* 1649–1654.

Livesey, F. J., & Cepko, C. L. (2001). Vertebrate neural cell-fate determination: lessons from the retina. *Nature Reviews. Neuroscience, 2,* 109–118.

Lo, L., Dormand, E., Greenwood, A., & Anderson, D. J. (2002). Comparison of the generic neuronal differentiation and neuron subtype specification functions of mammalian achaete-scute and atonal homologs in cultured neural progenitor cells. *Development, 129,* 1553–1567.

Logan, M. A., Steele, M. R., Van Raay, T. J., & Vetter, M. L. (2005). Identification of shared transcriptional targets for the proneural bHLH factors Xath5 and XNeuroD. *Developmental Biology, 285,* 570–583.

Matter-Sadzinski, L., Puzianowska-Kuznicka, M., Hernandez, J., Ballivet, M., & Matter, J. M. (2005). A bHLH transcriptional network regulating the specification of retinal ganglion cells. *Development, 132,* 3907–3921.

McConnell, S. K. (1988). Fates of visual cortical neurons in the ferret after isochronic and heterochronic transplantation. *The Journal of Neuroscience, 8,* 945–974.

McConnell, S. K. (1995). Strategies for the generation of neuronal diversity in the developing central nervous system. *The Journal of Neuroscience, 15,* 6987–6998.

McDonald, J. A., & Doe, C. Q. (1997). Establishing neuroblast-specific gene expression in the Drosophila CNS: huckebein is activated by Wingless and Hedgehog and repressed by Engrailed and Gooseberry. *Development, 124,* 1079–1087.

Molyneaux, B. J., Arlotta, P., Menezes, J. R., & Macklis, J. D. (2007). Neuronal subtype specification in the cerebral cortex. *Nature Reviews. Neuroscience, 8,* 427–437.

Mu, X., Fu, X., Sun, H., Beremand, P. D., Thomas, T. L., & Klein, W. H. (2005). A gene network downstream of transcription factor Math5 regulates retinal progenitor cell competence and ganglion cell fate. *Developmental Biology, 280,* 467–481.

Muhr, J., Andersson, E., Persson, M., Jessell, T. M., & Ericson, J. (2001). Groucho-mediated transcriptional repression establishes progenitor cell pattern and

neuronal fate in the ventral neural tube. *Cell, 104,* 861–873.

Noctor, S. C., Martinez-Cerdeno, V., & Kriegstein, A. R. (2008). Distinct behaviors of neural stem and progenitor cells underlie cortical neurogenesis. *The Journal of Comparative Neurology, 508,* 28–44.

Oh, E. C., Khan, N., Novelli, E., Khanna, H., Strettoi, E., & Swaroop, A. (2007). Transformation of cone precursors to functional rod photoreceptors by bZIP transcription factor NRL. *Proceedings of the National Academy of Sciences of the United States of America, 104,* 1679–1684.

Ohnuma, S., Philpott, A., Wang, K., Holt, C. E., & Harris, W. A. (1999). p27Xic1, a Cdk inhibitor, promotes the determination of glial cells in Xenopus retina. *Cell, 99,* 499–510.

Pattyn, A., Morin, X., Cremer, H., Goridis, C., & Brunet, J. F. (1999). The homeobox gene Phox2b is essential for the development of autonomic neural crest derivatives. *Nature, 399,* 366–370.

Perron, M., & Harris, W. A. (2000). Determination of vertebrate retinal progenitor cell fate by the Notch pathway and basic helix-loop-helix transcription factors. *Cellular and Molecular Life Sciences, 57,* 215–223.

Poggi, L., Vitorino, M., Masai, I., & Harris, W. A. (2005). Influences on neural lineage and mode of division in the zebrafish retina in vivo. *The Journal of Cell Biology, 171,* 991–999.

Ready, D. F. (1989). A multifaceted approach to neural development. *Trends in Neurosciences, 12,* 102–110.

Ready, D. F., Hanson, T. E., & Benzer, S. (1976). Development of the Drosophila retina, a neurocrystalline lattice. *Developmental Biology, 53,* 217–240.

Reh, T. A., & Kljavin, I. J. (1989). Age of differentiation determines rat retinal germinal cell phenotype: induction of differentiation by dissociation. *The Journal of Neuroscience, 9,* 4179–4189.

Reinke, R., & Zipursky, S. L. (1988). Cell-cell interaction in the Drosophila retina: the bride of sevenless gene is required in photoreceptor cell R8 for R7 cell development. *Cell, 55,* 321–330.

Reissmann, E., Ernsberger, U., Francis-West, P. H., Rueger, D., Brickell, P. M., & Rohrer, H. (1996). Involvement of bone morphogenetic protein-4 and bone morphogenetic protein-7 in the differentiation of the adrenergic phenotype in developing sympathetic neurons. *Development, 122,* 2079–2088.

Rubin, G. M. (1991). Signal transduction and the fate of the R7 photoreceptor in Drosophila. *Trends in Genetics, 7,* 372–377.

Sarin, S., O'Meara, M. M., Flowers, E. B., Antonio, C., Poole, R. J., Didiano, D., et al. (2007). Genetic screens for Caenorhabditis elegans mutants defective in left/right asymmetric neuronal fate specification. *Genetics, 176,* 2109–2130.

Schneider, C., Wicht, H., Enderich, J., Wegner, M., & Rohrer, H. (1999). Bone morphogenetic proteins are required in vivo for the generation of sympathetic neurons. *Neuron, 24,* 861–870.

Schweisguth, F., Gho, M., & Lecourtois, M. (1996). Control of cell fate choices by lateral signaling in the adult peripheral nervous system of Drosophila melanogaster. *Developmental Genetics, 18,* 28–39.

Shen, Q., Wang, Y., Dimos, J. T., Fasano, C. A., Phoenix, T. N., Lemischka, I. R., et al. (2006). The timing of cortical neurogenesis is encoded within lineages of individual progenitor cells. *Nature Neuroscience, 9,* 743–751.

Sidman, R. (1961). Histogenesis of the mouse retina studied with thymidine 3H. In G. Smelse (Ed.), *The structure of the eye* (pp. 487–506). New York: Academic Press.

Siegrist, S. E., & Doe, C. Q. (2006). Extrinsic cues orient the cell division axis in Drosophila embryonic neuroblasts. *Development, 133,* 529–536.

Sulston, J. E., Schierenberg, E., White, J. G., & Thomson, J. N. (1983). The embryonic cell lineage of the nematode Caenorhabditis elegans. *Dev Biol, 100,* 64–119.

Turner, D. L., & Cepko, C. L. (1987). A common progenitor for neurons and glia persists in rat retina late in development. *Nature, 328,* 131–136.

Vitorino, M., Jusuf, P. R., Maurus, D., Kimura, Y., Higashijima, S., & Harris, W. A. (2009). Vsx2 in the zebrafish retina: restricted lineages through derepression. *Neural Dev, 4,* 14.

Waid, D. K., & McLoon, S. C. (1998). Ganglion cells influence the fate of dividing retinal cells in culture. *Development, 125,* 1059–1066.

Wallace, V. A. (2008). Proliferative and cell fate effects of Hedgehog signaling in the vertebrate retina. *Brain Research, 1192,* 61–75.

Wang, J. C., & Harris, W. A. (2005). The role of combinational coding by homeodomain and bHLH transcription factors in retinal cell fate specification. *Developmental Biology, 285,* 101–115.

Watanabe, T., & Raff, M. C. (1990). Rod photoreceptor development in vitro: intrinsic properties of proliferating neuroepithelial cells change as development proceeds in the rat retina. *Neuron, 4,* 461–467.

Wetts, R., & Fraser, S. E. (1988). Multipotent precursors can give rise to all major cell types of the frog retina. *Science, 239,* 1142–1145.

Wirtz-Peitz, F., Nishimura, T., & Knoblich, J. A. (2008). Linking cell cycle to asymmetric division: Aurora-A phosphorylates the Par complex to regulate Numb localization. *Cell, 135,* 161–173.

Zhang, Y., Ma, C., Delohery, T., Nasipak, B., Foat, B. C., Bounoutas, A., et al. (2002). Identification of genes expressed in C. elegans touch receptor neurons. *Nature, 418,* 331–335.

Zhou, Q., Choi, G., & Anderson, D. J. (2001). The bHLH transcription factor Olig2 promotes oligodendrocyte differentiation in collaboration with Nkx2.2. *Neuron, 31,* 791–807.

Zigman, M., Cayouette, M., Charalambous, C., Schleiffer, A., Hoeller, O., Dunican, D., et al. (2005). Mammalian inscuteable regulates spindle orientation and cell fate in the developing retina. *Neuron, 48,* 539–545.

Zirlinger, M., Lo, L., McMahon, J., McMahon, A. P., & Anderson, D. J. (2002). Transient expression of the bHLH factor neurogenin-2 marks a subpopulation of neural crest cells biased for a sensory but not a neuronal fate. *Proceedings of the National Academy of Sciences of the United States of America, 99,* 8084–8089.

Axon growth and guidance

5

Newborn neurons send out single threadlike axons that carry information to target cells and they also send out a set of multiply branched dendritic processes that receive inputs from other neurons. Local interneurons have short axons and make connections to cells in their immediate vicinity, while projection neurons send long axons to distant targets. There is tremendous divergence and convergence of wiring in the brain. On the sensory side, peripheral neurons send axons into the CNS where they usually diverge to project to several distinct targets. Each of these targets contains neurons that also diverge to various targets of their own, and so on. Tracing pathways from the motor side backwards yields a similar complexity in convergence, with each motor neuron being innervated by many presynaptic neurons, and each of these having its own multitude of inputs. Thus, with thousands of target nuclei and billions of axons, the interweaving of axonal pathways is a remarkable and complex tapestry of axonal and dendritic processes. This complexity can be appreciated by looking at almost any region of the brain microscopically with a technique that labels individual axons (**Figure 5.1**). When looking at such images of an adult brain, it is difficult to imagine how the precise patterns of connections were ever made. However, by looking at early embryonic brains, there is the possibility

of seeing the very first axons and dendrites arising at a stage in development when the connectivity was much less complicated (Wilson et al., 1990; Ross et al., 1992; Easter et al., 1994) (**Figure 5.2**). These pioneer axons navigate in a simpler environment. But as the brain matures, more axons are added, and the weave becomes more intricate. The navigation of later axons is aided by trails blazed by earlier pioneers, but additions of layers upon layers of new fibers add complexity stage by stage, and the three-dimensional neuritic meshwork of brain is thus knitted thread by thread.

How do the first pioneer axons find their way? A good example of pioneer axonal navigation is that of the sensory neurons that arise in the distal part of a grasshopper leg (Keshishian and Bentley, 1983a, b, c). These Ti cells pioneer the tract that later developing sensory axons will follow to their targets in the central nervous system. So, how do the Tis find their way? Part of the answer is that the Tis use local cues on their journey. Spaced at short distances from one another are a series of spaced "guidepost" cells for which the Ti growth cones show a particular affinity (Caudy and Bentley, 1986). The distances between guidepost cells are small enough that a growth cone can reach out to a new guidepost while still contacting the previous one. Ti axons thus use these guidepost cells as stepping-stones into the CNS (**Figure 5.3**). Some of the guideposts are clearly critical for pathfinding because when they are ablated with a laser microbeam, the Ti axons get stuck and are unable to make it from one segment of the journey to the next (Bentley and Caudy, 1983).

Another important insight into axonal navigation came from a simple experiment performed by Hibbard (Hibbard, 1965) on another class of early arising neuron called Mauthner neurons in the hindbrain of salamander. He rotated a piece of the embryonic salamander hindbrain that contained the pair of giant Mauthner neurons before these cells had begun to send out their axons. Mauthner neurons are responsible for rapid escape response, and they send large diameter axons, easily visible with silver staining, caudally down the spinal cord. In the rotated piece of hindbrain, Hibbard saw the axons of Mauthner cells initially grew rostrally instead of caudally, as though they were guided by local cues within the transplant. However, when the axons of these rotated neurons reached the rostral boundary of the transplanted tissue and entered the rostral unrotated territory, they made dramatic U-turns and headed caudally down toward the spinal cord. This proved that axon navigation relies on cues provided by the external environment and is not just an intrinsic program of directions (**Figure 5.4**). The growth and orientation of dendrites is also guided by such local cues. In this

105

Fig. 5.1 Confocal image of axons and synapses in the hindbrain of an adult brainbow mouse. The brainbow technique is an ingenious genetic recombination strategy using a palette of genetically engineered fluorescent proteins that can be used to randomly label neurons in the brain with myriad colors *(Livet et al., 2007).*

chapter we will learn about these guidance cues and how they are used by axons and dendrites, but it is also worth pointing out that the same evolutionarily conserved molecules are used by a variety of migrating cells, such as cells in the neural crest and in the developing vasculature of the body, which may explain to a large extent why blood vessels and peripheral nerve fibers often grow close to each other along parallel pathways.

The axon of a projection neuron makes a journey to connect to its distant target much like a driver makes a journey from a particular address in one populated city to another address in another city (**Figure 5.5**). As he or she pulls out from home on a particular street, the driver knows which turns to make, which roads to get on, the signs to follow, what to avoid, which exits to take, and how to recognize and stop at the correct destination. Unlike humans, most axons make these long journeys without errors. Growing axons are able to recognize various molecules on the surfaces of other axons and cells, and to use these molecules as cues to navigate the sometimes circuitous pathways to their particular destinations. They can also respond to diffusible molecules such as morphogens (see Chapter 2) that percolate through the embryonic brain and provide cues about overall orientation. Intermediate targets are important for drivers and growth cones. A trip may involve finding a particular motel or restaurant en route to a final destination. Once this is found, new maps are brought out and the next phase of the journey is planned. Similarly, axons may dramatically modulate responses to cues as these intermediate targets are reached and passed.

Axons, like cars, also need motor abilities, as they must be able to move forward, make turns, and put on the brakes when they reach their target. They may also need to integrate information, for cues that have a particular significance during an early phase of their journey may have a different significance later on in the context of other signals, or they may need to adapt their responsiveness so that they remain sensitive as background levels of particular guidance cues change. These functions, the sensory, the motor, the integrative and the adaptive, are all contained within the specialized tip of a growing axon, called the growth cone. The fact that growth cones are autonomously capable of all these functions is demonstrated in experiments in which growth cones surgically isolated from their cell bodies continue to grow (Bray et al., 1978) and even navigate correctly (Harris et al., 1987) (**Figure 5.6**).

THE GROWTH CONE

In 1890, the Spanish microscopist, Ramón y Cajal, studying the nervous system of fixed embryos, saw what he referred to as "concentrations of protoplasm of conical form" and from their various shapes he considered them to be "endowed with amoeboid movements." This was the first description

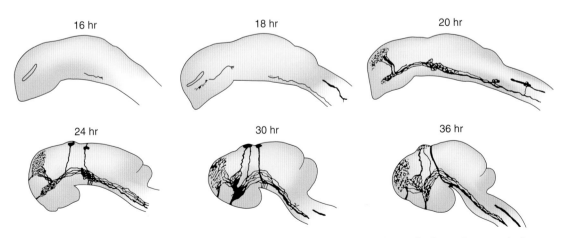

Fig. 5.2 The increasing complexity of fiber tracts in the developing vertebrate brain. Antibodies against axons in the embryonic zebrafish brain at successive stages of development over the course of just 20 hours reveal that a variety of new axons are added at each stage. *(After Wilson et al., 1990; Ross et al., 1992)*

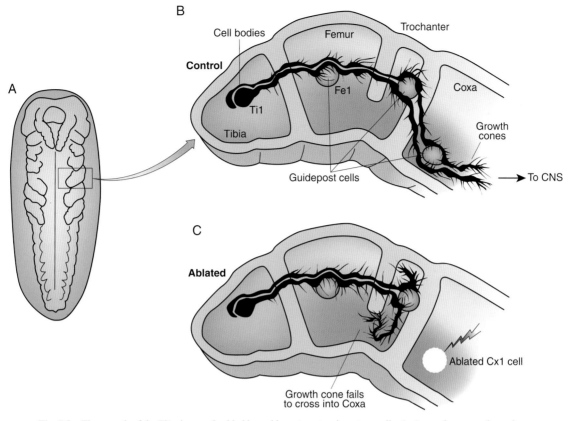

Fig. 5.3 The growth of the Ti1 pioneers is aided by guidepost or stepping-stone cells. A. A grasshopper embryo showing the developing legs. B. The Ti1 pioneers in black reach from one of the guidepost neurons to the next, successively contacting Fe1, Tr1, and Cx1 on their way to the CNS. C. When the Cx1 cells are ablated, the Ti1 cells lose their way and do not cross into the coxal segment of the embryonic leg. *(After Bentley and Caudy, 1983)*

of growth cones and the first insight into their function. He imagined the growth cone as a sort of "battering ram" that extending axons used to punch their way through the packed cells of the embryonic brain (Ramón y Cajal, 1890) (**Figure 5.7**A). Several years later, the American embryologist Ross Harrison invented a way to look at live tissue in a microscope using culture techniques. When he put pieces of embryonic neural tube into culture (See **Box 5.1**), he was amazed to see live growth cones crawling across a microscope slide, moving and changing their form in real time (Harrison, 1910). They seemed to feel their way along the surface, sending out long, thin filopodia and forming veils between the filopodia (Figure 5.7B). Speidel was the first to observe live growth cones *in vivo* at the ends of growing sensory axons in the transparent growing tail fin of a frog (Speidel, 1941). He followed single-growth cones over days, even weeks, and he watched how they responded to obstacles and injuries. He was impressed with the ability of these growth cones to change directions, branch, and respond to stimulation. Speidel described the occasional growth rates as high as 40 µm/hr (Figure 5.7C). Such growth cones *in vivo* advanced in an amoeboid fashion and showed a number of delicate transient processes, consistent with Harrison's earlier *in vitro* observations.

Growth cones assume several morphologies and may travel at different speeds as they navigate through different parts of their pathways (Tosney and Landmesser, 1985; Bovolenta and Mason, 1987). The growth cones of pioneer axons that are growing straight ahead have several active filopodia and

a few lamellipodia. The growth cones of follower axons that grow along earlier pioneer axons tend to be more simple, bullet-shaped with few filopodia. Growth cones get particularly complex when they arrive at choice points along the pathway such as the midline. A good example of this is at a commissure when axon tracts from each side of the brain meet, and the axons that were pioneers on one side become followers on the other. Using time-lapse imaging in zebrafish embryos, this transition can be appreciated (Bak and Fraser, 2003) (**Figure 5.8**). Static observations from fixed tissue provide neither the rate of growth nor the sampling strategy that growth cones employ as they make decisions. While Cajal's remarkable visual memory and artistry allowed him to produce accurate drawings of growth cones nearly a century ago, simple video microscopy systems now permit most laboratories to view and measure growth cones in the act of navigating and making decisions.

Low light video cameras, scanning laser illumination, and specialized image processing software coupled with the recent introduction of fluorescent labels that are rapidly transported along living axons, enable researchers to produce time-lapse movies of process outgrowth and innervation (Glover et al., 1986; Honig and Hume, 1986; Harris et al., 1987). Vital fluorescent dyes, such as the lipophilic dyes, DiI and DiO, intercalate into the neuron membrane and diffuse rapidly down axonal processes; they are then visualized with epifluorescent illumination. Fluorescent proteins of different colors such as Green Fluorescent Protein, (GFP), RFP (Red),

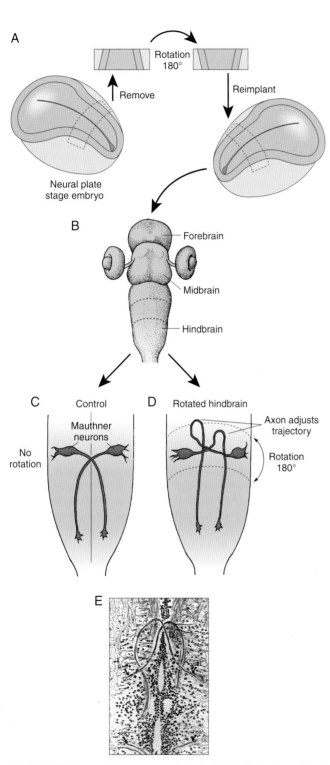

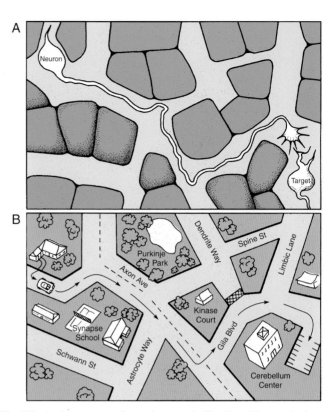

Fig. 5.5 A. An axon growing to its target is like B. a driver navigating through city streets. See text for details.

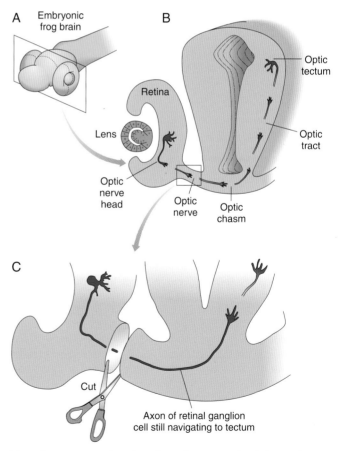

Fig. 5.6 Axons grow from the retina to the tectum using their growth cones to guide them. A. A dorso-lateral view of the embryonic frog brain. B. Images of single retinal ganglion cell axons through the plane of section indicated in (A) shows that as they grow to the tectum, they are always tipped by active growth cones. C. When the axon is separated from the cell body by cutting the optic stalk, time-lapse imaging shows that isolated growth cones still grow along the correct pathway. *(After Harris et al., 1987)*

Fig. 5.4 Mauthner cells grow posteriorly in the hindbrain due to local cues. A. At the neural plate stage, a segment of the hindbrain region of a salamander embryo is removed, rotated 180°, and reimplanted. B. shows a dorsal view of a larval brain of such an animal. C. The bilaterally symmetric giant Mauthner neurons in the normal unoperated larval brain. D. The trajectory of Mauthner axons in an experimental animal in which a segment of the hindbrain containing the Mauthner primordia is rotated. E. Photo of host and graft Mauthner cell axons in same animal. *(After Hibbard, 1965)*

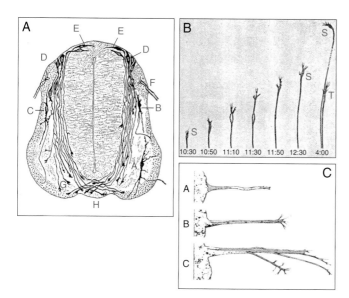

Fig. 5.7 Early observations of growth cones. A. In the late 1800s, Ramón y Cajal saw expansions of axons near the ventral midline of the chick neural tube. B. In the 1930s, Speidel observed growing nerve fibers tipped with axons in the frog tail. C. In the early 1900s, Harrison grew neural explants in culture and watched them extend axons, or leader axons of the opposite side, tipped with motile growth cones. *(From Ramón y Cajal, 1890; Harrison, 1910; Speidel, 1941)*

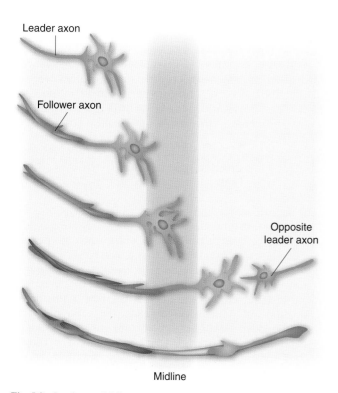

Fig. 5.8 Leaders and followers. A schematic drawing shows a leader axon (pink) and a number of follower axons (blue and black) growing through the midline. The leading axon, being the first, is completely exposed to the guidance cues in the environment. Its growth cone must sense all the positive and negative midline cues and interpret them accordingly, which results in slow progress and complex morphology of leader growth cones at the midline where these cues are found. By growing along the leader, follower axons, or leader axons of the opposite side, are less exposed to midline cues, have more simply shaped growth cones, and grow faster. *(From Bak and Fraser, 2003)*

and YFP (Yellow), are engineered variants of natural proteins found in certain types of jellyfish and corals. When a growing neuron is transfected with such a GFP gene, it creates its own fluorescent protein. By genetically engineering chimeric genes, combining GFP with a protein potentially involved in axon growth, one can see what particular molecules are doing in the growth cone and test the effect of misexpressing mutant versions of the protein on growth and navigation. By putting the GFP gene under the control of a promoter that is active in a subset of developing nerve cells in a transgenic animal, it is possible to monitor a whole class of growing axons. One problem, often encountered in thick specimens, is the excessive level of out-of-focus fluorescence. Various forms of microscopy, such as deconvolution or confocal microscopy, can overcome this obstacle and allow one to build sharp images of three dimensional objects such as growth cones even deep in the developing nervous system. Using a combination of these techniques, it has become possible to watch growth cones and key molecules within them as axons grow out and innervate their targets in the CNS of live embryos (**Figure 5.9**).

As growth cones crawl forward, they leave axons behind. This means that new material must be continually incorporated into the elongating axon. In culture, when a particle is attached to a growing axon, it remains relatively stationary compared to the distal tip of the axon. Thus new material is assembled distally in the growing axon, that is near the growth cone. Indeed, new glycoproteins are added preferentially at the distal tips of axons in the growth cone region (Hollenbeck and Bray, 1987), and the incorporation of this new material is calcium dependent, suggesting that new membrane is added by calcium-mediated fusion of internal vesicles to the growth cone's surface. The addition of cytoskeletal components such as actin filaments and microtubules also takes place at the tips of growing axons. This was shown by labeling neurons with fluorescent tubulin and actin, and then illuminating part of the axon with a bright spot to bleach the fluorescence at a particular location (**Figure 5.10**). The result is that the bleached spot stays relatively still as the growth cone continues to advance, suggesting that these components are assembled distally. If one looks carefully at the middle of the axon, one can see cargo moving up and down the axon. Some of the cargo heading out is destined for incorporation at the distal tips. While

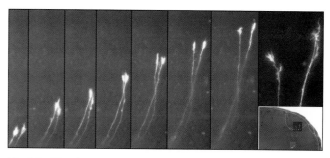

Fig. 5.9 Time-lapse images of two GFP-expressing retinal axons growing in the optic tract of a *Xenopus* brain. The images on the left are successive time points spaced about 20 minutes apart showing the two elongating axons, tipped with growth cones, rearranging their relative positions in the optic tract. The image in the top right shows the initial branching of the same two axons in the tectum, the target structure for these axons. The image at the bottom right is a low-power view of the preparation at the beginning and at the end of the time-lapse. *(Courtesy of Sonia Witte and Christine Holt)*

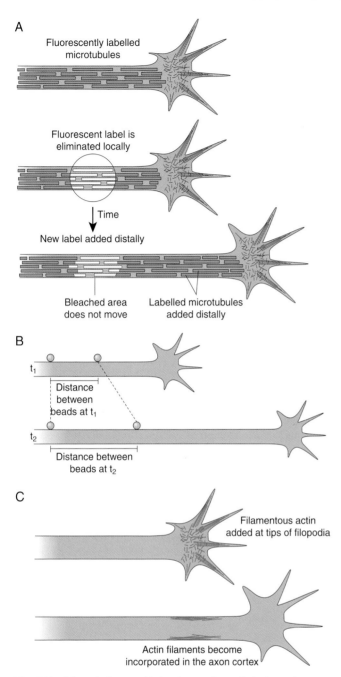

A

Fluorescently labelled microtubules

Fluorescent label is eliminated locally

↓ Time

New label added distally

Bleached area does not move

Labelled microtubules added distally

B

t_1

Distance between beads at t_1

t_2

Distance between beads at t_2

C

Filamentous actin added at tips of filopodia

Actin filaments become incorporated in the axon cortex

Fig. 5.10 Microtubules are added at the growing end. A. A growing axon is labeled with fluorescent tubulin, and then some of this fluorescence is bleached by a beam of light (circle) focused on the axon near the growth cone. As the axon elongates distally, the bleached spot stays in approximately the same place (bottom panel), implying that the microtubules along the axon shaft do not move forward but rather that new microtubules are assembled at the distal tip. B. Two beads placed on an axon move further apart from each other as the axon grows, with the front bead moving further forward than the rear bead. C. Filamentous actin is assembled in the filopodia of the growth cone, and some of this is left behind in the submembranous cortex of the axon as part of the axon's cytoskeleton.

most growth happens in the growth cone region, it is also true that if two beads are attached along the shaft of the axon, one can notice a slowly increasing separation between these two beads, suggesting that there is some interstitial growth in the axon, which is crucial as the brain or body enlarges after initial connections are made.

THE DYNAMIC CYTOSKELETON

If we want to know how a growth cone navigates, it might help to understand how it moves, and this means looking into its dynamic cytoskeleton. The cytoskeleton of a growth cone is filled with molecules that are involved in cell movements (Heidemann, 1996; Letourneau, 1996; Dent and Gertler, 2003; Gordon-Weeks, 2004). The two most important elements are the microtubules that extend along the axon and splay out in the central domain, C-domain, of the growth cone, and the actin fibers, which are more prominent in the peripheral, P-domain (**Figure 5.11**). In the P-domain, actin forms the basic cytoskeleton of the lamellipodia, the veils between filopodia, and the filopodia themselves, which are filled with thick bundles of filamentous F-actin. Actin filaments are assembled by the addition of G-actin momoners at the tips of the filopodia. Many other cytoskeletal-associated proteins in the growth cones do a variety of jobs such as anchoring actin and microtubules to the cell membrane, to each other, and to other cytoskeletal components including the molecular motors that generate force (**Figure 5.12**).

Drugs such as the actin-depolymerizing agent, cytochalasin, have been used to investigate the functions of actin in the growth cone. Treatment of growth cones with cytochalsin inhibits filopodia formation, and treated growth cones slow down dramatically, showing that actin-rich fibers are important for the forward progress of a growth cone. At low doses of cytochalasin that do not completely block actin polymerization, growth cones have few or no filopodia and display a narrow rather than expanded morphologies. Such growth cones often lose their way in the developing organism. Thus, when the Ti1 pioneer neuron in the grasshopper limb is treated with cytochalasin, its axon meanders off course and often does not make the turns necessary to grow to its targets in the CNS (Bentley and Toroian-Raymond, 1986) (**Figure 5.13**A). In the amphibian brain, retinal axons must make a posterior turn in the diencephalon to get to their targets in the midbrain. If these growing axons are treated with a low dose of cytochalasin, they grow past the turning point (Chien et al., 1993) (Figure 5.13B). Thus, actin filaments are critical for growth cone navigation.

The actin filaments in filopodia are bundled and oriented so that their fast-growing barbed or plus ends are pointing away from the growth cone center out toward the periphery. Time-lapse observations in tissue culture show that, as the growth cone advances, the actin bundles in the filopodia move rearward at almost exactly the same rate as they elogate distally through the addition of new actin monomers at the tip. At their minus ends, the actin filaments in the filopodia are disassembled into monomers which can cycle up to the front and be used again. The filopodia thus remain pretty much the same length although they are constantly flowing rearward like the tread of a tank (**Figure 5.14**) (Lin et al., 1996). The clutch hypothesis (Mitchison and Kirschner, 1988) suggests that adhesion complexes composed of transmembrane receptors interact intracellularly with actin binding molecules so that actin filaments themselves become anchored to the substrate. This engages the clutch at the growing front. The disassembly of actin filaments at the base of the filopodium, releases the clutch there. The force that pulls the growing filopodia rearward is generated by myosin molecules located at the base

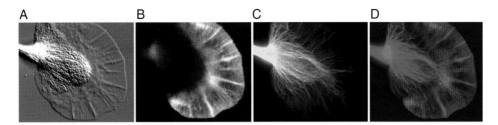

Fig. 5.11 Views of an *Aplysia* growth cone. A. Nomarski image showing the growth cone. The bulging central domain and the thin peripheral domain containing actin cables are visible. B. Labeling the actin filaments with a fluorescent probe reveals they are concentrated in the peripheral domain and the filopodia. C. Labeling the microtubules reveals that these structures are in the central domain. D. Pseudo-colored merged image of actin filaments (red) and microtubules (green). *(From Paul Forscher)*

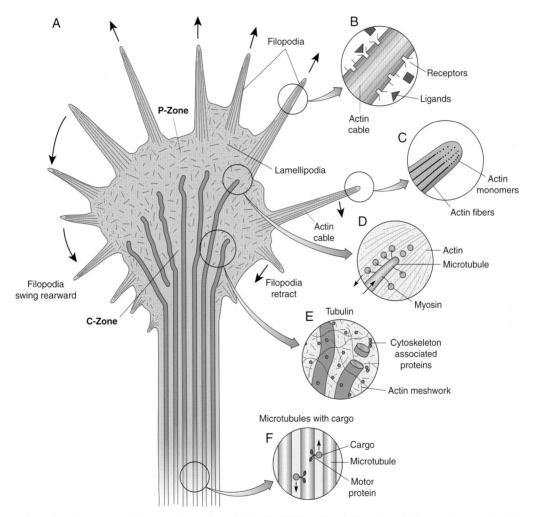

Fig. 5.12 The structure of the growth cone. A. Actin bundles fill filopodia, which are bounded by membranes with cell adhesion molecules and various receptors, poke out at the advancing edge, and are retracted at the trailing edge of the growth cone. Between the filopodia are sheets of lamellipodia that extend forward. They are filled with an actin meshwork that is continuous with that in the main body of the growth cone. Here also microtubules push forward and carry cargo to and from the cell body along the axon shaft as they enter the growth cone and fan out toward the filopodia. B–F. Close-ups of various regions show some of the molecular components of the cytoskeletal network that are localized in the growth cone.

of filopodia (**Figure 5.15**). Indeed, when myosin function in growth cones is blocked, forward progress is slowed, while the filopodia themselves tend to lengthen as if a force that pulled them rearward into the growth cone is attenuated (Lin et al., 1996). This rearward directed actomyosin-generated force against the substrate pulls on the main body of the growth cone, which then slides forward.

Depending on the molecular nature of the substrate, filopodia may attach firmly or weakly, thus generating more or less tension. This tension can be appreciated when a single filopodium

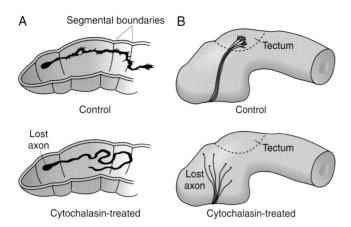

Fig. 5.13 Actin filaments are necessary to guide growth cones. A. In the grasshopper limb, the Ti1 growth cones are hairy with active filopodia (top). If the growth cones are treated with the actin-depolymerizing agent cytochalasin, the axon fails to navigate (bottom). B. In the vertebrate visual system, axons enter the brain from the optic nerve and grow toward the tectum by growing dorsally and turning posteriorly (top). When these axons are treated with cytochalasin, the axons fail to make the appropriate posterior turn, and most axons miss the tectum (bottom). *(After Bentley and Toroian-Raymond, 1986; Chien et al., 1993)*

from a growth cone contacts an axon lying in its path and pulls it back like the string on an archer's bow (Bray, 1979) (Figure 5.15). Similarly, a single filopodium that makes contact with a more adhesive substrate in tissue culture is able to steer the entire growth cone by pulling it toward the attachment point (Letourneau, 1996). The importance of single filopodia in directing growth cones *in vivo* has been demonstrated in the Ti1 pioneer axons of the grasshopper limb. Here it can be seen that when a single filopodium makes contact with a guidepost cell, it attaches firmly and pulls the growth cone in that direction while other less firmly attached filopodia retract into the growth cone (O'Connor et al., 1990) (Figure 5.15).

In culture, growth cones move straight ahead, and there are approximately equal numbers of filopodia on the left and right sides. Some conditions cause more actin polymerization or depolymerization on one side than the other, leading to an imbalance of filopodial number and resulting traction force, causing the growth cone to turn. In fact, this is suspected to be a main mechanism of growth cone reorientation. Indeed, if actin filaments are destabilized on one side of the growth by using a depolymerizing agent locally, the axon will turn in the other direction (**Figure 5.16**A) (Yuan et al., 2003). When a single filopodium is detached from the surface of a culture dish using a fine glass needle, the growth cone snaps into a new direction that is consistent with the tension exerted by the remaining filopodia (Wessells and Nuttall, 1978).

Microtubules are the other main cytoskeletal elements of the growth cone. Pools of unassembled tubulin are concentrated in the growth cone, which is the most sensitive part of the axon to the effects of microtubule depolymerizing agents such as nocodozole (Brown et al., 1992). Natural microtubule stabilizing proteins, such as Tau, are also highly concentrated near the growth cone, suggesting that this is where unpolymerized tubulin is fashioned into microtubules and stabilized. Microtubules run

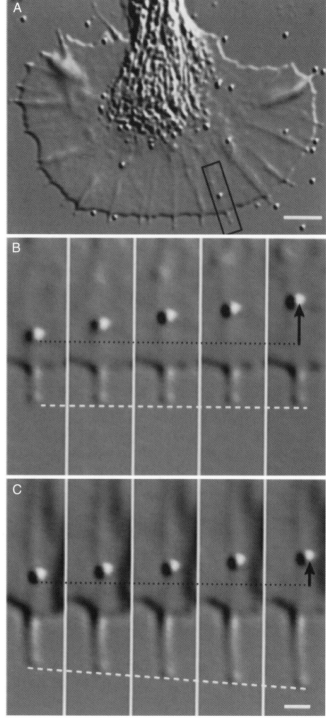

Fig. 5.14 Filopodium growth and retrograde f-actin flow are inversely proportional. A. To compare the rates of F-actin flow before and after treatments that inhibited myosin activity, 200 nm beads were positioned at the same location (box) using the laser trap. Bar represents 5 mm. B. Robust retrograde F-actin flow (black dashed line) and little, if any, filopodium outgrowth (white dashed line) were observed under control conditions. C. After application of 10 mM BDM (which inhibits myosin), filopodium elongation (white dashed line) occurred along with slowing of retrograde F-actin flow (black dashed line). Bar represents 1 mm. *(From Lin et al., 1996)*

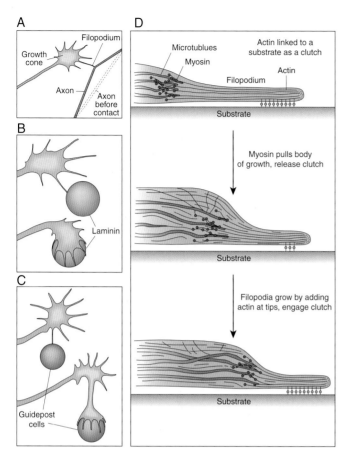

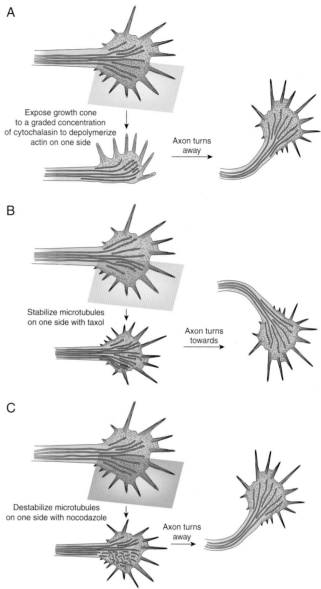

Fig. 5.15 Single filopodia can direct growth cones. A. A single filopodium from a growth cone exerts tension and pulls on an axon it contacts in culture. B. A single filopodium touches a laminin-coated spot in a culture dish and reorients. C. A single filopodium of a Ti1 cell contacts a guidepost cell, and by the process of microtubule invasion becomes the new leading edge. D. The clutch mechanism: myosin at the base of filopodia pulls on actin cables that are attached to the substrate through a transmembrane clutch and so pulls the main body of the growth cone forward. *(After Mitchison and Kirschner, 1988)*

Fig. 5.16 Actin and microtubules steer growth cones. A. Local depolymerization of actin on one side of a growth cone causes it to turn the other way. B. Local stabilization of microtubules on one side of a growth cone causes it to turn toward that side. C. Destabilization of microtubules on one side causes it to turn the other way. *(After Buck and Zheng, 2002; Yuan et al., 2003)*

straight and parallel inside of the axon, but when they enter the base of the growth cone, they splay out and bend like soft spokes and sometimes appear as broken fragments (Figure 5.11). Like the actin filaments in the filopodia, the microtubules of the axons have a "plus" end where polymerization takes place, and this is positioned at the growing tip. The fact that axons treated with cytochalasin B continue to grow, albeit slowly, is probably due to the distal growth of microtubules, as depolymerization of microtubules inhibits axon elongation completely (Marsh and Letourneau, 1984). Microtubule assembly is partially controlled by post-translational modifications that affect their stability. A carboxyl terminal tyrosine is added to alpha-tubulin by the enzyme, tubulin tyrosine ligase, inside of the growth cone. The tyrosinated form of tubulin is sensitive to depolymerizing agents so the new microtubules with the growth cone are very dynamic. In contrast, microtubules in which tubulin loses the tyrosine and becomes acetylated instead, as it does when it enters the axon, makes these axonal microtubules more stable (Brown et al., 1992). Dynamic microtubules in the growth cones can rapidly polymerize and extend transiently into the P-zone.

Dynamic microtubules can also go through catastrophies, that is, rapid disassembly. This dynamic instability of non-acetylated microtubules is critical for normal growth cone motility. If the dynamics are altered with a reagent like taxol which stabilizes microtubules, growth cones advance slowly, relentlessly, and become incapable of turning (Williamson et al., 1996).

Experiments on the local microtubule dynamics show that microtubules may be involved in growth cone turning in a similar way to actin (Figure 5.16). Thus, if the microtubule stabilizing agent taxol is delivered just on one side of a growth cone, the growth cone will turn in that direction, while if a depolymerizing agent such as nocodazole is delivered to one side, the growth cone will turn the other way (Buck and Zheng, 2002). The similarity of the results with actin and microtubule destabilizers suggests that actin bundles and microtubules interact in ways that are critical for proper growth cone navigation. The actin fibers in the central domain may restrict the dynamic microtubules from invading peripheral domains, but the actin cables in the filopodia may provide channels for dynamic microtubules to invade. These actin cables become linked to the microtubules through cross-linking molecules. Then, the myosin-driven retrograde flow of actin at the leading edges bends and breaks the dynamic microtubules, revealing new plus ends which can be elongated and new minus ends which can be depolymerized (Schaefer et al., 2002) (**Figure 5.17**). New growing dynamic microtubules thus continue to polymerize and probe the periphery of the growth cone, trying to get a foothold. Those microtubules that do successfully invade the peripheral domain can be stabilized there, and thus promote directional axon growth (Zhou and Cohan, 2004).

The keys to making the growth cone cytoskeleton dynamic in this way are the numerous proteins that are associated with microtubules and actin. Several microtubule-associated proteins (MAPs) are found in growing neurites, and these proteins may regulate the growth process. Some MAPs seem to be involved in stabilizing microtubules and inducing them to form bundles. Different MAPs are differentially localized: MAP2 is mostly in dendrites, and Tau is in axons. Antisense mediated depletion or knockouts of many of these MAPs diminish neurite outgrowth (Letourneau, 1996). Some MAPs, like the plus-end tracking protein Orbit/Clasp may help microtubules probe the actin rich peripheral domain of the growth cone where guidance cues operate (Lee et al., 2004) or participate in cross linking of actin fibers to microtubules, thus leading to the stabilization of actin-microtubule complexes. Microtubule destabilizing proteins such as SCG10 and stathmin are also important for growth cone function. For example, SCG10 overexpression in growth cones leads to enhanced dynamic instability of microtubules and increased neurite outgrowth (Grenningloh et al., 2004).

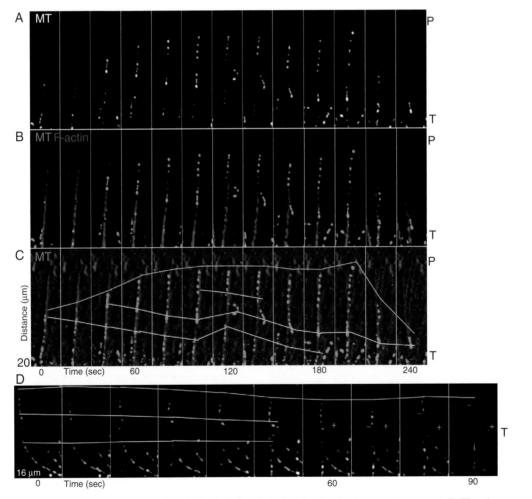

Fig. 5.17 MTs coupling to retrograde actin flow in P-domain leads to looping, breakage, and turnover. A–D. Time-lapse montages showing an example of an MT loop formation. A. MT. B. MT/F-actin overlay. C. MT/DIC overlay. Green, MT channel; Red, F-actin channel. MT is aligned along the prominent F-actin bundle and the loop forms as a proximal part of the MT translocates anterogradely (6.5 µm min-1) and dissociates from the actin bundle while the distal part of the MT is still aligned with the cable (100–140 s). MT grows at a rate of 6.6 µm min-1 and then experiences a catastrophe and shortens at 14.6 µm min-1. D. MT loop formation and breakage sequence in the T-zone. Note minus-end catastrophe and plus-end growth immediately after break. *(From Schaefer et al., 2002)*

There is also an array of actin-associated proteins (more than 50 have been identified) involved in growth cone function. Cofilin is an actin-binding protein found in the growth cone that depolymerizes F-actin into G-actin monomers, working a bit like SCG10 does on increasing dynamic instability, but on actin filaments rather than microtubules. Arps promote the branching of actin filaments, and the Ena/Vasp proteins that localize to filopodial tips act as anticapping agents, encouraging straight actin filaments to grow at the leading edge (Lanier and Gertler, 2000; Lebrand et al., 2004) (**Figure 5.18**). Without Ena/Vasp, growth cones have lammelopodia but no filopodia, yet surprisingly they can still navigate *in vivo* though have trouble making terminal arbors (Dwivedy et al., 2007). All these actin and microtubule-associated proteins and many others have to be regulated properly as the growth cone navigates, and there are a large and growing number of identified kinases and phosphatases that add and remove phosphate groups from these proteins, thus activating and inactivating them. These enzymes are in turn regulated by receptors on the growth cone surface that sense the substrates and guidance molecules that the growth cone encounters in its journey.

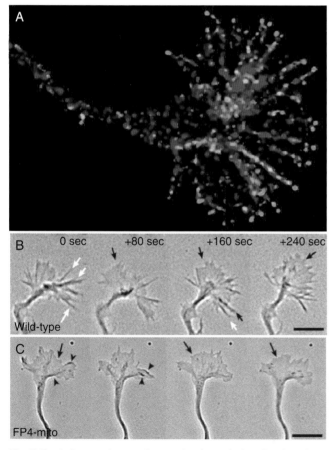

Fig. 5.18 A. Image of a growth cone showing actin in red and Ena/Vasp in green located at the tips of the filopodia where these proteins act as anti-capping agents, preventing the binding of actin capping proteins and thus encouraging plus end elongation. B. In wild-type neurons, growth cones are highly dynamic structures. Numerous filopodia and lamellipodia extended and retracted quickly at the surface of the growth cone. C. Growth cones of neurons expressing FP4-Mito, which sequestors Ena/Vasp to the mitochondria instead of the growth cone periphery, nearly lost their capacity to generate filopodia yet frequently developed lamellipodia and ruffles. *(From Lanier and Gertler, 2000 and Lebrand et al., 2004)*

DENDRITE FORMATION

Dendrites are more complicated than axons, but we know less about their development. Dendrites have to grow properly and be guided to the appropriate locations to make synaptic contact with the incoming axons. Dendrites are not only the main receivers of synaptic input, but they are the most distinctive features of neuronal morphology. Dendritic trees differ from axonal arbors in a variety of ways, the most obvious being that dendrites have mostly postsynaptic specializations while axons have mostly presynaptic specializations. While a neuron generally has one axon, it may have several distinct dendrites and each gets specific inputs. Single dendrites can often be further compartmentalized into regions that take particular types of synapses. The cytoskeleton of a dendrite is also different from that of an axon, usually having a higher ratio of microtubules to actin filaments and more rough endoplasmic reticulum and polyribosomes. Axonal microtubules, as mentioned earlier, have their plus ends pointed distally, while the dendritic microtubules have a mixture of plus- and minus-ends leading. Microtubule associated proteins are also differentially distributed in axons and dendrites. For example, MAP2 is located in dendrites, while Tau is located mainly in axons. Dendrites also grow somewhat differently to axons (McAllister, 2000; Whitford et al., 2002). Treatment of cultured neurons with antisense constructs that reduce MAP2 or Tau expression have the expected specific effects on the formation of dendrites or axons, indicating that these proteins are particularly critical in the formation or stabilization of these structures (Liu et al., 1999; Yu et al., 2000). Certain membrane proteins are also differentially distributed among axons and dendrites; for instance, transmitter receptors are more common on dendrites, whereas certain cell adhesion molecules and GAP-43 are found mainly on axons (Craig and Banker, 1994). While the molecular mechanisms for sorting different proteins to different neuronal processes are not yet well understood, the developmental origins of neuronal polarity are beginning to be.

When isolated embryonic neurons are plated in culture dishes, hippocampal neurons often start as spherical cells, but given time in culture they tend to grow into the typical neuronal shape with a single axon and multiple dendrites. Neurons thus have an intrinsic ability to break symmetry. *In vivo*, axons and dendrites are often oriented in respect to the apicobasal axis of the neuroepithelium in which they were born. For example, cortical pyramidal cells have apically emerging axons and basally extending dendrites, while retinal ganglion cells have just the reverse. This argues that the factors that are responsible for setting up the orientation of the neuroepithelium, like the basal lamina on the basal side and the primary cilium and protein components of the apical junction and the Par complex on the apical side, initiate the polarization of neurons (Shi et al., 2003; Barnes et al., 2008). Indeed, some external influences recreated in culture also affect the axonal versus dendritic decision perhaps by affecting cytoskeletal-associated proteins or the location of organelles. For instance, if a hippocampal cell is plated at the interface between a laminin and polylysine, the axon almost invariably grows on the laminin substrate. Other external factors favor dendrite formation. Superior cervical ganglion neurons grown in serum-free medium in the absence of nonneuronal cells were unipolar and only grew axons. When the same neurons were exposed to serum, they became multipolar and developed dendrites (Bruckenstein and Higgins, 1988a,b). The bone morphogenetic proteins, BMP2 and BMP6, were subsequently found to selectively induce the formation of dendrites and the expression of microtubule-associated protein-2 (MAP2) in sympathetic neurons in a concentration-dependent manner (Lein et al., 1995; Guo et al., 1998).

Hippocampal neurons differentiating in culture are very good models for exploring the question why neurons tend to have only one axon yet several dendrites. Initially, these neurons put out several short neurites, each tipped with small growth cones (**Figure 5.19**). All of these processes appear to be identical and they all have GAP-43 at their tips. One of these processes, however, begins to extend more rapidly than the others, and as it does so, it gathers axonal specific markers so that soon only the forming axon has a GAP-43 tipped growth cone and the other processes begin to assume dendrite-specific markers. Interestingly, if the emerging axon is selectively cut off, the longest of the short processes starts to grow faster than the others and it becomes the axon. Thus, neurons have an axon-to-dendrite ratio that is, to a certain extent, internally regulated through a feedback mechanism by which the axon inhibits the other neurites from assuming the axonal identity they would attain by default (Goslin and Banker, 1989). If, at an early stage of polarization when all processes are equal, the actin depolymerizing agent cytochalasin is transiently applied locally to just one neurite, that neurite will become the axon. If, however, cytochalasin is applied uniformly to all the neurites, then surprisingly, they all become axons (Bradke and Dotti, 1999). This suggests that actin instability, possibly allowing microtubule invasion, may be a key to the decision of a process to become an axon or a dendrite. This model is supported by direct manipulation of molecules that control microtubule stabilization, such as collapsin response mediator protein (CRMP-2, a MAP) and a GEF called Tiam1 (Kunda et al., 2001; Fukata et al., 2002). When either of these proteins is overexpressed in a developing hippocampal cell, all the processes become axons; when their function is reduced, all the processes become dendrites. It is thought that these proteins help microtubules invade actin networks.

What are the factors that control the growth and shape of dendrites? Dendritic tree growth is influenced by external factors. For example, the growth of dendrites in mouse cortical neurons is specifically enhanced by astrocytes derived from the forebrain (Le Roux and Reh, 1994). Similar results were obtained with glial conditioned medium. Dermatan sulfate also specifically enhances dendritic growth. Dendritic retraction occurs in many regions of the developing brain. Leukemia inhibitory factor (LIF) and ciliary neurotrophic factor (CNTF) specifically cause dendritic retraction in SCG cells (Guo et al., 1997; Guo et al., 1999). Axon growth is unaffected by these factors. Taken together, these results suggest the existence of separate but extensive molecular mechanisms for promoting and inhibiting dendrite versus axon outgrowth.

The Rho family of GTPases is critical for proper dendritic morphology. These GTPases are key regulators of actin polymerization. Cdc42 is required for multiple aspects of dendritic morphogenesis (Luo, 2000, 2002). For example, in neurons that are mutant for Cdc42, dendrites are longer than normal, branch abnormally, and have a reduced number of spines. Extra activation of RhoA, via experimental expression of a constitutively active form of the molecule, leads to a dramatic simplification of dendritic branch patterns. Such experiments suggest that different Rho family members have distinct roles in regulating dendritic morphogenesis.

Each type of neuron has a characteristic dendritic tree. In some neurons, like the Purkinje cell, the dendritic tree is enormously complex and supports synaptic input from thousands of presynaptic fibers. In other neurons, like some sensory neurons, the dendritic tree is simple, consisting of a single postsynaptic process. In the central nervous system of a cockroach or a leech, the dendritic tree of each identified cell has a unique signature branching pattern recognizable from individual to individual. What drives these particular morphologies? Transcription factors, identified in *Drosophila*, have a clear role in regulating the cell specific autonomous character dendritic trees (Jan and Jan, 2003; Parrish et al., 2007). The *hamlet* gene is expressed in external sensory neurons that have very simple monopolar dendrites, and when these neurons are mutant for *hamlet*, their dendritic trees become highly branched so they look very much like another class of sensory neurons called multidendritic neurons, which normally do not express *hamlet*. Conversely, multidendritic neurons express high levels of another transcription factor called *cut*, which is expressed at low levels in the external sensory neurons. Overexpression of *cut* in the latter causes them to acquire multiple dendrites, while loss of *cut* function in multidendritic neurons causes them to lose their dendritic complexity (Grueber et al., 2003). A genetic screen in *Drosophila*

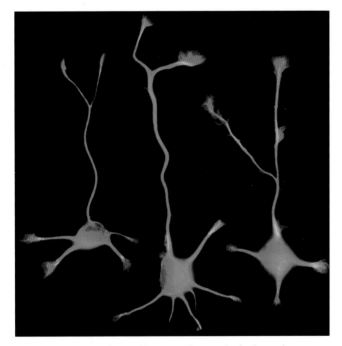

Fig. 5.19 In tissue culture, a hippocampal neuron begins by putting out several minor processes that are basically equivalent. One of these, the future axon, then begins to grow faster than the other process and collects axon-specific components like GAP43 and Tau. After the axon has elongated, dendrites begin to grow and express dendrite-specific components such as MAP2. This figure shows three young hippocampal neurons in culture stained for microtubules (red) and actin (green). At this stage, one process is elongating while the shorter processes are not yet definitive dendrites. If, at this stage, the emerging axon is cut, then a minor process, which would have otherwise become a dendrite, begins to grow more rapidly and becomes the axon. *(From Ruthel and Hollenbeck, 2003)*

found many more such transcription factors that influence dendritic shape. In vertebrates, the bHLH proneural transcription factor Neurogenin 2 (which is important for the specification of cortical neurons; see Chapter 4) also plays a role in dendritic morphology postmitotically, and misexpression of Neurogenin-2 in cultured neurons promote pyramidal dendrite morphology (Hand et al., 2005).

Dendrite to dendrite interactions are important for tiling the epidermis or the retina with nerve endings and to spread dendritic trees out so that dendritic branches of the same neuron do not overlap with each other. The remarkable way that this is done in *Drosophila* through the molecule called Dscam is shown in Chapter 6, Box 6.1.

It has been thought that mature dendrite formation is somehow dependent on the axon making proper connections to its target. The situation, however, may not be so one-sided, as it seems that dendrites can also "search" for their inputs (Jan and Jan, 2003). In some cases, growing dendrites are tipped with dendritic growth cones that appear as miniature equivalents to axonal growth cones. Dendritic growth cones express receptors for several classes of guidance factors and, indeed, factors such as Slits, Netrins, and Semaphorins can influence the growth of dendrites (Whitford et al., 2002). Interestingly the same factors that do one thing in axons can do a different thing in dendrites. For example, Sema3A can attract the apical dendrites of cortical neurons toward the pial surface, while the same factor can repel the axons of the same pyramidal cells. Where, however, there is specific matching between axons and dendrites, they both may be guided by the same factors to a molecularly specified region of a developing neuropil where the two might meet and synapse on each other. For example, in the embryonic *Drosophila* CNS the dendrites of motor neurons that innervate muscles in a particular region of the body are likely guided by the same guidance cues that lead the central projections of sensory axons, which innervate that part of the body, so that these axons meet the appropriate dendrites at a specific CNS location (Landgraf and Thor, 2006).

Experimental conditions have been worked out in which pyramidal neurons from the cortex or hippocampus, principal neurons of the SCG, and even cerebellar Purkinje cells are able to develop a dendritic tree that looks at least cell type specific in dissociated cell culture. Yet the way the dendritic tree grows and the final size, shape, and complexity of the dendritic tree is sensitive to activity and innervation. Purkinje cell dendrites elongate in culture until the cells become electrically active, at which point the dendrites stop growing and branch. If activity is inhibited by tetrodotoxin, the dendrites continue to elongate, as if they were still immature. Intracellular calcium seems to be a key component in this process (Schilling et al., 1991). Dendrites of the aCC neuron in *Drosophila* also respond to changes in input, growing when there is not enough activity, and these studies suggest that dendrites are homeostatic devices that adapt structurally in response to alterations in synaptic input to normalize activity levels in the brain (Tripodi et al., 2008).

If activity is important to the way dendrites develop, then it makes sense that axonal inputs are also important for dendrite growth and branching. The complexity of the dendritic tree is, in many systems, proportional to the amount of innervation. In *weaver* mutant mice, the granule cells do not migrate properly into the cortex of the cerebellum and thus fail to make synapses on the Purkinje cells. In response to this lack of innervation, Purkinje cells make a dendritic tree that is smaller

and less well-formed than the trees of properly innervated Purkinje cells (Bradley and Berry, 1978). Similarly, the dendritic trees of the principal sympathetic neurons of the superior cervical ganglion (SCG) are larger and more complex in larger mammals that have more inputs onto these cells. Tectal dendrites are innervated by the axons of RGCs. Time-lapse recordings of tectal dendrites show that terminal dendritic branches are very dynamic and can appear or disappear within minutes. Active visually driven input on dendrites tends to stabilize new branches through the activation of NMDA receptors, CAM kinase II, and Rho GTPases (Cline, 2001). Many dendritic trees are dynamic throughout the life of the animal. The plasticity of these branches in response to experience and neuronal activity will be discussed further in Chapter 9.

WHAT DO GROWTH CONES GROW ON?

Growth cones have the ability to sense their environment and make choices based on extracellular information. These choices are made as the growth cone passes through a complex chemical and physical terrain. On a dried cracked collagen surface, growing axons often follow the pattern of the stress fractures (**Figure 5.20**A). When tracts or commissures are wounded or experimentally severed, axons may be physically impeded from growing as they normally would. In such cases, it is sometimes possible to provide axons with an artificial mechanical pathway across the wound (Silver and Ogawa, 1983) (Figure 5.20B). So, mechanical support is necessary and influential, but axon growth and guidance appear to be based

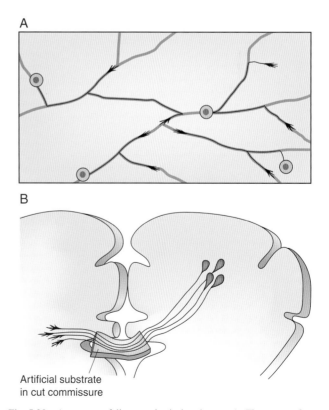

Artificial substrate
in cut commissure

Fig. 5.20 Axons may follow mechanical pathways. A. The axons of neurons on a dried collagen matrix growing through the cracks. B. Axons of the corpus callosum can use an artificial sling to grow from one side of the brain to the other.

more on molecular mechanisms. To discover these mechanisms, factors are often tested in tissue culture to see whether they influence the rate and direction of axonal growth. Many molecules that influence growth cone extension have been identified in this way, and such studies show that a single type of growth cone can respond to a variety of different cues. For example, embryonic spinal neurons from frog embryos respond to at least a dozen different factors tested, and this implies that growth cones have rich arrays of receptor molecules to sample various aspects of their environment. Genetic approaches in *Drosophila* and nematodes have led to the identification of many genes for guidance cues and their receptors which, when disrupted, lead to growth and pathfinding errors.

When neurons are plated on plain glass or tissue culture plastic, the cells often attach but rarely put out long neurites with active growth cones. However, when they are plated on polycationic substrates, such as polylysine, that stick well to negatively charged biological membranes, the neurons are much more likely to initiate axonal outgrowth. The growth cones of such neurons are flattened against the substrate as though they adhere very strongly to it (**Figure 5.21**). When axons are grown on a patterned dish that offers them the choice between a nonadhesive substrate versus an adhesive substrate, the growth cones follow the adhesive trail (Letourneau, 1975; Hammarback et al., 1985) (Figure 5.21). These findings led to the idea that growth cones might simply follow gradients of adhesion in the developing organism. This may indeed be the case for some neurons. For example, in the moth, sensory axons of the wing grow in the distal to proximal direction along a basal lamina of epithelial cells (Nardi and Vernon, 1990). Examination of this epithelium microscopically shows that it becomes increasingly complex toward the base. Transplantation of the epithelium suggests that this change corresponds to a gradient of adhesion. Axons readily cross onto a transplant that has been moved in the proximal to distal direction, but avoid distal transplants that have been moved proximally, suggesting that axons readily grow onto more adhesive membranes but will not grow onto less adhesive ones (Nardi, 1983).

To measure growth cone attachment to various cell adhesion molecules, culture media can be squirted at the growth cones through a pipette at a consistent force in an attempt to "blast" them off the substrate (**Figure 5.22**). The longer the growth cone stays attached to the surface, the stronger its adhesion must be. Neurite growth rate can then be measured on these same substrates for comparison (Lemmon et al., 1992).

Hierarchical adhesion decisions are critical for correct axonal growth *in vivo*. A good example of this comes from the ocellar axons of the simple eyes in *Drosophila*. These axons normally grow on the basal lamina of an epithelial pathway into the brain, but in mutants for a subunit of the laminin, these axons choose instead to extend upon the axons of mechanosensory neurons, and the result is that the ocellar axons do not make the correct connections (Garcia-Alonso et al., 1996). When axons are given a choice between pairs of naturally occurring cell adhesion molecules, they do not necessarily grow preferentially on the more adhesive substrate. Interestingly, it turns out that the most adhesive substrates, such as the lectin concanavalin A, do not support axon outgrowth. In fact, growth cones tend to get stuck on such a surface. They seem unable to retract their filopodia efficiently. So, they grow very slowly. The ability to detach is just as important as the ability to attach, and the molecules that have just the right amount of adhesion, not too little and not too much, are best able to support axon growth.

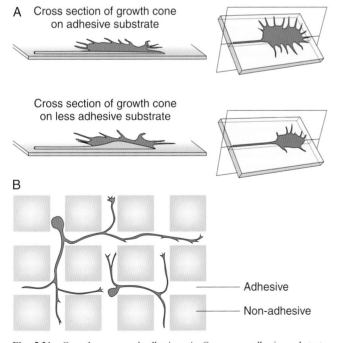

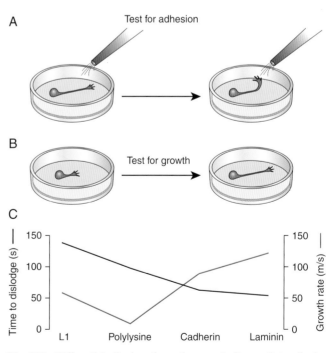

Fig. 5.21 Growth cones and adhesion. A. On a very adhesive substrate growth cones are flattened, have lots of filopodia, and do not move rapidly (top). On a less adhesive substrate, growth cones are more compact, rounded, have fewer processes, and often move more quickly. B. Neurites in culture given a choice between an adhesive and a nonadhesive substrate will tend to follow the adhesive trails.

Fig. 5.22 Differential adhesion of growth cones. A. To quantitate adhesivity, a measured blast of culture medium is directed at the growth cone. At a particular time, the growth cone becomes detached. B. Growth is quantified by axon length increase over an interval time. C. By using such tests, it can be shown that the neurons tested show a particular adhesion profile and tend to grow more slowly on more adhesive substrates. *(After Lemmon et al., 1992)*

Many molecules that are excellent supporters of axon growth have been isolated from the extracellular matrix (ECM) (Bixby and Harris, 1991; Tessier-Lavigne and Goodman, 1996). These factors were initially purified from culture media that was conditioned by cells known to support axonal outgrowth. Some of the most abundant proteins in ECM, such as laminin, fibronectin, vitronectin, and various forms of collagen, all promote axon outgrowth. Many of these ECM proteins are large and have many different functional domains for cell attachment, for collagen attachment, and for protein interactions. Different neurons seem to show a preference for particular ECM molecules. Vertebrate CNS cells grow particularly well on laminin, while peripheral neurons often seem to grow better on fibronectin. In experiments where retinal neurons are given a choice between laminin and fibronectin laid down in alternate stripes, retinal axons clearly prefer laminin, though they will grow on either substrate if given no choice.

Integrin is a receptor for many different extracellular matrix proteins, and it is composed of two subunits, alpha and beta (**Figure 5.23**). The extracellular matrix molecule that an axon will respond to is largely a matter of which integrin molecules are found at the growth cone as the specificity of integrin for particular ECM proteins depends on the combination of α and β subunits that are expressed (McKerracher et al., 1996). There are about 20 different α and about 10 different β subunits. Different tissues, including different neural tissues, use different subunit combinations. The $\alpha5$ subunit is particularly good at binding to fibronectin, while the $\alpha6$ subunit is better at binding to laminin. Over the course of development, axons may change which integrin subunits they express and thus change their sensitivity to a particular ECM molecule. For instance, chick retinal ganglion cells express $\alpha6$ and grow well on laminin when they are young. As they mature, they stop expressing this subunit and lose their ability to respond

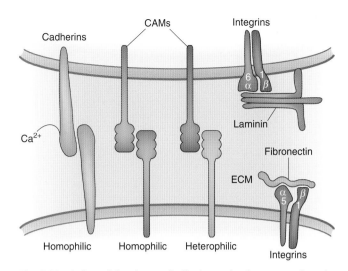

Fig. 5.23 A few of the classes of adhesion molecules expressed on the growth cone. Cadherins are calcium-dependent homophilic adhesion molecules. Some members of the IgG superfamily of cell adhesion molecules, CAMs, can bind homophilically; others are heterophilic. Integrins composed of various alpha and beta subunits bind to a variety of different extracellular matrix components with distinct affinity profiles.

to laminin just as their axons make contact with the tectum (Cohen and Johnson, 1991).

In this section, we have talked mainly about the ECM but as more and more neurons are added to the nervous system, the surfaces of other neurons become increasingly available as substrates. Neuronal and axonal surfaces can, through the expression of cell adhesion molecules (CAMs) provide good traction for growth cones and provide guidance information. For example, growth cones of Aplysia neurons preferentially

---

BOX 5.1   Tissue Culture

The tissue culture technique, a mainstay of all biological research in the last century, has continuously embraced innovative solutions to address neurobiological questions (Bunge, 1975; Banker and Goslin, 1991a). In fact, Ross Harrison invented tissue culture to study axon outgrowth (Harrison, 1907, 1910). His original preparations consisted of pieces of tissue, now termed *organotypic cultures*. Such cultures may now be obtained from vibratome sections of neural tissue and grown under conditions that promote thinning to a monolayer, thus providing greater access and visibility of individual neurons (Gšhwiler et al., 1991). Slices are attached to a coverglass and placed in rotating tissue culture tubes (hence the term *roller-tube culture*) such that the tissue culture media transiently washes over them. If one requires a slice of tissue with somewhat greater depth, then the cultures can be grown statically at the gas–liquid interface by using tissue culture plate inserts that provide a porous stage for the tissue and a reserve of media below (Stoppini et al., 1991). The relative simplicity of modern organotypic preparations has resulted in a wealth of data on the interaction between afferent and target populations, as described in the text.

In the arena of primary dissociated cell cultures, it has become feasible to isolate particular cell types. This may be performed by an immunoselection technique in which a cell-specific antibody is adsorbed to a plastic Petri dish, creating a surface on which one cell type will selectively attach. This approach has led to a 99% pure retinal ganglion cell preparation (Barres et al., 1988). A different means of separating cells relies upon selective prelabeling with a fluorescent dye and subsequently performing fluorescent-activated cell sorting (FACS). When passed through such a device, single cells are sequentially monitored for fluorescence and then selectively diverted to a receiver tube if they are labeled. This approach led to the isolation of retrogradely labeled spinal motor and preganglionic neurons (Calof and Reichardt, 1984; Clendening and Hume, 1990). Finally, it is possible to isolate large and small cell fractions following centrifugation on a Percoll density gradient, and then further enrich the cells with a short-duration plating step, which allows the more adhesive cells (e.g., astrocytes) to be retained on a treated surface. This approach led to the isolation of a > 95% pure granule cell population from cerebellar tissue (Hatten, 1985). Once specific cell types have been isolated, they may be mixed together in known ratios, or plated on two surfaces that are subsequently grown opposite one another as a sandwich. This technique allows one to produce a "feeder layer" of astrocytes on one surface that promotes survival of low-density neuronal cultures. It may also allow the experimenter to discriminate between contact-dependent and contact-independent phenomena.

Continued

Having obtained the neurons and glia of interest, tissue culture offers the opportunity to perform insightful manipulations. For example, it is possible to produce a nonuniform distribution of growth substrates to test the role of specific molecules in axon guidance (Letourneau, 1975) or stripes of membranous material as assays for the identification and characterization of axon guidance factors (Walter et al., 1987a; Walter et al., 1987b). The technique has been extended to create gradients of laminin or neuronal membrane on a surface that subsequently serves as the tissue culture substrate (McKenna and Raper, 1988; Baier and Bonhoeffer, 1992). Molecular micropatterns can even be stamped onto culture surfaces using a printing process (**Figure 5.24**). The gradients can be visualized and quantified by including a fluorescent or radioactive marker along with the intended substrate. Gradients of soluble molecules can be produced in vitro with repetitive pulsatile ejection of picoliter volumes from a micropipette tip into the tissue culture media (Lohof et al., 1992). The concentration gradient is quantified by ejecting a fluorescein-conjugated dextran and measuring the fluorescent signal at increasing distances from the pipette tip. Three-dimensional tissue culture is also possible by embedding neurons or explants in gelatinizing collagen or a mixture of ECM material. Such gels not only provide a more "realistic" substrate for axons to grow through, they also allow for the formation of relatively stable gradients of soluble factors that can percolate through the gel undisturbed by flows and currents that happen when the experimenter moves the culture dish. It was in such gels that evidence for diffusible guidance factors released from targets such as Max Factor and Netrin was first obtained (Lumsden and Davies, 1986). In such gels, it is also possible to inject diffusible factors in known quantities to create designer gradients of particular concentrations and steepnesses. In such designer gradients, it was possible to show that a growth cone can sense a difference across its width of one molecule per thousand (Rosoff et al., 2004). Microfluidic chambers and other devices made by nanofabrication techniques are being introduced into culture to get even finer spatial and temporal control of what neurons or parts of neurons are exposed to.

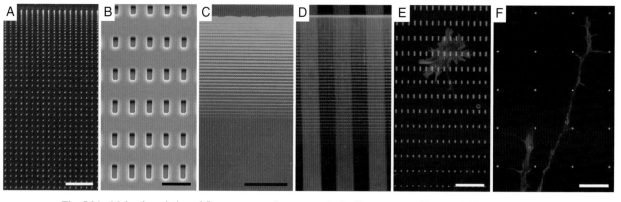

Fig. 5.24   Molecular printing of fluorescent protein patterns. A. A silicon master with a graded pattern built up by dots seen with the light microscope. B. Dots etched into a silicon master (approximately 650 nm deep) seen with the scanning electron microscope. C,D. Graded protein patterns seen with the fluorescence microscope. The printed protein (EphrinA5) is shown in green; laminin covering the pattern is shown in red. In D, laminin was applied in stripes. E,F. Growth cones navigating in protein patterns. The protein (EphrinA5) is shown in green, phalloidin-stained axons are shown in red. Scale bars: A,C,D, 100 mm; B, 5 mm; E,F, 15 mm. *(From von Philipsborn et al., 2006)*

stick to beads coated with the Aplysia CAM apCAM which becomes connected to the growth cytoskeleton based on a tyrosine kinase pathway, and experimentally tugging on such attached beads with laser tweezers is a way of steering these growth cones in a dish (Suter and Forscher, 2001). In the following sections, we will talk much more about steering and the role of CAMs.

WHAT PROVIDES DIRECTIONAL INFORMATION TO GROWTH CONES?

It has been suggested that four types of molecular cues influence the direction in which growth cones will travel (Tessier-Lavigne and Goodman, 1996). These are divided into short-range and long-range cues, each of which may be either attractive or repulsive (**Figure 5.25**). (1) Contact adhesion: a

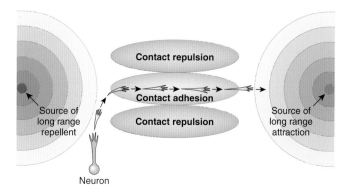

Fig. 5.25   Short-range (or contact) and long-range attractant and repellents guide an axon. This axon is pushed from the left and pulled from the right by long-range cues, while hemmed in along a narrow pathway by contact adhesive and repulsive cues. *(After Tessier-Lavigne and Goodman, 1996)*

sudden increase in the adhesivity of one cellular substrate compared to another may cause axons to switch pathways; (2) Contact repulsion: when growth cones bump into cells with repulsive membrane molecules, they may collapse or turn away and grow in a different direction; (3) Long-range attraction: growth cones may exhibit positive chemotaxis behavior to certain diffusible molecules originating at distant sources and so grow toward these chemoattractants; and (4) Long-range repulsion: growth cones may exhibit negative chemotaxis to diffusible molecules and so orient away from the sources of such factors. There are numerous cues and receptors expressed in the nervous system that provide and receive guidance information. Some of these are listed in **Figure 5.26**. In the following sections, we will discuss many of the specific guidance cues that work according to these basic mechanisms. In many systems, it has become clear that growth cones respond to a combination of these cues simultaneously or sequentially.

CELL ADHESION AND LABELED PATHWAYS

Growth cones, in addition to using ECM, make contact with the membranes of other cells and axons. In this, they are supported by a set of cell adhesion molecules (CAMs) (Walsh and Doherty, 1997). There are a host of such molecules, and they come in several classes. The most prominent class is the IgG superfamily. Members of this class have extracellular repeat domains similar to those found in antibodies, reflecting perhaps an ancient adhesive function for the IgG superfamily. Another class is the calcium-dependent cadherins. One property that many CAMs share is their ability to bind homophilically. Homophilic binding means that proteins of the same type bind to each other. Thus, if two cells express the same homophilic CAM on their surfaces, the CAM on the one cell will act as receptor for the CAM on the other cell and vice versa, causing the two cells to adhere to each other. To test whether a particular CAM is homophilic, nonadherent cells are transfected with the CAM and assayed with respect

Ligands	Receptors	Primary function
Netrin	DCC/frazzled	Attraction
	UNC5	Repulsion
Ephrin A	EphA	Repulsion
EphA	Ephrin A	Attraction
Ephrin B	EphBs	Repulsion
EphB	Ephrin B	Attraction
CXCL12/Sdf1	CXCR4	Attraction
Slit	Robo	Repulsion
Semaphorin	Plexin	Repulsion
	Neuropilin	Repulsion
Wnt	Frazzled	Attraction
	Ryk	Repulsion
Hedgehog	Bok	A and R
	Patched	A and R
IgCAM	IgCAM	Homophilic cell adhesion
Cadherin	Cadherin	Homophilic cell adhesion
ECM SAM	Integrin	Substrate adhesion
Growth factors	GFRs	Attraction
Trophic factors	Trks	Attraction
Sphingolipids	S1P	Homophilic cell adhesion
GAGs/carbohydrates	Various receptors	Modulation

Fig. 5.26 A partial list of some of the main classes of ligands and their receptors that affect the growth and guidance of axons. There are many molecules in each class, sometimes huge families of them, and so it is impossible to list them all here. The primary or first described function for each of these classes is given, although it must be understood that attraction and repulsion can flip depending on the state of the growth cone (see text for some details).

to whether they then form aggregates. During outgrowth, axons expressing the same CAMs often fasciculate with one another. Thus, single pioneer axons may use a specific CAM to guide the follower axons that express the same CAM, and these axons can attract the fasciculation of still later axons of the same type. In this way, pioneer axons can become the founders for large axonal tracts within the CNS.

Monoclonal antibodies raised against axonal membranes have been used to search for cell adhesion molecules that are expressed on particular fascicles of fibers during neural pathway formation. Several molecules were discovered in this manner. Some CAMs show a particularly restricted expression and result in very specific defects in axon growth. For example, a *Drosophila* CAM, called fasciclinII (FasII), is expressed on a subset of longitudinal tracts or fascicles in the CNS (Bastiani et al., 1987). Loss-of-function *fasII* mutations lead to defasciculation of these tracts, and this defasciculation can be rescued when FasII levels are specifically increased in these pathways. Indeed, in gain-of-function experiments, if FasII levels are kept too high, this can prevent normal defasciculation which is necessary for axons to reach their final targets (**Figure 5.27**) (Lin et al., 1994) (see also Chapter 6). Interestingly, in *fasII* mutants, the defasciculated axons still eventually find their appropriate targets, showing that there is manifest redundancy in neuronal wiring mechanisms, and illustrating how robust the system really is.

Some CAMs are first expressed in many cells of a neural circuit, so when these cells first contact each other, they attach and grow appropriately. For example, in mammals, limbic-associated membrane protein, LAMP, is a CAM expressed by neurons throughout the limbic system, a cortical and subcortical area of the brain that functions in emotion and memory (Horton and Levitt, 1988). Administration of LAMP antibodies to developing mouse brains results in abnormal growth of the fiber projections in the limbic system, suggesting that LAMP is an essential recognition molecule for the formation of limbic connections (Pimenta et al., 1995). LAMP has three IgG domains, and these domains appear to participate in different ways to enhance local wiring. LAMP enhances neurite outgrowth through homophilic interactions with other neurons in the limbic circuit using the first of its IgG domains, but inhibits neurite outgrowth of non-LAMP-expressing neurons using heterophilic interactions involving the second IgG domain (Eagleson et al., 2003). The function of the third domain is not yet known.

Neural cell adhesion molecule (NCAM), the earliest identified of the CAMs, appears to be expressed on virtually all vertebrate neurons and glia (Edelman, 1984). NCAM exists in many different forms, some with an intracellular domain that may interact with the cytoskeleton, and some without. The extracellular portion of NCAM can be highly modified by the addition of carbohydrates, particularly sialic acid residues. Nonsialylated forms are very adhesive compared to the sialylated forms. In the course of a neuron's development, there may be changes in the sialylation state of its NCAM, thus adjusting its adhesion (Walsh and Doherty, 1997). For example, developing motor neurons of the chick grow out of the spinal cord and enter a complicated plexus region, where they cross in many directions and eventually segregate into distinct nerve roots leading to their appropriate muscles. During the time when these axons are growing in the plexus, the NCAM they express is highly sialylated. This keeps the axons that are headed toward different muscles from fasciculating indiscriminately with one another. If the sialic acid is digested away with endoneuraminidase (Endo-N), errors in pathfinding occur in the plexus region, and motor axons exit into the wrong peripheral nerves (Tang et al., 1994) (**Figure 5.28**).

Growing axons respond to a variety of cues as they navigate, so the loss of just one of these may affect axon growth and navigation in a limited way. In many gene deletion studies in *Drosophila*, the majority of axons usually grow along the correct pathway. For instance, in *fasII* mutants mentioned above, the longitudinal tracts are defasciculated, but the axons are still able to grow in the correct directions. In some CAM knockouts, for instance the NCAM knockout in the mouse, the phenotype is surprisingly subtle, suggesting that it is just one of a number of molecules that are used for this purpose. This idea of molecular redundancy for pathfinding in the developing nervous system is supported by the finding that sometimes doubly mutant embryos, which have deletions of more than one particular adhesion molecule, exhibit severe axonal growth defects, whereas each of the single mutants is relatively normal. For example, *Drosophila* embryos that lack FasI, a CAM that is expressed in commissural fascicles, have a normal looking CNS. To test whether FasI is part of a redundant system, a series of double mutants were made with *fasI* and several other putative guidance or growth cone function mutants, each of which also had no striking phenotype on its own. One of the double mutants tested showed pathfinding defects in combination with *fasI*. This was the *fasI/abl* double mutant (Elkins et al., 1990). *abl* codes for a tyrosine kinase, which probably functions in a signal transduction pathway in the growth cone. Since both genes had to be knocked out to cause axon disorientation, the two genes are probably part of two distinct molecular pathways, either of which may suffice

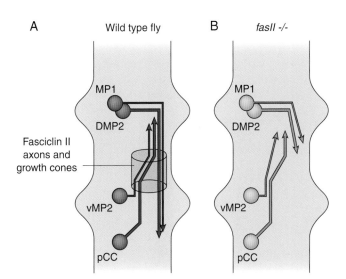

Fig. 5.27 The *fasII* loss-of-function and gain-of-function lead complementary phenotypes. A. In the wild-type the posteriorly directed axons of dMP2 and MP1 fasciculate with the anteriorly directed axons of pCC and vMP2. These axons all express FasII on their membranes. B. Loss-of-function mutants of *fasII* leads to a defasciculation phenotype. (Lin et al., 1994)

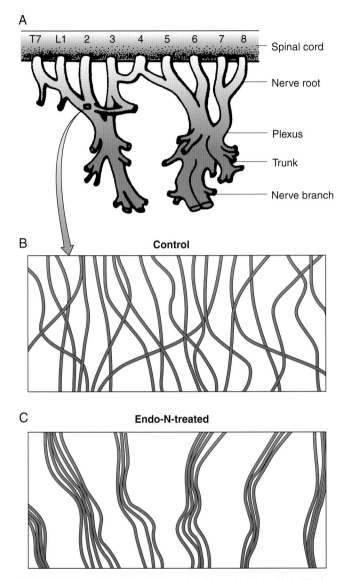

A

T7 L1 2 3 4 5 6 7 8 — Spinal cord

— Nerve root

— Plexus

— Trunk

— Nerve branch

B **Control**

C **Endo-N-treated**

Fig. 5.28 Homophilic adhesion is regulated by polysialic acid. A. The brachial plexus region in the chick where motor axons destined for particular muscles sort out into their correct nerve roots. B. Higher magnification of the plexus region showing fascicles breaking up and axons regrouping with other axons. C. After treatment with Endo-N to remove sialic acid residues from N-CAM, the axons do not defasciculate properly and stay in large fascicles. As a result, innervation errors are made. *(After Tang et al., 1994)*

pioneer axons because it expresses complementary CAMs. Particularly strong evidence for labeled pathways comes from the embryonic grasshopper central nervous system. Here, the axon of the G-neuron extends across the posterior commissure along a pathway pioneered by the Q-neurons (Bastiani et al., 1984; Raper et al., 1984). Once it has crossed the midline, the growth cone of the G-neuron pauses for a few hours. During this time, its filopodia seem to explore a number of different longitudinal fascicles in the near vicinity. In the electron microscope, it can be seen that filopodia from the G-neuron growth cone preferentially stick to the P-axons of the A/P-fascicle. The G-growth cone then joins the A/P fascicle and follows it anteriorly to the brain. If the P-axons are ablated before the G-growth cone crosses the midline, the G-growth cone upon reaching the other side does not show a high affinity for any other longitudinal bundle or even the A-axons. As a result, it often stalls and does not turn at all (**Figure 5.29**). Thus, the P-axons seem to have an important label on their surface that the G-growth cone can recognize, possibly because of a specific receptor on the G-cell membrane. Examination of CAM expression with the electron microscope confirms that specific CAMs are distributed on the surface of axons in particular fascicles (Bastiani et al., 1987). In vertebrates, too, there is evidence that some axons use a labeled pathway mechanism. The tract of the postoptic commissure (TPOC), for example, is a pioneering tract for axons from the pineal. Pineal axons fasciculate with the TPOC axons as the former turn posteriorly at the boundary between forebrain and midbrain. If the TPOC is ablated, pineal axons often fail to make the appropriate turn (Chitnis et al., 1992).

Just as you may have to change lines at a subway stop to reach your final destination, so axons may have to change pathways. To do so, an axon must change the CAM on its surface. Such a change in the expression of particular CAMs has now been seen in a number of systems at places where axons switch directions. For example, when axons in the embryonic central nervous system of the grasshopper travel on a longitudinal tract, they express FasII, but when they leave the longitudinal tract and turn onto a horizontal commissure,

for axon guidance. Similarly, in vertebrate tissue culture, motor axon growth over muscle fibers is not seriously impaired unless two or more adhesion molecules are simultaneously disabled (Tomaselli et al., 1986; Bixby et al., 1987). Axonal growth is such an important part of building an organism that such failsafe molecular mechanisms often operate to help ensure that the nervous system is properly wired.

Early axon tracts can be thought of as subway lines: the Orange Line, the Red Line, and the Green Line. As the pioneer axons grow, they express particular CAMs on their surfaces, creating a labeled "line" that other growth cones can follow. Guidance along previously pioneered tracts by selective adhesion is referred to as the labeled pathways hypothesis (Goodman et al., 1983). When a new neuron sends out its axon, the growth cone is able to choose between specific

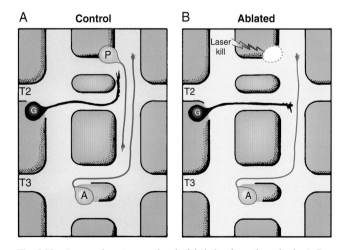

A **Control** **B** **Ablated**

Fig. 5.29 An experiment supporting the labeled pathway hypothesis. A. In a control embryo, the G-growth cone, after crossing the midline, fasciculates with P-axons and not A-axons. B. When the P-neuron is ablated, the G-growth cone stalls and does not fasciculate with the A-axons. *(After Raper et al., 1984)*

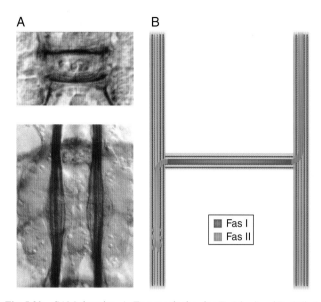

Fig. 5.30 CAM changing. A. Two panels showing FasI (top) and FasII (bottom) distribution in the embryonic CNS of *Drosophila* as revealed with specific antibodies. B. Axons express different CAMs on different segments. A commissural axon in an embryonic *Drosophila* CNS. This axon expresses FasII in the longitudinal pathway to help it fasciculate with other FasII expressing axons in this pathway, switches to FasI while it is in the commissure and fasciculating with other FasI expressing axons, and then switches back again to FasII once it has reached the other side. *(After Zinn et al., 1988; Harrelson and Goodman, 1988; Lin et al., 1994; Goodman, 1996)*

they stop expressing FasII and express FasI (**Figure 5.30**). In addition to following a scaffold of CAM-expressing axons, a new axon may also pioneer a new route during the last leg of its journey and add a new CAM to help future axons reach the same site. Thus, the simple scaffold of the first pioneers with a small number of CAMs becomes increasingly complex as more axons and more CAMs are added to the network.

Homophilic CAMs may balance other guidance cues, particularly repulsive cues. For example in *Drosophila* motor axons, Sema1a functions as an axonal repellent and mediates motor axon defasciculation, which is necessary for proper innervation of target muscles. As we have seen above with FasII overexpression, too much homophilic cell adhesion can cause axons to bundle abnormally, and so there should be an appropriate balance between attractive and repulsive guidance cues expressed on axons. Using genetic techniques it is possible to manipulate the levels of Sema1a and FasII on these axons independently, and change the balance of adhesive and repulsive cues to favor fasciculation or defasciculation. Not all CAMs are homophilic. Some are involved in heterophilic interactions with other CAMs. For example, the CAM, TAG-1, which is expressed on commissural interneurons of the spinal cord, binds to a different CAM, called NrCAM which is expressed on glial cells in the floorplate (Stoeckli and Landmesser, 1995). Some CAMs with heterophilic interactions are even involved in avoidance rather than attraction. In such cases they may be members of heteromultimeric receptor complexes that mediate signaling for repulsive guidance factors such as Semaphorins (Castellani et al., 2002).

REPULSIVE GUIDANCE

In addition to adhesion molecules and extracellular matrix molecules that promote axon growth, there are factors that do just the opposite. These are the inhibitory or repulsive factors (Kolodkin, 1996). Observations of mixtures of different tissues in culture first indicated that neural tissues produce substances that repel axons. In these sorts of experiments, the trajectories of axons are observed when they are cultured in the presence of tissues they normally avoid. For example, the axons of dorsal root ganglia (DRG) innervate the dorsal spinal cord and generally do not enter the ventral regions of the spinal cord. When these DRGs are grown in tissue culture alongside pieces of dorsal and ventral spinal cord, most DRG axons preferentially invade the dorsal cord and avoid the ventral cord even if they have to grow in a circuitous fashion around it (Peterson and Crain, 1981) (**Figure 5.31**A,B). Similar results are obtained by co-culturing pieces of olfactory bulb with septum, a medial structure of the forebrain. Axons from the bulb run laterally in the forebrain appearing to grow away from the septum. When an explant of bulb tissue is placed near the explant of septal tissue, the olfactory axons emerge from the side of the explant opposite the septum (Pini, 1993) (Figure 5.31C,D). These data provide indications that a chemorepulsive mechanism could play a role in axon guidance.

When two explants are co-cultured, there is usually an intermingling of axons, but in the case of retinal and sympathetic co-cultures, it is apparent that the axons avoid one another. Time-lapse video films made of growth cones from one explant as they approach the axons of the other show that these growth cones collapse when they make contact with the foreign axon. They lose their filopodia, retract, and become temporarily paralyzed (Kapfhammer and Raper, 1987a) (**Figure 5.32**). Often, after a few minutes, a new growth cone is formed which advances once more until it again encounters the unlike axon in its path and again collapses. These studies demonstrate that just the briefest contact from a single filopodium is all that is necessary to elicit this aversive behavior, strongly suggesting that growth cones sense a repulsive signal on the surface of the other axon. By pairing different types of explants in such cultures and observing the growth cone interactions, a variety of different collapsing activities effective on different types of growth cones were discovered, showing that there is a rich heterogeneity of repulsive interactions between neurons (Kapfhammer and Raper, 1987b).

Attempts to purify collapsing factors biochemically were aided by a bioassay in which reconstituted membrane vesicles were added to cultures of axons growing on laminin substrates. When vesicles enriched in collapsing activity from the CNS were added to cultures, they caused the immediate collapse and paralysis of all the sensory ganglion cell growth cones on the plate. A 100kd glycoprotein that could cause growth cone collapse, initially called Collapsin, was eventually purified sufficiently to obtain a partial protein sequence (Luo et al., 1993). Through use of the sequence data, the gene was obtained. Collapsin turned out to be a member of a large molecular family, the Semaphorins. Collapsin became known as Sema3A. The first member of the Semaphorin family to be identified, SemaI, came from grasshoppers, and was originally named FasciclinIV because it is expressed on particular axon fascicles (Kolodkin et al., 1993). SemaI is also expressed at segment borders on the limb bud epithelium, and antibodies that neutralize SemaI

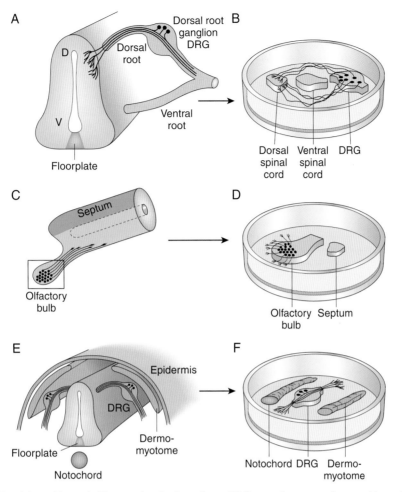

Fig. 5.31 Repulsive guidance. A. The central projections of most DRG axons do not enter the ventral horn of the spinal cord, but rather make synapses in the dorsal horn. B. When cultured together, DRG neurons avoid ventral spinal cord explants to grow to dorsal targets. C. The telencephalon shows olfactory tract fibers originating from the olfactory bulb traveling in the lateral region, far away from the medial septum. D. When cultured together, olfactory bulb axons travel away from the septum indicating the existence of a diffusible chemorepellent. E. Surround repulsion. DRG axons outside the spinal cord elongate in a bipolar fashion between the dermomyotome and the ventral spinal cord and notochord. Many surrounding tissues, including the epidermis, the dermomyotome, the floorplate, and the notochord, secrete diffusible repellents. F. When placed in a collagen gel between a piece of notochord and dermomyotome, DRG axons extend in a bipolar fashion, similar to their pattern in vivo. *(After Peterson and Crain, 1981; Pini, 1993; Keynes et al., 1997)*

function in the limb allow the Ti1 pioneers to cross the segment borders where they normally would not, suggesting that this molecule normally serves a repellent function *in vivo*. In vertebrates, many peripheral target tissues express Semas, particularly Sema3A, and when Sema3A is knocked out, or when the Sema-receptors, Neuropilin, or Plexins, are knocked out, there is strong overgrowth of projections from both motor and sensory nerves into peripheral targets (Taniguchi et al., 1997; Kitsukawa et al., 1997; Waimey et al., 2008) (**Figure 5.33**).

These results suggest that Semaphorin signals are specifically arranged to repel specific axons, presumably those that have receptors to Sema family members. The ectoderm, dermomyotome, and notochord, it turns out, also repel DRG axons (Figure 5.31E,F). If a DRG is placed between a dermomyotome and a piece of notochord in a collagen gel, mimicking its *in situ* position between these tissues, the result is bipolar axon growth that looks very much like the in vivo trajectory of these neurons and suggests that "surround repulsion" may

be a key mechanism for shaping the process trajectories of the developing neurons (Keynes et al., 1997).

An interesting wrinkle concerning repulsive guidance is that the receptor for this cue must first bind the repellent molecule before turning away from it. Since the specificity is often high, this means that the affinity between the receptor and the repellent molecule is also strong. This may not be a problem if the repulsive cue is in the form of a diffusible gradient, but many repellent molecules are attached to cell membranes or the ECM. In such cases, one might imagine that, if the affinity is high enough, the axon would attach rather than be repelled. Two mechanisms have been discovered that appear to resolve this problem. The first is extracellular protein clipping in which the ectodomains of activated receptors are attacked by metalloproteases and cleaved, breaking the attachment between the growth cone and its substrate (Hattori et al., 2000). The second is endocytosis in which the entire receptor-ligand complex is internalized into the growth

125

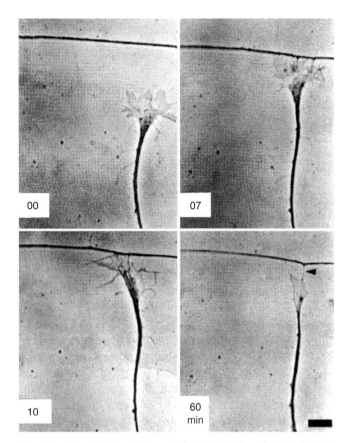

Fig. 5.32 Growth-cone collapse. A time-lapse series of a growth cone from a retinal ganglion cell encountering an axon of a sympathetic axon in culture. Upon first contact, the growth cone retracts and collapses. *(From Kapfhammer and Raper, 1987a)*

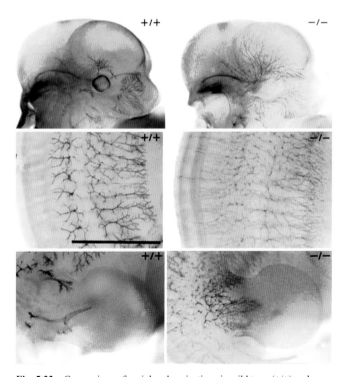

Fig. 5.33 Comparison of peripheral projections in wild type (+/+) and neuropilin knockout (-/-) mice. Top, trigeminal projections. Middle, intersomitic projections of spinal nerves, and bottom, projections into the limb, are all overgrown in mutant mice. *(From Kitsukawa et al., 1997)*

cone (Zimmer et al., 2003). In each case, the molecular bonds holding the growth cone to the repellent surface are neutralized, allowing the growth cone to retract.

The repulsive factor responsible for the guidance of olfactory tract axons away from the septum has been identified as the vertebrate homolog of a *Drosophila* protein called Slit, which is the ligand for the receptor, Robo (Li et al., 1999). The olfactory bulb axons express Robo, which enables them to sense the Slit. Motor neurons of the vertebrate spinal cord also express Robo and grow away from the ventral midline, which expresses Slit (Brose et al., 1999). The axons of motor neurons and olfactory bulb neurons also grow away from cells transfected with Slit in culture. Slit and Robo will become more important to us later in the chapter when we deal with the interesting subject of midline crossing.

CHEMOTAXIS, GRADIENTS, AND LOCAL INFORMATION

In a process termed *chemotaxis*, growth cones claw their way up concentration gradients of diffusible attractants to their source, or in the case of negative chemotaxis turn away from concentration gradients of repellents. For chemotaxis to occur, the growth cone must be positioned in the gradient such that one side is exposed to a higher level of the factor than the other. Gradients of different molecules can be experimentally produced by ejecting solution from the tip of an electrode and allowing the concentration to dissipate as it spreads out into the tissue culture media. By using this method, it was demonstrated that chick dorsal root axons turn toward a source of nerve growth factor (**Figure 5.34**) (Gundersen and Barrett, 1979; Zheng et al., 1994). Since this first demonstration, a large number of molecules have been shown to have chemotactic activity in such pipette-based turning assays. These experiments indicate that growing processes have a mechanism for recognizing concentration differences across a relatively small distance. Using a controlled gradient in a collagen gel, Rosoff et al. (2004) have been able to show that growth cones can sense a difference in concentration of one molecule

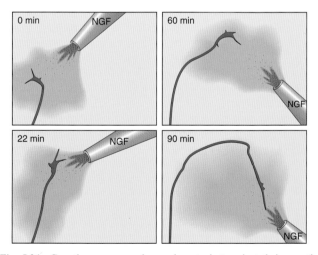

Fig. 5.34 Growth cones can rely on chemotaxis to orient their growth. A sensory neuron turns toward a pipette that is ejecting nerve growth factor (NGF) and thus producing a diffusible gradient. Each time the pipette is moved, the axon reorients its growth. *(After Gundersen and Barrett, 1979)*

in a thousand from one side of the growth cone to the other, making the growth cone the most sensitive reader of chemical gradients known in biology. In the case of pipette-based turning assays, growth cones grow forward at a reasonable pace and turn toward or away from the pipette without speeding up, suggesting that these guidance factors are not simply acting as a general growth-promoting or retarding substances but as directional cues. Growth cones that turn toward attractants also send out more filopodia on the side where the concentration is higher and fewer filopodia on the other side, and the opposite is true for diffusible chemorepellents. This presumably creates differential traction forces toward the attractant or away from the repellent.

Do axons use similar gradient-based mechanisms of growth cone guidance *in vivo*? One of the first examples of an *in vivo* gradient attracting axons by chemotaxis was that of trigeminal ganglion sensory axons in the mouse growing to the maxillary pad epithelium at the base of the whiskers. This is the most heavily innervated skin in the entire body. When the maxillary pad and trigeminal ganglion are removed and placed near each other in a three-dimensional collagen gel, the trigeminal axons preferentially grow toward the explant of the whisker pad, even when there are competing target explants of neighboring pieces of epidermis even closer (Lumsden and Davies, 1986) (**Figure 5.35**). Thus, it was hypothesized that the maxillary pad emits a tropic agent for axon growth. This factor, originally nicknamed "max factor," was later identified as the combination of trophic factors BDNF and NT-3 (O'Connor and Tessier-Lavigne, 1999). Trophic factors will be discussed further in Chapter 7 in connection with cell survival.

An *in vivo* diffusible guidance mechanism was also recognized in nematodes. Here a gene called *unc-6* is expressed at the ventral midline, and in *unc-6* mutants some axons show defects in their growth toward the ventral midline

(Hedgecock et al., 1990; Culotti, 1994). Nematodes with a deletion of the *unc-40* gene show defects in the orientation of those neuronal processes that normally extend toward the ventral midline. Unlike *unc-6* mutants, the *unc-40* phenotype is cell autonomous, with *unc-40* being expressed in the navigating axons. This suggests that Unc-40 might be a receptor for Unc-6. Indeed, the suggestion that Unc-6 is a diffusible ligand was robustly confirmed when the first diffusible axon attractant purified turned out to be a homolog of Unc-6. In the spinal cord of vertebrates, dorsal commissural interneurons grow to the ventral midline, cross the floorplate, and then turn 90° and grow in a longitudinal tract beside the floorplate. Two factors were biochemically purified from embryonic chick brains using a bioassay for directed outgrowth from commissural interneurons in explants of dorsal spinal cord. They were partially sequenced, and the genes encoding these proteins were pulled out of a cDNA library (Kennedy et al., 1994; Serafini et al., 1994). They were called Netrin-1 and Netrin-2 after the Sanskrit "netr," meaning "one who guides." In the spinal cord Netrin-1 is expressed in the floorplate (**Figure 5.36**). The Netrins are secreted molecules found largely associated with cell membranes, but they are diffusible and can clearly reorient growing commissural axons toward a local source of netrin over a distance of hundreds of microns (Colamarino and Tessier-Lavigne, 1995) (Figure 5.36).

In *netrin-1* knockout mice, the dorsal commissural interneurons of the spinal cord do not grow all the way toward the ventral midline. Netrins have also been found in *Drosophila* where they serve a similar role in guiding commissural axons to the ventral midline. The sequence of the *netrin* gene showed it to be a homolog of the nematode gene *unc-6* (discussed below). In vertebrates, DCC, which binds Netrin, is expressed in commissural interneurons, and is essential for the attraction of their axons to the floorplate (Keino-Masu et al., 1996). The DCC protein, which is a transmembrane receptor, is a homolog of the nematode Unc-40 (Chan et al., 1996). A mutation in the Drosophila Netrin receptor homolog, which is called Frazzled, shows similar defects in commissural guidance (Kolodziej et al., 1996). These results suggest a strongly conserved function of Netrins and Netrin receptors in chemoattraction toward the ventral midline.

Although guidance to a distant target through the target's release of a diffusible cue would appear to be a reasonable way to guide axons to their targets, it seems that this solution is used rather rarely in the nervous system, and target tissues do not generally put out very long-range attractants. Rather, as we shall see, axons tend to use a number of intermediate targets and several distinct local cues and gradients on the way to their destinations. The early neuroepithelium is like a patchwork quilt of various guidance cues that act as attractants and repellents and operate in context with particular ECM molecules and CAMs. This rich, detailed molecular array covers the entire developing brain. Axons read the terrain and respond appropriately. Consistent with this analogy are the results of various embryonic perturbations. If the tectum is removed, retinal axons grow toward the missing tissue, suggesting that optic axons use these local cues rather than a long-range diffusible attractant from the tectum as they grow along the optic tract (Taylor, 1990). If a small piece of the optic tract neuroepithelium is rotated 90° before the axons enter it, then they become misoriented when they

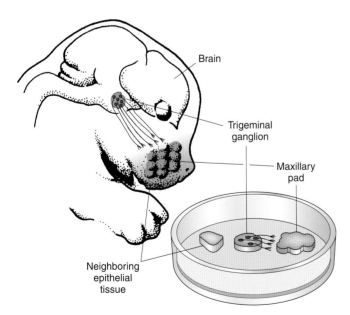

Fig. 5.35 Chemotactic agents from target tissues. Sensory axons from the trigeminal ganglion heavily innervate the maxillary pad of the mouse face, the site of the whisker field. When the trigeminal ganglion is placed into a three-dimensional collagen gel with the maxillary pad tissue and another piece of epithelium, the axons leaving the ganglion grow toward their appropriate target, suggesting that it is releasing a chemotropic agent. *(After Lumsden and Davies, 1986)*

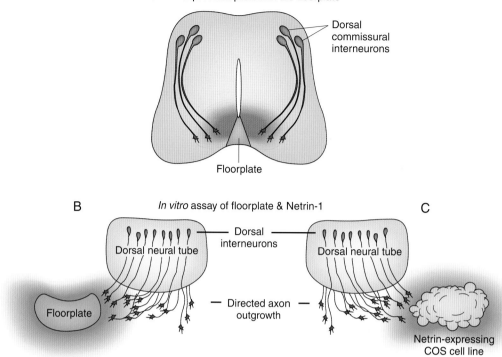

Fig. 5.36 Dorsal commissural interneurons are attracted by a gradient of Netrin. A. Dorsal commissural interneurons grow directly to the ventral midline of the spinal cord along a gradient of Netrin that is released by floorplate neurons. B. In collagen gels, dorsal interneurons are attracted at a distance and orient to the floorplate. C. They are also attracted to Netrin released from a pellet of COS cells which have been transfected with the *netrin* gene. *(After Kennedy et al., 1994)*

enter the rotated transplant (Harris, 1989) (**Figure 5.37**) and correct their course of growth when they exit. These results confirm that the neuroepithelium contains local information to which growing axons respond and that they are not simply following gradients of attractants released by their targets.

As a growth cone migrates through the embryonic nervous system, it encounters new molecular cues every 10 to 50 μms or so. How does such a molecularly complex terrain get established in the neuroepithelium? It appears that the early molecular events that pattern the embryo play a role in setting up the domains of these guidance cues. The homeobox genes that provide segmental identity (see Chapter 1) also direct the expression of various axon guidance cues, and pioneer axons are often found at the boundaries of brain territories that express different homeobox positional markers. Moreover, axonal tracts in embryonic brains exhibit unusual patterns when the expression of homeobox genes is perturbed (Wilson et al., 1997). Recent evidence suggests that homeobox transcription factors, which are usually nuclear, may themselves sometimes be released from cells and act as guidance factors for growth cones (Brunet et al., 2005).

Are the molecules that are involved in the initial patterning of the nervous system also used as cues in pathfinding? We learned in Chapter 2, for example, that Wnts are expressed in caudal high to rostral low gradient during gastrulation. The same Wnt gradient appears to be involved in giving cues to spinal axons about whether they are traveling toward or away from the brain (Lyuksyutova et al., 2003) (**Figure 5.38**). Those commissural axons that cross the midline and extend anteriorly toward the brain respond to Wnt through a conventional Wnt receptor called Frazzled. Thus, axons expressing Frazzled see Wnt as attractive. But what about axons that descend the spinal cord, like the corticospinal neurons that we discussed in the previous chapter; do they see Wnt as a repulsive guidance cue? Indeed, a different type Wnt receptor, called Ryk, originally identified in *Drosophila* as a tyrosine kinase that mediates axonal repulsion to Wnt is expressed in corticospinal axons and is necessary for their posterior growth in the spinal cord (Yoshikawa et al., 2003; Liu et al., 2005). Sonic hedgehog (Shh), which as we have seen participates in patterning the ventral neural tube, also acts as an attractive guidance cue at the ventral midline of the spinal cord (Charron et al., 2003). Conversely, BMPs that pattern the dorsal spinal cord are repulsive to the growth cones of dorsal interneurons that grow away from the dorsal midline (Augsburger et al., 1999). The receptor for Shh signaling as an axon guidance factor is a transmembrane kinase receptor called Boc (Okada et al., 2006). It is worth pointing out that the canonical Wnt and Shh pathways, which we discussed in Chapters 2 and 4, which regulate transcription in response to these patterning cues, operate through quite different, transcription independent, pathways in growth cones that respond to these factors as guidance cues.

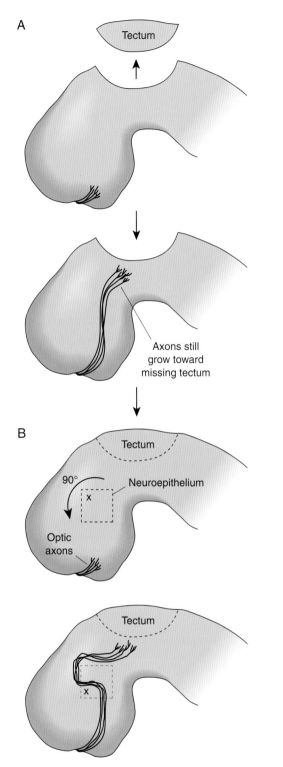

A

Tectum

Axons still
grow toward
missing tectum

B

Tectum

90°

Neuroepithelium

x

Optic
axons

Tectum

x

Fig. 5.37 Retinal axons follow local guidance cues in the neuroepithelium. A. When the tectum is removed, the axons still grow correctly to the tectum, indicating that the tectum is not the source of a diffusible attractant. B. A piece of neuroepithelium in front of the retinal axons is rotated 90° (top). When the retinal axons enter the rotated piece, they are deflected in the direction of the rotation, but they correct their trajectories when they exit the rotated piece, showing that these axons pay attention to localized cues within the epithelium. *(After Harris, 1989, and Taylor, 1990)*

SIGNAL TRANSDUCTION

Signal transduction in growth cones is the process by which guidance cues exert their effects on the dynamic cytoskeleton. CAMs, integrins, repulsive factors, attractive factors, and growth factors are received by receptors on the surface of the growth cone. Many such receptors have intracellular domains that have enzymatic activity when an extracellular ligand is bound and are thus able to amplify the signal (Strittmatter and Fishman, 1991). The Robo receptor has an intracellular domain that mediates a repulsive response to Slit, while Frazzled (the *Drosophila* homolog of DCC) has a different intracellular domain that mediates attraction to Netrin. The action of these intracellular domains is sufficient to drive attraction or repulsion, as is clearly shown in experiments where domains are switched, as in Robo-Frazzled fusion proteins (Bashaw and Goodman, 1999), where the intracellular domain of Frazzled is fused with the extracellular domain of Robo. When this Robo-Frazzled fusion protein is expressed in neurons, their growth cones are attracted to rather than repelled by Slit. Similarly, neurons expressing Frazzled-Robo fusion proteins are repelled by, rather than attracted to Netrin.

Many guidance receptors have intracellular tyrosine kinase (RTKs) or tyrosine phosphatase (RTPs) activities (Goldberg and Wu, 1996; Goodman, 1996). Some receptors, however, do not possess intracellular enzymatic domains on their own. Yet, they may be able to recruit other RTKs, such as the FGF receptor, to do the work. Thus, there is a CAM binding site on the FGF receptor, and NCAM binding can stimulate phosphorylation of growth cone proteins via FGF receptor activity (Williams et al., 1994). In addition to RTKs, some receptors are thought to signal through cytosolic tyrosine kinases and phosphatases. Among the nonreceptor tyrosine kinases (NRTK) are Src, Yes, and Fyn. Neurons from single-gene mouse knockouts that have no src are specifically unable to grow on the CAM, L1, while neurons from single-gene knockout mice that have no *fyn* are unable to grow on NCAM (Beggs et al., 1994). A number of RTPs are found predominantly in growth cones, such as DLAR, which is expressed in *Drosophila* motor axons. Mutants in these genes can also lead to growth and pathfinding defects (Desai et al., 1996).

These enzymatic activities (phosphorylation and dephosphorylation) eventually act on proteins involved in cytoskeletal dynamics. Well-known among these are the small GTPases of the Rho-family (Luo et al., 1997; Dickson, 2001). These exist in two states: an active GTP bound state and an inactive GDP bound state. Rho-family molecules are involved in various actin rearrangements, which are in turn regulated by a host of effector molecules like the Guanidine Exchange Factors (GEFs) that exchange GDP for GTP. Some well-known Rho-family members are RhoA, Rac1, and Cdc42. RhoA is involved in actin mediated neurite retraction, while Rac1 and Cdc42 promote filopodia and lamella formation and thus neurite extension. When Slit binds to Robo, an intracellular adaptor protein called Dock joins the intracellular domain of Robo. Dock then recruits and activates the GEF, Sos (Son of sevenless), to the membrane leading to the local regulation of Rac (Yang and Bashaw, 2006). Ephrins, Plexins and other receptors also associate with GEFs either directly or through adaptor proteins

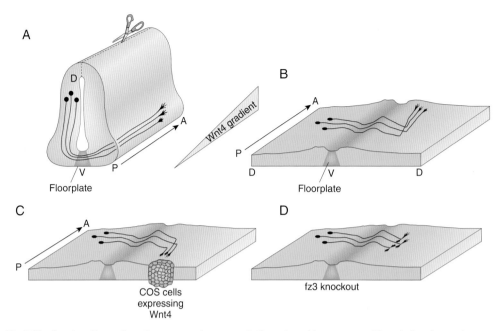

Fig. 5.38 Local gradients of morphogens can orient axons. A. Commissural interneurons of the spinal cord, once they cross the ventral midline, grow anteriorly toward the brain, and up the Wnt4 concentration gradient. B. These axons can be seen well in a filleted preparation grown in culture. The neural tube is sliced open at the dorsal midline and flattened out. Label is applied to the commissural interneurons. C. Commissural interneurons grow posteriorly if a ball of COS cells expressing Wnt4 is placed on the posterior side of such an explant. D. In an *fz3* knockout, lacking the Wnt4 receptor, commissural interneurons do not grow either anteriorly or posteriorly once they cross the midline. *(After Lyuksyutova et al., 2003)*

to activate and deactivate Rho-family members (Schmandke et al., 2007).

Growth cone filopodia are long, motile, and covered with receptors, so they are able to sample and compare different parts of their local environment. They also have a very high surface-to-volume ratio, which can help convert membrane signals into large changes in intracellular messengers such as calcium. Filopodia can show localized transient elevations of intracellular calcium. These transients reduce filopodial motility. Calcium transients can be directly artificially activated in growth cone filopodia by loading neurons with a calcium caging agent that releases calcium upon stimulation with a pulse of light (Gomez et al., 2001). Experiments in which calcium is uncaged on one side of a growth cone generally cause the growth cone to turn toward the side that has the elevated calcium (Zheng, 2000). Calcium may also be released from internal stores in response to a signal from the cell surface or enter the growth cone through calcium channels, and may stimulate or inhibit neurite outgrowth. Serotonin, which stops growth cone advancement in certain neurons of the snail *Heliosoma*, appears to work by increasing calcium levels locally. It is possible in these neurons to cut single filopodia off an active growth cone and study its behavior in isolation. Such isolated filopodia react to the application of serotonin by showing a marked increase in calcium along with a shortening response (Kater and Mills, 1991), giving an excellent insight into just how localized sensory and motor responses in growth cones are. It is not entirely clear how calcium triggers these responses; a likely possibility is that calcium stimulates the activity of actomyosin or certain cytoplasmic kinases, such as CAM kinaseII or PKC, which then go on to affect the cytoskeleton (Nishiyama et al., 2008).

Local protein synthesis also happens within the growth cone because of the presence of mRNAs and a full complement of protein synthetic machinery. Protein endocytosis and degradation machinery is also present in growth cones, including the ubiquitinating enzymes that target proteins to the proteasome for degradation (Campbell and Holt, 2001). When a guidance molecule stimulates protein synthesis or degradation, it does so through MAP kinase pathways. There are several different MAP kinases, some of which eventually activate a protein called Target of Rapamycin, mTOR (rapamycin being a natural toxin of stimulated protein synthesis in various cell types). TOR is a kinase that phosphorylates a protein called eIF4E-BP, a negative regulator of the translation initiation factor eIF4E (Campbell and Holt, 2001). Other MAP kinases activate specific ubiquitin ligases that attach ubiquitin molecules to particular proteins targeting them for degradation. In fact, it seems likely that there are a variety of downstream kinases that regulate different aspects of the growth cone response when stimulated by a single guidance cue.

THE MIDLINE: TO CROSS OR NOT TO CROSS?

One could say that the basic axon scaffold of the embryonic CNS starts with a set of decisions that exploring axons must make about the midline. To cross or not to cross? To grow up or down the nerve cord at what distance from the midline? When we investigate how axons make these decisions, we encounter several intriguing problems in axonal guidance. To get to the midline, axons use attractive guidance factors like Netrin and Shh, but why then if the midline is so attractive, do these axons ever leave it, and once they emerge on the opposite side of the midline, why do they turn longitudinally into specific lateral pathways without ever being attracted to the midline again (Dickson and Gilestro, 2006)? Insights into this problem came first from the *Drosophila* mutant *roundabout*

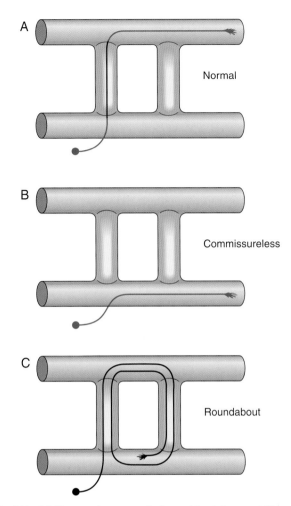

Fig. 5.39 Midline crossing mutants in *Drosophila*. A. In normal flies, many neurons cross the midline once in a commissure and then travel in longitudinal fascicles on the other side. B. In *commissureless* mutants, the axons do not cross but travel in longitudinal tracts on the same side. C. In *roundabout* mutants, the longitudinal tracts do not form properly because the axons keep crossing back and forth. *(After Seeger et al., 1993)*

(*robo*). Commissural axons in *robo* mutants cross the midline and then recross it at the next commissure. They do this over and over, sometimes going in circles (**Figure 5.39**). Robo is the receptor for a midline repellent Slit, and *slit* mutants show a similar phenotype. Interestingly, there is another class of mutant in *Drosophila* called *commissureless* that has the opposite phenotype (Seeger et al., 1993; Tear et al., 1993). In these mutants, commissural neurons are unable to cross the midline and remain ipsilateral (Figure 5.39). When the *comm* gene was cloned and its expression was studied in wildtype flies, it was found that only those axons that normally cross the midline normally express the Commissureless protein (Comm); those that never express Comm, never cross the midline. In those axons that do cross, Comm, a sorting protein, binds Robo as soon as it is made in cell body and sorts it for destruction in the endosome, so that Robo never gets to the cell membrane of the growth cone and the axons are not repelled by Slit (Keleman et al., 2002; Myat et al., 2002). Once they cross the midline, however, a switch is thrown, Comm is downregulated and Robo returns to the surface, so the axons, though still attracted by Netrin, are more strongly repelled by Slit, and they never recross the midline.

This illustrative mechanism is conserved, though with an interesting twist in molecular detail for vertebrates (**Figure 5.40**). Instead of Comm, vertebrate commissural axons use Rig1. Rig1 is an unusual Robo in that it does not lead to repulsion from Slit. Rather it appears to desensitize axons to Slit (Marillat et al., 2004). There are two splice variants of Rig1. The first splice variant, which has the activity described above, is localized in the growth cones of axons before they cross the midline. Once across the midline, and possibly as a result of crossing it, the first splice form is replaced by the second, which does not have the ability to desensitize axons to the repulsive effects of Slit, the result of which is that they are repelled rather than attracted to the midline (Chen et al., 2008). Experience of the floorplate also causes axons to lose their attraction for Netrin (**Figure 5.41**) (Shirasaki et al., 1998). This effect appears to be mediated by a direct interaction between Robo1 and the Netrin receptor (DCC) that happens as a result of exposure to Slit at the midline so once these axons have crossed the midline they no longer sense the attractive activity of Netrin (Stein and Tessier-Lavigne, 2001). Thus, postcrossing axons, while having regained sensitivity to Slit and lost sensitivity to Netrin, are at the same time repulsed from and no longer attracted to the midline, and so they never recross it.

Now that axons have decided to cross or not, the next decision is how far away from the midline to grow. In *Drosophila*, there are three basic longitudinal tracts—medial, intermediate and lateral—and three different *robo* genes. The axons that grow in the most medial tract closest to the midline express Robo1. The axons that grow in the most lateral tract, farthest away from the ventral midline, express Robo1, Robo2, and Robo3. Finally, the axons that grow in the intermediate tract express Robo1 and Robo3 (Rajagopalan et al., 2000; Simpson et al., 2000). All the Robo receptors appear to depend on Slit, yet an extensive analysis in which different Robos or parts of Robos replaced each other (Evans and Bashaw, 2010; Spitzweck et al., 2010) revealed that each Robo receptor acts in a distinct way. Robo1 is critical for midline repulsion, and neither Robo2 nor Robo3 can replace this function, which is dependent on a unique cyctoplasmic domain. Robo2 is critical for the lateral fascicle choice, and this maps to the distal region of the extracellular domain. Finally, in *robo3* mutants, axons that would normally grow in the intermediate tract grow in the medial one instead, and when misexpressed Robo3 shifts axons from the medial to the intermediate fascicles. Robo3, however, in this context is not doing anything unique among the Robos as misexpression of either the Robo1 or Robo2 has exactly the same effects. The intermediate position may thus be determined by the total level of Robo receptors that are expressed.

The third question, to turn up or down the cord, we dealt with earlier in the chapter in the context of the vertebrate spinal cord, with Wnt acting as a repellant to Ryk expressing descending axons and as an attractant to Frazzled expressing ascending ones.

ATTRACTION AND REPULSION: DESENSITIZATION AND ADAPTATION

The responses of growth cones to the same cues may change as axons progress toward their final destinations. We have just seen examples of this at the ventral midline, which can be considered an intermediate target. These axons must not get stuck

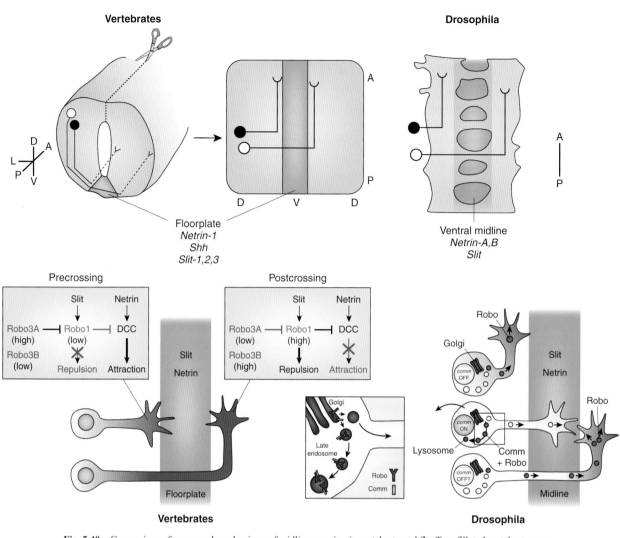

Fig. 5.40 Comparison of conserved mechanisms of midline crossing in vertebrate and fly. Top: filleted vertebrate preparation, obtained by dissecting the embryonic spinal cord of the chick and the embryonic *Drosophila* ventral nerve cord. Homologous guidance molecules (particularly Netrins and Slits) are expressed at the ventral midline. Bottom right. In flies, Comm in precrossing axons binds to newly made Robo coming from the Golgi apparatus and sorts it to the late endosome for degradation. Bottom left. In vertebrates, Robo3 antagonizes Robo1 to allow crossing in mice. In precrossing commissural axons, Robo3 levels are high, and Robo1 levels low. Robo3A inhibits Robo1-mediated repulsion in these axons so that they are instead attracted to the floorplate by Netrin-1 and Shh. After crossing, Robo3A is replaced by Robo3B, and Robo1 activity is high. Axons are now repelled by signaling of Slit through Robo1. In addition, attraction to Netrin is downregulated, owing to a Slit-dependent interaction between Robo1 and DCC. *(From Dickson and Gilestro, 2006)*

where the Netrin concentration is highest, but should move on and take the next portion of their journey. One way to do this is to neutralize the signalling of an attractant or a repellant as happens when commissural neurons cross the midline. But can one make attractants downright repulsive and vice versa? Netrin (or Unc-6 in worms) is attractive for axons that express DCC (or Unc-40). Mutants in DCC (or Unc-40) disrupt the migration of axons toward Netrin (or Unc-6). The *unc-5* mutant, however, has the opposite phenotype. It disrupts dorsal migration (Culotti, 1994), but the *unc-5* phenotype is dependent on wild-type *unc-6* function. Interestingly, when neurons whose axons normally grow toward Unc-6 at the ventral midline, are made to misexpress Unc-5, these axons now grow dorsally instead, again depending on the normal expression of Unc-6 to do so. Unc-5, like Unc-40, codes for a transmembrane protein,

possibly another Unc-6 receptor, but is involved in chemorepulsion rather than chemoattraction (**Figure 5.42**).

This means that Netrins can act as chemorepulsive agents for neurons that express Unc-5 type receptors and chemoattractive factors for neurons that express DCC-type Netrin receptors. That Netrin can act as a long-range chemorepulsive factor in vertebrates has been demonstrated in trochlear motor neurons whose axons grow dorsally away from the ventral midline (Colamarino and Tessier-Lavigne, 1995) (Figure 5.42). In collagen gels, these axons grow away from explanted floorplate tissues or from cells transfected with a Netrin-1 expressing gene. The switch from attraction versus repulsion to Netrin can be mediated by the misexpression of an Unc-5 in single neurons (Hong et al., 1999). Surprisingly, Netrin repulsion can also be mediated by DCC alone, depending on the

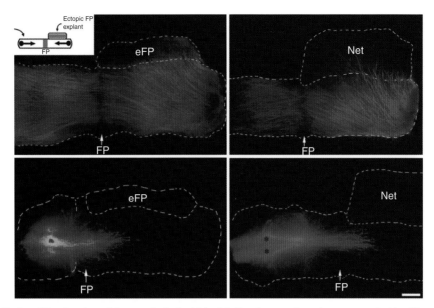

Fig. 5.41 In a filleted explant of the neural tube in chicks, labeling the commissural axons on each side with crystals of DiI shows that precrossing axons are attracted to either an ectopic floorplate or aggregate of Netrin expressing COS cells, whereas postcrossing axons that have experienced the floorplate are no longer attracted to either the ectopic floorplate or Netrin. *(From Shirasaki et al., 1998)*

level of intracellular cAMP (Ming et al., 1997). When cAMP inside the growth cone is high, Netrin acts like an attractant to spinal and retinal axons, but when cAMP is experimentally lowered using drugs that block adenylate-cyclase, then the response to Netrin by the very same axons is one of avoidance and repulsion (**Figure 5.43**). Similarly, many other attractants can be switched to repellents by pharmacologically down-modulating cAMP in growth cones. One hypothesis for how this works is that the signal from the guidance receptor catalyzes cytoskeletal polymerization/depolymerization reactions. In high cAMP this reaction favors polymerization, while in low cAMP it favors depolymerization (Ming et al., 1997; Song et al., 1997). Although some guidance cues are primarily modulated in this way by cAMP, others such as NGF and Sema3A are modulated by cGMP (Song et al., 1998). Neurotransmitters like serotonin can also work to convert attraction to repulsion in vertebrates, as in the case of thalamocortical afferants which become repelled from Netrin on exposure to serotonin (Bonnin et al., 2007).

As axons grow past intermediate targets, such as the optic nerve head, they are also getting older and maturing. This intrinsic aging process may also affect their internal cAMP levels and thus the way an axon responds to a guidance cue like Netrin. By the time retinal axons enter the brain, they are no longer responsive to Netrin (even if they are grown on low levels of laminin), and by the time they reach the tectum Netrin is repulsive. This change in responsiveness happens in a stage-dependent way even in isolated retina explants where the axons are not exposed to the optic pathway (Shewan et al., 2002). Another way that growth cones can change their sensitivity to a particular guidance cue is to make or degrade proteins that are critical for responding to the cue, such as a receptor. Retinal growth cones do not "see" Sema3A in their pathway until they grow along the optic tract. At early stages of pathway navigation, retinal axons do not express Neuropilin, the Sema3A receptor, and so Sema3A is neither attractive nor repulsive to them. However, by the time

these axons have entered the optic tract, they express Neuropilin and are repelled by Sema3A (Campbell et al., 2001).

It was initially thought that all proteins in the growth cone were made in the cell body and shipped to the growth cone, but as mentioned above, growth cones have the local machinery to make and degrade proteins rapidly in response to guidance cues, and in many cases, if either translation or degradation is pharmacologically inhibited, a number of guidance cues become ineffective, suggesting that growth cone navigation may depend on the local manufacture and degradation of proteins. mRNAs, such as the mRNA for beta-actin, which initiates F-actin polymerization, and the mRNA for Cofilin, which depolymerizes F-actin, are found in growth cones (Bassell et al., 1998). Attractive turning to a gradient like Netrin is associated with an increase in beta-actin synthesis preferentially on the side of the growth cone closest to the Netrin source, while repulsion is likely to involve an asymmetric rise in cofilin closest to the repellant (Lin and Holt, 2008). Local protein synthesis might also be involved in the manufacture of new receptors once growth cones have reached intermediate targets; this would allow them to become sensitive to new guidance cues on the next leg of the journey (Brittis et al., 2002).

Various studies have shown that growth cones adapt to guidance cues. For example, if a growth cone bumps into an axon with a repulsive guidance cue (as mentioned above), it will collapse and retract. When it regrows, it may run into the same axon again and repeat the collapse to regrowth cycle. However, after several such cycles, it has been observed that growth cones generally become desensitized to the collapse-inducing factor and are able to grow over repulsive axons (Kapfhammer et al., 1986). In another study, axons were tested to see how far they would crawl along a membrane on which there was an increasing gradient of the repulsive guidance factor, EphrinA. Axons were started on a platform of different levels of EphrinA, and those that started on higher concentrations ended up growing the farthest, suggesting that they

133

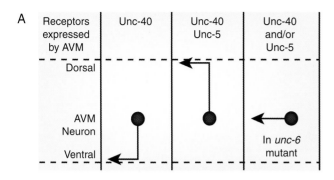

A

| Receptors expressed by AVM | Unc-40 | Unc-40 Unc-5 | Unc-40 and/or Unc-5 |

Dorsal

AVM Neuron

Ventral

In *unc-6* mutant

B

Netrin-1 expression pattern in the floorplate

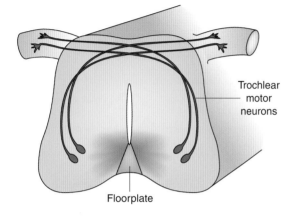

Trochlear motor neurons

Floorplate

C

In vitro assay of floorplate and Netrin-1

Ventral hindbrain

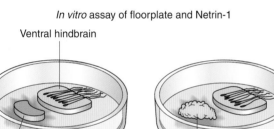

Floorplate

COS cells

Fig. 5.42 Repulsive guidance by Netrin through Unc-5. A. In *C. elegans*, guidance of the AVM neuron GC by the Unc-6/netrin guidance cue and the Unc-40 and Unc-5 receptor subtypes. AVM neurons normally express the UNC-40 but not the Unc-5 receptor subtype. AVM GCs migrate ventrally to the ventral nerve cord, then turn anteriorly and migrate within the ventral nerve cord to the nerve ring. The normal ventralward migration phase depends on ventrally expressed Unc-6 and other, unknown guidance cues. Ectopic expression of the Unc-5 receptor in AVM causes GCs to migrate in a dorsalward direction to the dorsal nerve cord, and this is also dependent on Unc-6. (From Merz and Culotti, 2000) B. Trochlear motor neurons in the vertebrate embryo are repelled by netrin. Trochlear motor neurons arise in the ventral neural tube at the midbrain/hindbrain region. They grow away from the ventral midline to decussate and leave the brain dorsally. C. Trochlear neurons in a collagen gel explant culture grow away from the floorplate, and in from COS cells expressing Netrin. *(After Colamarino and Tessier-Lavigne, 1995)*

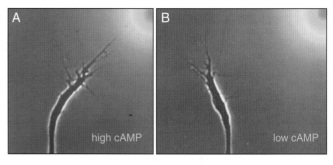

A high cAMP

B low cAMP

Fig. 5.43 cAMP modulates growth-cone turning. A. When internal cAMP is high, the growth cone of embryonic spinal neurons grows toward a source of Netrin ejected by a pipette. B. When cAMP is pharmacologically lowered, the same neurons are repelled by Netrin. *(After Ming et al., 1997)*

new proteins that counteract this process (Ming et al., 2002; Piper et al., 2005). Growth cones can thus adjust the levels of the proteins that are critical for axon navigation. When this is added to the ability of growth cones to regulate cyclic nucleotides that can rapidly mediate a switch between attraction and repulsion, the picture that emerges of the growth cone is that of a very autonomous machine capable of continually redefining itself as it navigates through the embryonic brain.

THE OPTIC PATHWAY: GETTING THERE FROM HERE

The optic pathway, composed of the axons of retinal ganglion cells (RGCs), leads from the retina to the tectum and is one of the most extensively studied examples of axonal guidance. Work on this pathway over many years by many laboratories shows how various cues and mechanisms, which we have discussed in this chapter, are combined in taking a single axon from its origin to its target (Erskine and Herrera, 2007; Johnson and Harris, 2000). The journey starts within the retina itself (Bao, 2008) (**Figure 5.44**). All RGCs initiate axons on the basal pole of their cell bodies, because the cells are polarized in this direction by ECM components of the basal lamina, to which all neuroepithelial cells are attached and which stimulates the emergence of growth cone (Zolessi et al., 2006). Retinal axons become confined to the optic fiber layer, where they grow with the support of laminin on the basal side and the cell bodies of retinal ganglion cells on the other. The pioneer retinal ganglion cells, the first to be born in the central retina, make a spoke-like radial pattern on the retinal surface as they head toward the optic fissure. They seem to prefer to grow on the surfaces of some RGCs rather than others. In chicks, Slit works as an adhesive guidance molecule on these RGCs which thus act as stepping-stones. As in the grasshopper limb, follower axons of RGCs that are born later need to fasciculate with the axons of pioneers to find the optic nerve head (Pittman et al., 2008). To do this they use a variety of homophilic CAMs including L1, NrCAM and DM-GRASP a.k.a. Neurolin, which are used in the fasciculation of these retinal axons along the entire pathway. If the pioneers are missing, the axons from more peripheral RGCs cannot find their way to the nerve head.

Slit also acts as a repellent in the inner retina so that retinal axons remain in the optic fiber layer. If the *slit* gene is knocked out in mice, RGC axons fasciculate abnormally and dive into

had partially adapted to EphrinA (Rosentreter et al., 1998). Adaptation can be subdivided into two parts, desensitization followed by resensitization. Desensitization involves the internalization and degradation of receptors to guidance factors (Piper et al., 2005), while resensitization involves making

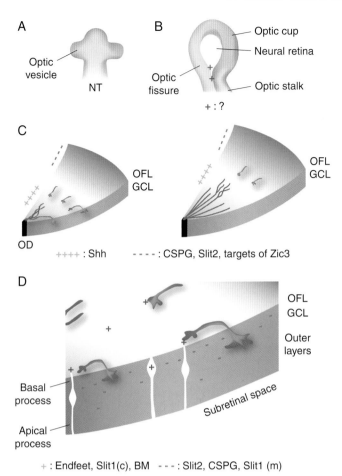

A
Optic vesicle
NT

B
Optic cup
Neural retina
Optic fissure
Optic stalk
+ : ?

C
OFL GCL
OFL GCL
OD
++++ : Shh   - - - - : CSPG, Slit2, targets of Zic3

D
OFL GCL
Outer layers
Basal process
Apical process
Subretinal space
+ : Endfeet, Slit1(c), BM   - - - : Slit2, CSPG, Slit1 (m)

Fig. 5.44 Schematic diagrams of intraretinal axon pathfinding. Proper guidance of RGC axons inside the retina requires cooperation of a multitude of factors. A. Optic vesicles are derived from the evagination of the neural tube. NT: neural tube. B. Invagination of optic vesicles results in formation of the optic cup and also a ventral groove. The ventral groove is called the "optic fissure" that serves as the passageway for the exiting of the first axons and inward migration of the mesodermal cells for formation of retinal artery. The optic fissure fuses and the inner layer of the optic cup becomes the neural retina. Signals emitting from the optic fissure/disc may be required for guiding the early RGC axons to the optic fissure/disc. C. At later stages, when RGC axons are projected at a substantial distance away from the optic fissure/disc, a combination of positive guidance factors (shown in green) and negative factors (shown in pink) ensure the correct orientation of the RGC axons. Once the axons are oriented correctly, they can grow along with the other axons of the RGC cells at more central positions. OD: optic disc. D. Mechanisms are also present to restrict the RGC axonal growth within the OFL layer of the retina. The undifferentiated neuroepithelial cells are shown as white cells with bipolar processes, whereas the RGCs are shown as gray cells with their soma in the GCL. Glial/neuroepithelial endfeet and basal lamina (BM) provide positive/permissive substrates whereas Slit2 provides inhibitory signals to the RGC axons. Slit1 is a positive (or permissive) factor in chick (c), not in mouse (m). *(From Bao, 2008)*

Chondroitin Sulfate (CSPG) that are expressed at higher levels peripherally. If these cues are missing, retinal axons become disoriented in the retina and often fail to converge at the optic nerve head.

When RGC axons get to within about 50 mm of the optic nerve head, they encounter a high concentration of Netrin, which acts as an attractant (Deiner et al., 1997). If retinal neurons are grown on low levels of laminin or fibronectin, cAMP is high in the growth cone, and Netrin is attractive, but Netrin can be switched to a repulsive guidance cue to these axons by experimentally lowering cAMP. Similarly, if the cAMP level is artificially raised in retinal growth cones, they are once again attracted to Netrin. Laminin dramatically reduces the cAMP level in retinal growth cones, which turns Netrin from an attractant to a repellant. So for retinal axons growing on the laminin rich basal lamina at the optic nerve head, Netrin is repulsive rather than attractive, and so these retinal axons leave the basal lamina, turning away from the surface of the retina to dive into the optic nerve where they travel until they enter the brain near the chiasm (Hopker et al., 1999). This is a nice example of how environmental cues, such as ECM molecules, can by modulating whether a guidance cue is attractive or repulsive, contribute to axonal guidance.

The optic nerves enter the brain at the optic chiasm, where RGC axons either cross to the contralateral side of the brain or remain ipsilateral. Molecularly this is a complex decision (Petros et al., 2008). At the chiasm, all axons are exposed to Slit and Shh again, although by this time the RGC axons have changed their sensitivity and see these same molecules as repulsive rather than attractive. Slit and Shh are expressed anterior of and posterior to the chiasm, but not in the chiasm itself (Erskine et al., 2000; Trousse et al., 2001). Thus, these molecules corral the RGC axons into the chiasm proper. Zebrafish mutants called *astray* do not have a functional Robo receptor in their RGCs, and in these mutants fewer RGC axons find the chiasm, and many get lost when they enter the brain (Fricke et al., 2001; Hutson and Chien, 2002). At the chiasm itself, there is a high concentration of another repulsive guidance molecule called EphrinB. The RGCs from the ventrotemporal part of the mouse retina express the transcription factor Zic2 which upregulates EphB, the receptor for EphrinB, in this part of the retina. These ventrotemporal axons are repelled by the EphrinB at the chiasm and take the ipsilateral optic tract. The dorsal and nasal axons, however, because they do not express Zic2 and therefore EphB, remain insensitive to EphrinB, and so cross the chiasm to the optic tract on the other side of the brain (Nakagawa et al., 2000; Williams et al., 2003).

Once in the optic tract, retinal axons are influenced by the repulsive guidance cue Sema3A and the ECM proteoglycan heparan sulfate so that they are guided toward the tectum (Walz et al., 1997; Campbell et al., 2001; Irie et al., 2002). At the front of the tectum, RGC axons encounter a sudden drop of FGF, which signals that they have entered the target area (McFarlane et al., 1995). In the target area they encounter orthogonal gradients of Ephs, Ephrins, and Wnts that signify tectal coordinates (see Chapter 6). At this point, the process of adaptation and resensitization allow axons to crawl to the appropriate position in these gradients. In the frog or fish embryo, retinal growth cones encounter all these cues and several others, which space does not allow us to discuss, during their journey of no more than about 800 μm (**Figure 5.45**).

the depths of the retina (Jin et al., 2003; Thompson et al., 2006). Confined to the optic fiber layer, the pioneer axons must navigate to the optic nerve head near the center of the retina. In *slit* mutants, deep axons still grow toward the central retina demonstrating that Slit is not essential for this centripedal growth. It seems that these retinal pioneers are attracted to positive cues, such as Shh, percolating from the optic fissure and repelled by factors such as Sdf-1 and the proteoglycan

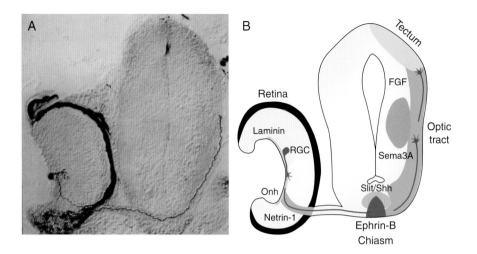

Fig. 5.45 Local cues for retinal ganglion cells. A. A photograph of a single retinal ganglion cell growing toward the tectum. The cell has been filled with a dye for visualization. B. Various guidance cues that the growth cone of the RGC uses to orient toward its target in the tectum are artificially colored in, although in close correspondence to the known distribution of these guidance cues. *(Courtesy of Christine Holt)*

BOX 5.2   Axon Regeneration

In the adult mammalian CNS, axons fail to regenerate following injury (Ramón y Cajal, 1928; Aguayo et al., 1990). Axons that are cut find it difficult or impossible to cross the lesion site, and many neuronal cells whose axons are cut die as the result of the injury. This means that injuries that break axons in the spinal cord of an adult human can lead to permanent paraplegia or quadriplegia. Work on axonal regeneration is therefore of intense medical interest. The inability of adult central axons to regenerate is in stark contrast to the situation in the peripheral nervous system where regeneration is possible and the situation in lower vertebrates such as fish and amphibia. In these animals CNS axons can regrow. For example, salamander retinal ganglion cells are fully capable of regeneration (Piatt, 1955), and severing the optic nerve in such an animal, an insult that would lead to permanent blindness in an adult human, is followed by the regrowth of these axons and the restoration of vision. The failure of regeneration in the adult mammalian nervous system is also in contrast to the ability of the developing nervous system to send out long axons. The capacity of central axons to regenerate is therefore lost during the early stages of mammalian development (Kalil and Reh, 1982). It is as though there were a connection between evolution and development in the ability of axons to regenerate central axons. Perhaps the key to central regeneration is to find a way of making the damaged tissue act more like it did during the time when it was developing or when vertebrates were more primitive.

For both intrinsic and extrinsic reasons central neurons of mammals are challenged by regeneration. Let us look at the extrinsic factors first. The importance of extracellular cues *in vivo* is clearly illustrated by the ability of peripheral nerve grafts to support central axonal regrowth (Richardson et al., 1980; David and Aguayo, 1981; Aguayo et al., 1990). In a set of classic studies, it was shown that while transected central axons were unable to grow within the CNS, they could grow for many centimeters through a sheath of nonneuronal cells that ordinarily provide insulation to motor axons in the periphery (**Figure 5.46**). The axons that are able to regenerate following a pyramidal tract lesion in neonatal hamsters or cats are the ones that grow *around* the lesion site. It appears that regenerating axons are not able to penetrate the injury site because there are inhibitory factors at the lesion site (Bregman

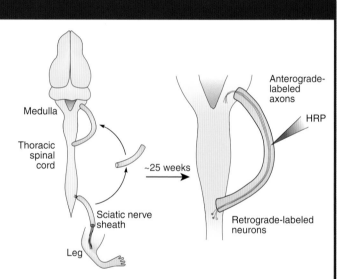

Fig. 5.46 Central neurons in spinal tracts do not regrow long axons after they are transected, but if they are allowed to innervate a sheath of peripheral nerve, they can regrow over substantial distances. *(After David and Aguayo, 1981)*

and Goldberger, 1983). Indeed, while embryonic and peripheral glial cells support neurite outgrowth, adult astrocytes and oligodendrocytes appear to inhibit neurite outgrowth. Part of the problem for injured CNS axons in mammals is the invasion of the wound site with various glia, which produce repulsive cues for axon growth. By X-irradiating mouse spinal cords during neonatal development, it was possible to create mice that are deficient in glial cells. In these animals, spinal axons regenerated past a transection point, a behavior that they never display in normal animals (Schwab and Bartholdi, 1996).

To find the molecular components involved in inhibiting central regeneration, the system was brought into culture where it was found that CNS neurons stop, and sometimes collapse, when they touch oligodendrocytes. Liposomes from these cells and preparations of myelin were used to identify an inhibitory factor

BOX 5.2   Axon Regeneration—Cont'd

that causes CNS growth cones to collapse. A monoclonal antibody to this factor, which is now called Nogo, was then made and tested in culture for its ability to block the collapsing activity. In the presence of antibody, axons grew over oligodendrocytes without stopping or collapsing. The antibody was then tested *in vivo*, using mice with partially severed spinal cords. In the presence of Nogo antibodies, many more axons were able to regenerate beyond the crush than in control animals, and there was considerable functional recovery, suggesting that Nogo is a critical component of the failure of spinal regeneration (Schnell and Schwab, 1990; Bregman et al., 1995), although there has been some debate as to whether knocking out Nogo in mice leads to enhanced regeneration. The Nogo receptor (NgR) was identified and found to be a receptor for other myelin-derived regeneration inhibitory factors such as myelin associate glycoprotein (MAG) and oligodendrocyte myelin glycoprotein (OMgp) indicating that this receptor may provide an insight into where the signals that inhibit regeneration converge (Fournier et al., 2001). NgR is part of a receptor complex that activates RhoA to inhibit axonal growth (**Figure 5.47**).

Astrocytes often accumulate around CNS wounds, forming complex scars. These cells produce an extracellular matrix that is also inhibitory to axon regeneration, and one of the key components of this inhibitory material appears to be chondroitin sulfate glycosaminoglycan chains, which are found on many proteoglycans in the astroglial scar (Asher et al., 2001). Even when plated on the growth-promoting ECM component, laminin, spinal neurons stop growing when they confront a stripe of chondroitin sulfate. In culture, the inhibitory component can be digested away with chondroitinase, rendering the matrix more permissive to axon growth and regeneration. To see if chondroitinase could be used to treat models of CNS injury *in vivo*, rats whose spinal cords had been transected, were treated locally with the enzyme. Such treatment restored synaptic activity below the lesion after electrical stimulation of corticospinal neurons and promoted functional recovery of locomotor activity (Bradbury et al., 2002).

Many repellents that are known to function during neural development become re-expressed in the adult following injury: including Wnts, Semas, and Ephrins and their receptors. These factors negatively influence regeneration, perhaps because they are not arrayed in the same kind of useful geometries for guidance as they were when the embryo was much smaller and neurons were pioneering the first pathways in the brain. In order for mammalian nerve regeneration to become more of a reality, it may be necessary to control the expression of or sensitivity to these guidance cues as well as to override the inhibitory factors at the site of the glial scar (Benowitz and Yin, 2007).

Issues with the regenerating growth cone's environment, however, are not the only impediment to regeneration in the mammalian CNS. There is still the problem that central axons simply do not regenerate very well even when the conditions are good. Adult axons can grow for a short distance (<500 mm) in many central locations (Liu and Chambers, 1958; Raisman and Field, 1973a). The growth of very young neurons does not appear to be so restricted in an adult nervous system. Human neuroblasts are able to form long axon pathways when transplanted into excitotoxin-lesioned adult rat striatum (Wictorin et al., 1990). Similarly, transplanted mouse embryonic retinal ganglion cells are able to grow long distances within the rostral midbrain of neonatal rats and selectively innervate normal targets (Radel et al., 1990). Indeed, myelin inhibits regeneration from old but not young neurons. What do these young axons have that older

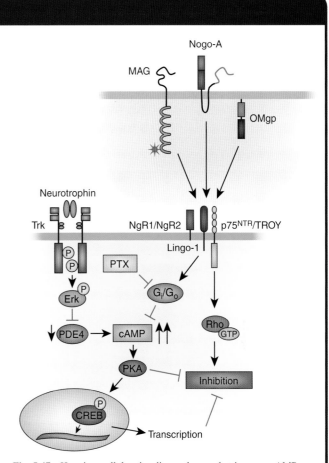

Fig. 5.47   How intracellular signaling pathways that increase cAMP may help in overcoming inhibition by CNS regeneration. Priming with neurotrophins activates Trk receptors and Erk, which in turn produces a transient inhibition of PDE4 activity. This causes intracellular cAMP levels to rise and, upon reaching a threshold level, cAMP will activate PKA and initiate transcription by CREB. These events allow primed neurons to overcome inhibition mediated by such inhibitors of regeneration as MAG, Nogo-A, and OMgp binding to the receptor complex consisting of NgR1 or NgR2, p75NTR or TROY and LINGO-1. Myelin inhibitors also activate Gi/Go, which inactivates adenylate cyclase and inhibits cAMP synthesis. When neurotrophins are added directly to neurons in the presence of myelin, intracellular cAMP rises, but the threshold level cannot be reached due to the activity of Gi/Go, and as a result inhibition persists. Administration of PTX in conjunction with neurotrophins blocks the effects of Gi/Go and allows neurons to overcome inhibition in the absence of priming. *(From Hannila and Filbin, 2008)*

axons do not? It was found that the levels of cAMP in young growth cones are much higher than in older axons (Cai et al., 2001). By increasing the cAMP levels, one can turn old neurons into neurons that behave more like young ones in terms of their regenerative potential (Qiu et al., 2002). Remembering, from the section of this chapter on how repulsive cues can be interpreted as attractive ones by increased cAMP, and how other signals can regulate cAMP levels in growth cones, there is excellent potential for using cAMP based strategies to aid in CNS regeneration (Figure 5.47) (Hannila and Filbin, 2008). The difference between the ability of neonatal versus mature CNS neurons to regenerate led researchers to check the activity of genes that are expressed selectively in differently aged

Continued

BOX 5.2   Axon Regeneration—Cont'd

neurons. One group of genes that is differentially expressed in this way is the Kruppel-Like Factor (KLF) family of transcription factors. Some KLFs (e.g., KLF4/9) are expressed more in adult neurons and inhibit regeneration, whereas others (e.g., KLF6/7) are expressed more in neonatal neurons and promote regeneration (Moore et al., 2009). Another key intrinsic difference between old and young axons may have to do with protein synthesis. Young growth cones are full of protein synthetic machinery, but the axons of older neurons have less of such machinery. There is a good correlation between the ability of

a growth cone to make new proteins and its ability to regenerate *in vitro* (Verma et al., 2005). Knocking out a key negative regulator of protein synthesis, known as PTEN, in mature retinal ganglion cells in mice can dramatically improve the regenerative ability of the optic nerve (Park et al., 2008). A new challenge will be to find ways to crank up the protein synthetic machinery in the growth cones of damaged CNS neurons to see if this can aid recovery. When considering all these data, it seems that full recovery after an injury will require a strategy for dealing with both intrinsic and extrinsic factors.

SUMMARY

We began this chapter by comparing axonal navigation with human navigation. We mentioned the need for a motor, and we have seen that the growth cone by virtue of its dynamic cytoskeleton is able to locomote forward, turn, stop, and even retract. We mentioned the need for guidance cues, and we have seen a variety of cues attached to the extracellular matrix and cell surfaces, and other guidance molecules that appear to act through diffusive gradients. Some of these guidance factors promote growth and adhesion, while others inhibit growth and adhesion. These various signals are integrated and communicated

to the dynamic growth cone cyctoskeleton. We are, however, still a long way from understanding how axons grow to their targets. The molecules mentioned in this chapter are used only in some neurons, and it is fair to say that for even the best studied neurons, such as retinal ganglion cells, we understand only parts of their navigation, but not their entire route. Many more factors than mentioned here are involved, and many remain to be discovered. Our understanding of how these cues regulate growth and guidance either alone or in combination is still rudimentary but it is possible that further insight into these issues will bring us closer to understanding how to rejuvenate adult neurons so that they can regenerate (see **Box 5.2**).

REFERENCES

Aguayo, A. J., Bray, G. M., Rasminsky, M., Zwimpfer, T., Carter, D., & Vidal-Sanz, M. (1990). Synaptic connections made by axons regenerating in the central nervous system of adult mammals. *Journal of Experimental Biology, 153,* 199–224.

Asher, R. A., Morgenstern, D. A., Moon, L. D., & Fawcett, J. W. (2001). Chondroitin sulphate proteoglycans: inhibitory components of the glial scar. *Progress in Brain Research, 132,* 611–619.

Augsburger, A., Schuchardt, A., Hoskins, S., Dodd, J., & Butler, S. (1999). BMPs as mediators of roof plate repulsion of commissural neurons. *Neuron, 24,* 127–141.

Baier, H., & Bonhoeffer, F. (1992). Axon guidance by gradients of a target-derived component. *Science, 255,* 472–475.

Bak, M., & Fraser, S. E. (2003). Axon fasciculation and differences in midline kinetics between pioneer and follower axons within commissural fascicles. *Development, 130,* 4999–5008.

Banker, G., & Goslin, K. (1991a). *Culturing nerve cells.* Cambridge: MIT Press.

Banker, G., & Goslin, K. (1991b). Rat hippocampal neurons in low density culture. In G. Banker, & K. Goslin (Eds.), *Culturing nerve cells* (pp. 251–281). Cambridge: MIT Press.

Bao, Z. Z. (2008). Intraretinal projection of retinal ganglion cell axons as a model system for studying axon navigation. *Brain Research, 1192,* 165–177.

Barnes, A. P., Solecki, D., & Polleux, F. (2008). New insights into the molecular mechanisms specifying neuronal polarity in vivo. *Current Opinion in Neurobiology, 18,* 44–52.

Barres, B. A., Silverstein, B. E., Corey, D. P., & Chun, L. L. (1988). Immunological, morphological, and

electrophysiological variation among retinal ganglion cells purified by panning. *Neuron, 1,* 791–803.

Bashaw, G. J., & Goodman, C. S. (1999). Chimeric axon guidance receptors: the cytoplasmic domains of slit and netrin receptors specify attraction versus repulsion. *Cell, 97,* 917–926.

Bassell, G. J., Zhang, H., Byrd, A. L., Femino, A. M., Singer, R. H., Taneja, K. L., et al. (1998). Sorting of beta-actin mRNA and protein to neurites and growth cones in culture. *The Journal of Neuroscience, 18,* 251–265.

Bastiani, M. J., Harrelson, A. L., Snow, P. M., & Goodman, C. S. (1987). Expression of fasciclin I and II glycoproteins on subsets of axon pathways during neuronal development in the grasshopper. *Cell, 48,* 745–755.

Bastiani, M. J., Raper, J. A., & Goodman, C. S. (1984). Pathfinding by neuronal growth cones in grasshopper embryos. III. Selective affinity of the G growth cone for the P cells within the A/P fascicle. *The Journal of Neuroscience, 4,* 2311–2328.

Beggs, H. E., Soriano, P., & Maness, P. F. (1994). NCAM-dependent neurite outgrowth is inhibited in neurons from Fyn-minus mice. *The Journal of Cell Biology, 127,* 825–833.

Benowitz, L. I., & Yin, Y. (2007). Combinatorial treatments for promoting axon regeneration in the CNS: strategies for overcoming inhibitory signals and activating neurons' intrinsic growth state. *Dev Neurobiol, 67,* 1148–1165.

Bentley, D., & Caudy, M. (1983). Pioneer axons lose directed growth after selective killing of guidepost cells. *Nature, 304,* 62–65.

Bentley, D., & Toroian-Raymond, A. (1986). Disoriented pathfinding by pioneer neurone growth cones deprived of filopodia by cytochalasin treatment. *Nature, 323,* 712–715.

Bixby, J. L., & Harris, W. A. (1991). Molecular mechanisms of axon growth and guidance. *Annual Review of Cell Biology, 7,* 117–159.

Bixby, J. L., Pratt, R. S., Lilien, J., & Reichardt, L. F. (1987). Neurite outgrowth on muscle cell surfaces involves extracellular matrix receptors as -dependent and -independent cell adhesion moleculeswell as Ca2. *Proceedings of the National Academy of Sciences of the United States of America, 84,* 2555–2559.

Bonnin, A., Torii, M., Wang, L., Rakic, P., & Levitt, P. (2007). Serotonin modulates the response of embryonic thalamocortical axons to netrin-1. *Nature Neuroscience, 10,* 588–597.

Bovolenta, P., & Mason, C. (1987). Growth cone morphology varies with position in the developing mouse visual pathway from retina to first targets. *The Journal of Neuroscience, 7,* 1447–1460.

Bradbury, E. J., Moon, L. D., Popat, R. J., King, V. R., Bennett, G. S., Patel, P. N., et al. (2002). Chondroitinase ABC promotes functional recovery after spinal cord injury. *Nature, 416,* 636–640.

Bradke, F., & Dotti, C. G. (1999). The role of local actin instability in axon formation. *Science, 283,* 1931–1934.

Bradley, P., & Berry, M. (1978). The Purkinje cell dendritic tree in mutant mouse cerebellum. A quantitative Golgi study of Weaver and Staggerer mice. *Brain Research, 142,* 135–141.

Bray, D. (1979). Mechanical tension produced by nerve cells in tissue culture. *Journal of Cell Science, 37,* 391–410.

Bray, D., Thomas, C., & Shaw, G. (1978). Growth cone formation in ultures of sensory neurons. *Proceedings of the National Academy of Sciences of the United States of America, 75,* 5226–5229.

Bregman, B. S., & Goldberger, M. E. (1983). Infant lesion effect: III. Anatomical correlates of sparing and recovery of function after spinal cord damage in newborn and adult cats. *Brain Research, 285,* 137–154.

Bregman, B. S., Kunkel-Bagden, E., Schnell, L., Dai, H. N., Gao, D., & Schwab, M. E. (1995). Recovery from spinal cord injury mediated by antibodies to neurite growth inhibitors. *Nature, 378,* 498–501.

Brittis, P. A., Lu, Q., & Flanagan, J. G. (2002). Axonal protein synthesis provides a mechanism for localized regulation at an intermediate target. *Cell, 110,* 223–235.

Brose, K., Bland, K. S., Wang, K. H., Arnott, D., Henzel, W., Goodman, C. S., et al. (1999). Slit proteins bind Robo receptors and have an evolutionarily conserved role in repulsive axon guidance. *Cell, 96,* 795–806.

Brown, A., Slaughter, T., & Black, M. M. (1992). Newly assembled microtubules are concentrated in the proximal and distal regions of growing axons. *The Journal of Cell Biology, 119,* 867–882.

Bruckenstein, D. A., & Higgins, D. (1988a). Morphological differentiation of embryonic rat sympathetic neurons in tissue culture. II. Serum promotes dendritic growth. *Developmental Biology, 128,* 337–348.

Bruckenstein, D. A., & Higgins, D. (1988b). Morphological differentiation of embryonic rat sympathetic neurons in tissue culture. I. Conditions under which neurons form axons but not dendrites. *Developmental Biology, 128,* 324–336.

Brunet, I., Weinl, C., Piper, M., Trembleau, A., Volovitch, M., Harris, W., et al. (2005). The transcription factor Engrailed-2 guides retinal axons. *Nature, 438,* 94–98.

Buck, K. B., & Zheng, J. Q. (2002). Growth cone turning induced by direct local modification of microtubule dynamics. *The Journal of Neuroscience, 22,* 9358–9367.

Bunge, R. P. (1975). Changing uses of nerve tissue culture. In D. B. Tower (Ed.), *The Nervous System, Vol. 1: The Basic Neurosciences* (pp. 31–42). New York: Raven Press.

Cai, D., Qiu, J., Cao, Z., McAtee, M., Bregman, B. S., & Filbin, M. T. (2001). Neuronal cyclic AMP controls the developmental loss in ability of axons to regenerate. *The Journal of Neuroscience, 21,* 4731–4739.

Calof, A. L., & Reichardt, L. F. (1984). Motoneurons purified by cell sorting respond to two distinct activities in myotube-conditioned medium. *Developmental Biology, 106,* 194–210.

Campbell, D. S., & Holt, C. E. (2001). Chemotropic responses of retinal growth cones mediated by rapid local protein synthesis and degradation. *Neuron, 32,* 1013–1026.

Campbell, D. S., Regan, A. G., Lopez, J. S., Tannahill, D., Harris, W. A., & Holt, C. E. (2001). Semaphorin 3A elicits stage-dependent collapse, turning, and branching in Xenopus retinal growth cones. *The Journal of Neuroscience, 21,* 8538–8547.

Castellani, V., De Angelis, E., Kenwrick, S., & Rougon, G. (2002). Cis and trans interactions of L1 with neuropilin-1 control axonal responses to semaphorin 3A. *The EMBO Journal, 21,* 6348–6357.

Caudy, M., & Bentley, D. (1986). Pioneer growth cone steering along a series of neuronal and non-neuronal cues of different affinities. *The Journal of Neuroscience, 6,* 1781–1795.

Chan, S. S., Zheng, H., Su, M. W., Wilk, R., Killeen, M. T., Hedgecock, E. M., et al. (1996). UNC-40, a C. elegans homolog of DCC (Deleted in Colorectal Cancer), is required in motile cells responding to UNC-6 netrin cues. *Cell, 87,* 187–195.

Charron, F., Stein, E., Jeong, J., McMahon, A. P., & Tessier-Lavigne, M. (2003). The morphogen sonic hedgehog is an axonal chemoattractant that collaborates with netrin-1 in midline axon guidance. *Cell, 113,* 11–23.

Chen, Z., Gore, B. B., Long, H., Ma, L., & Tessier-Lavigne, M. (2008). Alternative splicing of the Robo3 axon guidance receptor governs the midline switch from attraction to repulsion. *Neuron, 58,* 325–332.

Chien, C. B., Rosenthal, D. E., Harris, W. A., & Holt, C. E. (1993). Navigational errors made by growth cones without filopodia in the embryonic Xenopus brain. *Neuron, 11,* 237–251.

Chitnis, A. B., Patel, C. K., Kim, S., & Kuwada, J. Y. (1992). A specific brain tract guides follower growth cones in two regions of the zebrafish brain. *Journal of Neurobiology, 23,* 845–854.

Clendening, B., & Hume, R. I. (1990). Cell interactions regulate dendritic morphology and responses to neurotransmitters in embryonic chick sympathetic preganglionic neurons in vitro. *The Journal of Neuroscience, 10,* 3992–4005.

Cline, H. T. (2001). Dendritic arbor development and synaptogenesis. *Current Opinion in Neurobiology, 11,* 118–126.

Cohen, J., & Johnson, A. R. (1991). Differential effects of laminin and merosin on neurite outgrowth by developing retinal ganglion cells. *Journal of Cell Science. Supplement, 15,* 1–7.

Colamarino, S. A., & Tessier-Lavigne, M. (1995). The role of the floor plate in axon guidance. *Annual Review of Neuroscience, 18,* 497–529.

Craig, A. M., & Banker, G. (1994). Neuronal polarity. *Annual Review of Neuroscience, 17,* 267–310.

Culotti, J. G. (1994). Axon guidance mechanisms in Caenorhabditis elegans. *Current Opinion in Genetics & Development, 4,* 587–595.

David, S., & Aguayo, A. J. (1981). Axonal elongation into peripheral nervous system bridges after central nervous system injury in adult rats. *Science, 214,* 931–933.

Deiner, M. S., Kennedy, T. E., Fazeli, A., Serafini, T., Tessier-Lavigne, M., & Sretavan, D. W. (1997). Netrin-1 and DCC mediate axon guidance locally at the optic disc: loss of function leads to optic nerve hypoplasia. *Neuron, 19,* 575–589.

Dent, E. W., & Gertler, F. B. (2003). Cytoskeletal dynamics and transport in growth cone motility and axon guidance. *Neuron, 40,* 209–227.

Desai, C. J., Gindhart, J. G., Goldstein, L. S., Jr., & Zinn, K. (1996). Receptor tyrosine phosphatases are required for motor axon guidance in the Drosophila embryo. *Cell, 84,* 599–609.

Dickson, B. J. (2001). Rho GTPases in growth cone guidance. *Current Opinion in Neurobiology, 11,* 103–110.

Dickson, B. J., & Gilestro, G. F. (2006). Regulation of commissural axon pathfinding by slit and its Robo receptors. *Annual Review of Cell and Developmental Biology, 22,* 651–675.

Dwivedy, A., Gertler, F. B., Miller, J., Holt, C. E., & Lebrand, C. (2007). Ena/VASP function in retinal axons is required for terminal arborization but not pathway navigation. *Development, 134,* 2137–2146.

Eagleson, K. L., Pimenta, A. F., Burns, M. M., Fairfull, L. D., Cornuet, P. K., Zhang, L., et al. (2003). Distinct domains of the limbic system-associated membrane protein (LAMP) mediate discrete effects on neurite outgrowth. *Molecular and Cellular Neurosciences, 24,* 725–740.

Easter, S. S., Jr., Burrill, J., Marcus, R. C., Ross, L. S., Taylor, J. S., & Wilson, S. W. (1994). Initial tract formation in the vertebrate brain. *Progress in Brain Research, 102,* 79–93.

Edelman, G. M. (1984). Modulation of cell adhesion during induction, histogenesis, and perinatal development of the nervous system. *Annual Review of Neuroscience, 7,* 339–377.

Elkins, T., Zinn, K., McAllister, L., Hoffmann, F. M., & Goodman, C. S. (1990). Genetic analysis of a Drosophila neural cell adhesion molecule: interaction of fasciclin I and Abelson tyrosine kinase mutations. *Cell, 60,* 565–575.

Erskine, L., & Herrera, E. (2007). The retinal ganglion cell axon's journey: insights into molecular mechanisms of axon guidance. *Developmental Biology, 308,* 1–14.

Erskine, L., Williams, S. E., Brose, K., Kidd, T., Rachel, R. A., Goodman, C. S., et al. (2000). Retinal ganglion cell axon guidance in the mouse optic chiasm: expression and function of robos and slits. *The Journal of Neuroscience, 20,* 4975–4982.

Evans, T. A., & Bashaw, G. J. (2010). Functional diversity of Robo receptor immunoglobulin domains promotes distinct axon guidance decisions. *Current Biology: CB, 20,* 567–572.

Fournier, A. E., GrandPre, T., & Strittmatter, S. M. (2001). Identification of a receptor mediating Nogo-66 inhibition of axonal regeneration. *Nature, 409,* 341–346.

Fricke, C., Lee, J. S., Geiger-Rudolph, S., Bonhoeffer, F., & Chien, C. B. (2001). Astray, a zebrafish roundabout homolog required for retinal axon guidance. *Science, 292,* 507–510.

Fukata, Y., Kimura, T., & Kaibuchi, K. (2002). Axon specification in hippocampal neurons. *Neuroscience Research, 43,* 305–315.

GŠhwiler, B. H., Thompson, S. M., Audinat, E., & Robertson, R. T. (1991). Organotypic slice cultures of neural tissue. In G. Banker, & K. Goslin (Eds.), *Culturing nerve cells* (pp. 379–411). Cambridge: MIT Press.

Garcia-Alonso, L., Fetter, R. D., & Goodman, C. S. (1996). Genetic analysis of Laminin A in Drosophila: extracellular matrix containing laminin A is required for ocellar axon pathfinding. *Development, 122,* 2611–2621.

Glover, J. C., Petursdottir, G., & Jansen, J. K. S. (1986). Fluorescent dextran-amines used as axonal tracers in the nervous system of the chicken embryo. *Journal of Neuroscience Methods, 18,* 243–254.

Goldberg, D. J., & Wu, D. Y. (1996). Tyrosine phosphorylation and protrusive structures of the growth cone. *Perspectives on Developmental Neurobiology, 4,* 183–192.

Gomez, T. M., Robles, E., Poo, M., & Spitzer, N. C. (2001). Filopodial calcium transients promote substrate-dependent growth cone turning. *Science, 291,* 1983–1987.

Goodman, C. S. (1996). Mechanisms and molecules that control growth cone guidance. *Annual Review of Neuroscience, 19,* 341–377.

Goodman, C. S., Raper, J. A., Chang, S., & Ho, R. (1983). Grasshopper growth cones: divergent choices and labeled pathways. *Progress in Brain Research, 58,* 283–304.

Gordon-Weeks, P. R. (2004). Microtubules and growth cone function. *Journal of Neurobiology, 58,* 70–83.

Goslin, K., & Banker, G. (1989). Experimental observations on the development of polarity by hippocampal neurons in culture. *The Journal of Cell Biology, 108,* 1507–1516.

Grenningloh, G., Soehrman, S., Bondallaz, P., Ruchti, E., & Cadas, H. (2004). Role of the microtubule destabilizing proteins SCG10 and stathmin in neuronal growth. *Journal of Neurobiology, 58,* 60–69.

Grueber, W. B., Jan, L. Y., & Jan, Y. N. (2003). Different levels of the homeodomain protein cut regulate distinct dendrite branching patterns of Drosophila multidendritic neurons. *Cell, 112,* 805–818.

139

Gundersen, R. W., & Barrett, J. N. (1979). Neuronal chemotaxis: chick dorsal-root axons turn toward high concentrations of nerve growth factor. *Science, 206,* 1079–1080.

Guo, X., Chandrasekaran, V., Lein, P., Kaplan, P. L., & Higgins, D. (1999). Leukemia inhibitory factor and ciliary neurotrophic factor cause dendritic retraction in cultured rat sympathetic neurons. *The Journal of Neuroscience, 19,* 2113–2121.

Guo, X., Metzler-Northrup, J., Lein, P., Rueger, D., & Higgins, D. (1997). Leukemia inhibitory factor and ciliary neurotrophic factor regulate dendritic growth in cultures of rat sympathetic neurons. *Brain Research. Development Brain Research, 104,* 101–110.

Guo, X., Rueger, D., & Higgins, D. (1998). Osteogenic protein-1 and related bone morphogenetic proteins regulate dendritic growth and the expression of microtubule-associated protein-2 in rat sympathetic neurons. *Neuroscience Letters, 245,* 131–134.

Hammarback, J. A., Palm, S. L., Furcht, L. T., & Letourneau, P. C. (1985). Guidance of neurite outgrowth by pathways of substratum-adsorbed laminin. *Journal of Neuroscience Research, 13,* 213–220.

Hand, R., Bortone, D., Mattar, P., Nguyen, L., Heng, J. I., Guerrier, S., et al. (2005). Phosphorylation of Neurogenin2 specifies the migration properties and the dendritic morphology of pyramidal neurons in the neocortex. *Neuron, 48,* 45–62.

Hannila, S. S., & Filbin, M. T. (2008). The role of cyclic AMP signaling in promoting axonal regeneration after spinal cord injury. *Experimental Neurology, 209,* 321–332.

Harrelson, A. L., & Goodman, C. S. (1988). Growth cone guidance in insects: fasciclin II is a member of the immunoglobulin superfamily. *Science, 242,* 700–708.

Harris, W. A. (1989). Local positional cues in the neuroepithelium guide retinal axons in embryonic Xenopus brain. *Nature, 339,* 218–221.

Harris, W. A., Holt, C. E., & Bonhoeffer, F. (1987). Retinal axons with and without their somata, growing to and arborizing in the tectum of Xenopus embryos: a time-lapse video study of single fibres in vivo. *Development, 101,* 123–133.

Harrison, R. (1907). Observations on the living developing nerve fiber. *The Anatomical Record, 1,* 116–118.

Harrison, R. (1910). The outgrowth of the nerve fiber as a mode of protoplasmic movement. *Journal of Experimental Zoology, 9,* 787–846.

Hatten, M. E. (1985). Neuronal regulation of astroglial morphology and proliferation in vitro. *The Journal of Cell Biology, 100,* 384–396.

Hattori, M., Osterfield, M., & Flanagan, J. G. (2000). Regulated cleavage of a contact-mediated axon repellent. *Science, 289,* 1360–1365.

Hedgecock, E. M., Culotti, J. G., & Hall, D. H. (1990). The unc-5, unc-6, and unc-40 genes guide circumferential migrations of pioneer axons and mesodermal cells on the epidermis in C. elegans. *Neuron, 4,* 61–85.

Heidemann, S. R. (1996). Cytoplasmic mechanisms of axonal and dendritic growth in neurons. *International Review of Cytology, 165,* 235–296.

Hibbard, E. (1965). Orientation and directed growth of Mauthners cell axons form duplicated vestibular nerve roots. *Experimental Neurology, 13,* 289–301.

Hollenbeck, P. J., & Bray, D. (1987). Rapidly transported organelles containing membrane and cytoskeletal components: their relation to axonal growth. *The Journal of Cell Biology, 105,* 2827–2835.

Hong, K., Hinck, L., Nishiyama, M., Poo, M. M., Tessier-Lavigne, M., & Stein, E. (1999). A ligand-gated association between cytoplasmic domains of UNC5 and DCC family receptors converts netrin-induced growth cone attraction to repulsion. *Cell, 97,* 927–941.

Honig, M. G., & Hume, R. I. (1986). Fluorescent carbocyanine dyes allow living neurons of identified origin to be studied in long-term cultures. *The Journal of Cell Biology, 103,* 171–187.

Hopker, V. H., Shewan, D., Tessier-Lavigne, M., Poo, M., & Holt, C. (1999). Growth-cone attraction to netrin-1 is converted to repulsion by laminin-1. *Nature, 401,* 69–73.

Horton, H. L., & Levitt, P. (1988). A unique membrane protein is expressed on early developing limbic system axons and cortical targets. *The Journal of Neuroscience, 8,* 4653–4661.

Hutson, L. D., & Chien, C. B. (2002). Pathfinding and error correction by retinal axons: the role of astray/robo2. *Neuron, 33,* 205–217.

Irie, A., Yates, E. A., Turnbull, J. E., & Holt, C. E. (2002). Specific heparan sulfate structures involved in retinal axon targeting. *Development, 129,* 61–70.

Jan, Y. N., & Jan, L. Y. (2003). The control of dendrite development. *Neuron, 40,* 229–242.

Jin, Z., Zhang, J., Klar, A., Chedotal, A., Rao, Y., Cepko, C. L., et al. (2003). Irx4-mediated regulation of Slit1 expression contributes to the definition of early axonal paths inside the retina. *Development, 130,* 1037–1048.

Johnson, K. G., & Harris, W. A. (2000). Connecting the eye with the brain: the formation of the retinotectal pathway. *Results and Problems in Cell Differentiation, 31,* 157–177.

Kalil, K., & Reh, T. (1982). A light and electron microscopic study of regrowing pyramidal tract fibers. *The Journal of Comparative Neurology, 211,* 265–275.

Kapfhammer, J. P., Grunewald, B. E., & Raper, J. A. (1986). The selective inhibition of growth cone extension by specific neurites in culture. *The Journal of Neuroscience, 6,* 2527–2534.

Kapfhammer, J. P., & Raper, J. A. (1987a). Collapse of growth cone structure on contact with specific neurites in culture. *The Journal of Neuroscience, 7,* 201–212.

Kapfhammer, J. P., & Raper, J. A. (1987b). Interactions between growth cones and neurites growing from different neural tissues in culture. *The Journal of Neuroscience, 7,* 1595–1600.

Kater, S. B., & Mills, L. R. (1991). Regulation of growth cone behavior by calcium. *The Journal of Neuroscience, 11,* 891–899.

Keino-Masu, K., Masu, M., Hinck, L., Leonardo, E. D., Chan, S. S., Culotti, J. G., et al. (1996). Deleted in Colorectal Cancer (DCC) encodes a netrin receptor. *Cell, 87,* 175–185.

Keleman, K., Rajagopalan, S., Cleppien, D., Teis, D., Paiha, K., Huber, L. A., et al. (2002). Comm sorts robo to control axon guidance at the Drosophila midline. *Cell, 110,* 415–427.

Kennedy, T. E., Serafini, T., de la Torre, J. R., & Tessier-Lavigne, M. (1994). Netrins are diffusible chemotropic factors for commissural axons in the embryonic spinal cord. *Cell, 78,* 425–435.

Keshishian, H., & Bentley, D. (1983a). Embryogenesis of peripheral nerve pathways in grasshopper legs. II. The major nerve routes. *Developmental Biology, 96,* 103–115.

Keshishian, H., & Bentley, D. (1983b). Embryogenesis of peripheral nerve pathways in grasshopper legs. I. The initial nerve pathway to the CNS. *Developmental Biology, 96,* 89–102.

Keshishian, H., & Bentley, D. (1983c). Embryogenesis of peripheral nerve pathways in grasshopper

legs. III. Development without pioneer neurons. *Developmental Biology, 96,* 116–124.

Keynes, R., Tannahill, D., Morgenstern, D. A., Johnson, A. R., Cook, G. M., & Pini, A. (1997). Surround repulsion of spinal sensory axons in higher vertebrate embryos. *Neuron, 18,* 889–897.

Kitsukawa, T., Shimizu, M., Sanbo, M., Hirata, T., Taniguchi, M., Bekku, Y., et al. (1997). Neuropilin-semaphorin III/D-mediated chemorepulsive signals play a crucial role in peripheral nerve projection in mice. *Neuron, 19,* 995–1005.

Kolodkin, A. L. (1996): Growth cones and the cues that repel them. *Trends in Neurosciences, 19,* 507–513.

Kolodkin, A. L., Matthes, D. J., & Goodman, C. S. (1993). The semaphorin genes encode a family of transmembrane and secreted growth cone guidance molecules. *Cell, 75,* 1389–1399.

Kolodziej, P. A., Timpe, L. C., Mitchell, K. J., Fried, S. R., Goodman, C. S., Jan, L. Y., et al. (1996). Frazzled encodes a Drosophila member of the DCC immunoglobulin subfamily and is required for CNS and motor axon guidance. *Cell, 87,* 197–204.

Kunda, P., Paglini, G., Quiroga, S., Kosik, K., & Caceres, A. (2001). Evidence for the involvement of Tiam1 in axon formation. *The Journal of Neuroscience, 21,* 2361–2372.

Landgraf, M., & Thor, S. (2006). Development of Drosophila motoneurons: specification and morphology. *Seminars in Cell & Developmental Biology, 17,* 3–11.

Lanier, L. M., & Gertler, F. B. (2000). From Abl to actin: Abl tyrosine kinase and associated proteins in growth cone motility. *Current Opinion in Neurobiology, 10,* 80–87.

Le Roux, P. D., & Reh, T. A. (1994). Regional differences in glial-derived factors that promote dendritic outgrowth from mouse cortical neurons in vitro. *The Journal of Neuroscience, 14,* 4639–4655.

Lebrand, C., Dent, E. W., Strasser, G. A., Lanier, L. M., Krause, M., Svitkina, T. M., et al. (2004). Critical Role of Ena/VASP Proteins for Filopodia Formation in Neurons and in Function Downstream of Netrin-1. *Neuron, 42,* 37–49.

Lee, H., Engel, U., Rusch, J., Scherrer, S., Sheard, K., & Van Vactor, D. (2004). The microtubule plus end tracking protein Orbit/MAST/CLASP acts downstream of the tyrosine kinase Abl in mediating axon guidance. *Neuron, 42,* 913–926.

Lein, P., Johnson, M., Guo, X., Rueger, D., & Higgins, D. (1995). Osteogenic protein-1 induces dendritic growth in rat sympathetic neurons. *Neuron, 15,* 597–605.

Lemmon, V., Burden, S. M., Payne, H. R., Elmslie, G. J., & Hlavin, M. L. (1992). Neurite growth on different substrates: permissive versus instructive influences and the role of adhesive strength. *The Journal of Neuroscience, 12,* 818–826.

Letourneau, P. C. (1975). Cell-to-substratum adhesion and guidance of axonal elongation. *Developmental Biology, 44,* 92–101.

Letourneau, P. C. (1996). The cytoskeleton in nerve growth cone motility and axonal pathfinding. *Perspectives on Developmental Neurobiology, 4,* 111–123.

Li, H. S., Chen, J. H., Wu, W., Fagaly, T., Zhou, L., Yuan, W., et al. (1999). Vertebrate slit, a secreted ligand for the transmembrane protein roundabout, is a repellent for olfactory bulb axons. *Cell, 96,* 807–818.

Lin, A. C., & Holt, C. E. (2008). Function and regulation of local axonal translation. *Current Opinion in Neurobiology, 18,* 60–68.

Lin, C. H., Espreafico, E. M., Mooseker, M. S., & Forscher, P. (1996). Myosin drives retrograde F-actin flow in neuronal growth cones. *Neuron, 16,* 769–782.

Lin, D. M., Fetter, R. D., Kopczynski, C., Grenningloh, G., & Goodman, C. S. (1994). Genetic analysis of Fasciclin II in Drosophila: defasciculation, refasciculation, and altered fasciculation. *Neuron, 13,* 1055–1069.

Liu, C. N., & Chambers, W. W. (1958). Intraspinal sprouting of dorsal root axons; development of new collaterals and preterminals following partial denervation of the spinal cord in the cat. *A.M.A. Archives of Neurology and Psychiatry, 79,* 46–61.

Liu, C. W., Lee, G., & Jay, D. G. (1999). Tau is required for neurite outgrowth and growth cone motility of chick sensory neurons. *Cell Motility and the Cytoskeleton, 43,* 232–242.

Liu, Y., Shi, J., Lu, C. C., Wang, Z. B., Lyuksyutova, A. I., Song, X. J., et al. (2005). Ryk-mediated Wnt repulsion regulates posterior-directed growth of corticospinal tract. *Nature Neuroscience, 8,* 1151–1159.

Livet, J., Weissman, T. A., Kang, H., Draft, R. W., Lu, J., Bennis, R. A., et al. (2007). Transgenic strategies for combinatorial expression of fluorescent proteins in the nervous system. *Nature, 450,* 56–62.

Lohof, A. M., Quillan, M., Dan, Y., & Poo, M. M. (1992). Asymmetric modulation of cytosolic cAMP activity induces growth cone turning. *The Journal of Neuroscience, 12,* 1253–1261.

Lumsden, A. G., & Davies, A. M. (1986). Chemotropic effect of specific target epithelium in the developing mammalian nervous system. *Nature, 323,* 538–539.

Luo, L. (2000). Rho GTPases in neuronal morphogenesis. *Nature Reviews. Neuroscience, 1,* 173–180.

Luo, L. (2002). Actin cytoskeleton regulation in neuronal morphogenesis and structural plasticity. *Ann Rev Cell Dev Biol, 18,* 601–635.

Luo, L., Jan, L. Y., & Jan, Y. N. (1997). Rho family GTP-binding proteins in growth cone signalling. *Current Opinion in Neurobiology, 7,* 81–86.

Luo, Y., Raible, D., & Raper, J. A. (1993). Collapsin: a protein in brain that induces the collapse and paralysis of neuronal growth cones. *Cell, 75,* 217–227.

Lyuksyutova, A. I., Lu, C. C., Milanesio, N., King, L. A., Guo, N., Wang, Y., et al. (2003). Anterior-posterior guidance of commissural axons by Wnt-frizzled signaling. *Science, 302,* 1984–1988.

Marillat, V., Sabatier, C., Failli, V., Matsunaga, E., Sotelo, C., Tessier-Lavigne, M., et al. (2004). The slit receptor Rig-1/Robo3 controls midline crossing by hindbrain precerebellar neurons and axons. *Neuron, 43,* 69–79.

Marsh, L., & Letourneau, P. C. (1984). Growth of neurites without filopodial or lamellipodial activity in the presence of cytochalasin B. *The Journal of Cell Biology, 99,* 2041–2047.

McAllister, A. K. (2000). Cellular and molecular mechanisms of dendrite growth. *Cerebral Cortex, 10,* 963–973.

McFarlane, S., McNeill, L., & Holt, C. E. (1995). FGF signaling and target recognition in the developing Xenopus visual system. *Neuron, 15,* 1017–1028.

McKenna, M. P., & Raper, J. A. (1988). Growth cone behavior on gradients of substratum bound laminin. *Developmental Biology, 130,* 232–236.

McKerracher, L., Chamoux, M., & Arregui, C. O. (1996). Role of laminin and integrin interactions in growth cone guidance. *Molecular Neurobiology, 12,* 95–116.

Merz, D. C., & Culotti, J. G. (2000). Genetic analysis of growth cone migrations in Caenorhabditis elegans. *Journal of Neurobiology, 44,* 281–288.

Ming, G. L., Song, H. J., Berninger, B., Holt, C. E., Tessier-Lavigne, M., & Poo, M. M. (1997). cAMP-dependent growth cone guidance by netrin-1. *Neuron, 19,* 1225–1235.

Ming, G. L., Wong, S. T., Henley, J., Yuan, X. B., Song, H. J., Spitzer, N. C., et al. (2002). Adaptation in the chemotactic guidance of nerve growth cones. *Nature, 417,* 411–418.

Mitchison, T., & Kirschner, M. (1988). Cytoskeletal dynamics and nerve growth. *Neuron, 1,* 761–772.

Moore, D. L., Blackmore, M. G., Hu, Y., Kaestner, K. H., Bixby, J. L., Lemmon, V. P., et al. (2009). KLF family members regulate intrinsic axon regeneration ability. *Science, 326,* 298–301.

Myat, A., Henry, P., McCabe, V., Flintoft, L., Rotin, D., & Tear, G. (2002). Drosophila Nedd4, a ubiquitin ligase, is recruited by Commissureless to control cell surface levels of the roundabout receptor. *Neuron, 35,* 447–459.

Nakagawa, S., Brennan, C., Johnson, K. G., Shewan, D., Harris, W. A., & Holt, C. E. (2000). Ephrin-B regulates the ipsilateral routing of retinal axons at the optic chiasm. *Neuron, 25,* 599–610.

Nardi, J. B. (1983). Neuronal pathfinding in developing wings of the moth Manduca sexta. *Developmental Biology, 95,* 163–174.

Nardi, J. B., & Vernon, R. A. (1990). Topographical features of the substratum for growth of pioneering neurons in the Manduca wing disc. *Journal of Neurobiology, 21,* 1189–1201.

Nishiyama, M., von Schimmelmann, M. J., Togashi, K., Findley, W. M., & Hong, K. (2008). Membrane potential shifts caused by diffusible guidance signals direct growth-cone turning. *Nature Neuroscience, 11,* 762–771.

O'Connor, R., & Tessier-Lavigne, M. (1999). Identification of maxillary factor, a maxillary process-derived chemoattractant for developing trigeminal sensory axons. *Neuron, 24,* 165–178.

O'Connor, T. P., Duerr, J. S., & Bentley, D. (1990). Pioneer growth cone steering decisions mediated by single filopodial contacts in situ. *The Journal of Neuroscience, 10,* 3935–3946.

Okada, A., Charron, F., Morin, S., Shin, D. S., Wong, K., Fabre, P. J., et al. (2006). Boc is a receptor for sonic hedgehog in the guidance of commissural axons. *Nature, 444,* 369–373.

Park, K. K., Liu, K., Hu, Y., Smith, P. D., Wang, C., Cai, B., et al. (2008). Promoting axon regeneration in the adult CNS by modulation of the PTEN/mTOR pathway. *Science, 322,* 963–966.

Parrish, J. Z., Emoto, K., Kim, M. D., & Jan, Y. N. (2007). Mechanisms that regulate establishment, maintenance, and remodeling of dendritic fields. *Annual Review of Neuroscience, 30,* 399–423.

Peterson, E. R., & Crain, S. M. (1981). Preferential growth of neurites from isolated fetal mouse dorsal root ganglia in relation to specific regions of co-cultured spinal cord explants. *Brain Research, 254,* 363–382.

Petros, T. J., Rebsam, A., & Mason, C. A. (2008). Retinal axon growth at the optic chiasm: to cross or not to cross. *Annual Review of Neuroscience, 31,* 295–315.

Piatt, J. (1955). Regeneration in the spinal cord of the salamander. *Journal of Experimental Zoology, 129,* 177–207.

Pimenta, A. F., Zhukareva, V., Barbe, M. F., Reinoso, B. S., Grimley, C., Henzel, W., et al. (1995). The limbic system-associated membrane protein is an Ig superfamily member that mediates selective neuronal growth and axon targeting. *Neuron, 15,* 287–297.

Pini, A. (1993). Chemorepulsion of axons in the developing mammalian central nervous system. *Science, 261,* 95–98.

Piper, M., Salih, S., Weinl, C., Holt, C. E., & Harris, W. A. (2005). Endocytosis-dependent desensitization and protein synthesis-dependent resensitization in retinal growth cone adaptation. *Nature Neuroscience, 8*(2), 179–186 Epub 2005 Jan 9.

Pittman, A. J., Law, M. Y., & Chien, C. B. (2008). Pathfinding in a large vertebrate axon tract: isotypic interactions guide retinotectal axons at multiple choice points. *Development, 135,* 2865–2871.

Qiu, J., Cai, D., Dai, H., McAtee, M., Hoffman, P. N., Bregman, B. S., et al. (2002). Spinal axon regeneration induced by elevation of cyclic AMP. *Neuron, 34,* 895–903.

Radel, J. D., Hankin, M. H., & Lund, R. D. (1990). Proximity as a factor in the innervation of host brain regions by retinal transplants. *The Journal of Comparative Neurology, 300,* 211–229.

Raisman, G., & Field, P. M. (1973a). A quantitative investigation of the development of collateral reinnervation after partial deafferentation of the septal nuclei. *Brain Research, 50,* 241–264.

Rajagopalan, S., Vivancos, V., Nicolas, E., & Dickson, B. J. (2000). Selecting a longitudinal pathway: Robo receptors specify the lateral position of axons in the Drosophila CNS. *Cell, 103,* 1033–1045.

Ramón y Cajal, S. (1890). A quelle epoque aparaissent les expansions des cellule neurveuses de la moelle epinere du poulet. *Anatomischer Anzeiger, 5,* 609–613.

Ramón y Cajal, S. (1928). *Degeneration and regeneration of the nervous system.* New York: Hafner.

Raper, J. A., Bastiani, M. J., & Goodman, C. S. (1984). Pathfinding by neuronal growth cones in grasshopper embryos. IV. The effects of ablating the A and P axons upon the behavior of the G growth cone. *The Journal of Neuroscience, 4,* 2329–2345.

Richardson, P. M., McGuinness, U. M., & Aguayo, A. J. (1980). Axons from CNS neurons regenerate into PNS grafts. *Nature, 284,* 264–265.

Rosentreter, S. M., Davenport, R. W., Loschinger, J., Huf, J., Jung, J., & Bonhoeffer, F. (1998). Response of retinal ganglion cell axons to striped linear gradients of repellent guidance molecules. *Journal of Neurobiology, 37,* 541–562.

Rosoff, W. J., Urbach, J. S., Estrick, M. A., McAllister, R. G., Richards, L. J., & Goodhill, G. J. (2004). A novel chemotaxis assay reveals the extreme sensitivity of axons to molecular gradients. *Nature Neuroscience, 7,* 678–682.

Ross, L. S., Parrett, T., Easter, S. S., Jr. (1992). Axonogenesis and morphogenesis in the embryonic zebrafish brain. *The Journal of Neuroscience, 12,* 467–482.

Ruthel, G., & Hollenbeck, P. J. (2003). Response of mitochondrial traffic to axon determination and differential branch growth. *The Journal of Neuroscience, 23,* 8618–8624.

Schaefer, A. W., Kabir, N., & Forscher, P. (2002). Filopodia and actin arcs guide the assembly and transport of two populations of microtubules with unique dynamic parameters in neuronal growth cones. *The Journal of Cell Biology, 158,* 139–152.

Schilling, K., Dickinson, M. H., Connor, J. A., & Morgan, J. I. (1991). Electrical activity in cerebellar cultures determines Purkinje cell dendritic growth patterns. *Neuron, 7,* 891–902.

Schmandke, A., Schmandke, A., & Strittmatter, S. M. (2007). ROCK and Rho: biochemistry and neuronal functions of Rho-associated protein kinases. *The Neuroscientist: A Review Journal Bringing Neurobiology, Neurology and Psychiatry, 13,* 454–469.

Schnell, L., & Schwab, M. E. (1990). Axonal regeneration in the rat spinal cord produced by an antibody against myelin-associated neurite growth inhibitors. *Nature, 343,* 269–272.

Schwab, M. E., & Bartholdi, D. (1996). Degeneration and regeneration of axons in the lesioned spinal cord. *Physiological Reviews, 76,* 319–370.

Seeger, M., Tear, G., Ferres-Marco, D., & Goodman, C. S. (1993). Mutations affecting growth cone guidance in Drosophila: genes necessary for guidance toward or away from the midline. *Neuron, 10,* 409–426.

Serafini, T., Kennedy, T. E., Galko, M. J., Mirzayan, C., Jessell, T. M., & Tessier-Lavigne, M. (1994). The netrins define a family of axon outgrowth-promoting proteins homologous to C. elegans UNC-6. *Cell, 78,* 409–424.

Shewan, D., Dwivedy, A., Anderson, R., & Holt, C. E. (2002). Age-related changes underlie switch in netrin-1 responsiveness as growth cones advance along visual pathway. *Nature Neuroscience, 5,* 955–962.

Shi, S. H., Jan, L. Y., & Jan, Y. N. (2003). Hippocampal neuronal polarity specified by spatially localized mPar3/mPar6 and PI 3-kinase activity. *Cell, 112,* 63–75.

Shirasaki, R., Katsumata, R., & Murakami, F. (1998). Change in chemoattractant responsiveness of developing axons at an intermediate target. *Science, 279,* 105–107.

Silver, J., & Ogawa, M. Y. (1983). Postnatally induced formation of the corpus callosum in acallosal mice on glia-coated cellulose bridges. *Science, 220,* 1067–1069.

Simpson, J. H., Bland, K. S., Fetter, R. D., & Goodman, C. S. (2000). Short-range and long-range guidance by Slit and its Robo receptors: a combinatorial code of Robo receptors controls lateral position. *Cell, 103,* 1019–1032.

Song, H., Ming, G., He, Z., Lehmann, M., McKerracher, L., Tessier-Lavigne, M., et al. (1998). Conversion of neuronal growth cone responses from repulsion to attraction by cyclic nucleotides. *Science, 281,* 1515–1518.

Song, H. J., Ming, G. L., & Poo, M. M. (1997). cAMP-induced switching in turning direction of nerve growth cones. *Nature, 388,* 275–279.

Speidel, C. (1941). Adjustments of nerve endings. *Harvey Lectures, 36,* 126–158.

Spitzweck, B., Brankatschk, M., & Dickson, B. J. (2010). Distinct protein domains and expression patterns confer divergent axon guidance functions for Drosophila Robo receptors. *Cell, 140,* 409–420.

Stein, E., & Tessier-Lavigne, M. (2001). Hierarchical organization of guidance receptors: silencing of netrin attraction by slit through a Robo/DCC receptor complex. *Science, 291,* 1928–1938.

Stoeckli, E. T., & Landmesser, L. T. (1995). Axonin-1, Nr-CAM, and Ng-CAM play different roles in the in vivo guidance of chick commissural neurons. *Neuron, 14,* 1165–1179.

Stoppini, L., Buchs, P. A., & Muller, D. (1991). A simple method for organotypic cultures of nervous tissue. *Journal of Neuroscience Methods, 37,* 173–182.

Strittmatter, S. M., & Fishman, M. C. (1991). The neuronal growth cone as a specialized transduction system. *BioEssays: News and Reviews in Molecular, Cellular and Developmental Biology, 13,* 127–134.

Suter, D. M., & Forscher, P. (2001). Transmission of growth cone traction force through apCAM-cytoskeletal linkages is regulated by Src family tyrosine kinase activity. *The Journal of Cell Biology, 155,* 427–438.

Tang, J., Rutishauser, U., & Landmesser, L. (1994). Polysialic acid regulates growth cone behavior during sorting of motor axons in the plexus region. *Neuron, 13,* 405–414.

Taniguchi, M., Yuasa, S., Fujisawa, H., Naruse, I., Saga, S., Mishina, M., et al. (1997). Disruption of semaphorin III/D gene causes severe abnormality in peripheral nerve projection. *Neuron, 19,* 519–530.

Taylor, J. S. (1990). The directed growth of retinal axons towards surgically transposed tecta in *Xenopus*: An examination of homing behaviour by retinal ganglion cells. *Development, 108,* 147–158.

Tear, G., Seeger, M., & Goodman, C. S. (1993). To cross or not to cross: a genetic analysis of guidance at the midline. *Perspectives on Developmental Neurobiology, 1,* 183–194.

Tessier-Lavigne, M., & Goodman, C. S. (1996). The molecular biology of axon guidance. *Science, 274,* 1123–1133.

Thompson, H., Camand, O., Barker, D., & Erskine, L. (2006). Slit proteins regulate distinct aspects of retinal ganglion cell axon guidance within dorsal and ventral retina. *The Journal of Neuroscience, 26,* 8082–8091.

Tomaselli, K. J., Reichardt, L. F., & Bixby, J. L. (1986). Distinct molecular interactions mediate neuronal process outgrowth on non-neuronal cell surfaces and extracellular matrices. *The Journal of Cell Biology, 103,* 2659–2672.

Tosney, K. W., & Landmesser, L. T. (1985). Growth cone morphology and trajectory in the lumbosacral region of the chick embryo. *The Journal of Neuroscience, 5,* 2345–2358.

Tripodi, M., Evers, J. F., Mauss, A., Bate, M., & Landgraf, M. (2008). Structural homeostasis: compensatory adjustments of dendritic arbor geometry in response to variations of synaptic input. *PLoS Biology, 6,* e260.

Trousse, F., Marti, E., Gruss, P., Torres, M., & Bovolenta, P. (2001). Control of retinal ganglion cell axon growth: a new role for Sonic hedgehog. *Development, 128,* 3927–3936.

Verma, P., Chierzi, S., Codd, A. M., Campbell, D. S., Meyer, R. L., Holt, C. E., et al. (2005). Axonal protein synthesis and degradation are necessary for efficient growth cone regeneration. *The Journal of Neuroscience, 25,* 331–342.

von Philipsborn, A. C., Lang, S., Bernard, A., Loeschinger, J., David, C., Lehnert, D., et al. (2006). Microcontact printing of axon guidance molecules for generation of graded patterns. *Nat Protoc, 1,* 1322–1328.

Waimey, K. E., Huang, P. H., Chen, M., & Cheng, H. J. (2008). Plexin-A3 and plexin-A4 restrict the migration of sympathetic neurons but not their neural crest precursors. *Developmental Biology, 315,* 448–458.

Walsh, F. S., & Doherty, P. (1997). Neural cell adhesion molecules of the immunoglobulin superfamily: role in axon growth and guidance. *Annual Review of Cell and Developmental Biology, 13,* 425–456.

Walter, J., Henke-Fahle, S., & Bonhoeffer, F. (1987a). Avoidance of posterior tectal membranes by temporal retinal axons. *Development, 101,* 909–913.

Walter, J., Kern-Veits, B., Huf, J., Stolze, B., & Bonhoeffer, F. (1987b). Recognition of position-specific properties of tectal cell membranes by retinal axons in vitro. *Development, 101,* 685–696.

Walz, A., McFarlane, S., Brickman, Y. G., Nurcombe, V., Bartlett, P. F., & Holt, C. E. (1997). Essential role of heparan sulfates in axon navigation and targeting in the developing visual system. *Development, 124,* 2421–2430.

Wessells, N. K., & Nuttall, R. P. (1978). Normal branching, induced branching, and steering of cultured parasympathetic motor neurons. *Experimental Cell Research, 115,* 111–122.

Whitford, K. L., Dijkhuizen, P., Polleux, F., & Ghosh, A. (2002). Molecular control of cortical dendrite development. *Annual Review of Neuroscience, 25,* 127–149.

Wictorin, K., Brundin, P., Gustavii, B., Lindvall, O., & Bjorklund, A. (1990). Reformation of long axon pathways in adult rat central nervous system by human forebrain neuroblasts. *Nature, 347,* 556–558.

Williams, E. J., Furness, J., Walsh, F. S., & Doherty, P. (1994). Activation of the FGF receptor underlies neurite outgrowth stimulated by L1, N-CAM, and N-cadherin. *Neuron, 13,* 583–594.

Williams, S. E., Mann, F., Erskine, L., Sakurai, T., Wei, S., Rossi, D. J., et al. (2003). Ephrin-B2 and EphB1 mediate retinal axon divergence at the optic chiasm. *Neuron, 39,* 919–935.

Williamson, T., Gordon-Weeks, P. R., Schachner, M., & Taylor, J. (1996). Microtubule reorganization is obligatory for growth cone turning. *Proceedings of the National Academy of Sciences of the United States of America, 93,* 15221–15226.

Wilson, S. W., Brennan, C., Macdonald, R., Brand, M., & Holder, N. (1997). Analysis of axon tract formation in the zebrafish brain: the role of territories of gene expression and their boundaries. *Cell and Tissue Research, 290,* 189–196.

Wilson, S. W., Ross, L. S., Parrett, T., Easter, S. S., Jr. (1990). The development of a simple scaffold of axon tracts in the brain of the embryonic zebrafish, Brachydanio rerio. *Development, 108,* 121–145.

Yang, L., & Bashaw, G. J. (2006). Son of sevenless directly links the Robo receptor to rac activation to control axon repulsion at the midline. *Neuron, 52,* 595–607.

Yoshikawa, S., McKinnon, R. D., Kokel, M., & Thomas, J. B. (2003). Wnt-mediated axon guidance via the Drosophila Derailed receptor. *Nature, 422,* 583–588.

Yu, W., Cook, C., Sauter, C., Kuriyama, R., Kaplan, P. L., & Baas, P. W. (2000). Depletion of a microtubule-associated motor protein induces the loss of dendritic identity. *The Journal of Neuroscience, 20,* 5782–5791.

Yuan, X. B., Jin, M., Xu, X., Song, Y. Q., Wu, C. P., Poo, M. M., et al. (2003). Signalling and crosstalk of Rho GTPases in mediating axon guidance. *Nature Cell Biology, 5,* 38–45.

Zheng, J. Q. (2000). Turning of nerve growth cones induced by localized increases in intracellular calcium ions. *Nature, 403,* 89–93.

Zheng, J. Q., Felder, M., Connor, J. A., & Poo, M. M. (1994). Turning of nerve growth cones induced by neurotransmitters. *Nature, 368,* 140–144.

Zhou, F. Q., & Cohan, C. S. (2004). How actin filaments and microtubules steer growth cones to their targets. *Journal of Neurobiology, 58,* 84–91.

Zimmer, M., Palmer, A., Kohler, J., & Klein, R. (2003). EphB-ephrinB bi-directional endocytosis terminates adhesion allowing contact mediated repulsion. *Nature Cell Biology, 5,* 869–878.

Zinn, K., McAllister, L., & Goodman, C. S. (1988). Sequence analysis and neuronal expression of fasciclin I in grasshopper and Drosophila. *Cell, 53,* 577–587.

Zolessi, F. R., Poggi, L., Wilkinson, C. J., Chien, C. B., & Harris, W. A. (2006). Polarization and orientation of retinal ganglion cells in vivo. *Neural Dev, 1,* 2.

Target selection

6

As a growth cone nears the end of its journey, it must find appropriate target cells with which to synapse. Because a properly functioning nervous system depends on precise patterns of neural connectivity, this should be done with immense accuracy, but the actual degree of specificity that is achieved is different in different systems and so may be case-specific. At one extreme, every connection is genetically specified. At the other, connections start out only roughly specified, but then sort themselves out through functional validation. In either case, the task of making all these precise connections is daunting because the growth cone may be confronted with thousands or even millions of roughly similar cells in the general target area, from which it will have to choose only one or a few as postsynaptic partners.

The process of target selection can be broken down into a number of steps (Holt and Harris, 1998) (**Figure 6.1**). As they near the target area, axons tend to defasciculate from the tracts or nerves that they are growing along. When they enter the target area, they slow down, and begin searching for their postsynaptic partners who might also be searching, via dendritic growth, for the axons that will innervate them. As they search, the axon terminals begin to branch in the target area. Molecular barriers may also be erected around the borders of the target area so that the incoming axons are corralled until they find the most suitable partners. In large target areas, there is often a mapping strategy so that the axon and

its postsynaptic partner can meet at particular molecular coordinates. Some targets are multilayered, and it is important for axons to find the appropriate synaptic layers within the correct topographic region. Having finally arrived at the correct location, the growing axons choose postsynaptic cells, depending on the degree of genetically programmed specificity, perhaps particular dendrites or regions of these cells, and begin to make synapses.

Almost all of the above phases of targeting and synapse formation can happen very accurately in the complete absence of neural activity. When action potentials are blocked from the earliest stages, retinal axons can still navigate to their correct topographic connections in the tectum and begin to make synapses on tectal cells (Harris, 1984). Experiments in the mouse brain, using a mutant that completely inhibits the release of neurotransmitter, have shown that the brain can be put together such that it is anatomically normal in the complete absence of synaptic activity (Verhage et al., 2000). Finally, the very specific connections that *Drosophila* photoreceptor axons make with their synaptic partners in the optic lobes of the brain form normally at the ultrastructural level in mutants that block the generation of electrical potentials, neurotransmitter release and vesicle endocytosis. (Hiesinger et al., 2006).

The final phase of target selection, however, often entails the refinement of connections. This phase is dependent on neural and synaptic activity, for as synapses are formed, the nervous system can begin to function for the first time. During this phase, the developing organism can begin to test out the wiring and if necessary eliminate connections that are misplaced or poorly functioning. This functional verification is done so that the product of neural development, the brain, works as well as possible when it faces the real world. The refinement of connections is dealt with more thoroughly in Chapter 9.

DEFASCICULATION

In order to enter a target area or to find a target cell with which to make synapses, it is often first necessary to exit from a tight bundle. Nerves, tracts, columns, bundles, and fasciculi often travel past a variety of potential targets. As they do so, specific axons or groups of axons peel off of these common pathways, so that they can enter the target. In the last chapter, we saw that homophilic adhesion molecules such as N-CAM and its homologs can cause similar axons to fasciculate together into nerves or tracts in

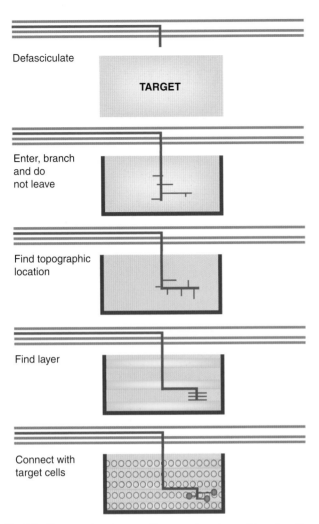

Fig. 6.1 Conceptual, not necessarily successive, stages of targeting. From top to bottom, an axon defasciculates in the region of the target. It enters the target and begins to branch, but is prevented from exiting by a repulsive border. The axon responds to a topographic gradient that promotes branching at the correct location. It then selects a particular layer and finally homes in on particular target cells. *(After Holt and Harris, 1998)*

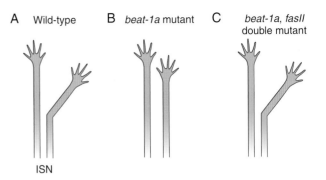

Fig. 6.2 Defasciculation is regulated by Beat proteins. A. Two motor axons growing off the intersegmental nerve (ISN) are shown, one that branches off the nerve in the region of its target. B. In a *beat-1a* mutant, the motor axon at right does not defasciculate. C. The *beat-1a* mutant phenotype is rescued in a *fasII* mutant background. *(After Vactor et al., 1993; Fambrough and Goodman, 1996; Pipes et al., 2001)*

the nervous system. In the motor system of the fruitfly, *Drosophila*, overexpression of the homophilic cell adhesion molecule and NCAM homolog FasciclinII (FasII) leads to enhanced fasciculation and loss of nerve branch formation in the periphery, suggesting that normally such factors might have to be downregulated or deactivated if axons are to defasciculate near a target properly. A failure to defasciculate from the main nerve and resultant bypassing of target muscles is found in a mutant called *beaten path* (Van Vactor et al., 1993). The protein encoded by this gene, Beat-1a, seems required for selective defasciculation of motor axons at particular choice points. Indeed, Beat-1a appears to be an anti-adhesion factor that is secreted by growth cones. Defasciculation can be rescued in *beaten path* mutants by down-regulating other neural cell adhesion molecules such as FasII (Fambrough and Goodman, 1996; Holmes and Heilig, 1999) (**Figure 6.2**).

In vertebrates, too, branching patterns of nerves are mediated by defasciculation. One way to reduce fasciculation is

to secrete an anti-adhesive factor, like Beat-1a. Another is to post-translationally modify adhesion molecules before putting them into the membranes. In the previous chapter, we looked at the role of polysialic acid (PSA) in decreasing the adhesivity of N-CAM in the nerve plexus, where defasciculation was necessary to get different motor neurons sorted into the correct mixed nerves. The relative PSA levels of L1 and N-CAM are also important in balancing axon–axon versus axon–muscle adhesion during target innervation (Landmesser et al., 1990). If PSA is removed, the result is increased axon fasciculation, reduced nerve branching, and reduced target innervation.

The targets themselves may contribute to the defasciculation of the axons that will innervate them. In *Drosophila* mutants lacking mesoderm, the main motor nerves form, but motor axons fail to defasciculate from these bundles. Experiments by Landgraf et al. (1999) have shown that founder myoblasts are the source of defasciculation cues and that a single founder myoblast can trigger the defasciculation of an entire nerve branch. One muscle-derived signal required for the defasciculation of motor nerve branches and their attraction to target muscles is the Ig-Superfamily protein Sidestep, which is expressed by all muscles (Sink et al., 2001). A parallel process that facilitates nerve branch defasciculation is mediated by the matrix metalloprotease Tolloid related 1 (Meyer and Aberle, 2006). Sidestep and Tolloid related 1 work together because double mutants have more severe defects than either mutant alone, and this suggests that the pattern of nerve branching into appropriate muscle target areas may be achieved through combinations of target-specific defasciculation factors.

TARGET RECOGNITION AND TARGET ENTRY

Targets can be internal organs, sensory cells, muscles, or other groups of neurons in CNS or PNS. What makes these tissues the targets for specific innervation? The sympathetic nervous system, with its diversity of end organs, provides an excellent opportunity to study this question. For sympathetic neurons, neurotrophins are the key to initial innervation. There are several different neurotrophins, originally named for their effects on neuronal growth and survival. In Chapter 7, we will see that the survival of sympathetic and other neurons is critically dependent on receiving enough of these target-derived trophic factors. However, the axons must first enter the targets to gain access to a supply of the neurotophins. Neurotrophins, therefore, have roles in target entry that are distinct from their roles in survival (Glebova and Ginty, 2004). Many sympathetic neurons grow along vasculature to reach their various somatic targets. These blood vessels are a source of the neurotrophin, NT-3. In the absence of NT-3, sympathetic cells often fail to invade their targets (Kuruvilla et al., 2004). Take, for example, the epidermis of the external ear. It is innervated by sympathetic neurons, and in NT-3 knockout mice, sympathetic fibers fail to invade the external ear. Exogenously administered NT-3 into the ear rescues the sympathetic innervation in the knockout mice (ElShamy et al., 1996) (**Figure 6.3**).

While NT-3 is required for target entry, axon tips upon entering the target are then presented with another target-derived neurotrophin, called nerve growth factor (NGF). NGF binds the receptor TrkA on these axons and leads to its internalization, which then retrogradely signals to the

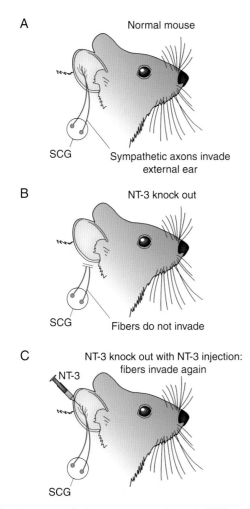

Fig. 6.3 Some sympathetic neurons use a change in NT-3 expression to innervate their targets in the ear. A. Some SCG neurons project to and arborize in the pinna of a normal mouse. B. In NT-3 knockout mice, these fibers do not invade the pinna. C. Restoration of targeting by injection of NT-3 into the ear. *(Adapted from ElShamy et al., 1996)*

cell nucleus promoting both survival and the expression of another neurotrophin receptor, called p75. p75 expression then leads to a reduction in the sensitivity of axons to NT-3, which become more reliant on NGF. This is a sequential hierarchy of neurotrophin signaling that coordinates distinct stages of axon growth, target innervation, and survival (Kuruvilla et al., 2004). The pancreas and other internal organs, which are innervated by sympathetic fibers, also express NGF. If NGF is overexpressed under the control of a beta-cell specific promoter in the islets of the pancreas in transgenic mice, there is a dramatic increase in the innervation of the islets (Hoyle et al., 1993). The role of NGF in attracting innervation has medical implications. For example, pancreatic cancer often metastasizes to neural tissue, and this may be because the pancreatic tumors attract innervation (Schneider et al., 2001). Similarly, pancreatic transplants may benefit by the local application of NGF to help attract innervation (Reimer et al., 2003). Another interesting example of target recognition by sympathetic axons concerns another neurotrophin, glial-derived neurotrophic factor (GDNF) (Ledda et al., 2002). GDNF is recognized by the c-Ret receptor on the growth cones of innervating axons, and it is also recognized by a GPI-linked secreted receptor called GFRal. The peripheral targets of the c-Ret expressing axons, such as the epidermis, secretes GFRa1, which captures circulating GDNF and holds it to the target sites, thus creating a very high level of GDNF right around the target, which attracts innervation from c-Ret expressing axons.

Neurotrophins are also involved in the innervation of neuronal targets by nonsympathetic neurons. A particularly interesting example is the innervation of the inner ear, which includes the vestibular organs of balance and the cochlear organs of hearing (Ernfors et al., 1995), (**Figure 6.4**). Neurotrophins are first expressed in the otocyst during the time at which ganglion cells with neurotrophin receptors send their processes toward this structure. Indeed, BDNF appears to be expressed by the hair cells, which are the cellular targets of this innervation, well before the hair cells have fully differentiated (Hallbook and Fritzsch, 1997). NT-3 is also expressed in the developing inner ear, and all the innervating fibers possess receptors for both neurotrophins: TrkB for BDNF and TrkC for NT-3. Knockout experiments of the ligands show that BDNF is necessary for the innervation of the vestibular hair cells, whereas NT-3 is more important in the innervation of the cochlear hair cells. Indeed, the phenotypes match where these factors are most heavily expressed (Fritzsch et al., 1997). In mice that lack both BDNF and NT-3, or both TrkB and TrkC, there is a complete loss of innervation to the inner ear (Fritzsch et al., 1997). Interestingly, if the BDNF coding sequence is inserted into the NT-3 gene in a transgenic mouse, the result is the expression of BDNF throughout the inner ear, and all the fibers that normally innervate the NT-3 rich areas survive and innervate the cochlea as usual (Tessarollo et al., 2004). So the neurotrophic factors can substitute for each other in a way. However, these transgenic mice show excessive innervation of the cochlea from neurons that would normally innervate the vestibular regions, a miswiring that probably occurs because of the changed spatiotemporal expression pattern of BDNF. This incorrect projection can be enhanced by knocking out the normal expression of BDNF in the vestibular region. These results suggest that correct temporal and spatial pattern of neurotrophin expression may be critical for the correct innervation of these inner-ear targets.

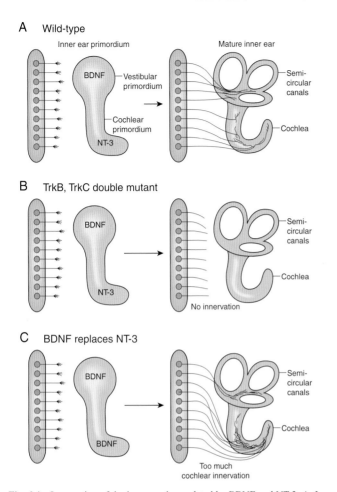

Fig. 6.4 Innervation of the inner ear is regulated by BDNF and NT-3. A. In the wild-type animal, the vestibulo-cochlear ganglion, all of whose neurons express TrkB and TrkC, grow toward the developing inner ear, which has a vestibular and a cochlear primordium. As the system develops and the primordia develop into semicircular canals and a cochlea, the ingrowing axons innervate both parts of the inner ear. B. In BDNF, NT-3 or TrkB, TrkC double mutants, the inner ear remains uninnervated. C. In transgenic mice in which BDNF has been knocked into the NT-3 coding region, the cochlear region becomes innervated by the vestibular part of the ganglion. *(After Ernfors et al., 1995; Fritzsch et al., 1997; Fritzsch et al., 2004; Tessarollo et al., 2004)*

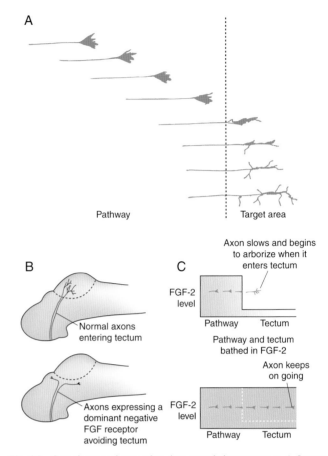

Fig. 6.5 Growth cones change when they enter their target zones. A. Images from a time-lapse movie of a retinal ganglion cell growing in the optic tract and then crossing (at the dotted line) into the tectum. The simple growth cone becomes much more complex and slows down dramatically as it enters the target. B. Tectal innervation by control retinal axons (top) and tectal avoidance by retinal axons that misexpress a dominant negative FGF receptor. C. Retinal axons slow down and branch when they reach the tectum in control animals (top), but when the pathway is exposed to high levels of FGF-2, the axons keep going and do not innervate the tectum. *(Adapted from Harris et al., 1987; McFarlane et al., 1995; McFarlane et al., 1996)*

SLOWING DOWN AND BRANCHING IN THE TARGET REGION

Time-lapse observations of fluorescently labeled RGC axons in *Xenopus* embryos show that they grow at a rate of about 60μm/hr in the optic tract but slow to about 16μm/hr when they enter the optic tectum (Harris et al., 1987) (**Figure 6.5**). Once within the optic tectum, these terminals may advance in a saltatory, stop and start, manner. Why do retinal growth cones slow down when they reach the tectum? Retinal axons grow toward their target on a pathway that is rich in FGF, and this molecule has been found to promote axonal growth in the tract and *in vitro*. As retinal axons enter the tectum, they encounter a sudden drop in external FGF because the tectum expresses very little of it (McFarlane et al., 1995). Therefore, one cue that decreases the growth rate of retinal axons at the target is a drop in FGF levels. If excess FGF is added, or if the retinal axons are made insensitive to FGF, the retinal axons

do not respond to the target and often grow by it, so they may read the drop in FGF as a target entry signal (McFarlane et al., 1995; McFarlane et al., 1996; Webber et al., 2003).

As retinal growth cones slow down in the tectum, time-lapse images show that they also become much more complex. Branches begin to form, and many of these arise at some distance behind the axonal tip (Harris et al., 1987). Thus, axonal arbors are built in a way that is reminiscent of the way the arbor of a tree develops, with new branches arising along the stems of older branches. Many of these new branches are not tipped with growth cones themselves but appear like worms wriggling out of the parent branch. What causes axons to branch in this way? The tectum expresses the repulsive guidance cue Sema3A. Sema3A, when applied to retinal growth cones *in vitro*, causes them to collapse, but this collapse is transient, and recovery from collapse is often associated with branching (Campbell et al., 2001). Thus, Sema3A and perhaps other repulsive guidance factors such as Ephrins may stimulate terminal branching in the tectum. In addition, the tectum is a source of BDNF, which also promotes branching of RGC

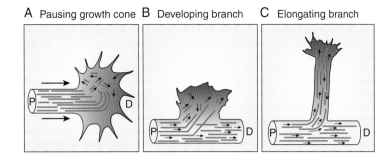

Fig. 6.6 Axons branch at pause points. A. A growth cone pauses, microtubules splayed. B. The growth cone moves on but leaves behind it a zone where the cytoskeleton remains somewhat disorganized. C. A branch forms at this zone. *(After Dent and Kalil, 2001)*

axons. Injection of BDNF into the optic tectum of live *Xenopus laevis* tadpoles increases the branching of RGC terminals, whereas blocking BDNF reduces axon arborization (Cohen-Cory and Fraser, 1995). Altering levels of BDNF in the retina had no effect on RGC branching in the tectum, indicating that the branch-promoting effects of BDNF are local on axon terminals (Lom and Cohen-Cory, 1999; Lom et al., 2002).

The sensory neurons of the dorsal root ganglion (DRG) provide another useful example of branching in the target region. These axons enter the spinal cord through the dorsal root and then bifurcate to grow in the anterior and posterior directions. Collaterals then sprout from these branches to innervate the gray matter of the cord (Ozaki and Snider, 1997). Using an *in vitro* assay, it was found that Slit2 promotes the formation and elongation of branches in DRG neurons (Wang et al., 1999). The identification of repulsive guidance molecules such as Slits and Semas, which can control growth cone guidance on the one hand and promote branching on the other, suggests that there may be a link between repulsion and branching. Indeed, *in vitro* observations have indicated that branches form behind the tip of the growth cone whenever the growth cone collapses in response to any collapse-inducing agent, even a mechanical one (Davenport et al., 1999). These observations fit well with the finding that the growth cones of callosal axons in the cortex pause for many hours beneath their targets prior to the development of branches (Kalil et al., 2000). Imaging of dissociated living cortical neurons shows that wherever a growth cone undergoes a lengthy pause, and then advances again, cytoskeletal remnants of the paused growth cone are left behind on the axon shaft (Szebenyi et al., 1998). Here, the cytoskeleton of the axon appears more splayed apart and fragmented, and it is from these regions that new branches form (Dent and Kalil, 2001) (**Figure 6.6**). Such results demonstrate that growth cone pausing is closely related to axon branching.

BORDER PATROL: THE PREVENTION OF INAPPROPRIATE TARGETING

Once they have recognized and entered a target area, slowed down, and started to branch, axons may be prevented from exiting the target area by repulsive cues at the perimeters. Sema3a, which we discussed in the last chapter, repels the growth cones of cutaneous sensory neurons. Analysis of knockout mice supports a critical role for Sema3a as an exclusion factor confining the peripheral ends of these axons to the correct target areas

of the skin (Taniguchi et al., 1997). In these Sema3a knockouts mice, axons that are normally restricted from innervating skin now enter these territories. Sema3a is also expressed at the posterior boundary of the olfactory bulb, where it seems to act to restrict olfactory axons to the bulb, preventing them from entering the telencephalon (Kobayashi et al., 1997). The repulsive molecule Ephrin-A5, which we will discuss in more detail below with respect to topography, reaches its highest concentration just posterior to the superior colliculus or tectum, the target of retinal axons, suggesting that this ligand may also serve as a factor that confines these axons to the target. Indeed, retinal axons extend beyond the posterior border of the superior colliculus in Ephrin-A5 knockout mice (Frisen et al., 1998).

This raises the question of whether this type of mechanism is used to help segregate neural circuits that carry different kinds of information to different, but nearby, centers of the brain, thus preventing inappropriate targeting. Reinnervation and cross-innervation experiments show that, when the normal targets of axons have been surgically removed, functional synapses can indeed be made on the wrong targets. Similarly, when the brain is injured, the normal targets of some axons may die and nearby regions may become denervated. In these cases, axons that originally innervated the injured areas may sprout new growth cones to invade denervated but inappropriate targets. To test how promiscuous axons are, and whom they will synapse with if given the chance, one can test a variety of foreign targets with different axonal populations. For example, to know how determined retinal ganglion cell axons are to invade a specific target, one of their normal targets, the visual thalamus, was left to degenerate by ablating the visual cortex and superior colliculus while the ventrobasal nucleus (VBN) of the somatosensory thalamus was denervated (Metin and Frost, 1989). In this case, retinal axons innervated the somatosensory VBN (**Figure 6.7**). In a similar experiment, retinal ganglion cell axons innervated the medial geniculate nucleus (MGN) of the auditory thalamus (Roe et al., 1992). The thalamocortical connections have not been changed and are basically normal in these animals, giving rise to the weird condition that these animals process visual information in the somatosensory or auditory cortex, thus perhaps having the conscious sensation of feeling or hearing the visual world (von Melchner et al., 2000) (Figure 6.7). Normally, of course, the nuclei of the thalamus have modality-specific innervation. The question is whether segregation is a result of molecular barriers that normally separate these brain areas. It is interesting to note, then, that high levels of Ephrin-A2 and Ephrin-A5 define a distinct border between the visual and auditory thalamus. If the normal input to the auditory thalamus is denervated and

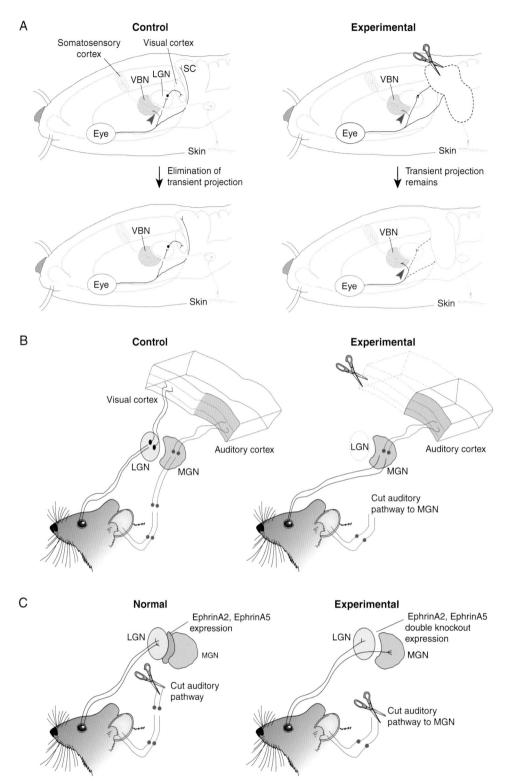

Fig. 6.7 Cross wiring of visual signals to the somatosensory and auditory cortex. A. In a normal control animal, the somatosensory input goes to the ventrobasal nucleus (VBN) of the thalamus and from there to the somatosensory cortex. Retinal input is to the lateral geniculate nucleus (LGN) and the superior colliculus (SC), although neonatally, there is a transient projection to the VBN that is normally eliminated. The visual input to the LGN then goes on to the visual cortex. In an experimental animal in which the superior colliculus, the visual cortex, and the input to the VBM are removed neonatally, the transient retinal projection to the VBN is stabilized and the visual information is thus provided to the somatosensory cortex via the VBN. B. When similar visual cortical and SC ablations are done and the auditory pathway is cut, the retinal input sprouts into the medial geniculate nucleus MGN (the normal thalamic target of the auditory input), which projects as usual giving visual physiological properties to the auditory cortex. C. EphrinA2 and EphrinA5 are expressed at the border between the LGN and the MGN. Even though the auditory pathway to the MGN has been cut, the retinal fibers do not invade the MGN, but they do so in EphrinA2, EphrinA5 double knockouts. *(After Metin and Frost, 1989; Roe et al., 1992; Lyckman et al., 2001)*

the visual thalamus is spared, retinal axons mostly remain in their uninjured normal targets. However, when this experiment is done in knockout mice that lack both Ephrin-A2 and Ephrin-A5, there is extensive rewiring and retinal axons invade and innervate the deafferented auditory thalamus (Lyckman et al., 2001) (Figure 6.7). These findings suggest that signals that induce innervation may compete with barriers, such as repulsive guidance molecules, that serve to contain axons within the normal targets.

Border patrol is not the only mechanism for maintaining appropriate targeting. In cross-innervation experiments in amphibians, in which extensor motor nerves are forced to innervate flexor muscles and flexor motor nerves are forced to innervate extensor muscles, the animals develop expected inappropriate motor behaviors after surgery. Interestingly, however, normal behavior usually recovers after a rather long period of time. This was first interpreted as the animals learning to use these muscles in a new way, but detailed anatomical investigations of these animals showed that the crossed nerves, over the course of time, had managed to uncross themselves and find their original muscles again (Mark, 1969). Competition experiments between original and foreign nerves for the innervation of particular muscles show that the original nerves always have an advantage (Dennis and Yip, 1978). Thus axons, although they will innervate inappropriate denervated targets when their own targets are unavailable, seem to have a natural preference for their own original targets.

TOPOGRAPHIC MAPPING

In many targets there is a topographical relationship between the position of the innervating neuron and the position of its terminal arbor in the target field. A good example is the visual system. RGCs in a particular position in this retina are maximally stimulated from a region of the visual world, and these cells send their axons to a particular region in the tectum. Neurons in neighboring retinal positions send their projections to neighboring regions in the tectum. This orderly projection preserves visual topography in the brain and provides a neuroanatomical basis for the contiguity of perceived visual space. Similarly, most central auditory nuclei have a representation of the cochlea's frequency axis. Such maps may be referred to by the anatomical substrates that they preserve (e.g., retinotopic, cochleotopic) or by more perceptual terms (e.g., visuotopic or tonotopic). Even simple animals like the nematode, with only 301 neurons, have ordered arrays of sensory receptors that make somatopically organized central projections. These help them respond appropriately to stimuli that strike the animal from different directions. There is a second type of neural map, a computational map, which can be revealed by recording from neighboring single nerve cells *in vivo*. What is represented in such maps is not obvious from the anatomy of the connections, yet these maps may also display orderly representations of a physical parameter. For example, in the auditory system, we find nuclei that display topography of sound source location, even though the ear contains only a one-dimensional array of spiral ganglion cells representing sound frequency. Such maps are constructed from cells that extract information and are referred to by the functional characteristic that they encode (e.g., map of auditory space). There

are also less intuitively obvious maps, such as maps of smell that we will discuss below.

What is the developmental basis for the establishing topographic projections in the nervous system? In the late 1800s John Langley discovered that superior cervical ganglion (SCG) neurons mediate reflexes in a topographic manner (Langley, 1897, 1985). When Langley stimulated the first or top thoracic root to the ganglion in the rat, this activated ganglion cells that caused dilation of the pupil. When he stimulated the fourth thoracic root to the ganglion, the blood vessels of the ear constricted. This suggested that there was some sort of topographical organization within the SCG. All reflexes were immediately lost when the preganglionic nerve to the SCG was cut, but the fibers reinnervated the SCG in several weeks, as peripheral nerves often do in mammals, and the autonomic reflexes recovered. The surprising discovery was that the connections reformed with such precision that all reflexes were re-established accurately (**Figure 6.8**). This result suggested that individual SCG neurons have some mechanism that enables the regenerating preganglionic fibers to distinguish one SCG neuron from another in a topographically appropriate manner.

The sympathetic chain ganglia provide a simple system in which to examine somatotopic specificity because the relative positions of ganglia along the A-P axis are innervated topographically through nerve roots from the spinal cord from corresponding A-P regions of the CNS. Each ganglion is selectively innervated by afferents from a limited number of spinal cord segments from a specific A-P domain. Thus,

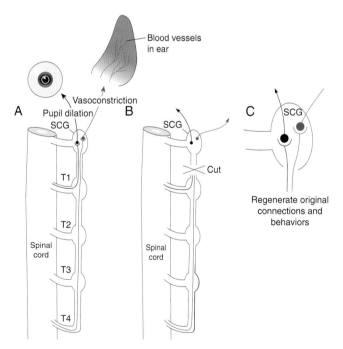

Fig. 6.8 Regeneration of topographic specificity. A. Langley's classic study showed that stimulation of preganglionic root T4 relayed through ganglion cells in the SCG that caused vasoconstriction of the ear pinna vessels, whereas stimulation of root T1 excited other SCG neurons that caused dilation of the pupil. B. When Langley cut the sympathetic tract above T1, all these sympathetic reflexes were abolished, but with time they recovered. C. Shows the specificity associated with this regeneration, such that the axons that enter the chain at T4 reinnervate the SCG cells that cause ear vasoconstriction, and the axons that enter at T1 reinnervate the cells that cause pupil dilation. *(After Langley, 1897, 1985)*

the superior cervical ganglion (SCG) is primarily innervated by preganglionic afferents from thoracic segments T1–T4, whereas the more caudally located fifth thoracic ganglion (T5) is primarily innervated by afferents from T4–T7 (Nja and Purves, 1977). In one experiment, a T5 ganglion was transplanted to different locations along the sympathetic chain, exposing this target to afferents from a large range of spinal cord segments (Purves et al., 1981). Selective reinnervation was then assessed electrophysiologically. The sympathetic chain ganglia were dissected out along with the ventral nerve roots through which all preganglionic fibers course from the ventral spinal cord. Stimulating electrodes were then placed on the ventral roots from each spinal cord segment, and an intracellular recording was obtained from the reinnervated T5 ganglion. The spinal segments that innervate each T5 neuron were thus recorded. The results clearly indicate that T5 neurons are selectively reinnervated by their original spinal segments (**Figure 6.9**). This is not merely an artifact of the host transplantation site, because when the SCG was placed in the same location, it, too, becomes reinnervated by its original set of afferents. These experiments strongly suggest that axons from different rostrocaudal levels can distinguish individual sympathetic ganglion cells, which must also carry some label of their rostrocaudal origin.

CHEMOSPECIFICITY AND EPHRINS

In the early 1940s, Roger Sperry cut the optic nerve of a newt, rotated the detached eye 180° in its orbit, and assayed the visuomotor behavior of the animal after its nerve had regenerated. The newts, and in subsequent studies, frogs, behaved as if their visual world were back-to-front and upside-down: when a lure was presented in front of them, "they wheeled rapidly to the rear instead of striking forward" and when the lure was presented above "the animals struck downward in front of them and got a mouthful of mud and moss" (Sperry, 1943) (**Figure 6.10**). This led him to propose that topographic nerve connections between the retina and its main central target, the optic tectum, were the result of anatomical rather than experiential features of the nervous system, as Sperry's unlucky frogs never did learn to snap in the correct direction. Sperry reasoned that the retinal fibers mapped onto the tectum according to original anatomical coordinates of the eye. The explanation he gave was the possible existence of biochemical tags across the retina and tectum. He postulated the existence of two or more cytochemical gradients "that spread across and through each other with their axes roughly perpendicular" (Sperry, 1963). These separate gradients successively superimposed on the retinal and tectal fields would stamp each cell with its appropriate

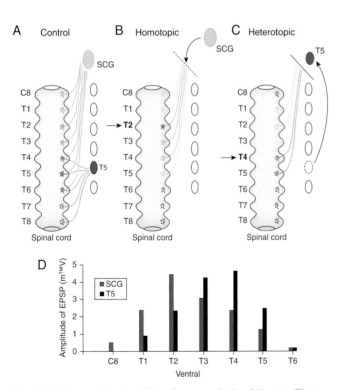

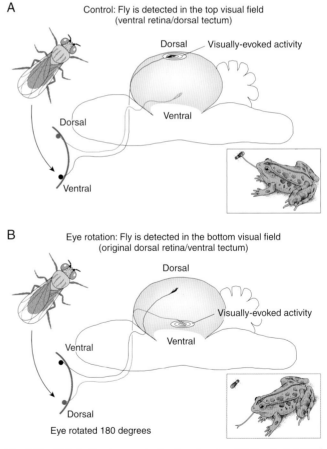

Fig. 6.9 Topographic input into the sympathetic chain. A. Electrophysiological studies show that the SCG receives input from many roots but primarily the more anterior ones. The ganglion at T5 receives its primary inputs from more posterior roots. B. When the SCG is removed and replaced with another SCG, the axons that reinnervate it tend to be from more anterior roots. C. When a T5 ganglion is put in place of the SCG, its neurons still tend to get innervated by more posterior roots even though the ganglion is in an anterior position. D. This topographic specificity of reinnervation is reflected in the shape of the histogram of EPSP amplitudes as a function of nerve root stimulation for the homototopic and heterotopic transplants. *(After Purves et al., 1981)*

Fig. 6.10 Maladaptive topography implies chemospecificity. A. A normal frog sees a fly above on its ventral retina, which projects to the dorsal tectum, leading to a snap in the appropriate upwards direction. B. A frog with a rotated eye sees the same fly on what used to be the dorsal retina, which projects as ever to the ventral tectum, leading to a snap in the wrong downward direction. *(After Sperry, 1943)*

latitude and longitude in a kind of chemical code with matching values between the retinal and tectal maps. This became known as the chemoaffinity hypothesis.

The chemoaffinity hypothesis inspired many biologists and biochemists to try to find the molecules that were responsible for topographic targeting in the retinotectal system. Such studies often took an *in vitro* approach, and for over 20 years, not much progress was made. Friedrich Bonhoffer and colleagues made a breakthrough when they used membranes from anterior and posterior parts of the tectum to make a striped carpet. When retinal tissue was cultured on such striped carpets, they found that temporal retinal axons, but not nasal axons, grow preferentially on anterior tectal membranes (Walter et al., 1987b) (**Figure 6.11**). Surprisingly, when the posterior membranes were heated, treated with formaldehyde, or exposed to an enzyme (PI-PLC) that removes PI-linked membrane molecules, temporal

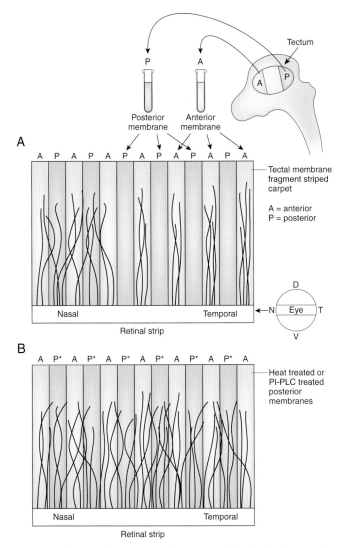

Fig. 6.11 The striped carpet assay. A. An equatorial strip of retina spanning the nasal (N) temporal (T) extent is positioned on a striped carpet of alternating anterior (A) and posterior (P) tectal membranes. The nasal fibers from the retinal explant grow on both A and P tectal membranes, but the temporal fibers grow only on the A membranes. B. If the tectal membranes are denatured or treated with PI-PLC, which releases PI-linked membrane proteins, the temporal axons also grow on both types of membranes, suggesting that the P membranes normally have a PI-linked repulsive guidance molecule. *(After Walter et al., 1987a; Walter et al., 1987b; Walter et al., 1990)*

axons no longer showed such a preference (Walter et al., 1987a; 1990). This suggests that the relevant activity is a membrane-linked protein that is repulsive to temporal axons, and to which nasal axons are rather insensitive. By examining the choices that temporal axons make between membranes extracted from successive rostrocaudal sixths of the tectum, it became clear that this inhibitory activity is graded across the tectum, highest at the caudal (posterior) pole and lowest at the rostral (anterior) pole. Two repulsive factors, now called Ephrin-A5 and Ephrin-A2, were then identified by the Bonhoffer and Flanagan laboratories as the inhibitory molecules involved (Cheng et al., 1995; Drescher et al., 1995). Ephrins, of which several are now known, come in two subfamilies, a GPI-linked or A-type, and a transmembrane or B-type. Their receptors, known as Ephs, also divide into two A- and B-type families. The Ephrin-As generally activate Eph-As, while the Ephrin-Bs generally activate Eph-Bs (Flanagan and Vanderhaeghen, 1998). Both Ephrin-A5 and Ephrin-A2 are expressed in a posterior (high) to anterior (low) gradient in the tectum (**Figure 6.12**). The retina, as expected, shows a gradient of Eph-As, the receptors for these ligands. Temporal axons that have high levels of Eph-A avoid the posterior pole of the tectum that has the highest level of the Ephrin-A ligands. When Ephrin-A2 is misexpressed by transfection across the entire tectum in chick embryos, temporal axons find it difficult even to enter the tectum. When membrane stripes are made from the transfected anterior tectal cells, temporal axons will not grow on them.

These results predict that when the Ephrins are knocked out, there will be mapping errors. In mutant mice, in which Ephrin-A5 is knocked out, temporal axons map more posteriorly (Frisen et al., 1998), but the mapping phenotype is even more striking in double knockouts of both Ephrin-A2 and Ephrin-A5 (**Figure 6.13**). In these mice, the anteroposterior order of both nasal and temporal axons is largely, though not totally, lost. Temporal axons terminate all over the tectum and freely invade the posterior poles (Feldheim et al., 2000). The fact that there is still some order left in this projection suggests that there may be as yet other undiscovered chemospecificity factors that are involved. Knocking out Eph-As, the receptors to the Ephrins, corroborates these findings (Feldheim et al., 2004). If Eph-A3 is knocked out in temporal axons in the chick, they project more posteriorly than they normally would within the tectum. A gene disruption of mouse Eph-A5 receptors caused similar map abnormalities. A very interesting experiment with Eph-A receptors involves gene targeting to elevate Eph-A receptor expression in just one subset of retinal ganglion cell in the mouse, while leaving the neighboring ganglion cells to express normal levels of Eph-A. The effect of this manipulation is to produce two intermingled ganglion cell populations, one that expresses more receptor and thus should be more sensitive to Ephrin-A, and another that is normally sensitive. The results are that the two populations of ganglion cells from the same eye form separate shifted maps in the tectum, leading to a kind of double vision (Brown et al., 2000) (Figure 6.13). The RGCs that express higher levels of Eph-A map more rostrally than those that express normal levels. This finding clearly favors a model in which retinal growth cones, by the levels of Eph-A they express, read the levels of Ephrin-A in the tectum to establish a graded map.

The data described above, however, do not fully explain the problem of topographic mapping across the anterior to the posterior axis of the tectum. Since Ephrin-As are noted as

151

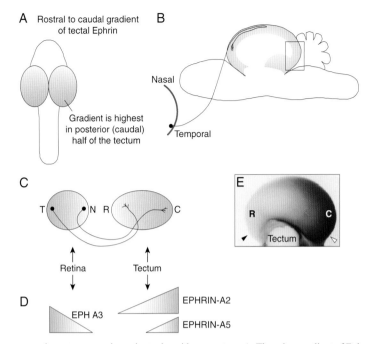

Fig. 6.12 Nasotemporal to anteroposterior retinotopic guidance system. A. There is a gradient of Ephrins in the tectum, high in the posterior or caudal (C) pole and low in the anterior or rostral (R) pole. B. A retinal ganglion cell in the temporal retina expresses active receptors for these tectal Ephrins and avoids the posterior tectum. C and D. The opposing gradients of active Eph-A receptor expressed in the retina and the gradients of A-type Ephrins in the tectum. This system can at least partially account for topographic mapping in this axis. E. The Ephrin-A gradient is shown in the tectum of a chick that uses a soluble Eph-A receptor fused to alkaline phosphatase to reveal the distribution of the Ephrin-A ligands in the tectum. *(After Cheng et al., 1995; Drescher et al., 1995)*

axon repellents, one of the key questions that remains is why any axons bother to go to the posterior tectum, especially as all axons express at least some Eph-As and should prefer to map to the anterior tectum. Is there a counterbalancing attractive mechanism as some theoretical work predicts (Honda, 1998)? One idea in this regard is that there might be interaxonal competition. For instance, the tectum might be a source of a limited supply of neurotrophin, for which retinal axons compete (Wilkinson, 2000). The nasal axons that have the least Eph-A would find the competition less fierce in the posterior tectum, which is why they go there. This idea may explain the otherwise puzzling observation that removal of the Ephrin-As from the tectum not only causes a posterior shift for temporal axons, but also causes an anterior shift for nasal axons, as if the two populations were competing. There is, however, another, though not mutually exclusive, explanation that has to do with the finding that Ephrins are not always repulsive. A systematic *in vitro* analysis shows that Ephrin-A2, while capable of inhibiting the growth of temporal axons at high concentrations, actually promotes the growth of these axons at low concentrations (Hansen et al., 2004) (**Figure 6.14**). Moreover, the transition from growth inhibition to growth promotion varies across the retina according to how much Eph-A is expressed in RGCs; so nasal axons with low levels of Eph-As may actually be attracted to the posterior tectum.

In addition to the repulsive activity of the Ephrin-As, there are other potential graded cues in the tectum such as repulsive guidance molecule (RGM) (Monnier et al., 2002) that work in concert with the Ephrin-As to repel axons that arise from the temporal retina, which express the RGM receptor neogenin. Such remarkable gradients of these guidance molecules and their receptors in the retina and the tectum presuppose that both tissues have polarity, and this results from the action of patterning molecules like FGF8 (as we have seen in Chapter 2). In the tectum, one of the first read-outs of positional information is the graded expression of the transcription factor Engrailed 2 which itself is expressed in a gradient in the tectum, high posteriorly and low anteriorly. Engrailed then regulates the graded expression of Ephrin-As in the tectum. The transcription factor Engrailed is thus a key in patterning the tectum along the anterior-posterior axis. Surprising, Engrailed in its own right can serve as a mapping cue as it is secreted by tectal cells and taken up by the axons of retinal ganglion cells, causing repulsion of temporal axons and attraction of nasal ones (Brunet et al., 2005).

How is the second axis of the retinotectal map formed, in which axons from the dorsal retina map to the lateral tectum while axons from the ventral retina map to the medial tectum? Eph-B receptors have been found in a ventral (high) to dorsal (low) gradient in the retina, whereas Ephrin-B ligands are found in a medial (high) to lateral (low) gradient in the tectum (Braisted et al., 1997; Holash et al., 1997) (**Figure 6.15**). Interestingly, Ephrin-Bs are also expressed in the retina, in a dorsal (high) to ventral (low) gradient, whereas Eph-Bs are expressed in the tectum in a lateral (high) to dorsal (low) gradient. These expression patterns suggest that attraction rather than repulsion is the overriding mechanism at work in this dimension. Here Eph-B expressing axons of the ventral retina are attracted to Ephrin-Bs in the medial tectum, and Ephrin-B expressing RGC axons of the dorsal retina are attracted to Eph-B expressing cells in the lateral tectum via backwards signaling from the receptor to the ligand. In *Xenopus*, the prevention of all Ephrin-B/Eph-B interactions causes dorsal axons to project medially rather than laterally (Mann et al., 2002a).

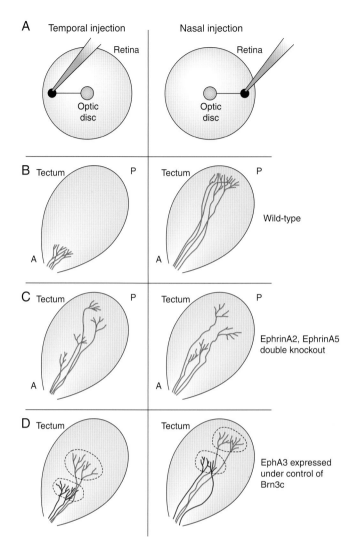

Fig. 6.13 Topographic mapping in Ephrin-A2 and Ephrin-A5 double knockouts and with mosaic Eph-A3 misexpression. A. Label is injected into the temporal (left) or nasal (right) retinas. B. The result in the normal mice is a projection to the anterior (left) or posterior (right) colliculus. C. In the Ephrin-A2, Ephrin-A5 double knockouts (tectum is gray), termination zones are all over the AP extent of the colliculus for both nasal and temporal injections. D. Two separate overlapping maps form from the subset of RGCs that express Isl2 and thus extra Eph-A3 (blue) and those that express the normal amount of Eph-A3 (red). *(After Brown et al., 2000; Feldheim et al., 2000)*

This effect depends on Ephrin-B function in the axons because the same phenotype occurs if retinal axons express a dominant negative form of Ephrin-B that lacks an intracellular domain. Thus, reverse signaling via the Ephrin-B intracellular domain seems to attract Ephrin-B expressing dorsal retinal axons to Eph-B-expressing cells in the lateral tectum (Figure 6.15). But then why does the ventral retina project to the medial tectum? In mice, when the Eph-B2 and Eph-B3 receptors are knocked out, there is an ectopic projection to the lateral tectum and the phenotype is even stronger if the Eph-B receptors are replaced with receptors that are unable to signal (Hindges et al., 2002). This result suggests that forward signaling through the intracellular domain of the Eph-B receptor is critical for ventral axons to map to the medial tectum (Figure 6.15). In the medial to lateral axis, there should also be countergradients that reinforce the action of the Ephrin-B system, and indeed, the morphogen Wnt3 is expressed in a medial high

to lateral low gradient in tectum. Wnt3 repels RGCs through the Wnt receptor Ryk, which is expressed in a ventral high to dorsal low gradient in the retina. Dorsal RGC axons, however, express Frizzled, another Wnt receptor, which causes them to be attracted to Wnt3.

These studies on Ephrin-As and Ephrin-Bs and their receptors strongly verify Sperry's chemospecificity hypothesis about retinotectal mapping by providing the molecular identities of at least some cytochemical tags of the kind that he proposed. The next question is whether the Ephs and Ephrins are involved in setting up topographic projections in other regions of the nervous system. Certainly, the fact that there are many of these ligands and receptors is consistent with such a possibility, as are the histological findings that the CNS is painted with a rich pattern of these ligands and receptors often in reciprocal graded arrangements of A-type ligands with A-type receptors and B-type ligands with B-type receptors (Zhang et al., 1996). Work in a number of systems has now established that this is the case. For example, there is evidence that Ephrin/Eph signaling is used in establishing the visuotopic projection from the retina to the visual thalamus (Feldheim et al., 1998) and from the thalamus to the visual cortex (Cang et al., 2005), the somatotopic map of the body surface on the primary somatosensory area of the cortex (Vanderhaeghen et al., 2000), and the tonotopic projection from the cochlea onto the nucleus magnocellularis in the hindbrain (Person et al., 2004).

THE THIRD DIMENSION, LAMINA-SPECIFIC TERMINATION

Many parts of the nervous system, like the tectum or the cerebral cortex, are layered structures, and innervating axons must not only map to their correct topographic position in two dimensions, but they must also find the appropriate layer in which to synapse, making targeting a three-dimensional problem. Lamina-specific targeting involves a variety of different issues. In many cases, a laminated target is composed of layers of physiologically and molecularly distinct cell types, and the innervating axons must therefore choose between different cell types, possibly based on chemical differences. In some cases, the layers are composed of essentially similar cells, but the innervating axons, through an activity-based competition, segregate them into layers. This latter case can be considered as an example of the refinement of synaptic connections and will be discussed more fully in Chapter 9.

We have already encountered the first kind of the laminated structure in the central projections of DRG fibers in the spinal cord. These axons enter the spinal cord and make synapses in various laminae of the dorsal horn or ventral gray matter depending on their modality. For instance, stretch receptors make monosynaptic contact with motor neurons in the ventral horn. In contrast, pain and temperature sensory fibers innervate neurons in dorsal laminae of the spinal cord (**Figure 6.16**). The result of this laminar arrangement by different types of input is that somatosensory modalities sort out in the spinal cord and so make a multilayered registered map, such that a column of cells in the spinal cord represents one area of the body with different modalities at different depths. Multimodal, layered maps are used in several places in the nervous system.

Why do only stretch receptors penetrate the more ventral layers of the spinal neuropil? In the previous chapter, we

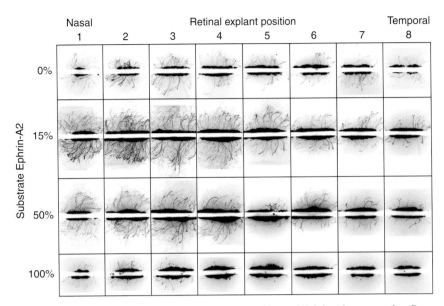

Fig. 6.14 Retinal axon outgrowth, with variation in both retinal position and Ephrin-A2 concentration. Representative photographs showing outgrowth from the eight contiguous explant positions (numbered 1 through 8) across the nasal-temporal axis of the retina, grown on substrates containing different proportions of membranes from Ephrin-A2 DNA-transfected and untransfected cells. Outgrowth varies with both retinal position and Ephrin concentration. Responses to Ephrin-A2 membranes vary from total outgrowth inhibition (at higher concentrations and more temporal positions) to several-fold outgrowth promotion (at lower concentrations and nasal positions). *(After Hansen et al., 2004)*

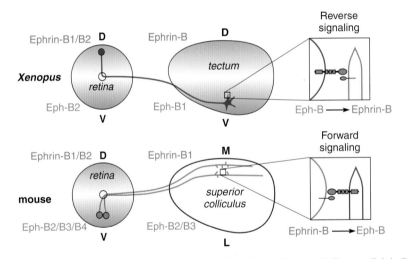

Fig. 6.15 Topographic mapping of the dorsal–ventral retina onto the mediolateral tectum. In *Xenopus*, Ephrin-B expressing retinal ganglion cells from the dorsal retina via reverse-signaling are attracted to the high levels of Eph-B in the lateral tectum, while in chicks Eph-B expressing retinal ganglion cells are attracted via forward signaling to the high levels of Ephrin-B in the medial tectum. Abbreviations: D, dorsal; V, ventral; M, medial; L, lateral. *(From Pittman and Chien, 2002)*

described the varying sensitivity of different classes of neurons to the repulsive effects of Sema3A, which is expressed in the ventral layers only. Sema3A repulses pain receptors and thermoreceptors, which therefore map to dorsal layers, while stretch receptors ignore Sema3A and map to ventral layers (Messersmith et al., 1995). In mice in which the *sema3A* gene is knocked out, the terminals of the pain and thermoreceptive axonal terminals appear to extend into the ventral regions of the spinal cord, similar to the stretch receptors and this layer-specific targeting is abolished (Taniguchi et al., 1997; Catalano et al., 1998) (Figure 6.16).

The cerebral cortex is another example of a highly laminated structure, composed of different cell types in different layers.

The different layers of the cortex are innervated by different inputs. *In vitro* studies using membrane fractions of cells in the different layers suggest that the targeting cues are membrane-associated (Castellani and Bolz, 1997). Somatosensory thalamic neurons, for example, innervate layer 4 of the somatosensory cortex and cross through layer 5 without branching (Bolz et al., 1996) (**Figure 6.17**). Ephrin-A5 is expressed on the membranes of cells in layer 4 but not layer 5, and *in vitro* studies show that membrane-bound Ephrin-A5 increases the branching of thalamic axons and that there are as yet unidentified repulsive activities to these neurons on the membranes of layers 2/3 and 5 cells (Mann et al., 2002b). Layer 6 neurons in the cortex, like thalamic cells, also arborize in layer 4, and

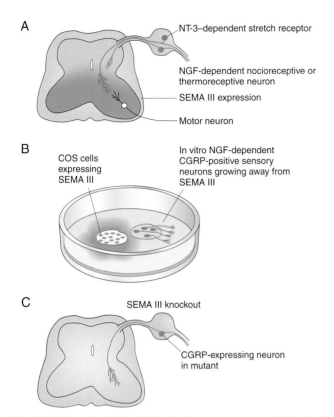

Fig. 6.16 The role of Sema3A in keeping some axons out of a target region. A. Stretch receptors project their central axons into the ventral horn of the spinal cord where they synapse on motor neuron dendrites. Nocioreceptive and thermoreceptive axons, which are NGF dependent, however, terminate in the dorsal gray matter of the spinal cord. Sema3A is expressed only in the ventral cord. B. COS cells in culture repel axons from DRG cultures treated with NGF but not NT-3, thereby preserving nocioceptive and thermoreceptive axons but not stretch receptors. C. In Sema3A knockout mice, these sensory neurons extend into the ventral horn. *(After Messersmith et al., 1995)*

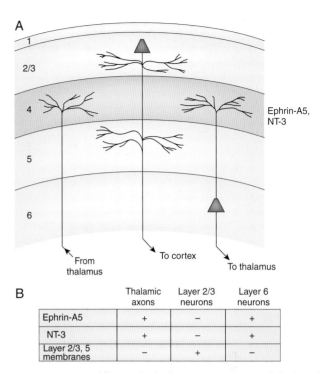

Fig. 6.17 Layer-specific targeting in the sensory cortex. A. Thalamic and Layer 6 axons invade the cortex and selectively branch in Layer 4, which expresses Ephrin-A5 and NT-3. The branches of Layer 2/3 neurons avoid Layer 4. B. In vitro studies show that Ephrin-A5 and NT-3 enhance growth and branching of thalamic and Layer 6 neurons, but inhibit the growth of Layer 2/3 neurons. There is also in vitro evidence to suggest that the membranes of Layer 2/3 and Layer 5 cells are specifically inhibitory to the growth of thalamic and Layer 6 axons.

the axons of these neurons respond to Ephrin-A5, as do thalamic axons (Castellani et al., 1998). Unlike the thalamic cells and layer 6 cells, layer 2/3 cells in the cortex send out axons that do not branch in layer 4. For the axons of layer 2/3 cells, *in vitro* experiments show that Ephrin-A5 inhibits rather than promotes branching. Interestingly, NT-3, which is expressed in layer 4, also promotes axonal branching of layer 6 axons while it inhibits branching of layer 2/3 axons (Castellani and Bolz, 1999). In summary, these studies demonstrate that many familiar factors are at work, leading to laminated projections, and that some factors may have bifunctional roles in promoting the branching of some axons while inhibiting the branching of others.

Laminar-selective growth is even more impressive in the chick tectum where there are 16 layers that receive input from at least 10 different sources. The retinal ganglion cells contribute input to just three of these layers, and each retinal ganglion cell sends its terminals to just one of these three layers. These retinorecipient layers express a number of molecules including one unidentified factor known only because it binds a particular plant lectin, known as VVA, which labels all three retinorecipient layers, but none of the other layers (Yamagata et al., 1995). To study the components of lamina-specific termination in this system, sections of formaldehyde-fixed

tectum are put into a culture dish with live retina. Amazingly, the retinal axons grow into the correct layers in this situation. Yet if VVA is added to this preparation, retinal axons become unable to map to the correct laminae. Several different cadherin molecules, such as N-cadherin, R-cadherin, and T-cadherin, are also expressed in different combinations of the tectal laminae, with N-cadherin being selectively present in the retinorecipient layers. Antibodies to N-cadherin when added to these cultures also cause lamination errors, though of a different type, with some axons stopping in the retinorecipient layers but not extending in them. If BDNF is added to the *in vitro* preparation, it does not affect the appropriate targeting or retinal axons to the retinorecipient layers, but it does cause excessive growth and branching in these layers. These studies suggest that different molecules regulate different aspects of laminar-specific innervation, including recognition, innervation, and branching (Inoue and Sanes, 1997; Sanes and Yamagata, 1999).

The retina is a multilayered structure, and the layers where synapses occur, such as the inner plexiform layer (IPL), are refined into several functionally specialized sublaminae, for example, the ON sublaminae that contain the synaptic terminals of bipolar and amacrine cells that fire when light is turned on and transmit this signal to ON-type retinal ganglion cells whose dendrites are in the same sublamina. The OFF sublamina does the same for lights off. Four closely related membrane associated molecules of the Immunoglobin superfamily, Dscam (see **Box 6.1**), DscamL, Sidekick-1, and Sidekick-2 are expressed in the chick

BOX 6.1 *Drosophila* Dscam: Protein isoforms for neuronal individuality

Dscam is an IgG-superfamily member. In flies, as opposed to vertebrates, the *dscam* gene includes 95 variable exons that code for protein variants with ten Ig and six fibronectin domains and single transmembrane and cytoplasmic domains (**Figure 6.18**) (Schmucker et al., 2000).

Splice variants of Dscam have the potential to produce 38,000 distinct protein isoforms through alternative splicing. Each isoform binds homophilically to itself but does not bind to other, even closely related, isoforms (Wojtowicz et al., 2004). Many of these variants are

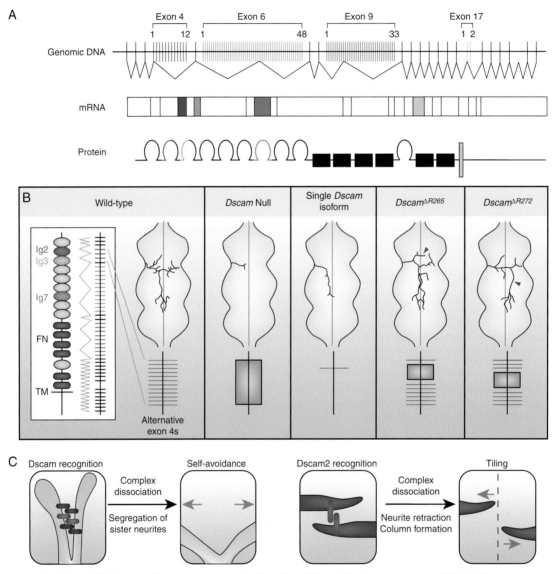

Fig. 6.18 Multiple forms of Dscam are generated by alternative splicing. A. The *Dscam* gene spans 61.2 kb of genomic DNA. Dscam mRNA extends 7.8 kb and comprises 24 exons. Mutually exclusive alternative splicing occurs for exons 4, 6, 9, and 17. 1 of 12 exon 4 alternatives, 1 of 48 exon 6 alternatives, 1 of 33 exon 9 alternatives, and 1 of 2 exon 17 alternatives are retained in each mRNA, as deduced from cDNA sequence. Variable exons are shown in color: exon 4, red; exon 6, blue; exon 9, green; and exon 17, yellow. Constant exons are represented by gray lines in genomic DNA and white boxes in mRNA. The splicing pattern shown (4.1, 6.28, 9.9, 17.1) corresponds to that obtained in the initial cDNA clone. The alternatively spliced exons 4, 6, 9, and 17 encode the N-terminal half of Ig2 (red), the N-terminal half of Ig3 (blue), the entire Ig7 (green), and the transmembrane domain (yellow), respectively (from Schmucker et al., 2000). B. Schematic of mechanosensory projections into the adult fly CNS. The first panel shows the wild-type trajectory. The remaining panels show the same mechanosensory neuron's projection in Dscam deletion flies with exon 4 isoform diversity indicated by an expanded view of the exons present in each mutant. The light gray boxes indicate deletions. Ectopic or misrouted branches are highlighted in red, branches prevalent in either Dscam deletion mutant (but not both) are highlighted by blue and green arrowheads, and the blue line denotes the CNS midline (from Bharadwaj and Kolodkin, 2006). C. The molecular mechanism of self-avoidance involves self-recognition and repulsion. As a neurite branches, each sister branch expresses the same set of Dscam1 isoforms. These isoforms bind homophilically and this transduces a signal to dissociate the receptor complex and initiate a repulsive response. *(From Millard and Zipursky, 2008)*

expressed differentially in neurons and each neuron in the *Drosophila* CNS seems to express about a dozen different isoforms, the choice of which in some classes of neurons, such as those of the mushroom bodies in the adult and the body wall sensory cells of the larva, seems stochastic. If one calculates how many combinations of isoforms are possible, it is astronomical, far more than enough to give every single neuron a unique identity. To investigate how this tremendous diversity from a single gene is used in axonal branching and targeting, different Dscam deletion lines were generated to take out some of the alternative splice forms from Exon 4 (Chen et al., 2006). In these deletion mutants, the axons of individual identified mechanosensory neurons show a variety of reproducible targeting defects in the CNS, including the presence of ectopic branches, misrouting of branches, and absence of certain branches normally observed in wild-type flies (**Figure 6.18**).

One thing that is clear from looking at a variety of neurons throughout the nervous system of any animal is that the dendritic branches of a neuron do not fasciculate with each other. As each neuron may be an individual based on the expression of its particular Dscam varients, the dendritic branches of the same neuron can recognize each other and distinguish themselves from the dendritic branches of all their sister neurons. Deletion and overexpression studies have shown that Dscam is required for the dendritic branches of single neurons to segregate from each other through a self-recognition based avoidance mechanism (Wang et al., 2002; Millard and Zipursky, 2008). The amazing diversity generated by this one gene is already of enormous significance to neuronal development in the fly, and one suspects that we may only have discovered the tip of this iceberg.

retina by nonoverlapping subsets of bipolar cells, amacrine cells, and RGCs that form synapses in distinct IPL sublaminae and each mediates homophilic adhesion (Yamagata et al., 2002; Yamagata and Sanes, 2008). These molecules mediate sublaminar specific termination, as shown, for example, by ectopic expression of Sdk-1 in Sdk-negative cells, which redirects their processes to the Sdk-1 positive sublamina. Similar experiments with Sdk-2, Dscam1, and Dscam2 show that they each direct processes to the appropriate sublaminae (**Figure 6.19**) and loss-of-function studies indicate that these molecules are also essential for correct sublaminar termination. That homophilic adhesion molecules bind axonal terminals to the dendritic processes of cells that are destined to synapse onto each other makes a good deal of sense. As we will see in the next section, this kind of process is used a great deal when we consider targeting at the cellular or synaptic level. Retinal ganglion cells, although they express this CAM code, are not absolutely critical for the formation of these sublaminae. In the zebrafish mutant *lak*, ganglion cells are never born, and yet the ON and OFF sublaminae form, though in a delayed and slightly disarrayed way (Kay et al., 2004).

CELLULAR AND SYNAPTIC TARGETING

A critical step in targeting occurs when axonal terminals choose specific cells with which they will form synapses. (The process of synapse formation itself is discussed in Chapter 8.) A good example of how individual axons choose particular postsynaptic partners is the neuromuscular system of the *Drosophila* larva. In each segment of the larva, the growth cones of about 40 motor neurons touch about 30 different muscles before they select those one or two onto which they will synapse (Nose et al., 1992b; Broadie et al., 1993). The differences between the muscles are subtle since they are, by and large, similar to one another. Each muscle has a variety of molecules on its surface, many of which are the same as the molecules expressed on the membranes of all its neighbors. What strategy do these *Drosophila* motor neurons use to make the correct cellular connections? Defasciculation of motor nerve bundles at topographically appropriate exit points leads groups of axons to subsets of target cells, but the next stage involves the actual adhesion of specific terminals to specific postsynaptic target cells.

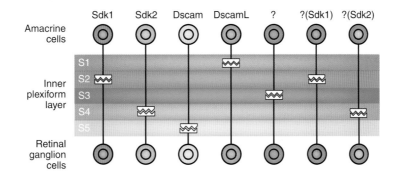

Fig. 6.19 The CAM code of sublaminar targeting. In the inner plexiform layer in the retina, neurons expressing the same sidekick or Dscam homophilic adhesion molecules send processes to the same sublamina, thereby establishing lamina-specific synaptic connections. Five of the approximately ten sublaminae are shown in this figure, and only four of these CAM code molecules have so far been identified, suggesting that there may be other CAMs for other sublaminae, indicated by the "?" for the neurons that project to sublamina 3. Overexpression of any of the known CAMs in neurons shifts their connectivity to the specified sublamina, and is shown when sidekick 1 or sidekick 2 are expressed in the "? neurons." *(After Yamagata et al., 2003, and Yamagata and Sanes, 2008)*

Individual muscle cells, although largely similar to each other, also express cell surface molecules that are shared with only some of its neighbors, and each particular muscle cell expresses different concentrations of the same molecules as its neighbors (Winberg et al., 1998). It is on this basis that highly cell specific connections can be made. For example, some motor neurons normally innervate muscles that express high levels of Netrin. In embryos in which Netrin is not expressed, the axons that would normally innervate these muscles go to inappropriate muscles (Mitchell et al., 1996). However, two other muscle cells that normally express Netrin remain inner-vated in these mutants, suggesting that additional recognition molecules must be involved in these muscles. Connectin is a second (homophilic cell adhesion) molecule that plays a role in nerve–muscle specificity in this system (Nose et al., 1992a; Meadows et al., 1994; Nose et al., 1994, 1997; Raghavan and White, 1997). There are few neuromuscular innervation defects in *connectin* mutants. However, when Connectin is expressed ectopically on all muscles in transgenic flies, motor axons frequently make targeting errors and invade nontarget muscles adjacent to their normal targets that do not normally express Connectin. This ectopic innervation depends on Connectin expression in the motor neurons that usually inner-vate Connectin positive muscles, suggesting this molecule contributes to neuromuscular specificity by homophilic adhe-sion. The defects seen with Connectin overexpression may, however, to some extent be attributed to increased adhesion between muscles that do not normally adhere to each other, making it difficult for the axon to take its usual pathway through the muscle field (Raghavan and White, 1997).

A third factor that plays a role in this system is the homophilic adhesion molecule, FasII, which is also expressed on muscle fibers (Schuster et al., 1996a,b; Davis et al., 1997). Since FasII is expressed on many muscle cells, a more subtle experiment was done, which was to use various cell-type spe-cific promoters to switch the relative levels of FasII expressed on specific muscles. The result is that extra synapses form on muscles that express higher levels of FasII at the expense of synapses formed on neighboring muscles that do not have increased FasII. This is true when the level of FasII on the less innervated muscles is high or low, so it is the relative and not the absolute level of FasII that is important. FasIII, another homophilic adhesion molecule, and SemaII, a secreted growth cone repulsive factor, are also expressed on overlapping spe-cific muscle subsets in *Drosophila*. As with *connectin* mutants, loss of function mutants in these molecules display no serious effects on neuromuscular targeting (Winberg et al., 1998). But as for the other molecules described, misexpression of FasIII or SemaII in inappropriate muscles leads to dramatic targeting effects. For example, the motor neuron RP3, which expresses FasIII, often incorrectly innervates neighboring nontarget muscle cells when these cells misexpress FasIII, suggesting that while it acts as a target recognition molecule, its absence can be compensated for by other molecules (Chiba et al., 1995). The change in probability of particular motor neurons targeting particular muscles caused by experimentally chang-ing the levels of the single-cell adhesion molecule is consis-tent with the idea that growth cones are able to distinguish targets by relative changes in the concentrations of a number of such molecules. Furthermore, targeting errors caused by the increase in an attractive or adhesive factor can be com-pensated by a simultaneous increase in a repulsive factor such

as Sema, showing that indeed the combination of amounts of various such factors is what counts. In summary, the results with FasII, FasIII, Connectin, Sema, and Netrin suggest that cellular targeting in the *Drosophila* neuromuscular system is based on a combinatorial code involving all these molecules and probably others as well (**Figure 6.20**).

In nematodes, a very intriguing case of cellular targeting is provided by the hermaphroditic specific motor neurons that innervate the vulva. In *syg-1* and *syg-2* mutants, vulval muscles remain uninnervated, and the neurons make ecto-pic synapses on inappropriate targets (Shen and Bargmann, 2003; Shen et al., 2004). The Syg-1 and Syg-2 proteins are adhesion molecules of the IgG superfamily. Syg-1 is expressed in the neuron, and its binding partner Syg-2 is nor-mally expressed transiently not on the postsynaptic targets but on a vulval epithelial guidepost cell. This interaction is critical for the positioning and maturation of the axonal terminal in preparation for synapse formation on the later arriving partner neuron. Interestingly, if Syg-2 is expressed under the control of a promoter that causes it to be localized on other epithelial cells, the hermaphrodite-specific motor neurons begin to make presynaptic specializations at these ectopic sites. Thus, heterophilic binding between Syg-1 and Syg-2 is involved in localizing the formation of synapses to appropriate sites.

SynCAMS and cadherins (especially protocadherins) are large families of homophilic adhesion molecules that may add another level of specificity by helping presynaptic termi-nals make contacts at the correct subcellular locations (Abbas, 2003; Yamagata et al., 2003). There are more than 50 differ-ent cadherins, and many of them are expressed at subsets of synapses. As many as 20 different genes of the cadherin superfamily genes are expressed in restricted patterns in the developing tectum (Miskevich et al., 1998), and recent specu-lation is that these types of molecules, like the Sdks and Sygs discussed above, are involved in a synaptic targeting code (Redies and Takeichi, 1996). As an example of how this might work, consider the N- and E-cadherins, which are distributed at synaptic junctions in a mutually exclusive pattern along the dendritic shafts of single pyramidal neurons (Fannon and Colman, 1996). Ultrastructural examination of double immu-nolabeled material revealed the existence of many unlabeled synapses on these cells, raising the possibility that synapses at these other synapses are linked by other cadherins or CAMs. In the hypothalamus, it has been found that some protocad-herins are found just at excitatory synapses (Phillips et al., 2003). Of course, these molecules are likely to do more than simply glue particular synapses together; they are also likely to be involved in the maturation, structural organization, and stabilization of synapses, topics that will be covered in more detail in Chapter 8.

SNIFFING OUT TARGETS

Visual and somatosensory maps in the brain can be seen as topographic and continuous. These are maps in which neigh-boring points in the environment impinge on neighboring receptors which project to neighboring postsynaptic cells, so a continuous map of space is preserved in the brain. There are, however, other kinds of maps in the brain that are discontinous

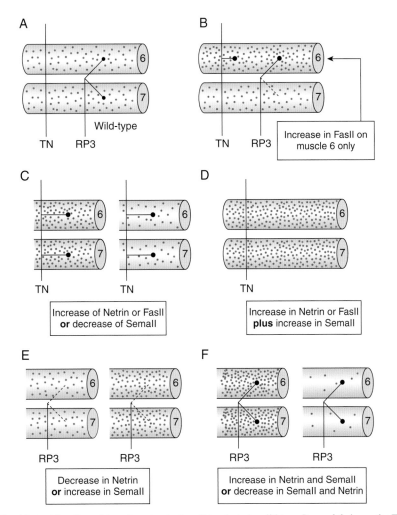

Fig. 6.20 Combinatorial coding of targeting at a single-cell level. A. In wild-type *Drosophila* larva, the TN nerve does not innervate muscle fibers 6 and 7, whereas the RP3 nerve innervates both. B. When FasII is increased on muscle 6 only, both the TN nerve and the RP3 nerve innervate this muscle differentially. C. When Netrin expression or FasII expression is increased on both muscles, or when SemaII is decreased on both, the TN nerve innervates both 6 and 7. D. When Netrin or FasII is increased but SemaII is also increased simultaneously, then the TN nerve does not innervate 6 and 7. E. When there is a decrease in Netrin or an increase SemaII, the RP3 nerve does not innervate muscles 6 and 7. F. However, when there is either an increase or a decrease in both SemaII and Netrin, the RP3 nerve innervates as normal. *(Adapted from Winberg et al., 1998)*

or discrete. Discrete maps are maps of qualities rather than spatial organization. A good example of such a map is the map of smell. In seminal work, Buck and Axel (1991) found that the nasal epithelium of the mammal expresses thousands of different seven transmembrane receptor molecules, and that each individual primary olfactory sense cell expresses only a single type of receptor. Olfactory sensory neurons send axons into the bulb where they make connections with second-order cells in synaptic complexes called *glomeruli*. Physiological studies reveal that distinct odorants cause activity in distinct glomeruli, and in the zebrafish, a careful anatomical study showed a reproducible pattern of about 80 glomeruli that have the same position and size from individual to individual (Baier and Korsching, 1994). Surprisingly, the receptors projecting to a single glomerulus are scattered all over the olfactory epithelium in a fairly random pattern, suggesting that there is no regionalization of odorant receptors on the sensory epithelium (**Figure 6.21**). Olfactory axons therefore follow a wiring

principle of "one receptor—one glomerulus" such that all the olfactory receptor neurons that express the same odorant receptors, and therefore smell the same odorants, project to the same glomerulus, creating a map of odor information. Although they appear quite differently organized (Figure 6.21), many similar strategies are used in making continuous and discrete maps (Luo and Flanagan, 2007).

The olfactory epithelium is a convoluted sheet of tissue, but it can be theoretically flattened out. If one then looks at all the sensory neurons that express a particular odorant receptor, one finds they are not spread across the entire width of the epithelium but are restricted to a band that covers approximately one quarter of the dorsomedial ventrolateral extent. Different receptors are restricted to slightly different bands and sensory cells from one band project to a correspondingly restricted dorsoventral domain within the olfactory bulb (**Figure 6.22**A) (Miyamichi et al., 2005). The continuous but rough topography in this axis, note Luo and Flanagan (2007),

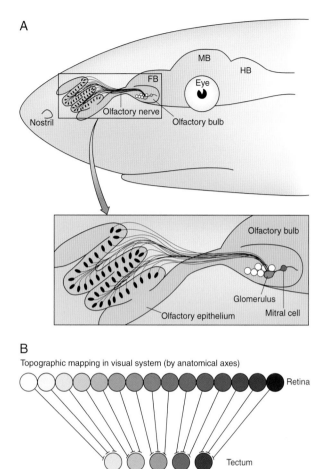

A

B

Topographic mapping in visual system (by anatomical axes)

Retina

Tectum

Topographic mapping in olfactory system (by odorant specificity)

Fig. 6.21 Comparison of topographic mapping in the visual system (B. top), where neighboring cells project to neighboring targets, creating a central representation of visual space; and the olfactory system (A.) where cells of the same type are intermingled and yet their axons sort out and converge forming an odor representation map (B. bottom).

"begins to 'smell' like the visual map." Within each band, however, all the receptors that express a particular odorant receptor gene are dispersed widely in the anteroposterior axis, so if there is any topography in this axis, it is not obvious. Amazingly, though different types of olfactory sensory cells are each scattered in a salt and pepper fashion across about a quarter of the entire epithelium, all the axons of each sensory cell type converge in highly reproducible manner to single glomeruli in specific regions of the olfactory bulb (Vassar et al., 1994; Mombaerts, 1996).

The rough mapping of the olfactory neurons onto the dorsoventral axis of the bulb uses, it seems, guidance cues and topography cues that we have already discussed, such as cell adhesion molecules and repulsive guidance factors.

But what concerns us in this case is the question of how the axons of a single class of receptor converge onto topographically fixed glomeruli so that a consistent olfactory map is created in the brain (Vassar et al., 1994; Mombaerts, 1996). There are two major possibilities. The first is that the expression of a particular receptor gene is somehow linked to the expression or activation of a particular set of guidance molecules. The other, more radical, possibility is that the odorant receptor molecules themselves are expressed on axonal growth cones and are involved in sniffing out the correct postsynaptic target area. Let us look at this more interesting possibility first. The first question is whether odorant receptor molecules are expressed in axons and growth cones. The answer is yes, both odorant receptor mRNA and protein are expressed in growth cones (Barnea et al., 2004) (**Figure 6.23**). The next question is what happens when a receptor is knocked out. Are the axons unable to find their targets? To answer this question, marker genes such as *lacZ* have been knocked in to replace specific receptor loci. Thus, *lacZ* is expressed in the cells that would have normally expressed a particular receptor and by examining the distribution of *lacZ,* which is transported down the axons of these cells, one can see that the axons appear disoriented and do not converge on their targets (Wang et al., 1998). This suggests that the olfactory receptors are critical for accurate targeting in the bulb. Interestingly, however, the correct targeting of these axons occurs if a constitutively active G protein is misexpressed in these cells (Imai et al., 2006; Chesler et al., 2007). The correct interpretation of the apparently disordered growth of the axons that do not express their normal receptor is thus complicated by the fact that when an olfactory neuron is prevented from expressing a particular odorant receptor which activates a G protein, it will randomly choose one of the thousand other receptors to express. Thus in the receptor knockout experiment, each of the cells that normally express the deleted oderant receptor, choose to express another receptor instead. Thus, these axons had not necessarily lost their way to their normal glomerulus, but had instead been carefully targeting a variety of different glomeruli that matched their new receptors.

Swapping receptors is perhaps a better way to test whether the odorant receptors themselves have a role in targeting (Mombaerts et al., 1996; Wang et al., 1998). Thus, in another set of experiments, a specific odorant receptor gene was replaced by a fusion gene driving not only *lacZ,* but also the cDNA for a different receptor, so that the axons misexpressing this receptor are easy to visualize (**Figure 6.24**). When olfactory neurons that target distinct zones of the bulb have their receptors swapped, they target neither their normal glomeruli nor the glomeruli typical of their new odorant receptor. Instead, they map to a new specific glomerulus somewhere in-between, suggesting that odorant receptors do have some role in targeting but suggesting also that there must be other factors that guide these axons to their particular targets. In fact, there is accumulating evidence that this is so. Like the muscles of *Drosophila* larvae discovered above, different combinations of cell adhesion and guidance molecules appear to be expressed by various olfactory receptor neurons. These factors may take an olfactory axon to an appropriate vicinity and then the axons may rely on their receptors for the final targeting. This idea

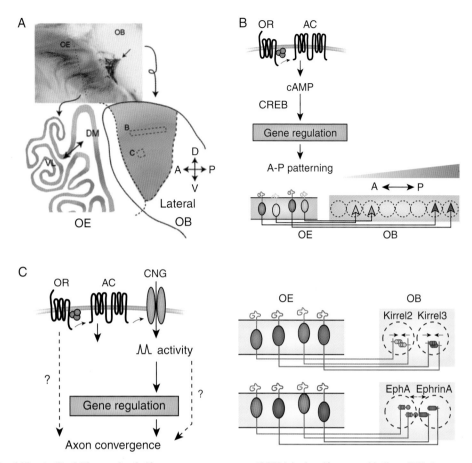

Fig. 6.22 A. (Top) Thousands of olfactory receptor neurons (ORNs) in the olfactory epithelium (OE) that express a common odorant receptor (OR) converge their axonal projections onto the same glomerulus (arrow) in the olfactory bulb (OB). *(From Mombaerts et al. (1996)* (Bottom) ORNs that express a given OR are distributed within a band along the dorsomedial (DM) to ventrolateral (VL) axis in the OE and project their axons to corresponding color-matched positions along the dorsal–ventral (D–V) axis of the OB. According to data from Miyamichi et al. (2005). Dotted rectangles correspond to OB schematics in (B) and (C) as indicated. A, anterior; P, posterior; D, dorsal; V, ventral. The A-P axis in bottom left (nasal epithelium) is orthogonal to the plane shown. B. Basal level G protein (three purple subunits) coupling with individual ORs, through the adenylate cyclase activation of cAMP and CREB, induces gene expression that contributes to ORN axon targeting. Specifically, putative axon guidance receptors have been found to be targets of this signaling pathway and exhibit graded distribution along the A-P axis in the OB. According to data from Imai et al. (2006). C. OR-correlated and activity-dependent expression of homophilic cell adhesion molecules Kirrel2 and Kirrel3, and repulsive axon guidance ligand/receptor pair Ephrin-A and EphA, could contribute to local axonal convergence and sorting. According to data from Serizawa et al. (2006). *(From Luo and Flanagan 2007)*

is supported by the finding that when receptors are swapped between sensory neurons that have nearby targets in the same region, the axons do target to the precise vicinity of the glomeruli of the switched receptors (Figure 6.24).

Even a minor change in the coding region of an olfactory receptor gene causes a change in the target destination of the axons that express this gene (Feinstein and Mombaerts, 2004). The critical amino acid residues that affect guidance tend to be clustered in the transmembrane domains. When these are interchanged between two parent receptors, the axons expressing the hybrid receptors may map to the same glomerulus as the axons that express one of the parent molecules or they may map to a different glomerulus altogether. Whatever they do, all the axons that express the same chimeric proteins seem to behave the same way. Moreover, axons that express the same receptor fasciculate with each

other before they enter the glomerular neuropil (Potter et al., 2001). If the odorant receptor molecule is used in axon fasciculation and target recognition, are the odorant receptors on the growth cones of these sensory cells "smelling" their way, or is there a different mechanism at play here? For instance, could it be that the odorant receptor somehow activates a targeting mechanism? As seven transmembrane receptor molecules, odorant receptors activate G proteins when they bind odorants. The G proteins when active, mediate cAMP signaling inside the cell and this results in changes in gene expression mediated by the CREB transcription factor. It is thought that CREB might differentially regulate the expression of guidance cue receptors such as neuropilin (the Sema receptor) on the growth cones of sensory neurons, which might help them target appropriately in the bulb. (Figure 6.22B).

161

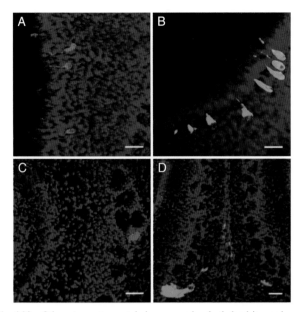

Fig. 6.23 Odorant receptor protein is expressed on both dendrites and axons of olfactory sensory neurons. A and B. Staining of mouse olfactory epithelium with antibodies to two different particular odorant receptors (one labelled in red, the other in green). C. and D. Staining of mouse olfactory bulb with the same antibodies. Scale bars, 10mm. *(From Barnea et al., 2004)*

SHIFTING AND FINE TUNING OF CONNECTIONS

In frog and the fish embryos, retinal axons project to roughly the correct topographic site in the tectum, and then begin to branch at this location (Harris et al., 1983; Stuermer and Raymond, 1989). The earliest terminal arbors that these axons make cover a substantial fraction of the tectum. Thus the retinotectal map is orderly, but there is a great deal of overlap among the individual retinal arbors. In fact, physiological recording of any tectal neuron at these early stages of map formation in the frogs indicates that each tectal neuron receives input from a large fraction of the RGCs in the embryonic retina (Zhang et al. 1998). Over the course of development, the terminal arbors become refined to a smaller and smaller fraction of the tectal area, and the topographic wiring becomes more and more precise (Sakaguchi and Murphey, 1985). In the chick and the mouse, retinal axons overshoot their termination points to some extent and begin to make interstitial branches at roughly the correct topographic position behind the growth cone (Nakamura and O'Leary, 1989; Simon and O'Leary, 1990, 1992) (**Figure 6.25**). The topographic specificity is then enhanced through the preferential arborization of appropriately positioned branches and elimination of ectopic branches.

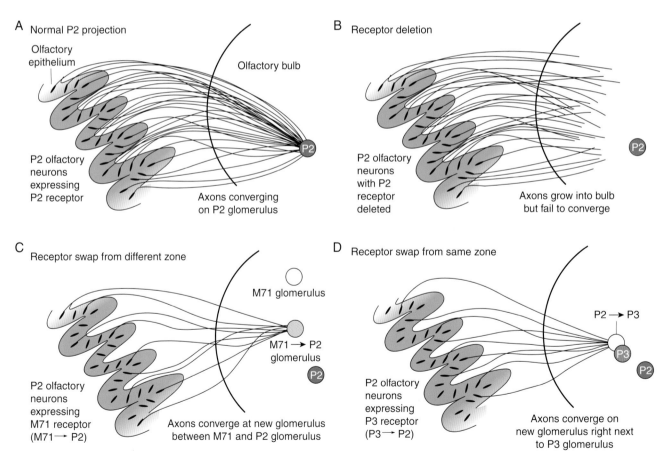

Fig. 6.24 Olfactory receptors are involved in central targeting. A. Axons from olfactory neurons expressing the P2 odorant receptor converge on the P2 glomerulus. B. If the P2 receptor is deleted, these axons do not converge on any glomerulus. C. If these neurons are made to express the M71 odorant receptor instead of the P2 receptor, they converge on a glomerulus somewhere in-between the normal P2 and the normal M71 glomerulus. D. If they are forced to express the P3 odorant receptor which is relatively near the P2 glomerulus, they converge on a glomerulus right next to P3. *(After Mombaerts, 1996; Wang et al., 1998)*

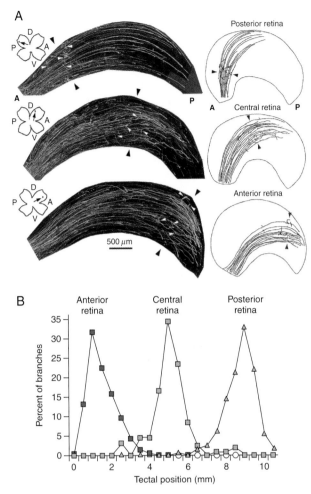

Fig. 6.25 RGC axons overshoot their correct termination zone in a position-dependent manner in chick embryos. A. RGC axons in tectal whole mounts labeled by a small focal DiI injection into peripheral temporal retina (top), central retina (middle), or peripheral nasal retina (bottom) on E11. The injection sites are shown in drawings of the retinal whole mounts and marked by arrows. The relative positioning of the labeled axons and branches within the tectum is shown to the right, with drawings of the outline of each tectum on which the labeled axons and branches are traced. Axons overshoot their topographically correct TZ (the predicted locations of the TZs are marked with black arrowheads), but the distribution of interstitial branches along the axon shafts (white arrowheads) is strongly biased for the location of the future TZ along the anterior (A)-posterior (P) tectal axis. Peripheral temporal axons exhibit the greatest overshoot and peripheral nasal axons the least. B. Distribution of interstitial branches along the axon shaft is expressed in percentage. The A-P tectal axis was divided into 500 mm bins, and the number of branches in each bin is graphed as the percentage of total branches for each of the three groups of injections. *(After Yates et al., 2001)*

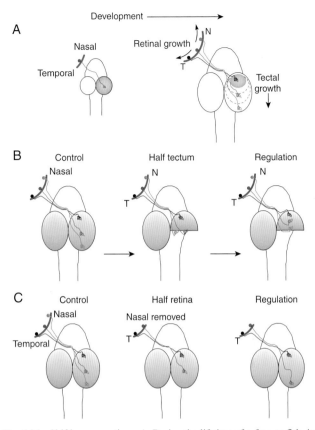

Fig. 6.26 Shifting connections. A. During the lifetime of a frog or fish, its eye and brain continue to grow. The retina grows circumferentially like a tree, but the tectum grows in expanding posterior crescents. As a result, new retina that is added temporally must send axons to the anterior primordial tectum, while fibers from the central primordial retina must shift posteriorly, and new nasal fibers map to the new posterior tectum in order to keep the map in topographic order. B. If half the tectum is removed from a fish, after about a month the retinotopic map will regulate and compress, mapping out evenly over the remaining half tectum. C. Similar regulation occurs when half the retina is removed. The remaining projection eventually expands over the whole tectum. *(After Schmidt and Easter, 1978; Gaze et al., 1979; Schmidt and Coen, 1995)*

Topographic refinement may occur throughout life. Consider the case of a goldfish. It hatches as a tiny 1 mg animal and over the course of its life may attain a weight of 1 kg or more. It has increased in volume a millionfold. As the animal grows, the retina grows in proportion by adding cells circumferentially at the rim or margin. The tectum grows as well but mostly at the caudal end. In order for the retinotectal map to remain evenly distributed, the retinal axons must continually retract anterior branches and send out new branches more posteriorly in the tectum (Gaze et al., 1979). Axons from the center of a large adult retina are from the oldest retinal ganglion cells that were born when the fish was just a small larva (**Figure 6.26**). Initially these axons projected to the center of the larval tectum. These larval tectal cells remain at the anterior pole of the tectum as new tectal cells are added caudally throughout out life. So, by the time the fish has grown to adult proportions, axons from retinal ganglion cells at the center of the retina project to cells in the middle of the large adult tectum, perhaps a millimeter or so away (Easter and Stuermer, 1984). Thus these axons have continued to switch their preferred targets to more posterior cells throughout their lifetimes. A similar type of shifting reorganization of connections is evident when half of the retina or half of the tectum of a fish is ablated. When half the retina is ablated, the projection of the remaining half of the retina expands to cover the entire tectum. When half the tectum is removed, the retina's projection compresses to cover the remaining half (Schmidt and Easter, 1978; Schmidt and Coen, 1995). This sort of regulation is also observed in neonatal hamsters with a partially deleted superior colliculus (Figure 6.17). This form of topographic expansion or compression, like the natural shift that is a consequence of the asymmetric growth of the tectum, does not depend on the activity patterns in retinal fibers. The regulation can occur in the dark or even in the continuous presence of tetrodotoxin (TTX), which blocks action potentials (Meyer and Wolcott, 1987).

These shifting connections are part of larger developmental phenomenon whereby, once a topographic map is roughly established, it is adjusted, modified, and fine-tuned. Part of this refinement may be based on growth patterns or injury, as above, but refinement also has an activity or experience-dependent aspect. Without impulse activity, the retinotectal map of the goldfish is topographic, but the sizes of the receptive fields recorded in the tectum are larger and less precise than normal. Analysis of individual retinal axonal arbors shows that they are up to four times as large as normal ones (Schmidt and Buzzard, 1990). Repeated examination of single retinal arbors over time shows the effects that activity has on branching and topography. When retinal activity is abolished by tetrodotoxin (TTX), the result is increased branch addition and elimination, or in other words decreased branch stability and continued exploration of a larger territory (Schmidt and Buzzard, 1990).

There is a simple saying in the field of development neurobiology: "Neurons that fire together, wire together." In the retinotectal system of the fish, it is possible to use this rule, by adjusting firing patterns, to shift the receptive fields of tectal neurons in any direction in visual space under experimental control (Vislay-Meltzer et al., 2006). The mechanisms by which simultaneous activity may have such effects on the fine-tuning of connections are dealt with in detail in Chapter 9. Here, we would simply like to point out how activity affects topographic maps in the nervous system. A particularly interesting example in this regard is the somatosensory cortex. The neurosurgeon, Wilder Penfield (Penfield, 1954a), while performing operations to remove epileptic foci in the brains of fully conscious patients, took the opportunity to study the organization of the cortex by locally stimulating different regions with an electrode. When he stimulated points in the postcentral gyrus, patients reported the sensation of touch in specific areas of their bodies. Stimulation of neighboring points caused the patients to experience sensations in neighboring parts of their body surface although there were occasional jumps, such as between the hand and the face. By mapping these sensations on the cortex of different patients, Penfield was able to come up with a consistent somatosensory homunculus and in the precentral gyrus a matching topographic homunculus where stimulation caused movements of specific body parts (**Figure 6.27**). One striking feature of the homunculus that Penfield noticed immediately is the relative magnification of parts of the map. This appears to be a consistent feature of many maps in the CNS. The largest features of the human homunculus are the lips, tongue, and tips of the fingers. In contrast, the representation of the upper back is quite small. In other animals, the somatosensory cortex has an expanded representation of different body parts: the hands of the raccoon, the snout of a star nose mole, and the whiskers of the mouse, for example, are particularly enlarged. The differential magnification of certain body parts in the cortical representation of the body is probably due to the density of peripheral innervation. Thus, in humans, each fingertip has almost as many sensory receptors as the whole of the upper back.

In humans, the loss of sensation in one area of the body, as happens when a peripheral nerve is cut through accident or medical intervention, often leads to the cortical representation of that area to be invaded by representation from neighboring parts. This is thought to be one reason why people who

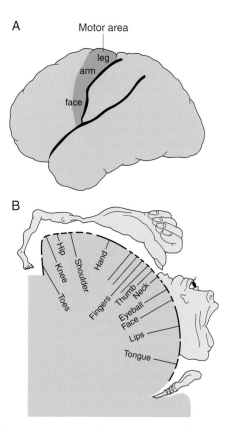

Fig. 6.27 Somatotopic representation in the cortex. A. The motor area of the precentral gyrus of the cerebral cortex was stimulated electrically in human patients during neurosurgery. B. A "homunculus" of the body on the motor cortex illustrates the sequence of representation as well as the disproportionate representation given to the various muscles involved in skilled movements. *(After Penfield, 1954)*

have lost a limb may report sensations in the phantom limb, especially when a part of the body is touched whose cortical representation is adjacent to the missing limb (Ramachandran and Rogers-Ramachandran, 2000). A touch to the face in such a person can be experienced as a touch on the missing hand. The explanation is that nerve fibers that carry information about touch on the face invade the neighboring cortical area that used to receive such information from the lost limb. The rest of the brain, however, has not yet "learned" the change in the meaning of the input to this part of the cortex, and still interprets it as a touch to the hand (**Figure 6.28**). Experiments with monkeys, in which a cuff of TTX on the nerve temporarily paralyzes a single finger, have shown that there is a rapid reorganization of the somatosensory map in the cortex. Within days, the representation of the insensitive finger shrinks, and the representation of the neighboring fingers expands (Merzenich and Jenkins, 1993). These cortical changes in the representation of somatotopy can be extremely large even in normal animals, as was revealed by an unusual experiment at the National Institutes of Health. Antivivisectionists stole a set of experimental monkeys after their somatosensory cortex was first mapped, and they were not recovered until about 10 years later. When the scientists remapped their somatosensory cortex, they found that the extent of the rearrangement was dramatic, a matter of tens of millimeters (Palca, 1991). Thus, minor reorganization of the cortical somatosensory map is happening throughout

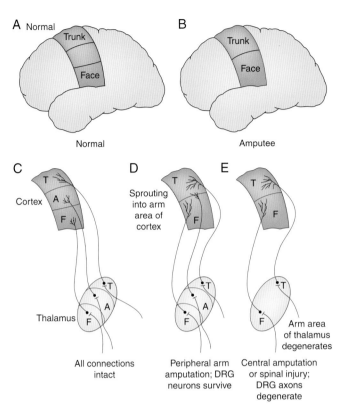

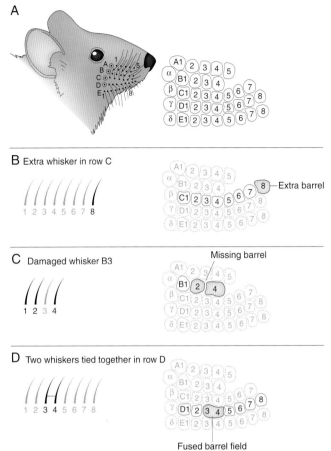

Fig. 6.28 Large-scale plasticity in the somatosensory cortex. A. The normal organization of the cortical topography in the somatosensory system. B. Damage to the arm may result in large-scale reorganization of the cortical map. C. There is an isomorphic mapping of the somatosensory thalamus onto the primary cortex. D. When the arm is damaged peripherally and the sensory neurons in the DRG survive, the reorganization is cortical rather than thalamic. E. When the damage is more central causing the DRG cells to degenerate, both the thalamus and cortex get reorganized. *(After Merzenich, 1998)*

Fig. 6.29 Plasticity of the mouse barrel field in the somatosensory cortex. A. The correspondence between bristles and barrel field in the cortex of a normal mouse. B. An extra whisker in row C leads to the formation of an extra barrel in the appropriate location in the cortex. C. Neonatal damage to the B3 whisker causes the shrinkage of this barrel and the expansion of neighboring ones. D. Tying two whiskers together causes their barrel field to coalesce. *(After Woolsey and Van der Loos, 1970)*

life, presumably influenced by experience and activity. Even normal use can change topographic representations in an impressive way. Monkeys trained to use just one finger to feel textural differences for a few hours a day over a period of months have a hugely expanded cortical representation of that finger (Recanzone et al., 1992).

Central representations of the somatosensory system are flexible and may depend on sensory stimulation, especially during early life. In the mouse, single cortical areas, called *barrels,* are devoted to each vibrissa. The barrel fields of the cortex are almost equal in size to the somatosensory cortex devoted to the rest of the body (Woolsey and Van der Loos, 1970). There are five rows of barrels that correspond to the five rows of vibrissae. When a bristle is destroyed by cauterization in early life, the cortical barrel that represents it shrinks, while the neighboring barrels expand into the territory of the cortex originally devoted to the cauterized whisker (Dietrich et al., 1981; Simons et al., 1984) (**Figure 6.29**). When two whiskers are glued together, their cortical barrels fuse. Perhaps the most surprising finding is the case of a mouse that was born with an extra whisker, as sometimes happens. This mouse had an extra barrel in its cortex (Van der Loos et al., 1984). From these results, it is clear that the neural representation of the body surface has flexibility in its structure. The sensory fields themselves and their activity guide this flexibility.

Activity may play a critical role in the refinement of the target selection of odorant sensory axons. If axons that are active at the same time tend to terminate together, then olfactory neurons with the same receptor molecule would respond the same way to odorants. If olfactory sensory neurons express a mutation that makes them electrically unresponsive to odorants, though they initially map correctly to their target glomeruli, they then fail to compete with active axons expressing the same receptor and are eliminated (Zhao and Reed, 2001). Interestingly, these axons can be rescued by nose plugs, which cause the potentially active axons to be odorant deprived. These studies suggest that competitive activity is necessary to stabilize specific connections in the bulb. The idea of competition is strengthened by the discovery that very early in the innervation of the bulb, some glomeruli are innervated by more than one receptor type of axon, but this sorts out postnatally so that eventually each glomerulus is innervated by only a single receptor type (Zou et al., 2004). The importance of successful physiological activity in mapping is also supported by experiments showing that inhibition of all spontaneous activity in a subset of olfactory sensory neurons that express a single odorant receptor causes a selective unraveling of the target glomerulus, with the axons that originally invaded this glomerulus now making inappropriate contacts in various regions of the bulb (Yu et al., 2004). The impulse activity of olfactory axons, like G-protein activity, regulates the expression of repulsive guidance molecules such as Ephrin-A5 and Eph-A5 and homophilic adhesion molecules such as Kirrel2 and Kirrel3 (Serizawa et al., 2006). The activity-dependent expression of such molecules can help axons expressing the same odorant receptor fasciculate together and segregate from axons belonging to other odorant classes that map nearby (Figure 6.22c).

SUMMARY

Pathfinding to the vicinity of a correct target is only the first step in the process of selecting appropriate postsynaptic cells on which to synapse. Having come to the doorstep of the target, axons use a variety of signals, such as relative changes in growth factors, to enter the target and begin to arborize. Growing axons are often encouraged to enter the target at a particular gateway and discouraged from exiting the target through repulsive barriers that surround it elsewhere. There are a variety of molecules within the target zone, including growth factors neurotrophins and CAMs, which conspire, often in combination, to encode different possible target cells along various axes and layers. In continuous maps, the incoming afferents are distinguished from each other by the graded amounts of various receptors such as Eph receptors on their surfaces so that they respond in a topographic way to the gradients of guidance cues such as Ephrins across the target cells. These gradients of ligands and receptors are often the result of very early patterning events in the embryo, for example, those that lay down rostrocaudal and dorsoventral patterns (see Chapter 2). Discrete maps, although they represent information qualities instead of space, use many similar mechanisms of targeting. Molecular cues, in the form of specific receptors, homophilic or heterophilic adhesion molecules, and especially molecules such as Dscam that can show great diversity, are often involved in the final cellular and synaptic levels of targeting. The molecular nature of target recognition leads one to appreciate how a nervous system can wire up to a fairly high degree of precision in the absence of function. The precision of neural connections, however, is usually dependent on neuronal activity, which comes into play once synapses are made (see Chapter 8). This is a time when the wiring of the nervous system can be tested out and fine-tuned to make sure it is in good working order. In Chapter 9, we shall learn more about how neural activity works to alter synaptic connectivity.

REFERENCES

Abbas, L. (2003). Synapse formation: let's stick together. *Current Biology*, 13, R25–R27.

Baier, H., & Korsching, S. (1994). Olfactory glomeruli in the zebrafish form an invariant pattern and are identifiable across animals. *The Journal of Neuroscience*, 14, 219–230.

Barnea, G., O'Donnell, S., Mancia, F., Sun, X., Nemes, A., Mendelsohn, M., et al. (2004). Odorant receptors on axon termini in the brain. *Science*, 304, 1468.

Bharadwaj, R., & Kolodkin, A. L. (2006). Descrambling Dscam diversity. *Cell*, 125, 421–424.

Bolz, J., Castellani, V., Mann, F., & Henke-Fahle, S. (1996). Specification of layer-specific connections in the developing cortex. *Progress in Brain Research*, 108, 41–54.

Braisted, J. E., McLaughlin, T., Wang, H. U., Friedman, G. C., Anderson, D. J., & O'Leary, D. D. (1997). Graded and lamina-specific distributions of ligands of EphB receptor tyrosine kinases in the developing retinotectal system. *Developmental Biology*, 191, 14–28.

Broadie, K., Sink, H., Van Vactor, D., Fambrough, D., Whitington, P. M., Bate, M., et al. (1993). From growth cone to synapse: the life history of the RP3 motor neuron. *Development Supplement*, 227–238.

Brown, A., Yates, P. A., Burrola, P., Ortuno, D., Vaidya, A., Jessell, T. M., et al. (2000). Topographic mapping from the retina to the midbrain is controlled by relative but not absolute levels of EphA receptor signaling. *Cell*, 102, 77–88.

Brunet, I., Weinl, C., Piper, M., Trembleau, A., Volovitch, M., Harris, W., et al. (2005). The transcription factor Engrailed-2 guides retinal axons. *Nature*, 438, 94–98.

Buck, L., & Axel, R. (1991). A novel multigene family may encode odorant receptors: a molecular basis for odor recognition. *Cell*, 65, 175–187.

Campbell, D. S., & Holt, C. E. (2001). Chemotropic responses of retinal growth cones mediated by rapid local protein synthesis and degradation. *Neuron*, 32, 1013–1026.

Campbell, D. S., Regan, A. G., Lopez, J. S., Tannahill, D., Harris, W. A., & Holt, C. E. (2001). Semaphorin 3A elicits stage-dependent collapse, turning, and branching in Xenopus retinal growth cones. *The Journal of Neuroscience*, 21, 8538–8547.

Cang, J., Kaneko, M., Yamada, J., Woods, G., Stryker, M. P., & Feldheim, D. A. (2005). Ephrin-as guide the formation of functional maps in the visual cortex. *Neuron*, 48, 577–589.

Castellani, V., & Bolz, J. (1997). Membrane-associated molecules regulate the formation of layer-specific cortical circuits. *Proceedings of the National Academy of Sciences of the United States of America*, 94, 7030–7035.

Castellani, V., & Bolz, J. (1999). Opposing roles for neurotrophin-3 in targeting and collateral formation of distinct sets of developing cortical neurons. *Development*, 126, 3335–3345.

Castellani, V., Yue, Y., Gao, P. P., Zhou, R., & Bolz, J. (1998). Dual action of a ligand for Eph receptor tyrosine kinases on specific populations of axons during the development of cortical circuits. *The Journal of Neuroscience*, 18, 4663–4672.

Catalano, S. M., Messersmith, E. K., Goodman, C. S., Shatz, C. J., & Chedotal, A. (1998). Many major CNS axon projections develop normally in the absence of semaphorin III. *Molecular and Cellular Neurosciences*, 11, 173–182.

Chen, B. E., Kondo, M., Garnier, A., Watson, F. L., Puettmann-Holgado, R., Lamar, D. R., et al. (2006). The molecular diversity of Dscam is functionally required for neuronal wiring specificity in Drosophila. *Cell*, 125, 607–620.

Cheng, H. J., Nakamoto, M., Bergemann, A. D., & Flanagan, J. G. (1995). Complementary gradients in expression and binding of ELF-1 and Mek4 in development of the topographic retinotectal projection map. *Cell*, 82, 371–381.

Chesler, A. T., Zou, D. J., Le Pichon, C. E., Peterlin, Z. A., Matthews, G. A., Pei, X., et al. (2007). A G protein/cAMP signal cascade is required for axonal convergence into olfactory glomeruli. *Proceedings of the National Academy of Sciences of the United States of America*, 104, 1039–1044.

Chiba, A., Snow, P., Keshishian, H., & Hotta, Y. (1995). Fasciclin III as a synaptic target recognition molecule in Drosophila. *Nature*, 374, 166–168.

Cohen-Cory, S., & Fraser, S. E. (1995). Effects of brain-derived neurotrophic factor on optic axon branching and remodelling in vivo. *Nature*, 378, 192–196.

Davenport, R. W., Thies, E., & Cohen, M. L. (1999). Neuronal growth cone collapse triggers lateral extensions along trailing axons. *Nature Neuroscience*, 2, 254–259.

Davis, G. W., Schuster, C. M., & Goodman, C. S. (1997). Genetic analysis of the mechanisms controlling target selection: target-derived Fasciclin II regulates the pattern of synapse formation. *Neuron*, 19, 561–573.

Dennis, M. J., & Yip, J. W. (1978). Formation and elimination of foreign synapses on adult salamander muscle. *The Journal of Physiology*, 274, 299–310.

Dent, E. W., & Kalil, K. (2001). Axon branching requires interactions between dynamic microtubules and actin filaments. *The Journal of Neuroscience*, 21, 9757–9769.

Dietrich, W. D., Durham, D., Lowry, O. H., & Woolsey, T. A. (1981). Quantitative histochemical effects of whisker damage on single identified cortical barrels in the adult mouse. *The Journal of Neuroscience*, 1(9), 929–935.

Drescher, U., Kremoser, C., Handwerker, C., Loschinger, J., Noda, M., & Bonhoeffer, F. (1995). In vitro guidance of retinal ganglion cell axons by RAGS, a 25 kDa tectal protein related to ligands for Eph receptor tyrosine kinases. *Cell*, 82, 359–370.

Easter, S. S., Jr., & Stuermer, C. A. (1984). An evaluation of the hypothesis of shifting terminals in goldfish optic tectum. *The Journal of Neuroscience*, 4, 1052–1063.

ElShamy, W. M., Linnarsson, S., Lee, K. F., Jaenisch, R., & Ernfors, P. (1996). Prenatal and postnatal requirements of NT-3 for sympathetic neuroblast survival and innervation of specific targets. *Development*, 122(2), 491–500.

Ernfors, P., Van De Water, T., Loring, J., & Jaenisch, R. (1995). Complementary roles of BDNF and NT-3 in vestibular and auditory development. *Neuron*, *14*, 1153–1164.

Fambrough, D., & Goodman, C. S. (1996). The Drosophila beaten path gene encodes a novel secreted protein that regulates defasciculation at motor axon choice points. *Cell*, *87*, 1049–1058.

Fannon, A. M., & Colman, D. R. (1996). A model for central synaptic junctional complex formation based on the differential adhesive specificities of the cadherins. *Neuron*, *17*, 423–434.

Feinstein, P., & Mombaerts, P. (2004). A contextual model for axonal sorting into glomeruli in the mouse olfactory system. *Cell*, *117*, 817–831.

Feldheim, D. A., Kim, Y. I., Bergemann, A. D., Frisen, J., Barbacid, M., & Flanagan, J. G. (2000). Genetic analysis of ephrin-A2 and ephrin-A5 shows their requirement in multiple aspects of retinocollicular mapping. *Neuron*, *25*, 563–574.

Feldheim, D. A., Nakamoto, M., Osterfield, M., Gale, N. W., DeChiara, T. M., Rohatgi, R., et al. (2004). Loss-of-function analysis of EphA receptors in retinotectal mapping. *The Journal of Neuroscience*, *24*, 2542–2550.

Feldheim, D. A., Vanderhaeghen, P., Hansen, M. J., Frisen, J., Lu, Q., Barbacid, M., et al. (1998). Topographic guidance labels in a sensory projection to the forebrain. *Neuron*, *21*, 1303–1313.

Flanagan, J. G., & Vanderhaeghen, P. (1998). The ephrins and Eph receptors in neural development. *Annual Review of Neuroscience*, *21*, 309–345.

Frisen, J., Yates, P. A., McLaughlin, T., Friedman, G. C., O'Leary, D. D., & Barbacid, M. (1998). Ephrin-A5 (AL-1/RAGS) is essential for proper retinal axon guidance and topographic mapping in the mammalian visual system. *Neuron*, *20*, 235–243.

Fritzsch, B., Silos-Santiago, I., Bianchi, L. M., & Farinas, I. (1997). The role of neurotrophic factors in regulating the development of inner ear innervation. *Trends in Neurosciences*, *20*, 159–164.

Fritzsch, B., Tessarollo, L., Coppola, E., & Reichardt, L. F. (2004). Neurotrophins in the ear: their roles in sensory neuron survival and fiber guidance. *Prog Brain Res*, *146*, 265–278.

Gaze, R. M., Keating, M. J., Ostberg, A., & Chung, S. H. (1979). The relationship between retinal and tectal growth in larval Xenopus: implications for the development of the retino-tectal projection. *Journal of Embryology and Experimental Morphology*, *53*, 103–143.

Glebova, N. O., & Ginty, D. D. (2004). Heterogeneous requirement of NGF for sympathetic target innervation in vivo. *The Journal of Neuroscience*, *24*, 743–751.

Hallbook, F., & Fritzsch, B. (1997). Distribution of BDNF and trkB mRNA in the otic region of 3.5 and 4.5 day chick embryos as revealed with a combination of in situ hybridization and tract tracing. *The International Journal of Developmental Biology*, *41*, 725–732.

Hansen, M. J., Dallal, G. E., & Flanagan, J. G. (2004). Retinal axon response to ephrin-as shows a graded, concentration-dependent transition from growth promotion to inhibition. *Neuron*, *42*, 717–730.

Harris, L., McKenna, W. J., Rowland, E., Holt, D. W., Storey, G. C., & Krikler, D. M. (1983). Side effects of long-term amiodarone therapy. *Circulation*, *67*, 45–51.

Harris, W. A. (1984). Axonal pathfinding in the absence of normal pathways and impulse activity. *The Journal of Neuroscience*, *4*, 1153–1162.

Harris, W. A., Holt, C. E., & Bonhoeffer, F. (1987). Retinal axons with and without their somata, growing to and arborizing in the tectum of Xenopus embryos: a time-lapse video study of single fibres in vivo. *Development*, *101*, 123–133.

Hiesinger, P. R., Zhai, R. G., Zhou, Y., Koh, T. W., Mehta, S. Q., Schulze, K. L., et al. (2006). Activity-independent prespecification of synaptic partners in the visual map of Drosophila. *Current Biology*, *16*, 1835–1843.

Hindges, R., McLaughlin, T., Genoud, N., Henkemeyer, M., & O'Leary, D. D. (2002). EphB forward signaling controls directional branch extension and arborization required for dorsal-ventral retinotopic mapping. *Neuron*, *35*, 475–487.

Holash, J. A., Soans, C., Chong, L. D., Shao, H., Dixit, V. M., & Pasquale, E. B. (1997). Reciprocal expression of the Eph receptor Cek5 and its ligand(s) in the early retina. *Developmental Biology*, *182*, 256–269.

Holmes, A. L., & Heilig, J. S. (1999). Fasciclin II and beaten path modulate intercellular adhesion in Drosophila larval visual organ development. *Development*, *126*, 261–272.

Holt, C. E., & Harris, W. A. (1998). Target selection: invasion, mapping and cell choice. *Current Opinion in Neurobiology*, *8*, 98–105.

Honda, H. (1998). Topographic mapping in the retinotectal projection by means of complementary ligand and receptor gradients: a computer simulation study. *Journal of Theoretical Biology*, *192*, 235–246.

Hoyle, G. W., Mercer, E. H., Palmiter, R. D., & Brinster, R. L. (1993). Expression of NGF in sympathetic neurons leads to excessive axon outgrowth from ganglia but decreased terminal innervation within tissues. *Neuron*, *10*, 1019–1034.

Imai, T., Suzuki, M., & Sakano, H. (2006). Odorant receptor-derived cAMP signals direct axonal targeting. *Science*, *314*, 657–661.

Inoue, A., & Sanes, J. R. (1997). Lamina-specific connectivity in the brain: regulation by N-cadherin, neurotrophins, and glycoconjugates. *Science*, *276*, 1428–1431.

Kalil, K., Szebenyi, G., & Dent, E. W. (2000). Common mechanisms underlying growth cone guidance and axon branching. *Journal of Neurobiology*, *44*, 145–158.

Kay, J. N., Roeser, T., Mumm, J. S., Godinho, L., Mrejeru, A., Wong, R. O., et al. (2004). Transient requirement for ganglion cells during assembly of retinal synaptic layers. *Development*, *131*, 1331–1342.

Kobayashi, H., Koppel, A. M., Luo, Y., & Raper, J. A. (1997). A role for collapsin-1 in olfactory and cranial sensory axon guidance. *The Journal of Neuroscience*, *17*, 8339–8352.

Kuruvilla, R., Zweifel, L. S., Glebova, N. O., Lonze, B. E., Valdez, G., Ye, H., et al. (2004). A neurotrophin signaling cascade coordinates sympathetic neuron development through differential control of TrkA trafficking and retrograde signaling. *Cell*, *118*, 243–255.

Landgraf, M., Baylies, M., & Bate, M. (1999). Muscle founder cells regulate defasciculation and targeting of motor axons in the Drosophila embryo. *Current Biology*, *9*, 589–592.

Landmesser, L., Dahm, L., Tang, J. C., & Rutishauser, U. (1990). Polysialic acid as a regulator of intramuscular nerve branching during embryonic development. *Neuron*, *4*, 655–667.

Langley, J. N. (1897). On the regeneration of pre-ganglionic and post-ganglionic visceral nerve fibres. *The Journal of Physiology*, *22*, 215–230.

Langley, J. N. (1985). Note on regeneration of pre-ganglionic fibres of the sympathetic. *The Journal of Physiology*, *18*, 280–284.

Ledda, F., Paratcha, G., & Ibanez, C. F. (2002). Target-derived GFRalpha1 as an attractive guidance signal for developing sensory and sympathetic axons via activation of Cdk5. *Neuron*, *36*, 387–401.

Lom, B., Cogen, J., Sanchez, A. L., Vu, T., & Cohen-Cory, S. (2002). Local and target-derived brain-derived neurotrophic factor exert opposing effects on the dendritic arborization of retinal ganglion cells in vivo. *The Journal of Neuroscience*, *22*, 7639–7649.

Lom, B., & Cohen-Cory, S. (1999). Brain-derived neurotrophic factor differentially regulates retinal ganglion cell dendritic and axonal arborization in vivo. *The Journal of Neuroscience*, *19*, 9928–9938.

Luo, L., & Flanagan, J. G. (2007). Development of continuous and discrete neural maps. *Neuron*, *56*, 284–300.

Lyckman, A. W., Jhaveri, S., Feldheim, D. A., Vanderhaeghen, P., Flanagan, J. G., & Sur, M. (2001). Enhanced plasticity of retinothalamic projections in an ephrin-A2/A5 double mutant. *The Journal of Neuroscience*, *21*, 7684–7690.

Mann, F., Ray, S., Harris, W., & Holt, C. (2002a). Topographic mapping in dorsoventral axis of the Xenopus retinotectal system depends on signaling through ephrin-B ligands. *Neuron*, *35*, 461–473.

Mann, F., Peuckert, C., Dehner, F., Zhou, R., & Bolz, J. (2002b). Ephrins regulate the formation of terminal axonal arbors during the development of thalamocortical projections. *Development*, *129*, 3945–3955.

Mark, R. F. (1969). Matching muscles and motoneurons. A review of some experiments on motor nerve regeneration. *Brain Research*, *14*, 245–254.

McFarlane, S., Cornel, E., Amaya, E., & Holt, C. E. (1996). Inhibition of FGF receptor activity in retinal ganglion cell axons causes errors in target recognition. *Neuron*, *17*, 245–254.

McFarlane, S., McNeill, L., & Holt, C. E. (1995). FGF signaling and target recognition in the developing Xenopus visual system. *Neuron*, *15*, 1017–1028.

Meadows, L. A., Gell, D., Broadie, K., Gould, A. P., & White, R. A. (1994). The cell adhesion molecule, connectin, and the development of the Drosophila neuromuscular system. *Journal of Cell Science*, *107*(Pt 1), 321–328.

Merzenich, M. (1998). Long-term change of mind. *Science*, *282*, 1062–1063.

Merzenich, M. M., & Jenkins, W. M. (1993). Reorganization of cortical representations of the hand following alterations of skin inputs induced by nerve injury, skin island transfers, and experience. *Journal of Hand Therapy*, *6*, 89–104.

Messersmith, E. K., Leonardo, E. D., Shatz, C. J., Tessier-Lavigne, M., Goodman, C. S., & Kolodkin, A. L. (1995). Semaphorin III can function as a selective chemorepellent to pattern sensory projections in the spinal cord. *Neuron*, *14*, 949–959.

Metin, C., & Frost, D. O. (1989). Visual responses of neurons in somatosensory cortex of hamsters with experimentally induced retinal projections to somatosensory thalamus. *Proceedings of the National Academy of Sciences of the United States of America*, *86*, 357–361.

Meyer, F., & Aberle, H. (2006). At the next stop sign turn right: the metalloprotease Tolloid-related 1 controls defasciculation of motor axons in Drosophila. *Development*, *133*, 4035–4044.

Meyer, R. L., & Wolcott, L. L. (1987). Compression and expansion without impulse activity in the retinotectal projection of goldfish. *Journal of Neurobiology*, *18*, 549–567.

Millard, S. S., & Zipursky, S. L. (2008). Dscam-mediated repulsion controls tiling and self-avoidance. *Current Opinion in Neurobiology*, *18*, 84–89.

Miskevich, F., Zhu, Y., Ranscht, B., & Sanes, J. R. (1998). Expression of multiple cadherins and catenins in the chick optic tectum. *Molecular and Cellular Neurosciences*, *12*, 240–255.

Mitchell, K. J., Doyle, J. L., Serafini, T., Kennedy, T. E., Tessier-Lavigne, M., Goodman, C. S., et al. (1996). Genetic analysis of Netrin genes in Drosophila: Netrins guide CNS commissural axons and peripheral motor axons. *Neuron*, *17*, 203–215.

Miyamichi, K., Serizawa, S., Kimura, H. M., & Sakano, H. (2005). Continuous and overlapping expression domains of odorant receptor genes in the olfactory epithelium determine the dorsal/ventral positioning of glomeruli in the olfactory bulb. *The Journal of Neuroscience, 25*, 3586–3592.

Mombaerts, P. (1996). Targeting olfaction. *Current Opinion in Neurobiology, 6*, 481–486.

Mombaerts, P., Wang, F., Dulac, C., Chao, S. K., Nemes, A., Mendelsohn, M., et al. (1996). Visualizing an olfactory sensory map. *Cell, 87*, 675–686.

Monnier, P. P., Sierra, A., Macchi, P., Deitinghoff, L., Andersen, J. S., Mann, M., et al. (2002). RGM is a repulsive guidance molecule for retinal axons. *Nature, 419*, 392–395.

Nakamura, H., & O'Leary, D. D. (1989). Inaccuracies in initial growth and arborization of chick retinotectal axons followed by course corrections and axon remodeling to develop topographic order. *The Journal of Neuroscience, 9*, 3776–3795.

Njå, A., & Purves, D. (1977). Specific innervation of guinea-pig superior cervical ganglion cells by preganglionic fibres arising from different levels of the spinal cord. *The Journal of Physiology, 264*, 565–583.

Nose, A., Mahajan, V. B., & Goodman, C. S. (1992a). Connectin: a homophilic cell adhesion molecule expressed on a subset of muscles and the motoneurons that innervate them in Drosophila. *Cell, 70*, 553–567.

Nose, A., Van Vactor, D., Auld, V., & Goodman, C. S. (1992b). Development of neuromuscular specificity in Drosophila. *Cold Spring Harbor Symposia on Quantitative Biology, 57*, 441–449.

Nose, A., Takeichi, M., & Goodman, C. S. (1994). Ectopic expression of connectin reveals a repulsive function during growth cone guidance and synapse formation. *Neuron, 13*, 525–539.

Nose, A., Umeda, T., & Takeichi, M. (1997). Neuromuscular target recognition by a homophilic interaction of connectin cell adhesion molecules in Drosophila. *Development, 124*, 1433–1441.

Ozaki, S., & Snider, W. D. (1997). Initial trajectories of sensory axons toward laminar targets in the developing mouse spinal cord. *The Journal of Comparative Neurology, 380*, 215–229.

Palca, J. (1991). Famous monkeys provide surprising results [news]. *Science, 252*(5014), 1789.

Penfield, W. (1954). *The excitable cortex in conscious man.* Liverpool: Liverpool University Press.

Person, A. L., Cerretti, D. P., Pasquale, E. B., Rubel, E. W., & Cramer, K. S. (2004). Tonotopic gradients of Eph family proteins in the chick nucleus laminaris during synaptogenesis. *Journal of Neurobiology, 60*, 28–39.

Phillips, G. R., Tanaka, H., Frank, M., Elste, A., Fidler, L., Benson, D. L., et al. (2003). Gamma-protocadherins are targeted to subsets of synapses and intracellular organelles in neurons. *The Journal of Neuroscience, 23*, 5096–5104.

Pipes, G. C., Lin, Q., Riley, S. E., & Goodman, C. S. (2001). The beat generation: a multigene family encoding IgSF proteins related to the beat axon guidance molecule in Drosophila. *Development, 128*, 4545–4552.

Pittman, A., & Chien, C. B. (2002). Understanding dorsoventral topography: backwards and forwards. *Neuron, 35*, 409–411.

Potter, S. M., Zheng, C., Koos, D. S., Feinstein, P., Fraser, S. E., & Mombaerts, P. (2001). Structure and emergence of specific olfactory glomeruli in the mouse. *The Journal of Neuroscience, 21*, 9713–9723.

Purves, D., Thompson, W., & Yip, J. W. (1981). Re-innervation of ganglia transplanted to the neck from different levels of the guinea-pig sympathetic chain. *The Journal of Physiology, 313*, 49–63.

Raghavan, S., & White, R. A. (1997). Connectin mediates adhesion in Drosophila. *Neuron, 18*, 873–880.

Ramachandran, V. S., & Rogers-Ramachandran, D. (2000). Phantom limbs and neural plasticity. *Archives of Neurology, 57*, 317–320.

Recanzone, G. H., Merzenich, M. M., & Schreiner, C. E. (1992). Changes in the distributed temporal response properties of SI cortical neurons reflect improvements in performance on a temporally based tactile discrimination task. *Journal of Neurophysiology, 67*(5), 1071–1091.

Redies, C., & Takeichi, M. (1996). Cadherins in the developing central nervous system: an adhesive code for segmental and functional subdivisions. *Developmental Biology, 180*, 413–423.

Reimer, M. K., Mokshagundam, S. P., Wyler, K., Sundler, F., Ahren, B., & Stagner, J. I. (2003). Local growth factors are beneficial for the autonomic reinnervation of transplanted islets in rats. *Pancreas, 26*, 392–397.

Roe, A. W., Pallas, S. L., Kwon, Y. H., & Sur, M. (1992). Visual projections routed to the auditory pathway in ferrets: receptive fields of visual neurons in primary auditory cortex. *The Journal of Neuroscience, 12*, 3651–3664.

Sakaguchi, D. S., & Murphey, R. K. (1985). Map formation in the developing Xenopus retinotectal system: an examination of ganglion cell terminal arborizations. *The Journal of Neuroscience, 5*, 3228–3245.

Sanes, J. R., & Yamagata, M. (1999). Formation of lamina-specific synaptic connections. *Current Opinion in Neurobiology, 9*, 79–87.

Schmidt, J., & Coen, T. (1995). Changes in retinal arbors in compressed projections to half tecta in goldfish. *Journal of Neurobiology, 28*, 409–418.

Schmidt, J. T., & Buzzard, M. (1990). Activity-driven sharpening of the regenerating retinotectal projection: effects of blocking or synchronizing activity on the morphology of individual regenerating arbors. *Journal of Neurobiology, 21*, 900–917.

Schmidt, J. T., & Easter, S. S. (1978). Independent biaxial reorganization of the retinotectal projection: a reassessment. *Experimental Brain Research, 31*, 155–162.

Schmucker, D., Clemens, J. C., Shu, H., Worby, C. A., Xiao, J., Muda, M., et al. (2000). Drosophila Dscam is an axon guidance receptor exhibiting extraordinary molecular diversity. *Cell, 101*, 671–684.

Schneider, M. B., Standop, J., Ulrich, A., Wittel, U., Friess, H., Andren-Sandberg, A., et al. (2001). Expression of nerve growth factors in pancreatic neural tissue and pancreatic cancer. *Journal of Histochemistry and Cytochemistry, 49*, 1205–1210.

Schuster, C. M., Davis, G. W., Fetter, R. D., & Goodman, C. S. (1996a). Genetic dissection of structural and functional components of synaptic plasticity. I. Fasciclin II controls synaptic stabilization and growth. *Neuron, 17*, 641–654.

Schuster, C. M., Davis, G. W., Fetter, R. D., & Goodman, C. S. (1996b). Genetic dissection of structural and functional components of synaptic plasticity. II. Fasciclin II controls presynaptic structural plasticity. *Neuron, 17*, 655–667.

Serizawa, S., Miyamichi, K., Takeuchi, H., Yamagishi, Y., Suzuki, M., & Sakano, H. (2006). A neuronal identity code for the odorant receptor-specific and activity-dependent axon sorting. *Cell, 127*, 1057–1069.

Shen, K., & Bargmann, C. I. (2003). The immunoglobulin superfamily protein SYG-1 determines the location of specific synapses in C. elegans. *Cell, 112*, 619–630.

Shen, K., Fetter, R. D., & Bargmann, C. I. (2004). Synaptic specificity is generated by the synaptic guidepost protein SYG-2 and its receptor, SYG-1. *Cell, 116*, 869–881.

Simon, D. K., & O'Leary, D. D. (1990). Limited topographic specificity in the targeting and branching of mammalian retinal axons. *Developmental Biology, 137*, 125–134.

Simon, D. K., & O'Leary, D. D. (1992). Development of topographic order in the mammalian retinocollicular projection. *The Journal of Neuroscience, 12*, 1212–1232.

Simons, D. J., Durham, D., & Woolsey, T. A. (1984). Functional organization of mouse and rat SmI barrel cortex following vibrissal damage on different postnatal days. *Somatosensory Research, 1*, 207–245.

Sink, H., Rehm, E. J., Richstone, L., Bulls, Y. M., & Goodman, C. S. (2001). sidestep encodes a target-derived attractant essential for motor axon guidance in Drosophila. *Cell, 105*, 57–67.

Sperry, R. W. (1943). Effect of 180 degree rotation of the retinal field on visuomotor coordination. *J Exp Zool, 92*, 263–279.

Sperry, R. W. (1963). Chemoaffinity in the orderly growth of nerve fiber patterns and connections. *Proceedings of the National Academy of Sciences of the United States of America, 50*, 703–710.

Stuermer, C. A., & Raymond, P. A. (1989). Developing retinotectal projection in larval goldfish. *The Journal of Comparative Neurology, 281*, 630–640.

Szebenyi, G., Callaway, J. L., Dent, E. W., & Kalil, K. (1998). Interstitial branches develop from active regions of the axon demarcated by the primary growth cone during pausing behaviors. *The Journal of Neuroscience, 18*, 7930–7940.

Taniguchi, M., Yuasa, S., Fujisawa, H., Naruse, I., Saga, S., Mishina, M., et al. (1997). Disruption of semaphorin III/D gene causes severe abnormality in peripheral nerve projection. *Neuron, 19*, 519–530.

Tessarollo, L., Coppola, V., & Fritzsch, B. (2004). NT-3 replacement with brain-derived neurotrophic factor redirects vestibular nerve fibers to the cochlea. *The Journal of Neuroscience, 24*, 2575–2584.

Van der Loss, H., Dorfl, J., & Welker, E. (1984). Variation in pattern of mystacial vibrissae in mice. A quantitative study of ICR stock and several inbred strains. *J Hered, 75*, 326–336.

Van Vactor, D. V., Sink, H., Fambrough, D., Tsoo, R., & Goodman, C. S. (1993). Genes that control neuromuscular specificity in Drosophila. *Cell, 73*, 1137–1153.

Vanderhaeghen, P., Lu, Q., Prakash, N., Frisen, J., Walsh, C. A., Frostig, R. D., et al. (2000). A mapping label required for normal scale of body representation in the cortex. *Nature Neuroscience, 3*, 358–365.

Vassar, R., Chao, S. K., Sitcheran, R., Nunez, J. M., Vosshall, L. B., & Axel, R. (1994). Topographic organization of sensory projections to the olfactory bulb. *Cell, 79*, 981–991.

Verhage, M., Maia, A. S., Plomp, J. J., Brussaard, A. B., Heeroma, J. H., Vermeer, H., et al. (2000). Synaptic assembly of the brain in the absence of neurotransmitter secretion. *Science, 287*, 864–869.

Vislay-Meltzer, R. L., Kampff, A. R., & Engert, F. (2006). Spatiotemporal specificity of neuronal activity directs the modification of receptive fields in the developing retinotectal system. *Neuron, 50*, 101–114.

von Melchner, L., Pallas, S. L., & Sur, M. (2000). Visual behaviour mediated by retinal projections directed to the auditory pathway. *Nature, 404*, 871–876.

Walter, J., Henke-Fahle, S., & Bonhoeffer, F. (1987a). Avoidance of posterior tectal membranes by temporal retinal axons. *Development, 101*, 909–913.

Walter, J., Kern-Veits, B., Huf, J., Stolze, B., & Bonhoeffer, F. (1987b). Recognition of position-specific properties of tectal cell membranes by retinal axons in vitro. *Development, 101*, 685–696.

Walter, J., Muller, B., & Bonhoeffer, F. (1990). Axonal guidance by an avoidance mechanism. *Journal De Physiologie, 84*, 104–110.

Wang, F., Nemes, A., Mendelsohn, M., & Axel, R. (1998). Odorant receptors govern the formation of a precise topographic map. *Cell, 93*, 47–60.

Wang, J., Zugates, C. T., Liang, I. H., Lee, C. H., & Lee, T. (2002). Drosophila Dscam is required for divergent segregation of sister branches and suppresses ectopic bifurcation of axons. *Neuron, 33*, 559–571.

Wang, K. H., Brose, K., Arnott, D., Kidd, T., Goodman, C. S., Henzel, W., et al. (1999). Biochemical purification of a mammalian slit protein as a positive regulator of sensory axon elongation and branching. *Cell, 96*, 771–784.

Webber, C. A., Hyakutake, M. T., & McFarlane, S. (2003). Fibroblast growth factors redirect retinal axons in vitro and in vivo. *Developmental Biology, 263*, 24–34.

Wilkinson, D. G. (2000). Topographic mapping: organising by repulsion and competition? *Current Biology, 10*, R447–R451.

Winberg, M. L., Mitchell, K. J., & Goodman, C. S. (1998). Genetic analysis of the mechanisms controlling target selection: complementary and combinatorial functions of netrins, semaphorins, and IgCAMs. *Cell, 93*, 581–591.

Wojtowicz, W. M., Flanagan, J. J., Millard, S. S., Zipursky, S. L., & Clemens, J. C. (2004). Alternative splicing of Drosophila Dscam generates axon guidance receptors that exhibit isoform-specific homophilic binding. *Cell, 118*, 619–633.

Woolsey, T. A., & Van der Loos, H. (1970). The structural organization of layer IV in the somatosensory region (SI) of mouse cerebral cortex. The description of a cortical field composed of discrete cytoarchitectonic units. *Brain Research, 17*, 205–242.

Yamagata, M., Herman, J. P., & Sanes, J. R. (1995). Lamina-specific expression of adhesion molecules in developing chick optic tectum. *The Journal of Neuroscience, 15*, 4556–4571.

Yamagata, M., & Sanes, J. R. (2008). Dscam and Sidekick proteins direct lamina-specific synaptic connections in vertebrate retina. *Nature, 451*, 465–469.

Yamagata, M., Sanes, J. R., & Weiner, J. A. (2003). Synaptic adhesion molecules. *Current Opinion in Cell Biology, 15*, 621–632.

Yamagata, M., Weiner, J. A., & Sanes, J. R. (2002). Sidekicks: synaptic adhesion molecules that promote lamina-specific connectivity in the retina. *Cell, 110*, 649–660.

Yates, P. A., Roskies, A. L., McLaughlin, T., & O'Leary, D. D. (2001). Topographic-specific axon branching controlled by ephrin-As is the critical event in retinotectal map development. *The Journal of Neuroscience, 21*, 8548–8563.

Yu, C. R., Power, J., Barnea, G., O'Donnell, S., Brown, H. E., Osborne, J., et al. (2004). Spontaneous neural activity is required for the establishment and maintenance of the olfactory sensory map. *Neuron, 42*, 553–566.

Zhang, J. H., Cerretti, D. P., Yu, T., Flanagan, J. G., & Zhou, R. (1996). Detection of ligands in regions anatomically connected to neurons expressing the Eph receptor Bsk: potential roles in neuron-target interaction. *The Journal of Neuroscience, 16*, 7182–7192.

Zhang, L. I., Tao, H. W., Holt, C. E., Harris, W. A., & Poo, M. m. (1998). A critical window for cooperation and competition among developing retinotectal synapses. *Nature, 395*, 37–44.

Zhao, H., & Reed, R. R. (2001). X inactivation of the OCNC1 channel gene reveals a role for activity-dependent competition in the olfactory system. *Cell, 104*, 651–660.

Zou, D. J., Feinstein, P., Rivers, A. L., Mathews, G. A., Kim, A., Greer, C. A., et al. (2004). Postnatal refinement of peripheral olfactory projections. *Science, 304*, 1976–1979.

Naturally-occurring neuron death

7

Nervous system differentiation is accompanied by tremendous growth; neuron cell bodies and dendrites expand, glial cells and myelin are added, blood vessels arborize, and extracellular matrix is secreted. Even after the period of neurogenesis has largely ended, the human brain continues to increase in size from approximately 400 grams at birth to 1400 grams in adulthood (Dekaban and Sadowsky, 1978). Surprisingly, neurogenesis and this later period of growth overlap with a tremendous loss of neurons and glia, both of which die from "natural causes." At first, it was difficult to accept the concept that neurons were born, only to die a short time later. Although

there were reports of neuron death following the removal of their target of innervation (see below), it was not clear that postmitotic neurons were lost in any significant number during normal development (Clarke and Clarke, 1996). We now understand that nerve cells participate actively in their own demise through gene transcription and protein synthesis, and this process is called *programmed cell death* (PCD).

Depending on the brain region, 20 to 80% of differentiated cells degenerate during development (Oppenheim, 1991; Oppenheim and Johnson, 2003). Whether or not a neuron survives depends on many factors (**Figure 7.1**). Soluble survival factors may be supplied by the postsynaptic target, by neighboring nerve and glial cells, or by the circulatory system. Neurons also depend upon the synaptic contacts that they receive, and deafferentation early in life can cause a postsynaptic neuron to atrophy or die. These diverse signals are referred to as *trophic factors* because one cell is nourished or sustained by another. The first part of this chapter will describe the characteristics of naturally-occurring cell death in the developing nervous system. Relatively little will be said of injury-evoked cell death. We then discuss the trophic factors and intracellular signals that regulate this process. Finally, we will learn that electrical activity and synaptic transmission can have an important influence on neuron survival.

WHAT DOES NEURON DEATH LOOK LIKE?

Naturally-occurring neuron death was discovered over a century ago by John Beard (1896), who followed the fate of a very large, easily recognized neuron found at the surface of the skate spinal cord. He found that these Rohon-Beard cells were born in the neural crest and differentiated in the spinal cord, sending out processes to the ectoderm before degenerating. To the trained eye, a dying neuron looks quite different from a healthy one (**Figure 7.2**, left side).

PCD is usually accomplished through a process called *apoptosis*. During apoptosis, the chromatin becomes very condensed and aggregates in crescent-shaped figures at the nuclear membrane, a process called *pyknosis*. The plasma membrane remains intact, but the neuron gradually shrinks (**Figure 7.3**, top). As proteins become cross linked at the membrane, small protuberances, called *apoptotic* bodies, pinch off from the cell body and are phagocytosed by *macrophages* (see *The Removal of Dying Neurons*, page 196).

171

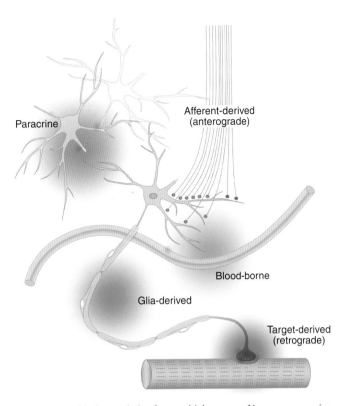

Fig. 7.1 Trophic factors derive from multiple sources. Neurons can receive survival signals from the cells that they innervate (target-derived), from their synaptic inputs (afferent-derived), from neighboring neuron cell bodies (paracrine), from distant sources via the circulatory system (blood-borne), and from nonneuronal cells (glia-derived).

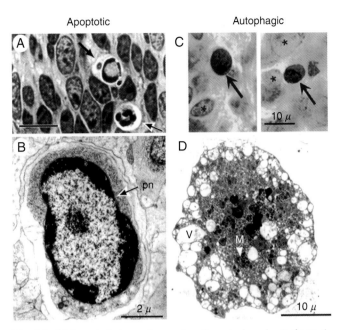

Fig. 7.2 Light and electron micrographs of apoptosis and autophagy. A. A high-power photomicrograph of the kitten retina at embryonic day 57 shows two neurons with condensed chromatin in their nuclei (arrows). B. An electron micrograph shows a degenerating retinal ganglion cell with a clearly pyknotic nucleus (pn). (Reprinted from Wong and Hughes, 1987) C. High-power photomicrographs of E16.5 spinal cord shows two neurons from caspase-deficient mice (arrows). These dying cells do not display a prominent pyknosis or apoptotic bodies, but have an increased cytoplasmic density. Healthy neurons are indicated with asterisks. (Reprinted from Oppenheim et al., 2001) D. An electron micrograph shows a degenerating motor neuron in the developing moth with characteristic autophagic morphology. The nucleus is missing, the cytoplasm has increased in electron density, mitochondria (M) are electron dense and clustered, and there is massive accumulation of autophagic bodies and vacuoles (V). *(Reprinted from Kinch et al., 2003)*

As the large, crescent-shaped aggregates of nuclear material form, enzymes are activated that cleave the DNA, producing fragments of about 180 base pairs. Although this process is too small to see anatomically, it is possible to stain the broken ends of DNA strands with molecular markers. One technique, called *TUNEL* (for **T**erminal transferase **U**TP **N**ick **E**nd **L**abeling), employs an enzyme that attaches labeled nucleotides to the exposed ends of the DNA fragments (Figure 7.3, bottom). This approach is useful when studying cell death in a large population of cells that has no clear boundaries, such as an area of cerebral cortex. However, TUNEL is not a foolproof assay for cell death. Unlabeled cells may, in fact, enter a cell death pathway in which chromatin break down and condensation are not featured (Oppenheim et al., 2001a).

PCD may also follow a second path in which the neuron essentially digests itself (Figure 7.2, right side). Autophagy is a normal catabolic process by which cells recycle cytoplasm and dispose of organelles. During normal autophagy, cytoplasmic proteins are sequestered in vesicles (called autophagosomes), and are then degraded when the vesicles fuse with lysosomes (called autolysosomes). However, a more extreme form of this process can lead to cell death when all of the cell's constituents are degraded. For example, a set of moth motor neurons undergo PCD in response to a steroid hormone signal, and the cytoplasm of these neurons fills with autophagosomes and autolysosomes before there is any sign of pyknosis (Kinch et al., 2003). It is thought that later stages of PCD, occurring after synaptic connections have been established, may employ an autophagic process (Oppenheim et al., 2001a, 2008).

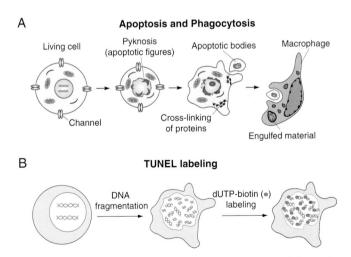

Fig. 7.3 Features of apoptosis. A. Naturally-occurring cell death is usually accomplished through a process called *apoptosis*. During apoptosis, the neuron begins to shrink, and the nuclear matter becomes condensed, forming crescent-shaped figures. As proteins become cross linked at the membrane, small apoptotic bodies break off and are phagocytized by macrophages. B. During apoptosis, DNA is broken down by endonuclease activity to produce double-stranded, low-molecular-weight fragments. These DNA fragments can be detected by a labeling technique called TUNEL. A modified nucleotide, such as dUTP-biotin, is catalytically attached to the free 3'-hydroxyl end of each DNA fragment by the enzyme, terminal deoxynucleotidyl transferase. Thus, the nuclei of dying neurons can be detected before the cells break up and are phagocytosed.

A third set of degenerative changes, called *necrosis*, are observed in many instances of traumatic injury. In this case, the mitochondria stop producing energy, and the neuron becomes unable to regulate ionic content and hydrostatic pressure. The neuron and its organelles begin to swell, lysosomal enzymes become activated, and cytoplasmic components are broken down. Finally, the soma bursts open. Clearly, a neuron that dies gracefully by budding off neat little packages of membrane (as is the case for PCD) is unlikely to injure healthy neighboring neurons, and serves as an efficient means to eliminate a subset of cells. Many additional forms of cell death have been described (Clarke, 1990), but for the most part we will consider only the mechanisms that accompany PCD in the absence of injury or insult to the nervous system.

EARLY ELIMINATION OF PROGENITOR CELLS

The first episode of PCD occurs among the cells that are responsible for populating the nervous system. During late embryonic development, neural progenitors begin to die in great numbers. This is clearly shown for cells in the proliferative zone of the mouse cerebral cortex which display heavy labeling with a variant of the TUNEL technique, indicating a high level of DNA fragmentation (**Figure 7.4**A). Furthermore, counts of pyknotic and TUNEL-positive nuclei are consistent

with one another (Blaschke et al., 1996; Blaschke et al., 1998; Rakic and Zecevic, 2000).

To determine whether the dying cells are still in the proliferative pool, animals were injected with BrdU for several hours to label cells going through the synthesis phase of mitosis, and the tissue was then examined for TUNEL labeling (Thomaidou et al., 1997). About 70% of the TUNEL-labeled cells were colabeled with BrdU, suggesting that cell death occurred immediately after the synthesis phase (**Figure 7.4**B), and before differentiation occurred. Using these methods, it has been estimated that the percentage of newly generated cells undergoing PCD gradually increases to 37% by birth.

Little is known about the signals that lead neural progenitors to enter the apoptotic pathway. However, evidence from cultured quail neural crest cells indicates that cell–cell contact is involved. When neural crest clusters were grown on a nonadhesive substrate to prevent them from dispersing, the cells displayed a marked increase in TUNEL-labeling as compared to dissociated crest cells that were permitted to disperse (Maynard et al., 2000).

The presence of PCD in the proliferative zone suggests that the total number of neurons in the brain is regulated, in part, by the elimination of stem cells. It also raises the interesting possibility that these two stages of development—birth and death—share certain molecular pathways, a concept that is discussed below (see *Intracellular Signaling Pathways That Mediate Death*, page 191).

HOW MANY DIFFERENTIATED NEURONS DIE?

It might seem a straightforward matter to determine how many neurons are being added or removed from a population: Simply count the neurons in a young animal, and subtract this number from an identical count obtained in an adult. If the number is positive, then neurons must have been added. If the number is negative, then neurons must have been eliminated. Unfortunately, obtaining an accurate neuron count is trickier than one might suppose. For example, if neurogenesis and cell death overlap in time, then cell counts can remain relatively stable, concealing the existence of both cell birth and death. A second difficulty revolves around the counting strategy itself. Since it is often too laborious to count each cell in a neuronal structure, estimates are made from tissue sections, and changes in the size or packing density of cell bodies can each influence the final counts. Finally, neurons are not the only type of cell to die during development. For example, about 50% of oligodendrocytes in the rat optic nerve die naturally during development, and their survival depends upon the presence of retinal axons (Barres et al., 1992). Therefore, those who tally up cell bodies must be careful to discriminate neurons from glia.

Once it became clear that cell death was a general feature of the developing nervous system, its magnitude was characterized by rigorous cell counting studies. One of the most convincing ways to demonstrate that neurons are dying is to count both the healthy cells and the pyknotic cells in the same tissue section (Hughes, 1961). Counts of spinal motor neuron (MN) cell bodies in the chick and frog demonstrate that the decrease in number of healthy-looking MNs is well-correlated with the appearance of pyknotic cells (**Figure 7.5**). At first glance, it

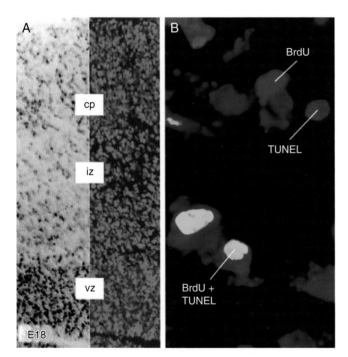

Fig. 7.4 Dying cells in the proliferative zone. A. Two micrographs of the embryonic day 18 mouse cortex. To the left, apoptotic cells are labeled with a technique similar to TUNEL, called ISEL. To the right, all cells are revealed with fluorescent labeling of nuclei. Note the prominent ISEL labeling in the ventricular zone where proliferation occurs (cp, cortical plate; iz, intermediate zone; vz, ventricular zone) (Blaschke et al., 1996). B. To demonstrate that proliferating cells are dying, the subventricular zone is colabeled with BrdU (green) and TUNEL (red). This confocal image shows colocalization (yellow) of both labels, although not all TUNEL-positive cells are dividing. *(Thomaidou et al., 1997)*

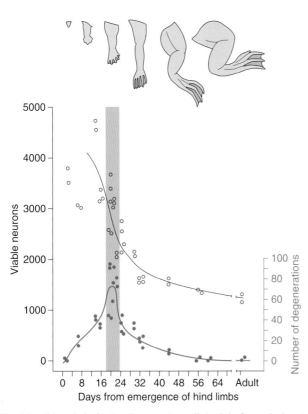

Fig. 7.5 The period of naturally-occurring cell death in frog spinal motor neurons. The graph shows the total number of viable (black) and pyknotic (red) motor neurons that innervate the frog hind limb during each of several developmental stages. At the top are pictures of leg size during this period. The number of degenerating neurons reaches a peak at precisely the time when the loss of viable neurons is most rapid, as indicated by the shaded bar. *(Adopted from Hughes, 1961)*

may not be clear why so few pyknotic neurons are observed during the period of maximal neuron elimination. This is due to the rapid removal of cell debris, which has been estimated to occur in as little as three hours.

The magnitude of naturally-occurring neuron death is striking. The data in Figure 7.5 show that the motor neuron population innervating frog hind limbs declines from an initial size of about 4000 to a final size of 1200. Thus, over 60% of differentiated motor neurons are eliminated. A similar amount of cell death has been detected at all levels of the nervous system. About 50% of rat retinal ganglion cells die within two weeks of birth, and 50% of embryonic chick ciliary ganglion neurons are lost after they project to the iris and choroid (Potts et al., 1982; Landmesser and Pilar, 1974). While it is impractical to count the total number of neurons in any one area of cerebral cortex, it is estimated that 20 to 50% of postmigratory neurons are eliminated. However, this may depend on the type of nerve cell, the region of cortex, and the time of birth (Miller, 1995; Finlay and Slattery, 1983).

SURVIVAL DEPENDS ON THE SYNAPTIC TARGET

One possibility is that cell death serves to match the number of afferents to the size of the target population. In this scenario, less motor neuron PCD would occur in spinal cord segments projecting to the ample limb musculature than would occur for

segments projecting to thoracic muscle. This theory makes the rational assumption, recognized by every woodworker, that it is easier to trim off the excess than to paste on a bit more. There are many interesting examples of this principle. The limbless lizard, *Anguis fragilis*, produces a set of motor neurons in the limb region of the spinal cord which then proceed to die during development (Raynaud et al., 1977). However, there are certain limitations to the relationship between afferent pool and target size, particularly across animal species of different size (Purves, 1988). For example, the number of superior cervical ganglion neurons innervated by each projecting spinal afferent increases from 14 in mice to 110 in man. Thus, neurogenesis is the first regulatory mechanism for setting neuron number (see Chapter 3), and cell death is the second.

Even before naturally-occurring neuron death was well-accepted, it was known that developing nerve cells died when their target was removed. A common manipulation was to remove a limb bud prior to the time of innervation (i.e., no direct damage to the axons) and then examine the motor neurons or DRG cells that would have made synapses there. These studies were performed on amphibian or chick embryos because it was relatively easy to carry out the surgeries. In the salamander, *Amblystoma*, the sensory ganglia that normally innervate a limb are much smaller when the limb is excised. In contrast, sensory ganglia that normally innervate axial musculature are much larger than normal if provided with a transplanted limb (Detwiler, 1936).

A similar relationship exists for DRG or motor neurons and target size in the chick. As greater and greater amounts of muscle are removed from developing chick embryos, the ventral horn of the spinal cord becomes smaller and fewer neurons are found in the DRG (**Figure 7.6**, left). When an extra limb bud is transplanted next to the original one, providing a larger than normal target region, developing processes grow into the added target, and the population of motor and DRG neurons is found to be much larger than normal (Figure 7.6, right). The addition of an extra limb bud saves up to 25% of the motor neurons that would otherwise die. Experimental results of this sort led to the hypothesis that the target provides a survival factor (Hamburger, 1934; Hollyday and Hamburger, 1976).

A complementary experiment can be performed by reducing the number of neurons projecting to a target, and determining whether the remaining cells die off anyway. In one such experiment, about two-thirds of the ciliary ganglion neurons innervating one eye are killed off by cutting their axons. This reduction in the number of competing neurons results in rescue of 40% of the neurons that would normally have died. Furthermore, many of the surviving axons sprout into territory vacated by the elimination of axotomized ciliary neurons. Therefore, particular neurons are not preordained to die, but do so during some sort of competition for a feature of the target (Pilar et al., 1980).

If PCD is a developmental mechanism for matching the size of a presynaptic population to its target, then it should be more prominent among neuronal populations that project to small targets. This was explored in developing chicks, where about 30% of dorsal root ganglia (DRG) neurons die between embryonic day 4.5 to 7, after they have made central and peripheral connections (**Figure 7.7**). However, the amount of cell death varies greatly. Naturally-occurring cell death occurs primarily amongst DRGs that innervate axial musculature. In contrast, the DRGs that project to wings and legs

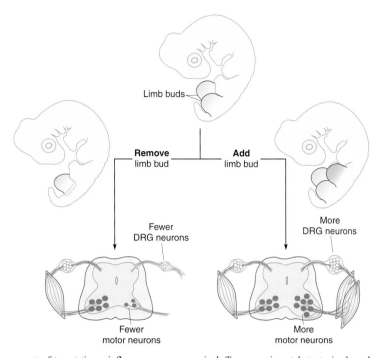

Fig. 7.6 The amount of target tissue influences neuron survival. Two experimental strategies have been used to test whether target tissue provides neurons with a survival factor. In the embryonic chick, a limb bud can be surgically removed, or an extra limb bud can be grafted nearby. The animal is then permitted to progress through the time when cell death would normally occur. (Left) When a limb bud is removed, the process of cell death is enhanced, and there are fewer motor neurons and DRG cells. (Right) When an extra limb bud is grafted on, the process of cell death is decreased, and there are a greater number of motor neurons and DRG cells. *(Adapted from Hamburger, 1934; Hollyday and Hamburger, 1976)*

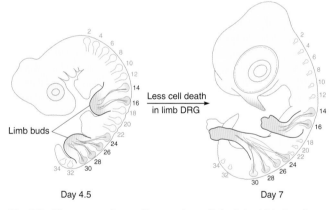

Fig. 7.7 The pattern of naturally-occurring cell death in chick dorsal root ganglia. There is less cell death among neurons in the DRGs that innervate limb buds (blue filled), compared to those that innervate the axial musculature (unfilled). The numbers refer to the somitic segment of the spinal cord. Thus, the size of DRGs at segments 14–16 decrease only slightly during development, whereas DRGs at segments 18–22 become much smaller. This result shows that neuron survival is correlated with the amount of target tissue. *(Adapted from Hamburger and Levi-Montalcini, 1949)*

have more muscle to innervate, and this reduces the amount of PCD. Therefore, naturally-occurring cell death does appear to be an important mechanism for thinning out neuron populations with less target to innervate (Ernst, 1926; Hamburger and Levi-Montalcini, 1949).

A basic question that arises from these studies is whether the target influences neuron proliferation or cell death. By carefully studying the pattern of DRG neuron degeneration following wing bud removal, Viktor Hamburger and Rita Levi-Montalcini (1949) demonstrated that target reduction leads to an increase in the number of dying neurons. One wing bud was removed at about embryonic day 3, and the DRGs were examined ipsilateral and contralateral to the ablation. Within 2–3 days, the ganglia ipsilateral to the extirpated wing buds were much smaller than normal and contained a large number of darkly stained pyknotic cells (**Figure 7.8**). Subsequent studies showed that target removal had little or no effect on the amount of ^{3}H-thymidine incorporated into DRG neurons (Carr and Simpson, 1978).

There is only one system in which cell death has been quantified in both the pre- and postsynaptic neuronal population. This provides a natural situation in which to ask whether a decrease in the target population precedes and brings about cell death in the afferent population. The presynaptic population of cells, called *nucleus magnocellularis* (NM), is the first region of the chick brain to receive input from the ear, and these cells project to a second-order auditory nucleus, called *nucleus laminaris* (NL) (**Figure 7.9**, left). The percentage of neurons that die is nearly identical in these two nuclei (Rubel et al., 1976; Solum et al., 1997). However, PCD progresses more rapidly in NM than in NL, as highlighted by the shaded bar in Figure 7.9. Since the presynaptic NM neurons largely complete their PCD days before the target population has, it seems unlikely that target size is the sole determinant of NM survival. In fact, it raises the possibility that afferent innervation may influence the survival of target neurons (see *Afferent Regulation of Neuron Survival*, page 198).

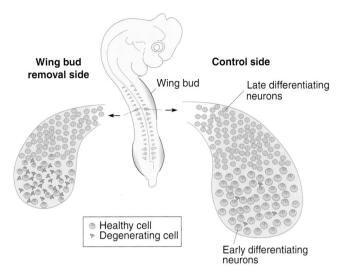

Wing bud removal side

Wing bud

Control side

Late differentiating neurons

Healthy cell
Degenerating cell

Early differentiating neurons

Fig. 7.8 The target influences neuron survival. Following unilateral limb bud removal in a 3-day chick embryo (left side), there is an increase in the number of degenerating cells in the DRG ipsilateral to the ablated limb, as compared to the control side (right). The number of viable DRG neurons is reduced dramatically. *(Adapted from Hamburger and Levi-Montalcini, 1949)*

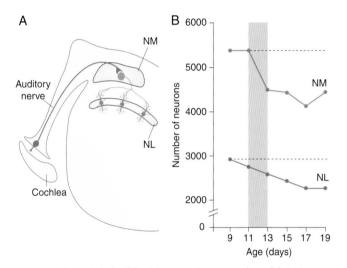

A

NM

Auditory nerve

NL

Cochlea

B

Number of neurons

6000

5000 — NM

4000

3000 — NL

2000

0

9 11 13 15 17 19
Age (days)

Fig. 7.9 The period of cell death in pre- and postsynaptic nuclei. A. A transverse hemisection through the chick auditory brainstem. The *nucleus magnocellularis* (NM) is a central auditory nucleus that is innervated by auditory nerve terminals. It projects to a second-order nucleus, called nucleus laminaris (NL). B. The graph shows the total number of cell bodies in NM and NL during the latter period of embryonic development. In both nuclei, about 20% of the neurons are lost between embryonic day 9 and 17. However, most of the NM neurons have died by day 13, several days before cell death is complete in NL. *(Adapted from Rubel et al., 1976 and Solum et al., 1997)*

Neuron survival is also influenced by transient interactions that occur during axon outgrowth, well before synaptic connections are made with the target. During embryonic development, commissural neurons in the dorsal spinal cord are attracted to the floorplate where they eventually cross the midline (see Chapter 5). The floorplate region also appears to provide a survival signal. When E13 commissural neurons are grown *in vitro*, they all die within two days, but they can survive for several days when grown in the presence of floorplate-conditioned medium (Wang and Tessier-Lavigne, 1999).

Transient growth cone interactions can also serve the opposite roll, preventing afferents from reaching their target and surviving. In the zebrafish, the death of a ventral primary motor neuron (VaP) depends on contact with a transient population of muscle cells. If these muscle cells are ablated before the VaP growth cone arrives, then the VaP motor neuron grows on to innervate the ventral musculature and survives (Eisen and Melancon, 2001).

NGF: A TARGET-DERIVED SURVIVAL FACTOR

Neuron survival clearly depends on the presence of target tissue, but what is being procured? One simple hypothesis is that target cells secrete a chemical that presynaptic neurons require for their survival. In fact, an extraordinary series of experiments, coupled with a few strokes of serendipity, led to the first endogenous neurotrophic substance to be discovered, the nerve growth factor (NGF). NGF has since been shown to control the survival of sympathetic neurons, and contribute to the survival of DRG neurons during development. Since NGF is the best understood survival factor, we are going to first consider the experiments that led to its discovery and an understanding of its mechanism of action.

Viktor Hamburger (1934) first suggested that the target produces a factor that is retrogradely transported by the innervating neurons and that influences their subsequent development. As described above, careful observations strongly suggested that the hypothetical substance worked by maintaining the survival of differentiating neurons. By modern standards, the next step would have been to harvest the target tissue and isolate soluble substances that enhance survival. Since most of the necessary biochemical tools did not yet exist in the 1950s, the isolation of NGF took a few decades to achieve, and it began with a curious set of observations. In an effort to identify a source tissue that could substitute for target muscle, various mouse tumors were implanted into the chick hind limb (Bueker, 1948). One tumor, a connective tissue cell line called sarcoma, grew rapidly and was invaded by nerve fibers. Within five days of the transplant, there was a dramatic increase in the survival of sensory and sympathetic neurons, while motor neurons were unaffected (**Figure 7.10**A). When the tissue was examined in more detail, a key observation was made: ganglia with no apparent physical connection to the tumor were also greatly enlarged. This provided the first indication that cell survival was mediated by a diffusible chemical (Levi-Montalcini and Hamburger, 1951; Levi-Montalcini and Hamburger, 1953).

A more direct demonstration that the survival factor was diffusible came from experiments in which tumor cells were placed on a vascularized respiratory structure in the chick embryo called the chorioallantoic membrane. In this case, the tumor was not in contact with sympathetic and sensory ganglia, but it did share the same blood supply. Even though the tumor was physically isolated from the nervous system, it was able to elicit a strong growth-promoting effect (Figure 7.10B). Thus, sarcoma tumor cells must have released a soluble factor that could be transported to the neurons through the circulatory system.

As a first step towards isolating the putative survival factor found in mouse sarcoma, an *in vitro* assay system was

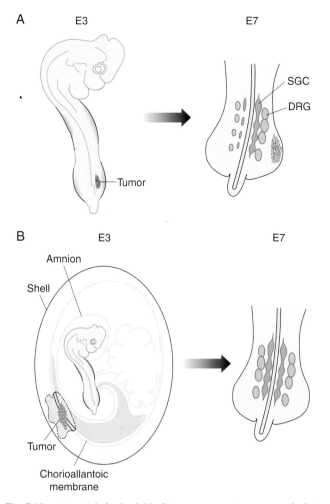

A

E3   E7

SGC

DRG

Tumor

B

E3   E7

Amnion

Shell

Tumor

Chorioallantoic
membrane

Fig. 7.10 A target-derived soluble factor can support neuron survival. A. When a tumor cell line is placed in the chick embryo at E3, the size of sympathetic ganglia and DRGs is much larger ipsilateral to the tumor by E7. B. When the same tumor cell line is placed on the chorioallantoic membrane at E3, such that nerve fibers have no direct access, all of the sympathetic and dorsal root ganglia are much increased in size by E7. Thus, the tumor must have secreted a soluble factor that enhanced neuron survival. *(Adapted from Bueker, 1948; Levi-Montalcini and Hamburger, 1951; Levi-Montalcini and Hamburger, 1953)*

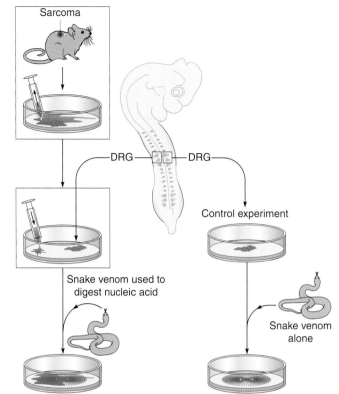

Sarcoma

DRG — DRG

Control experiment

Snake venom used to
digest nucleic acid

Snake venom
alone

Fig. 7.11 A soluble factor that supports the survival and growth of DRG neurons is discovered in a mouse sarcoma and, later, in snake venom. (Left) DRG neurons obtained from chick embryos were placed in a tissue culture dish, and conditioned medium from a mouse sarcoma was added. The venom of a snake was added to the culture to determine whether nucleic acids mediate the trophic effects. The DRG neurons survived and grew processes under these conditions. (Right) When the control experiment was performed, in which only snake venom was added to the DRG neuron cultures, a surprising discovery was made. The DRG neurons survived and grew, indicating that the snake venom must also have contained a soluble survival factor. *(Adapted from Cohen and Levi-Montalcini, 1956; Levi-Montalcini and Cohen, 1956)*

developed. Sympathetic ganglia were obtained from chick embryos and placed in a tissue culture dish, either by themselves or next to mouse sarcoma tumor cells. When grown next to tumor, the neurons survived and grew a dense halo of axons within hours, providing a simple and convenient assay system. Although biochemical isolation was a slow process, it was possible to obtain a tumor cell fraction that had only proteins and nucleic acids. In order to determine whether either of these components contained the growth factor, a biochemical trick was employed. Snake venom was known to contain high levels of an enzyme that breaks down nucleic acids (phosphodiesterase). Therefore, it was added to the extract to determine whether this class of molecules mediated the trophic effect (Cohen and Levi-Montalcini, 1956; Levi-Montalcini and Cohen, 1956). If the biological activity remained, then one could conclude that growth factor contained protein. Of course, control cultures containing only sympathetic ganglia and snake venom were also prepared. Surprisingly, the tumor fraction containing the snake venom was even more potent

than the origin protein-nucleic acid fraction. Even more peculiar, the snake venom itself was found to support nerve growth (**Figure 7.11**).

As it turned out, the discovery of a growth-promoting effect in snake venom was extremely fortunate. It suggested that growth-promoting activity would also be found in a mammalian analog, the salivary gland. In fact, the mouse submaxillary gland proved to be a superb source for the nerve growth factor, and this eventually led to its complete isolation and sequencing.

Once the NGF was purified, it was possible to perform two critical experiments *in vivo* to determine whether this protein is both necessary and sufficient to keep sensory and sympathetic neurons alive during development. First, the NGF-containing fraction that was purified from snake venom or salivary gland was injected directly into neonatal mammals, and it did produce a dramatic increase in the size of sensory and sympathetic ganglia (Levi-Montalcini and Cohen, 1956; Levi-Montalcini and Booker, 1960a). Sympathetic ganglia from control and treated animals are shown in **Figure 7.12**A. In addition to keeping neurons alive, NGF promotes process outgrowth. Thus, when NGF is injected into neonatal rodents, sympathetic nerve fibers

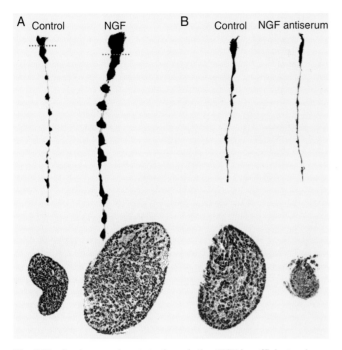

Fig. 7.12 In vivo experiments testing whether NGF is sufficient and necessary for sympathetic neuron survival. A. Whole mounts of the sympathetic thoracic chain ganglia from a P19 control mouse (top left) and an animal injected with NGF-containing salivary extract since birth (top right). Transverse sections through the stellate ganglion of each animal (dotted red line) is shown below. (*Levi-Montalcini and Booker, 1960a*) B. Whole mounts of stellate and sympathetic thoracic chains from a control P3.5 rabbit (top left) and an animal injected with NGF antiserum since birth (top right). Transverse sections through the superior cervical ganglion of a P9 control mouse (bottom left) and an animal injected with NGF antiserum since birth (bottom right). (*Levi-Montalcini and Booker, 1960b*)

are no longer restricted to their normal synaptic target, but grow widely in the peripheral field and can even invade blood vessels or the central nervous system.

To determine whether endogenous NGF is necessary for survival, an antibody directed against the NGF protein was injected into neonatal rodents (Levi-Montalcini and Booker, 1960b). This treatment leads to the loss of almost all sympathetic neurons (Figure 7.12B). It was later found that DRG cells are no longer dependent on NGF at the age when antibody was administered, but they can be destroyed by prenatal exposure to NGF antibody (Johnson et al., 1978). In fact, not all sensory neurons are dependent on NGF for survival. Those neurons that derive from sensory placodes (e.g., nodose ganglion), rather than the neural crest, are unresponsive to NGF treatment. In the DRG, only small peptidergic neurons that carry nociceptive signals to the spinal cord are killed following loss of the NGF signal. It is possible to reproduce the effects of antibody treatment in genetically engineered mice. When a deletion is made in the coding sequence of the NGF gene, homozygous animals display profound cell loss in both sympathetic and sensory ganglia (Crowley et al., 1994).

If NGF is the endogenous survival factor, then it is important to verify its presence at the sympathetic and sensory ganglion target regions at an appropriate time during development. Although NGF levels are extremely low (except in the fortuitous case of the male mouse salivary gland, from which it was purified), it has been possible to localize the protein with immunohistochemical staining, and the NGF mRNA with *in situ* hybridization. For example, trigeminal axons arrive at their cutaneous target just before the NGF mRNA and protein are manufactured, and the initial outgrowth of trigeminal axons is NGF-independent (Davies et al., 1987), suggesting that the maintenance of trigeminal neurons depends on NGF derived from their target. The success with NGF was achieved by 1960, and the expectation was that many other neurotrophic substances would quickly be found in the central nervous system. While several growth and survival factors were discovered in nonneuronal systems, the search for another bona fide neurotrophic substance was, at first, somewhat frustrating.

THE NEUROTROPHIN FAMILY

The full amino acid sequence of NGF was obtained by 1971, yet a decade elapsed before a second neurotrophic factor was identified. The search began with the simple observation that, in contrast to its effect on the retinae from lower vertebrates, NGF does not stimulate neurite outgrowth from cultured rat retina. Working under the assumption that there must be a growth factor for mammalian retina, a soluble extract was prepared from the entire pig brain. This extract did, in fact, stimulate retinal process outgrowth in a dose-dependent manner. When the active substance, named brain derived growth factor (BDNF), was purified and its amino acid sequence determined, its structure displayed a striking similarity to that of NGF (Turner et al., 1982; Leibrock et al., 1989).

Several members of the neurotrophin family have now been isolated, and they are found in both the peripheral and central nervous systems. The more recent additions to the family have been given the less colorful names: neurotrophin-3 (NT-3), NT-4/5, NT-6, and NT-7; the last two are found only in fish (**Figure 7.13**). In each case, a precursor protein of about 250 amino acids, called the *pro-peptide*, is processed post-translationally to produce active peptides of about 120 amino acids. These peptides form homodimers and become biologically active. The family members share about 50% sequence homology with one another, particularly within six hydrophobic regions that are responsible for linking the two protomers together. Each neurotrophin also contains a unique amino acid sequence, and it is this variable region that is responsible for binding to a specific receptor (Ibanez, 1994).

Each of the neurotrophins has been shown to play a role in the survival of specific peripheral neuron populations. As with NGF, two general classes of experiment have been performed: One can provide excess neurotrophin (*in vivo* or *in vitro*), or decrease the amount of endogenous neurotrophin, typically by single-gene knockout experiments (Chapter 2). Experiments of this sort indicate that BDNF is a necessary endogenous signal for the survival of vestibular ganglia, while NT-3 is an endogenous survival signal for the cochlear ganglion. Both BDNF and NT-3 also contribute to the survival of neurons in the sensory, trigeminal, and nodose ganglia. Even though increasing exogenous growth factor may promote survival, this does not necessarily reflect a role for the endogenous factor. For example, exogenous BDNF is able to save chick motor neurons when administered during the period of naturally-occurring cell death, but there is no effect on motor neuron survival in BDNF knockout mice (Oppenheim et al., 1992; Ernfors et al., 1994, 1995).

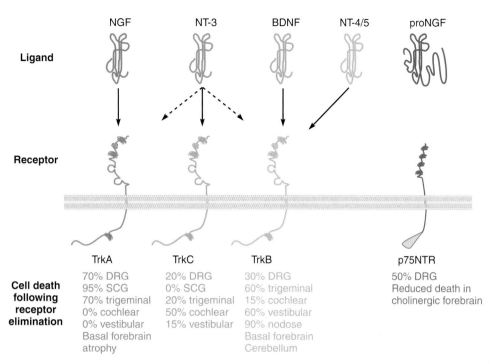

Fig. 7.13 Neurotrophins and their receptors. Following the discovery of NGF, several homologous proteins were found, including NT-3, BDNF, and NT-4. Each of these proteins binds selectively to a member of the Trk receptor tyrosine kinase family, as illustrated. In addition, there is a low-affinity receptor, called P75NTR. The effect of eliminating a neurotrophin receptor in mice is shown beneath each pair.

Neurotrophins may promote survival even before an axon has reached its target. The development of NT-3 expression can be followed with a *lacZ* reporter gene during the embryonic period when DRG neurons first extend their axons. NT-3 is expressed heavily along the path of axon growth (**Figure 7.14**), and this may explain why loss of NT-3 expression can affect neuron survival before target innervation has occurred (Farinas et al., 1996).

THE TRK FAMILY OF NEUROTROPHIN RECEPTORS

Even before a receptor for NGF was discovered, it was known that NGF binds to a site on the axon terminal with very high affinity. The β subunit of NGF can be labeled with ^{125}I, and used to perform binding studies on freshly dissociated chick sensory neurons. These experiments reveal two types of binding sites (Sutter et al., 1979). The first displays a lower affinity for NGF (e.g., nanomolar concentrations saturate the binding sites), while the second displays a higher affinity for NGF (e.g., picomolar concentrations saturate the binding sites). In fact, there are two different types of NGF receptor that are associated with these binding kinetics, and each one has now been isolated.

The high-affinity receptor was discovered through a series of interesting observations (**Figure 7.15**). Initially, it was found that NGF exposure induces rapid phosphorylation of proteins on their tyrosine residues, suggesting that the receptor might be a kinase (Maher, 1988). Soon after, an oncogene was discovered in human colon carcinoma cells, and this turned out to contain a transmembrane protein with a tyrosine kinase on its cytoplasmic tail (Martin-Zanca et al., 1986). The oncogene apparently results from a genetic rearrangement that

fuses a tyrosine kinase with part of a nonmuscle tropomyosin sequence, leading to its name: tropomyosin related kinase or Trk (later called TrkA). The *trk* proto-oncogene was cloned, and the distribution of its mRNA was examined *in vivo*. The highest levels of expression are confined to the cranial sensory, dorsal root, and sympathetic ganglia (Martin-Zanca et al., 1990). Most importantly, the TrkA protein is a high-affinity binding site for NGF, and the binding event induces tyrosine kinase activity (Kaplan et al., 1991a).

Three high-affinity neurotrophin Trk receptors have now been isolated: TrkA, TrkB, and TrkC (Figure 7.13). The last two were discovered by taking the *trk* sequence and performing low-stringency binding screens with cDNA libraries. In this manner, two sequences were isolated that encoded for 145kD receptor tyrosine kinases, named TrkB and TrkC (Barbacid, 1994). Each Trk receptor has two immunoglobulin-like repeats in the extracellular domain and a tyrosine kinase with auto-phosphorylation sites in the cytoplasmic domain. The extracellular domains are about 50% homologous, but each Trk displays a specific affinity for one or two of the neurotrophins: TrkA is specifically activated by NGF, Trk B is specifically activated by BDNF or NT-4, and TrkC is specifically activated by NT-3. Trk receptors may also be activated in the absence of neurotrophins from within the neurons, a process called *transactivation.* For example, the small neurotransmitter, adenosine, can produce Trk phosphorylation through its G protein-coupled receptor. The transactived Trk is then able to promote survival of PC12 cells and hippocampal neurons in vitro (Lee and Chao, 2001).

Each Trk receptor gene is differentially spliced, resulting in the expression of many isoforms. In fact, the *trkC* genes may encode for up to eight different TrkC receptor proteins. There are a number of truncated Trk receptors (those missing the tyrosine

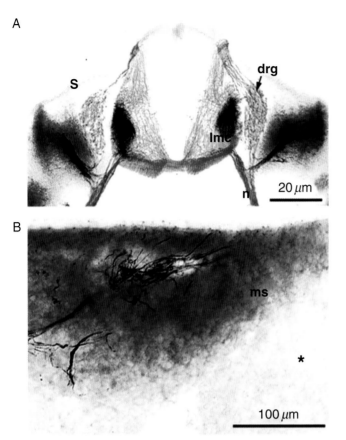

Fig. 7.14 The pattern of NT-3 expression is revealed by a *lacZ* reporter gene, and nerves are counterstained with a neurofilament antibody. A. A transverse section through the thoracic region of an E11 mouse embryo shows DRGs (drg), and peripheral nerves (n). NT-3 expression (blue) is prominently around the DRG and sensory-motor projection. B. Peripheral axons within the fore-limb growing through mesenchyme (ms) are surrounded by NT-3 expression. There is almost no NT-3 expression at the end of the limb (asterisk), which is not yet innervated. Lmc = lateral motor column, s = skin. *(Reprinted from Farinas et al., 1996)*

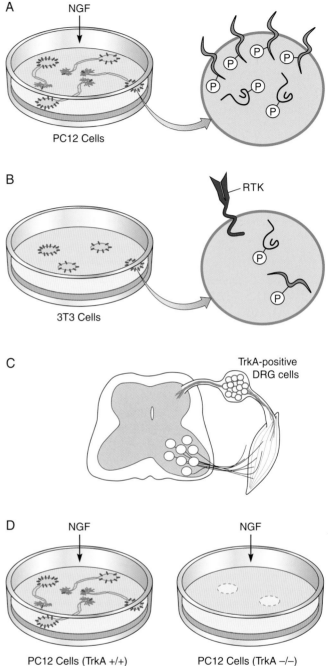

Fig. 7.15 The high-affinity NGF receptor was discovered through a series of disparate observations. A. NGF was found to elicit protein phosphorylation in PC12 cells. B. The oncogene in a cancer cell line was found to be a transmembrane receptor tyrosine kinase (RTK). C. The messenger RNA for this RTK, known as TrkA, was found in extremely high levels in DRG neurons. D. When TrkA was eliminated from PC12 cells, they became unresponsive to NGF. *(Adapted from Maher, 1988; Martin-Zanca et al., 1986; Martin-Zanca et al., 1990; Loeb et al., 1991)*

kinase domain) that are able to bind to their cognate ligand, and these are generally expressed only by glial cells during development. Differential splicing also leads to differences in the extracellular domains, and these can influence ligand binding.

Several lines of evidence indicate that Trk receptors do mediate a survival signal when bound to neurotrophin. A mutant line of PC12 cells that lack TrkA protein are unresponsive to NGF (Figure 7.15), but they can be rescued if they are transfected with expression vectors encoding a full-length rat *trk* cDNA. Perhaps the most compelling evidence is that transgenic mice lacking Trk receptors display extensive death in specific neuron populations (Figure 7.13). TrkA−/− mice exhibit large-scale cell death in sympathetic and dorsal root ganglia, in agreement with earlier experiments that eliminated NGF with function-blocking antibodies (Smeyne et al., 1994). Targeted disruption of the *trkB* gene in mice seems to be particularly devastating in that all mice die within two days of birth. Several peripheral populations are affected, such as the trigeminal, nodose, and vestibular ganglia. In contrast, TrkC receptor deletion leads to a 50% loss of cochlear ganglion neurons but spares most nodose and trigeminal neurons. The developmental effects of eliminating a neurotrophin or

its cognate Trk receptor are not necessarily identical (Klein et al., 1993; Smeyne et al., 1994). For example, the amount of cell death that results from TrkB deletion can be greater than for BDNF disruption because TrkB also serves as a receptor for NT-4/5.

The expression of TrkB and TrkC is widely distributed in the CNS, remaining quite high into adulthood, and some evidence suggests that central neurons may depend on neurotrophin signaling for survival (Barbacid, 1994). Cholinergic cells of the basal forebrain are NGF-sensitive, and exogenous NGF is able to keep them alive when their axons are cut (Gage et al., 1988). A projection from the chick midbrain to its retina also requires target-derived BDNF during development (Von Bartheld and Johnson, 2001). In addition, *trkA−/−* mice have fewer axonal projections from cholinergic basal forebrain neurons to the hippocampus and cortex, suggesting a role in process outgrowth (see Chapter 4). In *trkB−/−* or *trkB/trkC−/−* mice, increased PCD has been reported in the developing hippocampus, cerebellum, cortex, striatum, and thalamus (Minichiello and Klein, 1996; Alcantara et al., 1997; Lotto et al., 2001). However, it is not yet clear whether this PCD is associated with the period of naturally-occurring cell death.

HOW DOES THE NEUROTROPHIN SIGNAL REACH THE SOMA?

Exposure of the growing tips of axons to NGF is sufficient to prevent cell death. This was demonstrated in an elegant tissue culture study where sympathetic neuron cell bodies were placed in a central chamber that was physically isolated from the growth media that bathed the neuritic processes. When grown without NGF, most of the neurons die (**Figure 7.16**A). However, when NGF is provided only to neurites that grow out and reach one of the isolated side chambers, 95% of the neurons survive (Figure 7.16B) (Campenot, 1977, 1982). Many studies demonstrate that NGF and activated TrkA receptors

are internalized, and transported retrogradely to neuron cell bodies in vivo (Hendry et al., 1974; Johnson et al., 1978; Bhattacharyya et al., 1997; Ehlers et al., 1995; Tsui-Pierchala and Ginty, 1999; Watson et al., 1999). In fact, sympathetic neurons die when retrograde axonal transport is disrupted, and the cells can be saved by supplying exogenous NGF (Johnson, 1978).

To test whether retrograde transport of NGF is necessary for survival, rat sympathetic neurons were grown in the presence of NGF that was covalently linked to 1 μm beads to prevent internalization (Figure 7.16C). Once again, only the neuronal processes had access to the NGF. The bead-linked NGF is almost as effective as free NGF: 81% of the neurons survive. In a second test to determine whether retrogradely transported NGF influences survival, a function blocking antibody was introduced in the somata of sympathetic neurons. In this case, only 60% of the neurons survive. Together, these experiments emphasize the importance of local signaling at the distal axon membrane (MacInnis and Campenot, 2002; Ye et al., 2003).

How, then, is the survival signal relayed back to the cell body? The signaling endosome hypothesis suggests that neurotrophin binding stimulates Trk internalization through endocytosis (**Figure 7.17**). The activated Trk recruits certain effector molecules to the endosome, and together they are transported retrogradely to the cell body where they activate transcription (Howe and Mobley, 2005). Trk may be internalized by more than one mechanism; there is evidence both for clathrin-mediated endocytosis, and a pinocytotic mechanism that employs a membrane trafficking protein, called Pincher (Howe et al., 2001; Valdez et al., 2005).

There are several lines of evidence that the retrogradely transported endosome provides an important signal in the soma which is required for cell survival. First, anatomical studies

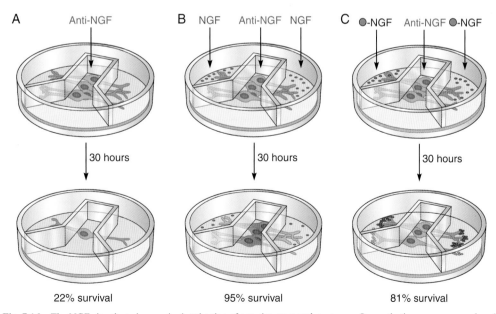

Fig. 7.16 The NGF signal can be acquired at the tips of growing neuronal processes. Sympathetic neurons were placed in a special tissue culture system that permitted the cell bodies and neurites to be bathed in different media. A. Most neurons die when grown in the absence of NGF for 30 hours. B. Neurons can be kept alive by adding NGF only to the compartments with growing neurites. In both cases, an antibody against NGF is added to the central compartment to prevent activation of TrkA. C. To test whether internalized NGF contributes to sympathetic neuron survival, NGF is covalently bound to beads, preventing internalization but permitting local activation of the TrkA receptor. In this case, 81% of the neurons survive for 30 hours. *(Adapted from Campenot, 1977, 1982; MacInnis and Campenot, 2002)*

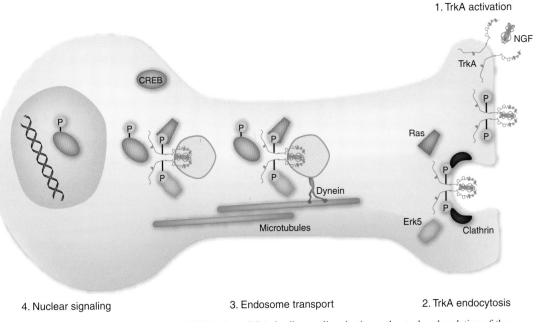

Fig. 7.17 The signaling endosome. (1) NGF binds to TrkA, leading to dimerization and autophosphorylation of the receptor. (2) The ligand-receptor complex then undergoes clathrin-dependent endocytosis. During this period, downstream effector proteins can be recruited to the endocytotic vesicle and activated. (3) The endosome is transported retrogradely along microtubules by a dynein motor. (4) When the endosome reaches the cell body, it provides a signal that can enter the nucleus and control transcription (in this case, phosporylation of CREB).

show that NGF, TrkA, and an effector molecule are colocalized with endosomal markers in the cell body (Delcroix et al., 2003). Second, neuron death increases when retrograde transport of activated TrkA is prevented by interfering with the dynein motor that transports the endosome to the soma (Heerssen et al., 2004). Third, neurons were grown in the chamber illustrated in Figure 7.16, and a specific kinase inhibitor was used to block TrkA-mediated phosphorylation in the cell body compartment (Riccio et al., 1997). This treatment reduces the phosphorylation of an effector kinase and a transcription factor that mediate cell survival (for a description of the molecular pathway see *Intracellular Signaling Pathways that Mediate Survival*, page 188). In fact, TrkA activity can be blocked transiently en route to the soma, and neurons still survive (Ye et al., 2003). This means that TrkA catalytic activity is generated anew when the endosome arrives in the cell body. Therefore. the signaling endosome provides a mechanism for long-range retrograde signaling from the terminal of a process back to the soma.

THE P75 NEUROTROPHIN RECEPTOR CAN INITIATE CELL DEATH

A second neurotrophin receptor was originally described as a low-affinity binding site for NGF. Known as *P75NTR* (75kD neurotrophin receptor), this glycoprotein is a member of the tumor necrosis factor (TNFR) family of receptors (Johnson et al., 1986; Locksley et al., 2001). Unlike Trk receptors, the P75NTR does not have an intracellular catalytic domain, but it can activate several intracellular signaling pathways (**Figure 7.18**A). One intriguing clue about the function of P75NTR comes from its homology to other members of the TNFR family. Each of these proteins has a "death domain"

on the cytoplasmic tail which is similar to the *reaper* gene product in *Drosophila*. The deletion of *reaper* blocks most cell death in the embryonic fly nervous system, and its overexpression in the retina leads to the complete loss of cells (White et al., 1996). As suggested by its homology to the *reaper* gene product, P75NTR activation has been shown to promote PCD in several developing neuronal populations. In P75NTR knockout mice there is less cell death in the retina, the cholinergic brainstem, and the spinal cord mantle zone (Frade and Barde, 1999; Naumann et al., 2002).

Since P75NTR can initiate cell death, there must be a specific receptor to activate pro-apoptotic activity. As it turns out, neurotrophins can be released as uncleaved pro-peptides which are biologically active. Proneurotrophins bind with a high-affinity to the P75NTR and a coreceptor called sortilin. Thus, when sympathetic neurons are grown in the presence of mature NGF, there is virtually no cell death, but the neurons display a significant increase in TUNEL labeling when grown in the presence of proNGF (Figure 7.18B). In a similar manner, blockade of proNGF binding to sortilin decreases cell death in symapthatic neurons (Lee et al., 2001; Nykjaer et al., 2004).

TNFR family members also contribute to neuronal cell death during development. In the developing cortex, some neuroblasts express a TNFR, and neighboring cells express its ligand. Activation of this receptor produces cell death in the primary cultures of cortex neuroblasts. Spinal motor neurons also express a TNFR and its ligand during the embryonic period of normal cell death. When primary cultures of motor neurons are deprived of trophic factor, they increase expression of a TNF (FasL) which results in their death. Furthermore, cultures can be kept alive when grown in the presence of a compound that blocks TNFR activation (Cheema et al., 1999;

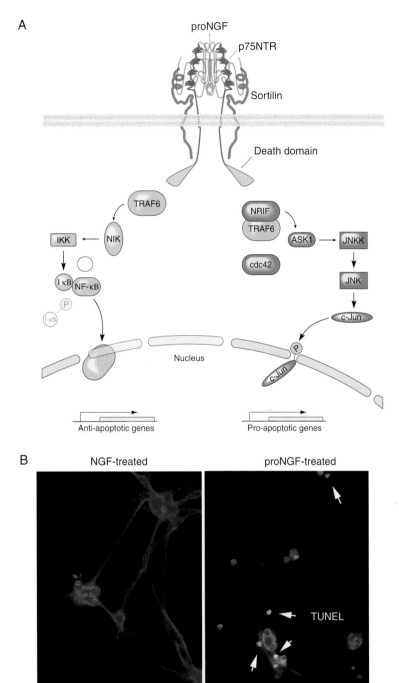

Fig. 7.18 The P75NTR influences neuron survival. A. The schematic shows two intracellular pathways activated by the P75NTR. Proneurotrophins (e.g., proNGF) bind with high affinity to the P75NTR and its coreceptor, sortilin, and this can mediate either survival or PCD. Each is mediated by adaptor proteins that bind to the p75NTR cytoplasmic tail which contains a death domain. Survival involves the recruitment a cytoplasmic adaptor protein that transduces signals from several members of the TNFR family (TRAF6), which leads to the release of a transcription activator, nuclear factor kB (NF-kB). NF-kB enters the nucleus and regulates the transcription of genes that promote survival. PCD involves the recruitment of TRAF6 and neurotrophin receptor interacting factor (NRIF) which lead to the activation of the c-Jun kinase (JNK). JNK phosphorylates the transcription factor, c-Jun, which enters the nucleus to promote the transcription of pro-apoptotic genes. B. The images show two sympathetic neuron cultures grown either in the presence of NGF (left) or proNGF (right). The NGF promotes survival and axon outgrowth, whereas the proNGF induces more PCD, as assessed with TUNEL labeling (green dots) through activation of the P75NTR. *(Adapted from Lee et al., 2001)*

Raoul et al., 1999; Barthélémy et al., 2004). Even NGF-dependent neurons have been shown to express TNF which participates in their death in vivo. Thus, TNFα-/- mice display fewer pyknotic neurons and more living cells in sympathetic ganglia (Barker et al., 2001).

Alternate signaling pathways can be activated following neurotrophin binding to P75NTR and sortilin. Each is mediated by adaptor proteins that bind to the P75NTR cytoplasmic tail. As a cell death signal, P75NTR employs the c-Jun N-terminal kinase (JNK) and its target (the transcription factor, c-Jun) to activate pro-apoptotic genes (Figure 7.18A). This pathway is mediated by two adaptor proteins, TRAF6 and NRIF (Gentry et al., 2004; Linggi et al., 2005). The survival-promoting influence of P75NTR may employ a transcription activator called nuclear factor κB (NF-κB). The phosphorylation of a NF-κB inhibitor, IκB, results in its degradation, liberating NF-κB which enters the nucleus. There, NF-κB activates gene transcription that promotes sensory neuron survival (Hamanoue et al. 1999) (Figure 7.18A).

The relative amount of ligand type (proneurotrophin and neurotrophin), and receptors (Trks and P75NTR) may determine whether life or death occurs. Rat brain oligodendrocytes grown in culture express the P75NTR receptor, but not TrkA, and NGF treatment kills the majority of cells (Casaccia-Bonnefil et al., 1996). In the developing retina, the depletion of endogenous NGF with antibody results in better survival of retinal neurons. The ability of NGF to kill retinal neurons is apparently mediated by P75NTR because antibodies against this receptor prevent the cell death (Frade et al., 1996). In cultured sympathetic neurons that express both TrkA and P75NTR, BDNF exposure causes cell death because it selectively activates the P75NTR receptor, but cannot activate TrkA (Bamji et al., 1998).

Many experiments have examined the effect of mature neurotrophins only because it was not yet known that propeptide was a high affinity ligand. These studies suggest that the P75NTR can also improve the chance for neuron survival during development. For example, P75NTR collaborates with Trk receptors to enhance ligand binding and phosphorylation (Chao, 1994). Cutaneous sensory trigeminal neurons cultured from P75NTR knockout mice require a fourfold greater concentration of NGF in order to survive (Davies et al., 1993). Consistent with this survival-promoting role, the complete loss of P75NTR function in mice leads to a 50% reduction in lumbar DRG neuron number (von Schack et al., 2001).

Finally, P75NTR may be involved in adult degenerative disorders. For example, the beta amyloid peptide also binds to the P75NTR, and this may contribute to degeneration of cholinergic brainstem neurons in Alzheimer's disease (Yaar et al., 1997). Furthermore, proNGF expression is increased twofold in the parietal cortex of Alzheimer's patients (Fahnestock et al., 2001).

CYTOKINES ACT AS NEURON SURVIVAL FACTORS

As research on the NGF and its receptors accelerated, it was difficult to understand why other survival factors had not been found for the many other cells that die during development. Whereas scientists were vigorously searching for even a single endogenous neuron survival factor in 1980, there is now evidence that many families of factors and receptors influence survival. Unfortunately, the trophic influence of most factors has been tested in relatively few brain regions, and these tests are seldom confirmed with in vivo manipulations. Of the neurons that have been investigated, most are influenced by more than one trophic factor, and the array of factors (or receptors) can vary during the course of development.

Several cytokines have been found to keep neurons alive in dissociated primary culture. Cytokines are a diverse family of secreted proteins that were originally discovered as growth factors in lymphocyte cultures, and many of these have turned out to have a primary role in neuron survival. The names of individual cytokines derive from the first biological activity that they were discovered to have, such as killing tumors (TNF, tumor necrosis factor) or promoting mitosis of hematopoietic stem cells (CSF, colony stimulating factor).

One of the most thoroughly studied cytokines is ciliary neurotrophic factor (CNTF), which binds to an intrinsic membrane protein, called CNTFRα, and recruits two other transmembrane proteins (gp130 and LIFRβ) to form the β subunit of the receptor complex (**Figure 7.19**A). The α subunit provides specificity to the trimeric receptor, while the β subunits are responsible for signal transduction (Sleeman et al., 2000). When the receptor complex forms, a tyrosine kinase (member of the Jak family) that is associated with the cytoplasmic tail of each β subunit becomes activated, and phosphorylates a DNA-binding protein that translocates to the nucleus and activates transcription (Bonni et al., 1993, p. 91). CNTF has been shown to support the in vitro survival of autonomic, DRG, hippocampal, and motor neurons. However, CNTF can also act as a pro-apoptotic signal. Blockade of CNTF signaling in vivo can reduce naturally-occurring PCD in the retinal layer containing rods (Elliott et al., 2006).

The TGF-β family of cytokines and their receptors seem to be involved primarily in promoting the death of sympathetic, sensory, motor, and retinal neurons. When all three isoforms of TGF-β were neutralized with antibody treatment in vivo, virtually all of the normally occurring cell death was prevented. Mice with a deficiency in two TGF-β receptors also displayed less retinal PCD (Krieglstein et al., 2000; Dünker and Krieglstein, 2003). TGF-β signaling also controls the number of glial cells in the peripheral nervous system. When the TGF-β receptor is conditionally deleted only in Schwann cells, there is a significant reduction in their developmental PCD (D'Antonio et al., 2006).

Motor neuron survival factors are of particular interest because naturally-occurring cell death is well-characterized in this population, and is closely linked to the target (Figures 7.5 and 7.6). When chick embryos are treated with human recombinant CNTF, half of the naturally-occurring motor neuron death is prevented (Oppenheim et al., 1991). Surprisingly, parasympathetic, sympathetic, and sensory neuron cell death is unaffected. Although CNTF knockout mice display little effect on cell survival during development, including motor neurons, the functional loss of CNTF receptors does increase cell death. Loss of CNTFRα increases motor neuron death by about a third, and similar observations have been made on LIFRβ and gp130 knockout mice (DeChiara et al., 1995; Li et al., 1995; Nakashima et al., 1999). This result implies that there is at least one target-derived cytokine that supports motor neuron survival through activation of the CNTFRα. In fact, cardiotrophin-1 (CT-1), a cytokine that is expressed

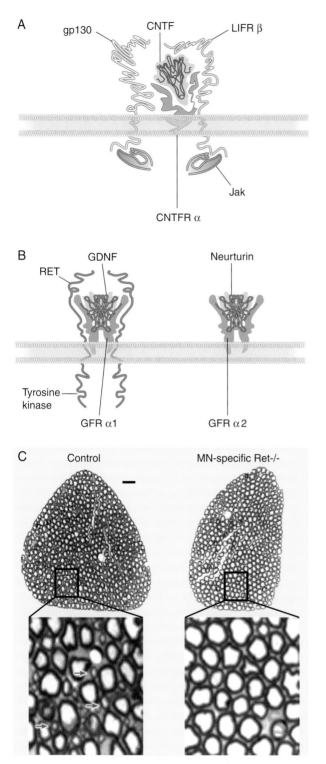

Fig. 7.19 Survival factors and receptors in the nervous system. A. Cytokine signaling: CNTF binds first to an intrinsic membrane protein called CNTFRα. This event causes two other transmembrane proteins, gp130 and LIFR, to form the β subunit of the receptor complex. The activated receptor complex signals via a tyrosine kinase (Jak) that is associated with the cytoplasmic tail of each β subunit. B. GDNF signaling: GDNF and Neurturin bind first to their cognate GFRα receptor. This complex then recruits the receptor tyrosine kinase, RET, to homodimerize and to become autophosphorylated, leading to intracellular signaling. C. Cross section of a lumbar ventral nerve root showing small (red arrows) and large caliber motor axons in control mice (left). A conditional deletion of Ret in motor neurons (right) during development led to a dramatic and specific loss of the small myelinated motor axons. scale bar=20 μm *(Adapted from Gould et al., 2008)*

in embryonic skeletal muscle, contributes to the survival of embryonic motor neurons. CT-1 deficient mice display 20–40% greater motor neuron loss during the normal period of cell death. However, CT-1 does not bind to CNTFRα, suggesting that an array of factors and receptors keep motor neurons alive during development (Oppenheim et al., 2001; Gould and Oppenheim, 2004).

A second substance, glial cell line-derived neurotrophic factor (GDNF), has been identified as preventing naturally-occurring motor neuron death in vivo. GDNF was initially characterized by its ability to keep midbrain dopaminergic cells alive in vitro. This assay was chosen because Parkinson's disease involves the death of these dopaminergic neurons, and a survival factor may have important therapeutic value (Lin et al., 1993). Four members of the GDNF ligand family have now been isolated, and they all belong to the TGF-β superfamily. Each ligand binds to a specific ligand recognition α subunit (GFRα1 through 4). The GFRα subunits are attached to the membrane by a glycosyl phosphatidylinositol anchor (Figure 7.19B). The ligand-receptor complex becomes associated with a transmembrane tyrosine kinase, called RET, and leads to its activation (Airaksinen and Saarma, 2002).

There is strong evidence that GDNF is an endogenous, target-derived survival factor for at least some motor neurons. GDNF mRNA is found in the limb, and GFRα and RET are expressed by subsets of chick motor neurons during the period of normal cell death (Homma et al., 2003). GDNF treatment prevents naturally-occurring motor neuron death in chick embryos, and over-expression of GDNF in the musculature of mice increased survival in most motor neuron populations. Conversely, motor neuron death was increased in GDNF-, RET-, or GFRα-deficient mice. However, this effect is largely due to the loss of GDNF within the motor axon pathway; that is, GDNF provides a target-independent survival signal (Oppenheim et al., 1995; Oppenheim et al., 2000a; Gould et al., 2008). Furthermore, when cell counts were made on identified motor neuron pools, the effects of disrupting GDNF signaling are highly restricted; only small caliber γ-motor neurons that innervate muscle spindles die (Figure 7.19C). The GDNF signaling pathway influences more than just motor neuron survival. Mice lacking either GDNF, Neurturin (a second ligand), GFRα1, GFRα2, or RET also exhibit specific loss of parasympathetic and enteric neurons (Huang and Reichardt, 2001). A third motor neuron trophic factor has been reported, called hepatocyte growth factor (HGF) which also influences the survival of only a subset of motor neurons. In the chick, only lumbar motor neurons are dependent on HGF for their survival (Ebens et al., 1996; Yamamoto et al., 1997; Novak et al., 2000).

Despite the increase in the number of candidate growth factors, their elimination during development has surprisingly minor effects on the survival of CNS neurons. One hypothesis is that central neurons, unlike peripheral ganglion cells, have multiple targets and afferents, perhaps giving them access to many different growth factors during development. The prediction from this hypothesis is that one must eliminate two or more growth factors or receptors in order to disrupt survival. It is also likely that many survival factors have yet to be identified. For example, the survival of embryonic retinal ganglion cells is enhanced by tectal cell-conditioned media in a manner that cannot be duplicated by CNTF or the neurotrophins (Meyer-Franke et al., 1995).

185

A final consideration is that many trophic factors also promote certain aspects of differentiation (see Chapters 2–4), and progress along these pathways may be entangled with the decision to live or die. An interesting example of this occurs in the developing fly eye. The epidermal growth factor receptor (EGFR) mediates a survival signal from a nearby cluster of postmitotic cells in the fly retina. In its absence, the number of ommatidial cells declines significantly. However, the EGFR is also playing an important role in progression through the cell cycle during this same period (Baker and Yu, 2001).

HORMONAL CONTROL OF NEURON SURVIVAL

Endocrine signaling controls many aspects of development, and this includes PCD. The most intensively studied endocrine signals are the steroid hormones because of their role in sexual differentiation. Several brain structures are quantitatively different in males and females of the same species, referred to as *sexual dimorphism*. These sexual dimorphisms are thought to arise from regional differences in the amount of steroid hormones or their receptors during development. Steroid hormones (e.g., estrogens and androgens) are lipid-soluble molecules that are released by the gonads. They bind to cytoplasmic receptors, and these receptors can translocate to the nucleus where they regulate gene transcription (see *Hormonal Signals* and *Hormonal Control of Brain Gender*, Chapter 10, page 337).

The first brain structure shown to differ anatomically between males and females was the sexual dimorphic nucleus of the preoptic area (SDN-POA). The nucleus is much larger in males, and this depends on intact gonads during early development. The number of cells can be greatly reduced by castrating genetic males within a few days of birth, and testosterone-treated females develop an enlarged nucleus (Gorski et al., 1978). To measure programmed cell death, the SDN-POA was labeled for fragmented DNA (Figure 7.3B). Although both the male and female SDN-POA display a period of cell death postnatally, it persists for a longer time in females (Davis et al., 1996). Since testosterone can be converted enzymatically to estradiol, female pups treated with testosterone display fewer intracellular effectors of apoptosis and less DNA fragmentation (Arai et al., 1996; Tsukahara et al., 2008). Therefore, the sexual dimorphism appears to arise from the protective influence of estradiol in the male SDN-POA. Not surprisingly, other sexual dimorphisms have been reported amongst hypothalamic nuclei that are associated with sex-specific behaviors, and many of these are generated by hormonal regulation of cell death (Bleier et al., 1982; Chung et al., 2000; Forger et al., 2004).

A particularly well-studied system is the lumbar spinal cord of male rats, where there is a motor nucleus that innervates striated muscle of the penis: the spinal nucleus of the bulbocavernosus (SNB). As one might expect, this nucleus and the muscles that it innervates are present in adult male rats, but are nearly absent females. This sexual dimorphism arises from the selective loss of motor neurons in female rats during development. During the first 10 postnatal days, the total number of SNB neurons decreases by nearly 70% in females, but only by 30% in males (**Figure 7.20**). When neonatal females are treated with the steroid hormone, testosterone,

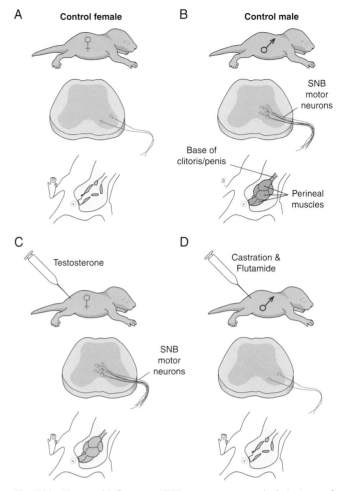

Fig. 7.20 Hormonal influence on SNB motor neuron survival. A. Areas of the lumbar spinal cord that innervate perineal muscles in male rats are nearly absent in female rats, owing to motor neuron death during development. B. Male rats exhibit a greater number of SNB motor neurons. C. When neonatal female rats are treated with the androgen steroid hormone, testosterone, SNB motor neuron death is decreased. D. In contrast, when neonatal males are castrated and reared with an androgen antagonist, flutamide, SNB motor neurons display increased cell death. The target muscle size is affected by each treatment and appears to explain the effect of hormone. *(Adapted from Nordeen et al., 1985; Breedlove and Arnold, 1983)*

the amount of cell death is decreased and resembles the pattern seen in males (Breedlove et al., 1982; Nordeen et al., 1985). When males are castrated and reared with an androgen antagonist, flutamide, their SNB neurons are as scarce as in female rats (Breedlove and Arnold, 1983). Similarly, male rats with a mutation of the androgen receptor display the female pattern of cell loss (Breedlove and Arnold, 1981; Sengelaub et al., 1989). SNB motor neurons do not express androgen receptors during the time when androgens spare them from death. In fact, androgen signal probably keeps male SNB neurons alive by promoting survival of the target muscles, and preserving target-derived motor neuron survival factors. Thus, when females are treated with androgen, but also receive a treatment that blocks trophic factor signaling, there is very little sparing of SNB motor neuron number (Freeman et al., 1997, Xu et al., 2001).

A role for endocrine signaling in sexual dimorphism is also well-studied in the telencephalic nuclei of songbirds: those species where males learn to produce mating calls, while females vocalize little, if at all. In canaries and zebra finches, at least three areas of the brain that support song production are much larger in males than females (Nottebohm and Arnold, 1976). In some nuclei such as the robustus nucleus of the archistriatum (RA), which shares some features with motor cortex, the differences in neuron number arise from a selective loss of cells in the female (Konishi and Akutagawa, 1985; Nordeen and Nordeen, 1988; Kirn and DeVoogd, 1989). What is the evidence that steroid hormones influence cell survival in males? If developing females are treated with testosterone or its active metabolite, estradiol, then the number of neurons in RA becomes masculinized, and the birds acquire male-like vocalizations (Gurney, 1981). This effect is mediated by androgen receptor since it can be prevented with a specific antagonist, flutamide (Grisham et al., 2002). More recently, the idea that steroid hormones can account for sexual dimorphism of songbird vocal nuclei has been challenged. For example, when genetic females are "engineered" to grow testicular tissue that secretes androgens, their vocal nuclei do not become masculinized (Wade and Arnold, 1996). Therefore, there are strong reasons to think that steroid hormones play a role in control of cell number in male and female songbirds, but these hormones may be produced locally within the sexually dimorphic brain regions.

Although the impact of steroid hormones on naturally-occurring cell death has not been examined quantitatively in the majority of brain regions, it may turn out to be relatively common. In the peripheral nervous system, cell death in the sympathetic ganglion that innervates the face is greater in female rats than in males. Furthermore, castration of neonatal male rats significantly increases the number of dying neurons, and treatment of neonatal females with testosterone improves survival (Wright and Smolen, 1987). Male-female differences in cell death have also been found at the level of cortex. The period of normal cell death in rat primary visual cortex ends at about P15 for rats, but extends to P28 for females; males end up with 19% more neurons. The hormonal mechanism that is responsible for this difference appears to be ovarian estrogen. Removal of the female ovaries at P20 is sufficient to produce the male phenotype (Reid and Juraska, 1992; Nuñez et al., 2001; Nuñez et al., 2002). Reports indicate that cell death in the hippocampus, locus coeruleus, and medial amygdala may also be regulated by circulating steroid hormones during development (Guillamon et al., 1988; Morris et al., 2005; Spritzer and Galea, 2007; Morris et al., 2008).

Several nonsteroidal hormones can also influence brain development. One essential signal is thyroid hormone (TH), and a deficiency during development can produce a range of sensory and cognitive deficits (Zoeller and Rovet, 2004). The circulating TH, thyroxine, crosses cell membranes where it is converted to a ligand that binds to members of the superfamily of ligand-dependent transcription factors, which include receptors for steroids and retinoids. Depending on the developmental time and anatomical location, thyroid hormone receptors (TR) regulate a variety of genes (Dong et al., 2005; Takahashi et al., 2008). Therefore, it is not too surprising that thyroxine has been reported to act as both a

pro- and anti-apoptotic signal. The surge of thyroxine that initiates metamorphosis in tadpoles causes lysosomal activity to increase in motor neurons that innervate regressing tail musculature, leading to their death (Decker, 1976; 1977). Thyroxine exposure also elicits cell death in a region of the adult zebra finch forebrain that is involved in song production (Tekumalla et al., 2002).

Circulating thyroxine may also be necessary for cell survival in the mammalian cerebellum, hippocampus, and cortex. When hypothyroidism was induced chemically in neonatal rats, the level of PCD was found to increase in somatosensory cortex using TUNEL labeling. Interestingly, the loss of thyroxine was associated with increased expression of pro-NGF, p75NTR, and other pro-apoptotic factors (Kumar et al., 2006). Despite the profound effects of severe hypothyroidism, TR knockout mice have a surprisingly mild phenotype. This led to the theory that ligand-free TRs interact with corepressors to inhibit transcription while TH binding facilitates interaction with coactivators and stimulates transcription. To test this idea, transgenic mice were generated with a TR that could not bind ligand, and nervous system development was found to be disrupted, similar to TH deficiency models (Hashimoto et al., 2001).

Two other hormones have been implicated as survival factors in the central nervous system. Growth hormone (GH) released from the pituitary promotes growth and differentiation, but it can also be synthesized in nonpituitary sites. The developing chick retina expresses GH during the period of normal cell death, and immunoneutralization of this endogenous hormone increases the number of TUNEL-positive cells in the ganglion cell layer (Sanders et al., 2005). Circulating GH acts directly on many organs to stimulate Insulin-like growth factor 1 (IGF-1) production, with the liver providing the main source of circulating IGF-1. However, IGF-1 can also be synthesized locally in the brain, and may modulate cell survival. By genetically increasing the expression of IGF-1 in mice, it was possible to inhibit naturally-occurring neuron death in cerebral cortex, and these surviving cells persisted into adulthood (Hodge et al., 2007).

Finally, endocrine signals are responsible for the extensive remodeling of the insect nervous system that occurs during metamorphosis. Many neurons are required only at specific times during insect development; some are necessary for larvae, while others are required for the process of metamorphosis from caterpillar to pupa to moth. A surge of the steroid hormone, 20-hydroxecdysone (20E), triggers each molt, during which the larva sheds its cuticle and grows. The 20E also acts as a trigger for normal cell death. For example, a small increase of 20E triggers pupa formation in moths, and simultaneously brings about the death of specific abdominal motor neurons (Hoffman and Weeks, 1998). Later in development there is another prolonged surge of 20E that causes the adult moth to emerge from its pupa. In this case, the survival of many neurons becomes dependent on 20E, and as the level falls nearly 40% of abdominal neurons are lost (Truman and Schwartz, 1984; Zee and Weeks, 2001). In fruit flies, there are about 300 neurons in the ventral cord that express much higher levels of the 20E receptor (EcR) which is similar to the thyroid receptor. As in moths, these cells become extremely dependent on 20E and die at metamorphosis when 20E levels drop (Robinow et al., 1993). This period of cell death can be delayed by treatment with 20E in both the moth and fruit fly. The molecular signals that mediate ecdysone-induced cell death are quite similar to those found in other systems (Hoffman and Weeks, 2001; Buszczak and Segraves, 2000; Choi et al., 2006). The EcR also plays an important role in remodeling of the larval fly eye as it turns into the adult "eyelet." During metamorphosis, all the green-sensitive photoreceptors are killed by EcR activation, while blue-sensitive photoreceptors are induced to switch fates (Sprecher and Desplan, 2008).

CELL DEATH REQUIRES PROTEIN SYNTHESIS

One might suppose that when a neuron is deprived of a trophic factor, it fails to maintain normal levels of synthesis and metabolism, and simply expires. In fact, developing neurons participate actively in their own death by expressing genes and synthesizing proteins that damage cell structure and function. That is, they "commit suicide." Is it possible that neuron cell death can actually be prevented by blocking RNA or protein synthesis?

As described above, embryonic sympathetic neurons can survive in vitro when grown in the presence of NGF. When NGF is removed from the culture media, few neurons remain after two days. Therefore, the first experiment asked whether inhibitors of RNA or protein synthesis could save NGF-deprived neurons (**Figure 7.21**). Actinomycin-D blocks transcription by binding to DNA and preventing the movement of RNA polymerase, while cycloheximide prevents translation by blocking the peptidyl transferase reaction on ribosomes. Each of these treatments completely rescue sympathetic neurons following NGF deprivation, demonstrating that new RNA and proteins must be manufactured to bring about cell death. To determine when the harmful phase of translation occurs, cycloheximide was delivered at several times after NGF deprivation, and it was found that the cell death-promoting proteins are produced at about 18 hours (Martin et al., 1988). If all the molecular machinery for cell death was present in the cytoplasm, one would expect that neurons would be committed to die within a few hours.

To determine whether mRNA and protein synthesis are general features of cell death in vivo, animals were treated with synthesis inhibitors at the age when neurons are normally lost (Figure 7.21). When chick embryos are treated with either cycloheximide or actinomycin D on embryonic day 8, the time of maximum motor neuron and DRG cell death, they exhibit a striking reduction in the number of dying neurons (Oppenheim et al., 1990). Similarly, the cell death that occurs in response to declining levels of 20-HE in moths can be reduced by RNA or protein synthesis inhibitors (Fahrbach et al., 1994). These studies suggest that in the absence of a trophic signal harmful proteins may be synthesized. The discovery of these proteins is discussed below.

Perhaps the most compelling evidence for gene transcription came from genetic studies of cell death in the nematode, *Caenorhabditis elegans* (Metzstein et al., 1998). About 10% of cells die during development in *C. elegans*, most of them being neurons, but inactivation of two specific genes (called *ced-3* and *ced-4*) rescues all of these cells, including neurons (discussed in more detail below).

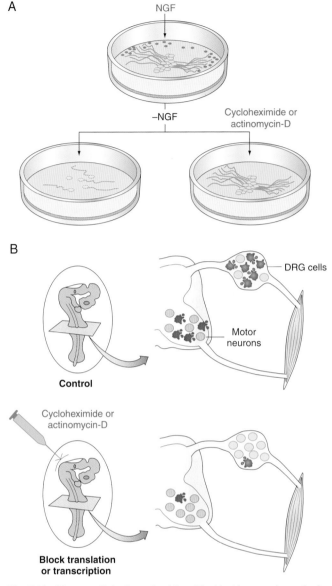

Fig. 7.21 Neuron cell death can be delayed by blocking protein synthesis. A. Sympathetic neurons die within two days, when NGF is removed from the culture media (left). When either a translation blocker (cycloheximide) or a transcription blocker (actinomycin-D) was added to NGF-deprived cultures, sympathetic neurons were rescued (right). B. The synthesis of mRNA and protein is also required for naturally-occurring cell death in vivo. In control chick embryos, pyknotic motor neurons and DRG neurons are counted during normal development (top). When chick embryos are treated with either cycloheximide or actinomycin-D during the time when motor neuron and DRG cell death is at its greatest, the number of pyknotic neurons is decreased (bottom). *(Adapted from Martin et al., 1988; Oppenheim et al., 1990).*

INTRACELLULAR SIGNALING PATHWAYS THAT MEDIATE SURVIVAL

A full description of the molecular pathways that lead to PCD can quickly overpower the reader with a long list of acronyms. In simplified form, the developing neuron contains two types of proteins in its cytoplasm: those that maintain survival and those that mediate death. When a survival factor binds to its receptor, many anti-apoptotic proteins are activated by phosphorylation, and their expression may increase. When survival factors are withdrawn, pro-apoptotic proteins are activated by phosphorylation, and their expression increases. Therefore, the vital purpose of a survival factor is to upregulate the function and/or expression of pro-apoptotic proteins and suppress the anti-apoptotic proteins. Many of the intracellular signaling pathways induced by growth receptors have now been identified, in large measure from studies of cancer cell growth. To make this discussion somewhat manageable, we will focus on the cytoplasmic signals that have been associated with the survival of neurons during normal development.

Neurotrophin signal transduction involves the sequential recruitment and activation of several intracellular proteins, leading to the modification of existing proteins and regulation of gene transcription. Many components of this intracellular pathway are identical to those that are recruited by receptor tyrosine kinases that were discussed earlier (e.g., *sevenless*, see Figure 4.9). When NGF binds to TrkA, it first induces receptor dimerization, followed by the rapid phosphorylation of five tyrosine residues on the cytoplasmic tail by neighboring Trk receptors (Kaplan et al., 1991b). The phosphorylation sites on the Trk receptors serve as docking sites for adaptor proteins, such as Shc, which are themselves phosphorylated by TrkA (**Figure 7.22**). In fact, the Shc adapter protein is critical to the survival of NT-4-dependent sensory neurons in vivo (Minichiello et al., 1998).

As shown in **Figure 7.23**, one intracellular pathway becomes activated upon the recruitment of a second adaptor protein (Grb2, growth factor receptor-bound protein-2) and docking protein (Gab1, Grb2-associated Binder-1). In this pathway, a phosphotidylinositol 3-kinase (PI3K) activates a serine/threonine kinase, called Akt.

The PI3K-Akt pathway provides a crucial intracellular signal for keeping NGF-dependent neurons alive. Specific blockade of either PI3K or Akt activity kills primary cultures of sympathetic neurons even in the presence of NGF, while constitutive activation of either kinase results in neuron survival even in the absence of NGF (Crowder and Freeman, 1998).

There is a second major pathway that also leads to phosphorylation of cyoplasmic components and that participates in cell survival. In this case, the adaptor and docking proteins

recruit a guanine nucleotide exchange factor (SOS) to activate a membrane-associated GTPase called Ras (Figure 7.23). Ras is activated when it exchanges a GDP for a GTP, whereupon it binds to and activates a serine/threonine kinase, called Raf. Raf activation leads to activation of the mitogen-activated protein kinase cascade (MAPK), eventually resulting in transcriptional regulation (Figure 7.23). Activated Ras can also engage the PI3K-Akt pathway, discussed above, to promote neuron survival (Vaillant et al. 1999).

The potential importance of Ras-Raf signaling to neuron survival is demonstrated in experiments where the Ras protein is injected directly into cultured chick DRG neurons. Although DRG neurons depend on NGF for their survival, the Ras protein is sufficient to prevent cell death (Borasio et al., 1989). Furthermore, the same treatment promotes the survival of BDNF-dependent nodose ganglion neurons and CNTF-dependent ciliary ganglion neurons.

Which MAPKs are responsible for the positive effects of NGF binding? NGF promotes the survival of PC12 cells, and this is accompanied by activation of several MAPKs. Within six hours of NGF withdrawal, PC12 cells begin to die in great numbers, and there is a prominent decrease in the activity of an extracellular signal-regulated kinase (ERK). To test whether ERK is responsible for the positive effect of NGF, PC12 cells were engineered to produce constitutively high levels of ERK activity (**Figure 7.24**). These cells survived much better following NGF withdrawal (Xia et al., 1995).

How does activation of the two major kinase pathways, PI3K-Akt and MAPK, prevent cell death? Akt can promote cell survival by inactivating pro-apoptotic proteins in the cytoplasm and increasing transcription of anti-apoptotic proteins. For example, RSK phosphorylates, and inactivates, one of the pro-apoptotic regulatory proteins (see *Bcl-2 Proteins: Regulators of Programmed Cell Death*, page 194). The Ras-MAPK pathway promotes survival in the fly using a similar strategy. When active Ras is ectopically expressed in the developing eye, it decreases the expression of pro-apoptotic proteins and prevents normal cell death (Kurada and White, 1998).

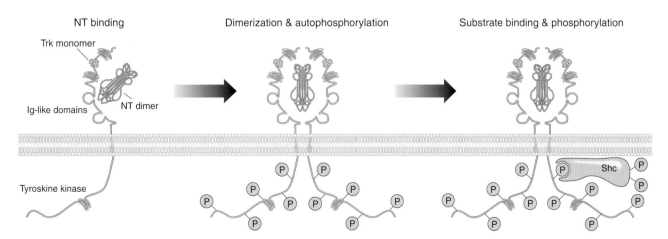

Fig. 7.22 Neurotrophin-Trk receptor interaction. The biologically active forms of neurotrophins are dimers of identical 13 kDa peptide chains. The neurotrophin dimer binds to the Trk protein (left). The binding induces 2 Trk receptors to dimerize. The ligand-receptor complex leads to transphosphorylation of tyrosine residues on the cytoplasmic tail of neighboring receptors (middle). Cytoplasmic adaptor proteins (Shc) bind to specific phosphorylation sites on the cytoplasmic tail, and these substrates are then phosphorylated (right).

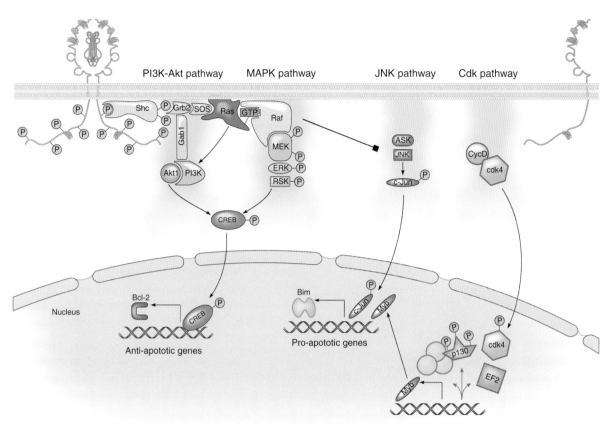

Fig. 7.23 The intracellular signaling pathways associated with the presence (left) or absence (right) of NGF. The first pathway leads to activation of the Akt kinase. Two proteins (Grb-2, Gab-1) are recruited to the receptor complex, resulting in the activation of a phosphatidylinositol-3 kinase (PI3K). PI3K generates phosphoinositide phophases that activate the serine/threonine kinase Akt. The second pathway leads to activation of the mitogen-activated protein kinase (MAPK) cascade. In this pathway, a membrane-associated G protein, Ras, first becomes activated through GDP-GTP exchange, mediated by a guanine nucleotide exchange factor (Sos). The activated Ras phosphorylates several substrates, including a serine/threonine kinase (Raf). Active Raf initiates the MAPK cascade, concluding with the activation of a ribosomal S6 protein kinase (RSK). Both Akt and RSK can phosphorylate the transcription factor, cyclic AMP response element binding protein (CREB). The phospho-CREB then enters the nucleus where it can increase the expression of anti-apoptoptic proteins, such as Bcl-2. In the absence of NGF, two pro-apoptotic pathways can be activated. The first one involves the sequential activation of two kinases (ASK, JNK), and the phosphorylation of a transcription factor (c-Jun) which then enters the nucleus where it can increase the expression of pro-apoptoptic proteins, such as Bim. The second pathway involves the activation of cyclin-dependent kinase 4 which enters the nucleus and hyperphosphorylates p130, which tethers proteins (including the promotor binding factor, E2F) to a promoter and represses transcription. When the protein complex dissociates, the transcription factor, Myb, is expressed. Along with c-Jun, Myb can induce the transcription of pro-apoptotic proteins, such as Bim.

The PI3K-Akt and MAPK pathways also promote survival at the level of gene transcription. The transcription factor, cyclic AMP response element binding protein (CREB), is phosphorylated by RSK2. The activated phospho-CREB enters the nucleus and binds to DNA where it can increase the expression of both neurotrophins and anti-apoptotic regulatory protein that are discussed below (Tao et al., 1998). This mechanism supports the in vitro survival of both sympathetic neurons through TrkA activation, and cerebellar granule cells through TrkB activation (Xing et al., 1996; Riccio et al., 1999; Bonni et al., 1999).

The extent to which CREB-dependent transcription contributes to neuron survival in vivo is not entirely clear. In a CREB knockout mouse, sensory neurons die in greater numbers during the period when they depend on neurotrophins, while sympathetic neuron death increases prior to neurotrophin dependence because the cells do not migrate properly (Lonze

et al., 2002). In fact, when CREB inactivation was restricted to noradrenergic sympathetic neurons, their migration was normal. Surprisingly, their survival was enhanced. To reconcile these findings, it was suggested that the loss of CREB signaling within sympathetic neurons led to the down-regulation of a pro-apoptotic protein such as p75NTR (Parlato et al., 2007). In any event, other transcription pathways are likely to influence cell survival. For example, Akt phosphorylates a transcription factor, called forkhead (FOXO), preventing it from translocating to the nucleus and upregulating Fas ligand (Brunet et al., 1999).

The intracellular signals that promote survival are also able to suppress the pathway that leads to death, discussed below. For example, the Ras-Raf pathway can block the activation of a transcription factor (c-Jun) that promotes cell death (Figure 7.23). Thus, when sympathetic neurons are transfected with constitutively active Ras, the level of c-Jun declines, and the

A

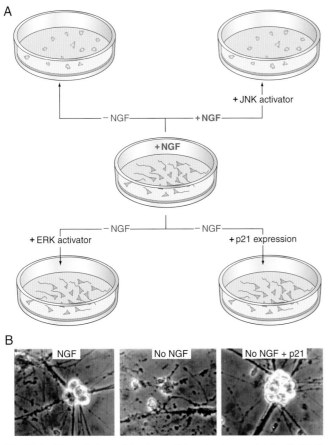

B

Fig. 7.24 Manipulations of the intracellular signals associated with NGF. A. When NGF is withdrawn from the culture media, neurons die (top left). NGF is not able to support neuron survival when the MAP kinase, JNK, is constitutively activated (top right). In contrast, when ERK is constitutively activated, neurons are able to survive even in the absence of NGF (bottom left). Similarly, neurons are able to survive without NGF when p21 is expressed with a viral vector. B. Phase contrast images of sympathetic neuron cultures with NGF (left), without NGF (middle), or with expression of the cyclin-dependent kinase inhibitor, p21, in the absence of NGF (right). *(Adapted from Xia et al., 1995; Park et al., 1997)*

neurons survive in the absence of NGF. This pathway appears to play a role in the survival mechanism during in vivo maturation. The genetic elimination of p53, which is known to be a downstream effector of c-Jun, results in increased survival of SCG neurons (Mazzoni et al., 1999; Aloyz et al., 1998; Kanamoto et al., 2000).

Although we have tried to present a unified picture of cell survival signaling, it is likely that different neurons will employ specific cytoplasmic mechanisms. For example, Ras appears to be a principal cytoplasmic signal for NGF signaling in DRG neurons. However, it is also known that NGF-dependent sympathetic neurons from caudal regions of the nervous system are *not* saved by Ras injection, whereas sympathetic neurons from the rostral SCG are saved (Markus et al., 1997a).

The many molecular pathways between receptor and nucleus interact with one another, producing another level of complexity. For example, Ras is able to activate both the PI3K-Akt and MAPK pathways (Figure 7.23). Therefore, it will be crucial to understand how receptors are activated in vivo if we are to evaluate the contribution of each of the cytoplasmic effectors.

INTRACELLULAR SIGNALING PATHWAYS THAT MEDIATE DEATH

When neurotrophin-dependent neurons are deprived of NGF, a different kinase signal, called JNK, becomes activated; this pathway leads to the phosphorylation of a transcription factor, c-Jun, which enters the nucleus and increases the transcription of pro-apoptotic genes, ultimately leading to cell death (Figure 7.23). In fact, PC12 cell death will occur in the presence of NGF if JNK is constitutively activated (Figure 7.24), and rat spinal motor neurons can be kept alive in vitro by blocking JNK activity (Maroney et al., 1998). Similarly, experimental reduction of c-Jun is sufficient to rescue NGF-deprived neurons, and constitutive overexpression of c-Jun is sufficient to kill neurons in the presence of NGF (Ham et al., 1995).

It appears that c-Jun is required for sympathetic neuron PCD that attends neurotrophin withdrawal, but not the cell death associated with p75NTR activation. When c-Jun was conditionally deleted in sympathetic neurons, they were able to survive NGF withdrawal, but not the activation of p75NTR (Palmada et al., 2002). Thus, the loss of NGF can lead to the up-regulation of proteins needed for cell death via JNK signaling.

A similar death pathway is recruited when cultured motor neurons are deprived of trophic factors. In this case, removal of trophic factors leads to reduced signaling through the PI3K-Akt pathway and the activation of a FOXO transcription factor. The active FOXO causes motor neurons to express the death-promoting TNF, called FasL which is sufficient to kill them (Barthélémy et al., 2004).

A second intracellular pathway that is initiated by NGF withdrawal makes use of the machinery that is normally associated with generation of new cells: cell cycle regulatory proteins. In neurons deprived of trophic support, Cyclin D levels rise, and lead to the activation of Cdk4 (Freeman et al., 1994; Herrup and Busser, 1995). During neurogenesis, these proteins cause neuroblasts to enter the synthesis phase of mitosis (see *What Determines the Number of Cells Produced by the Progenitors?* Chapter 3).

The importance of cell cycle regulatory proteins to PCD has been demonstrated in a host of developing neuronal populations. One approach makes use of Cdk antagonists, and a second takes advantage of the endogenous proteins that bind to and inhibit the Cdks, known as cyclin-dependent kinase inhibitors (CKI). The in vitro blockade of Cdks has been shown to rescue motor, sensory, sympathetic, retinal, and cerebellar neurons in primary culture (Farinelli and Greene, 1996; Park et al., 1996; Park et al., 1997; Markus et al., 1997b; Maas et al., 1998; Padmanabhan et al., 1999; Appert-Collin et al., 2006). Figure 7.24B shows that the expression of a CKI called p21 can rescue sympathetic neurons after NGF withdrawal.

Activated Cdk enters the nucleus and hyperphosphorylates one protein in a complex that acts to repress transcription (Figure 7.23). One of the genes that is derepressed is the transcription factor, Myb, which promotes the expression of Bim, the same pro-apoptotic protein that was induced by the JNK-c-Jun pathway. It is possible to demonstrate that each stage of this pathway is involved in cell death. For example, neurons can survive NGF withdrawal if p130 is eliminated with RNA interference, or if they express p130 protein that is resistant to phosphorylation (Liu et al., 2001; Liu et al., 2005; Biswas et al., 2007). The Bim protein which is expressed in response to Cdk activation does play a role in cell death. Thus, sympathetic neurons cultured from Bim-deficient mice can survive in the absence of NGF (Biswas et al., 2005; Putcha et al., 2001).

CASPASES: AGENTS OF DEATH

The discovery of specific genes that are directly involved in PCD was made in experiments on the nematode, *Caenorhabditis elegans*. In 2002, Robert Horvitz shared the Nobel Prize for Physiology or Medicine for discovering the stereotyped series of intracellular events that constitute apoptosis, the most characteristic form of PCD. Two genes, *ced-3* and *ced-4, must* be expressed by each *C. elegans* cell if it is to die during development (*ced* stands for cell death abnormal) (Yuan and Horvitz, 1990). When either of these genes is mutated, almost all of the PCD is prevented (**Figure 7.25**A). Analysis of mosaic animals (i.e., animals in which the *ced* gene is expressed in only a few identified cells) indicates that the gene product acts within the cell that produces it, showing that PCD proceeds in a cell autonomous manner, rather than a destructive neighbor. What are these gene products, and how do they control the life or death decision?

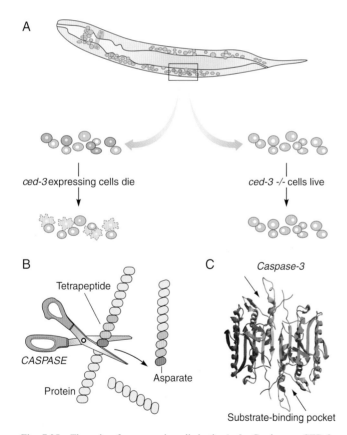

Fig. 7.25 The role of caspases in cell death. A. In *C. elegans*, CED-3-expressing cells (yellow cytoplasm) die during development, but almost all cell death is prevented when the *ced-3* gene is mutated. B. CED-3 is a member of the caspase family, enzymes that recognize a tetrapeptide sequence and specifically cleave proteins after an aspartate residue. C. The structure of Caspase-3 shows that two heterodimers assemble into a single globular unit. The substrate binding pockets (arrows) are located at opposite ends of the assembled enzyme *(Panel A adapted from Yuan and Horvitz, 1990; Panel C from Ni et al., 2003)*

Proteases appear to be a common weapon of choice for PCD. Although it has been known for some time that one can rescue injured cells by blocking proteolytic enzymes, the evidence for protease involvement in naturally-occurring cell death is more recent. One of the gene products that kills *C. elegans* neurons, CED-3, turns out to be a cysteine protease, an enzyme that specifically cuts up proteins after an aspartate residue. Cysteine proteases have been identified in many species, and they are generally referred to as **C**ysteine requiring **ASP**artate prote**ASE**s, or caspases (Figure 7.25B). Each caspase recognizes a short tetrapeptide sequence and has an absolute requirement for aspartate in the first position. The structure of caspase-3 is shown in Figure 7.25C.

In *C. elegans*, the *ced-4* gene product is another necessary constituent of the death pathway. CED-4 binds to the inactive form of CED-3 and leads to its activation (**Figure 7.26**). In mammals and flies, the pathway involves *ced-3* and *ced-4* homologs, but the mechanism leading to caspase activation begins at the mitochondrion. Various stimuli, such as the withdrawal of growth factor, lead to increased permeability of the outer mitochondrial membrane. When this occurs, a member of the electron transport chain, called cytochrome *c* (cyt *c*), leaks out of the mitochondrion. As it enters the cytoplasm, cyt *c* binds tightly to an adaptor protein called apoptosis protease activating factor-1 (Apaf-1). This is the mammalian homolog of *ced-4*. Procaspase-9 is then recruited to this complex, called an *apoptosome*. This is the site where caspase-9 becomes activated autoproteolytically. The activated caspase-9, in turn, cleaves pro-caspase-3, resulting in its activation. Mutations of Apaf-1 lead to a decrease in cell death in vivo (Li et al., 1997), similar to the effects of *ced-4* inactivation in *C. elegans*.

There are 14 caspase members in the mammalian genome, 6 in flies, and only 3 in worms. Many lines of evidence indicate that caspase activity is required for apoptosis in many neuronal populations. When a caspase inhibitor (crmA) is microinjected into chick DRG neurons in vitro, they can survive the withdrawal of NGF (Gagliardini et al., 1994). Members of the caspase family may also mediate the death-promoting effect of "death domain" containing proteins, such as P75NTR or the *Drosophila* protein caller Reaper (discussed above). For example, Reaper overexpression in the *Drosophila* eye causes all the cells to die, but a caspase inhibitor is able to block this effect (White et al., 1996; Vernooy et al., 2000).

To determine whether caspases are involved in the normal period of cell death, chick embryos were treated with a synthetic peptide inhibitor of caspase on embryonic day 8, the peak of motor neuron death. After 24 hours, the number of pyknotic cell bodies was cut in half compared to animals treated with a less selective protease inhibitor (Milligan et al., 1995). Since the synthetic peptide inhibitor could have blocked several caspases, the effect of a single protease was examined in caspase-3 null mutant mice. The brains of these animals are disorganized, and there are few signs of the pyknotic cell clusters that accompany nervous system morphogenesis, suggesting a decrease in normal cell death. This is most apparent in the proliferative zone or immature populations in the forebrain (Kuida et al., 1996, Pompeiano et al., 2000). In contrast, cell death of motor, sensory, and sympathetic neurons proceeds unchecked in caspase-3-/- mice. Interestingly, electron micrographs of the dying neurons suggest that they die

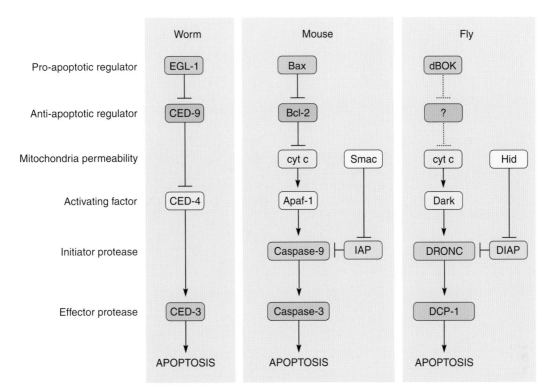

Fig. 7.26 Regulation of cell death machinery in three species. (Left) In *C. elegans*, the effector protease that leads to apoptosis (CED-3) is activated by CED-4. An anti-apoptotic regulator (CED-9) can complex with CED-4 and interfere with CED-3 activation. A pro-apoptotic regulator (EGL-1) can bind to CED-9, and facilitate the processing of CED-3. (Middle) Many of the same components are present in mammals, including pro- and anti-apoptotic regulators (Bax and Bcl-2). However, Caspase-3 is activated by an initiator protease (Caspase-9) that is processed in molecular complex with cytochrome c and Apaf-1. An additional regulatory component consists of IAP, which prevents Caspase-9 activation, and Smac, which can inhibit IAP and permit the apoptotic pathway to progress. (Right) In the fruit fly, the pathway is relatively similar to that described for mammals.

in a somewhat different manner than those in control mice; there is little sign of chromatin condensation or fragmentation into apoptotic bodies. Together, these results suggest that other caspases, or caspase-independent pathways, mediate cell death in many developing neuronal populations. They also suggest that TUNEL labeling alone does not characterize cell death. It is possible that cell death is occurring, but without DNA fragmentation, in cells that are not stained with TUNEL (Oppenheim et al., 2001).

For those nerve cells that die by destroying their own proteins, is caspase activity restricted to specific substrates? The evidence suggests great specificity, particularly for proteins that are involved in genome regulation, such as DNA repair, DNA replication, and RNA splicing enzymes (Lazebnik et al., 1994; Loetscher et al., 1997; Nicholson and Thornberry, 1997). Structural proteins of the nucleus and cytoskeleton, such as actin and fodrin, are also targets for cleavage. One example of a caspase target is the DNA repair enzyme called poly (ADP-ribose) polymerase (PARP), suggesting that cell death is achieved by compromising the neuron's transcription machinery. A second set of targets, of some interest to those studying Alzheimer's disease, are the transmembrane proteins called presenilins that are apparently involved in the Notch signaling pathway (see Chapter 2). Although NGF-deprived PC12 cells normally die, they can be rescued by transfection with presenilin 2 antisense mRNA.

Caspases are a primary contributor to PCD within some areas of the developing nervous system. However, conspiracy theorists can take comfort in the many other death mechanisms that underlie normal cell death. A caspase-independent pathway begins with the release of a mitochondrial flavoprotein, called apoptosis-inducing factor (AIF). In this case, the AIF enters the nucleus and initiates DNA cleavage (**Figure 7.27**). A second caspase-independent pathway involves the regulation of superoxide ($O_2.-$), which accumulates as a result of oxygen usage in the mitochondrial respiratory chain. Free radicals such as $O_2.-$ have unpaired electrons, making them an extremely reactive species. Excess $O_2.-$ can disrupt membrane integrity, inhibit pumps, and fragment DNA. Superoxide dismutase (SOD) is the endogenous enzyme that eliminates $O_2.-$ by catalyzing a reaction to O_2 and H_2O_2. Interestingly, sympathetic neurons can survive for a longer period of time after NGF deprivation if injected with SOD. Although a caspase mediates the cell death process initiated by trophic factor deprivation, it is *not* responsible for the death initiated by free radicals (Troy et al., 1997).

Thus, for each developing population of neurons, the naturally-occurring period of cell death may invoke a distinctive set of molecular mechanisms. In fact, one of the most varied features of cell death involves the regulatory proteins that determine whether or not caspases cross the threshold to their active state.

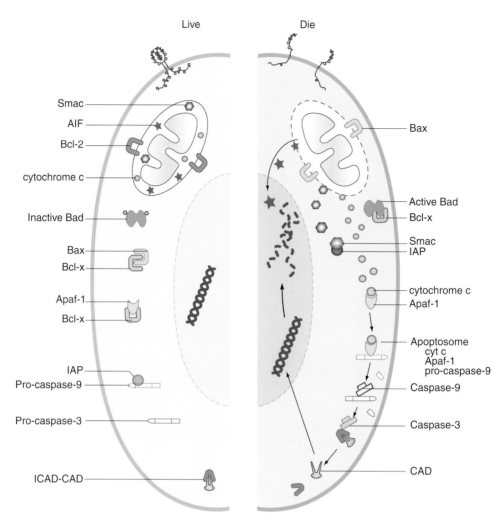

Fig. 7.27 The molecular state of a neuron that permits it to survive or leads it to death. (Left) The living cell contains mitochondria that are preserved in a nonpermeable state due to the presence of an anti-apoptotic regulator, Bcl-2. A second anti-apoptotic regulator, Bcl-x, complexes with Apaf-1, preventing the activation of caspase-9. Bcl-x also binds to a pro-apoptotic regulator, Bax, and prevents it from influencing the mitochondrion. A different pro-apoptotic regulator, Bad, is inactive, having been phosphoryated by neurotrophin-elicited kinase activity. IAP binds to pro-caspase-9, and this also serves to prevent activation. A nuclease that is responsible for DNA fragmentation, caspase-activated deoxyribonuclease (CAD), is bound by its inhibitor, ICAD. (Right) In dying neurons, Bad becomes active when it is dephosphorylated, and it binds to Bcl-x. This permits Bax to associate with the mitochondrion, leading to the release of cytochrome c, AIF, and Smac. It also permits Apaf-1 to form an apoptosome with cyt *c*, process caspase-9, and activate caspase-3. Smac binds to IAP, which also permits the processing of pro-caspase-9. One target of caspase-3 is ICAD, leading to the release of CAD, and the fragmentation of DNA. In a caspase-independent pathway, AIF also enters the nucleus and fragments DNA.

BCL-2 PROTEINS: REGULATORS OF PROGRAMMED CELL DEATH

The continuous presence of death-promoting molecules in the cytoplasm is rather like keeping several loaded guns scattered about the house; safety locks of some sort are essential. In fact, there are several important checks and balances to ensure that only the correct neurons die. There are two key sites of regulation. First, the mitochondrion outer membrane must remain relatively impermeable so that cytochrome *c* does not leak into the cytoplasm. Mitochondrion permeability is controlled by a large family of proteins that interact with the outer mitochondrial membrane as well as with one another. Second, inactive pro-caspases must be controlled such that

their threshold for activation remains relatively high. The activation of pro-caspases is regulated by two interacting molecules, one of which is released from mitochondria.

In *C. elegans*, activation of the *ced-9* gene prevents PCD in all cells. Mutations that inactivate *ced-9* lead to death among cells that would normally survive through development (Hengartner et al., 1992). CED-9 apparently blocks death by complexing with CED-4 and interfering with its ability to activate the protease, CED-3 (Figure 7.26). The mammalian homolog of CED-9 is a membrane-associated protein called Bcl-2 (named for its discovery in B Cell Lymphoma cells) that was originally discovered in studies of tumor formation.

The Bcl-2 family has since been found to include members that promote survival (anti-apoptotic), as well as those that promote cell death (pro-apoptotic). To date, 12 pro-apoptotic and 5 anti-apoptotic members have been described in mammals, although few of these have been evaluated as participants in naturally-occurring neuron cell death (Youle and Strasser, 2008). In healthy neurons, anti-apoptotic members of the Bcl-2 family, such as Bcl-2 itself, are closely associated with mitochondria (Figure 7.27). Other anti-apoptotic members, such as Bcl-x, associate with specific proteins to keep them inactive: a pro-apoptotic member of the Bcl-2 family (Bax) and the caspase activator (Apaf-1).

An impressive demonstration of Bcl-2 influence on neuron survival comes from transgenic mice that overexpress this protein. These mice have much bigger brains than controls, and cell counts in the facial nucleus and retina reveal a 40% increase in neuron number (Martinou et al., 1994). However, genetic studies suggest that there is some redundancy among survival-promoting members of the Bcl-2 family: Targeted disruption of the *bcl-2* gene does *not* have an effect on neuron survival, despite its high expression in the developing nervous system (Veis et al., 1993). In contrast, disruption of a different anti-apoptotic gene, *bcl-x*, does produce a clear increase of PCD in the nervous system (Motoyama et al., 1995).

In dying neurons, the pro-apoptotic members of the Bcl-2 family bind to members that maintain survival (i.e., Bcl-2 and Bcl-x), and then mount an assault on the mitochondria (Figure 7.27). One subset of pro-apoptotic proteins (those containing only a single Bcl-2 homology domain, such as Bad, Bid, and Bim) recruit a second subset into play (e.g., Bax and Bak). This latter set of pro-apoptotic proteins oligomerize, become associated with the mitochondrial membrane, and somehow permeabilize it. As a result, several mitochondrial proteins, such as cyt c, are released into the cytoplasm (Chipuk and Green, 2008; Youle and Strasser, 2008).

At the same time that the mitochondrial membrane is under attack, pro-apoptotic proteins are dimerizing with anti-apoptotic members and inactivating them. Bad binds to Bcl-x, and Bax binds to Bcl-2. As the caspases become activated, anti-apoptotic members of the Bcl-2 family are themselves a substrate. When Bcl-2 is cleaved by a caspase, its protective influence is eradicated (Cheng et al., 1997).

The pro-apoptotic Bcl-2 members have a clear influence on neuron cell death in vivo. PCD in sympathetic ganglia and motor neurons is virtually eliminated in *bax* knockout mice, and it is significantly reduced in many areas of the CNS (White et al., 1998; Fan et al., 2001). The functional interaction between pro- and anti-apoptotic members is illustrated in a mouse with deletions of one or both types of Bcl-2 protein (**Figure 7.28**). Mice deficient for *bcl-x* exhibit increased PCD, suggesting that pro-apoptotic proteins are no longer being suppressed. To test this idea, a double knockout mouse deficient for both *bcl-x* and the pro-apoptotic member, *bax*, was examined. When PCD was examined in the spinal cord of *bcl-x⁻/⁻/bax⁻/⁻* mice, the level had returned to normal (Shindler et al., 1997).

Elements of the neurotrophin withdrawal response have been tied to elevation of pro-apoptotic Bcl-2 family members. When sympathetic neurons were transfected with a virus that expresses a dominant negative form of c-Jun, the cells could survive NGF withdrawal. Without active c-Jun, there were two basic changes to the cells' physiology when NGF

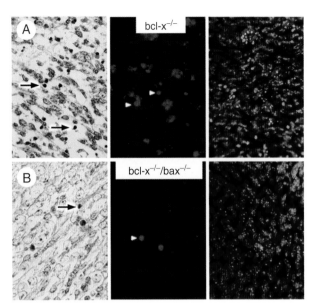

Fig. 7.28 In vivo regulation of neuron survival by pro- and anti-apoptotic regulatory proteins. A. A spinal cord tissue section from an E12 *bcl-x⁻/⁻* mouse shows many pyknotic nuclei (arrows), indicative of massive cell death (left). TUNEL-positive cells (red) are common in the E12 *bcl-x⁻/⁻* spinal cord (middle). A nuclear stain (bright blue) indicates condensed chromatin (right). B. A spinal cord tissue section from an E12 *bcl-x⁻/⁻/bax⁻/⁻* double mutant mouse shows few pyknotic nuclei, indicating that cell death is curtailed (left). In contrast to the *bcl-x-/-* cord, there are few TUNEL positive cells in *bcl-x⁻/⁻/bax⁻/⁻* mice (middle). In contrast to the *bcl-x-/-* cord, the nuclear stain reveals few cases of condensed chromatin in *bcl-x⁻/⁻/bax⁻/⁻* mice (right). *(Adopted from Shindler et al., 1997, by permission of Soc. of Neuroscience)*

was withdrawn. First, the expression of a pro-apoptotic Bcl-2 family member, Bim, did not increase as it normally does. Second, cyt c was not released from the mitochondria. This result suggests that neurotrophins, at least, prevent cell death by suppressing the pathway that leads to pro-apoptotic protein expression (Whitfield et al., 2001).

The second major point of regulation is located at the caspases themselves. One family of regulators, called inhibitors of apoptosis proteins (IAPs), act directly on the caspases (Figure 7.26). IAPs have been identified that bind directly to caspase-3 or caspase-9, and block their proteinase activity. One striking indication that IAPs keep neurons alive comes from a clinical observation. A member of the IAP family is deleted in humans with a disorder called spinal muscular atrophy, a condition in which spinal cord motor neurons gradually die (Roy et al., 1995). There are also antagonists to the IAPs. A mitochondrial protein called Smac/DIABLO, which is also released when the outer membrane becomes permeable, can bind to the apoptosome and inhibit IAP. While IAPs and Smac have been shown to promote caspase activation and PCD in some systems, mice with null mutations of either gene do not display greater cell death (Harlin et al., 2001; Okada et al., 2002). This suggests that there are redundant pathways mediating naturally-occurring cell death.

Therefore, there are two safety latches on caspase activation. The first is control of cyt c release from mitochondria, and the second is IAP (Figure 7.27). In fact, both safety latches must be removed to kill sympathetic neurons grown in the presence of NGF. Injection of cyt c alone is not able to activate caspases and produce PCD, but if Smac is co-injected, then caspases are activated and the neurons die (Du et al., 2000; Verhagen et al., 2000; Deshmukh et al., 2002). In *Drosophila*, loss of IAP function alone leads to unrestrained activation of caspases and death. Conversely, the loss of all three gene products that inhibit IAP (*grim*, *reaper*, and *hid*) result in almost no PCD (White et al., 1994; Hay et al., 1995).

REMOVAL OF DYING NEURONS

The fate of developing nerve cells that contain activated caspases and pro-apoptotic proteins appears to be sealed. However, engulfment and phagocytosis by microglial cells may actually nail the coffin shut. Microglia derive from mesodermal cells in vertebrates, and they migrate into the CNS during development (Kaur et al., 2001). They protect the adult CNS against infection or injuries through phagocytosis,

antigen presentation, and cytokines secretion. During development, microglia gather during the period of naturally-occurring death, and are actively involved in this process by engulfing and digesting the dying nerve cells. **Figure 7.29** shows examples of dying neurons that are in the process of engulfment by microglia.

The role of engulfment and phagocytosis has been described most clearly in *C. elegans*. There are seven genes involved in this process, with some involved in the recognition of cell corpses and others controlling the cytoskeletal rearrangements that are required to engulf dying cells (Figure 7.29). Although null mutations of these engulfment genes do not result in greater survival, is possible that engulfment does participate actively in cell death. To determine this, studies were performed on mutant animals with decreased killer gene function, such as a *ced-3* allele that encodes a weak caspase-3. Fewer cells died in these mutants, and many showed signs of PCD but recovered. These animals were then crossed with null

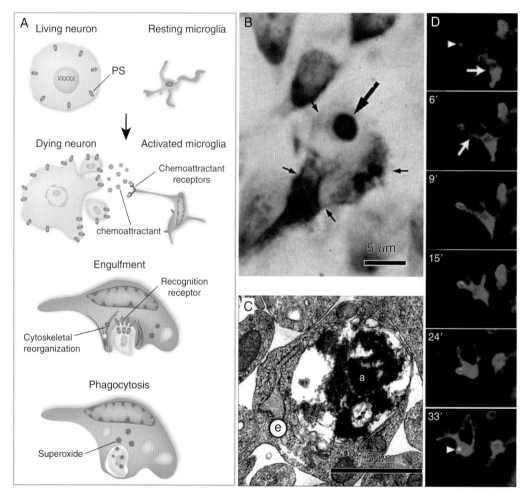

Fig. 7.29 Microglia engulf dying neurons. A. Schematic of engulfment and phagocytosis shows a living neuron with phosphatidylserine (PS), held in the inner leaflet of the plasma membrane, and resting microglial cell. When the neuron undergoes apoptosis, PS becomes exposed, serving to identify the corpse, and a microglia chemoattractant is released. When the microglial cell recognition receptors bind to PS, the cytoskeleton becomes reorganized and the apoptotic body is engulfed. Finally, phagocytosis occurs when microglia-derived effectors such as superoxide break down the corpse. B. A pyknotic nucleus (large arrow) in the mouse retina is phagocytosed by a microglial cell stained for the macrophage-specific antigen, F4/80. The boundary of the microglial cell is indicated by small arrows. C. An electron micrograph of an apoptotic retinal cell nucleus (a) engulfed by a microglial cell (e). D. Time lapse from hippocampal slice culture showing a macrophage (green) extending a process (arrow at 6′) to engulf a dead cell nucleus (red, arrowhead). *(Panel B adopted from Hume et al., 1983; Panel C adopted from Mellén et al., 2008; Panel D adopted from Peterson and Dailey, 2004)*

mutants for an engulfment gene. The double mutants display more surviving cells (than the weak ced-3 alone), suggesting that even in an advanced stage of degeneration, neurons can recover, unless they are engulfed and phagocytized (Hoeppner et al., 2001; Reddien et al., 2001).

Naturally-occurring cell death may also be influenced by engulfment in the developing mammalian CNS. In slice cultures of early postnatal cerebellum, more than half of the Purkinje neurons express caspase-3 and are engulfed by microglia. When the microglia were selectively eliminated with a drug, there was a profound increase in surviving Purkinje neurons (Marín-Teva et al., 2004).

The signal that first recruits microglia to the site of dying neurons is not entirely clear. It is known that microglia respond rapidly to injury in the adult nervous system, and this response can be elicited with ATP. In fact, microglia that are deficient for a purinergic receptor to ATP do not become activated by neuronal injury (Davalos et al., 2005; Haynes et al., 2006).

Having arrived at an area of cell death, microglia require a local signal to specify which cells are corpses that are to be engulfed. The most common alteration on the surface of cells undergoing apoptosis is the movement of phosphatidylserine (PS) from the inner to the outer leaflet of the plasma membrane (Figure 7.29). PS is held within healthy cells by an ATP-binding transporter. One of the *C. elegans* genes that must be expressed for corpse recognition to occur, *ced-7*, encodes just such a transporter, and is responsible for PS exposure. When PS expression is masked with a binding protein, engulfment does not occur (Venegas and Zhou, 2007). A similar scenario occurs in the developing chick retina; surface expression of PS decreases when the lysosomal degradation pathway is inhibited, and dying retinal neurons fail to be engulfed (Mellén et al., 2008).

Several cell surface proteins may play a role in recognition of apoptotic cells by macrophages or other glial cells. In *C. elegans*, CED-1 acts as a corpse recognition receptor which may bind to PS, along with a second protein called PS-binding receptor (PSR). The process of engulfment employs both of these receptors, although CED-1 appears to play the central role. If PS is not exposed at the surface due to the inactivation of *ced-7*, then CED-1 no longer clusters at the site of contact with dying cells (Zhou et al., 2001b; Wang et al., 2003). In Drosophila, the engulfment of dying neurons may involve at least two receptors, including a CED-1 homolog called Draper (Freeman et al., 2003; Kurant et al., 2008). In vertebrates, there is also evidence that PSR plays a role in engulfment. A greater number of corpses are found in the brains of PSR-deficient mouse and zebrafish embryos (Li et al., 2003; Hong et al., 2004). However, there is an expanding list of PS-binding proteins, and it is not clear how recognition receptors are involved in engulfment (Bratton and Henson, 2008).

In *C. elegans*, activation of the CED-1 receptor is associated with the reorganization of the cytoskeleton, particularly actin (Figure 7.29). Although some of the key molecules have been identified, the signaling cascade that leads to phagocytosis is not yet understood fully (Kinchen et al., 2005; Yu et al., 2006). The final stage, phagocytosis, may employ microglia-derived effectors such as superoxide ions. In fact, a free radical scavenger was able to rescue cultured Purkinje cells from microglia-dependent death (Marín-Teva et al., 2004).

SYNAPTIC TRANSMISSION AT THE TARGET

The earliest experimental manipulations of target size suggested that functional synaptic contacts are correlated with survival. Removal of the nasal placodes in salamander embryos do not, at first, decrease the size of the innervating forebrain region. It is only after the system becomes functional that loss of the target results in hypoplasia (Burr, 1916). Similarly, there is a very strong correlation between the onset of neuromuscular activity in chicks and the onset of normal motor neuron cell death. Therefore, the formation of functional synapses may regulate the trophic support that is provide by the target.

If synapse activity at the target is necessary for survival, then one would predict that more neurons would die when synaptic transmission is blocked. To test this hypothesis, chick embryos were treated with an acetylcholine receptor (AChR) antagonist (curare or bungarotoxin) during a four-day period that overlapped the normal period of motor neuron death. Curare was quite effective at blocking neuromuscular transmission, as spontaneous movements were virtually eliminated for much of the treatment period. Rather than increasing cell death, the surprising result was that synapse blockade saved motor neurons (**Figure 7.30**). Over 90% of the motor neurons that would have died were still alive after the period of normal cell death ended and the curare had been removed. This effect is due to the interaction at the neuromuscular junction because agents that selectively block muscle AChRs produce the effect, while those that

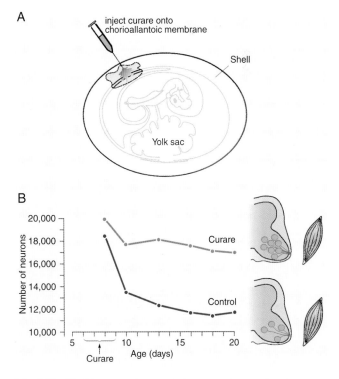

Fig. 7.30 Blocking synaptic transmission prevents normal motor neuron cell death. A. Neuromuscular transmission can be blocked by applying curare onto the chorioallantoic membrane of chick embryos. B. In control animals, over 30% of motor neurons die after embryonic day 5. When animals are treated with curare from E6-9, the magnitude of normal cell death is greatly diminished. *(Panel B adopted from Pittman and Oppenheim, 1979)*

block AChRs in the CNS do not prevent cell death (Pittman and Oppenheim, 1979; Oppenheim et al., 2000b; Oppenheim et al., 2008). A similar decrease in normal cell death has been observed in the isthmo-optic nucleus when activity is blocked in its target, the eye, by injecting TTX during development (Péquignot and Clarke, 1992).

Why would a loss of functional innervation improve survival? One clue comes from the anatomy of the motor neurons' axonal projections. There was a threefold increase in the number of motor axon branches and synaptic contacts during the period when normal cell death should have occurred (Dahm and Landmesser, 1991). A similar phenomenon is observed in mice that lack the signals to form proper post-synaptic specializations, including the aggregation of ACh receptors. These mice display increased axonal branching and improved survival of motor neurons during the normal period of cell death (Banks et al., 2001; Terrado et al., 2001; Banks et al., 2003). Synaptic activity could also regulate the expression of trophic factor by the target. To account for increased survival of motor neurons, one would suppose that decreased synaptic activity would result in an increased level of muscle-derived trophic factor; at present, the evidence does not favor this theory (Tanaka, 1987; Bartlett et al., 2001). Taken together, these findings have led to the hypothesis that neuron survival depends on their level of access to the target-derived survival factor (Oppenheim, 1989). Thus, the increased branching may improve the capacity of motor neurons to obtain muscle-derived trophic factors.

AFFERENT REGULATION OF NEURON SURVIVAL

Many observations suggest that synaptic activation of a neuron can regulate its survival. The role of afferent connections was demonstrated originally by surgically removing an axonal projection and observing whether its target developed properly in their absence. For example, Larsell (1931) removed one eye in tree frog larvae and found that its target, the contralateral optic tectum, had many fewer cells than expected. Modern experiments suggest that if the amount of synaptic transmission is too low during development, then postsynaptic neurons can cease protein synthesis, become atrophic, and may even die. At the other extreme, too much excitatory activation has been shown to kill neurons by opening voltage-gated channels and loading the cytoplasm with calcium (Nicholls and Ward, 2000; Duchen, 2000).

One of the best studied cases of afferent regulated survival is the *nucleus magnocellularis* (NM) in the chick central auditory system (**Figure 7.31**). Just before taking up her studies of NGF, Rita Levi-Montalcini had been studying the effect of cochlear nerve fibers on the survival of NM neurons and other brainstem nuclei. These studies have fascinated students of biology because they were performed with very little equipment in the countryside of Italy while World War II raged around her. In spite of these privations, Levi-Montalcini (1949) was able to show that the period of normal cell death is elevated when the cochlea is removed. Although there was little sign of degeneration at the age when auditory nerve fibers first activate NM neurons, there was a dramatic loss of cells soon thereafter. Subsequent studies showed that about 30% of NM neurons are lost following cochlear ablation, and the effect of denervation

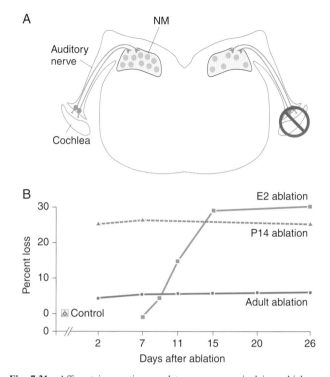

Fig. 7.31 Afferent innervation regulates neuron survival in a chick central auditory nucleus. A. Auditory neurons from the cochlea innervate the nucleus magnocellularis (NM) in the chick auditory brainstem. The removal of a cochlea (right) completely denervates NM neurons on the ipsilateral side. B. When a cochlea is removed at embryonic day 2 (E2), about 30% of NM neurons are lost during the ensuing two weeks, although cell death does not begin until E10. When the cochlea is removed at posthatch day 14, about 25% of neurons die within two days. In adults, cochlear ablation results in the loss of only about 5% of NM neurons. *(Adapted from Parks, 1979; Born and Rubel, 1985)*

is much reduced in adult animals (Parks, 1979; Born and Rubel, 1985). In fact, neuron survival can change from being dependent upon afferent innervation to being completely independent over the course of a few days (Tierney et al., 1997; Mostafapour et al., 2000). When the cochlea is removed in P7 gerbils, about 50% of the postsynaptic cochlear nucleus neurons are lost. However, when the cochlea is removed just two days later, at P9, there is no neuronal cell loss (**Figure 7.32**). Microarray analysis shows that cochlear nucleus neurons from mice constitutively express higher levels of transcript for pro-apoptotic genes during the age range when they depend on afferent innervation for survival. After this age, cochlear nucleus neurons undergo a dramatic switch to express higher levels of transcript for anti-apoptotic genes. Somewhat surprisingly, the age of deafferentation does not have a large effect on the expression of pro- or anti-apoptotic genes. This leads to the conclusion that deafferentation at an early age is more likely to result in cell death because of the constitutive expression of pro-apoptotic genes, rather than their induction by the manipulation (Harris et al., 2005; Harris et al., 2008).

Survival in other peripheral and central neurons also depends, in part, on afferent connections during development (Linden, 1994). However, surgical removal of the afferent population does not really tell us much about the trophic signal. Does the synapse provide a survival factor such as NGF? Does the neurotransmitter itself enhance survival? To address this

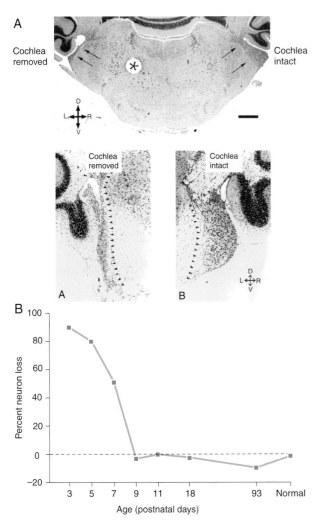

Fig. 7.32 The survival of neurons in the gerbil cochlear nucleus depends on afferent innervation from the cochlea only through the first postnatal week. A. Cross section through the brainstem showing the ventral cochlear nucleus that was deafferented by cochlea removal (left) and the cochlear nucleus that received normal innervation (right). The cochlea was removed at P3. Shown below are higher magnification images of a deafferented (left) and control (right) cochlear nucleus from an animal in which the cochlea was removed at P5. There are very few neurons remaining on the cochlea removal side. B. The plot shows the percentage of nerve cells lost on the cochlea removal side, as compared to the control side, for different ages of surgical removal. The number of neurons lost is 50% or greater when the cochlea is removed by postnatal day 7. However, there is no cell death when the cochlea is removed after P9. *(Adapted from Tierney et al., 1997)*

question, intact afferent pathways were treated with agents that block neuronal activity (Maderdrut et al., 1988; Born and Rubel, 1988). In the chick ciliary ganglion, cell death is increased when transmission is blocked, whereas neurogenesis and migration proceed normally. Similarly, action potential blockade in the cochlea for two days is sufficient to increase normal cell death in the chick NM. Therefore, synaptic activity seems to play a critical role in postsynaptic neuron survival. To determine whether glutamatergic synaptic transmission provides a positive survival signal, as opposed to the action potentials elicited by transmission, brain slices containing the chick NM and its auditory nerve afferents were placed in vitro and provided with two different stimulation protocols

(**Figure 7.33**). Although the experimental period was too brief to observe dying neurons, denervation of NM neurons leads to a rapid decrease in protein synthesis that is thought to be a condition preceding cell death in this and many other neural systems. When the auditory nerve is stimulated, NM neurons receive synaptically evoked activity, and protein synthesis is maintained. In contrast, when NM axons are stimulated to produce antidromic action potentials in their cell body, protein synthesis is not maintained (Hyson and Rubel, 1989). Thus, the preservation of postsynaptic neuron metabolism, and presumably its survival, depend on the release of something from the synaptic terminal.

It is also possible that afferents release trophic factors, along with neurotransmitters, during development. Neurotrophins, such as NT-3, are produced in the developing retina, and they are transported anterogradely down retinal ganglion cell axons to the optic tectum (von Bartheld et al., 1996). Since the survival of optic tectum neurons depends on both axonal transport and electrical activity by retinal axons, it seems possible that NT-3 mediates this afferent regulation (Catsicas et al., 1992). Of course, if neurotrophins provided an anterograde signal, then the target neurons would be expected to have Trk receptors at the synapse. In fact, an electron microscopic study of TrkB and TrkC receptors shows that they are located at postsynaptic profiles in the developing (and adult) central nervous system (Hafidi et al., 1996).

INTRACELLULAR CALCIUM MEDIATES BOTH SURVIVAL AND DEATH

Intracellular free calcium often plays a decisive role in neuron survival, serving as a mediator for both survival and cell death signals (Orrenius et al., 2003). In fact, the activity of afferent connections appears to be an important survival signal because synaptic transmission can raise the level of postsynaptic calcium. This can occur when depolarizing postsynaptic potentials activate voltage-gated calcium channels, or when the pore of a neurotransmitter receptor is permeable to calcium. An example of the latter is the N-methyl D-aspartate (NMDA) receptor which is activated by the neurotransmitter, glutamate. NMDA receptors are expressed throughout the developing nervous system, and are known to influence many events, including survival and synapse plasticity (see Chapter 9).

Calcium influx through synaptic NMDA receptors has now been shown to support neuron survival in many different CNS populations (Gould et al., 1994; Ikonomidou et al., 1999; Monti and Contestabile, 2000; Fiske and Brunjes, 2001). For example, naturally occurring neuron cell death increases dramatically in mouse brainstem and thalamus when NMDA receptors are deleted transgenically, or when they are blocked pharmacologically (Adams et al., 2004; de Rivero Vaccari et al., 2006). **Figure 7.34** shows anatomical sections through the mouse somatosensory thalamus in control mice and in animals with a genetic deletion of the NMDA receptor 1 subunit. Staining for degeneration and caspase-3 expression are each greater in the knock-out NMDAR1 null mice.

The location at which calcium enters a neuron may determine its effect on survival (Hardingham et al., 2002, Papadia et al., 2005; Zhang et al., 2009). When calcium enters the neuron through NMDA receptors located at synapses, it leads to an increase in nuclear calcium and phosphorylation of the

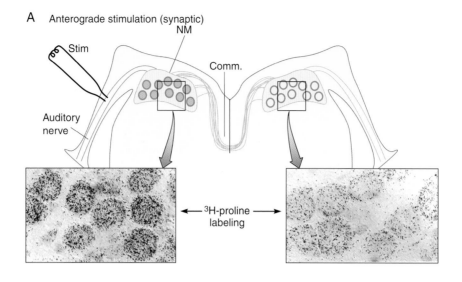

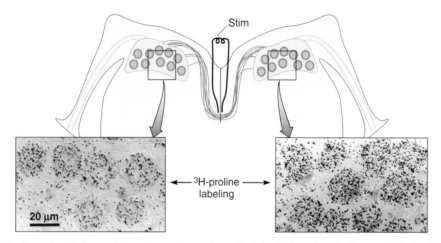

Fig. 7.33 Synaptic activity regulates postsynaptic protein synthesis in chick auditory brainstem. A. Brain slices containing the chick nucleus *magnocellularis* (NM) and its auditory nerve afferents were incubated in an oxygenated solution containing ³H-proline. Synaptic transmission in one NM was elicited by electrically stimulating the auditory nerve. When the tissue was processed for autoradiography, it was found that synaptically stimulated NM neurons incorporated far more proline (black dots) into newly synthesized proteins compared to the control side. B. When the axons of NM neurons were stimulated at the commissure, evoking a retrograde action potential in the cell body, protein synthesis was not maintained. *(Adapted from Hyson and Rubel, 1989)*

cAMP response element-binding protein (CREB) through a calcium-calmodulin dependent process. The activated pCREB then induces transcription of many survival-promoting factors such as BDNF. Synaptic NMDA receptor activity can also suppress the transcription of pro-apoptotic genes, such as Procaspase-9 and Apaf-1. Consistent with these findings, activation of synaptic NMDA receptors promotes neuron survival in vitro. However, when nonsynaptic NMDA receptors are selectively activated, there is a tremendous dephosphorylation of CREB, called CREB shut-off, and this results in increased cell death. The mechanism whereby NMDA receptors communicate with nuclear CREB is unclear, but at least one protein, called Jacob, may lead to CREB shut-off when it is transported to the nucleus (Dieterich et al., 2008).

There is some evidence calcium-dependent neuron survival is due, in part, to increased expression of the neurotrophin, BDNF.

The activity-dependent expression of BDNF has been shown to support the survival of embryonic cortex neurons in vitro (**Figure 7.35**). BDNF expression and survival depend on the entry of calcium into the neurons when they are depolarized in a medium containing a high potassium ion concentration. When function-blocking antibodies directed against BDNF are added to these cultures, the trophic effect of depolarization is eliminated (Ghosh et al., 1994). Similarly, NMDA receptor activation promotes survival of retinal neurons, and this effect can be eliminated by blocking BDNF signaling (Martins et al., 2005).

In the chick cochlear nucleus, deafferentation initially results in calcium elevation, and it is this calcium that appears to be responsible for killing 30% of the neurons (Zirpel et al., 1998). One proapoptotic signaling pathway that is recruited by high intracellular calcium levels is the transcription factor NFAT. Activated NFAT translocates to the cell nucleus

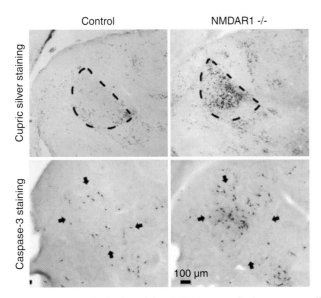

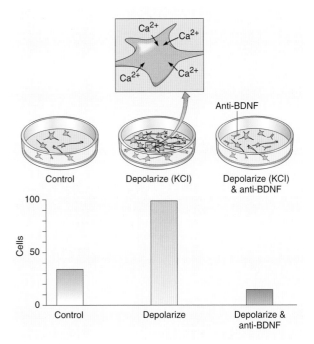

Fig. 7.34 Genetic elimination of the NMDA receptor leads to greater cell death in the developing somatosensory thalamus. (Top) Degenerating thalamic neurons are revealed by cupric silver staining at postnatal day 0 in control (left) and NMDAR1-/- (right) mice. (Bottom) Immunostaining for caspase-3 shows greater expression in NMDAR1-/- mice (right) as compared to a control (left). Arrows indicate the boundary of the thalamus. *(Adopted from Adams et al., 2004)*

Fig. 7.35 Electrical activity enhances the survival of embryonic cortex neurons by way of a neurotrophic signal. When the cultures are depolarized by adding KCl to the culture media, calcium enters the neurons, and the level of BDNF expression increases, leading to greater neuron survival compared to control media. The trophic influence of depolarization is eliminated by adding a function-blocking anti-BDNF antibody to the growth medium. *(Adopted from Ghosh et al., 1994)*

where it can increase the transcription of death receptor ligand, FasL. When NFAT activation is blocked, deafferentation of the cochlear nucleus does not lead to an increase in FasL expression and cell death is largely prevented (Luoma and Zirpel, 2008). Although deafferentation does lead to calcium-dependent cell death, the majority of cochlear nucleus neurons survive the insult. This may well depend on CREB, which is phosphorylated within one hour of deafferentation. In this case, the source of calcium appears to be AMPA-type glutamate receptors. The survival pathway may be quite similar to that described for NMDA receptors above. Within six hours of deafferentation, the expression of an anti-apoptotic gene, *bcl-2*—known to be regulated by CREB—has increased (Zirpel et al., 2000; Wilkinson et al., 2002).

SUMMARY

Naturally-occurring cell death claims up to 80% of the neuroblasts and differentiating neurons and glia in the developing brain. Depending on the particular group of neurons, survival may rely on target-derived trophic factors, synaptic activity, hormonal signals, and other cues. The diversity of survival factors is accompanied by a variety of cytoplasmic mechanisms for dying. All forms of normal cell death require either the production or activation of proteins that can do damage to the neuron, such as caspases. Given the inherent risk of constitutively expressing the pro-apoptotic machinery, developing neurons also express a broad array of regulatory proteins which ensure that PCD occurs only under the appropriate conditions. Small perturbations of synaptic activity can have a profound impact on the number of surviving neurons and, therefore, the amount of postsynaptic membrane that is available for innervation (see Chapter 9). It is not too difficult to imagine that these mechanisms are necessary to optimize the diverse kinds of neural circuitry found within each animal. Despite the wealth of candidate mechanisms that are known to mediate cell death, these have been studied in a relatively small number of neuronal populations. Thus, the process remains poorly understood for most regions of the CNS.

REFERENCES

Adams, S. M., de Rivero Vaccari, J. C., & Corriveau, R. A. (2004). Pronounced cell death in the absence of NMDA receptors in the developing somatosensory thalamus. *The Journal of Neuroscience, 24,* 9441–9450.

Airaksinen, M. S., & Saarma, M. (2002). The GDNF family: Signaling, biological functions and therapeutic value. *Nat Neuro Rev, 3,* 383–394.

Alcantara, S., Frisen, J., del Rio, J. A., Soriano, E., Barbacic, M., & Silos-Santiago, I. (1997). TrkB signaling is required for postnatal survival of CNS neurons and protects hippocampal and motor neurons from axotomy-induced cell death. *The Journal of Neuroscience, 17,* 3623–3633.

Aloyz, R. S., Bamji, S. X., Pozniak, C. D., Toma, J. G., Atwal, J., Kaplan, D. R., et al. (1998). P53 is essential

for developmental neuron death as regulated by the TrkA and p75 neurotrophin receptors. *The Journal of Cell Biology, 143,* 1691–1703.

Appert-Collin, A., Hugel, B., Levy, R., Niederhoffer, N., Coupin, G., Lombard, Y., et al. (2006). Cyclin dependent kinase inhibitors prevent apoptosis of postmitotic mouse motoneurons. *Life Sciences, 79,* 484–490.

Arai, Y., Sekine, Y., & Murakami, S. (1996). Estrogen and apoptosis in the developing sexually dimorphic preoptic area in female rats. *Neuroscience Research, 25*, 403–407.

Baker, N. E., & Yu, S. Y. (2001). The EGF receptor defines domains of cell cycle progression and survival to regulate cell number in the developing drosophila eye. *Cell, 104*, 699–708.

Bamji, S. X., Majdan, M., Pozniak, C. D., Belliveau, D. J., Aloyz, R., Kohn, J., et al. (1998). The p75 neurotrophin receptor mediates neuronal apoptosis and is essential for naturally occurring sympathetic neuron death. *The Journal of Cell Biology, 140*, 911–923.

Banks, G. B., Chau, T. N., Bartlett, S. E., & Noakes, P. G. (2001). Promotion of motoneuron survival and branching in rapsyn-deficient mice. *The Journal of Comparative Neurology, 429*, 156–165.

Banks, G. B., Choy, P. T., Lavidis, N. A., & Noakes, P. G. (2003). Neuromuscular synapses mediate motor axon branching and motoneuron survival during the embryonic period of programmed cell death. *Developmental Biology, 257*, 71–84.

Barbacid, M. (1994). The Trk family of neurotrophin receptors. *Journal of Neurobiology, 25*, 1386–1403.

Barker, V., Middleton, G., Davey, F., & Davies, A. M. (2001). TNFalpha contributes to the death of NGF-dependent neurons during development. *Nature Neuroscience, 4*, 1194–1198.

Barres, B. A., Hart, I. K., Coles, H. S., Burne, J. F., Voyvodic, J. T., Richardson, W. D., et al. (1992). Cell death and control of cell survival in the oligodendrocyte lineage. *Cell, 70*, 31–46.

Barthélémy, C., Henderson, C. E., & Pettmann, B. (2004). Foxo3a induces motoneuron death through the Fas pathway in cooperation with JNK. *BMC Neuroscience, 5*, 48.

Bartlett, S. E., Banks, G. B., Reynolds, A. J., Waters, M. J., Hendry, I. A., & Noakes, P. G. (2001). Alterations in ciliary neurotrophic factor signaling in rapsyn deficient mice. *Journal of Neuroscience Research, 64*, 575–581.

Beard, J. (1986). On the disappearance of a transient nervous apparatus in the series: Scyllium, Acanthias, Mustelus and Torpedo. *Anat Anz, 19*, 371–374.

Bhattacharyya, A., Watson, F., Bradlee, T., Pomeroy, S., Stiles, C., & Segal, R. (1997). Trk receptors function as rapid retrograde signal carriers in the adult nervous system. *The Journal of Neuroscience, 17*, 7007–7016.

Biswas, S. C., Liu, D. X., & Greene, L. A. (2005). Bim is a direct target of a neuronal E2F-dependent apoptotic pathway. *The Journal of Neuroscience, 25*, 8349–8358.

Biswas, S. C., Shi, Y., Sproul, A., & Greene, L. A. (2007). Pro-apoptotic Bim induction in response to nerve growth factor deprivation requires simultaneous activation of three different death signaling pathways. *The Journal of Biological Chemistry, 282*, 29368–29374.

Blaschke, A. J., Staley, K., & Chun, J. (1996). Widespread programmed cell death in proliferative and postmitotic regions of the fetal cerebral cortex. *Development, 122*, 1165–1174.

Blaschke, A. J., Weiner, J. A., & Chun, J. (1998). Programmed cell death is a universal feature of embryonic and postnatal neuroproliferative regions throughout the central nervous system. *The Journal of Comparative Neurology, 396*, 39–50.

Bleier, R., Byne, W., & Siggelkow, I. (1982). Cytoarchitectonic sexual dimorphisms of the medial preoptic and anterior hypothalamic areas in guinea pig, rat, hamster, and mouse. *The Journal of Comparative Neurology, 212*, 118–130.

Bonni, A., Brunet, A., West, A. E., Datta, S. R., Takasu, M. A., & Greenberg, M. E. (1999). Cell survival promoted by the Ras-MAPK signaling pathway by transcription-dependent and independent mechanisms. *Science, 286*, 1358–1362.

Bonni, A., Frank, D. A., Schindler, C., & Greenberg, M. E. (1993). Characterization of a pathway for ciliary neurotrophic factor signaling to the nucleus. *Science, 262*, 1575–1579.

Borasio, G. D., John, J., Wittinghofer, A., Barde, Y. A., Sendtner, M., & Heumann, R. (1989). Ras p21 protein promotes survival and fiber outgrowth of cultured embryonic neurons. *Neuron, 2*, 1087–1096.

Born, D. E., & Rubel, E. W. (1985). Afferent influences on brain stem auditory nuclei of the chicken: neuron number and size following cochlea removal. *The Journal of Comparative Neurology, 22*, 435–445.

Born, D. E., & Rubel, E. W. (1988). Afferent influences on brain stem auditory nuclei of the chicken: presynaptic action potentials regulate protein synthesis in nucleus magnocellularis neurons. *The Journal of Neuroscience, 8*, 901–919.

Bratton, D. L., & Henson, P. M. (2008). Apoptotic cell recognition: will the real phosphatidylserine receptor(s) please stand up? *Current Biology, 18*, R76–79.

Breedlove, S. M., & Arnold, A. P. (1981). Sexually dimorphic motor nucleus in the rat lumbar spinal cord: response to adult hormone manipulation, absence in androgen-insensitive rats. *Brain Research, 225*, 297–307.

Breedlove, S. M., & Arnold, A. P. (1983). Hormonal control of a developing neuromuscular system. I. Complete Demasculinization of the male rat spinal nucleus of the bulbocavernosus using the anti-androgen flutamide. *The Journal of Neuroscience, 3*, 417–423.

Breedlove, S. M., Jacobson, C. D., Gorski, R. A., & Arnold, A. P. (1982). Masculinization of the female rat spinal cord following a single neonatal injection of testosterone propionate but not estradiol benzoate. *Brain Research, 237*, 173–181.

Brunet, A., Bonni, A., Zigmond, M. J., Lin, M. Z., Juo, P., Hu, L. S., et al. (1999). Akt promotes cell survival by phosphorylating and inhibiting a Forkhead transcription factor. *Cell, 96*, 857–868.

Bueker, E. D. (1948). Implantation of tumors in the hind limb field of the embryonic chick and the developmental response of the lumbosacral nervous system. *The Anatomical Record, 102*, 369–390.

Burr, H. S. (1916). The effects of the removal of the nasal pits in Amblystoma embryos. *Journal of Experimental Zoology, 20*, 27–57.

Buszczak, M., & Segraves, W. A. (2000). Insect metamorphosis: out with the old, in with the new. *Current Biology, 10*, R830–R833.

Campenot, R. B. (1977). Local control of neurite development by nerve growth factor. *Proceedings of the National Academy of Sciences of the United States of America, 74*, 4516–4519.

Campenot, R. B. (1982). Development of sympathetic neurons in compartmentalized cultures. II. Local control of neurite survival by nerve growth factor. *Developmental Biology, 93*, 13–21.

Carr, V. M., & Simpson, S. B. (1978). Proliferative and degenerative events in the early development of chick dorsal root ganglia. II. Responses to altered peripheral fields. *The Journal of Comparative Neurology, 182*, 741–755.

Casaccia-Bonnefil, P., Carter, B. D., Dobrowsky, R. T., & Chao, M. V. (1996). Death of oligodendrocytes mediated by the interaction of nerve growth factor with its receptor p75. *Nature, 383*, 716–719.

Catsicas, M., Péquignot, Y., & Clarke, P. G. H. (1992). Rapid onset of neuronal death induced by blockade of either axoplasmic transport or action potentials in afferent fibers during brain development. *The Journal of Neuroscience, 12*, 4642–4650.

Chao, M. V. (1994). The p75 neurotrophin receptor. *Journal of Neurobiology, 25*, 1373–1385.

Cheema, Z. F., Wade, S. B., Sata, M., Walsh, K., Sohrabji, F., & Miranda, R. C. (1999). Fas/Apo [apoptosis]-1 and associated proteins in the differentiating cerebral cortex: induction of caspase-dependent cell death and activation of NF-kappaB. *The Journal of Neuroscience, 19*, 1754–1770.

Cheng, E. H., Kirsch, D. G., Clem, R. J., Ravi, R., Kastan, M. B., Bedi, A., et al. (1997). Conversion of Bcl-2 to a Bax-like death effector by caspases. *Science, 278*, 1966–1968.

Chipuk, J. E., & Green, D. R. (2008). How do BCL-2 proteins induce mitochondrial outer membrane permeabilization? *Trends in Cell Biology, 18*, 157–264.

Choi, Y. J., Lee, G., & Park, J. H. (2006). Programmed cell death mechanisms of identifiable peptidergic neurons in Drosophila melanogaster. *Development, 133*, 2223–2232.

Chung, W. C., Swaab, D. F., & De Vries, G. J. (2000). Apoptosis during sexual differentiation of the bed nucleus of the stria terminalis in the rat brain. *Journal of Neurobiology, 43*, 234–243.

Clarke, P. G. (1990). Developmental cell death: morphological diversity and multiple mechanisms. *Anatomy and Embryology, 181*, 195–213.

Clarke, P. G., & Clarke, S. (1996). Nineteenth century research on naturally occurring cell death and related phenomena. *Anatomy and Embryology, 193*, 81–99.

Cohen, S., & Levi-Montalcini, R. (1956). A nerve growth-stimulating factor isolated from snake venom. *Proceedings of the National Academy of Sciences of the United States of America, 42*, 571–574.

Crowder, R. J., & Freeman, R. S. (1998). Phosphatidylinositol 3-kinase and Akt protein kinase are necessary and sufficient for the survival of nerve growth factor-dependent sympathetic neurons. *The Journal of Neuroscience, 18*, 2933–2943.

Crowley, C., Spencer, S. D., Nishimura, M. C., Chen, K. S., Pitts-Meek, S., Armanini, M. P., et al. (1994). Mice lacking nerve growth factor display perinatal loss of sensory and sympathetic neurons yet develop basal forebrain cholinergic neurons. *Cell, 76*, 1001–1011.

Dahm, L. M., & Landmesser, L. T. (1991). The regulation of synaptogenesis during normal development and following activity blockade. *The Journal of Neuroscience, 11*, 238–255.

D'Antonio, M., Droggiti, A., Feltri, M. L., Roes, J., Wrabetz, L., Mirsky, R., et al. (2006). TGFbeta type II receptor signaling controls Schwann cell death and proliferation in developing nerves. *The Journal of Neuroscience, 26*, 8417–8827.

Davalos, D., Grutzendler, J., Yang, G., Kim, J. V., Zuo, Y., Jung, S., et al. (2005). ATP mediates rapid microglial response to local brain injury in vivo. *Nature Neuroscience, 8*, 752–758.

Davies, A. M., Bandtlow, C., Heumann, R., Korsching, S., Rohrer, H., & Thoenen, H. (1987). Timing and site of nerve growth factor synthesis in developing skin in relation to innervation and expression of the receptor. *Nature, 326*, 353–358.

Davies, A. M., Lee, K. F., & Jaenisch, R. (1993). p75-deficient trigeminal sensory neurons have an altered response to NGF but not to other neurotrophins. *Neuron, 11*, 565–574.

Davis, E. C., Popper, P., & Gorski, R. A. (1996). The role of apoptosis in sexual differentiation of the rat sexually dimorphic nucleus of the preoptic area. *Brain Research, 734*, 10–18.

DeChiara, T. M., Vejsada, R., Poueymirou, W. T., Acheson, A., Suri, C., Conover, J. C., et al. (1995). Mice lacking the CNTF receptor, unlike mice lacking CNTF, exhibit profound motor neuron deficits at birth. *Cell, 83*, 313–322.

Decker, R. S. (1976). Influence of thyroid hormones on neuronal death and differentiation in larval Rana pipiens. *Developmental Biology, 49*, 101–118.

Decker, R. S. (1977). Lysosomal properties during thyroxine-induced lateral motor column neurogenesis. *Brain Research, 132*, 407–422.

Dekaban, A. S., & Sadowsky, D. (1978). Changes in brain weights during the span of human life: relation of brain weights to body heights and body weights. *Annals of Neurology, 4*, 345–356.

Delcroix, J. D., Valletta, J. S., Wu, C., Hunt, S. V., Kowal, A. S., & Mobley, W. C. (2003). NGF signaling in sensory neurons: evidence that early endosomes carry NGF retrograde signals. *Neuron, 39*, 69–84.

de Rivero Vaccari, J. C., Casey, G. P., Aleem, S., Park, W. M., & Corriveau, R. A. (2006). NMDA receptors promote survival in somatosensory relay nuclei by inhibiting Bax-dependent developmental cell death. *Proceedings of the National Academy of Sciences of the United States of America, 103*, 16971–16976.

Deshmukh, M., Du, C., Wang, X., Johnson, E. M., Jr. (2002). Exogenous smac induces competence and permits caspase activation in sympathetic neurons. *The Journal of Neuroscience, 22*, 8018–8027.

Detwiler, S. R. (1936). *Neuroembryology. An Experimental Study*. New York: MacMillan.

Dieterich, D. C., Karpova, A., Mikhaylova, M., Zdobnova, I., König, I., Landwehr, M., et al. (2008). Caldendrin-Jacob: a protein liaison that couples NMDA receptor signalling to the nucleus. *PLoS Biology, 6*, e34.

Dong, H., Wade, M., Williams, A., Lee, A., Douglas, G. R., & Yauk, C. (2005). Molecular insight into the effects of hypothyroidism on the developing cerebellum. *Biochemical and Biophysical Research Communications, 330*, 1182–1193.

Du, C., Fang, M., Li, Y., Li, L., & Wang, X. (2000). Smac, a mitochondrial protein that promotes cytochrome c-dependent caspase activation by eliminating IAP proteins. *Cell, 102*, 33–42.

Duchen, M. R. (2000). Mitochondria and calcium: from cell signalling to cell death. *The Journal of Physiology, 529*, 57–68.

Dünker, N., & Krieglstein, K. (2003). Reduced programmed cell death in the retina and defects in lens and cornea of Tgfbeta2(-/-) Tgfbeta3(-/-) double-deficient mice. *Cell and Tissue Research, 313*, 1–10.

Ebens, A., Brose, K., Leonardo, E. D., Hanson, M. G., Jr, Bladt, F., Birchmeier, C., et al. (1996). Hepatocyte growth factor/scatter factor is an axonal chemoattractant and a neurotrophic factor for spinal motor neurons. *Neuron, 17*, 1157–1172.

Ehlers, M. D., Kaplan, D. R., Price, D. L., & Koliatsos, V. E. (1995). NGF-stimulated retrograde transport of TrkA in the mammalian nervous system. *The Journal of Cell Biology, 130*, 149–156.

Eisen, J. S., & Melancon, E. (2001). Interactions with identified muscle cells break motoneuron equivalence in embryonic zebrafish. *Nature Neuroscience, 4*, 1065–1070.

Elliott, J., Cayouette, M., & Gravel, C. (2006). The CNTF/LIF signaling pathway regulates developmental programmed cell death and differentiation of rod precursor cells in the mouse retina in vivo. *Developmental Biology, 300*, 583–598.

Ernfors, P., Lee, K. F., & Jaenisch, R. (1994). Mice lacking brain-derived neurotrophic factor develop with sensory deficits. *Nature, 368*, 147–150.

Ernfors, P., Van De Water, T., Loring, J., & Jaenisch, R. (1995). Complementary roles of BDNF and NT-3 in vestibular and auditory development. *Neuron, 14*, 1153–1164.

Ernst, M. (1926). Ueber Untergang von Zellen während der normalen Entwicklung bei Wirbeltieren. *Zeitschrift fur Anatomie und Entwicklungsgeschichte, 79*, 228–262.

Fahnestock, M., Michalski, B., Xu, B., & Coughlin, M. D. (2001). The precursor pro-nerve growth factor is the predominant form of nerve growth factor in brain and is increased in Alzheimer's disease. *Molecular and Cellular Neurosciences, 18*, 210–220.

Fahrbach, S. E., Choi, M. K., & Truman, J. W. (1994). Inhibitory effects of actinomycin D and cycloheximide on neuronal death in adult Manduca sexta. *Journal of Neurobiology, 25*, 59–69.

Fan, H., Favero, M., & Vogel, M. W. (2001). Elimination of Bax expression in mice increases cerebellar purkinje cell numbers but not the number of granule cells. *The Journal of Comparative Neurology, 436*, 82–91.

Farinas, I., Yoshida, C. K., Backus, C., & Reichardt, L. F. (1996). Lack of neurotrophin-3 results in death of spinal sensory neurons and premature differentiation of their precursors. *Neuron, 17*, 1065–1078.

Farinelli, S. E., & Greene, L. A. (1996). Cell cycle blockers mimosine, ciclopirox, and deferoxamine prevent the death of PC12 cells and postmitotic sympathetic neurons after removal of trophic support. *The Journal of Neuroscience, 16*, 1150–1162.

Finlay, B. L., & Slattery, M. (1983). Local differences in the amount of early cell death in neocortex predict adult local specializations. *Science, 219*, 1349–1351.

Fiske, B. K., & Brunjes, P. C. (2001). NMDA receptor regulation of cell death in the rat olfactory bulb. *Journal of Neurobiology, 47*, 223–232.

Forger, N. G., Rosen, G. J., Waters, E. M., Jacob, D., Simerly, R. B., & de Vries, G. J. (2004). Deletion of Bax eliminates sex differences in the mouse forebrain. *Proceedings of the National Academy of Sciences of the United States of America, 101*, 13666–13671.

Frade, J. M., & Barde, Y. A. (1999). Genetic evidence for cell death mediated by the nerve growth factor and the neurotrophin receptor p75 in the developing mouse retina and spinal cord. *Dev, 126*, 683–690.

Frade, J. M., Rodríguez-Tébar, A., & Barde, Y. A. (1996). Induction of cell death by endogenous nerve growth factor through its p75 receptor. *Nature, 383*, 166–168.

Freeman, L. M., Watson, N. V., & Breedlove, S. M. (1997). Androgen spares androgen-insensitive motoneurons from apoptosis in the spinal nucleus of the bulbocavernosus in rats. *Hormones and Behavior, 30*, 424–433.

Freeman, M. R., Delrow, J., Kim, J., Johnson, E., & Doe,, C. Q. (2003). Unwrapping glial biology. Gcm target genes regulating glial development, diversification, and function. *Neuron, 38*, 567–580.

Freeman, R. S., Estus, S., Johnson, E. M., Jr. (1994). Analysis of cell cycle-related gene expression in postmitotic neurons: selective induction of Cyclin D1 during programmed cell death. *Neuron, 12*, 343–355.

Gage, F. H., Armstrong, D. M., Williams, L. R., & Varon, S. (1988). Morphological response of axotomized septal neurons to nerve growth factor. *The Journal of Comparative Neurology, 269*, 147–155.

Gagliardini, V., Fernandez, P. A., Lee, R. K., Drexler, H. C., Rotello, R. J., Fishman, M. C., et al. (1994). Prevention of vertebrate neuronal death by the crmA gene. *Science, 263*, 826–828.

Gentry, J. J., Rutkoski, N. J., Burke, T. L., & Carter, B. D. (2004). A functional interaction between the p75 neurotrophin receptor interacting factors, TRAF6 and NRIF. *Journal of Biological Chemistry, 279*, 16646–16656.

Ghosh, A., Carnahan, J., & Greenberg, M. E. (1994). Requirement for BDNF in activity-dependent survival of cortical neurons. *Science, 263*, 1618–1623.

Gorski, R. A., Gordon, J. H., Shryne, J. E., & Southam, A. M. (1978). Evidence for a morphological sex difference within the medial preoptic area of the rat brain. *Brain Research, 148*, 333–346.

Gould, E., Cameron, H. A., & McEwen, B. S. (1994). Blockade of NMDA receptors increases cell death and birth in the developing rat dentate gyrus. *The Journal of Comparative Neurology, 340*, 551–565.

Gould, T. W., & Oppenheim, R. W. (2004). The function of neurotrophic factor receptors expressed by the developing adductor motor pool in vivo. *The Journal of Neuroscience, 24*, 4668–4682.

Gould, T. W., Yonemura, S., Oppenheim, R. W., Ohmori, S., & Enomoto, H. (2008). The neurotrophic effects of glial cell line-derived neurotrophic factor on spinal motoneurons are restricted to fusimotor subtypes. *J Neurosci, 28*, 2131–2146.

Grisham, W., Lee, J., McCormick, M. E., Yang-Stayner, K., & Arnold, A. P. (2002). Antiandrogen blocks estrogen-induced masculinization of the song system in female zebra finches. *Journal of Neurobiology, 51*, 1–8.

Guillamón, A., de Blas, M. R., & Segovia, S. (1988). *Effects of sex steroids on the development of the locus coeruleus in the rat*.

Gurney, M. E. (1981). Hormonal control of cell form and number in the zebra finch song system. *The Journal of Neuroscience, 1*, 658–673.

Hafidi, A., Moore, T., & Sanes, D. H. (1996). Regional distribution of neurotrophin receptors in the developing auditory brainstem. *The Journal of Comparative Neurology, 367*, 454–464.

Hamanoue, M., Middleton, G., Wyatt, S., Jaffray, E., Hay, R. T., & Davies, A. M. (1999). p75-mediated NF-kappaB activation enhances the survival response of developing sensory neurons to nerve growth factor. *Molecular and Cellular Neurosciences, 14*, 28–40.

Hamburger, V. (1934). The effects of wing bud extirpation on the development of the central nervous system in chick embryos. *Journal of Experimental Zoology, 68*, 449–494.

Hamburger, V., & Levi-Montalcini, R. (1949). Proliferation, differentiation and degeneration in the spinal ganglia of the chick embryo under normal and experimental conditions. *Journal of Experimental Zoology, 111*, 457–501.

Ham, J., Babij, C., Whitfield, J., Pfarr, C. M., Lallemand, D., Yaniv, M., et al. (1995). A c-Jun dominant negative mutant protects sympathetic neurons against programmed cell death. *Neuron, 14*, 927–939.

Hardingham, G. E., Fukunaga, Y., & Bading, H. (2002). Extrasynaptic NMDARs oppose synaptic NMDARs by triggering CREB shut-off and cell death pathways. *Nature Neuroscience, 5*, 405–414.

Harlin, H., Reffey, S. B., Duckett, C. S., Lindsten, T., & Thompson, C. B. (2001). Characterization of XIAP-deficient mice. *Molecular and Cellular Biology, 21*, 3604–3608.

Harris, J. A., Hardie, N. A., Bermingham-McDonogh, O., & Rubel, E. W. (2005). Gene expression differences over a critical period of afferent-dependent neuron survival in the mouse auditory brainstem. *The Journal of Comparative Neurology, 493*, 460–474.

Harris, J. A., Iguchi, F., Seidl, A. H., Lurie, D. I., & Rubel, E. W. (2008). Afferent deprivation elicits a transcriptional response associated with neuronal survival after a critical period in the mouse cochlear nucleus. *The Journal of Neuroscience, 28*, 10990–11002.

Hashimoto, K., Curty, F. H., Borges, P. P., Lee, C. E., Abel, E. D., Elmquist, J. K., et al. (2001). An unliganded thyroid hormone receptor causes severe neurological dysfunction. *Proceedings of the National Academy of Sciences of the United States of America, 98*, 3998–4003.

Hay, B. A., Wassarman, D. A., & Rubin, G. M. (1995). Drosophila homologs of baculovirus inhibitor of apoptosis proteins function to block cell death. *Cell, 83*, 1253–1262.

Haynes, S. E., Hollopeter, G., Yang, G., Kurpius, D., Dailey, M. E., Gan, W. B., et al. (2006). The P2Y12 receptor regulates microglial activation by extracellular nucleotides. *Nature Neuroscience, 9*, 1512–1519.

Heerssen, H. M., Pazyra, M. F., & Segal, R. A. (2004). Dynein motors transport activated Trks to promote survival of target-dependent neurons. *Nature Neuroscience, 7*, 596–604.

Hendry, I. A., Stöckel, K., Thoenen, H., & Iversen, L. L. (1974). The retrograde axonal transport of nerve growth factor. *Brain Research, 68*, 103–121.

Hengartner, M. O., Ellis, R. E., & Horvitz, H. R. (1992). Caenorhabditis elegans gene ced-9 protects cells from programmed cell death. *Nature, 356*, 494–499.

Herrup, K., & Busser, J. C. (1995). The induction of multiple cell cycle events precedes target-related neuronal death. *Development, 121*, 2385–2395.

Hodge, R. D., D'Ercole, A. J., & O'Kusky, J. R. (2007). Insulin-like growth factor-I (IGF-I) inhibits neuronal apoptosis in the developing cerebral cortex in vivo. *International Journal of Developmental Neuroscience, 25*, 233–241.

Hoeppner, D. J., Hengartner, M. O., & Schnabel, R. (2001). Engulfment genes cooperate with ced-3 to promote cell death in Caenorhabditis elegans. *Nature, 412*, 202–206.

Hoffman, K. L., & Weeks, J. C. (1998). Programmed cell death of an identified motoneuron in vitro: Temporal requirements for steroid exposure and protein synthesis. *Journal of Neurobiology, 35*, 300–322.

Hoffman, K. L., & Weeks, J. C. (2001). Role of caspases and mitochondria in the steroid-induced programmed cell ceath of a motoneuron during metamorphosis. *Developmental Biology, 229*, 517–536.

Hollyday, M., & Hamburger, V. (1976). Reduction of the naturally occurring motor neuron loss by enlargement of the periphery. *The Journal of Comparative Neurology, 170*, 311–320.

Homma, S., Yaginuma, H., Vinsant, S., Seino, M., Kawata, M., Gould, T., et al. (2003). Differential expression of the GDNF family receptors RET and GFRalpha1, 2, and 4 in subsets of motoneurons: a relationship between motoneuron birthdate and receptor expression. *The Journal of Comparative Neurology, 456*, 245–259.

Hong, J. R., Lin, G. H., Lin, C. J., Wang, W. P., Lee, C. C., Lin, T. L., et al. (2004). Phosphatidylserine receptor is required for the engulfment of dead apoptotic cells and for normal embryonic development in zebrafish. *Development, 131*, 5417–5427.

Howe, C. L., & Mobley, W. C. (2005). Long-distance retrograde neurotrophic signaling. *Current Opinion in Neurobiology, 15*, 40–48.

Howe, C. L., Valletta, J. S., Rusnak, A. S., & Mobley, W. C. (2001). NGF signaling from clathrin-coated vesicles: evidence that signaling endosomes serve as a platform for the Ras–MAPK pathway. *Neuron, 32*, 801–814.

Huang, E. J., & Reichardt, L. F. (2001). Neurotrophins: Roles in neuronal development and function. *Ann Rev Neurosci, 24*, 677–736.

Hughes, A. F. (1961). Cell degeneration in the larval ventral horn of Xenopus laevis. *Journal of Embryology and Experimental Morphology, 9*, 269–284.

Hume, D. A., Perry, V. H., & Gordon, S. (1983). Immunohistochemical localization of a macrophage-specific antigen in developing mouse retina: phagocytosis of dying neurons and differentiation of microglial cells to form a regular array in the plexiform layers. *The Journal of Cell Biology, 97*, 253–257.

Hyson, R. L., & Rubel, E. W. (1989). Transneuronal regulation of protein synthesis in the brain-stem auditory system of the chick requires synaptic activation. *The Journal of Neuroscience, 9*, 2835–2845.

Ibanez, C. F. (1994). Structure-function relationships in the neurotrophin family. *Journal of Neurobiology, 25*, 1349–1361.

Ikonomidou, C., Bosch, F., Miksa, M., Bittigau, P., Vockler, J., Dikranian, K., et al. (1999). Blockade of NMDA receptors and apoptotic neurodegeneration in the developing brain. *Science, 283*, 70–74.

Johnson, D., Lanahan, A., Buck, C. R., Sehgal, A., Morgan, C., Mercer, E., et al. (1986). Expression and structure of the human NGF receptor. *Cell, 47*, 545–554.

Johnson, E. M., Jr (1978). Destruction of the sympathetic nervous system in neonatal rats and hamsters by vinblastine: prevention by concomitant administration of nerve growth factor. *Brain Research, 141*, 105–118.

Johnson, E. M., Jr, Andres, R. Y., & Bradshaw, R. A. (1978). Characterization of the retrograde transport of nerve growth factor (NGF) using high specific activity [125I] NGF. *Brain Research, 150*, 319–331.

Kanamoto, T., Mota, M. A., Takeda, K., Rubin, L. L., Miyazopo, K., Ichijo, H., et al. (2000). Role of apoptosis signal-regulating kinase in regulation of the c-Jun N-terminal kinase pathway and apoptosis in sympathetic neurons. *Molecular and Cellular Biology, 20*, 196–204.

Kaplan, D. R., Hempstead, B. L., Martin-Zanca, D., Chao, M. V., & Parada, L. F. (1991a). The trk proto-oncogene product: a signal transducing receptor for nerve growth factor. *Science, 252*, 554–558.

Kaplan, D. R., Martin-Zanca, D., & Parada, L. F. (1991b). Tyrosine phosphorylation and tyrosine kinase activity of the trk proto-oncogene product induced by NGF. *Nature, 350*, 158–160.

Kaur, C., Hao, A. J., Wu, C. H., & Ling, E. A. (2001). Origin of microglia. *Microscopy Research and Technique, 54*, 2–9.

Kinch, G., Hoffman, K. L., Rodrigues, E. M., Zee, M. C., & Weeks, J. C. (2003). Steroid-triggered programmed cell death of a motoneuron is autophagic and involves structural changes in mitochondria. *The Journal of Comparative Neurology, 457*, 384–403.

Kinchen, J. M., Cabello, J., Klingele, D., Wong, K., Feichtinger, R., Schnabel, H., et al. (2005). Two pathways converge at CED-10 to mediate actin rearrangement and corpse removal in C. elegans. *Nature, 434*, 93–99.

Kirn, J. R., & DeVoogd, T. J. (1989). Genesis and death of vocal control neurons during sexual differentiation in the zebra finch. *The Journal of Neuroscience, 9*, 3176–3187.

Klein, R., Smeyne, R. J., Wurst, W., Long, L. K., Auerbach, B. A., Joyner, A. L., et al. (1993). Targeted disruption of the trkB neurotrophin receptor gene results in nervous system lesions and neonatal death. *Cell, 8*, 113–122.

Konishi, M., & Akutagawa, E. (1985). Neuronal growth, atrophy and death in a sexually dimorphic song nucleus in the zebra finch brain. *Nature, 315*, 145–147.

Krieglstein, K., Richter, S., Farkas, L., Schuster, N., Dunker, N., Oppenheim, R. W., et al. (2000). Reduction of endogenous transforming growth factors beta prevents ontogenetic neuron death. *Nature Neuroscience, 3*, 1085–1090.

Kuida, K., Zheng, T. S., Na, S., Kuan, C., Yang, D., Karasuyama, H., et al. (1996). Decreased apoptosis in the brain and premature lethality in CPP32-deficient mice. *Nature, 384*, 368–372.

Kumar, A., Sinha, R. A., Tiwari, M., Pal, L., Shrivastava, A., Singh, R., et al. (2006). Increased pro-nerve growth factor and p75 neurotrophin receptor levels in developing hypothyroid rat cerebral cortex are associated with enhanced apoptosis. *Endocrinology, 147*, 4893–4903.

Kurada, P., & White, K. (1998). Ras promotes cell survival in Drosophila by downregulating hid expression. *Cell, 95*, 319–329.

Kurant, E., Axelrod, S., Leaman, D., & Gaul, U. (2008). Six-microns-under acts upstream of Draper in the glial phagocytosis of apoptotic neurons. *Cell, 133*, 498–509.

Landmesser, L., & Pilar, G. (1974). Synaptic transmission and cell death during normal ganglionic development. *The Journal of Physiology, 241*, 737–749.

Larsell, O. (1931). The effect of experimental excision of one eye on the development of the optic lobe and opticus layer in larvae of the tree-frog. *Journal of Experimental Zoology, 58*, 1–20.

Lazebnik, Y. A., Kaufmann, S. H., Desnoyers, S., Poirier, G. G., & Earnshaw, W. C. (1994). Cleavage of poly(ADP-ribose) polymerase by a proteinase with properties like ICE. *Nature, 371*, 346–347.

Lee, F. S., & Chao, M. V. (2001). Activation of Trk neurotrophin receptors in the absence of neurotrophins. *Proceedings of the National Academy of Sciences of the United States of America, 98*, 3555–3560.

Lee, R., Kermani, P., Teng, K. K., & Hempstead, B. L. (2001). Regulation of cell survival by secreted proneurotrophins. *Science, 294*, 1945–1948.

Leibrock, J., Lottspeich, F., Hohn, A., Hofer, M., Gengerer, B., Masiakowski, P., et al. (1989). Molecular cloning and expression of brain-derived neurotrophic factor. *Nature, 341*, 149–152.

Levi-Montalcini, R. (1949). The development of the acoustico-vestibular centers in the chick embryo in the absence of the afferent root fibers and of descending fiber tracts. *The Journal of Comparative Neurology, 91*, 209–241.

Levi-Montalcini, R., & Booker, B. (1960a). Excessive groth of the sympathetic ganglia evoked by a protein isolated from mouse salivary glands. *Proceedings of the National Academy of Sciences of the United States of America, 46*, 373–384.

Levi-Montalcini, R., & Booker, B. (1960b). Destruction of the sympathetic ganglia in mammals by an antiserum to a nerve growth protein. *Proceedings of the National Academy of Sciences of the United States of America, 46*, 384–391.

Levi-Montalcini, R., & Cohen, S. (1956). In vitro and in vivo effects if a nerve growth-stimulating agent isolated from snake venom. *Proceedings of the National Academy of Sciences of the United States of America, 42*, 695–699.

Levi-Montalcini, R., & Hamburger, V. (1951). Selective growth stimulating effects of mouse sarcoma on the sensory and sympathetic nervous system of the chick embryo. *Journal of Experimental Zoology, 116*, 321–361.

Levi-Montalcini, R., & Hamburger, V. (1953). A diffusible agent of mouse sarcoma, producing hyperplasia of sympathetic ganglia and hyperneurotization of viscera on the chick embryo. *Journal of Experimental Zoology, 123*, 233–287.

Li, M., Sendtner, M., & Smith, A. (1995). Essential function of LIF receptor in motor neurons. *Nature*, *378*, 724–727.

Li, P., Nijhawan, D., Budihardjo, I., Srinivasula, S. M., Ahmad, M., Alnemri, E. S., et al. (1997). Cytochrome c and dATP-dependent formation of Apaf-1/caspase-9 complex initiates an apoptotic protease cascade. *Cell*, *91*, 479–489.

Li, M. O., Sarkisian, M. R., Mehal, W. Z., Rakic, P., & Flavell, R. A. (2003). Phosphatidylserine receptor is required for clearance of apoptotic cells. *Science*, *302*, 1560–1563.

Linden, R. (1994). The survival of developing neurons: A review of afferent control. *Neurosci*, *58*, 671–682.

Linggi, M. S., Burke, T. L., Williams, B. B., Harrington, A., Kraemer, R., Hempstead, B. L., et al. (2005). Neurotrophin receptor interacting factor (NRIF) is an essential mediator of apoptotic signaling by the p75 neurotrophin receptor. *The Journal of Biological Chemistry*, *280*, 13801–13808.

Lin, L. F. H., Doherty, D. H., Lile, J. D., Bektesh, S., & Collins, F. (1993). GDNF: A glial cell line-derived neurotrophic factor for midbrain dopaminergic neurons. *Science*, *260*, 1130–1132.

Liu, D. X., Nath, N., Chellappan, S. P., & Greene, L. A. (2005). Regulation of neuron survival and death by p130 and associated chromatin modifiers. *Genes & Development*, *19*, 719–732.

Liu, J. P., Laufer, E., & Jessell, T. M. (2001). Assigning the positional identity of spinal motor neurons: rostrocaudal patterning of Hox-c expression by FGFs, Gdf11, and retinoids. *Neuron*, *32*(6), 997–1012.

Locksley, R. M., Killeen, N., & Lenardo, M. J. (2001). The TNF and TNF receptor superfamilies: integrating mammalian biology. *Cell*, *104*, 487–501.

Loeb, D. M., Maragos, J., Martin-Zanca, D., Chao, M. V., Parada, L. F., & Greene, L. A. (1991). The trk proto-oncogene rescues NGF responsiveness in mutant NGF-nonresponsive PC12 cell lines. *Cell*, *66*, 961–966.

Loetscher, H., Deuschle, U., Brockhaus, M., Reinhardt, D., Nelboeck, P., Mous, J., et al. (1997). Presenilins are processed by caspase-type proteases. *Journal of Biological Chemistry*, *272*, 20655–20659.

Lonze, B. E., Riccio, A., Cohen, S., & Ginty, D. D. (2002). Apoptosis, axonal growth defects, and degeneration of peripheral neurons in mice lacking CREB. *Neuron*, *34*, 371–385.

Lotto, R. B., Asavaritikrai, P., Vali, L., & Price, D. J. (2001). Target-derived neurotrophic factors regulate the death of developing forebrain neurons after a change in their trophic requirements. *The Journal of Neuroscience*, *21*, 3904–3910.

Luoma, J. I., & Zirpel, L. (2008). Deafferentation-induced activation of NFAT (nuclear factor of activated T-cells) in cochlear nucleus neurons during a developmental critical period: a role for NFATc4-dependent apoptosis in the CNS. *The Journal of Neuroscience*, *28*, 3159–3169.

Maas, J. W., Horstmann, S., Borasio, G. D., Anneser, J. M., Shooter, E. M., & Kahle, P. J. (1998). Apoptosis of central and peripheral neurons can be prevented with cyclin-dependent kinase/mitogen-activated protein kinase inhibitors. *Journal of Neurochemistry*, *70*, 1401–1410.

MacInnis, B. L., & Campenot, R. B. (2002). Retrograde support of neuronal survival without retrograde transport of nerve growth factor. *Science*, *295*, 1536–1539.

Maderdrut, J. L., Oppenheim, R. W., & Prevette, D. (1988). Enhancement of naturally occurring cell death in the sympathetic and parasympathetic ganglia of the chicken embryo following blockade of ganglionic transmission. *Brain Research*, *444*, 189–194.

Maher, P. A. (1988). Nerve growth factor induces protein tyrosine-phosphorylation. *Proceedings of the National Academy of Sciences of the United States of America*, *85*, 6788–6791.

Marin-Teva, J. L., Dusart, I., Colin, C., Gervais, A., van Rooijen, N., & Mallat, M. (2004). Microglia promote the death of developing Purkinje cells. *Neuron*, *41*, 535–547.

Markus, A., von Holst, A., Rohrer, H., & Heumann, R. (1997a). NGF-mediated survival depends on p21ras in chick sympathetic neurons from the superior cervical but not from lumbosacral ganglia. *Developmental Biology*, *191*, 306–310.

Markus, M. A., Kahle, P. J., Winkler, A., Horstmann, S., Anneser, J. M., & Borasio, G. D. (1997b). Survival-promoting activity of inhibitors of cyclin-dependent kinases on primary neurons correlates with inhibition of c-Jun kinase-1. *Neurobiology of Disease*, *4*, 122–133.

Maroney, A. C., Glicksman, M. A., Basma, A. N., Walton, K. M., Knight, E., Jr, Murphy, C. A., et al. (1998). Motoneuron apoptosis is blocked by CEP-1347 (KT 7515), a novel inhibitor of the JNK signaling pathway. *The Journal of Neuroscience*, *18*, 104–111.

Martin, D. P., Schmidt, R. E., DiStefano, P. S., Lowry, O. H., Carter, J. G., Johnson, E. M., Jr (1988). Inhibitors of protein synthesis and RNA synthesis prevent neuronal death caused by nerve growth factor deprivation. *The Journal of Cell Biology*, *106*, 829–844.

Martinou, J. C., Dubois-Dauphin, M., Staple, J. K., Rodriguez, I., Frankowski, H., Missotten, M., et al. (1994). Overexpression of BCL-2 in transgenic mice protects neurons from naturally occurring cell death and experimental ischemia. *Neuron*, *13*, 1017–1030.

Martins, R. A., Silveira, M. S., Curado, M. R., Police, A. I., & Linden, R. (2005). NMDA receptor activation modulates programmed cell death during early post-natal retinal development: a BDNF-dependent mechanism. *Journal of Neurochemistry*, *95*, 244–253.

Martin-Zanca, D., Barbacid, M., & Parada, L. F. (1990). Expression of the trk proto-oncogene is restricted to the sensory cranial and spinal ganglia of neural crest origin in mouse development. *Genes & Development*, *4*, 683–694.

Martin-Zanca, D., Hughes, S. H., & Barbacid, M. (1986). A human oncogene formed by the fusion of truncated tropomyosin and protein tyrosine kinase sequences. *Nature*, *319*, 743–748.

Maynard, T. M., Wakamatsu, Y., & Weston, J. A. (2000). Cell interactions within nascent neural crest cell populations transiently promote death of neurogenic precursors. *Development*, *127*, 4561–4572.

Mazzoni, I. E., Said, F. A., Aloyz, R., Miller, F. D., & Kaplan, D. (1999). Ras regulates sympathetic neuron survival by suppressing the p53-mediated cell death pathway. *The Journal of Neuroscience*, *19*, 9716–9727.

Mellén, M. A., de la Rosa, E. J., & Boya, P. (2008). The autophagic machinery is necessary for removal of cell corpses from the developing retinal neuroepithelium. *Cell Death and Differentiation*, *15*, 1279–1290.

Metzstein, M. M., Stanfield, G. M., & Horvitz, H. R. (1998). Genetics of programmed cell death in C. elegans: past, present and future. *Trends in Genetics*, *14*, 410–416.

Meyer-Franke, A., Kaplan, M. R., Pfrieger, F. W., & Barres, B. A. (1995). Characterization of the signaling interactions that promote the survival and growth of developing retinal ganglion cells in culture. *Neuron*, *15*, 805–819.

Miller, M. W. (1995). Relationship of the time of origin and death of neurons in rat somatosensory cortex: barrel versus septal cortex and projection versus local circuit neurons. *The Journal of Comparative Neurology*, *355*, 6–14.

Milligan, C. E., Prevette, D., Yaginuma, H., Homma, S., Cardwell, C., Fritz, L. C., et al. (1995). Peptide inhibitors of the ICE protease family arrest programmed cell death of motoneurons in vivo and in vitro. *Neuron*, *15*, 385–393.

Minichiello, L., Casagranda, F., Tatche, R. S., Stucky, C. L., Postigo, A., Lewin, G. R., et al. (1998). Point mutation in trkB causes loss of NT4-dependent neurons without major effects on diverse BDNF responses. *Neuron*, *21*, 335–345.

Minichiello, L., & Klein, R. (1996). TrkB and TrkC neurotrophin receptors cooperate in promoting survival of hippocampal and cerebellar granule neurons. *Genes & Development*, *10*, 2849–2858.

Monti, B., & Contestabile, A. (2000). Blockade of the NMDA receptor increases developmental apoptotic elimination of granule neurons and activates caspases in the rat cerebellum. *European Journal of Neuroscience*, *12*, 3117–3123.

Morris, J. A., Jordan, C. L., & Breedlove, S. M. (2008). Sexual dimorphism in neuronal number of the posterodorsal medial amygdala is independent of circulating androgens and regional volume in adult rats. *The Journal of Comparative Neurology*, *506*, 851–859.

Morris, J. A., Jordan, C. L., Dugger, B. N., & Breedlove, S. M. (2005). Partial demasculinization of several brain regions in adult male (XY) rats with a dysfunctional androgen receptor gene. *The Journal of Comparative Neurology*, *487*, 217–226.

Mostafapour, S. P., Cochran, S. L., Del Puerto, N. M., & Rubel, E. W. (2000). Patterns of cell death in mouse anteroventral cochlear nucleus neurons after unilateral cochlea removal. *The Journal of Comparative Neurology*, *426*, 561–571.

Motoyama, N., Wang, F., Roth, K. A., Sawa, H., Nakayama, K., Nakayama, K., et al. (1995). Massive cell death of immature hematopoietic cells and neurons in Bcl-x-deficient mice. *Science*, *267*, 1506–1510.

Nakashima, K., Wiese, S., Yanagisawa, M., Arakawa, H., Kimura, N., Hisatsune, T., et al. (1999). Developmental requirement of gp130 signaling in neuronal survival and astrocyte differentiation. *The Journal of Neuroscience*, *19*, 5429–5434.

Naumann, T., Casademunt, E., Hollerbach, E., Hofmann, J., Dechant, G., Frotscher, M., et al. (2002). Complete deletion of the neurotrophin receptor p75NTR leads to long-lasting increases in the number of basal forebrain cholinergic neurons. *The Journal of Neuroscience*, *22*, 2409–2418.

Ni, C. Z., Li, C., Wu, J. C., Spada, A. P., & Ely, K. R. (2003). Conformational restrictions in the active site of unliganded human caspase-3. *Journal of Molecular Recognition*, *16*, 121–124.

Nicholls, D. G., & Ward, M. W. (2000). Mitochondrial membrane potential and neuronal glutamate excitotoxicity: mortality and millivolts. *Trends in Neurosciences*, *23*, 166–174.

Nicholson, D. W., & Thornberry, N. A. (1997). Caspases: killer proteases. *Trends Biochem*, *22*, 299–306.

Nordeen, E. J., & Nordeen, K. W. (1988). Sex and regional differences in the incorporation of neurons born during song learning in zebra finches. *The Journal of Neuroscience*, *8*, 2869–2874.

Nordeen, E. J., Nordeen, K. W., Sengelaub, D. R., & Arnold, A. P. (1985). Androgens prevent normally occurring cell death in a sexually dimorphic spinal nucleus. *Science*, *229*, 671–673.

Nottebohm, F., & Arnold, A. P. (1976). Sexual dimorphism in vocal control areas of the songbird brain. *Science, 194,* 211–213.

Novak, K. D., Prevette, D., Wang, S., Gould, T. W., & Oppenheim, R. W. (2000). Hepatocyte growth factor/scatter factor is a neurotrophic survival factor for lumbar but not for other somatic motoneurons in the chick embryo. *The Journal of Neuroscience, 20,* 326–337.

Nuñez, J. L., Lauschke, D. M., & Juraska, J. M. (2001). Cell death in the development of the posterior cortex in male and female rats. *The Journal of Comparative Neurology, 436,* 32–41.

Nuñez, J. L., Sodhi, J., & Juraska, J. M. (2002). Ovarian hormones after postnatal day 20 reduce neuron number in the rat primary visual cortex. *Journal of Neurobiology, 52,* 312–321.

Nykjaer, A., Lee, R., Teng, K. K., Jansen, P., Madsen, P., Nielsen, M. S., et al. (2004). Sortilin is essential for proNGF-induced neuronal cell death. *Nature, 427,* 843–848.

Okada, H., Suh, W. K., Jin, J., Woo, M., Du, C., Elia, A., et al. (2002). Generation and characterization of Smac/DIABLO-deficient mice. *Molecular and Cellular Biology, 22,* 3509–3517.

Oppenheim, R. W. (1989). The neurotrophic theory and naturally occurring motoneuron death. *Trends in Neurosciences, 12,* 252–255.

Oppenheim, R. W. (1991). Cell death during development in the nervous system. *Ann Rev Neurosci, 14,* 453–501.

Oppenheim, R. W., Calderó, J., Cuitat, D., Esquerda, J., McArdle, J. J., Olivera, B. M., et al. (2008). The rescue of developing avian motoneurons from programmed cell death by a selective inhibitor of the fetal muscle-specific nicotinic acetylcholine receptor. *Dev Neurobiol, 68,* 972–980.

Oppenheim, R. W., Flavell, R. A., Vinsant, S., Prevette, D., Kuan, C. Y., & Rakic, P. (2001a). Programmed cell death of developing mammalian neurons after genetic deletion of caspases. *The Journal of Neuroscience, 21,* 4752–4760.

Oppenheim, R. W., Wiese, S., Prevette, D., Armanini, M., Wang, S., Houenou, L. J., et al. (2001b). Cardiotrophin-1, a muscle-derived cytokine, is required for the survival of subpopulations of developing motoneurons. *The Journal of Neuroscience, 21,* 1283–1291.

Oppenheim, R. W., Houenou, L. J., Johnson, J. E., Lin, L. F., Li, L., Lo, A. C., et al. (1995). Developing motor neurons rescued from programmed and axotomy-induced cell death by GDNF. *Nature, 373,* 344–346.

Oppenheim, R. W., Houenou, L. J., Parsadanian, A. S., Prevette, D., Snider, W. D., & Shen, L. (2000a). Glial cell line-derived neurotrophic factor and developing mammalian motoneurons: regulation of programmed cell death among motoneuron subtypes. *The Journal of Neuroscience, 20,* 5001–5011.

Oppenheim, R. W., Prevette, D., D'Costa, A., Wang, S., Houenou, L. J., & McIntosh, J. M. (2000b). Reduction of neuromuscular activity is required for the rescue of motoneurons from naturally occurring cell death by nicotinic-blocking agents. *The Journal of Neuroscience, 20,* 6117–6124.

Oppenheim, R. W., & Johnson, J. E. (2003). Programmed cell death and neurotrophic factors. In L. E. Squire, F. E. Bloom, S. K. McConnell, J. L. Roberts, N. C. Spitzer & M. J. Zigmond (Eds.), *Fundamental Neuroscience.* New York: Academic Press.

Oppenheim, R. W., Prevette, D., Tytell, M., & Homma, S. (1990). Naturally occurring and induced neuronal death in the chick embryo in vivo requires protein

and RNA synthesis: evidence for the role of cell death genes. *Developmental Biology, 138,* 104–113.

Oppenheim, R. W., Prevette, D., Yin, Q. W., Collins, F., & MacDonald, J. (1991). Control of embryonic motoneuron survival in vivo by ciliary neurotrophic factor. *Science, 29,* 1616–1618.

Oppenheim, R. W., Yin, Q. W., Prevette, D., & Yan, Q. (1992). Brain-derived neurotrophic factor rescues developing avian motoneurons from cell death. *Nature, 360,* 755–757.

Orrenius, S., Zhivotovsky, B., & Nicotera, P. (2003). Regulation of cell death: the calcium-apoptosis link. *Nature Reviews. Molecular Cell Biology, 4,* 552–565.

Padmanabhan, J., Park, D. S., Greene, L. A., & Shelanski, M. L. (1999). Role of cell cycle regulatory proteins in cerebellar granule neuron apoptosis. *The Journal of Neuroscience, 19,* 8747–8756.

Palmada, M., Kanwal, S., Rutkoski, N. J., Gustafson-Brown, C., Johnson, R. S., Wisdom, R., et al. (2002). c-jun is essential for sympathetic neuronal death induced by NGF withdrawal but not by p75 activation. *The Journal of Cell Biology, 158,* 453–461.

Papadia, S., Stevenson, P., Hardingham, N. R., Bading, H., & Hardingham, G. E. (2005). Nuclear Ca2+ and the cAMP response element-binding protein family mediate a late phase of activity-dependent neuroprotection. *The Journal of Neuroscience, 25,* 4279–4287.

Park, D. S., Farinelli, S. E., & Greene, L. A. (1996). Inhibitors of cyclin-dependent kinases promote survival of post-mitotic neuronally differentiated PC12 cells and sympathetic neurons. *Journal of Biological Chemistry, 271,* 8161–8169.

Park, D. S., Levine, B., Ferrari, G., & Greene, L. A. (1997). Cyclin dependent kinase inhibitors and dominant negative cyclin dependent kinase 4 and 6 promote survival of NGF-deprived sympathetic neurons. *The Journal of Neuroscience, 17,* 8975–8983.

Parks, T. N. (1979). Afferent influences on the development of the brain stem auditory nuclei of the chicken: otocyst ablation. *The Journal of Comparative Neurology, 183,* 665–678.

Parlato, R., Otto, C., Begus, Y., Stotz, S., & Schütz, G. (2007). Specific ablation of the transcription factor CREB in sympathetic neurons surprisingly protects against developmentally regulated apoptosis. *Development, 134,* 1663–1670.

Péquignot, Y., & Clarke, P. G. (1992). Changes in lamination and neuronal survival in the isthmo-optic nucleus following the intraocular injection of tetrodotoxin in chick embryos. *The Journal of Comparative Neurology, 321,* 336–350.

Petersen, M. A., & Dailey, M. E. (2004). Diverse microglial motility behaviors during clearance of dead cells in hippocampal slices. *Glia, 46,* 195–206.

Pilar, G., Landmesser, L., & Burstein, L. (1980). Competition for survival among developing ciliary ganglion cells. *Journal of Neurophysiology, 43,* 233–254.

Pittman, R., & Oppenheim, R. W. (1979). Cell death of motoneurons in the chick embryo spinal cord. IV. Evidence that a functional neuromuscular interaction is involved in the regulation of naturally occurring cell death and the stabilization of synapses. *The Journal of Comparative Neurology, 187,* 425–446.

Pompeiano, M., Blaschke, A. J., Flavell, R. A., Srinivasan, A., & Chun, J. (2000). Decreased Apoptosis in Proliferative and Postmitotic Regions of the Caspase 3-Deficient Embryonic Central Nervous System. *The Journal of Comparative Neurology, 423,* 1–12.

Potts, R. A., Dreher, B., & Bennett, M. R. (1982). The loss of ganglion cells in the developing retina of the rat. *Dev Brain Res, 3,* 481–486.

Purves, D. (1988). *Body and Brain: A Trophic Theory of Neural Connections.* Cambridge, MA: Harvard University Press.

Putcha, G. V., Moulder, K. L., Golden, J. P., Bouillet, P., Adams, J. A., Strasser, A., et al. (2001). Induction of BIM, a proapoptotic BH3-only BCL-2 family member, is critical for neuronal apoptosis. *Neuron, 29,* 615–628.

Rakic, S., & Zecevic, N. (2000). Programmed cell death in the developing human telencephalon. *European Journal of Neuroscience, 12,* 2721–2734.

Raoul, C., Henderson, C. E., & Pettmann, B. (1999). Programmed cell death of embryonic motoneurons triggered through the Fas death receptor. *The Journal of Cell Biology, 147,* 1049–1062.

Raynaud, A., Clairambault, P., Renous, S., & Gasc, J. P. (1977). Organisation des cornes ventrales de la moelle epiniere, dans les regions brachiale et lombaire, chez les embryons de Reptiles serpentiformes et de Reptiles a membres bien developpes. *Comptes Rendus Hebdomadaires Des Seances de l'Academie Des Sciences. Serie D, 19,* 1507–1509.

Reddien, P. W., Cameron, S., & Horvitz, H. R. (2001). Phagocytosis promotes programmed cell death in C. elegans. *Nature, 412,* 198–202.

Reid, S. N., & Juraska, J. M. (1992). Sex differences in neuron number in the binocular area of the rat visual cortex. *The Journal of Comparative Neurology, 321,* 448–455.

Riccio, A., Ahn, S., Davenport, C. M., Blendy, J. A., & Ginty, D. D. (1999). Mediation by a CREB family transcription factor of NGF-dependent survival of sympathetic neurons. *Science, 286,* 2358–2361.

Riccio, A., Pierchala, B. A., Ciarallo, C. L., & Ginty, D. D. (1997). An NGF-TrkA–mediated retrograde signal to transcription factor creb in sympathetic neurons. *Science, 277,* 1097–1100.

Robinow, S., Talbot, W. S., Hogness, D. S., & Truman, J. W. (1993). Programmed cell death in the Drosophila CNS is ecdysone-regulated and coupled with a specific ecdysone receptor isoform. *Development, 119,* 1251–1259.

Roy, N., Mahadevan, M. S., McLean, M., Shutler, G., Yaraghi, Z., Farahani, R., et al. (1995). The gene for neuronal apoptosis inhibitory protein is partially deleted in individuals with spinal muscular atrophy. *Cell, 80,* 167–178.

Rubel, E. W., Smith, D. J., & Miller, L. C. (1976). Organization and development of brain stem auditory nuclei of the chicken: Ontogeny of n. magnocellularis and n laminaris. *The Journal of Comparative Neurology, 166,* 469–490.

Sanders, E. J., Parker, E., Arámburo, C., & Harvey, S. (2005). Retinal growth hormone is an anti-apoptotic factor in embryonic retinal ganglion cell differentiation. *Exp Eye Res, 81,* 551–560.

Sengelaub, D. R., Jordan, C. L., Kurz, E. M., & Arnold, A. P. (1989). Hormonal control of neuron number in sexually dimorphic nuclei in the rat spinal cord. II. Development of the spinal nucleus of the bulbocavernosus in androgen insensitive (Tfm) rats. *The Journal of Comparative Neurology, 280,* 630–636.

Shindler, K. S., Latham, C. B., & Roth, K. A. (1997). Bax deficiency prevents the increased cell death of immature neurons in bcl-x-deficient mice. *The Journal of Neuroscience, 17,* 3112–3119.

Sleeman, M. W., Anderson, K. D., Lambert, P. D., Yancopoulos, G. D., & Wiegand, S. J. (2000). The ciliary neurotrophic factor and its receptor, CNTFR alpha. *Pharmaceutica Acta Helvetiae, 74,* 265–272.

Smeyne, R. J., Klein, R., Schnapp, A., Long, L. K., Bryant, S., Lewin, A., et al. (1994). Severe sensory and sympathetic neuropathies in mice carrying a disrupted Trk/NGF receptor gene. *Nature, 368*, 246–249.

Solum, D., Hughes, D., Major, M. S., & Parks, T. N. (1997). Prevention of normally occurring and deafferentation-induced neuronal death in chick brainstem auditory neurons by periodic blockade of AMPA/kainate receptors. *The Journal of Neuroscience, 17*, 4744–4751.

Sprecher, S. G., & Desplan, C. (2008). Switch of rhodopsin expression in terminally differentiated Drosophila sensory neurons. *Nature, 454*, 533–537.

Spritzer, M. D., & Galea, L. A. (2007). Testosterone and dihydrotestosterone, but not estradiol, enhance survival of new hippocampal neurons in adult male rats. *Developmental Neurobiology, 67*, 1321–1333.

Sutter, A., Riopelle, R. J., Harris-Warrick, R. M., & Shooter, E. M. (1979). Nerve growth factor receptors. Characterization of two distinct classes of binding sites on chick embryo sensory ganglia cells. *Journal of Biological Chemistry, 254*, 5972–5982.

Takahashi, M., Negishi, T., & Tashiro, T. (2008). Identification of genes mediating thyroid hormone action in the developing mouse cerebellum. *Journal of Neurochemistry, 104*, 640–652.

Tanaka, H. (1987). Chronic application of curare does not increase the level of motoneuron survival-promoting activity in limb muscle extracts during the naturally occurring motoneuron cell death period. *Developmental Biology, 124*, 347–357.

Tao, X., Finkbeiner, S., Arnold, D. B., Shaywitz, A. J., & Greenberg, M. E. (1998). Ca2+ influx regulates BDNF transcription by a CREB family transcription factor-dependent mechanism. *Neuron, 20*, 709–726.

Tekumalla, P. K., Tontonoz, M., Hesla, M. A., & Kirn, J. R. (2002). Effects of excess thyroid hormone on cell death, cell proliferation, and new neuron incorporation in the adult zebra finch telencephalon. *Journal of Neurobiology, 51*, 323–341.

Terrado, J., Burgess, R. W., DeChiara, T., Yancopoulos, G., Sanes, J. R., & Kato, A. C. (2001). Motoneuron survival is enhanced in the absence of neuromuscular junction formation in embryos. *The Journal of Neuroscience, 21*, 3144–3150.

Thomaidou, D., Mione, M. C., Cavanagh, J. F., & Parnavelas, J. G. (1997). Apoptosis and its relation to the cell cycle in the developing cerebral cortex. *The Journal of Neuroscience, 17*, 1075–1085.

Tierney, T. S., Russell, F. A., & Moore, D. R. (1997). Susceptibility of developing cochlear nucleus neurons to deafferentation-induced death abruptly ends just before the onset of hearing. *The Journal of Comparative Neurology, 378*, 295–306.

Troy, C. M., Stefanis, L., Greene, L. A., & Shelanski, M. L. (1997). Nedd2 is required for apoptosis after trophic factor withdrawal, but not superoxide dismutase (SOD1) downregulation, in sympathetic neurons and PC12 cells. *The Journal of Neuroscience, 17*, 1911–1918.

Truman, J. W., & Schwartz, L. M. (1984). Steroid regulation of neuronal death in the moth nervous system. *The Journal of Neuroscience, 4*, 274–280.

Tsui-Pierchala, B. A., & Ginty, D. D. (1999). Characterization of an NGF-P-TrkA retrograde signaling complex and age-dependent regulation of TrkA phosphorylation in sympathetic neurons. *The Journal of Neuroscience, 19*, 8207–8218.

Tsukahara, S., Hojo, R., Kuroda, Y., & Fujimaki, H. (2008). Estrogen modulates Bcl-2 family protein expression in the sexually dimorphic nucleus of the preoptic area of postnatal rats. *Neuroscience Letters, 432*, 58–63.

Turner, J. E., Barde, Y. A., Schwab, M. E., & Thoenen, H. (1982). Extract from brain stimulates neurite outgrowth from fetal rat retinal explants. *Brain Research, 282*, 77–83.

Vaillant, A. R., Mazzoni, I., Tudan, C., Boudreau, M., Kaplan, D. R., & Miller, F. D. (1999). Depolarization and neurotrophins converge on the phosphatidylinositol 3-kinase-Akt pathway to synergistically regulate neuronal survival. *The Journal of Cell Biology, 146*, 955–966.

Valdez, G., Akmentin, W., Philippidou, P., Kuruvilla, R., Ginty, D. D., & Halegoua, S. (2005). Pincher-mediated macroendo-cytosis underlies retrograde signaling by neurotrophin receptors. *The Journal of Neuroscience, 25*, 5236–5247.

Veis, D. J., Sorenson, C. M., Shutter, J. R., & Korsmeyer, S. J. (1993). Bcl-2-deficient mice demonstrate fulminant lymphoid apoptosis, polycystic kidneys, and hypopigmented hair. *Cell, 75*, 229–240.

Venegas, V., & Zhou, Z. (2007). Two alternative mechanisms that regulate the presentation of apoptotic cell engulfment signal in Caenorhabditis elegans. *Molecular Biology of the Cell, 18*, 3180–3192.

Verhagen, A. M., Ekert, P. G., Pakusch, M., Silke, J., Connolly, L. M., Reid, G. E., et al. (2000). Identification of DIABLO, a mammalian protein that promotes apoptosis by binding to and antagonizing IAP proteins. *Cell, 102*, 43–54.

Vernooy, S. Y., Copeland, J., Ghaboosi, N., Griffin, E. E., Yoo, S. J., & Hay, B. A. (2000). Cell death regulation in Drosophila: conservation of mechanism and unique insights. *The Journal of Cell Biology, 150*, F69–F76.

von Bartheld, C. S., Byers, M. R., Williams, R., & Bothwell, M. (1996). Anterograde transport of neurotrophins and axodendritic transfer in the developing visual system. *Nature, 379*, 830–833.

Von Bartheld, C. S., & Johnson, J. E. (2001). Target-derived BDNF (brain-derived neurotrophic factor) is essential for the survival of developing neurons in the isthmo-optic nucleus. *The Journal of Comparative Neurology, 433*, 550–564.

von Schack, D., Casademunt, E., Schweigreiter, R., Meyer, M., Bibel, M., & Dechant, G. (2001). Complete ablation of the neurotrophin receptor p75NTR causes defects both in the nervous and the vascular system. *Nature Neuroscience, 4*, 977–978.

Wade, J., & Arnold, A. P. (1996). Functional testicular tissue does not masculinize development of the zebra finch song system. *Proceedings of the National Academy of Sciences of the United States of America, 93*, 5264–5268.

Wang, H., & Tessier-Lavigne, M. (1999). En passant neurotrophic action of an intermediate axonal target in the developing mammalian CNS. *Nature, 401*, 765–769.

Wang, X., Wu, Y. C., Fadok, V. A., Lee, M. C., Gengyo-Ando, K., Cheng, L. C., et al. (2003). Cell corpse engulfment mediated by C. elegans phosphatidylserine receptor through CED-5 and CED-12. *Science, 302*, 1563–1566.

Watson, F. L., Heerssen, H. M., Moheban, D. B., Lin, M. Z., Sauvageot, C. M., Bhattacharyya, A., et al. (1999). Rapid nuclear responses to target-derived neurotrophins require retrograde transport of ligand-receptor complexes. *The Journal of Neuroscience, 19*, 7889–7900.

White, F. A., Keller-Peck, C. R., Knudson, C. M., Korsmeyer, S. J., & Snider, W. D. (1998). Widespread elimination of naturally occurring neuronal death in Bax-deficient mice. *The Journal of Neuroscience, 18*, 1428–1439.

White, K., Grether, M. E., Abrams, J. M., Young, L., Farrell, K., & Steller, H. (1994). Genetic control of programmed cell death in Drosophila. *Science, 264*, 677–683.

White, K., Tahaoglu, E., & Steller, H. (1996). Cell killing by the Drosophila gene reaper. *Science, 271*, 805–807.

Whitfield, J., Neame, S. J., Paquet, L., Bernard, O., & Ham, J. (2001). Dominant-negative c-jun promotes neuronal survival by reducing BIM expression and inhibiting mitochondrial cytochrome c release. *Neuron, 29*, 629–643.

Wilkinson, B. L., Sadler, K. A., & Hyson, R. L. (2002). Rapid deafferentation-induced upregulation of bcl-2 mRNA in the chick cochlear nucleus. *Mol Brain Res, 99*, 67–74.

Wong, R. O. L., & Hughes, A. (1987). Role of cell death in the topogenesis of neuronal distributions in the developing cat retinal ganglion cell layer. *The Journal of Comparative Neurology, 262*, 496–511.

Wright, L. L., & Smolen, A. J. (1987). The role of neuron death in the development of the gender difference in the number of neurons in the rat superior cervical ganglion. *International Journal of Developmental Neuroscience, 5*, 305–311.

Xia, Z., Dickens, M., Raingeaud, J., Davis, R. J., & Greenberg, M. E. (1995). Opposing effects of ERK and JNK-p38 MAP kinases on apoptosis. *Science, 270*, 1326–1331.

Xing, J., Ginty, D. D., & Greenberg, M. E. (1996). Coupling of the RAS-MAPK pathway to gene activation by RSK2, a growth factor-regulated CREB kinase. *Science, 273*, 959–963.

Xu, J., Gingras, K. M., Bengston, L., Di Marco, A., & Forger, N. G. (2001). Blockade of endogenous neurotrophic factors prevents the androgenic rescue of rat spinal motoneurons. *The Journal of Neuroscience, 21*, 4366–4372.

Yaar, M., Zhai, S., Pilch, P. F., Doyle, S. M., Eisenhauer, P. B., Fine, R. E., et al. (1997). Binding of beta-amyloid to the p75 neurotrophin receptor induces apoptosis. A possible mechanism for Alzheimer's disease. *The Journal of Clinical Investigation, 100*, 2333–2340.

Yamamoto, Y., Livet, J., Pollock, R. A., Garcés, A., Arce, V., deLapeyriére, O., et al. (1997). Hepatocyte growth factor (HGF/SF) is a muscle-derived survival factor for a subpopulation of embryonic motoneurons. *Development, 124*, 2903–2913.

Ye, H., Kuruvilla, R., Zweifel, L. S., & Ginty, D. D. (2003). Evidence in support of signaling endosome-based retrograde survival of sympathetic neurons. *Neuron, 39*, 57–68.

Youle, R. J., & Strasser, A. (2008). The BCL-2 protein family: opposing activities that mediate cell death. *Nature Reviews. Molecular Cell Biology, 9*, 47–59.

Yuan, J. Y., & Horvitz, H. R. (1990). The Caenorhabditis elegans genes ced-3 and ced-4 act cell autonomously to cause programmed cell death. *Developmental Biology, 138*, 33–41.

Yu, X., Odera, S., Chuang, C. H., Lu, N., & Zhou, Z. (2006). C. elegans Dynamin mediates the signaling of phagocytic receptor CED-1 for the engulfment and degradation of apoptotic cells. *Developmental Cell, 10*, 743–757.

Zee, M. C., & Weeks, J. C. (2001). Developmental change in the steroid hormone signal for cell-autonomous, segment-specific programmed cell death of a motoneuron. *Developmental Biology, 235*, 45–61.

Zhang, S. J., Zou, M., Lu, L., Lau, D., Ditzel, D. A., Delucinge-Vivier, C., et al. (2009). Nuclear calcium signaling controls expression of a large gene pool: identification of a gene program for acquired neuroprotection induced by synaptic activity. *PLoS Genetics, 5*, e1000604.

Zhou, Q., Choi, G., & Anderson, D. J. (2001a). The bHLH transcription factor Olig2 promotes oligodendrocyte differentiation in collaboration with Nkx2.2. *Neuron, 31,* 791–807.

Zhou, Z., Hartwieg, E., & Horvitz, H. R. (2001b). CED-1 is a transmembrane receptor that mediates cell corpse engulfment in C. elegans. *Cell, 104,* 43–56.

Zirpel, L., Janowiak, M. A., Veltri, C. A., & Parks, T. N. (2000). AMPA receptor-mediated, calcium-dependent CREB phosphorylation in a subpopulation of auditory neurons surviving activity deprivation. *The Journal of Neuroscience, 20,* 6267–6275.

Zirpel, L., Lippe, W. R., & Rubel, E. W. (1998). Activity-dependent regulation of [Ca2+]i in avian cochlear nucleus neurons: roles of protein kinases A and C and relation to cell death. *Journal of Neurophysiology, 79,* 2288–2302.

Zoeller, R. T., & Rovet, J. (2004). Timing of thyroid hormone action in the developing brain: Clinical observations and experimental findings. *Journal of Neuroendocrinology, 16,* 809–818.

Synapse formation and function

8

The emergence of functional connections between nerve cells distinguishes brain development from that of all other tissues. This process begins with axonal growth cones following a set of extracellular cues to reach a precise location within a distant target (see Chapters 5 and 6). When the growth cone finally comes in contact with an appropriate postsynaptic cell, a decision is made to stop growing and to differentiate into a presynaptic terminal. Simultaneously, the target neuron begins to create a minute specialization that will serve as the postsynaptic site. In fact, both the growth cone and postsynaptic neuron generate many of the components needed for neurotransmission well before innervation occurs, and the formation of a functional contact can be remarkably swift. However, the final adult phenotype may only be reached after weeks or months of further development.

Despite these unique properties, the precursors of synaptic transmission have been present for millions of years (**Figure 8.1**). This can be shown through a proteomic analysis of existing species from clades that diverged at different periods of evolution (Emes et al., 2008; Ryan and Grant, 2009). Unicellular eukaryotes, such as amoeba, have energy-dependent transporters to establish ion gradients, and cytoplasmic protein kinases to modify proteins in response to stimulation. In fact, when genes that encode postsynaptic proteins in mammals are compared to the genome of yeast, homologous genes can be identified for about 20% of them. Transmembrane proteins that are involved in synapse formation, such as the cadherin cell adhesion molecules and various intracellular anchoring proteins, are also present in unicellular eukaryotes. The emergence of some neurotransmitter receptors, such as ionotropic glutamate receptors, probably occurred in primitive metazoans (e.g., jellyfish). This suggests that glutamate receptors made use of a preexisting protosynaptic complex. Figure 8.1 shows the approximate time of divergence for representative species, and indicates what sort of synapse-associated genes are present in that species.

One general problem in studying synapses is that they are extremely small, often having a contact length of less than 1 μm. This makes them nearly impossible to see with a light microscope, and one might wonder how they were discovered in the first place. At the turn of the twentieth century, one group of biologists believed that neuronal processes fused with one another to produce long fibers with a continuous protoplasm, called a *syncytium* (**Figure 8.2**). Others believed that neurons remained separate, as had been shown for other cell types, and that they must be in contact with one another at the tips of their processes (Ramón y Cajal, 1905). The great interest that was then focused on the tips of neuronal processes led both to the discovery of growth cones in very young tissue (see Chapter 5) and to the first descriptions of presynaptic terminals in older animals (Held, 1897). Charles Sherrington, winner of the 1932 Nobel Prize in Physiology or Medicine, realized that a separation between nerve cells would allow for a new form of intercellular communication (cf. chemical transmission), and he popularized the term *synapse* (Sherrington, 1906).

The average mammalian neuron receives synapses along its soma and dendrites. The majority of these synapses release glutamate, which excites the postsynaptic neuron, or GABA, which acts to inhibit the neuron. Still others release neuromodulatory transmitters that regulate brain function. At a single glutamatergic synapse, there may be several types of receptors.

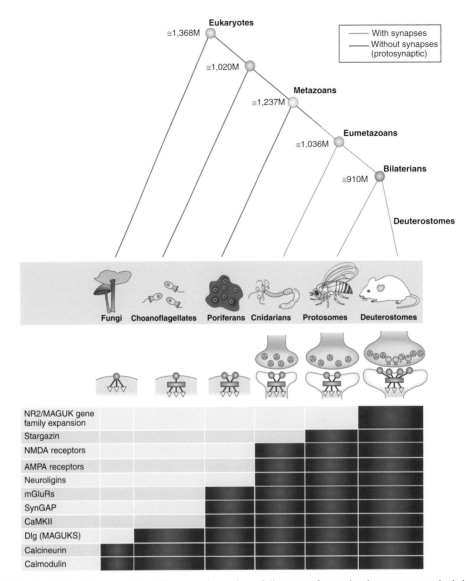

Fig. 8.1 Evolution of synapses. (Top) The approximate time of divergence for species that represent each clade are shown, in millions of years (M). Those without synapses (protosynaptic) are connected by a green line, and those with synapses are connected by a red line. (Bottom) The evolutionary appearance of several synaptic proteins is drawn schematically, and presented in tabular form, where a dark gray indicates that the molecule is present. Intracellular signaling components (yellow triangles: calcineurin, calmodulin) are found in Fungi, and intracellular proteins that cluster receptors and ion channels (red rectangle: Dlg) are found in unicellular Choanoflagellates. The appearance of glutamate receptors and synaptogenetic signals (blue circles: AMPA and NMDA receptors, neuroligin) are found in Cnidarians. Presynaptic vesicles (green) and neurotransmitter (yellow) are shown after the emergence of synaptic specializations. *(From Ryan and Grant, 2009)*

Glutamate binding causes some of these to gate open ion channels and leads others to activate a second messenger system. When glutamate causes ion channels to open, the change in membrane potential can rapidly recruit nearby ion channels (see Box: Maturation of Electrical Properties, next page). In the case of voltage-gated sodium channels, this leads to an action potential.

This simplified description of synaptic organization highlights many of the challenges to a developing nervous system. For example, each growth cone must identify an appropriate patch of membrane on which to differentiate. In the cortex, most glutamatergic synapses are located on postsynaptic specializations, called *dendritic spines*; GABAergic synapses tend to form on the cell body and proximal dendrite. A tight little cluster of GABA$_A$ receptors on a dendritic spine head would be of little use to the glutamate-releasing terminals that are located there. There must also be a mechanism to control the total number of synapses that can form on any one neuron. That number varies tremendously, from only a single excitatory synapse on neurons of the medial nucleus of the trapezoid body to over 10,000 for a cortical pyramidal neuron.

To make some sense of this complexity, we will consider separately how presynaptic terminals and postsynaptic specializations arise. Three general observations emerge from this section. First, neurons manufacture many of the synaptic

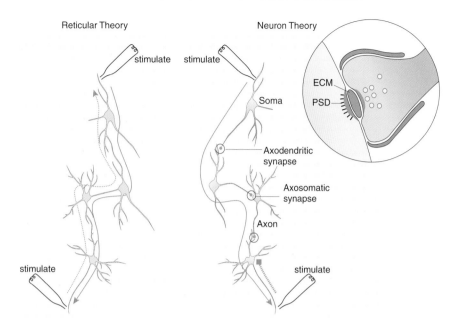

Fig. 8.2 Reticular versus neuron theory. Over a century ago, the nervous system was thought to be a syncytium (left) of cells that were joined together by their processes. This arrangement would permit electrical activity to travel through the syncytium (arrows) in either direction upon stimulation. As evidence mounted that neurons were separate cells (right), it was recognized that a chemical synapse (inset) would permit electrical activity to travel in only one direction. ECM, extracellular matrix; PSD, postsynaptic density.

building blocks even before making contact with one another. Second, intercellular signaling induces the differentiation of newly formed synapses from the moment of contact. Signals from glia, extracellular matrix, and neighboring neurons all participate in synaptogenesis. Third, synapses do not function in a mature manner for quite some time after they are fabricated. We will discuss how their functional properties change with development and how this might explain some of the behavioral limitations that young animals display (see Chapter 10).

Box 8.1   Maturation of Electrical Properties

Information processing by the central nervous system is based on electrical signals. The developmental regulation of each neuron's resting potential and voltage-gated ion channels is essential for the emergence of adult function. The resting membrane potential becomes more negative during development (Kullberg et al., 1977; Burgard and Hablitz, 1993; Tepper and Trent, 1993; Sanes, 1993; Ramoa and McCormick, 1994; Warren and Jones, 1997). This is due to regulation of extracellular K+ by glial cells which are proliferating and differentiating throughout the brain (Connors et al., 1982; Skoff et al., 1976; Syková et al., 1992). For example, extracellular K+ drops from about 35 mM in the cortex of newborn rabbits to 3 mM in adults (Mutani et al., 1974). This difference translates into a shift of almost 35 mV in membrane potential.

A few simple properties determine the size and speed of electrical events **(Figure 8.3)** . The first, membrane input resistance, determines how much the membrane potential will change for a given current pulse. The second, membrane time constant, determines how rapidly the membrane will reach a new potential when current is injected. Both of these properties tend to decrease with age, probably reflecting an increase in cell size (i.e., total membrane). Thus, input resistance decreases because the number of resistors (i.e., channels) increase as membrane is added to a cell. The mature neuron is able to

process information rapidly and accurately because the synaptic currents elicit brief changes in membrane potential.

The Action Potential: Sodium and Potassium Channels
When a neuron becomes slightly depolarized, perhaps owing to a synaptic potential, the opening of voltage-gated sodium channels permits a large depolarizing current due to the relatively high extracellular sodium concentration. As the neuron depolarizes, a second set of voltage-gated channels are activated that permit potassium to leave the cell, thus returning the membrane potential to rest. In many cases, the initial depolarization recruits a third type of voltage-gated channel that permits calcium to enter the neuron. When do these channels appear during development?

In many developing systems, the action potential is first carried by calcium ions. Since the calcium channels tend to remain open for a longer time, the action potentials can be very slow. Thus, *Xenopus* neurons begin life with 60–90 ms action potentials, although they quickly decrease to about 1 ms in duration **(Figure 8.4)**. Two basic changes explain this decrease. First, sodium channels become the primary conduit for inward current (Spitzer and Lamborghini, 1976; Baccaglini and Spitzer, 1977). Second, there is a 3.5-fold increase in a potassium channel current, called the *delayed rectifier*, that is

Box 8.1   Maturation of Electrical Properties—Cont'd

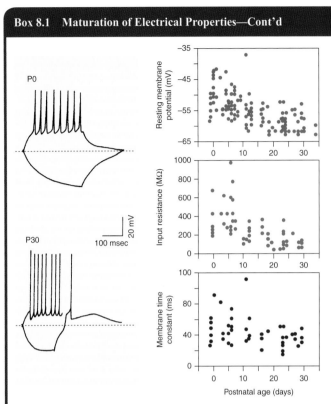

Fig. 8.3 Development of passive membrane properties. (Left) The intracellularly recorded voltage response to positive and negative current pulses in a P0 and a P30 neuron from ferret lateral geniculate nucleus (LGN) brain slices. The P0 neuron displayed a longer time constant and larger voltage deflection. (Right) Plots of membrane potential, input resistance, and time constant from LGN neurons during postnatal development. Membrane potential becomes about 10 mV more negative, input resistance decreases by about 200 MΩ, and the time constant decreases by about 10 ms. *(From Ramoa and McCormick, 1994)*

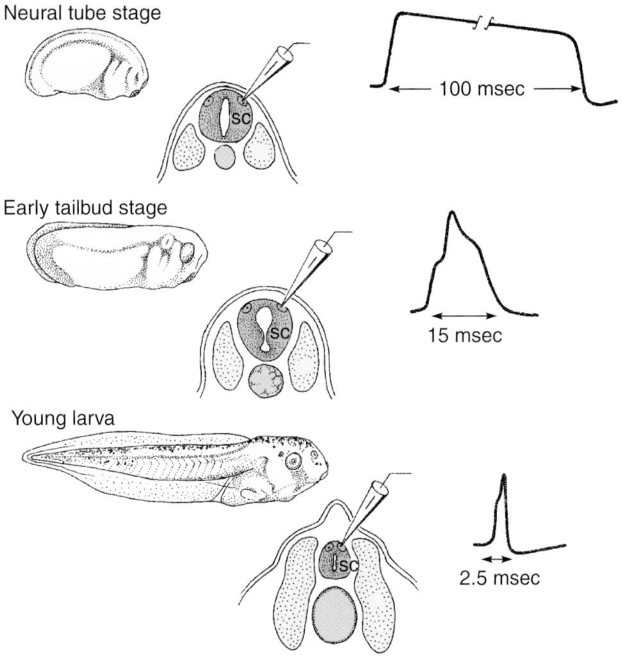

Fig. 8.4 Action potentials are initially calcium-dependent. (Top) When intracellular recordings are made from spinal cord (sc) Rohon-Beard neurons in neural tube stage *Xenopus* embryos, depolarizing current injection produces long-lasting calcium action potentials. (Middle) In early tailbud embryos, current injection evokes a mixed sodium/calcium response. (Bottom) In the young larva, current evokes a brief sodium-dependent action potential. *(Adopted from Spitzer, 1981)*

activated during membrane depolarization (Barish, 1986). The maturation of this large outward current brings the cell back to rest and limits the amount of calcium that enters during an action potential. Different kinds of sodium current can emerge at distinct times during development. For example, in cerebellar Purkinje neurons, the emergence of transient, resurgent, and persistent sodium currents occur at times corresponding with changes in action potential properties (Fry, 2006). At a molecular level, the eight genes that code for the Na(v)1 family of sodium channel pore-forming subunits display distinct expression patterns during zebrafish embryonic and larval development (Novak et al., 2006).

Sodium and potassium currents, as measured in dissociated *Xenopus* spinal neurons, increase dramatically within about 24 hours of the neuron's terminal mitosis (O'Dowd et al., 1988). Similar observations have been made in explants of chick cortex (Mori-Okamoto et al., 1983). In acutely dissociated rat cortical neurons, the sodium current density increases sixfold during the first two postnatal weeks (Huguenard et al., 1988). However, there is no uniform order of channel appearance in the nervous system. Chick motor neurons have significant sodium and delayed rectifying potassium currents from the outset, and there is a relatively late appearance of at least one type of potassium channel, and two types of calcium channels (McCobb et al., 1989; McCobb et al., 1990). Yet another potassium channel, which depends upon both

membrane potential *and* the intracellular calcium concentration to open, is generally expressed after the delayed rectifier (O'Dowd et al., 1988; Dourado and Dryer, 1992).

The localization of two sodium channels, Na(v)1.2 and Na(v)1.6, to the small gaps between myelin sheath segments (i.e., the nodes of Ranvier) requires a signal from oligodendrocytes. Clustering of Na(v)1.2 occurs first and is induced by a soluble factor released by oligodendrocytes. As myelination proceeds, Na(v)1.6 channels gradually replace Na(v)1.2 channels. In mice that lack compact myelin, Na(v)1.2 is found throughout adult axons, and little Na(v)1.6 is observed (Salzer et al., 2008; Vacher et al., 2008). The axonal cell adhesion molecule, neurofascin, is required for the accumulation of Na(v)1 channels at nodes of Ranvier, and deletion of the neurofascin gene spares myelination while eliminating Nav1 clusters (Sherman et al., 2005). Neurofascin recruits the cytoskeletal protein, ankyrin G, which serves to anchor Na(v)1 channels to the node (Dzhashiashvili et al., 2007).

Significance of Calcium Channel Expression

Calcium currents that are activated by small depolarizatons, called *low-voltage activated (LVA)* or *T currents*, are broadly expressed in developing tissue. As the nervous system matures, there is an increasing prominence of calcium channels that activate only when the cell is greatly depolarized **(Figure 8.5)**. These are referred to as *high-voltage activated (HVA)* or *N and L currents*. When hippocampal neurons from E19 rats are placed in a dissociated culture, only LVA currents are recorded at first. However, HVA currents appear over the next few

Box 8.1 Maturation of Electrical Properties—Cont'd

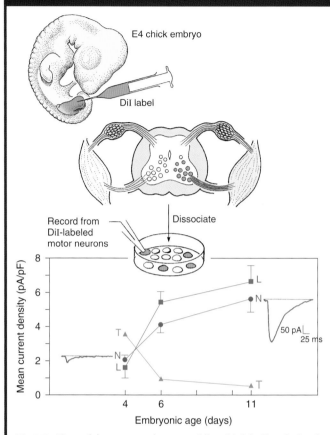

Fig. 8.5 Two calcium currents increase while a third declines in developing spinal motor neurons. A. To obtain identified motor neurons, a dye (DiI) was injected into the leg bud, and this dye was retrogradely transported by motor neurons. Thus, when the tissue was dissociated, it was possible to identify motor neurons because they carried the DiI label. B. When calcium currents were recorded from the dissociated motor neurons, it was found that T-type calcium channels declined with age, while N- and L-type channels increased. *(Adopted from McCobb et al., 1989)*

Regulation of Ionic Channel Expression

The addition of new channels to the membrane is necessary for most increases in current density. For example, when *Xenopus* neurons are grown in the presence of RNA or protein synthesis inhibitors, the transition from calcium- to sodium-dependent action potentials is prevented (Blair, 1983; O'Dowd, 1983). In like manner, transcription blockers prevent the normal increase in potassium current density (Ribera and Spitzer, 1989). The signal to increase production of potassium channels is present for only a brief period of time. A nine-hour exposure to an RNA synthesis inhibitor prevents the normal increase in potassium current density, even though RNA synthesis resumes upon withdrawal of the inhibitor. However, the appearance of A currents is not permanently blocked by transcription inhibitors. Regulation of channel expression within the neuron can occur at the level of both transcription and translation. Increasing the expression of the transcription factor, Even-skipped, is sufficient to decrease a fast K+ conductance in *Drosophila* motoneurons (Pym et al., 2006). In addition, a translational repressor, Pumilio, binds directly to the mRNA encoding a voltage-gated sodium channel, limiting its expression (Muraro et al., 2008).

If transynaptic signals regulate ion channel maturation, they remain largely unknown. However, the glycoprotein that stimulates AChR synthesis in muscle cells, neuregulin, can induce a twofold increase in sodium channels (Corfas and Fischbach, 1993). Other well-described growth factors, such as FGF, can upregulate the density of calcium channels in dissociated cultures of hippocampal neurons, an effect that requires protein synthesis (Shitaka et al., 1996).

The extrinsic signals that regulate ion channel expression are beginning to be understood in parasympathetic neurons of the chick ciliary ganglion. The expression of an A-type (I_A) and a calcium-activated potassium current ($I_{K[Ca]}$) is reduced when ciliary neurons are grown in dissociated culture, in the absence of their target or preganglionic afferents (Dourodo and Dryer, 1992). To determine whether synaptic connectivity influences potassium channel expression, in vivo manipulations were performed in which either the optic vesicle containing the target tissue, or a portion of the midbrain primordium containing the preganglionic nucleus was removed (Dourodo et al., 1994). The parasympathetic neurons were then acutely dissociated so that whole-cell voltage-clamp recordings could be obtained easily. The density of I_A was unaffected by either manipulation, although the channels did appear to open and close more rapidly than normal. In contrast, $I_{K[Ca]}$ was reduced by 90 to 100% following either target removal or deafferentation. In contrast, blocking the spontaneous activity of chick lumbar motor neurons does lead to a dramatic reduction in I_A expression, both in ovo and in vitro (Casavant et al., 2004).

A factor has been isolated from a target of the ciliary ganglion, the iris, that is able to upregulate the density of $I_{K[Ca]}$ (Subramony et al., 1996). When neurons are cultured in the presence of iris extract, the density of $I_{K[Ca]}$ reaches normal levels within seven hours **(Figure 8.6)**. This factor turns out to be a TGFβ1 family member. When an antibody directed against the TGFβ family is added to iris extract or injected into the eye, the expression of $I_{K[Ca]}$ on ciliary ganglion neurons is inhibited (Cameron et al., 1998). Interestingly, transcripts of the calcium-activated potassium channel are present in cultured ganglia before the current can be recorded, and the effect of iris extract does not require protein synthesis. However, brief exposure to TGFβ1 also elicits a persistent increase in $I_{K[Ca]}$ that depends on transcription (Lhuillier and Dryer, 2000). These results suggest that retrograde signals can affect

days and become a major contributor (Yaari et al., 1987). Similarly, it is the LVA calcium currents that are primarily observed when neurons from chick dorsal root ganglia, ciliary ganglia, or ventral horn are first recorded from (Gottmann et al., 1988; McCobb et al., 1989). These are overtaken by HVA currents within about 24–48 hours.

The initial appearance of LVA calcium channels can contribute greatly to a neuron's differentiation. For example, spontaneous calcium transients in developing *Xenopus* spinal neurons, largely carried by LVA calcium channels, have been implicated in the acquisition of GABAergic phenotype and process outgrowth (Spitzer, 1994). In fact, these calcium transients regulate the maturation of electrical properties, including a switch in potassium channel isoforms. The rate of activation for single potassium channels also increases by two to three times as *Xenopus* spinal neurons mature in vitro. This transition in channel kinetics is dependent upon calcium influx and can be induced by activation of a protein kinase C (Desarmenien and Spitzer, 1991).

Box 8.1 Maturation of Electrical Properties—Cont'd

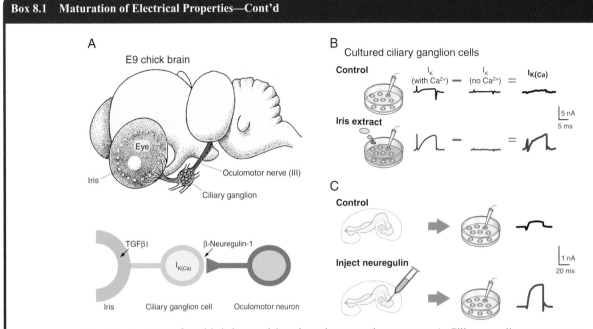

Fig. 8.6 An extract from iris induces calcium-dependent potassium currents. A. Ciliary ganglion neurons are innervated by occulomotor neuron axons and send their projections to innervate the iris (top). Two signaling systems regulate the expression of the calcium-activated potassium current ($I_{K[Ca]}$): target-derived TGFβ1 and afferent-derived β−Neuregulin-1 (bottom). B. Ciliary ganglion neurons were isolated from E9 chicks and placed in dissociated culture. When the cultures are grown in control culture medium, very little $I_{K[Ca]}$ is observed. This is shown by recording total current and subtracting the current in Ca_{2+}-free media. When an extract from the iris is added to the cultured neurons, a much larger $I_{K[Ca]}$ is recorded. This effect is also obtained with TGFβ1. C. When embryos were injected with β−Neuregulin-1 at E8, and dissociated neurons were recorded a day later, they displayed larger $I_{K[Ca]}$ as compared to controls. *(Adapted from Subramony et al., 1996; Cameron et al., 2001)*

the translation, insertion, or modification of potassium channels with a very short latency. The regulation of $I_{K[Ca]}$, and many other channels, is likely to depend on several signals. For example, there is a second isoform of TGFβ that inhibits the functional expression of $I_{K[Ca]}$, and an afferent-derived signal (neuregulin-1) participates in upregulating this channel (Cameron et al., 1999, 2001).

Electrical activity, itself, may affect the expression level of certain channels. Action potential blockade delays or prevents the normal increase in sodium and potassium current density in *Xenopus* myocytes in vitro (Linsdell and Moody, 1995). In fact, expression of $I_{K[Ca]}$ in the chick spinal cord is reduced significantly when activity is chronically inhibited by transfecting cells with a potassium channel that remains open and clamps the resting potential to a negative value (Yoon et al., 2008). In some

systems, the appearance of specific channels has been suggested to play a role in generating spontaneous activity that is essential for maturation of synaptic connections (Vasilyev and Barish, 2002; Picken et al., 2003). For example, spontaneous calcium spikes in the developing *Xenopus* system have been implicated directly in developmental events such as process outgrowth and channel expression. How are these spontaneous events initiated at the proper time during development? One mechanism involves control of resting membrane potential by the transporter that determines the potassium equilibrium potential (Na-K-ATPase). Neurons decrease the expression of a specific Na-K-ATPase subunit, thus inactivating the transporter and depolarizing the membrane potential. This apparently results in the activation of high voltage activated calcium channels and the initiation of calcium spikes (Chang and Spitzer, 2009).

WHAT DO NEWLY FORMED SYNAPSES LOOK LIKE?

Studies of the synapse began in earnest during the early 1950s with the arrival of two new techniques. Electron microscopy permitted neuroscientists to see the complex structure of synaptic contacts for the first time. Intracellular recordings allowed one to observe their electrical behavior (Palade and Palay, 1954; Fatt and Katz, 1951). Together, these techniques established the benchmarks by which we determine whether two nerve cells are, in fact, connected to one another.

For the sake of simplicity, we begin with an anatomical description; the molecular and physiological transformations that accompany synapse formation are considered below. At the ultrastructural level, there are three clear signs of a synaptic specialization: small vesicles accumulate at the presynaptic membrane, a narrow cleft filled with extracellular matrix is found between pre- and postsynaptic membranes, and the postsynaptic membrane appears thickened owing to the accumulation of membrane associated proteins and cytoskeletal elements, called the postsynaptic density (PSD). In contrast, newly formed synapses have few, if any, vesicles in the presynaptic terminal profile (**Figure 8.7**). In the rodent cortex,

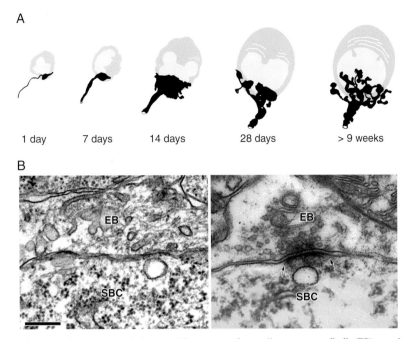

Fig. 8.7 Development of synapse morphology. A. The contact of an auditory nerve endbulb (EB) onto its postsynaptic target, the spherical bushy cell (SBC), is shown at several ages. The end bulb initially forms a small contact on the spherical bushy cell in newborn mice. In adults, the endbulb forms an extremely large ending on the spherical busy cell. B. When contacts from newborn animals are examined at the ultrastructural level, there is little evidence of synapse formation. (Right) In adults, these contacts (between arrows) display many signs of mature synapses, including presynaptic vesicles and a postsynaptic membrane density. (Limb and Ryugo, 2000)

the average number of vesicles found in a synaptic profile increases almost threefold during the first postnatal month (Dyson and Jones, 1980). A second characteristic of newly formed synapses is the close apposition of pre- and postsynaptic membranes, referred to as a tight junction (Figures 8.2 and 8.7). Finally, the postsynaptic membrane does not yet display a PSD. Therefore, newly formed synapses do not display any of the key adult structural features.

Even the highest power electron microscope cannot detect the onset of synaptogenesis between a growth cone and postsynaptic cell (Vaughn, 1989). There is simply not much to be seen. More importantly, the morphology does not tell us how the synapse functions. To get around these problems, many labs use the tissue culture technique where it is possible to monitor synaptic contacts and the onset of function in real time. One of the first in vitro systems used to monitor synapse formation consisted of a piece of fetal rat spinal cord plated next to dissociated superior cervical ganglion neurons, a target of autonomic motor neurons (Rees et al., 1976).

Within the first several hours of contact, there are only subtle changes in morphology to indicate that synaptogenesis is underway (**Figure 8.8**). At first, the growth cone loses its filopodia and forms a punctate contact that is unusually close to the postsynaptic cell membrane (about 7 nm, less than the diameter of a hemoglobin molecule). This suggests that an adhesive interaction may be involved in the initial stages of synaptogenesis, as discussed below. There are also many examples of presynaptic protrusions being engulfed by postsynaptic membrane, termed *coated pits*. These observations reveal an intense interaction at the initial site of contact. The first sign of differentiation is found below the postsynaptic membrane, where the Golgi apparatus accumu-

lates and coated vesicles proliferate, both of which probably contribute to the construction of the postsynaptic density. While these early in vitro studies suggested that *structural* maturation occurs over a relatively long period, more sensitive techniques suggest that the vesicular release machinery is recruited quite rapidly upon contact (see *The first signs of synapse function, page 217*).

WHERE DO SYNAPSES FORM ON THE POSTSYNAPTIC CELL?

The location of synapse formation has a critical impact on its efficacy and integration with other connections. Even motor axons form synapses at distinct central locations on the target myofibers. On the typical central neuron there are many more options. Synapses that form near the soma can have a greater influence on the postsynaptic neuron's determination to fire an action potential. Thus, inhibitory synapses are often found nestled up around the cell body so that they can halt activity effectively. In contrast, many excitatory synapses form on dendritic spines where their activity produces tiny potentials that must be summed together to produce a significant change in the neuron's membrane potential. When many different types of afferents synapse on a postsynaptic neuron, each with a distinct functional role, then the problem becomes quite difficult indeed. How does each synapse determine where to form on the postsynaptic cell?

In many systems, including the NMJ, the spinal cord, the hippocampus, and the cortex, contacts seem to form initially on postsynaptic processes rather than on the soma. In fact, early observations from Golgi stained spinal cord material showed

215

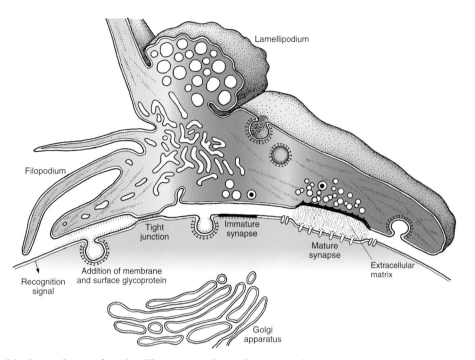

Fig. 8.8 Stages of synapse formation. When a presynaptic growth cone comes into contact with the postsynaptic membrane, its filipodia retract (left), and the membranes become tightly apposed to one another. Vesicles are found in both pre- and postsynaptic elements, possibly to add membrane and extracellular glycoproteins. The immature synapse may display a few vesicles presynaptically and a small postsynaptic density (center). The mature synapse (right) exhibits an accumulation of presynaptic vesicles, a dense extracellular matrix in the cleft, and a postsynaptic density. *(Adapted from Rees, 1978)*

axonal growth cones and dendritic growth cones seemingly reaching out for one another, suggesting that axodendritic synapses result from an early trophic interaction. In the cortex and hippocampus, axodendritic synapses are present in new-born tissue, while few axosomatic synapses are found until two to three weeks later (Pappas and Purpura, 1964; Schwartz et al., 1968). In fact, when a postsynaptic marker protein (PSD-95) is visualized in dendrites as they grow within either hippocampal organotypic slices or the zebrafish midbrain, the new postsynaptic sites appear first on dendritic filopodia (Marrs et al., 2001; Niell et al., 2004). This preference was quantified in the embryonic mouse spinal cord, where nearly 75% of axodendritic synapses are found on dendritic growth cones (Vaughn et al., 1974). Even muscle cells extend tiny processes, called *myopodia*, just before motor terminals arrive, and these postsynaptic processes are the preferred site of contact (Ritzenthaler et al., 2000). Therefore, it is the most recently generated postsynaptic processes that are first contacted by axonal growth cones (Fiala et al., 1998).

It is possible that some postsynaptic membrane is inaccessible for contact. For example, excitatory connections to an auditory brainstem nucleus called the *medial superior olive* (MSO) are at first restricted to the dendritic regions of the cell. At this stage, MSO cell bodies are completely surrounded by glial membrane. As the glial processes withdraw, synapses begin to form on the MSO cell bodies (Brunso-Bechtold et al., 1992). In this regard, it is interesting that elimination of a putative cell adhesion molecule at the neuromuscular junction (β2 laminin) permits glial processes to invade the synaptic region and impede synapse maturation (Noakes et al., 1995; Patton et al., 1998). Thus, glial processes may serve as gatekeepers, determining when and where synapses can be formed.

During the time when synapses are forming between nerve cells, it is quite common to see pre- or postsynaptic structures all by themselves: essentially, synapses to nowhere (**Figure 8.9**). From the amphibian spinal cord to the rodent olfactory cortex, presynaptic-like structures with an accumulation of vesicles can form in the absence of a postsynaptic cell. Similarly, postsynaptic densities that are not in contact with a presynaptic terminal have been found in the olfactory bulb and cortex. These lonesome structures indicate that growth cones and dendrites are poised on the brink of differentiating into synaptic specializations (Hayes and

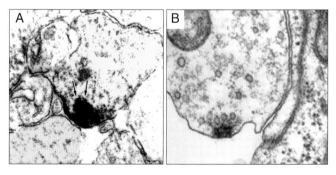

Fig. 8.9 Pre- and postsynaptic differentiation without a partner. A. An electron micrograph showing clustering of α2 adrenergic receptor (arrows) in a postnatal day 4 rat visual cortex neuron. Both of the red-tinted structures are dendrites. B. An electron micrograph showing an apparent presynaptic terminal adjacent to hemolymph. This terminal is made by a *Drosophila* motoneuron in a mutant strain, twisted (*twi*), that does not generate postsynaptic muscle cells. *(Panel A from C. Aoki, unpublished observations; panel B adapted from Prokop et al., 1996)*

Roberts, 1973; Newman-Gage et al., 1987; Westrum, 1975; Hinds and Hinds, 1976). Transient presynaptic-like terminals can also be found on glial cells during axon ingrowth, particularly in areas without dendritic processes (Hendrikson and Vaughn, 1974). In a *Drosophila* mutant that has no mesoderm, and therefore no muscle, motor axons continue to form presynaptic-like profiles on glia and other cells (Figure 8.9B). Thus, a growing neuronal process can display a synaptic morphology with little encouragement from its appropriate partner.

HOW RAPIDLY ARE SYNAPSES ADDED TO THE NERVOUS SYSTEM?

Ultrastructural studies have provided the most quantitative description of the time period when synapses are added. Bursts of synapse formation are found throughout the nervous system, but the timing and duration varies greatly (Vaughn, 1989). In the mouse olfactory bulb, synaptic profiles can first be recognized in electron micrographs at embryonic (E) day 14. The total number of synaptic profiles increases exponentially through the first postnatal week and then continues to increase at a lower rate over the next several weeks. As illustrated in **Figure 8.10A**, synapse density increases for over a month during the postnatal period in cat visual cortex (Cragg, 1975). In human cortex, synapses are first observed by the 7th gestational week and increase by more than 30% by 14 weeks (Zecevic, 1998). After birth, synapse number increases exponentially and reaches a peak at 1–4 years, depending on the region of cortex (Figure 8.10B; Huttenlocher, 1979; Huttenlocher and deCourten, 1987; Huttenlocher and Dabholkar, 1997).

There is a noticeable difference between in vivo descriptions of synapse formation and in vitro studies that focus on cellular and molecular mechanism (discussed below). The former demonstrate that synapses form over an extended period, while the latter emphasize that individual synapses can form within minutes of contact. One reason for this extended period of synaptogenesis observed in vivo is that dendrites continue to grow, and the addition of postsynaptic membrane may attract new contacts. It is also likely that certain afferent projections may arborize at different times. In the rat visual cortex, where the synaptic profiles of excitatory and inhibitory synapses can be recognized, their increase in number occurs at different times. Other areas display a steady increase in synapse number, such as the rat superior cervical ganglion, where the process occurs gradually from innervation at E14 to over one month after birth (Smolen, 1981).

An important caution is that neither anatomy nor physiology alone is sufficient to identify a developing synapse. Purely anatomical measures of synapse formation can be misleading because synaptic physiology can develop rapidly (see below), with little evidence of specialized morphology. On the other hand, an exclusively functional assay of synapse formation may create problems because there is evidence that "silent" synapses exist in the CNS, which nonetheless display normal structure and have the potential to become functional. Therefore, a precise chronology of synapse addition is still missing for most regions of the CNS.

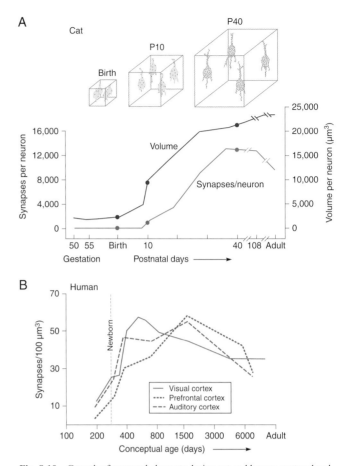

Fig. 8.10 Growth of neuronal elements during cat and human cortex development. A. Increase in volume and synapse density in cat visual cortex. The schematics (top) show that the density of neuronal cell bodies decreases as gliogenesis and angiogenesis occurs from birth until postnatal day 40 (P40). During this same period, dendritic arbors are expanding, and synaptic terminals (red dots) are accumulating on the postsynaptic membrane. The graph shows that the total volume of visual cortex occupied by each neuron increases by almost tenfold during the first postnatal month. When neuron packing density is taken into consideration, the accumulation of synapses can be expressed as synapses per neuron. As shown in the graph, there is a dramatic increase in synapses from P10 to P30, and significant decline after P108. B. The mean synaptic density is shown for three different cortical areas in the human brain. Synaptic density increases dramatically during the first few postnatal years and declines thereafter. *(Panel A adapted from Cragg, 1975; panel B adapted from Huttenlocher and Dabholkar, 1997)*

THE FIRST SIGNS OF SYNAPSE FUNCTION

At the moment a growth cone comes in contact with its postsynaptic target, it transforms from a spindly pathfinding organelle to a bulbous presynaptic terminal. One surprise is that the growth cone comes equipped with a rudimentary transmitter-releasing mechanism (Young and Poo, 1983a; Hume et al., 1983). This was first demonstrated in a primary culture of *Xenopus* spinal neurons and myocytes. During the first two days of culture, spinal neurons produce growth cones, extend neurites, and form functional cholinergic synapses with neighboring muscle cells. To detect the release of ACh, a special type of recording electrode is manufactured (**Figure 8.11**). The electrode has an excised piece of muscle cell membrane at its tip, and this membrane contains ACh receptors (see **Box 8.2**). These electrodes are brought within a few microns of a growth cone before it has

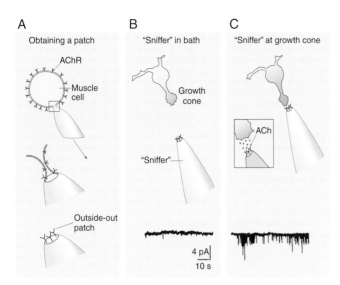

Fig. 8.11 Spontaneous release of neurotransmitter from growth cones. A. A biological sensor for ACh (a "sniffer") is created by excising a patch of membrane from a muscle cell with a recording pipette. The membrane contains AChRs that were facing outward. B. Recording of ACh-evoked currents (downward deflections) when the "sniffer" pipette is distant from the growth cone, and C. when it is within a few microns of the growth cone. The increased activity indicates that the growth cone is releasing ACh. *(From Young and Poo, 1983a)*

contacted a myocyte. When the growth cone releases ACh, then the ACh receptor-coupled channels open, and current flows across the electrode (Figure 8.11). The release of transmitter is probably a general property of all growth cones. For example, the neurotransmitter GABA is released from growth cones of mammalian CNS neurons (Gao and van den Pol, 2000). One study has suggested a role for transmitter release from growth cones in axon pathfinding: In *Drosophila*, the photoreceptor axon projection pattern depends on the release of ACh during the period of axon outgrowth (Yang and Kunes, 2004).

To demonstrate that vesicular release can occur from growth cone filopodia (see Chapter 5), an optical technique was employed. As shown in **Figure 8.12**A, neurons are incubated in the presence of a fluorescent dye (FM4-64) that does not cross the cell membrane. When the neuron is depolarized, the dye enters vesicles that have fused transiently with the membrane. Vesicles that are loaded with fluorescent dye in this manner will then release the dye when they next fuse with the membrane; that is, when the neuron is next depolarized. Using this technique, it can be demonstrated that growth cone filopodia can incorporate and release dye in response to depolarization, suggesting that a vesicular release mechanism is present (Figure 8.12B). Several synaptic vesicle proteins are colocalized within growth cone filopodia (Figure 8.12C), raising the possibility that vesicular release occurs prior to synapse formation (Sabo and McAllister, 2003).

Box 8.2 Biophysics: Nuts and Bolts of Functional Maturation

The study of neural tissue development is unique because the cells possess diverse electrical properties. These properties result from two essential components. First, the neuron must produce batteries by selectively transporting ions from one side of the membrane to the other. Second, the neuron must produce switches, commonly referred to as voltage- and ligand-gated channels, that allow the batteries to discharge across the membrane (that is, ionic currents flow due to an electrochemical gradient). To determine how transporters and channels operate, it is necessary to record from single neurons (or small pieces of membrane), and to control their environment. The most important parameters to control include ionic composition, voltage across the membrane, and the presence of ligands. The technical challenges presented by these requirements have largely been overcome in the past three decades, leading to fundamental discoveries about developing neurons.

To study voltage-gated channels, one must be able to adjust the membrane potential to different holding voltages (voltage-clamp), and then observe whether current flows across the membrane. Thus, if one depolarizes an axon, voltage-gated sodium channels will open at some criterion voltage, termed *threshold*, and Na+ will enter the cell (i.e., inward current). A novel set of recording techniques, called *patch-clamping*, was introduced to fully characterize different types of channels. Patch-clamp electrodes can form high-resistance seals ("gigaOhm-seals") with small areas of membrane, and these patches of membrane can then be excised from the cell (Hamill et al., 1981). This approach has several advantages. Small patches of membrane often contain single channels, they are relatively easy to voltage-clamp, and either side of the membrane may be exposed to a defined medium. Patch-clamping allows one to determine a channel's characteristic properties: the voltage at which activation and inactivation occur, the mean channel open time, the mean current amplitude, the relative permeability to different ions, and the pharmacological profile. Finally, when the excised patches of

tissue contain a known class of neurotransmitter receptor, then the recording pipette may be used to detect the release of neurotransmitter ("sniffer pipettes"). This approach has led to the discovery that growth cones release transmitter (Young and Poo, 1983a; Hume et al., 1983).

It is also possible to form a gigaOhm-seal with a neuron, and then rupture the membrane to form a whole-cell recording configuration. Although this technique is qualitatively similar to a standard sharp electrode intracellular recording, there are some benefits. The tip of the patch recording electrode is much larger than that of a sharp electrode, and this improves both the signal-to-noise ratio and the ability to inject current. The large tip diameter translates into a large hole in the membrane through which the patch pipette solution travels quite easily, allowing the intracellular composition to be controlled within a matter of minutes. In a more elegant form of this technique, a perfusion system is added to the recording pipette such that the intracellular composition can be altered during a recording session (Chen et al., 1990).

Although the patch-clamp technique offers rigorous biophysical measures, it does not, in itself, allow one to evaluate the movement of a single type of ion. Therefore, a common strategy utilizes several pharmacological agents or ions to block the contribution of specific ion channels (e.g., magnesium ions block the flow of calcium). A second approach makes use of a novel group of electrodes, each of which is responsive to changes in the concentration of a specific ion, such as potassium (Syková, 1992). The tips of these electrodes are filled with a liquid membrane that is selectively permeable to one species of ion, so that local changes in concentration result in the net movement of that ion across the membrane, resulting in a detectable potential difference. When employed in the central nervous system, these electrodes reveal substantial developmental changes in the regulation of extracellular potassium and pH (Connors et al., 1982; Jendelova and Syková, 1991).

The fields of electrophysiology and image processing have established a productive relationship in the area of ion channel function. The introduction of ion-sensitive fluorescent dyes provides a less invasive means of assessing functional properties, while providing a high degree of spatial resolution. Each of these dyes emits light at a specific wavelength when activated with a beam of light at a specific excitation wavelength. The amount of emitted light is proportional to the free concentration of a specific species of ion. That is because a dye's absorption or emission properties are altered when it binds to a specific ion. Selective indicator dyes now exist for a wide range of ions including Na^+, Ca^{2+}, Cl^-, and H^+. The calcium indicator dye, Oregon Green 488 BAPTA-1, has been used to demonstrate the elevation of Ca^{2+} on one side of the growth cone that has been exposed to a guidance signal (Nishiyama et al., 2008). Through molecular genetic techniques, it is also possible to express proteins that serve as ion indicators in specific neuron populations. For example, Clomeleon is a fusion protein that can serve as a chloride indicator, and this has been used to monitor the developmental reduction of intracellular chloride (Berglund et al., 2006).

A novel variation of this technology makes use of compounds that exist in a "caged" configuration and that only become activated when exposed to light of a specific wavelength. In this manner, one may elevate the concentration of a specific substance with great temporal and spatial resolution. Finally, it is now possible to induce the flow of ions across a neuron's membrane by exposure to a specific wavelength of light, a technology called optogenetics (Zhang et al., 2007). By genetically introducing a microbial opsin (channelrhodopsin-2) into neurons, the influx of sodium ions can be induced rapidly with light exposure. This technique has been used to selectively activate newly born neurons in the adult hippocampus to show that they form functional synapses with recorded postsynaptic neurons (Toni et al., 2008). Together, these techniques have permitted the measurement and manipulation of individual ion channels or receptors during development.

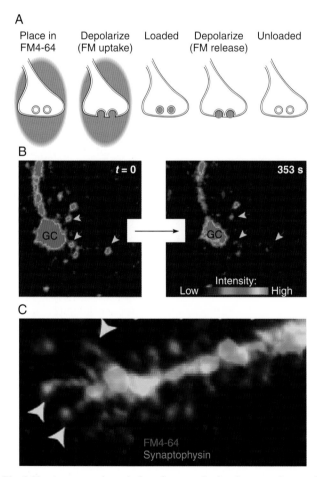

Fig. 8.12 A presynaptic vesicular release mechanism is present in growth cone filopodia. A. A schematic of the FM dye technique. The dye is taken up into vesicles as they fuse with the membrane. When the terminal depolarized again, the vesicles fuse with the membrane again and the terminal unloads the dye. B. Release of FM4–64 from growth cone filopodia in response to depolarization (yellow arrows). C. Colocalization of FM4–64 (red) and the synaptic vesicle protein, synaptophysin (green), in growth cone filopodia. *(Panels B and C from Sabo and McAllister, 2003)*

The timing of axon ingrowth and synaptic transmission has been followed with great precision in the peripheral nervous system and at the NMJ. In the rat superior cervical ganglion, axons first enter the target between embryonic (E) days 12 and 13, and afferent-evoked postsynaptic potentials are recorded by E13. Similarly, motor axons grow out of the *Xenopus* spinal cord and form functional synapses on the developing myotubes over a period of hours (Kullberg et al., 1977). In the fruit fly, it takes only eight hours for neuromuscular transmission to reach a mature level of function (Broadie and Bate, 1993a). However, it is nearly impossible to record from a cell at the exact moment that it is first contacted by a growth cone in vivo.

To study the appearance of synaptic transmission in real time, innervation has been explored with great accuracy in dissociated cultures. When intracellular recordings were obtained from isolated *Xenopus* muscle cells, and the formation of a neurite contact was visually monitored on a microscope, it was found that synaptic potentials could be elicited within minutes of lamellopodial contact (Kidokoro and Yeh, 1982). To provide even better temporal resolution, muscle cells were manipulated into contact with growing neurites while the muscles were being recorded from (**Figure 8.13**A). The tight seal between the large tip of a whole-cell recording electrode and the muscle membrane permits the recordings to continue while a small round muscle cell, called a *myoball*, is detached from the substrate and repositioned in the culture dish (see Box 8.2). By using this technique, it is possible to observe spontaneous synaptic events within seconds of contact (Figure 8.13B), and they continue to increase in both rate and amplitude over the first 10 to 20 minutes (Xie and Poo, 1986). Nerve-evoked synaptic transmission that is great enough to elicit an action potential can be found as early as 30 seconds after nerve–muscle contact. However, in most cases, evoked synaptic responses continue to increase during the first 15 minutes of contact (Figure 8.13C; Sun and Poo, 1987; Evers et al., 1989). Clearly, functional maturation proceeds briskly at the NMJ in vitro. However, most analyses of the mammalian CNS, both in vitro and in vivo, indicate that synaptic properties take days or weeks to reach maturity (see below).

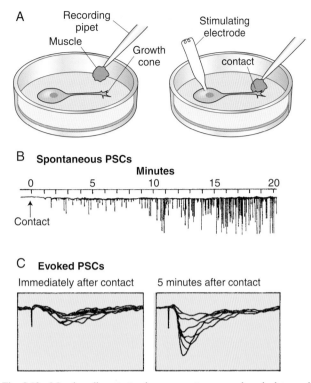

A

Recording pipet

Muscle

Growth cone

Stimulating electrode

contact

B Spontaneous PSCs

Minutes

0   5   10   15   20

Contact

C Evoked PSCs

Immediately after contact   5 minutes after contact

Fig. 8.13 Muscle cell contact enhances spontaneous and evoked transmission. A. Cultures of *Xenopus* spinal neurons were grown in culture, and whole cell pipettes were used to record from round muscle cells and to manipulate them into contact with the neuron. B. A continuous recording from a muscle cell shows spontaneous transmission (downward deflections) during the first 20 minutes after contact. C. Nerve evoked postsynaptic currents (downward deflections) increased in amplitude from the moment of contact to 5 minutes later. *(Adapted from Evers et al., 1989; Xie and Poo, 1986)*

In comparing the development of synaptic structure and function, it is interesting that the maturation of transmission seems to evolve far more rapidly. In the chick ciliary ganglion, synaptic potentials can be recorded before synapses are detected with an electron microscope (Landmesser and Pilar, 1972). Similarly, when a muscle is manipulated into contact with a growth cone in a *Xenopus* culture, the recorded synaptic currents can be quite large at contacts that show no appreciable differentiation (Buchanan et al., 1989). Therefore, a rapid phase of functional maturation occurs over minutes, and is due primarily to developmental events that have preceded contact: the expression of a neurotransmitter release mechanism by the growth cone and neurotransmitter receptors by the postsynaptic cell.

THE DECISION TO FORM A SYNAPSE

Growth cones usually slow down when they enter their target, and this may involve signals that reduce growth cone motility and promote synapse formation (see Chapter 6). Evidence for a target stop signal comes from a system in which newborn mouse basilar pontine nuclei are cocultured with their target neurons, the granule cells of the cerebellum (Baird et al., 1992). Pontine neurites grow rapidly on cerebellar glial cells (greater than 100 μm/hr), indicating that glia do not provide

the stop signal. However, when pontine nuclei were cultured on a bed of glia along with dissociated granule cells, the outgrowth of neurites was reduced. By closely examining individual neurites it was found that the reduced growth was due to contact with granule cells (Figure 8.14A). Neurites that did not come upon a granule cell during their outgrowth continued to grow for a normal distance. Moreover, when granule cells were suspended above the pontine explants, the neurites grew at a normal rate. Thus, growth cones can be terminated at the appropriate target by a contact-dependent mechanism.

Growth cones can display rather dramatic changes in membrane potential in response to known guidance factors (see Chapter 5). When exposed to the chemoattractant, Netrin-1, growth cones become depolarized, but when exposed to the chemorepellant, Sema3A, they become hyperpolarized (**Figure 8.14**A). Thus, it is not surprising that the initial dialogue between pre- and postsynaptic cells involves flow of calcium into the growth cone filopodia at the moment of contact (Dai and Peng, 1993; Zoran et al., 1993). This was determined for both frog and snail motor neurons that were grown in dissociated tissue culture and filled with a Ca^{2+}-sensitive indicator dye. When a muscle cell is manipulated

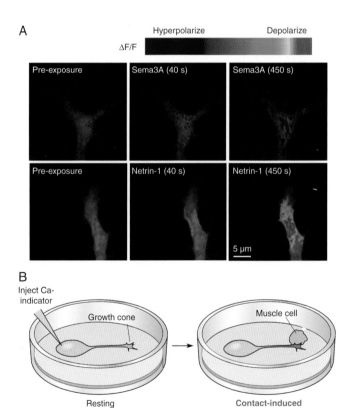

A

Hyperpolarize   Depolarize

ΔF/F

Pre-exposure   Sema3A (40 s)   Sema3A (450 s)

Pre-exposure   Netrin-1 (40 s)   Netrin-1 (450 s)

5 μm

B

Inject Ca-indicator

Growth cone

Muscle cell

Resting Ca²⁺ level   Contact-induced Ca²⁺ increase

Fig. 8.14 Induced current flow at the growth cone. A. Changes in growth cone membrane potential were monitored with a voltage sensitive dye during exposure to either of two guidance factors. Sema3A (top) causes a hyperpolarization of the growth cone, while Netrin-1 (bottom) leads to a depolarization. The color scale represents depolarization as green-red and hyperolarization as blue. B. Contact with target increases free calcium in the growth cone. Dissociated neurons were filled with a Ca^{2+}-sensitive dye (left), and the growth cone was imaged while a muscle cell was brought into contact (right). Intracellular free calcium increased during contact with the muscle (red). *(Panel A adapted from Nishiyama et al., 2008; panel B adapted from Dai and Peng, 1993)*

into contact with a growth cone, the Ca^{2+} increases locally within seconds (Figure 8.14B). This response exhibits some target-specificity, as the Ca^{2+} rise does not occur when a spinal neuron was manipulated into contact with the growth cone. It is not yet clear how calcium levels increase, but one possibility is that calcium is released from internal stores. In the rat central nervous system, IP_3 receptors, which transduce a signal leading to calcium release from endoplasmic reticulum, are upregulated during the period of intense synaptogenesis (Dent et al., 1996).

What is the evidence that a contact-evoked rise in Ca^{2+} provides a signal for growth cone differentiation? Intracellular Ca^{2+} can be manipulated in growth cones by exposing them to an ionophore such as A23187, a molecule that spontaneously inserts into a neuron membrane, allowing Ca^{2+} to pass freely into the cell (Mattson and Kater, 1987). As calcium rises, growth cones are often found to slow down and to assume a rounded appearance. In fact, growth cones tend to become stationary in all but a limited range of Ca^{2+} concentrations, from 200 to 300 nM (Lankford and Letourneau, 1991).

The rapid influx of calcium may have an immediate effect on actin polymerization, perhaps facilitating maturation. For example, silent presynaptic contacts can be converted rapidly into transmitter-releasing terminals by a brief electrical stimulus. This conversion is associated with an increase in actin polymerization at the newly functional terminals, and interfering with actin polymerization prevents the conversion (Shen et al., 2006; Yao et al., 2006).

By observing the giant growth cones produced by cultured *Aplysia* bag cells, one finds that microtubules extend toward the site of contact with a target, and filamentous actin begins to accumulate (Forscher et al., 1987). A similar transformation can be produced by raising cAMP levels within the growth cone. The cytoskeleton reorganizes and neurosecretory granules invade the growth cone's lamellapodia, resulting in a presynaptic-like morphology. Activation of a second intracellular signaling pathway, protein kinase C (PKC), results in the rapid appearance of new calcium channels at the edge of the growth cone (Knox et al., 1992). Thus, calcium, PKC, and cAMP may operate in tandem to support the accumulation of secretory vesicles and ion channels at the site of contact. In fact, the contact-evoked increase in calcium (Figure 8.14) may actually be a result of cAMP signaling. When an identified snail motor neuron is manipulated into contact with its normal target in vitro, it exhibits an increase in calcium (Funte and Haydon, 1993). This rise is mimicked by a membrane permeable analog of cAMP and is prevented by injecting an inhibitor of cAMP-dependent protein kinase (PKA) into the motor neuron.

An important extrinsic factor that regulates the formation of new synapses comes from astrocytes. When retinal ganglion cells are isolated and grown in a defined medium, they display little synaptic activity. However, the addition of astrocytes from their target region leads to a dramatic increase in the number and strength of synaptic contacts (Ullian et al., 2001). The glial cells need not be in contact with retinal neurons to elicit this response, suggesting that they release a soluble factor. However, the synapses induced without astrocyte contact were found to be presynaptically active but postsynaptically silent, indicating a second role for a contact-dependent mechanism. A similar effect has been found in hippocampal cultures; here, an astrocyte-derived soluble factor is

able to increase the number of inhibitory synapses (Elmariah et al., 2005). A search for the secreted synapse-inducing activity led to the discovery that an essential membrane constituent, cholesterol, plays an important role in synapse-formation. Apparently, developing neurons manufacture only enough cholesterol to survive and grow dendrites, but depend on the delivery of additional cholesterol from astrocytes to produce synapses (Mauch et al., 2001).

There are likely to be many other glial signals; a family of large extracellular matrix proteins, thrombospondins, can partly explain the ability of astrocytes to induce synapse formation. Thrombospondins can mimic the synapse-inducing effect of astrocytes, and the genetic deletion of two family members leads to a significant reduction in cortical synapses in vivo (Christopherson et al., 2005). Astrocytes also induce synapse formation through direct perisynaptic contact with postsynaptic neurons, and this interaction is mediated by members of the γ-protocadherins family of cell adhesion molecules (Garrett and Weiner, 2009). Many more glial-derived signals are likely to promote synapse formation, including molecules that have a primary role in earlier stages of development. For example, astrocyte-like cells induce presynaptic terminals in *C. elegans* by releasing Netrin, a molecule associated primarily with pathfinding (Colón-Ramos et al., 2007).

THE STICKY SYNAPSE

Adhesion between growth cones and target cells increases rapidly upon contact. To demonstrate this sort of adhesion in vitro, round muscle cells, known as myoballs, were lifted off the culture dish with an electrode and placed in contact with the growing tip of a *Xenopus* spinal neuron (**Figure 8.15**). At either 1.5 or 15 minutes after a contact was initiated, the intercellular adhesion was categorized by observing how much the neurite was deformed by the retracted myoball. A low level of adhesion is evident after 1.5 minutes, and the percentage of tightly adherent contacts more than doubles during the first 15 minutes of contact (Evers et al., 1989). The closely apposed contact sites of pre- and postsynaptic membranes on either side of the active zone have been called "puncta adherens," highlighting the fact that the synapse is a type of modified adherens junction.

What kinds of adhesion molecules are involved in the formation of early contacts? Several groups of cell adhesion molecules are localized at developing synapses, and many of these contain immunoglobulin-domains (Yamagata et al., 2003). For example, synaptic cell adhesion molecules (synCAMs) are brain-specific adhesion molecules that were discovered by searching for a mammalian homolog of FasII, the *Drosophila* cell adhesion molecule that contributes to nerve–muscle synaptogenesis. SynCAM expression gradually increases in rat brain during the first three postnatal weeks and is highly enriched at both pre- and postsynaptic plasma membrane. SynCAMs are found at both excitatory and inhibitory synapses, and tend to form heterophillic interactions. When nonneuronal cells are transfected with SynCAM1, they can induce presynaptic differentiation in cultured hippocampal neurons, and these terminals are able to release glutamate. Furthermore, overexpression of synCAM2 in postsynaptic neurons can promote the formation of presynaptic terminals (Biederer et al., 2002; Fogel et al., 2007).

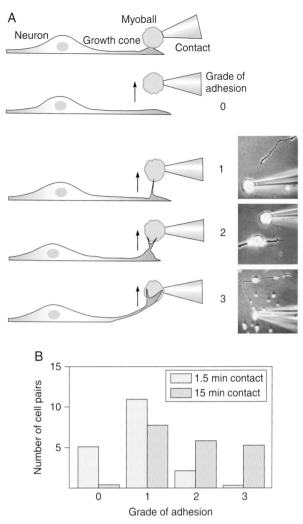

Fig. 8.15 Rapid adhesion between growth cone and postsynaptic muscle cell. A. A muscle cell was manipulated into contact with a growth cone in dissociated cultures of *Xenopus* spinal cord. After 1.5 or 15 minutes, the muscle cell was withdrawn, and the degree of adhesion was graded: (0) no attachment, (1) filamentous attachment, (2) deformation of growth cone, and (3) detachment of growth cone from substrate. B. After 1.5 minutes of contact, most pairs exhibited only grade 0–1 adhesion. However, after 15 minutes of contact, the level of adhesion shifted to grade 1–3. *(Adapted from Evers et al., 1989)*

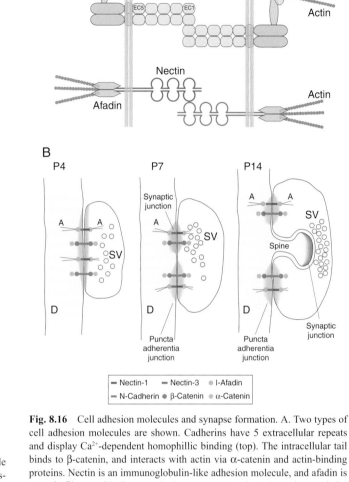

Fig. 8.16 Cell adhesion molecules and synapse formation. A. Two types of cell adhesion molecules are shown. Cadherins have 5 extracellular repeats and display Ca²⁺-dependent homophillic binding (top). The intracellular tail binds to β-catenin, and interacts with actin via α-catenin and actin-binding proteins. Nectin is an immunoglobulin-like adhesion molecule, and afadin is an actin filament–binding protein that connects nectin to the actin cytoskeleton (bottom). B. The schematic shows the nectin–afadin adhesion system during the formation of a synapse in developing hippocampal pyramidal neurons. The nectin–afadin system organizes adherens junctions cooperatively with the cadherin–catenin system. During development, nectin-1 and -3 localize at both the puncta adherentia junctions (i.e., mechanical anchoring sites) and at synaptic junctions. This changes during development such that nectin–afadin comes to be localized around the synaptic active zone. The cadherin–catenin system is likely to colocalize with the nectin–afadin system at each stage. D, dendritic trunks of pyramidal cells; SV, synaptic vesicles; A, actin filaments. *(Adapted from Mizoguchi et al., 2002)*

Multiple members of the large family of calcium-dependent cell adhesion molecules, called cadherins, are located at developing synapses throughout the CNS. There are about 20 "classical" cadherins expressed in the nervous system, as well as over 80 members of the cadherin superfamily subgroup known as protocadherins. The classical cadherins have an extracellular portion consisting of five extracellular cadherin (EC) domains, with calcium ions inserted between them. Cadherin adhesion is generally mediated by homophilic binding of the extracellular domains to one another, while conserved cytoplasmic tail sequences interact with actin filaments via the catenins (**Figure 8.16**A). The diversity of the cadherin superfamily implies that these molecules might mediate the specificity of synapse formation between individual types of partner neurons. Function-blocking antibodies against N-Cadherin can prevent axons from stopping at the appropriate target layer both in the chick optic tectum (Inoue and Sanes, 1997) and in thalamocortical

slice cocultures (Poskanzer et al., 2003). In the hippocampus, three members of the family (N-cadherin, E-cadherin, cadherin-8) may play dissimilar roles during target selection and contact formation and are restricted to separate synapses along the dendrite (Fannon and Colman, 1996; Bekirov et al. (2008). Two family members (cadherin-11 and -13) appear to be essential for both excitatory and inhibitory synapse formation in the hippocampus (Paradis et al., 2007), and when cadherin function is disrupted in hippocampal neurons, neither the postsynaptic spines nor the presynaptic terminals are as differentiated as in control cultures (Togashi et al., 2002). Similarly, deletion of the 22 member γ-protocadherin gene cluster results in a dramatic reduction of synapses in the developing spinal cord (Weiner et al., 2005). Cadherins not only bring pre- and postsynaptic membranes in close apposition, but also can bind to glutamate receptors and anchor them to the newly formed synapse (Saglietti et al., 2007). It has been difficult, however, to delineate clear roles for the cadherins in synaptogenesis per se, as opposed to their roles in axon guidance and presynaptic differentiation. Furthermore, transfection of non-neuronal cells with N-cadherin is not sufficient to induce presynaptic terminal formation in contacting axons (Scheiffele et al., 2000).

Synaptogenesis likely requires interactions between these two major superfamilies (Ig and Cadherin) of adhesion molecules. A distinct family of Ig-like CAMs, the Nectins, are connected to the cytoskeleton by the actin filament-binding protein, l-afadin (Figure 8.16A). The nectin-afadin system is colocalized and functionally linked with the cadherin-catenin system during synapse formation in the hippocampus. Initially, these proteins are found at the site of contact between growth cone and postsynaptic neuron, a close apposition of membrane called an *adherens junction* (see EM of neonatal synapse in Figure 8.7). As the synapse matures, the adhesion molecules gradually become localized adjacent to the synapse in *puncta adherens* (Figure 8.16B). As with the cadherin system, blockade of nectin-1 function has been shown to affect the size and number of synapses (Mizoguchi et al., 2002). Therefore, transmembrane adhesion molecules such as cadherin and nectin participate in the initiation of synaptogenesis.

CONVERTING GROWTH CONES TO PRESYNAPTIC TERMINALS

Soon after the initial contact is made, there is a dramatic improvement of neurotransmission (Figure 8.13C). In fact, the presynaptic release site (called the active zone) appears to be "preassembled" and shipped down the axon as a complex of multiple proteins (Ahmari et al., 2000; Shapira et al., 2003; Bresler et al., 2004). The movement of presynaptic transport packets was visualized in developing neuronal processes using a fluorescently labeled vesicular protein (VAMP-GFP). After fixation, VAMP (a.k.a. synaptobrevin) was found to colocalize with many other presynaptic proteins that are necessary for vesicular release (**Figure 8.17**A). These include presynaptic cytoskeleton-associated proteins, a regulator of vesicle fusion, and calcium channels. It is estimated that a new active zone can be established by the insertion of only two to three of these "presynaptic packet" vesicles.

To determine how quickly synaptic proteins accumulate at a contact site, new active zones were labeled with the FM4–64 method (illustrated in Figure 8.12A), and then counterstained for pre- and postsynaptic marker proteins (Friedman et al., 2000). Within 45 minutes of detecting a new release site, about 90% of them have already accumulated the presynaptic active zone protein, Bassoon. In contrast, less than 30% of the postsynaptic sites are labeled for the postsynaptic density protein, PSD-95 (Figure 8.17B). However, time-lapse imaging of labeled NMDA receptors in cultured cortical neurons indicate that they also move along neuronal processes in transport packets, and can arrive within minutes of contact by an axonal growth cone (Washbourne et al., 2002). In fact, preformed postsynaptic sites can apparently induce the conversion of a growth cone to presynaptic terminal (Gerrow

et al., 2006). By simultaneously imaging postsynaptic (PSD-95) and presynaptic (Synaptophysin) markers, it was found that the presynaptic transport packets can be stabilized at stationary postsynaptic sites (Figure 8.17C, top). Thus, new synapses can display rapid molecular development, and there is evidence that both pre- or postsynaptic membranes can initiate maturation of the other.

Since pre- and postsynaptic structure differs, a different set of instructions must be delivered to the growth cone and the postsynaptic membranes. Two signaling systems that exhibit this sort of asymmetry have been shown to recruit specific synaptic components to the site of contact (**Figure 8.18**). When pontine neurons are grown in vitro, their axons come to a halt and accumulate synaptic vesicles soon after contacting their postsynaptic target neurons, cerebellar granule cells (Figure 8.18A). The growth cones of these axons express neurexins (NRXN), transmembrane proteins that contain either one or six extracellular binding domains. Cerebellar granule cells express neuroligins (NLGN), a family of NRXN binding partners that are distinguished by an extracellular acetylcholinesterase-like domain. NLGN-NRXN signaling is sufficient to induce presynaptic differentiation of pontine axons. When pontine axons contact nonneuronal cells transfected with NLGN-1 or -2, they stop growing and accumulate presynaptic proteins (e.g., synapsin) and synaptic vesicles (Figure 8.18B). This effect can be blocked by the addition of soluble NRXN to the culture media (Scheiffele et al., 2000; Dean et al., 2003).

NLGN may induce presynaptic differentiation by stabilizing transport packets. For example, it is common to observe presynaptic transport packets becoming stabilized at postsynaptic loci containing a cluster of NLGN (Figure 8.17C, bottom). Excitatory and inhibitory synapses may be induced by different neuroligin family members. NLGN-2 is localized primarily to GABAergic terminals, suggesting that it induces inhibitory contacts, whereas NLGN-1 is found at excitatory synapses (Graf et al., 2004; Varoqueaux et al., 2004). It is possible that neuroligin-neurexin signaling is quite diverse; there are five NLGN genes and three NRXN family members, each with up to five canonical splice sites, and these may be distributed heterogeneously in the developing nervous system (Südhof, 2008).

Although NLGN-NRXN signaling is not required for the generation of anatomical synaptic contacts, the elimination of either protein leads to a dramatic loss of synaptic transmission *in vivo*. When three NLGN genes were deleted in mice, synapses appeared normal morphologically, yet the animals died at birth. In fact, synaptic transmission was reduced significantly, particularly at inhibitory contacts (Varoqueaux et al., 2006). Similarly, deletion of α-NRXNs genes in mice causes a loss of presynaptic Ca^{2+}-channel function which reduces transmitter release (Missler et al., 2003).

Cerebellar granule cells also release soluble factors, such as Wnt-7a, that can induce presynaptic differentiation in vitro. Pontine axons that express the *Wnt* receptor, *Frizzled*, accumulate synapsin when grown in the presence of *Wnt-7a*-transfected cells. There is now evidence that *Wnt* supports synapse differentiation in other systems, although the in vivo loss of this signaling system appears to delay, rather than prevent, differentiation (Hall et al., 2000; Krylova et al., 2002; Packard et al., 2002; Davis et al., 2008). A second soluble factor, FGF22, is also localized to granule neurons when synapse

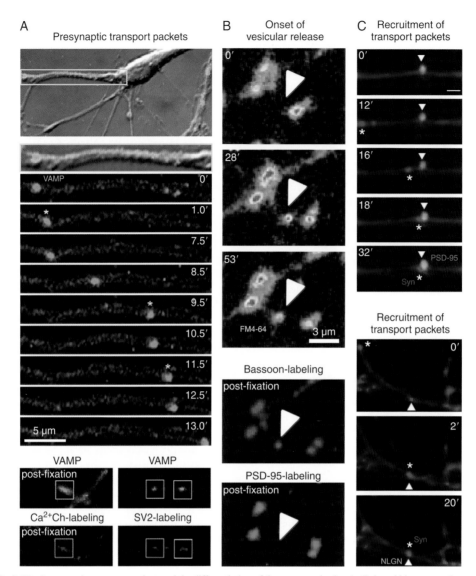

Fig. 8.17 Presynaptic transport packets and the differentiation of the presynaptic site. A. Cultered hippocampal neurons were transfected with a fusion protein containing GFP and the presynaptic vesicle protein, VAMP (top, yellow box). When imaged over the course of 13 mins, an individual VAMP-containing packet was found to move along the neuronal process. The tissue was fixed, and counter-stained for other presynaptic marker proteins (bottom), which were found to colocalize with VAMP. This indicates that a large fraction of presynaptic proteins are cotransported to the site of contact. B. Using a dissociated culture, the onset of vesicular release was monitored continuously with FM 4–64 (top). A new release site appeared sometime between 0 and 28 minutes (arrowhead), and this site was retained at 53 minutes. Following fixation, the cells were counterstained for pre- and postsynaptic markers (bottom). The new release site was associated with a presynaptic marker (Bassoon), but not with a postsynaptic marker (PSD-95). C. Presynaptic packets are stabilized at a stationary postsynaptic site. Cultured neurons were transfected with a presynaptic marker (Synaptophysin, Syn) and either of two postsynaptic markers (PSD-95 or NLGN). The top series shows that the presynaptic marker (Syn, asterisk) moved along the axon and became stabilized at a stationary postsynaptic site (PSD-95, arrowhead). The bottom series shows that the presynaptic marker (Syn, asterisk) moved along the axon and became stabilized at the site of a stationary signaling protein, neuroligin (NLGN, arrowhead). *(Adapted from Ahmari et al., 2000; Friedman et al., 2000; Gerrow et al., 2006)*

formation is at its peak. When FGF22 signaling was blocked with an antagonist, there was a significant reduction in the ability of pontine axons to form presynaptic varicosities, both in granule cell-containing cultures and in vivo (Umemori et al., 2004). Similarly, in vivo deletion of the receptor for FGF22 led to a loss of presynaptic differentiation in the cerebellum.

Together, these studies suggest that presynaptic differentiation is controlled by multiple factors, but the manner in which these signals are integrated is unclear. Genetic analysis of

synapse formation in *C. elegans* is likely to provide a more coherent description of the minimal complement of pathways involved (**Figure 8.19**). Two Ig CAMs, SYG-1 and -2, are necessary for the formation of specific synapses: mutation of either causes the HSNL identified neuron to mislocate its presynaptic terminals. It was found that this positioning is regulated by interactions between SYG-1 in the HSNL axon and SYG-2 in an epithelial cell that acts as a guidepost cell (Shen et al., 2003). SYG-1 then recruits two active zone proteins

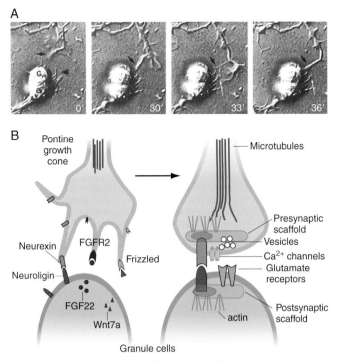

Fig. 8.18 Signals that promote differentiation of growth cones into presynaptic boutons. A. When imaged continuously in dissociated culture, pontine growth cones (arrows) were observed to stop when they contacted granule cells (arrowhead), their normal postsynaptic target. B. Pontine growth cones express three receptors, Frizzled, FGFR2, and neurexin, on their surface. Granule cells express their respective ligands: *Wnt7a, FGF22,* and Neuroligin. These signaling pathways recruit vesicles and synaptic proteins to the site of contact. *(Panel A from Baird et al., 1992)*

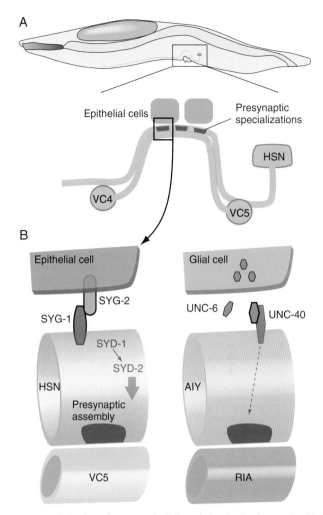

Fig. 8.19 Induction of presynaptic differentiation in *C. elegans*. A. HSN neurons control egg-laying by forming synapses onto the vulva muscles and the VC4 and 5 interneurons. The presynaptic specializations (box, red rectangle) form opposite the point where epithelial cells contact HSN. B. The signaling cascade that leads to presynaptic induction in HSN axons has been identified through genetic screens. The epithelial cells express an immunoglobulin superfamily member (SYG-2) that recruits a transmembrane molecule expressed by HSN cells (SYG-1) to the position where a presynaptic site will form. This signaling system recruits 2 active zone proteins (SYD-1 and -2) that are key regulators of presynaptic assembly and stability. Elsewhere in the organism, glial cells release UNC-6 (also known as Netrin in vertebrates), which activates its receptor (UNC-40) to promote the assembly of a presynaptic specialization. *(Adapted from Margeta et al., 2008)*

(SYD-1 and -2) which appear to be the key regulators of presynaptic assembly (Shen et al., 2003; Shen et al., 2004; Dai et al., 2006). Through genetic screens, a complete description of synapse assembly is possible, and many of the molecular signals identified in *C. elegans* are likely to be found in other species (Margeta et al., 2008).

Novel synaptogenic proteins can also be discovered by transfecting fibroblasts with members of a cDNA expression library obtained from the developing brain, and determining whether cocultured neurons are induced to form synapses. Using this strategy, it was found that the leucine rich repeat transmembrane neuronal protein family (LRRTM) members could induce excitatory presynaptic differentiation (Linhoff et al., 2009), and this role appears to be mediated by their interaction with NRXNs (Ko et al., 2009; de Wit et al., 2009). Another synaptogenic signaling system may employ a family of netrin-G ligands (NGLs) and either of two receptors: a family of cell adhesion proteins (netrin-Gs, proteins that resemble the axon guidance protein, Netrin, but do not bind to the DCC receptor) or a receptor protein tyrosine phosphatase (LAR). In vitro studies show that NGLs can induce excitatory presynaptic terminals by interaction with either netrin-G or LAR (Kim et al., 2006; Woo et al., 2009).

Many of the factors that support axon outgrowth and neuron survival (Chapters 5 and 7) also play an important role during synapse formation; these range from the induction of new synapses to the upregulation of neurotransmitter release. For example, Netrin-1 and BDNF signaling can increase the number of synapses and their differentiation, both in cultures and in vivo (Vicario-Abejon et al., 1998; Marty et al., 2000; Alsina et al., 2001; Manitt et al., 2009; Betley et al., 2009).

RECEPTOR CLUSTERING AND POSTSYNAPTIC DIFFERENTIATION AT THE NMJ

The aggregation of neurotransmitter receptors apposed to the presynaptic terminal is a principal feature of synaptogenesis. Since this process has been most thoroughly described at the vertebrate neuromuscular junction (NMJ), we will discuss this well-studied synapse, before turning to postsynaptic differentiation in the central nervous system.

Is receptor clustering a cell-autonomous process, or is it induced by the presynaptic terminal? At first inspection, the postsynaptic site appears to be produced in an autonomous fashion. Acetylcholine receptors (AChRs) form small clusters on the muscle cell membrane before the motor axons arrive (Fischbach and Cohen, 1973). Structures that resemble postsynaptic densities, but with no apparent presynaptic element, have also been found in the developing olfactory bulb and visual cortex during early development (Hinds and Hinds, 1976; Bahr and Wolff, 1985). In fact, most neurotransmitter receptors are expressed before innervation occurs, and they are often found in clusters, similar in appearance to a postsynaptic site (Figure 8.9A).

At the time of innervation, the muscle cell membrane still displays an immature distribution of AChRs. This was originally demonstrated by recording the response from rat muscle cells in vivo as ACh was applied at different places along the myofiber surface (Diamond and Miledi, 1962). Early in development, ACh application at each site evokes a similar shift in membrane potential. As the muscle is innervated, the ACh-evoked response becomes much larger at the site of innervation, and the response at extrasynaptic regions declines. Methods were subsequently developed to visualize the distribution of AChRs by labeling them with radioactive α-bungarotoxin (α-Btx), a high-affinity peptide from the venom of the Taiwanese cobra (Bevan and Steinbach, 1977; Burden, 1977a). Consistent with the electrophysiological measures, α-Btx labeling is broadly distributed at first and then becomes highly localized to the synapse. The process of clustering leads to a dramatic disparity in receptor concentration: there are >10,000 AChRs/μm^2 at the synaptic region but <10/μm^2 in extrasynaptic regions (Fertuck and Salpeter, 1976; Burden, 1977a; Salpeter and Harris, 1983).

While these observations implied that the motor axons induce receptor clustering, higher AChR concentrations occur autonomously at the center of developing skeletal muscle fibers (Yang et al., 2001). This was demonstrated in a mouse mutant where the diaphragm muscle remains uninnervated during embryonic development, yet AChRs nonetheless become concentrated in the central muscle (**Figure 8.20**). In fact, this initial localization of AChRs is caused by a postsynaptic clustering system that will ultimately become responsive to a motor nerve-released signal for stabilization (see next section, below, for discussion of Agrin-MuSK signaling).

Although the localization of AChRs is, at first, independent of motor axons, the nerve terminal does exert a potent clustering influence when it arrives. This was first suggested by observing fluorescently-labeled AChRs during the period of innervation (Anderson and Cohen, 1977; Cohen et al., 1979). Although small AChR clusters are present prior to innervation, they do not serve as preferential sites of innervation. Rather, the growing neurites induce the rapid accumulation of AChRs as they extended across the muscle (**Figure 8.21**). Furthermore, the ability to induce AChR clusters is specific to spinal cord neurons, which presumably include motor neurons (Cohen and Weldon, 1980; Kidokoro et al., 1980). However, the AChRs do not have to be activated in order to cluster. When myocytes and spinal cord are cultured in the presence of an AChR antagonist, clustering occurs normally at the site of neurite contact in the absence of cholinergic transmission (Cohen, 1972; Anderson and Cohen, 1977).

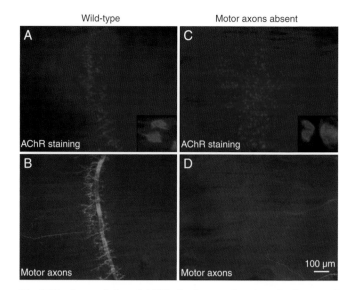

Wild-type   Motor axons absent

A   AChR staining   C   AChR staining

B   Motor axons   D   Motor axons   100 μm

Fig. 8.20 Accumulation of AChRs in the central region of uninnervated muscle. Diaphragm muscle from wild-type animals (A and B), and mutant embryos which lack motor axon ingrowth (C and D) were labeled with a probe for AChRs (Texas red-α-bungarotoxin, αBGT) and a second probe for axons and terminals (antibodies to neurofilament and synaptophysin, anti-NF/Syn). At embryonic day 18.5, there is dense innervation at the center of the wild-type muscle but no motor innervation in the mutants. (A few sensory and/or autonomic axons can be seen at the edge of mutant muscle in panel D.) The insets show individual AChR clusters at high magnification in wild-type (A) and mutant (C) embryos. The bar is 100 μm for the low magnification images. *(Adapted from Yang et al., 2001)*

AGRIN IS A TRANSYNAPTIC CLUSTERING SIGNAL AT THE NMJ

The studies discussed above suggest that a nerve terminal-derived signal initiates or stabilizes receptor clustering. In fact, the clustering signal was discovered to be a secreted proteoglycan that becomes incorporated into the extracellular matrix that ensheathes each muscle cell. When muscle cells were damaged in adult frogs, they degenerated, leaving behind the basal lamina (**Figure 8.22**). New myofibers then formed beneath this basal lamina, and AChR clusters were observed to reform at the original synaptic sites along the basal lamina, even if motor nerve terminals were absent (Burden et al., 1979). Even if the damaged muscle is irradiated so that new myofibers do not form and only basal lamina "ghosts" are left behind, these "ghosts" can be reinnervated by motor nerve terminals at the original sites (Sanes et al., 1978). A proteoglycan, named Agrin, was subsequently isolated from the electric organ of the marine ray *Torpedo californica*, a site rich in cholinergic synapses (Sanes et al., 1978; Godfrey et al., 1984; Nitkin et al., 1987). Monoclonal antibodies directed against Agrin have been used to localize this protein to motor neuron cell bodies, the synaptic basal lamina, and muscle cells (Reist et al., 1987; Magill-Solc and McMahon, 1988; Fallon and Gelfman, 1989).

The Agrin produced by neurons was shown to be responsible for AChR cluster formation by specifically blocking its function in vitro (Reist et al., 1992). A neuron-specific isoform of Agrin, generated by alternative splicing of the mRNA, has since been shown to have greater cluster-inducing activity than that found in muscles (Ruegg et al., 1992; Ferns et al., 1993).

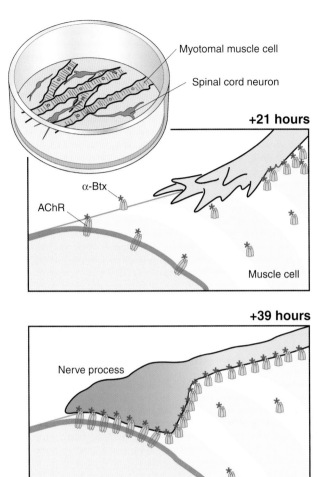

Fig. 8.21 ACh receptor clustering on muscle fibers is induced by contact with spinal neurites. A culture of spinal neurons and muscle cells was labeled with a fluorescent α-Btx (red) at 21 and 39 hours after plating. Soon after the spinal neurite grew across the muscle surface, fluorescent α-Btx appeared at the contact site, indicating that AChR aggregation is induced. *(Adapted from Anderson and Cohen, 1977)*

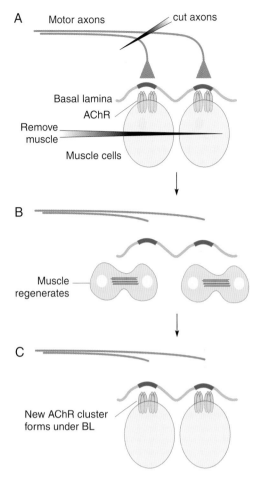

Fig. 8.22 The extracellular matrix contains a factor that induces AChR clustering. A. In the adult frog cutaneous pectoris muscle, the motor axons were cut and the muscle cells were destroyed, leaving only the basal lamina which contains extracellular matrix molecules. B. New myofibers are generated as a result of cell division. C. AChR clusters (green) form on the regenerated muscle fibers, directly beneath the synaptic portion (red) of the basal lamina. *(Adapted from Burden et al., 1979)*

Observations of Agrin knockout mice demonstrate that this signal is not required for the embryonic aggregation of AChRs to the central region of the muscle (Figure 8.20), but is necessary for their precise alignment with the motor nerve terminals. Furthermore, postsynaptic sites appear to be less mature than normal, suggesting that this signaling pathway regulates more than just receptor clustering (Gautam et al., 1996).

During a widespread search for the postsynaptic Agrin transducer, a novel receptor tyrosine kinase (muscle-specific kinase, or MuSK) was identified and localized to the synaptic junction (Valenzuela et al., 1995). MuSK is expressed when motor axons are first growing into the muscle and Agrin induces the tyrosine phosphorylation of MuSK within minutes (**Figure 8.23C**), which leads rapidly to increased MuSK clustering (Moore et al., 2001).

The MuSK protein is necessary both for the early prepatterning of AChRs that is observed in the absence of motor axons (Figure 8.20), as well as nerve-induced clustering. This was verified in a targeted disruption of the gene in mice (Figure 8.23B). These animals display a dramatic loss of postsynaptic maturation, including the absence of AChR clustering (DeChiara et al., 1996).

Although Agrin and MuSK display strong binding kinetics, this is only apparent when MuSK is expressed in muscle cells, suggesting that the protein must form a complex with one or more accessory proteins (Glass et al., 1996). A member of the low density lipoprotein receptor-related protein, Lrp4, probably serves as the primary Agrin binding site (Figure 8.23A). Lrp4 binds selectively to neural isoforms of Agrin and is required for Agrin to stimulate MuSK phosphorylation. Furthermore, mice carrying mutations in the *lrp4* gene display defects in AChR clustering that are similar to those found in MuSK mutant mice (Weatherbee et al., 2006; Kim et al., 2008; Zhang et al., 2008).

At least three additional cytoplasmic proteins are required to mediate AChR clustering (Figure 8.23A). The first, rapsyn, contains one domain that mediates self-association, and a second that associates with the main intracellular loop of AChRs (Ramarao et al., 2001). The co-expression of Rapsyn is sufficient to promote AChR clustering in vitro, and the phenotype of Rapsyn-deficient mice is consistent with a primary role in cluster formation (Frohner et al., 1990; Phillips et al., 1991; Gautam et al., 1995). The second protein, Dok-7, interacts directly with the cytoplasmic tail of MuSK, and is required for tyrosine kinase activity

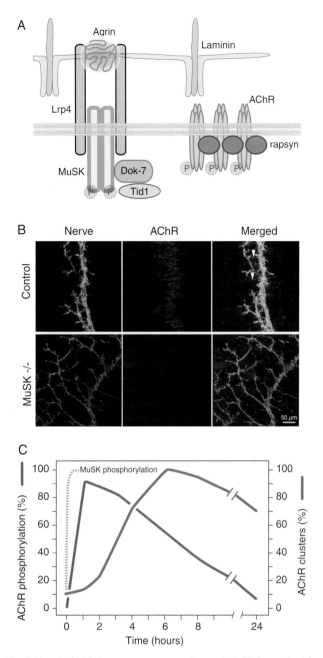

Fig. 8.23 Agrin binds to a receptor complex, and MuSK is required for AChR clustering. A. The schematic shows that Agrin activates a receptor complex that includes binding protein (Lrp4) and muscle-specific kinase (MuSK). The intracellular MuSK-interacting protein (Dok-7) is required for the Agrin-mediated MuSK activation that leads to AChR phosphorylation and clustering. The cytoskeletal adaptor protein rapsyn, which scaffolds AChRs, is also required for clustering. B. In wild-type mice, AChRs aggregate at the nerve-muscle junction (merged), but in MuSK-deficient mice (MuSK−/−) there is no clustering and the postsynaptic site does not differentiate. C. When myocyte cultures are exposed to Agrin, the Agrin receptor (MuSK) is phosphorylated within minutes (blue line), and the AChRs are maximally phosphorylated within an hour (red line). Receptor aggregation occurs over the next few hours (green line), but the level of phosphorylation displays a gradual decline over 24 hours. In contrast, AChR clustering is maximal by six hours after Agrin exposure and declines only slightly by 24 hours. *(Adapted from Ferns et al., 1996; Glass et al., 1996; Lin et al., 2001)*

(Inoue et al., 2009). The third protein, Tid1, activates the small GTPases and stimulates tyrosine phosphorylation of the AChR β subunit necessary for interactions with Rapsyn (Linnoila et al., 2008).

There is strong evidence that Agrin-MuSK signaling initiates or stabilizes AChR clustering at the developing NMJ, but the signal transduction mechanism is not fully resolved. Agrin induces tyrosine phosphorylation of AChR subunits, and inhibiting this process prevents AChR clustering (Wallace et al., 1991; Qu and Huganir, 1994; Wallace, 1994; Ferns et al., 1996). There is a close temporal relationship between β-subunit phosphorylation and receptor clustering. Receptor phosphorylation reaches a peak within one hour, and receptor clustering then occurs over the next six hours. By the time that clustering has reached a maximum, phosphorylation is in steep decline (Figure 8.23C). The tremendous impact of MuSK signaling is clearly seen in a group of patients suffering from the autoimmune disease myasthenia gravis. In most of these patients, autoantibodies against the AChR are made, leading to a decrease in synaptic transmission and muscle weakness. However, some patients make auto-antibodies against the MuSK protein, and this also leads to severe problems with neuromuscular transmission (Hoch et al., 2001). Similarly, mutations in Dok-7 underlie some cases of congenital myasthenic syndrome (Beeson et al., 2006).

The development of the *Drosophila* NMJ differs somewhat from that of vertebrates. Fly motor neurons make several glutamatergic contacts with a muscle fiber, somewhat like presynaptic boutons in the vertebrate CNS, and glutamate receptors become clustered at the postsynaptic muscle cell. Activity of the nerve terminal is required for glutamate receptor clustering, although release of glutamate is not required (Broadie and Bate, 1993c; Sweeney et al., 1995; Aravamudan et al., 1999; Saitoe et al., 2001). In fact, there are signs that a motor terminal does organize postsynaptic structure from the earliest stage. *Drosophila* muscle fibers produce tiny processes, called *myopodia*, that interact with motor nerve terminals. These myopodia gradually become restricted to the site of innervation. When the motor axons are delayed, as in the *prospero* mutant, the myopodia no longer become clustered (Ritzenthaler et al., 2000). Although Wnt and TGFβ signaling play a role in the number of boutons and their morphology, molecules other than the glutamate receptors themselves that initiate clustering have yet to be identified (Collins and DiAntonio, 2007).

RECEPTOR CLUSTERING SIGNALS IN THE CNS

Since receptor clustering is the most identifiable postsynaptic feature of central synapses, it is reasonable to ask whether Agrin-like molecules play a role. In fact, both Agrin and MuSK are present in the embryonic CNS, and contribute to the development of some excitatory synapses in cortex (Ksiazek et al., 2007). However, since many synapses develop normally in the absence of Agrin signaling, there must be additional receptor clustering signals in the CNS.

One system that is able to cluster glutamate receptors is the EphrinB-EphB signaling pathway which plays such an important role in axon pathfinding and target selection (Chapters 5 and 6). When cultured cortical neurons are exposed to ephrinB1, they rapidly display clusters of the ephrin receptor, EphB2, followed by the appearance of NMDA-type glutamate receptor clusters (**Figure 8.24**A) (Dalva et al., 2000). Furthermore, when hippocampal neurons are cultured from mice that lack three EphB receptors, there is a dramatic loss of normal spine morphology and glutamate receptor clustering (Figure 8.24B). However, this signaling pathway does not appear to disrupt inhibitory GABAergic synapse formation, presumably because these contacts are not made on dendritic spines (Henkemeyer et al., 2003).

A second system involves a family of proteins called pentraxins that have been implicated in the clustering of AMPA-type glutamate receptors (AMPAR), apparently through binding directly to a specific receptor subunit, GluR4. To determine whether an endogenous pentraxin, called Narp, participates in AMPAR clustering, cultured spinal neurons were transfected with a form of Narp that could not be transported to the synapse. This leads to a dramatic reduction in clustered AMPARs at axon–dendrite contacts (O'Brien et al., 1999; 2002). Similarly, when three neuronal pentraxins (Narp, NP-1, NRP) are genetically eliminated, GluR4 is no longer targeted to AMPAR clusters (Sia et al., 2007).

A third system that participates in receptor clustering is the neurotrophin-Trk signaling pathway, which also plays a fundamental role in target selection and neuron survival (Chapters 5 and 7). There is some indication that neurotrophins may be released by afferent terminals (Wang et al., 2002). A role for neurotrophins was assessed in a dissociated culture of embryonic hippocampal neurons which express the BDNF receptor, TrkB, diffusely along their dendrites and soma. When these

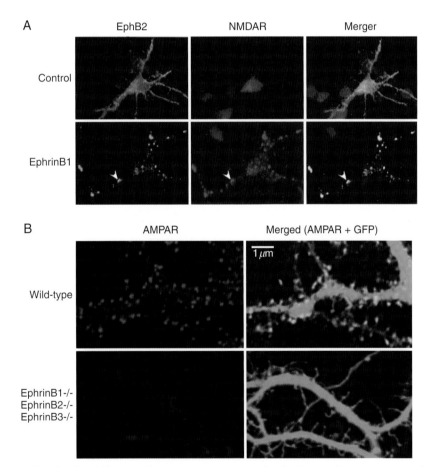

Fig. 8.24 EphB signaling influences glutamate receptor aggregation in the central nervous system. A. Cortical neurons were transfected with labeled EphB2 constructs (green, left images), and exposed to EphrinB1 or a control solution for one hour. The cultures were then fixed, and NMDA receptors (NMDAR) were stained with antibody (red, center images). The merged images (right) show that EphrinB1 induced NMDAR clusters colocalize with EphB2 (white arrowhead) on cortical neuron dendrites. B. Hippocampal neurons were cultured from wild-type or triple EphB-deficient mice (EphB1−/−, EphB2−/−, EphB3−/−). The cells were transfected with green fluorescent protein (GFP) so that the dendrite could be observed, and AMPA receptors (AMPAR) were stained with antibody (red, left images) after 21 days in vitro. Wild-type neurons exhibit punctate labeling of AMPAR clusters all along the dendrites (merged images, right). In contrast, AMPAR clustering is not observed in EphB-deficient neurons. *(Adapted from Dalva et al., 2000; Henkemeyer et al., 2003)*

neurons are exposed to BDNF, the number of GABA and NMDA receptor clusters double, and they are far more likely to be located adjacent to a presynaptic terminal (Elmariah et al., 2004). Conversely, decreasing endogenous BDNF in the cultures leads to a loss of these receptor clusters.

At least two additional systems, discussed above, contribute to receptor clustering: Neurexin can induce receptor clustering in a tissue culture system through specific interactions with postsynaptic neuroligins (NLGN). NLGN-1, -3, and -4 are found only at excitatory postsynaptic sites and mediate glutamate receptor clustering, whereas NLGN-2 is found at both excitatory and inhibitory postsynaptic sites and mediates both glutamate and GABA$_A$ receptor clustering (Graf et al., 2004). When NLGN-1, -2 and -3 were eliminated in mice, fewer GABA$_A$ receptors were recruited to the postsynaptic site and there was a substantial reduction of inhibitory currents (Varoqueaux et al., 2006).

Finally, it has been found that cell adhesion molecules in the cadherin family can induce postsynaptic differentiation, in part by interacting directly with an extracellular region of the glutamate receptor to induce their clustering (Paradis et al., 2007; Saglietti et al., 2007). To summarize, several candidate signals may be released by excitatory and inhibitory nerve terminals and may recruit the aggregation of postsynaptic receptors, similar to the role of Agrin at the nerve–muscle junction.

Several studies suggest that vesicular release is associated with receptor accumulation. For example, the largest glutamate receptor clusters occur opposite the presynaptic terminals that release the most glutamate at the fly neuromuscular junction (Marrus and DiAntonio, 2004). Furthermore, when spontaneous vesicular fusion is blocked during development, glutamate receptor clusters do not form (Saitoe et al., 2001). An effect of activity has also been observed for glycine receptors and glutamate receptors in tissue culture preparations (Kirsch and Betz, 1998; Shi et al., 1999; Liao et al., 2001; Lu et al., 2001). It is not yet clear whether activity-dependent receptor clustering is a developmental mechanism, or is used largely to adjust synaptic strength throughout the animal's life (Aoki et al., 2003).

It must be emphasized that the influence of synaptic transmission on receptor clustering remains an unsettled issue. In *C. elegans*, GABA is not required to obtain synaptically clustered GABA$_A$ receptors (Gally and Bessereau, 2003). A careful in vitro study showed that glutamate receptors are evenly distributed along the dendrite at a low density ($\approx 3/\mu m^2$) prior to innervation but form high-density ($\approx 10,000/\mu m^2$) aggregates at the site of presynaptic contacts (**Figure 8.25**). However, these glutamate receptor clusters appear even when synaptic or electrical activity is blocked (Cottrell et al., 2000). One potential explanation for these findings is that clustering of specific receptor family members may be either activity-dependent or -independent. For example, it has been found that neurexin-neuroigin signaling leads to the accumulation of the AMPAR protein, GluR1, but the clustering of another isoform, GluR2, requires activity (Heine et al., 2008). Activity at the nascent glutamatergic synapse may stabilize receptors in the membrane. A local increase of intracellular calcium can instantly reduce the lateral mobility of GluR2, and the activity of individual synapses can produce a similar effect (Borgdorff and Choquet, 2002; Ehlers et al., 2007). In a similar manner, AChR can aggregate at the neuromuscular junction through lateral diffusion in the membrane, and this process depends on

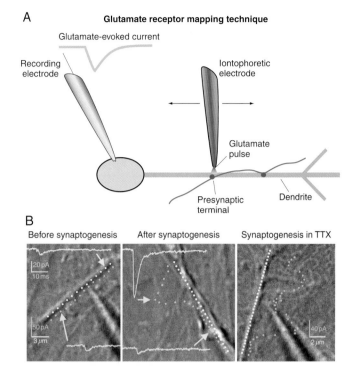

Fig. 8.25 Mapping glutamate receptor location during synaptogenesis. A. To map the location of glutamate receptors, an iontophoretic pipette (red) ejects glutamate focally, and the evoked response is recorded at the soma. B. Glutamate-evoked currents become restricted to the site of synaptic contacts (green) after the dendrites become innervated. White dots show the positions of glutamate application; the relative distance of the yellow dot indicates the evoked current magnitude for each position. Representative glutamate-evoked currents are shown (cyan). When neurons are cultured in the presence of TTX to block all action potentials, the synaptic localization of glutamate receptors occurs nonetheless. *(Adapted from Cottrell et al., 2000)*

activity (Young and Poo, 1983b; Akaaboune et al., 1999). The recovery of acetylcholine responses at the sites of local inactivation did not occur when receptor diffusion was perturbed by cross-linking surface receptors. Whether or not synaptic activity induces synapse formation, it has a major influence on their stability and function (see Chapter 9).

SCAFFOLD PROTEINS AND RECEPTOR AGGREGATION IN THE CNS

For each of the receptor clustering signals at central synapses, there must be a molecular mechanism to hold the receptors together, similar to the way that rapsyn restricts AChR mobility. In fact, there is an array of proteins located at the cytoplasmic surface of the synaptic membrane that can bind to both membrane receptors and cytoskeletal elements. These internal membrane proteins are generally referred to as scaffolding proteins (**Figure 8.26**).

The scaffolding proteins that cluster glutamate receptors comprise about 6% of the postsynaptic apparatus, and are relatively well-described (Cheng et al., 2006; Sheng and Hoogenraad, 2007; Feng and Zhang, 2009). In general, each excitatory postsynaptic site contains several interacting proteins with PDZ domains that bind to a short amino acid sequence at the C-terminal of glutamate receptors (Figure 8.26). Similar to the NMJ, cluster formation is regulated by

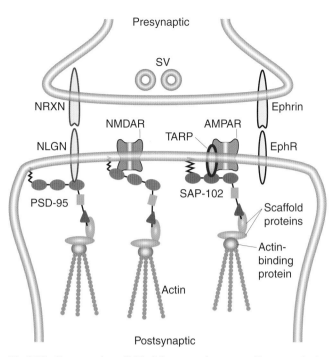

Fig. 8.26 Postsynaptic scaffold of glutamatergic synapses. Transynaptic signaling from presynaptic neurexin (NRXN) to postsynaptic neuroligin (NLGN) can influence receptor aggregation via NLGN interaction with PSD-95. NMDA receptors (NMDAR) can interact directly with a PDZ domain on PSD-95. AMPA receptors (AMPAR) first become anchored via an interaction with SAP-102, and transmembrane AMPAR regulatory proteins (TARPs). Ephrin-eph receptor signaling also facilitates NMDAR and AMPAR clustering. PSD-95 and SAP-102 interact with additional scaffold proteins such as GKAP and Shank. Actin-binding proteins connect the scaffold to actin filaments.

protein phosphorylation. There is a specific 4 amino acid sequence on the C-terminal tail of receptors and channels, called ET/SXV, that serves as an important phosphorylation site (Niethammer et al., 1996; Cohen et al., 1996).

One of the first scaffolding proteins to be isolated was postsynaptic density protein 95 (PSD-95). PSD-95 and other members of this family (PSD-93, SAP-102, SAP-97) are membrane-associated guanylate kinases (MAGUKs), and they may have distinct functional roles at different developmental stages. For example, a specific developmental role of individual scaffolding proteins was demonstrated in the hippocampus by manipulating expression level in vivo. When SAP-102 expression is decreased in the embryo, there is a dramatic reduction of synaptic AMPA and NMDA receptors in neonates; the opposite result is obtained when SAP-102 is over-expressed during development. On the other hand, manipulating PSD-95 or -93 expression does not alter the initial appearance of either glutamate receptor. PSD-95 is, however, required for a later stage of AMPAR accumulation and is specifically involved in clustering the mature NMDA receptor subunit, NR2A (Tao et al., 2003; Elias et al., 2006, Elias et al., 2008). Thus, SAP-102 is primary responsible for glutamate receptor clustering during synapse formation, but other scaffolding proteins participate in subsequent maturational events. Another member of the family, SAP-97, appears not to be required for glutamate synapse formation (Howard et al., 2010). However, when CaMKII is activated, SAP-97 is driven into the spine heads, and this is directly correlated with

an increase in AMPARs at these spines (Mauceri et al., 2004). This may provide a mechanism by which activity facilitates receptor clustering (see Chapter 9).

Many MAGUKs do not bind to glutamate receptors, suggesting the involvement of adapter proteins. A family of transmembrane AMPAR regulatory proteins (TARPs) have been shown to bind to AMPAR subunits, and may participate in clustering throughout the nervous system (Figure 8.26). For example, stargazin is a TARP that traffics AMPARs to the membrane surface, but must interact with PSD-95 in order to localize the AMPARs to the postsynaptic membrane. When stargazin is deleted, AMPARs are eliminated from cerebellar granule cells, while NMDA receptors are unaffected (Chen et al., 2000; Tomita et al., 2003).

Clustering proteins are conserved across species. In *Drosophila*, a scaffolding protein called discs-large (DLG) colocalizes with glutamate receptors at the nerve–muscle junction. When the *dlg* gene is inactivated, synaptic structure and function are profoundly altered (Woods and Bryant, 1991; Lahey et al., 1994; Budnik et al., 1996). The synaptic localization of DLG is regulated by CaMKII activity (Koh et al., 1999). DLG is necessary for the localization of FASII, a cell adhesion molecule that regulates synapse formation (see Chapter 9). As with the mammalian family members, the DLG protein has a PDZ binding domain, and the membrane proteins that it anchors have the conserved C-terminal motif. The developmental importance of PDZ-containing scaffolding proteins is underlined by studies showing moderate to severe mental retardation when one of the genes is truncated (Tarpey et al., 2004).

Many scaffold proteins and adapters have been identified and their function validated in at least one group of developing neurons (Feng and Zhang, 2009). Two important families that are present in excitatory synapses throughout the CNS are the guanylate kinase-associated proteins (GKAP) and SH3 and multiple ankyrin repeat domain proteins (Shank). Despite the great success in identifying these proteins, the specific developmental contribution of each one and the full range of protein–protein interactions will take many years to clarify.

The accumulation of glycine receptors at inhibitory synapses requires a protein called gephyrin that displays a high affinity for both polymerized tubulin and and the glycine receptor β subunit (Kirsch et al., 1993a,b; Schrader et al., 2004). Spinal neurons grown in dissociated culture normally display glycine receptor clusters. However, when the cells are grown in the presence of a gephyrin antisense nucleotide, which presumably prevents the translation of gephyrin mRNA, then clusters do not form at the membrane (**Figure 8.27**A). Similarly, glycine receptors do not aggregate within the membrane in gephyrin-deficient mice, indicating that they are required for normal clustering (Feng et al., 1998).

Gephyrin is also involved in the clustering of some GABA receptors. For example, gephyrin binds directly to the α2 subunit, and facilitates clustering of α2-containing GABA receptors (Tretter et al., 2008). For $GABA_A$ receptor clusters that are induced by neurexin-neuroligin signaling (*see above*), gephyrin may play an important role in coupling the clustering signal to receptor aggregation (Figure 8.27B). For example, once NRLG-2 is recruited to the postsynaptic site, it binds to gephyrin and then activates a GDP/GTP exchange factor, called collybistin. This leads to the recruitment of additional

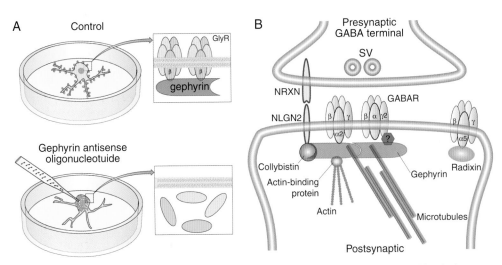

Fig. 8.27 Scaffold proteins support receptor clustering at inhibitory synapses. A. Gephyrin is required for glycine receptor (GlyR) clustering via a direct interaction with the β subunit. When spinal neurons are grown in culture, the peripheral membrane protein, gephyrin, colocalizes with glycine receptor clusters (top). However when translation of the gephyrin protein is blocked with an antisense oligonucleotide (bottom), the glycine receptors largely remain in the cytoplasm and clusters do not form. B. Postsynaptic scaffold of GABAergic synapses. Transynaptic signaling from presynaptic neurexin (NRXN) to postsynaptic neuroligin-2 (NLGN2) can influence GABA receptor (GABAR) aggregation via NLGN interaction with collybistin and gephyrin. GABARs that contain an α2 subunit can interact directly with gephyrin. GABARs that contain a γ2 subunit can cluster with gephyrin via an unidentified (?) binding protein. GABARs containing an α5 subunit are localized to extrasynaptic compartments through direct interaction with radixin. Gephyrin interacts directly with microtubules, and actin-binding proteins connect the gephyrin scaffold to actin filaments. *(Panel A adapted from Kirsch et al., 1993)*

gephyrin-collybistin complexes, and the establishment of a scaffold to which GABA$_A$ receptors become anchored. Deletion of collybistin during embryonic development prevents the clustering of gephyrin and GABA$_A$ receptors in only a few locations, such as the the hippocampus, suggesting that other clustering proteins exist (Papadopoulos et al., 2008; Poulopoulos et al., 2009).

Depending on the subunit composition, many GABA receptor-containing synapses form in neurons that lack gephyrin, leading to the conclusion that other scaffolding proteins must be involved (Levi et al., 2004; Fritschy et al., 2008). Although the inhibitory synaptic scaffold remains a work in progress, a novel scaffolding protein has been identified for GABA$_A$ receptor clusters that are located at nonsynaptic regions of the dendrite. These clusters are thought to play an important role in nervous system function by establishing a baseline level of inhibitory transmission. The actin-binding protein, radixin, anchors extrasynaptic GABA$_A$ receptor by interacting directly with the α5 subunit, and the loss of gephyrin does not disrupt their clustering (Loebrich et al., 2006).

INNERVATION INCREASES RECEPTOR EXPRESSION AND INSERTION

Even while receptor clustering is underway, the synthesis of new synaptic proteins increases dramatically. At the neuromuscular junction, the majority of AChRs within a cluster are newly inserted a short time after innervation (Salpeter and Harris, 1983; Ziskind-Conhaim et al., 1984; Role et al., 1985; Dubinsky et al., 1989). This was initially shown by labeling existing and newly inserted AChRs at newly formed synapses with separate markers (**Figure 8.28**). Before innervation, all

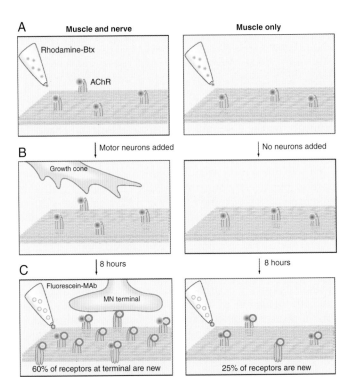

Fig. 8.28 Insertion of new ACh receptors occurs within hours of innervation. A. Cultures of muscle cells were prelabeled with rhodamine-conjugated bungarotoxin (red). B. In one set of cultures, motor neurons were added (left), while a second set of cultures remained without neurons (right). C. After eight hours, both cultures were labeled with a fluorescein-conjugated antibody against AChRs (yellow). The cultures with motor neurons contained many AChRs that were labeled with only antibody (yellow), indicating that they had been newly inserted after the addition of motor neurons. The muscle cell cultures had AChRs that were primarily labeled by both Rhod-Btx (red) and Fluor-MAb (yellow). *(Adapted from Role et al., 1985)*

of the AChRs on the muscle membrane surface were labeled with α-Btx ("old" AChRs). Following the addition of neurons, a monoclonal antibody directed against an extracellular AChR epitope was applied to label all AChRs ("old" plus "new" AChRs). In this way, it was possible to determine the percentage of "new" receptors (i.e., antibody-labeled minus α-Btx labeled). Within eight hours of neuron addition, more than 60% of the AChRs are newly inserted, indicating that synthesis or insertion is upregulated rapidly (Role et al., 1985). Similarly, when innervation of the *Drosophila* neuromuscular junction is delayed or prevented in *prospero* mutants, the normal increase in functional glutamate receptors fails to occur (Broadie and Bate, 1993b).

It is still not entirely clear how AChR transcription is controlled. Neuregulin-1 (NRG-1) signaling, which influences Schwann cell survival, was initially thought to be the neural factor that upregulates AChR transcription upon innervation. NRG-1 and its receptors (ErbB2 and 4) are localized to neuromuscular synapses, and NRG-1 can induce AChR transcription in cultured muscle cells (Falls, 2003). However, near-normal levels of AChR transcription occur at the neuromuscular synapses of mice lacking either NRG-1 in nerve or its ErbB receptors in muscle (Escher et al., 2005; Jaworski and Burden, 2006). In contrast, the loss of Agrin or MuSK leads to the loss of specialized AChR synthesis in subsynaptic nuclei (DeChiara et al., 1996; Gautam et al., 1996). Furthermore, Agrin-MuSK signaling can drive the transcription of synaptic genes, including AChR subunits independently of ErbBs when ectopically expressed in muscle in vivo (Lacazette et al., 2003). Therefore, Agrin appears to be the primary neuron-specific signal to increase transcription, and NRG-1 may contribute through its trophic support of Schwann cells which, themselves, serve as a source of NRG.

As new receptors are added, they also become more stable. This is shown by measuring how long the receptors remain in the membrane before being replaced. In vertebrate muscle, the rate of AChR turnover gradually increases from a half-life of ≈30 hours at the time of synaptogenesis to many days (Burden, 1977a, 1977b; Reiness and Weinberg, 1981). Receptor stability may come about as a result of an activity-dependent rise in second messengers (e.g., calcium or cAMP), and an accumulation of rapsyn (Rotzler et al., 1991; Shyng et al., 1991; Gervásio et al., 2007).

Innervation can also regulate the expression of receptors at neuron–neuron contacts. The ACh-evoked response recorded in sympathetic neurons increases almost tenfold after innervation by spinal afferents, and this can be mimicked with a soluble factor from spinal cord-conditioned media (Role, 1985; Gardette et al., 1991). The influence of innervation on receptor synthesis can be quite specific. When cultured in the absence of spinal interneurons, dissociated chick motoneurons exhibit much smaller glutamate-evoked currents, but the addition of interneurons serves to localize glutamate sensitivity to the motoneuron processes (O'Brien and Fischbach, 1986a, 1986b).

Are receptors synthesized only within the cell body and then transported to distant synapses? In fact, protein synthesis occurs in neuronal dendrites, often near synapses, and polyribosomal aggregates appear in dendritic spines during development (Miyashiro et al., 1994; Steward and Schuman, 2001). However, it has been difficult to determine whether the synthesis of new glutamate receptor subunits occurs subsynaptically in the dendrites of central neurons. To get around this problem, existing AMPARs were prelabeled in a culture of hippocampal neurons, and then counter-stained with a second label to identify the location of newly synthesized subunits. In concept, this approach is nearly identical to that employed on AChRs by Role et al. (see Figure 8.24). Hippocampal neurons were first transfected with a glutamate receptor subunit (GluR1 or GluR2) that was modified to contain a tetracysteine motif on its intracellular C-terminal (**Figure 8.29**). This permitted the use of

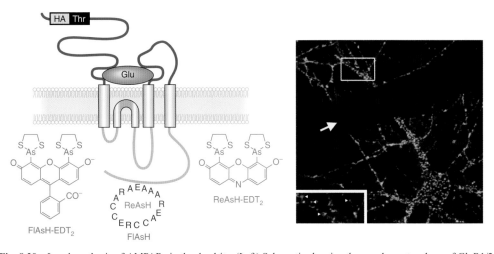

Fig. 8.29 Local synthesis of AMPARs in the dendrite. (Left) Schematic showing the membrane topology of GluR1/2 and the location of the intracellular tetracysteine and extracellular HA/thrombin tags. The tetracysteine binds to FlAsH-EDT2 or ReAsH-EDT2, yielding a green or red fluorescent signal, respectively. (Right) An example of a hippocampal neuron in which all GluR1 protein was prelabeled with red dye, followed by transection of the dendrite (white arrow) and labeling with green dye. Note that some GluR aggregates are labeled only with the green dye (inset, white arrowheads), indicating that they must have been synthesized after the dendrite was detached from the soma. Preexisting aggregates that are labeled with both red and green dye are orange. (Ju et al., 2004)

two different dyes that fluoresced (red or green) only when bound to tetracysteine. To show that some receptors are synthesized locally, the dendrite was transected from the soma after prelabeling all existing GluRs with the red dye. Thus, when GluR aggregates were subsequently labeled with only the green dye, they must have been synthesized within the dendrite (Figure 8.29). Finally, it was possible to show that the GluRs synthesized in the dendrite were inserted into the membrane (Ju et al., 2004).

SYNAPTIC ACTIVITY REGULATES RECEPTOR DENSITY

The increase in receptor synthesis that accompanies synapse formation suggests that the presynaptic terminal initiates this process. One simple possibility is that the transmitter itself can regulate expression of synaptic proteins. In fact, muscle cells are polynuclear, and motor neuron terminals selectively regulate the expression of AChR mRNA in the nuclei that lay directly beneath the synaptic cleft (Klarsfeld et al., 1991; Sanes et al., 1991; Simon et al., 1992; Merlie and Sanes, 1985). Activity-dependent regulation of receptor expression was first demonstrated in adult muscle: cutting the motor axon connection increases the response to ACh applied at any position along the surface. This is referred to as denervation supersensitivity, and it requires new receptor synthesis (Axelsson and Thesleff, 1959; Lømo and Rosenthal, 1972; Berg and Hall, 1975; Merlie et al., 1984). Denervation supersensitivity can also be produced by simply blocking presynaptic action potentials (**Figure 8.30**). The opposite manipulation, direct electrical stimulation of muscle cells in vitro, produces a decrease in AChR synthesis (Shainberg and Burstein, 1976). At least in muscle cells, synaptic activity limits receptor synthesis through increasing postsynaptic calcium, and may depend, in part, on the activation and phosphorylation of the muscle transcription factor, myogenin (Klarsfeld et al., 1989; Laufer et al., 1991; Huang et al., 1992; Tang et al., 2004).

Acetylcholine receptor aggregation is influenced by ACh-evoked activity almost immediately after synapse formation. As discussed above, genetic deletion of Agrin results in a loss of AChR clusters at the embryonic neuromuscular junction. However, relatively normal AChR clustering can be rescued in Agrin[-/-] mice by interfering with the synaptic transmission or the intracellular elements that it recruits (Misgeld et al., 2005; Lin et al., 2005). For example, deleting choline acetyltransferase (ChAT), the enzyme that synthesizes ACh, can rescue AChR aggregation in agrin-deficient mice. (**Figure 8.31**A, left). The ACh-AChR signal is likely mediated by the calcium-dependent activation of Cdk5, because its deletion enables AChR clusters to form in Agrin[-/-] mice (Figure 8.31A, right). The interpretation of this finding is that ACh-evoked synaptic activity acts to decrease the synthesis and accumulation of AChRs at the endplate. Therefore, receptor clustering is a competition between two forces: Agrin-MuSK signaling facilitates the synthesis and clustering of AChRs, while ACh-AChR signaling inhibits synthesis and acts to disperse existing receptor clusters (Figure 8.31B).

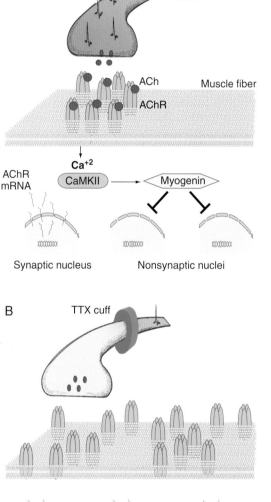

Fig. 8.30 Extrasynaptic ACh receptors accumulate when the nerve is inactive. A. At the control nerve–muscle junction, the electrically active terminal releases ACh and the receptors are clustered at the postsynaptic membrane. The activity-dependent signal that suppresses extrajunctional receptors involves calcium influx and activation of the calcium calmodulin-dependent protein kinase II (CamKII). A transcription factor found in muscle (myogenin) is phosphorylated and blocks transcription in extrasynaptic nuclei. B. When motor axon activity is blocked with the sodium channel blocker, tetrodotoxin (TTX), extrajunctional ACh receptors are distributed over the entire muscle surface. *(Adapted from Lømo and Rosenthal, 1972)*

A similar sort of regulation likely occurs at excitatory synapses in the CNS. For example, the functional expression of NMDA receptors decreases with age in the visual cortex of normal kittens, but when animals are reared in complete darkness to decrease visually driven activity, NMDAR-mediated transmission remains at an unusually high level (Fox et al., 1992).

When AMPAR synthesis is measured in the dendrites of cultured hippocampal neurons, it can be demonstrated that decreased neuronal activity leads to an upregulation of the GluR1 subunit (Ju et al., 2004). The converse is also true,

as illustrated in **Figure 8.32**. Activating NMDA receptors in these cultures leads to a significant decrease of the mRNA and protein for GluR1 and GluR2 in dendrites, and this is due to a calcium-dependent signal that decreases transcription at the nucleus (Grooms et al., 2006).

The ability of glutamate-evoked activity to regulate AMPAR density can also be mediated by a local pathway that controls protein internalization and degradation (**Figure 8.33**). In this case, synaptic activity leads to a local increase in the translation of a cytoskeleton-associated protein, called Arc, that is involved in AMPAR internalization (Bramhan et al., 2008). The ability of Arc to decrease surface expression of AMPARs is constrained by a second emzyme, E3 ubiquitin protein ligase (Ube3A), that marks Arc for degradation by attaching a polyubiquitin chain to it. When Ube3A function is disrupted in mice, Arc expression remains high and there is a steep loss of AMPARs from the synapse (Yashiro et al., 2009; Greer et al., 2010). In fact, a human disorder called Angelman syndrome, characterized by severe mental retardation and autistic symptoms, is caused by inheritance of a mutant copy of the Ube3A gene (Kishino et al., 1997; Matsuura et al., 1997).

Activity-dependent signals have also been shown to up-regulate surface expression of receptors. In fact, glutamatergic synaptic transmission often appears to be absent in young animals because excitatory synapses contain only NMDARs, and these receptors tend to remain closed at the resting membrane potential. These are sometimes referred to as "silent synapses." When NMDARs are actived by stimulating the synapse during depolarizing current pulses, the synapses are soon found to have functional AMPARs (Durand et al., 1996). This process depends on the influx of calcium when NMDARs are activated. When a calcium-sensitive kinase (CaMKII) is constitutively expressed in frog optic tectum neurons, the appearance of AMPAergic transmission is facilitated (Wu et al., 1996). Activation of "silent" synapses has also been observed at slightly later periods of development and may, in fact, underlie certain forms of learning or memory (see Chapter 9).

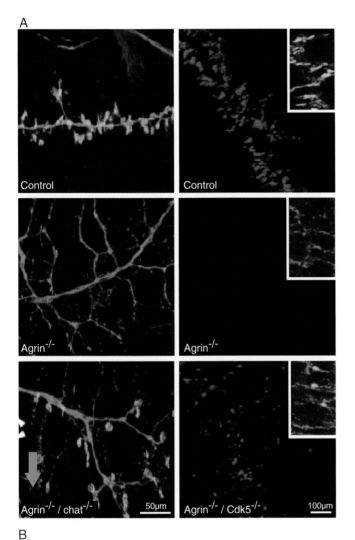

Fig. 8.31 Synaptic activity disperses AChRs at the neuromuscular junction. A. The developing mouse neuromuscular junction was stained for AChRs (red) and motor axons (green). Control animals (top) displayed aggregation of AChRs apposed to motor axons. Agrin-deficient animals (middle) displayed a loss of AChR aggregation. When cholinergic transmission was also eliminated by deleting the choline acetyltransferase gene (bottom, left), AChR aggregation was rescued. Similarly, when the cyclin-dependent kinase 5 (cdk5) was eliminated in Agrin-deficient mice, the AChR aggregation was rescued (bottom, right). B. The schematic illustrates that there are two competing pathways at the developing neuromuscular junction. Agrin-MuSK signaling increases the expression of the εAChR subunit and MuSK, and mediates AChR aggregation (thick black arrows). ACh-AChR signaling results in the calcium-dependent activation of Cdk5, and leads to the dispersal of AChR aggregation, possibly through endocytosis (thick red arrows). There is mutual antagonism between these two pathways: MuSK signaling inhibits the dispersal signal (thick black line) and AChR activity inhibits transcription (thick red line). *(Panel A adopted from Misgeld et al., 2005 and Lin et al., 2005)*

235

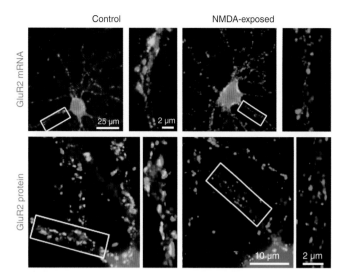

Fig. 8.32 Activity-dependent regulation of AMPAR surface expression at glutamatergic synapses. (Top) Hippocampal neurons from neonatal rats were grown under control conditions (left) or in the presence of NMDA (right), and then stained for a presynaptic marker (green) and GluR2 mRNA (red). When grown in NMDA, there is significantly less GluR2 mRNA in dendrites (insets show region within white box at higher magnification). (Bottom) The identical experiment was performed, but the neurons were stained for GluR2 protein (red). Again, NMDA exposure leads to significantly less GluR2 protein in dendrites. *(Adopted from Grooms et al., 2006)*

MATURATION OF TRANSMISSION AND RECEPTOR ISOFORM TRANSITIONS

Synapse formation is rapid, but adult functional properties emerge only gradually during development. One of the most common observations is that the duration of excitatory or inhibitory synaptic potentials declines over the course of days (**Figure 8.34**). This is observed from brainstem to cortex. For example, synapses in the brainstem display marked alterations during postnatal development. In the lateral superior olive, the maximum duration of both glutamatergic EPSPs and glycinergic inhibitory postsynaptic potentials (IPSPs) declines approximately tenfold during the first three postnatal weeks (Sanes, 1993). Similarly, in the rat neocortex, the duration of excitatory postsynaptic potentials (EPSPs) decreases

from approximately 400 to 100 ms during the first two postnatal weeks of development (Burgard and Hablitz, 1993), and synaptic potentials in the rat hippocampus display a similar schedule of maturation. The reduction in IPSP and EPSP duration has a similar rate of development, suggesting that some of the underlying mechanisms are the same. These long-lasting synaptic potentials probably limit the behavior capabilities of young animals (see Chapter 10).

Why are synaptic potentials of such long duration in developing neurons? One common difference is that young synapses usually express a unique form of the neurotransmitter receptor, called a *neonatal isoform*. These transiently expressed receptors have different functional properties than the receptor that is expressed by adult neurons. In particular, the receptor-coupled ion channels in young cells tend to remain open for a longer period of time, compared to those in mature cells. In mammalian muscle cells, recordings made from single channels with the patch-clamp recording technique (see Box 8.2), demonstrate that the mean channel open time declines from about 6 to 1 ms during development (Siegelbaum et al., 1984; Vicini and Schuetze, 1985). This functional change is due, in part, to a developmental change in AChR subunit composition (**Figure 8.35**). Neonatal AChRs are composed of four subunits ($\alpha_2\beta\gamma\delta$) but during the first two postnatal weeks in rat, the many nuclei beneath each synapse stop expressing the γ-subunit and increase their expression of ϵ-subunit transcripts. (Gu and Hall, 1988; Harris et al., 1988; Martinou et al., 1991).

Even though nerve cells limp along on one nucleus in their soma, it appears that they are able to respond to innervation by altering the receptor isoform expression. When chick sympathetic ganglion neurons are innervated, their sensitivity to ACh is enhanced, and this is correlated with increased expression of 5 AChR transcripts (Moss and Role, 1993; Corriveau and Berg, 1993). At E11, only 30% of neurons have significant AChR activity, and each individual patch of membrane contains a mixture of AChRs. At E17, the great majority of patches have a single functional type receptor. Similar changes in specific AChR subunits have been observed in rat brainstem, spinal cord, and dorsal route ganglia during prenatal development.

A developmental switch in receptor subunits has now been demonstrated in nearly every transmitter system in the central nervous system. For example, the adult form of the glycine receptor heteromer involves the substitution of a

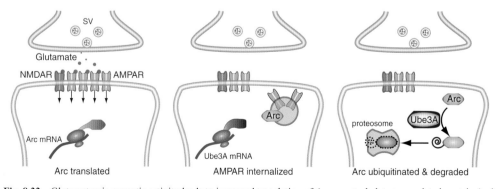

Fig. 8.33 Glutamatergic synaptic activity leads to increased translation of Arc, a cytoskeleton-associated protein (red, left). The Arc increases the endocytotic removal of AMPARs from the synaptic membrane (middle). During this time, the expression of E3 ubiquitin protein ligase (Ube3A) gradually rises (blue, middle). Ube3A attaches a polyubiquitin chain to Arc, marking it for degradation by a proteosome, and terminating AMPAR internalization.

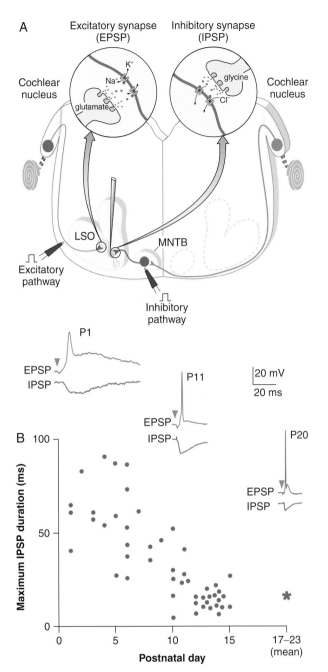

Fig. 8.34 The duration of synaptic potentials decreases during development. A. A schematic of a central auditory nucleus, the lateral superior olive (LSO), that receives excitatory synapses from the ipsilateral cochlear nucleus and inhibitory synapses from the medial nucleus of the trapezoid body (MNTB). The inset at left shows that excitatory terminals release glutamate and open receptors that are permeable to Na+ and K+. The inset at right shows that inhibitory terminals release glycine and open receptors that are permeable to Cl−. B. When intracellular recordings are made from LSO neurons during the first three postnatal weeks, the afferent-evoked EPSP and IPSP durations decline by approximately tenfold. Examples for postnatal day 1, 11, and 20 are shown at the top, and a summary of all IPSPs is plotted in the graph. *(Adapted from Sanes, 1993)*

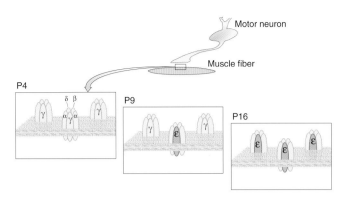

Fig. 8.35 Substitution of receptor subunits during development. In rat muscle, AChRs are composed of α, β, δ, and γ subunits at postnatal (P) day 4. By P9, there are a mix of receptors: some have the initial complement of subunits, and others have substituted the ε subunit in place of the γ subunit. At P16, all receptors contain the ε subunit.

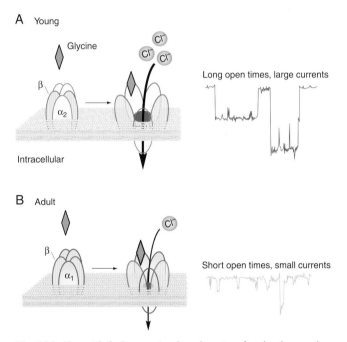

Fig. 8.36 Neonatal glycine receptors have immature functional properties. In neonatal mammalian neurons, the glycine receptor is composed of β and α2 subunits. When bound by glycine, the receptors remain open for a relatively long time and pass a relatively large current. In adult neurons, the glycine receptor contains a β subunit, but the neonatal isoform, α2, is replaced by the α1 subunit. These receptors open briefly and pass less current than the neonatal form. *(Panel A adapted from Gu and Hall, 1988; Panel B adapted from Takahashi et al., 1992)*

48kD ligand-binding subunit for a neonatal isoform (Becker et al., 1988). Recordings from rat dorsal spinal cord neurons during development showed that there is a complementary change in function (Takahashi et al., 1992). The glycine-gated channels from young animals (<P5) open for a much longer period of time and pass a greater amount of current, compared to older postnatal animals (**Figure 8.36**). By examining the properties of two different glycine receptor subunits in a *Xenopus* expression system, it was determined that a transition from the α2 to the α1 subunit could explain the functional change.

When a receptor family has many subunits, the type of receptor that is produced becomes a combinatorial problem. The temporal and regional expression of 13 different GABA$_A$ receptor subunits in the developing rat brain provides an interesting example. The expression patterns are determined by in situ hybridization, a technique in which radiolabeled antisense oligonucleotides are used as probes for each species of mRNA (Laurie et al., 1992). The onset of expression and the adult level of expression can vary greatly for a single subunit, depending on location. Moreover, there are a large number of subunits that are transiently expressed within a given structure. As one of many examples, there is little α1 in the brain of P6 rats, but it is heavily expressed in adults (**Figure 8.37**). In contrast, the onset of γ2 subunit expression occurs throughout the brain at embryonic day 17. Whereas the level of γ2 expression gradually increases in the hippocampus and cerebellum, it ceases to be expressed in the cortex and thalamus.

The long duration of excitatory synaptic events in many regions of the CNS is at least partly due to a neonatal form of the NMDA-gated glutamate receptor (**Figure 8.38**). The duration of afferent-evoked excitatory postsynaptic currents (EPSCs) in the rat superior colliculus that are mediated by NMDA receptors declines several-fold during the first three postnatal weeks (Hestrin, 1992). This transition is due, in part, to a developmental change in the subunit composition of NMDARs. In many areas of the CNS, the NMDARs initially contain the NR2B subunit, but the expression of NR2A rises gradually as the system matures, and this transition is associated with the shorter EPSCs (Sheng et al., 1994; Flint et al., 1997; Quinlan et al., 1999). NMDAR subunit composition may also affect other important properties of the receptor.

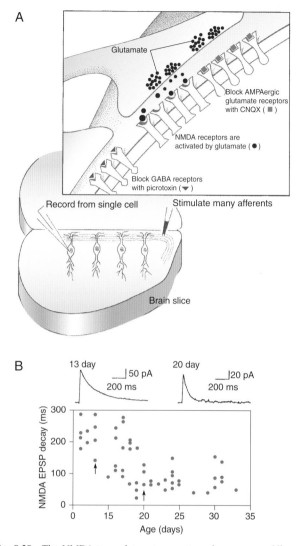

Fig. 8.38 The NMDA-type glutamate receptors close more rapidly with age. A. Intracellular recordings were obtained from rat hippocampal neurons in a brain slice preparation, and AMPARs and GABA receptors were blocked. Thus, stimulation of afferents evoked glutamate release, and only postsynaptic NMDA-type receptors were activated. B. The afferent-evoked EPSPs were longer lasting in neurons from young neurons due to the slow decay time. *(Adapted from Hestrin, 1992)*

For example, NMDARs are sensitive to the presence of both ligand (glutamate) and membrane depolarization in adults, but voltage sensitivity may be absent in the neonatal hippocampus (Ben-Ari et al., 1988). Apparently, the neonatal receptors are less sensitive to Mg^{2+}, the ion that must be expelled from the channel pore during depolarization, thus permitting Na^+ and Ca^{2+} to pass through (Bowe and Nadler, 1990; Kirson et al., 1999). Since resting membrane potential is generally more depolarized early in development, excitatory transmission through the NMDAR may contribute a larger fraction of the the synaptic current.

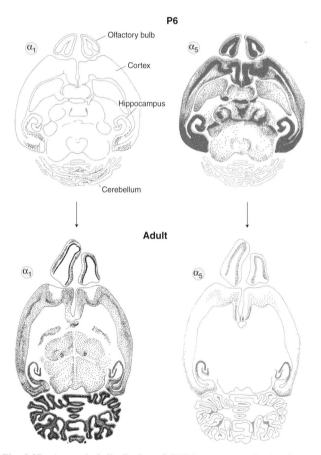

Fig. 8.37 Anatomical distribution of GABA_A receptor subunits changes dramatically during development. Each panel displays the staining pattern of an antibody directed against either the α1 or the α5 GABA_AR subunits. (Top) At P6, there is little α1 in the brain, whereas α5 is heavily expressed in the hippocampus and cortex. (Bottom) In adult, α1 is heavily expressed, and α5 expression is restricted. *(Adapted from Laurie et al., 1992)*

MATURATION OF TRANSMITTER REUPTAKE

The amount of time that the neurotransmitter remains in the synaptic cleft will also affect the duration of synaptic potentials, and the development of transmitter uptake systems is critical for the emergence of mature function. Neurotransmitter transporter protein development has been studied by expressing polyadenylated brain RNA (polyadenylation, or the addition of about 200 adenylate residues, is a common modification to transcripts in eukaryotic cells) in *Xenopus* oocytes (Blakely et al., 1991). Messenger RNA was obtained from animals of different ages and placed in a *Xenopus* oocyte expression system. The amount of transport was quantified by incubating the oocyte in a radiolabeled amino acid neurotransmitter, such as ³H-glycine, and the amount of ³H was quantified with a liquid scintillation counter. By using this assay, it was found that glutamate and GABA transporters first appear in the cortex at postnatal day 3 and increase to adult levels over the next two weeks. In the brainstem, the expression of a glycine transporter gradually increases to adult levels over the first three postnatal weeks.

A number of amino acid transporters have now been identified at the molecular level, and a few studies have traced their developmental appearance using in situ hybridization. The excitatory amino acid transporters, mEAAT1 and mEAAT2, are first found in the proliferative zone of mouse forebrain and midbrain during gliogenesis (E15–E19). However, mEAAT2 mRNA continues to increase in many areas of the CNS during the first two to three postnatal weeks (Sutherland et al., 1996). Transcripts for the Na⁺/Cl⁻-dependent glycine transporter (GlyT1), found almost exclusively in glial cells, achieve maximal levels in E13 mice, much earlier in neural development (Adams et al., 1995). There is also evidence for early transient expression of GlyT-1, as well as glycine, in the kitten retina (Pow and Hendrickson, 2000). The GABA transporters (GAT-1 and -3) are expressed in rat thalamic and cortical astrocytes at birth, achieving an adult pattern of expression between 2–3 weeks postnatal (Vitellaro-Zuccarello et al., 2003).

Although the presence of transporter mRNA suggests that the neurotransmitter could be efficiently cleared at the onset of synaptogenesis, studies of amino acid transporter function show that their physiology remains immature for some time (Blakely et al., 1991). For example, in the cerebellum, there is evidence that GAT-1 is first expressed by the GABAergic terminals during the period of inhibitory synaptogenesis, followed by the expression of GAT-3 in perisynaptic astrocytes (Takayama and Inoue, 2005). Therefore, the maturation of transporter proteins probably limits the kinetics of synaptic transmission.

SHORT-TERM PLASTICITY

To this point we have considered only the most basic functional properties of a synapse: the release of a transmitter elicited by a single action potential and the postsynaptic current that it produces. Of course, neurons will fire many times per second under realistic conditions, and the synaptic response may become facilitated or depressed over time. These changes in synaptic response are called short-term plasticity, and their maturation depends on the development of presynaptic release properties and the complement of postsynaptic receptors and ion channels.

A simple approach to examine short-term plasticity involves taking slices of brain tissue at increasing postnatal ages and recording the synaptic responses that are elicited when trains of stimuli are delivered to the afferent pathway. To examine short-term plasticity, synaptic currents were recorded from MNTB neurons in response to stimulation of excitatory afferents from the cochlear nucleus (see schematic in Figure 8.34). MNTB neurons are innervated by only a single glutamatergic afferent that makes a large synapse on the cell body, called the *endbulb of Held*. When these synapses are stimulated at 200 Hz just prior to the onset of hearing (P9), they display a rapid depression of the postsynaptic response, and there are complete failures where the transmitter is apparently not released by the endbulb of Held (**Figure 8.39**A). However, when the

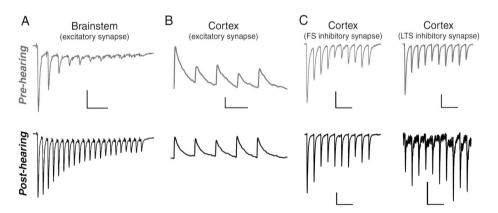

Fig. 8.39 Development of short-term synaptic plasticity in the auditory system before and after the onset of hearing. A. AMPAR-mediate excitatory postsynaptic currents are recorded in auditory brainstem MNTB neurons in response to a 200 Hz stimulus train. Examples are shown from neurons at P9 and P14. Before the onset of hearing, there is synaptic depression and failures, but the depression is much reduced after the hearing begins. (bars: 2 nA and 20 ms) B. Excitatory postsynaptic potentials (EPSP) are recorded in auditory cortex Layer 3 pyramidal neurons in response to stimulation of a second Layer 3 neuron. In P11 cortex, stimulation of the presynaptic neuron at 10 Hz evoked EPSPs that declined in amplitude, or short-term depression. At P28, stimulation at the same rate did not produce short-term depression. (bars: 0.5 mV and 100 ms) C. Inhibitory postsynaptic currents (IPSC) are recorded in auditory cortex Layer 3 pyramidal neurons in response to stimulation of either FS (left) or LTS (right) inhibitory interneurons. In P10 cortex, stimulation of either inhibitory neuron led to short-term depression, but after the onset of hearing the FS-evoked response did not depress. (FS bars: 20 pA, 100 pA, 200 ms; LTS bars: 50 pA, 10 pA, 200 ms). *(Panel A from Joshi et al., 2002; panel B from Oswald and Reyes, 2008; panel C from Takesian et al., 2010)*

same stimulus is delivered to afferents at P14, the response does not display as much depression, and there are no failures of transmitter release (Joshi and Wang, 2002). Several mechanisms may account for this maturation. In young animals, the endbulb of Held produces an action potential that lasts a relatively long time, and this prevents it from responding to each stimulus. It is also likely that the pool of vesicles available for release during rapid stimulation increases with development. Postsynaptic mechanisms could also contribute to this depression, including desensitization of glutamate receptors.

The developing auditory cortex displays an even greater transformation in short-term plasticity (Figure 8.39B). When two interconnected neurons are recorded in P11 auditory cortex, it is found that excitatory synapses display a depression when stimulated at 10 Hz. When a similar pair of neurons is recorded at P28, the excitatory connections no longer display depression (Oswald and Reyes, 2008). The transition of short-term plasticity is specific to the type of connection. For example, the connection between one type of auditory cortex inhibitory interneuron, fast spiking cells, displays short-term depression before and after the onset of hearing (Figure 8.39C). In contrast, the synaptic connections from another type of interneuron, low threshold spiking cells (LTS), transition from depressing to nondepressing during development (Figure 8.39D). The switch from depressing to nondepressing LTS-evoked inhibition is due, in part, to a developmental reduction of presynaptic $GABA_B$ receptors, and this transition depends on early auditory experience (Takesian et al., 2010). At excitatory synapses, a variety of factors are involved in the maturation of short-term plasticity, including regulation of presynaptic Ca^{+2} concentration, glutamate receptor desensitization, and an increase in presynaptic metabotropic glutamate receptors (Chen and Roper, 2004).

APPEARANCE OF SYNAPTIC INHIBITION

Up to this point, our discussion has been dominated by developmental studies performed on excitatory synapses; these connections have provided the great majority of information on synaptogenesis, and much of that from the cholinergic NMJ. Initially, it was thought that inhibitory synapses, those releasing GABA or glycine, matured after excitatory synapses. This is because inhibitory postsynaptic potentials (IPSPs) are difficult to observe in neonatal animals. For example, intracellular recordings from the kitten visual cortex demonstrate that afferent-evoked IPSPs are absent from over half the neurons during the first postnatal week, whereas all neurons display IPSPs by adulthood (Komatsu and Iwakiri, 1991). Similar observations have been made on the developing rat neocortex (Luhmann and Prince, 1991).

However, synaptic inhibition appears with a similar time course as synaptic excitation in diverse areas such as the spinal cord, cerebellar nuclei, olfactory bulb, lateral superior olive, and somatosensory cortex. Inhibitory events are probably more difficult to detect in young animals, both because they are concealed by excitatory events (Agmon et al., 1996) and measurements of their equilibrium potential indicate that they are close to the resting membrane potential (Zhang et al., 1991). Therefore, it is likely that inhibitory synapses are present from the outset, but their functional properties are immature.

IS INHIBITION REALLY INHIBITORY DURING DEVELOPMENT?

In adult animals, inhibitory synaptic potentials are generally hyperpolarizing. This is because the receptor is coupled to a Cl^- channel and the Cl^- equilibrium potential is more negative than the cell's resting potential. However, inhibitory synaptic transmission usually produces *depolarizing* potentials during the initial phase of development (Obata et al., 1978; Bixby and Spitzer, 1982; Mueller et al., 1983, 1984; Ben-Ari et al., 1989). For example, during the first postnatal week, rat hippocampal neurons display large spontaneous and evoked depolarizations that are blocked by the $GABA_A$ receptor antagonist, bicuculline (**Figure 8.40**A).

These depolarizing IPSPs are apparently large enough to open voltage-gated calcium channels. In dissociated cultures obtained from embryonic rat hypothalamus, intracellular free calcium is decreased by bicuculline during the first 10 days in vitro (Obrietan and van den Pol, 1995). As the cultures mature, bicuculline increases calcium, presumably by allowing excitatory synaptic acitivty to have a greater depolarizing influence (Figure 8.40B). Therefore, inhibitory synapses may provide a qualitatively different input to postsynaptic neurons during development.

The depolarizing inhibitory potentials seen in young animals are probably due to the outward flow of Cl^- through $GABA_A$ or glycine receptor-coupled channels (Reichling et al., 1994; Owens et al., 1996). Therefore, intracellular chloride must be elevated in young neurons, and it is important to understand how chloride is distributed across the membrane.

Intracellular chloride $[Cl^-]_i$ is regulated primarily by two cation-chloride cotransporter family members: a Na-K-2Cl cotransporter (NKCC1) leads to cytoplasmic accumulation of chloride, and a K-Cl cotransporter (KCC2) extrudes chloride (Payne et al., 2003). During early development, $[Cl^-]_i$ is relatively high due to NKCC1 activity (Clayton et al., 1998; Kanaka et al., 2001). In LSO neurons, NKCC1 transports Cl^- into the cell, particularly in immature neurons, and this contributes to the depolarizing IPSPs (Figure 8.40C; Kakazu et al., 1999). As KCC2 expression increases, $[Cl^-]_i$ drops below the electrochemical equilibrium (Lu et al., 1999; DeFazio et al., 2000; Hübner et al., 2001). This event plays an important role in the transition from inhibitory synapse-evoked depolarizations to hyperpolarizations (Zhang et al., 1991; Kandler and Friauf, 1995; Agmon et al., 1996; Owens et al., 1996; Ehrlich et al., 1999; Kakazu et al., 1999; Rivera et al., 1999; Stil et al., 2009). However, it has also been shown that transporter function is quite dependent on an energy source, and this may change during development. Thus, it can be shown that ketone bodies, an energy source utilized by developing animals, can rescue normal adult-like IPSP reversal potentials in neonatal cortical neurons (Rheims et al., 2009).

KCC2 is expressed at high levels in LSO neurons during the time when they display depolarizing IPSPs (Figure 8.40C and D), suggesting that a post-translational modification must be involved (Balakrishnan et al., 2003). For example, neuron-specific KCC2 has a tyrosine phosphorylation consensus site (Payne, 1997), and its function may be modulated during development. It has been found that cultured

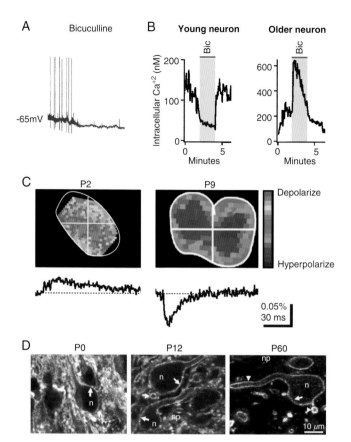

Fig. 8.40 The development of hyperpolarizing inhibition. GABAergic inhibition initially evokes membrane depolization and calcium entry. A. Intracellular recording from a neonatal rat hippocampal neuron shows that the GABA$_A$ receptor antagonist, bicuculline (BIC), blocks the spontaneous action potential and causes the cell to hyperpolarize. B. Intracellular free calcium was monitored in hypothalamic cultures during exposure to a GABA$_A$ receptor antagonist (Bic, red arrow and shaded bar). At 8 days in vitro (young neuron), Bic produced a *decrease* in calcium, suggesting that spontaneously active GABAergic transmission was depolarizing the neuron. At 33 days in vitro (older neuron), BIC increased calcium, suggesting that spontaneous inhibitory activity was now keeping the neuron hyperpolarized. C. Recordings were made from the lateral superior olivary nucleus (see schematic in Figure 8.34A) with a voltage sensitive dye, and a response was evoked by stimulating the presynaptic inhibitory projection. The heat maps show the full LSO nucleus at P2 and P9, and the evoked response from a single neuron is shown beneath each map. The evoked responses were primarily depolarizing at P2 (red color), and switched to hyperpolarizing responses at P9. D. Immunohistochemical staining of rat LSO neurons shows that the KCC2 protein labeling is widespread in the neuropil (np) at P0, but the signal becomes restricted to the plasma membrane surrounding the somata and proximal dendrites at P60 (arrows). *(Adopted from Ben-Ari et al., 1989; Obrietan and van den Pol, 1995; Löhrke et al., 2005; Blaesse et al., 2006)*

hippocampal neurons initially expressed an inactive KCC2 protein, which becomes activated during maturation (Kelsch et al., 2001). Activation of KCC2 in immature neurons can be induced by IGF-1 or a Src kinase, whereas membrane-permeable protein tyrosine kinase inhibitors deactivate KCC2. Therefore, endogenous protein tyrosine kinases may mediate the developmental switch of inhibitory responses by modifying KCC2. In fact, there is an increase in both KCC2 protein expression and tyrosine phosphorylation during normal development of the mouse cortex (Stein et al., 2004).

There is also evidence that KCC2 isoforms are initially expressed at the cell surface as monomers, and only become active as they begin to form oligomers (Blaesse et al., 2006). The presynaptic terminal can also influence chloride transporter expression or function. In hippocampal cultures, GABA$_A$ receptor activation facilitates KCC2 expression and the appearance of GABA$_A$-mediated hyperpolarizations (Ganguly et al., 2001). Finally, the neurotrophin signaling system has been implicated in regulating chloride transport. In transgenic embryonic mice that overexpress BDNF under the control of the *nestin* promoter, KCC2 expression increases dramatically (Aguado et al., 2003). Interestingly, it appears that BDNF may exert the opposite influence in older animals (Rivera et al., 2002).

Inhibitory synapses initially communicate with the postsynaptic neuron with multiple transmitters. The projection from MNTB to LSO (schematic in Figure 8.30A) releases both glycine and GABA during early development, but gradually comes to release mostly glycine. This is due to a reduction of presynaptic GABA as well as a postsynaptic reduction of GABA$_A$ receptors (Kotak et al., 1998; Nabekura et al., 2004). More surprisingly, the MNTB terminals release glutamate, and can elicit a postsynaptic response from NMDARs (Gillespie et al., 2005). Thus, inhibitory synapses have at their disposal a set of interesting signals that can be used during the stage of synapse formation when connections are strengthened or eliminated (see Chapter 9).

SUMMARY

The generic cortical neuron with which we began the chapter somehow manages to express just the right complement of receptors and channels, and place each of them in the correct compartments of the cell. This differentiation depends upon ongoing, bidirectional signaling between pre- and postsynaptic partners. Fortunately, we now have a basic understanding of synapse formation, including a few of these transynaptic signaling pathways. As extraordinary as these accomplishments are, it is important to recognize that we have ignored most of the neuromodulatory afferents, many of the neurotransmitter receptors, and multiple cytoplasmic signaling cascades. Furthermore, although we have begun to understand how the most basic attributes of transmission and electrogenesis emerge, there is scant information about how these functional building blocks shape the computational properties of a developing neuron and how these help to explain behavioral maturation. After all, that is the ultimate goal of neural development. Why is 10,000 synapses the correct number for a cortical pyramidal neuron, while an MNTB neuron receives only a single synapse? Would each of these neurons perform adequately with a different number of inputs? Similarly, how does the number of glutamatergic synapses influence the number of GABAergic or serotonergic synapses? How would a neuron operate if inhibitory synapses formed on dendritic spines, instead of excitatory synapses? As we start to understand how individual synapses are constructed, it becomes critical to explore the activity-dependent mechanisms that regulate their placement and strength (Chapter 9), and the relationship of synapse function to behavioral traits (Chapter 10).

REFERENCES

Adams, R. H., Sato, K., Shimada, S., Tohyama, M., Puschel, A. W., & Betz, H. (1995). Gene structure and glial expression of the glycine transporter GlyT1 embryonic and adult rodents. *The Journal of Neuroscience, 15*, 2524–2532.

Agmon, A., Hollrigel, G., & O'Dowd, D. K. (1996). Functional GABAergic synaptic connection in neonatal mouse barrel cortex. *The Journal of Neuroscience, 16*, 4684–4695.

Aguado, F., Carmona, M. A., Pozas, E., Aguilo, A., Martinez-Guijarro, F. J., Alcantara, S., et al. (2003). BDNF regulates spontaneous correlated activity at early developmental stages by increasing synaptogenesis and expression of the K+/Cl-co-transporter KCC2. *Development, 130*, 1267–1280.

Ahmari, S. E., Buchanan, J., & Smith, S. J. (2000). Assembly of presynaptic active zones from cytoplasmic transport packets. *Nature Neuroscience, 3*, 445–451.

Akaaboune, M., Culican, S. M., Turney, S. G., & Lichtman, J. W. (1999). Rapid and reversible effects of activity on acetylcholine receptor density at the neuromuscular junction in vivo. *Science, 286*, 503–507.

Alsina, B., Vu, T., & Cohen-Cory, S. (2001). Visualizing synapse formation in arborizing optic axons in vivo: dynamics and modulation by BDNF. *Nature Neuroscience, 4*, 1093–1101.

Anderson, M. J., & Cohen, M. W. (1977). Nerve-induced and spontaneous redistribution of acetylcholine receptors on cultured muscle cells. *The Journal of Physiology, 268*, 757–773.

Aoki, C., Fujisawa, S., Mahadomrongkul, V., Shah, P. J., Nader, K., & Erisir, A. (2003). NMDA receptor blockade in intact adult cortex increases trafficking of NR2A subunits into spines, postsynaptic densities, and axon terminals. *Brain Research, 963*, 139–149.

Aravamudan, B., Fergestad, T., Davis, W. S., Rodesch, C. K., & Broadie, K. (1999). Drosophila UNC-13 is essential for synaptic transmission. *Nature Neuroscience, 2*, 965–971.

Axelsson, J., & Thesleff, F. (1959). A study of supersensitivity in denervated mammalian skeletal muscle. *The Journal of Physiology, 147*, 178–193.

Baccaglini, P. I., & Spitzer, N. C. (1977). Developmental changes in the inward current of the action potential of Rohon-Beard neurones. *The Journal of Physiology, 271*, 93–117.

Bahr, S., & Wolff, J. R. (1985). Postnatal development of axosomatic synapses in the rat visual cortex: morphogenesis and quantitative evaluation. *The Journal of Comparative Neurology, 233*, 405–420.

Baird, D. H., Hatten, M. E., & Mason, C. A. (1992). Cerebellar target neurons provide a stop signal for afferent neurite extension in vitro. *The Journal of Neuroscience, 12*, 619–634.

Balakrishnan, V., Becker, M., Lohrke, S., Nothwang, H. G., Guresir, E., & Friauf, E. (2003). Expression and function of chloride transporters during development of inhibitory neurotransmission in the auditory brainstem. *The Journal of Neuroscience, 23*, 4134–4145.

Barish, M. E. (1986). Differentiation of voltage-gated potassium current and modulation of excitability in cultured amphibian spinal neurones. *The Journal of Physiology, 375*, 229–250.

Becker, C.-M., Hoch, W., & Betz, H. (1988). Glycine receptor heterogeneity in rat spinal cord during postnatal development. *The EMBO Journal, 7*, 3717–3726.

Beeson, D., Higuchi, O., Palace, J., Cossins, J., Spearman, H., Maxwell, S., et al. (2006). Dok-7 mutations underlie a neuromuscular junction synaptopathy. *Science, 313*, 1975–1978.

Bekirov, I. H., Nagy, V., Svoronos, A., Huntley, G. W., & Benson, D. L. (2008). Cadherin-8 and N-cadherin differentially regulate pre- and postsynaptic development of the hippocampal mossy fiber pathway. *Hippocampus, 18*, 349–363.

Ben-Ari, Y., Cherubini, E., Corradetti, R., & Gaiarsa, J. L. (1989). Giant synaptic potentials in immature rat CA3 hippocampal neurones. *The Journal of Physiology, 416*, 303–325.

Ben-Ari, Y., Cherubini, E., & Krnjevic, K. (1988). Changes in voltage dependence of NMDA currents during development. *Neuroscience Letters, 94*, 88–92.

Berg, D. K., & Hall, Z. W. (1975). Increased extrajunctional acetylcholine sensitivity produced by chronic postsynatpic neuromuscular blockade. *The Journal of Physiology, 244*, 659–676.

Berglund, K., Schleich, W., Krieger, P., Loo, L. S., Wang, D., Cant, N. B., et al. (2006). Imaging synaptic inhibition in transgenic mice expressing the chloride indicator, Clomeleon. *Brain Cell Biology, 35*, 207–228.

Betley, J. N., Wright, C. V., Kawaguchi, Y., Erdélyi, F., Szabó, G., Jessell, T. M., et al. (2009). Stringent specificity in the construction of a GABAergic presynaptic inhibitory circuit. *Cell, 139*, 161–174.

Bevan, S., & Steinbach, J. H. (1977). The distribution of alpha-bungarotoxin binding sites on mammalian skeletal muscle developing in vivo. *The Journal of Physiology, 267*, 195–213.

Biederer, T., Sara, Y., Mozhayeva, M., Atasoy, D., Liu, X., Kavalali, E. T., & Sudhof, T. C. (2002). SynCAM, a synaptic adhesion molecule that drives synapse assembly. *Science, 297*, 1525–1531.

Bixby, J. L., & Spitzer, N. C. (1982). The appearance and development of chemosensitivity in Rohon-Beard neurones of the Xenopus spinal cord. *The Journal of Physiology, 330*, 513–536.

Blaesse, P., Guillemin, I., Schindler, J., Schweizer, M., Delpire, E., Khiroug, L., et al. (2006). Oligomerization of KCC2 correlates with development of inhibitory neurotransmission. *The Journal of Neuroscience, 26*, 10407–10419.

Blair, L. A. C. (1983). The timing of protein synthesis required for the development of the sodium action potential in embryonic spinal neurons. *The Journal of Neuroscience, 3*, 1430–1436.

Blakely, R. D., Clark, J. A., Pacholczyk, T., & Amara, S. G. (1991). Distinct, developmentally regulated brain mRNAs direct the synthesis of neurotransmitter transporters. *Journal of Neurochemistry, 56*, 860–871.

Borgdorff, A. J., & Choquet, D. (2002). Regulation of AMPA receptor lateral movements. *Nature, 417*, 649–653.

Bowe, M. A., & Nadler, J. V. (1990). Developmental increase in the sensitivity to magnesium of NMDA receptors on CA1 hippocampal pyramidal cells. *Developmental Brain Research, 56*, 55–61.

Bramham, C. R., Worley, P. F., Moore, M. J., & Guzowski, J. F. (2008). The immediate early gene arc/arg3.1: regulation, mechanisms, and function. *The Journal of Neuroscience, 28*, 11760–11767.

Bresler, T., Shapira, M., Boeckers, T., Dresbach, T., Futter, M., Garner, C. C., et al. (2004). Postsynaptic density assembly is fundamentally different from presynaptic active zone assembly. *The Journal of Neuroscience, 24*, 1507–1520.

Broadie, K. S., & Bate, M. (1993a). Development of the embryonic neuromuscular synapse of Drosophila melanogaster. *The Journal of Neuroscience, 13*, 144–166.

Broadie, K. S., & Bate, M. (1993b). Innervation directs receptor synthesis and localization in Drosophila embryo synaptogenesis. *Nature, 361*, 350–353.

Broadie, K., & Bate, M. (1993c). Activity-dependent development of the neuromuscular synapse during Drosophila embryogenesis. *Neuron, 11*, 607–619.

Brunso-Bechtold, J. K., Henkel, C. K., & Linville, C. (1992). Ultrastructural development of the medial superior olive (MSO) in the ferret. *The Journal of Comparative Neurology, 324*, 539–556.

Buchanan, J., Sun, Y.-a., & Poo, M.-m. (1989). Studies of nerve-muslce interactions in Xenopus cell culture: fine structure of early functional contacts. *The Journal of Neuroscience, 9*, 1540–1554.

Budnik, V., Koh, Y.-H., Guan, B., Hartmann, B., Hough, C., Woods, D., et al. (1996). Regulation of synapse structure and function by the Drosophila tumor suppressor gene dlg. *Neuron, 17*, 627–640.

Burden, S. (1977a). Development of the neuromuscular junction in the chick embryo: The number, distribution, and stability of acetylcholine receptors. *Developmental Biology, 57*, 317–329.

Burden, S. (1977b). Acetylcholine receptors at the neuromuscular junction: Developmental change in receptor turnover. *Developmental Biology, 61*, 79–85.

Burden, S. J., Sargent, P. B., & McMahon, U. J. (1979). Acetylcholine receptors in regenerating muscle accumulate at original synaptic sites in the absence of the nerve. *The Journal of Cell Biology, 82*, 412–425.

Burgard, E. C., & Hablitz, J. J. (1993). Developmental changes in NMDA and Non-NMDA receptor-mediated synaptic potentials in rat neocortex. *Journal of Neurophysiology, 69*, 230–240.

Cameron, J. S., Dryer, L., & Dryer, S. E. (1999). Regulation of neuronal K(+) currents by target-derived factors: opposing actions of two different isoforms of TGFbeta. *Development, 126*, 4157–4164.

Cameron, J. S., Dryer, L., & Dryer, S. E. (2001). beta-Neuregulin-1 is required for the in vivo development of functional Ca2+-activated K+ channels in parasympathetic neurons. *Proceedings of the National Academy of Sciences of the United States of America, 98*, 2832–2836.

Cameron, J. S., Lhuillier, L., Subramony, P., & Dryer, S. E. (1998). Developmental regulation of neuronal K+ channels by target-derived TGF beta in vivo and in vitro. *Neuron, 21*, 1045–1053.

Casavant, R. H., Colbert, C. M., & Dryer, S. E. (2004). A-current expression is regulated by activity but not by target tissues in developing lumbar motoneurons of the chick embryo. *Journal of Neurophysiology, 2004 May 2* [Epub ahead of print].

Chang, L. W., & Spitzer, N. C. (2009). Spontaneous calcium spike activity in embryonic spinal neurons is regulated by developmental expression of the Na+, K+-ATPase beta3 subunit. *The Journal of Neuroscience, 29*, 7877–7885.

Chen, H. X., & Roper, S. N. (2004). Tonic activity of metabotropic glutamate receptors is involved in developmental modification of short-term plasticity in the neocortex. *Journal of Neurophysiology, 92*, 838–844.

Chen, L., Chetkovich, D. M., Petralia, R. S., Sweeney, N. T., Kawasaki, Y., Wenthold, R. J., et al. (2000). Stargazin regulates synaptic targeting of AMPA receptors by two distinct mechanisms. *Nature, 408*, 936–943.

Chen, Q. X., Stelzner, A., Kay, A. R., & Wong, R. K. S. (1990). GABAA receptor function is regulated by phosphorylation in acutely dissociated guinea-pig hippocampal neurones. *The Journal of Physiology, 420*, 207–221.

Cheng, D., Hoogenraad, C. C., Rush, J., Ramm, E., Schlager, M. A., Duong, D. M., et al. (2006). Relative and absolute quantification of postsynaptic density proteome isolated from rat forebrain and cerebellum. *Molecular & Cellular Proteomics: MCP*, 5, 1158–1170.

Christopherson, K. S., Ullian, E. M., Stokes, C. C., Mullowney, C. E., Hell, J. W., Agah, A., et al. (2005). Thrombospondins are astrocyte-secreted proteins that promote CNS synaptogenesis. *Cell*, 120, 421–433.

Clayton, G. H., Owens, G. C., Wolff, J. S., & Smith, R. L. (1998). Ontogeny of cation-Cl- cotransporter expression in rat neocortex. *Development Brain Research*, 109, 281–292.

Cohen, M. W. (1972). The development of neuromuscular connexions in the presence of D-tubocurarine. *Brain Research*, 41, 457–463.

Cohen, M. W., Anderson, M. J., Zorychta, E., & Weldon, P. R. (1979). Accumulation of acetylcholine receptors at nerve-muscle contacts in culture. *Progress in Brain Research*, 49, 335–349.

Cohen, M. W., & Weldon, P. R. (1980). Localization of acetylcholine receptors and synaptic ultrastructure at nerve-muscle contacts in culture: dependence on nerve type. *The Journal of Cell Biology*, 86, 388–401.

Cohen, N. A., Brenman, J. E., Snyder, S. H., & Bredt, D. S. (1996). Binding of the inward rectifier K+ channel Kir 2.3 to PSD-95 is regulated by protein kinase A phosphorylation. *Neuron*, 17, 759–767.

Collins, C. A., & DiAntonio, A. (2007). Synaptic development: insights from Drosophila. *Current Opinion in Neurobiology*, 17, 35–42.

Colón-Ramos, D. A., Margeta, M. A., & Shen, K. (2007). Glia promote local synaptogenesis through UNC-6 (netrin) signaling in C. elegans. *Science*, 318, 103–106.

Connors, B. W., Ransom, B. R., Kunis, D. M., & Gutnick, M. J. (1982). Activity dependent K+ accumulation in the developing rat optic nerve. *Science*, 216, 1341–1343.

Corfas, G., & Fischbach, G.D. (1993). The number of Na+ channels in cultured chick muscle is increased by ARIA, an acetylcholine receptor-inducing activity. *J Neurosci*, 13, 2118–2125.

Corriveau, R. A., & Berg, D. K. (1993). Coexpression of multiple acetylcholine receptor genes in neurons: quantification of transcripts during development. *The Journal of Neuroscience*, 13, 2662–2671.

Cottrell, J. R., Dube, G. R., Egles, C., & Liu, G. (2000). Distribution, density, and clustering of functional glutamate receptors before and after synaptogenesis in hippocampal neurons. *Journal of Neurophysiology*, 84, 1573–1587.

Cragg, B. G. (1975). The development of synapses in the visual system of the cat. *The Journal of Comparative Neurology*, 160, 147–166.

Dai, Z., & Peng, H. B. (1993). Elevation in presynaptic Ca2+ accompanying initial nerve-muscle contact in tissue culture. *Neuron*, 10, 827–837.

Dai, Y., Taru, H., Deken, S. L., Grill, B., Ackley, B., Nonet, M. L., et al. (2006). SYD-2 Liprin-alpha organizes presynaptic active zone formation through ELKS. *Nat Neurosci*, 9, 1479–1487.

Dalva, M. B., Takasu, M. A., Lin, M. Z., Shamah, S. M., Hu, L., Gale, N. W., et al. (2000). EphB receptors interact with NMDA receptors and regulate excitatory synapse formation. *Cell*, 103, 945–956.

Davis, E. K., Zou, Y., & Ghosh, A. (2008). Wnts acting through canonical and noncanonical signaling pathways exert opposite effects on hippocampal synapse formation. *Neural Development*, 3, 32.

de Wit, J., Sylwestrak, E., O'Sullivan, M. L., Otto, S., Tiglio, K., Savas, J. N., et al. (2009). LRRTM2 interacts with Neurexin1 and regulates excitatory synapse formation. *Neuron*, 64, 799–806.

Dean, C., Scholl, F. G., Choih, J., DeMaria, S., Berger, J., Isacoff, E., et al. (2003). Neurexin mediates the assembly of presynaptic terminals. *Nature Neuroscience*, 6, 708–716.

DeChiara, T. M., Bowen, D. C., Valenzuela, D. M., Simmons, M. V., Poueymirou, W. T., Thomas, S., et al. (1996). The receptor tyrosine kinase MuSK is required for neuromuscular junction formation in vivo. *Cell*, 85, 501–512.

DeFazio, R. A., Keros, S., Quick, M. W., & Hablitz, J. J. (2000). Potassium-coupled chloride cotransport controls intracellular chloride in rat neocortical pyramidal neurons. *The Journal of Neuroscience*, 20, 8069–8076.

Dent, M. A., Raisman, G., & Lai, F. A. (1996). Expression of type 1 inositol 1,4,5-trisphosphate receptor during axogenesis and synaptic contact in the central and peripheral nervous system of developing rat. *Development*, 122, 1029–1039.

Desarmenien, M. G., & Spitzer, N. C. (1991). Role of calcium and protein kinase C in development of the delayed rectifier potassium current in Xenopus spinal neurons. *Neuron*, 7, 797–805.

Diamond, J., & Miledi, R. (1962). A study of fetal and new-born rat muscle fibres. *The Journal of Physiology*, 162, 393–408.

Dourado, M. M., Brumwell, C., Wisgirda, M. E., Jacob, M. H., & Dryer, S. E. (1994). Target tissues and innervation regulate the characteristics of K+ currents in chick ciliary ganglion neurons developing in situ. *The Journal of Neuroscience*, 14, 3156–3165.

Dourado, M. M., & Dryer, S. E. (1992). Changes in the electrical properties of chick ciliary ganglion neurones during embryonic development. *The Journal of Physiology*, 449, 411–428.

Dubinsky, J. M., Loftus, D. J., Fischbach, G. D., & Elson, E. L. (1989). Formation of acetylcholine receptor clusters in chick myotubes: migration or new insertion? *The Journal of Cell Biology*, 109, 1733–1743.

Durand, G. M., Kovalchuk, Y., & Konnerth, A. (1996). Long-term potentiation and functional synapse induction in developing hippocampus. *Nature*, 381, 71–75.

Dyson, S. E., & Jones, D. G. (1980). Quantitation of terminal parameters and their interrelationships in maturing central synapses: A perspective for experimental studies. *Brain Research*, 183, 43–59.

Dzhashiashvili, Y., Zhang, Y., Galinska, J., Lam, I., Grumet, M., & Salzer, J. L. (2007). Nodes of Ranvier and axon initial segments are ankyrin G-dependent domains that assemble by distinct mechanisms. *The Journal of Cell Biology*, 177, 857–870.

Ehlers, M. D., Heine, M., Groc, L., Lee, M. C., & Choquet, D. (2007). Diffusional trapping of GluR1 AMPA receptors by input-specific synaptic activity. *Neuron*, 54, 447–460.

Ehrlich, I., Lohrke, S., & Friauf, E. (1999). Shift from depolarizing to hyperpolarizing glycine action in rat auditory neurones is due to age-dependent Cl-regulation. *The Journal of Physiology*, 520, 121–137.

Elias, G. M., Elias, L. A., Apostolides, P. F., Kriegstein, A. R., & Nicoll, R. A. (2008). Differential trafficking of AMPA and NMDA receptors by SAP102 and PSD-95 underlies synapse development. *Proceedings of the National Academy of Sciences of the United States of America*, 105, 20953–20958.

Elias, G. M., Funke, L., Stein, V., Grant, S. G., Bredt, D. S., & Nicoll, R. A. (2006). Synapse-specific and developmentally regulated targeting of AMPA receptors by a family of MAGUK scaffolding proteins. *Neuron*, 52, 307–320.

Elmariah, S. B., Crumling, M. A., Parsons, T. D., & Balice-Gordon, R. J. (2004). Postsynaptic TrkB-mediated signaling modulates excitatory and inhibitory neurotransmitter receptor clustering at hippocampal synapses. *The Journal of Neuroscience*, 24, 2380–2393.

Elmariah, S. B., Oh, E. J., Hughes, E. G., & Balice-Gordon, R. J. (2005). Astrocytes regulate inhibitory synapse formation via Trk-mediated modulation of postsynaptic GABAA receptors. *The Journal of Neuroscience*, 25, 3638–3650.

Emes, R. D., Pocklington, A. J., Anderson, C. N., Bayes, A., Collins, M. O., Vickers, C. A., et al. (2008). Evolutionary expansion and anatomical specialization of synapse proteome complexity. *Nature Neuroscience*, 11, 799–806.

Escher, P., Lacazette, E., Courtet, M., Blindenbacher, A., Landmann, L., Bezakova, G., et al. (2005). Synapses form in skeletal muscles lacking neuregulin receptors. *Science*, 308, 1920–1923.

Evers, J., Laser, M., Sun, Y.-a., Xie, Z.-p., & Poo, M.- m. (1989). Studies of nerve-muscle interactions in Xenopus cell culture: analysis of early synaptic currents. *The Journal of Neuroscience*, 9, 1523–1539.

Fallon, J. R., & Gelfman, C. E. (1989). Agrin-related molecules are concentrated at acetylcholin receptor clusters in normal and aneural developing muscle. *The Journal of Cell Biology*, 108, 1527–1535.

Falls, D. L. (2003). Neuregulins and the neuromuscular system: 10 years of answers and questions. *J Neurocytol*, 32, 619–647.

Fannon, A. M., & Colman, D. R. (1996). A model for central synaptic junctional complex formation based on the differential adhesive specificities of the cadherins. *Neuron*, 17, 423–434.

Fatt, P., & Katz, B. (1951). Spontaneous subthreshold activity at motor nerve endings. *The Journal of Physiology*, 117, 109–128.

Feng, G., Tintrup, H., Kirsch, J., Nichol, M. C., Kuhse, J., Betz, H., et al. (1998). Dual requirement for gephyrin in glycine receptor clustering and molybdoenzyme activity. *Science*, 282, 1321–1324.

Feng, W., & Zhang, M. (2009). Organization and dynamics of PDZ-domain-related supramodules in the postsynaptic density. *Nature Reviews. Neuroscience*, 10, 87–99.

Ferns, M., Deiner, M., & Hall, Z. (1996). Agrin-induced acetylcholine receptor clustering in mammalian muscle requires tyrosine phosphorylation. *The Journal of Cell Biology*, 132, 937–944.

Ferns, M. J., Campanelli, J. T., Hoch, W., Scheller, R. H., & Hall, Z. (1993). The ability of agrin to cluster AChRs depends on alternative splicing and on cell surface proteoglycans. *Neuron*, 11, 491–502.

Fertuck, H. C., & Salpeter, M. (1976). Quantitation of junctional and extrajunctional acetylcholine receptors by electron microscopic autoradiography after 125I–alpha-bungarotoxin binding at mouse neuromuscular junctions. *The Journal of Cell Biology*, 69, 144–158.

Fiala, J. C., Feinberg, M., Popov, V., & Harris, K. M. (1998). Synaptogenesis via dendritic filopodia in developing hippocampal area CA1. *The Journal of Neuroscience*, 18, 8900–8911.

Fischbach, G. D., & Cohen, S. A. (1973). The distribution of acetylcholine sensitivity over uninnervated muscle fibres grown in cell culture. *Developmental Biology*, 31, 147–162.

Flint, A. C., Maisch, U. S., Weishaupt, J. H., Kriegstein, A. R., & Monyer, H. (1997). NR2A subunit expression shortens NMDA receptor synaptic currents in developing neocortex. *The Journal of Neuroscience*, 17, 2469–2476.

Fogel, A. I., Akins, M. R., Krupp, A. J., Stagi, M., Stein, V., & Biederer, T. (2007). SynCAMs organize synapses through heterophilic adhesion. *The Journal of Neuroscience*, 27, 12516–12530.

Forscher, P., Kaczmarek, L. K., Buchanan, J. A., & Smith, S. J. (1987). Cyclic AMP induces changes in distribution and transport of organelles within growth cones of Aplysia bag cell neurons. *The Journal of Neuroscience, 7,* 3600–3611.

Fox, K., Daw, N., Sato, H., & Czepita, D. (1992). The effect of visual experience on development of NMDA receptor synaptic transmission in kitten visual cortex. *The Journal of Neuroscience, 12,* 2672–2684.

Friedman, H. V., Bresler, T., Garner, C. C., & Ziv, N. E. (2000). Assembly of new individual excitatory synapses: time course and temporal order of synaptic molecule recruitment. *Neuron, 27,* 57–69.

Fritschy, J. M., Harvey, R. J., & Schwarz, G. (2008). Gephyrin: where do we stand, where do we go? *Trends in Neurosciences, 31,* 257–264.

Frohner, S. C., Luetje, C. W., Scotland, P. B., & Patrick, J. (1990). The postsynaptic 43K protein clusters muscle nicotinic acetylcholine receptors in Xenopus oocytes. *Neuron, 5,* 403–410.

Fry, M. (2006). Developmental expression of Na+ currents in mouse Purkinje neurons. *The European Journal of Neuroscience, 24,* 2557–2566.

Funte, L. R., & Haydon, P. G. (1993). Synaptic target contact enhances presynaptic calcium influx by activating cAMP-dependent protein kinase during synaptogenesis. *Neuron, 10,* 1069–1078.

Gally, C., & Bessereau, J. L. (2003). GABA is dispensable for the formation of junctional GABA receptor clusters in Caenorhabditis elegans. *The Journal of Neuroscience, 23,* 2591–2599.

Ganguly, K., Schinder, A. F., Wong, S. T., & Poo, M.- m. (2001). GABA itself promotes the developmental switch of neuronal GABAergic responses from excitation to inhibition. *Cell, 105,* 521–532.

Gao, X. B., & van den Pol, A. N. (2000). GABA release from mouse axonal growth cones. *The Journal of Physiology, 523,* 629–637.

Gardette, R., Listerud, M. D., Brussaard, A. B., & Role, L. W. (1991). Developmental changes in transmitter sensitivity and synaptic transmission in embryonic chicken sympathetic neurons innervated in vitro. *Developmental Biology, 147,* 83–95.

Garrett, A. M., & Weiner, J. A. (2009). Control of CNS synapse development by {gamma}-protocadherin-mediated astrocyte-neuron contact. *The Journal of Neuroscience, 29,* 11723–11731.

Gautam, M., Noakes, P. G., Moscoso, L., Rupp, F., Scheller, R. H., Merlie, J. P., et al. (1996). Defective neuromuscular synaptogenesis in agrin-deficient mutant mice. *Cell, 85,* 525–535.

Gautam, M., Noakes, P. G., Mudd, J., Nichol, M., CHu, G. C., Sanes, J. R., et al. (1995). Failure of postsynaptic specialization to develop at neruomuscular junctions of rapsyn-deficient mice. *Nature, 377,* 232–236.

Gerrow, K., Romorini, S., Nabi, S. M., Colicos, M. A., Sala, C., & El-Husseini, A. (2006). A preformed complex of postsynaptic proteins is involved in excitatory synapse development. *Neuron, 49,* 547–562.

Gervásio, O. L., Armson, P. F., & Phillips, W. D. (2007). Developmental increase in the amount of rapsyn per acetylcholine receptor promotes postsynaptic receptor packing and stability. *Developmental Biology, 305,* 262–275.

Gillespie, D. C., Kim, G., & Kandler, K. (2005). Inhibitory synapses in the developing auditory system are glutamatergic. *Nature Neuroscience, 8,* 332–338.

Glass, D. J., Bowen, D. C., Stitt, T. N., Radziejewski, C., Bruno, J., Ryan, T. E., et al. (1996). Agrin acts via a MuSK receptor complex. *Cell, 85,* 513–523.

Godfrey, E. W., Nitkin, R. M., Wallace, B. G., Rubin, L. L., & McMahon, U. J. (1984). Components of Torpedo electric organ and muscle that cause aggregation of acetylcholine receptors in cultured muscle cells. *The Journal of Cell Biology, 99,* 615–627.

Gottmann, K., Dietzel, I. D., Lux, H. D., Huck, S., & Rohrer, H. (1988). Development of inward currents in chick sensory and autonomic neuronal precursor cells in culture. *The Journal of Neuroscience, 8,* 3722–3732.

Graf, E. R., Zhang, X., Jin, S. X., Linhoff, M. W., & Craig, A. M. (2004). Neurexins induce differentiation of GABA and glutamate postsynaptic specializations via neuroligins. *Cell, 119,* 1013–1026.

Greer, P. L., Hanayama, R., Bloodgood, B. L., Mardinly, A. R., Lipton, D. M., Flavell, S. W., et al. (2010). The Angelman Syndrome protein Ube3A regulates synapse development by ubiquitinating arc. *Cell, 140,* 704–716.

Grooms, S. Y., Noh, K. M., Regis, R., Bassell, G. J., Bryan, M. K., Carroll, R. C., et al. (2006). Activity bidirectionally regulates AMPA receptor mRNA abundance in dendrites of hippocampal neurons. *The Journal of Neuroscience, 26,* 8339–8351.

Gu, Y., & Hall, Z. W. (1988). Immunological evidence for a change in subunits of the acetylcholine receptor in developing and denervated rat muscle. *Neuron, 1,* 117–125.

Hall, A. C., Lucas, F. R., & Salinas, P. C. (2000). Axonal remodeling and synaptic differentiation in the cerebellum is regulated by WNT-7a signaling. *Ceil, 100,* 525–535.

Hamill, O. P., Marty, A., Neher, E., Sakmann, B., & Sigworth, F. J. (1981). Improved patch-clamp techniques for high-resolution current recording from cell and cell-free membrane patches. *Pflügers Archiv: European Journal of Physiology, 391,* 85–100.

Harris, D. A., Falls, D. L., Dill-Devor, R. M., & Fischbach, G. D. (1988). Acetylcholine receptor-inducing factor from chicken brain increases the level of mRNA encoding the receptor alpha subunit. *Proceedings of the National Academy of Sciences of the United States of America, 85,* 1983–1987.

Hayes, B. P., & Roberts, A. (1973). Synaptic junction development in the spinal cord of an amphibian embryo: An electron microscope study. *Z Zellforsch, 137,* 251–269.

Heine, M., Thoumine, O., Mondin, M., Tessier, B., Giannone, G., & Choquet, D. (2008). Activity-independent and subunit-specific recruitment of functional AMPA receptors at neurexin/neuroligin contacts. *Proceedings of the National Academy of Sciences of the United States of America, 105,* 20947–20952.

Held, H. (1897). Beiträge zur Structur der Nervenzellen und ihrer Fortädtze. *Arch Anat Physiol Anat Abt, 21,* 204–294.

Hendrikson, C. K., & Vaughn, J. E. (1974). Fine structural relationships between neurites and radial glial processes in developing mouse spinal cord. *J Neuroscytol, 3,* 659–675.

Henkemeyer, M., Itkis, O. S., Ngo, M., Hickmott, P. W., & Ethell, I. M. (2003). Multiple EphB receptor tyrosine kinases shape dendritic spines in the hippocampus. *The Journal of Cell Biology, 163,* 1313–1326.

Hestrin, S. (1992). Developmental regulation of NMDA receptor-mediated synaptic currents at a central synapse. *Nature, 357,* 686–689.

Hinds, J. W., & Hinds, P. A. (1976). Synapse formation in the mouse olfactory bulb. II. Morphogenesis. *The Journal of Comparative Neurology, 169,* 41–62.

Hoch, W., McConville, J., Helms, S., Newsom-Davis, J., Melms, A., & Vincent, A. (2001). Auto-antibodies to the receptor tyrosine kinase MuSK in patients with myasthenia gravis without acetylcholine receptor antibodies. *Nature Medicine, 7,* 365–368.

Howard, M. A., Elias, G. M., Elias, L. A., Swat, W., & Nicoll, R. A. (2010). The role of SAP97 in synaptic glutamate receptor dynamics. *Proceedings of the National Academy of Sciences of the United States of America, 107,* 3805–3810.

Huang, C. F., Tong, J., & Schmidt, J. (1992). Protein kinase C couples membrane excitation to acetylcholine receptor gene inactivation in chick skeletal muscle. *Neuron, 9,* 671–678.

Hübner, C. A., Stein, V., Hermans-Borgmeyer, I., Meyer, T., Ballanyi, K., & Jentsch, T. J. (2001). Disruption of KCC2 reveals an essential role of K-Cl cotransport already in early synaptic inhibition. *Neuron, 30,* 515–524.

Huguenard, J. R., Hamill, O. P., & Prince, D. A. (1988). Developmental changes in Na+ conductances in rat neocortical neurons: appearance of a slowly inactivating component. *Journal of Neurophysiology, 9,* 778–795.

Hume, R. I., Role, L. W., & Fischbach, G. D. (1983). Acetylcholine release from growth cones detected with patches of acetylcholine receptor rich membranes. *Nature, 305,* 632–634.

Huttenlocher, P. R. (1979). Synaptic density inhuman frontal cortex. Developmental changes and effects of aging. *Brain Research, 163,* 195–205.

Huttenlocher, P. R., & Dabholkar, A. S. (1997). Regional differences in synaptogenesis in human cerebral cortex. *The Journal of Comparative Neurology, 387,* 167–178.

Huttenlocher, P. R., & de Courten, C. (1987). The development of synapses in striate cortex of man. *Human Neurobiology, 6,* 1–9.

Inoue, A., & Sanes, J. R. (1997). Lamina-specific connectivity in the brain: regulation by N-cadherin, neurotrophins, and glycoconjugates. *Science, 276,* 1428–1431.

Inoue, A., Setoguchi, K., Matsubara, Y., Okada, K., Sato, N., Iwakura, Y., et al. (2009). Dok-7 activates the muscle receptor kinase MuSK and shapes synapse formation. *Sci Signal, 2,* ra7.

Jaworski, A., & Burden, S. J. (2006). Neuromuscular synapse formation in mice lacking motor neuron- and skeletal muscle-derived Neuregulin-1. *The Journal of Neuroscience, 26,* 655–661.

Jendelová, P., & Syková, E. (1991). The role of glia in K+ and pH homeostasis in the neonatal rat spinal cord. *Glia, 4,* 56–63.

Joshi, I., & Wang, L. Y. (2002). Developmental profiles of glutamate receptors and synaptic transmission at a single synapse in the mouse auditory brainstem. *The Journal of Physiology, 540,* 861–873.

Ju, W., Morishita, W., Tsui, J., Gaietta, G., Deerinck, T. J., Adams, S. R., et al. (2004). Activity-dependent regulation of dendritic synthesis and trafficking of AMPA receptors. *Nature Neuroscience, 7,* 244–253.

Kakazu, Y., Akaike, N., Komiyama, S., & Nabekura, J. (1999). Regulation of intracellular chloride by cotransporters in developing lateral superior olive neurons. *The Journal of Neuroscience, 19,* 2843–2851.

Kanaka, C., Ohno, K., Okabe, A., Kuriyama, K., Itoh, T., Fukuda, A., et al. (2001). The differential expression patterns of messenger RNAs encoding K-Cl cotransporters (KCC1,2) and Na-K-2Cl cotransporter (NKCC1) in the rat nervous system. *Neuroscience, 104,* 933–946.

Kandler, K., & Friauf, E. (1995). Development of glycinergic and glutamatergic synaptic transmission in the auditory brainstem of perinatal rats. *The Journal of Neuroscience, 15,* 6890–6904.

Kelsch, W., Hormuzdi, S., Straube, E., Lewen, A., Monyer, H., & Misgeld, U. (2001). Insulin-like growth factor 1 and a cytosolic tyrosine kinase activate chloride outward transport during maturation of hippocampal neurons. *The Journal of Neuroscience, 21*, 8339–8347.

Kidokoro, Y., Anderson, M. J., & Gruener, R. (1980). Changes in synaptic potential properties during acetylcholine receptor accumulation and neurospecific interactions in Xenopus nerve-muscle cell cultures. *Developmental Biology, 78*, 464–483.

Kidokoro, Y., & Yeh, E. (1982). Initial synaptic transmission at the growth cone in Xenopus nerve-muscle cultures. *Proceedings of the National Academy of Sciences of the United States of America, 79*, 6727–6731.

Kim, N., Stiegler, A. L., Cameron, T. O., Hallock, P. T., Gomez, A. M., Huang, J. H., et al. (2008). Lrp4 is a receptor for Agrin and forms a complex with MuSK. *Cell, 135*, 334–342.

Kim, S., Burette, A., Chung, H. S., Kwon, S. K., Woo, J., Lee, H. W., et al. (2006). NGL family PSD-95-interacting adhesion molecules regulate excitatory synapse formation. *Nature Neuroscience, 9*, 1294–1301.

Kirsch, J., & Betz, H. (1998). Glycine-receptor activation is required for receptor clustering in spinal neurons. *Nature, 392*, 717–720.

Kirsch, J., Malosio, M.-L., Wolters, I., & Betz, H. (1993a). Distribution of gephyrin transcripts in the adult and developing rat brain. *The European Journal of Neuroscience, 5*, 1109–1117.

Kirsch, J., Wolters, I., Triller, A., & Betz, H. (1993b). Gephyrin antisense oligonucleotides prevent glycine receptor clustering in spinal neurons. *Nature, 366*, 745–748.

Kirson, E. D., Schirra, C., Konnerth, A., & Yaari, Y. (1999). Early postnatal switch in magnesium sensitivity of NMDA receptors in rat CA1 pyramidal cells. *The Journal of Physiology, 521*, 99–111.

Kishino, T., Lalande, M., & Wagstaff, J. (1997). UBE3A/E6-AP mutations cause Angelman syndrome. *Nature Genetics, 15*, 70–73.

Klarsfeld, A., Laufer, R., Fontaine, B., Devillers-Thiery, A., Dubreuil, C., & Changeux, J. P. (1989). Regulation of muscle AChR alpha subunit gene expression by electrical activity: involvement of protein kinase C and Ca2+. *Neuron, 2*, 1229–1236.

Klarsfeld, A., Bessereau, J. L., Salmon, A. M., Triller, A., Babinet, C., & Changeux, J. P. (1991). An acetylcholine receptor alpha-subunit promoter conferring preferential synaptic expression in muscle of transgenic mice. *The EMBO Journal, 10*, 625–632.

Knox, R. J., Quattrocki, E. A., Connor, J. A., & Kaczmarek, L. K. (1992). Recruitment of Ca2+ channels by protein kinase C during rapid formation of putative neruopeptide release sites in isolated Aplysia neurons. *Neuron, 8*, 883–889.

Ko, J., Fuccillo, M. V., Malenka, R. C., & Südhof, T. C. (2009). LRRTM2 functions as a neurexin ligand in promoting excitatory synapse formation. *Neuron, 64*, 791–798.

Koh, Y. H., Popova, E., Thomas, U., Griffith, L. C., & Budnik, V. (1999). Regulation of DLG localization at synapses by CaMKII-dependent phosphorylation. *Cell, 98*, 353–363.

Komatsu, Y., & Iwakiri, M. (1991). Postnatal development of neuronal connections in cat visual cortex studied by intracellular recording in slice preparation. *Brain Research, 540*, 14–24.

Kotak, V. C., Korada, S., Schwartz, I. R., & Sanes, D. H. (1998). A developmental shift from GABAergic to glycinergic transmission in the central auditory system. *The Journal of Neuroscience, 18*, 4646–4655.

Krylova, O., Herreros, J., Cleverley, K. E., Ehler, E., Henriquez, J. P., Hughes, S. M., et al. (2002). WNT-3, expressed by motoneurons, regulates terminal arborization of neurotrophin-3-responsive spinal sensory neurons. *Neuron, 35*, 1043–1056.

Ksiazek, I., Burkhardt, C., Lin, S., Seddik, R., Maj, M., Bezakova, G., et al. (2007). Synapse loss in cortex of agrin-deficient mice after genetic rescue of perinatal death. *The Journal of Neuroscience, 27*, 7183–7195.

Kullberg, R. W., Lentz, T. L., & Cohen, M. W. (1977). Development of the myotomal neuromuscular junction in Xenopus laevis: An electrophysiological and fine-structural study. *Developmental Biology, 60*, 101–129.

Lacazette, E., Le Calvez, S., Gajendran, N., & Brenner, H. R. (2003). A novel pathway for MuSK to induce key genes in neuromuscular synapse formation. *The Journal of Cell Biology, 161*, 727–736.

Lahey, T., Gorczyca, M., Jia, X. X., & Budnik, V. (1994). The Drosophila tumor suppressor gene dlg is required for normal synaptic bouton structure. *Neuron, 13*, 823–1385.

Landmesser, L., & Pilar, G. (1972). The onset and development of transmission in the chick ciliary ganglion. *The Journal of Physiology, 222*, 691–713.

Lankford, K. L., & Letourneau, P. C. (1991). Roles of actin filaments and three second-messenger systems in short-term regulation of chick dorsal root ganglion neurite outgrowth. *Cell Motility and the Cytoskeleton, 20*, 7–29.

Laufer, R., Klarsfeld, A., & Changeux, J. P. (1991). Phorbol esters inhibit the activity of the chicken acetylcholine receptor alpha-subunit gene promoter. Role of myogenic regulators. *European Journal of Biochemistry/FEBS, 202*, 813–818.

Laurie, D. J., Wisden, W., & Seeburg, P. H. (1992). The distribution of 13 GABAA receptor subunit mRNAs in the rat brain. III. Embryonic and postnatal development. *The Journal of Neuroscience, 12*, 4151–4172.

Levi, S., Logan, S. M., Tovar, K. R., & Craig, A. M. (2004). Gephyrin is critical for glycine receptor clustering but not for the formation of functional GABAergic synapses in hippocampal neurons. *The Journal of Neuroscience, 24*, 207–217.

Lhuillier, L., & Dryer, S.E. (2000). Developmental regulation of neuronal KCa channels by TGFbeta 1: transcriptional and posttranscriptional effects mediated by Erk MAP kinase. *J Neurosci, 20*, 5616–5622.

Liao, D., Scannevin, R. H., & Huganir, R. (2001). Activation of silent synapses by rapid activity-dependent synaptic recruitment of AMPA receptors. *The Journal of Neuroscience, 21*, 6008–6017.

Limb, C. J., & Ryugo, D. K. (2000). Development of primary axosomatic endings in the anteroventral cochlear nucleus of mice. *Journal of the Association for Research in Otolaryngology: JARO, 1*, 103–119.

Lin, W., Burgess, R. W., Dominguez, B., Pfaff, S. L., Sanes, J. R., & Lee, K. F. (2001). Distinct roles of nerve and muscle in postsynaptic differentiation of the neuromuscular synapse. *Nature, 410*, 1057–1064.

Lin, W., Dominguez, B., Yang, J., Aryal, P., Brandon, E. P., Gage, F. H., et al. (2005). Neurotransmitter acetylcholine negatively regulates neuromuscular synapse formation by a Cdk5-dependent mechanism. *Neuron, 46*, 569–579.

Linhoff, M. W., Laurén, J., Cassidy, R. M., Dobie, F. A., Takahashi, H., Nygaard, H. B., et al. (2009). An unbiased expression screen for synaptogenic proteins identifies the LRRTM protein family as synaptic organizers. *Neuron, 61*, 734–749.

Linnoila, J., Wang, Y., Yao, Y., & Wang, Z. Z. (2008). A mammalian homolog of Drosophila tumorous imaginal discs, Tid1, mediates agrin signaling at the neuromuscular junction. *Neuron, 60*, 625–641.

Linsdell, P., & Moody, W. J. (1995). Electrical activity and calcium influx regulate ion channel development in embryonic Xenopus skeletal muscle. *The Journal of Neuroscience, 15*, 4507–4514.

Loebrich, S., Bahring, R., Katsuno, T., Tsukita, S., & Kneussel, M. (2006). Activated radixin is essential for GABAA receptor 5 subunit anchoring at the actin cytoskeleto. *The EMBO Journal, 25*, 987–999.

Löhrke, S., Srinivasan, G., Oberhofer, M., Doncheva, E., & Friauf, E. (2005). Shift from depolarizing to hyperpolarizing glycine action occurs at different perinatal ages in superior olivary complex nuclei. *The European Journal of Neuroscience, 22*, 2708–2722.

Lømo, T., & Rosenthal, J. (1972). Control of ACh sensitivity by muscle activity in the rat. *The Journal of Physiology, 221*, 493–513.

Lu, J., Karadsheh, M., & Delpire, E. (1999). Developmental regulation of the neuronal-specific isoform of K-Cl cotransporter KCC2 in postnatal rat brains. *Journal of Neurobiology, 39*, 558–568.

Lu, W., Man, H., Ju, W., Trimble, W. S., MacDonald, J. F., & Wang, Y. T. (2001). Activation of synaptic NMDA receptors induces membrane insertion of new AMPA receptors and LTP in cultured hippocampal neurons. *Neuron, 29*, 243–254.

Luhmann, H. J., & Prince, D. A. (1991). Postnatal maturation of the GABAergic system in rat neocortex. *Journal of Neurophysiology, 65*, 247–263.

Magill-Solc, C., & McMahon, U. J. (1988). Motor neurons contain agrin-like molecules. *The Journal of Cell Biology, 107*, 1825–1833.

Manitt, C., Nikolakopoulou, A. M., Almario, D. R., Nguyen, S. A., & Cohen-Cory, S. (2009). Netrin participates in the development of retinotectal synaptic connectivity by modulating axon arborization and synapse formation in the developing brain. *The Journal of Neuroscience, 29*, 11065–11077.

Margeta, M. A., Shen, K., & Grill, B. (2008). Building a synapse: lessons on synaptic specificity and presynaptic assembly from the nematode C. elegans. *Current Opinion in Neurobiology, 18*, 69–76.

Marrs, G. S., Green, S. H., & Dailey, M. E. (2001). Rapid formation and remodeling of postsynaptic densities in developing dendrites. *Nature Neuroscience, 4*, 1006–1013.

Marrus, S. B., & DiAntonio, A. (2004). Preferential localization of glutamate receptors opposite sites of high presynaptic release. *Current Biology: CB, 14*, 924–931.

Martinou, J. C., Falls, D. L., Fischbach, G. D., & Merlie, J. P. (1991). Acetylcholine receptor-inducing acitivity stimulates expression of the epsilon-subunit gene of the muscle acetylcholine receptor. *Proceedings of the National Academy of Sciences of the United States of America, 88*, 7669–7673.

Marty, S., Wehrle, R., & Sotelo, C. (2000). Neuronal activity and brain-derived neurotrophic factor regulate the density of inhibitory synapses in organotypic slice cultures of postnatal hippocampus. *The Journal of Neuroscience, 20*, 8087–8095.

Matsuura, T., Sutcliffe, J. S., Fang, P., Galjaard, R. J., Jiang, Y. H., Benton, C. S., et al. (1997). De novo truncating mutations in E6-AP ubiquitin-protein ligase gene (UBE3A) in Angelman syndrome. *Nature Genetics, 15*, 74–77.

Mattson, M. P., & Kater, S. B. (1987). Calcium regulation of neurite elongation and growth cone motility. *The Journal of Neuroscience, 7*, 4034–4043.

Mauceri, D., Cattabeni, F., Di Luca, M., & Gardoni, F. (2004). Calcium/calmodulin-dependent protein kinase II phosphorylation drives synapse-associated protein 97 into spines. *Journal of Biological Chemistry, 279*, 23813–23821.

Mauch, D. H., Nägler, K., Schumacher, S., Göritz, C., Müller, E. C., Otto, A., et al. (2001). CNS synaptogenesis promoted by glia-derived cholesterol. *Science, 294*, 1354–1357.

McCobb, D. P., Best, P. M., & Beam, K. G. (1989). Development alters the expression of calcium currents in chick limb motoneurons. *Neuron, 2*, 1633–1643.

McCobb, D. P., Best, P. M., & Beam, K. G. (1990). The differentiation of excitability in embryonic chick limb motoneurons. *Journal of Neuroscience, 10*, 2974–2984.

Merlie, J. P., Isenberg, K. E., Russell, S. D., & Sanes, J. R. (1984). Denervation supersensitivity in skeletal muscle: analysis with a cloned cDNA probe. *The Journal of Cell Biology, 99*, 332–335.

Merlie, J. P., & Sanes, J. R. (1985). Concentration of acetylcholine receptor mRNA in synaptic regions of adult muscle fibres. *Nature, 317*, 66–68.

Misgeld, T., Kummer, T. T., Lichtman, J. W., & Sanes, J. R. (2005). Agrin promotes synaptic differentiation by counteracting an inhibitory effect of neurotransmitter. *Proceedings of the National Academy of Sciences of the United States of America, 102*, 11088–11093.

Missler, M., Zhang, W., Rohlmann, A., Kattenstroth, G., Hammer, R. E., Gottmann, K., et al. (2003). α-Neurexins couple Ca2+ channels to synaptic vesicle exocytosis. *Nature, 423*, 939–948.

Miyashiro, K., Dichter, M., & Eberwine, J. (1994). On the nature and differential distribution of mRNAs in hippocampal neurites: implications for neuronal functioning. *Proceedings of the National Academy of Sciences of the United States of America, 91*, 10800–10804.

Mizoguchi, A., Nakanishi, H., Kimura, K., Matsubara, K., Ozaki-Kuroda, K., Katata, T., et al. (2002). Nectin: an adhesion molecule involved in formation of synapses. *The Journal of Cell Biology, 156*, 555–565.

Moore, C., Leu, M., Muller, U., & Brenner, H. R. (2001). Induction of multiple signaling loops by MuSK during neuromuscular synapse formation. *Proceedings of the National Academy of Sciences of the United States of America, 98*, 14655–14660.

Mori-Okamoto, J., Ashida, H., Maru, E., & Tatsuno, J. (1983). The development of action potentials in cultures of explanted cortical neurons from chick embryos. *Developmental Biology, 97*, 408–416.

Moss, B. L., & Role, L. W. (1993). Enhanced ACh sensitivity is accompanied by changes in ACh receptor channel properties and segregation of ACh receptor subtypes on sympathetic neurons during innervation in vivo. *The Journal of Neuroscience, 13*, 13–28.

Mueller, A. L., Chesnut, R. M., & Schwartzkroin, P. A. (1983). Actions of GABA in developing rabbit hippocampus: An in vitro study. *Neuroscience Letters, 39*, 193–198.

Mueller, A. L., Taube, J. S., & Schwartzkroin, P. A. (1984). Development of hyperpolarizing inhibitory postsynaptic potentials and hyperpolarizing responses to gamma-aminobutyric acid in rabbit hippocampus studied in vitro. *The Journal of Neuroscience, 4*, 860–867.

Muraro, N. I., Weston, A. J., Gerber, A. P., Luschnig, S., Moffat, K. G., & Baines, R. A. (2008). Pumilio binds para mRNA and requires Nanos and Brat to regulate sodium current in Drosophila motoneurons. *The Journal of Neuroscience, 28*(9), 2099–2109.

Mutani, R., Futamachi, K., & Prince, D. A. (1974). Potassium activity in immature cortex. *Brain Research, 75*, 27–39.

Nabekura, J., Katsurabayashi, S., Kakazu, Y., Shibata, S., Matsubara, A., Jinno, S., et al. (2004). Developmental switch from GABA to glycine release in single central synaptic terminals. *Nature Neuroscience, 7*, 17–23.

Newman-Gage, H., Westrum, L. E., & Bertrum, J. F. (1987). Stereological analysis of synaptogenesis in the molecular layer of piriform cortex in the prenatal rat. *The Journal of Comparative Neurology, 261*, 295–305.

Niell, C. M., Meyer, M. P., & Smith, S. J. (2004). In vivo imaging of synapse formation on a growing dendritic arbor. *Nature Neuroscience, 7*, 254–260.

Niethammer, M., Kim, E., & Sheng, M. (1996). Interaction between the C terminus of NMDA receptor subunits and multiple members of the PSD-95 family of membrane-associated guanylate kinases. *The Journal of Neuroscience, 16*, 2157–2163.

Nishiyama, M., von Schimmelmann, M. J., Togashi, K., Findley, W. M., & Hong, K. (2008). Membrane potential shifts caused by diffusible guidance signals direct growth-cone turning. *Nature Neuroscience, 11*, 762–771.

Nitkin, R. M., Smith, M. A., Magill, C., Fallon, J. R., Yao, M., Wallace, B. G., et al. (1987). Identification of agrin, a synaptic organizing protein from Torpedo electric organ. *The Journal of Cell Biology, 105*, 2471–2478.

Noakes, P. G., Gautam, M., Mudd, J., Sanes, J. R., & Merlie, J. P. (1995). Aberrant differentiation of neuromuscular junctions in mice lacking s-laminin/laminin beta 2. *Nature, 374*, 258–262.

Novak, A. E., Taylor, A. D., Pineda, R. H., Lasda, E. L., Wright, M. A., & Ribera, A. B. (2006). Embryonic and larval expression of zebrafish voltage-gated sodium channel alpha-subunit genes. *Developmental Dynamics, 235*, 1962–1973.

O'Brien, R. J., & Fischbach, G. D. (1986a). Isolation of embryonic chick motoneurons and their survival in vitro. *The Journal of Neuroscience, 6*, 3265–3274.

O'Brien, R. J., & Fischbach, G. D. (1986b). Modulation of embryonic chick motoneuron glutamate sensitivity by interneurons and agonists. *The Journal of Neuroscience, 6*, 3290–3296.

O'Brien, R. J., Xu, D., Petralia, R. S., Steward, O., Huganir, R. L., & Worley, P. (1999). Synaptic clustering of AMPA receptors by the extracellular immediate-early gene product Narp. *Neuron, 23*, 309–323.

O'Brien, R., Xu, D., Mi, R., Tang, X., Hopf, C., & Worley, P. (2002). Synaptically targeted narp plays an essential role in the aggregation of AMPA receptors at excitatory synapses in cultured spinal neurons. *The Journal of Neuroscience, 22*, 4487–4498.

O'Dowd, D. K. (1983). RNA synthesis dependence of action potential development in spinal cord neurones. *Nature, 303*, 619–621.

O'Dowd, D. K., Ribera, A. B., & Spitzer, N. C. (1988). Development of voltage-dependent calcium, sodium, and potassium currents in Xenopus spinal neurons. *The Journal of Neuroscience, 8*, 792–805.

Obata, K., Oide, M., & Tanaka, H. (1978). Excitatory and inhibitory actions of GABA and glycine on embryonic chick spinal neurons in culture. *Brain Research, 144*, 179–184.

Obrietan, K., & van den Pol, A. N. (1995). GABA neurotransmission in the hypothalamus: developmental reversal from Ca2+ elevating to depressing. *The Journal of Neuroscience, 15*, 5065–5077.

Oppenheim, R. W., & Reitzel, J. (1975). Ontogeny of behavioral sensitivity to strychnine in the chick embryo: Evidence for the early onset of CNS inhibition. *Brain, Behavior and Evolution, 11*, 130–159.

Oswald, A. M., & Reyes, A. D. (2008). Maturation of intrinsic and synaptic properties of layer 2/3 pyramidal neurons in mouse auditory cortex. *Journal of Neurophysiology, 99*, 2998–3008.

Owens, D. F., Boyce, L. H., Davis, M. B. E., & Kriegstein, A. R. (1996). Excitatory GABA responses in embryonic and neonatal cortical slices demonstrated by gramicidin perforated-patch recordings and calcium imaging. *The Journal of Neuroscience, 16*, 6414–6423.

Packard, M., Koo, E. S., Gorczyca, M., Sharpe, J., Cumberledge, S., & Budnik, V. (2002). The Drosophila Wnt, wingless, provides an essential signal for pre- and postsynaptic differentiation. *Cell, 111*, 319–330.

Palade, G. E., & Palay, S. L. (1954). Electron microscope observations of interneuronal and neuromuscular synapses. *The Anatomical Record, 118*, 335–336.

Papadopoulos, T., Eulenburg, V., Reddy-Alla, S., Mansuy, I. M., Li, Y., & Betz, H. (2008). Collybistin is required for both the formation and maintenance of GABAergic postsynapses in the hippocampus. *Molecular and Cellular Neurosciences, 39*, 161–169.

Pappas, G. D., & Purpura, D. P. (1964). Electron microscopy of immature human and feline hippocampus. *Progress in Brain Research, 4*, 176–186.

Paradis, S., Harrar, D. B., Lin, Y., Koon, A. C., Hauser, J. L., Griffith, E. C., et al. (2007). An RNAi-based approach identifies molecules required for glutamatergic and GABAergic synapse development. *Neuron, 53*, 217–232.

Patton, B. L., Chiu, A. Y., & Sanes, J. R. (1998). Synaptic laminin prevents glial entry into the synaptic cleft. *Nature, 393*, 698–701.

Payne, J. A. (1997). Functional characterization of the neuronal-specific K-Cl cotransporter: implications for [K+]o regulation. *The American Journal of Physiology, 273*, C1516–1525.

Payne, J. A., Rivera, C., Voipio, J., & Kaila, K. (2003). Cation-chloride co-transporters in neuronal communication, development, and trauma. *Trends Neurosci, 26*, 199–206.

Phillips, W. D., Kopta, C., Blount, P., Gardner, P. D., Steinbach, J. H., & Merlie, J. P. (1991). ACh receptor-rich domains organized in fibroblasts by recombinant 43-Kilodalton protein. *Science, 251*, 568–570.

Picken Bahrey, H. L., & Moody, W. J. (2003). Early development of voltage-gated ion currents and firing properties in neurons of the mouse cerebral cortex. *Journal of Neurophysiology, 89*, 1761–1773.

Poskanzer, K., Needleman, L. A., Bozdagi, O., & Huntley, G. W. (2003). N-cadherin regulates ingrowth and laminar targeting of thalamocortical axons. *The Journal of Neuroscience, 23*, 2294–2305.

Poulopoulos, A., Aramuni, G., Meyer, G., Soykan, T., Hoon, M., Papadopoulos, T., et al. (2009). Neuroligin 2 drives postsynaptic assembly at perisomatic inhibitory synapses through gephyrin and collybistin. *Neuron, 63*, 628–642.

Pow, D. V., & Hendrickson, A. E. (2000). Expression of glycine and the glycine transporter Glyt-1 in the developing rat retina. *Visual Neuroscience, 17*(3), 1R–9R.

Prokop, A., Landgraf, M., Rushton, E., Broadie, K., & Bate, M. (1996). Presynaptic development at the neuromuscular junction: Assembly and localization of presynaptic active zones. *Neuron, 17*, 617–626.

Pym, E. C., Southall, T. D., Mee, C. J., Brand, A. H., & Baines, R. A. (2006). The homeobox transcription factor Even-skipped regulates acquisition of electrical properties in Drosophila neurons. *Neural Development, 1*, 3.

Qu, Z., & Huganir, R. L. (1994). Comparison of innervation and agrin-induced tyrosine phosphorylation of the nicotinic acetylcholine receptor. *The Journal of Neuroscience, 14*, 6834–6841.

Quinlan, E. M., Olstein, D. H., & Bear, M. F. (1999). Bidirectional, experience-dependent regulation of N-methyl-D-aspartate receptor subunit composition in the rat visual cortex during postnatal development. *Proceedings of the National Academy of Sciences of the United States of America, 96*, 12876–12880.

Ramarao, M. K., Bianchetta, M. J., Lanken, J., & Cohen, J. B. (2001). Role of rapsyn tetratricopeptide repeat and coiled-coil domains in self-association and nicotinic acetylcholine receptor clustering. *Journal of Biological Chemistry, 276,* 7475–7483.

Ramoa, A. S., & McCormick, D. A. (1994). Developmental changes in electrophysiological properties of LGNd neurons during reorganization of retinogeniculate connections. *The Journal of Neuroscience, 14,* 2089–2097.

Ramón y Cajal, S. (1905). Genèse des fibres nerveuses de l'embryon et observations contraires à la thérie catenaire. *Trab Lab Invest Biol, Univ Madrid, 4,* 219–284.

Rees, R. P., Bunge, M. B., & Bunge, R. P. (1976). Morphological changes in the neuritic growth cone and target neuron during synaptic junction development in culture. *The Journal of Cell Biology, 68,* 240–263.

Rees, R. P. (1978). The morphology of interneuronal synaptogenesis: a review. *Federation Proceedings, 37,* 2000–2009.

Reichling, D. B., Kyrozis, A., Wang, J., & MacDermott, A. B. (1994). Mechanisms of GABA and glycine depolarization-induced calcium transients in rat dorsal horn neurons. *The Journal of Physiology, 476,* 411–421.

Reiness, C. G., & Weinberg, C. B. (1981). Metabolic stabilization of acetylcholine receptors at newly formed neuromuscular junctions in rat. *Developmental Biology, 84,* 247–254.

Reist, N. E., Magill, C., & McMahon, U. J. (1987). Agrin-like molecules at synaptic sites in normal, denervated, and damaged skeletal muscles. *The Journal of Cell Biology, 105,* 2457–2469.

Reist, N. E., Werle, M. J., & McMahon, U. J. (1992). Agrin released by motor neurons induces the aggregation of acetylcholine receptors at neuromuscular junctions. *Neuron, 8,* 865–868.

Rheims, S., Holmgren, C. D., Chazal, G., Mulder, J., Harkany, T., Zilberter, T., et al. (2009). GABA action in immature neocortical neurons directly depends on the availability of ketone bodies. *Journal of Neurochemistry, 110,* 1330–1338.

Ribera, A. B., & Spitzer, N. C. (1989). A critical period of transcription required for differentiation of the action potential of spinal neurons. *Neuron, 2,* 1055–1062.

Ritzenthaler, S., Suzuki, E., & Chiba, A. (2000). Postsynaptic filopodia in muscle cells interact with innervating motoneuron axons. *Nature Neuroscience, 3,* 1012–1017.

Rivera, C., Voipio, J., Payne, J. A., Ruusuvuori, E., Lahtinen, H., Lamsa, K., et al. (1999). The K+/Cl- co-transporter KCC2 renders GABA hyperpolarizing during neuronal maturation. *Nature, 397,* 251–255.

Rivera, C., Li, H., Thomas-Crusells, J., Lahtinen, H., Viitanen, T., Nanobashvili A, et al. (2002). BDNF-induced TrkB activation down-regulates the K+-Cl-cotransporter KCC2 and impairs neuronal Cl-extrusion. *J Cell Biol, 159,* 747–752.

Role, L. W. (1985). Neural regulation of acetylcholine sensitivity in embryonic sympathetic neurons. *Proceedings of the National Academy of Sciences of the United States of America, 85,* 2825–2829.

Role, L. W., Matossian, V. R., O'Brien, R. J., & Fischbach, G. D. (1985). On the mechanism of acetylcholine receptor accumulation at newly formed synapses on chick myotubes. *The Journal of Neuroscience, 5,* 2197–2204.

Rotzler, S., Schramek, H., & Brenner, H. R. (1991). Metabolic stabilization of endplate acetylcholine receptors regulated by Ca2+ influx associated with muscle activity. *Nature, 349,* 337–339.

Ruegg, M. A., Tsim, K. W. K., Horton, S. E., Kroger, S., Escher, G., Gensch, E. M., et al. (1992). The agrin gene codes for a family of basal lamina proteins that differ in function and distribution. *Neuron, 8,* 691–699.

Ryan, T. J., & Grant, S. G. (2009). The origin and evolution of synapses. *Nature Reviews. Neuroscience, 10,* 701–712.

Sabo, S. L., & McAllister, A. K. (2003). Mobility and cycling of synaptic protein-containing vesicles in axonal growth cone filopodia. *Nature Neuroscience, 6,* 1264–1269.

Saglietti, L., Dequidt, C., Kamieniarz, K., Rousset, M. C., Valnegri, P., Thoumine, O., et al. (2007). Extracellular interactions between GluR2 and N-cadherin in spine regulation. *Neuron, 54,* 461–477.

Saitoe, M., Schwarz, T. L., Umbach, J. A., Gundersen, C. B., & Kidokoro, Y. (2001). Absence of junctional glutamate receptor clusters in Drosophila mutants lacking spontaneous transmitter release. *Science, 293,* 514–517.

Salpeter, M. M., & Harris, R. (1983). Distribution and turnover rate of acetylcholine receptors throughout the junction folds at a vertebrate neuromuscular junction. *The Journal of Cell Biology, 96,* 1781–1785.

Salzer, J. L., Brophy, P. J., & Peles, E. (2008). Molecular domains of myelinated axons in the peripheral nervous system. *Glia, 56,* 1532–1540.

Sanes, D. H. (1993). The development of synaptic function and integration in the central auditory system. *The Journal of Neuroscience, 13,* 2627–2637.

Sanes, J. R., Johnson, Y. R., Kotzbauer, P. T., Mudd, J., Hanley, T., Martinou, J. C., et al. (1991). Selective expression of an acetylcholine receptor-lacZ transgene in synaptic nuclei of adult muscle fibers. *Development, 113,* 1181–1191.

Sanes, J. R., Marshall, L. M., & McMahan, U. J. (1978). Reinnervation of muscle fiber basal lamina after removal of myofibers. Differentiation of regenerating axons at original synaptic sites. *The Journal of Cell Biology, 78,* 76–198.

Scheiffele, P., Fan, J., Choih, J., Fetter, R., & Serafini, T. (2000). Neuroligin expressed in nonneuronal cells triggers presynaptic development in contacting axons. *Cell, 101,* 657–669.

Schrader, N., Kim, E. Y., Winking, J., Paulukat, J., Schindelin, H., & Schwarz, G. (2004). Biochemical characterization of the high affinity binding between the glycine receptor and gephyrin. *Journal of Biological Chemistry, 279,* 18733–18741.

Schwartz, I. R., Pappas, G. D., & Purpura, D. P. (1968). Fine structure of neruons and synapses in the feline hippocampus during postnatal ontogenesis. *Experimental Neurology, 22,* 394–407.

Shainberg, A., & Burstein, M. (1976). Decrease of acetylcholine receptor synthesis in muscle cultures by electrical stimulation. *Nature, 264,* 368–369.

Shapira, M., Zhai, R. G., Dresbach, T., Bresler, T., Torres, V. I., Gundelfinger, E. D., et al. (2003). Unitary assembly of presynaptic active zones from Piccolo-Bassoon transport vesicles. *Neuron, 38,* 237–252.

Shen, K., & Bargmann, C. I. (2003). The immunoglobulin superfamily protein SYG-1 determines the location of specific synapses in C. elegans. *Cell, 112,* 619–630.

Shen, K., Fetter, R. D., & Bargmann, C. I. (2004). Synaptic specificity is generated by the synaptic guidepost protein SYG-2 and its receptor, SYG-1. *Cell, 116,* 869–881.

Shen, W., Wu, B., Zhang, Z., Dou, Y., Rao, Z. R., Chen, Y. R., et al. (2006). Activity-induced rapid synaptic maturation mediated by presynaptic cdc42 signaling. *Neuron, 50,* 401–414.

Sheng, M., Cummings, J., Roldan, L. A., Jan, Y. N., & Jan, L. Y. (1994). Changing subunit composition of heteromeric NMDA receptors during development of rat cortex. *Nature, 368,* 144–147.

Sheng, M., & Hoogenraad, C. C. (2007). The postsynaptic architecture of excitatory synapses: a more quantitative view. *Annual Review of Biochemistry, 76,* 823–847.

Sherman, D. L., Tait, S., Melrose, S., Johnson, R., Zonta, B., Court, F. A., et al. (2005). Neurofascins are required to establish axonal domains for saltatory conduction. *Neuron, 48,* 737–742.

Sherrington, C. S. (1906). *The integrative action of the nervous system.* New York: Charles Scribner's Sons.

Shi, S. H., Hayashi, Y., Petralia, R. S., Zaman, S. H., Wenthold, R. J., Svoboda, K., et al. (1999). Rapid spine delivery and redistribution of AMPA receptors after synaptic NMDA receptor activation. *Science, 284,* 1811–1816.

Shitaka, Y., Matsuki, N., Saito, H., & Katsuki, H. (1996). Basic fibroblast growth factor increases functional L-type Ca2+ channels in fetal hippocampal neurons: Implications for neurite morphogenesis in vitro. *The Journal of Neuroscience, 16,* 6476–6489.

Shyng, S.-L., Xu, R., & Salpeter, M. M. (1991). cAMP stabilizes the degradation of original junctional acetylcholine receptors in denervated muscle. *Neuron, 6,* 469–475.

Sia, G. M., Béïque, J. C., Rumbaugh, G., Cho, R., Worley, P. F., & Huganir, R. L. (2007). Interaction of the N-terminal domain of the AMPA receptor GluR4 subunit with the neuronal pentraxin NP1 mediates GluR4 synaptic recruitment. *Neuron, 55,* 87–102.

Siegelbaum, S. A., Trautmann, A., & Koenig, J. (1984). Single acetylcholine-activated channel currents indeveloping muscle cells. *Developmental Biology, 104,* 366–379.

Simon, D. K., Prusky, G. T., O'Leary, D. D. M., & Constantine-Paton, M. (1992). N-methyl-D-aspartate receptor antagonists disrupt the formation of a mammalian neural map. *Proceedings of the National Academy of Sciences of the United States of America, 89,* 10593–10597.

Skoff, R. P., Price, D. L., & Stocks, A. (1976). Electron microscopic autoradiographic studies of gliogenesis in rat optic nerve. I. Cell proliferation. *The Journal of Comparative Neurology, 169,* 291–312.

Smolen, A. J. (1981). Postnatal development of ganglionic neurons in the absence of preganglionic input: Morphological observations on synapse formation. *Developmental Brain Research, 1,* 49–58.

Spitzer, N. C. (1981). Development of membrane properties in vertebrates. *TINS, 4,* 169–172.

Spitzer, N. C., & Lamborghini, J. E. (1976). The development of the action potential mechanism of amphibian neurons isolated in cell culture. *Proceedings of the National Academy of Sciences of the United States of America, 73,* 1641–1645.

Spitzer, N. C. (1994). Spontaneous Ca2+ spikes and waves in embryonic neurons: signaling systems for differentiation. *Trends Neurosci, 17,* 115–118.

Stein, V., Hermans-Borgmeyer, I., Jentsch, T. J., & Hubner, C. A. (2004). Expression of the KCl cotransporter KCC2 parallels neuronal maturation and the emergence of low intracellular chloride. *The Journal of Comparative Neurology, 468,* 57–64.

Steward, O., & Schuman, E. M. (2001). Protein synthesis at synaptic sites on dendrites. *Annual Review of Neuroscience, 24,* 299–325.

Stil, A., Liabeuf, S., Jean-Xavier, C., Brocard, C., Viemari, J. C., & Vinay, L. (2009). Developmental up-regulation of the potassium-chloride cotransporter type 2 in the rat lumbar spinal cord. *Neuroscience, 164,* 809–821.

Subramony, P., Raucher, S., Dryer, L., & Dryer, S. E. (1996). Posttranslational regulation of Ca2+-activated K + currents by a target-dervied factor in developing parasympathetic neurons. *Neuron, 17,* 115–124.

Südhof, T. C. (2008). Neuroligins and neurexins link synaptic function to cognitive disease. *Nature, 455,* 903–911.

247

Sun, Y.-a., & Poo, M. M. (1987). Evoked release of acetylcholine from the growing embryonic neuron. *Proceedings of the National Academy of Sciences of the United States of America, 84*, 2540–2544.

Sutherland, M. L., Delaney, T. A., & Noebels, J. L. (1996). Glutamate transporter mRNA expression in proliferative zones of the developing and adult murine CNS. *The Journal of Neuroscience, 16*, 2191–2207.

Sweeney, S. T., Broadie, K., Keane, J., Niemann, H., & O'Kane, C. J. (1995). Targeted expression of tetanus toxin light chain in Drosophila specifically eliminates synaptic transmission and causes behavioral defects. *Neuron, 14*, 341–351.

Syková, E. (1992). Ion-sensitive electrodes. In J. A. Stamford (Ed.), *Monitoring neuronal activity: a practical approach* (pp. 261–282). New York: IRL Press.

Syková, E., Jendelová, P., Simonová, Z., & Chvátal, A. (1992). K+ and pH homeostasis in the developing rat spinal chord is impaired by early postnatal X-irradiation. *Brain Research, 594*, 19–30.

Takahashi, T., Momiyama, A., Hirai, K., Hishinuma, F., & Akagi, H. (1992). Functional correlation of fetal and adult forms of glycine receptors with developmental changes in inhibitory synaptic receptor channels. *Neuron, 9*, 1155–1161.

Takayama, C., & Inoue, Y. (2005). Developmental expression of GABA transporter-1 and 3 during formation of the GABAergic synapses in the mouse cerebellar cortex. *Developmental Brain Research, 158*, 41–49.

Takesian, A. E., Kotak, V. C., & Sanes, D. H. (2010). Presynaptic GABA(B) receptors regulate experience-dependent development of inhibitory short-term plasticity. *The Journal of Neuroscience, 30*, 2716–2727.

Tang, H., Macpherson, P., Argetsinger, L. S., Cieslak, D., Suhr, S. T., Carter-Su, C., et al. (2004). CaM kinase II-dependent phosphorylation of myogenin contributes to activity-dependent suppression of nAChR gene expression in developing rat myotubes. *Cellular Signalling, 16*, 551–563.

Tao, Y. X., Rumbaugh, G., Wang, G. D., Petralia, R. S., Zhao, C., Kauer, F. W., et al. (2003). Impaired NMDA receptor-mediated postsynaptic function and blunted NMDA receptor-dependent persistent pain in mice lacking postsynaptic density-93 protein. *The Journal of Neuroscience, 23*, 6703–6712.

Tarpey, P., Parnau, J., Blow, M., Woffendin, H., Bignell, G., Cox, C., et al. (2004). Mutations in the DLG3 gene cause nonsyndromic X-linked mental retardation. *American Journal of Human Genetics, 75*, 318–324.

Tepper, J. M., & Trent, F. (1993). In vivo studies of the postnatal development of rat neostriatal neurons. *Progress in Brain Research, 99*, 35–50.

Togashi, H., Abe, K., Mizoguchi, A., Takaoka, K., Chisaka, O., & Takeichi, M. (2002). Cadherin regulates dendritic spine morphogenesis. *Neuron, 35*, 77–89.

Tomita, S., Chen, L., Kawasaki, Y., Petralia, R. S., Wenthold, R. J., Nicoll, R. A., et al. (2003). Functional studies and distribution define a family of transmembrane AMPA receptor regulatory proteins. *The Journal of Cell Biology, 161*, 805–816.

Toni, N., Laplagne, D. A., Zhao, C., Lombardi, G., Ribak, C. E., Gage, F. H., et al. (2008). Neurons born in the adult dentate gyrus form functional synapses with target cells. *Nature Neuroscience, 11*, 901–907.

Tretter, V., Jacob, T. C., Mukherjee, J., Fritschy, J. M., Pangalos, M. N., & Moss, S. J. (2008). The clustering of GABA(A) receptor subtypes at inhibitory synapses is facilitated via the direct binding of receptor alpha 2 subunits to gephyrin. *J Neurosci, 28*, 1356–1365.

Ullian, E. M., Sapperstein, S. K., Christopherson, K. S., & Barres, B. A. (2001). Control of synapse number by glia. *Science, 291*, 657–661.

Umemori, H., Linhoff, M. W., Ornitz, D. M., & Sanes, J. R. (2004). FGF22 and its close relatives are presynaptic organizing molecules in the mammalian brain. *Cell, 118*, 257–270.

Vacher, H., Mohapatra, D. P., & Trimmer, J. S. (2008). Localization and targeting of voltage-dependent ion channels in mammalian central neurons. *Physiological Reviews, 88*, 1407–1447.

Valenzuela, D. M., Stitt, T. N., DiStefano, P. S., Rojas, E., Mattsson, K., Compton, D. L., et al. (1995). Receptor tyrosine kinase specific for the skeletal muscle lineage: Expression in embryonic muscle, at the neuromuscular junction, and after injury. *Neuron, 15*, 573–584.

Varoqueaux, F., Aramuni, G., Rawson, R. L., Mohrmann, R., Missler, M., Gottmann, K., et al. (2006). Neuroligins determine synapse maturation and function. *Neuron, 51*, 741–754.

Varoqueaux, F., Jamain, S., & Brose, N. (2004). Neuroligin 2 is exclusively localized to inhibitory synapses. *European Journal of Cell Biology, 83*, 449–456.

Vasilyev, D. V., & Barish, M. E. (2002). Postnatal development of the hyperpolarization-activated excitatory current Ih in mouse hippocampal pyramidal neurons. *The Journal of Neuroscience, 22*, 8992–9004.

Vaughn, J. E. (1989). Review: Fine structure of synaptogenesis in the vertebrate central nervous system. *Synapse, 3*, 255–285.

Vaughn, J. E., Henrikson, C. K., & Grieshaber, J. A. (1974). A quantitative study of synapses on motor neuron dendritic growth cones in developing mouse spinal cord. *The Journal of Cell Biology, 60*, 664–672.

Vicario-Abejon, C., Collin, C., McKay, R. D., & Segal, M. (1998). Neurotrophins induce formation of functional excitatory and inhibitory synapses between cultured hippocampal neurons. *The Journal of Neuroscience, 18*, 7256–7271.

Vicini, S., & Schuetze, S. M. (1985). Gating properties of acetylcholine channels at developing rat endplates. *The Journal of Neuroscience, 5*, 2212–2224.

Vitellaro-Zuccarello, L., Calvaresi, N., & De Biasi, S. (2003). Expression of GABA transporters, GAT-1 and GAT-3, in the cerebral cortex and thalamus of the rat during postnatal development. *Cell and Tissue Research, 313*, 245–257.

Wallace, B. G., Qu, Z., & Huganir, R. L. (1991). Agrin induces phosphorylation of the nicotinic acetylcholine receptor. *Neuron, 6*, 869–878.

Wallace, B. G. (1994). Staurosporine inhibits agrin-induced acetylcholine receptor phosphorylation and aggregation. *The Journal of Cell Biology, 125*, 661–668.

Wang, X., Butowt, R., Vasko, M. R., & von Bartheld, C. S. (2002). Mechanisms of the release of anterogradely transported neurotrophin-3 from axon terminals. *The Journal of Neuroscience, 22*, 931–945.

Warren, R. A., & Jones, E. G. (1997). Maturation of neuronal form and function in a mouse thalamo-cortical circuit. *The Journal of Neuroscience, 17*, 277–295.

Washbourne, P., Bennett, J. E., & McAllister, A. K. (2002). Rapid recruitment of NMDA receptor transport packets to nascent synapses. *Nature Neuroscience, 5*, 751–759.

Weatherbee, S. D., Anderson, K. V., & Niswander, L. A. (2006). LDL-receptor-related protein 4 is crucial for formation of the neuromuscular junction. *Development, 133*, 4993–5000.

Weiner, J. A., Wang, X., Tapia, J. C., & Sanes, J. R. (2005). Gamma protocadherins are required for synaptic development in the spinal cord. *Proc Natl Acad Sci USA, 102*, 8–14.

Westrum, L. E. (1975). Electron microscopy of synaptic structures in olfactory cortex of early postnatal rats. *Journal of Neurocytology, 4*, 713–732.

Woo, J., Kwon, S. K., Choi, S., Kim, S., Lee, J. R., Dunah, A. W., et al. (2009). Trans-synaptic adhesion between NGL-3 and LAR regulates the formation of excitatory synapses. *Nature Neuroscience, 12*, 428–437.

Woods, D. F., & Bryant, P. J. (1991). The discs-large tumor suppressor gene of Drosophila encodes a guanylate kinase homolog localized at septate junctions. *Cell, 66*, 451–464.

Wu, G.-Y., Malinow, R., & Cline, H. T. (1996). Maturation of central glutamatergic synapse. *Science, 274*, 972–976.

Xie, Z.-p., & Poo, M. M. (1986). Initial events in the formation of neuromuscular synapse: Rapid induction of acetylcholine release from embryonic neuron. *Proceedings of the National Academy of Sciences of the United States of America, 83*, 7069–7073.

Yaari, Y., Hamon, B., & Lux, H. D. (1987). Development of two types of calcium channels in cultured mammalian hippocampal neurons. *Science, 235*, 680–682.

Yamagata, M., Sanes, J. R., & Weiner, J. A. (2003). Synaptic adhesion molecules. *Current Opinion in Cell Biology, 15*, 621–632.

Yang, H., & Kunes, S. (2004). Nonvesicular release of acetylcholine is required for axon targeting in the Drosophila visual system. *Proceedings of the National Academy of Sciences of the United States of America, 101*, 15213–15218.

Yang, X., Arber, S., William, C., Li, L., Tanabe, Y., Jessell, T. M., et al. (2001). Patterning of muscle acetylcholine receptor gene expression in the absence of motor innervation. *Neuron, 30*, 399–410.

Yao, J., Qi, J., & Chen, G. (2006). Actin-dependent activation of presynaptic silent synapses contributes to long-term synaptic plasticity in developing hippocampal neurons. *The Journal of Neuroscience, 26*, 8137–8147.

Yashiro, K., Riday, T. T., Condon, K. H., Roberts, A. C., Bernardo, D. R., Prakash, R., et al. (2009). Ube3a is required for experience-dependent maturation of the neocortex. *Nature Neuroscience, 12*, 777–783.

Yoon, Y. J., Kominami, H., Trimarchi, T., & Martin-Caraballo, M. (2008). Inhibition of electrical activity by retroviral infection with Kir2.1 transgenes disrupts electrical differentiation of motoneurons. *PLoS ONE, 3*, e2971.

Young, S. H., & Poo, M. M. (1983a). Spontaneous release of transmitter grom growth cones of embryonic neurones. *Nature, 305*, 634–637.

Young, S. H., & Poo, M. M. (1983b). Rapid lateral diffusion of extrajunctional acetylcholine receptors in the developing muscle membrane of Xenopus tadpole. *The Journal of Neuroscience, 3*, 225–231.

Zecevic, N. (1998). Synaptogenesis in layer I of the human cerebral cortex in the first half of gestation. *Cerebral Cortex, 8*, 245–252.

Zhang, B., Luo, S., Wang, Q., Suzuki, T., Xiong, W. C., & Mei, L. (2008). LRP4 serves as a coreceptor of agrin. *Neuron, 60*, 285–297.

Zhang, F., Aravanis, A. M., Adamantidis, A., de Lecea, L., & Deisseroth, K. (2007). Circuit-breakers: optical technologies for probing neural signals and systems. *Nature Reviews. Neuroscience, 8*, 577–581.

Zhang, L., Spigelman, I., & Carlen, P. L. (1991). Development of GABA-mediated, chloride-dependent inhibition in CA1 pyramidal neurones of immature rat hippocampal slices. *The Journal of Physiology, 444*, 25–49.

Ziskind-Conaim, L., Geffen, I., & Hall, Z. W. (1984). Redistribution of acetylcholine receptors on developing rat myotubes. *The Journal of Neuroscience, 4*, 2346–2349.

Zoran, M. J., Funte, L. R., Kater, S. B., & Haydon, P. G. (1993). Contact with a synaptic target causes an elevation in a presynaptic neuron's resting calcium set-point. *Developmental Biology, 158*, 163–171.

Refinement of synaptic connections

9

The process of development would appear to be complete once all neurons are born, differentiate, and connect with one another. Yet even as synaptogenesis proceeds (Chapter 8), a separate process is set in motion that leads to the *elimination* of some existing synapses and the *stabilization* of others. In many instances, this loss or stability can be influenced by an animal's sensory or motor experience. What, then, distinguishes developmental plasticity of this sort from learning and memory in adult animals? One important difference is that the developing nervous system can be altered permanently by some manipulations that have little effect during adulthood. As discussed below, there are developmental critical periods during which synaptic connections can be altered by manipulations to the sensory environment. These manipulations alter the amount or the pattern of synaptic transmission and action potentials, and this altered activity state influences the growth and differentiation of synaptic connections. Even though the immature nervous system is particularly sensitive to some manipulations, we will see that developmental plasticity and adult learning do share several molecular mechanisms.

Why are synapses being assembled and disbanded at the same time, particularly when the molecular cues that support axon guidance, and target selection (Chapters 5 and 6) lead to such accurate connections? One possibility is that synapse addition and loss do improve the specificity of neural connections beyond a level that can be obtained with molecular cues. The correct complement and strength of synaptic connections should optimize the computational properties of each neuron and, ultimately, improve behavioral performance in the animal's rearing environment (Chapter 10). In fact, the central nervous system is designed for continuous modification, from birth into adulthood, and we explore the earliest iteration of this process in this chapter.

THE EARLY PATTERN OF CONNECTIONS

Three general patterns of innervation distinguish the developing nervous system from that of the adult. First, individual axons that arborize in the correct topographic position (see Chapter 6) may spread out further than they do in the adult, perhaps a few tens of microns past what will become their mature boundary (**Figure 9.1**A). While this may seem to be a trivial distance, if billions of neurons make projections of this sort, then neural computations could be adversely affected. Second, an immature innervation pattern occurs when a developing neuron receives a greater number of inputs than will remain in adulthood (Figure 9.1B). The ratio of innervating afferent axons per postsynaptic neuron, called *convergence*, varies greatly in the nervous system. At the mammalian nerve–muscle junction there is one motor axon synapse per muscle cell. Similarly, each cerebellar Purkinje cell is innervated by a single climbing fiber axon (cf. convergence = 1), but is contacted by thousands of parallel fiber synapses on its dendritic tree (convergence $\approx$ 200,000). For many postsynaptic neurons the adult number of afferents is attained only after a fraction of their functional contacts are eliminated. A third way that innervation may become more specific during development is through the elimination of terminals from one region of the postsynaptic neuron (Figure 9.1C). In one auditory brainstem nucleus, the MSO, inhibitory terminals are eliminated from the dendrite and gradually become restricted to the cell body.

The addition or elimination of synapses during development is the most decisive way to modify a neural circuit. However, the postsynaptic response magnitude that is produced by an individual synapse, often called synaptic strength, can also be regulated. In the adult nervous system, the strength of synaptic transmission changes dramatically with use, and these

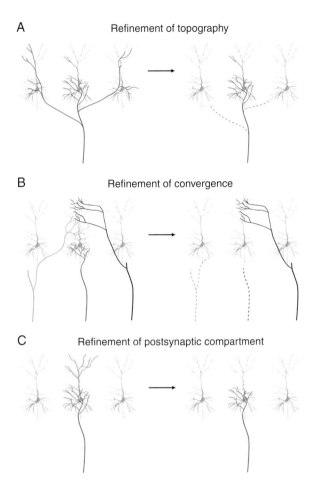

A   Refinement of topography

B   Refinement of convergence

C   Refinement of postsynaptic compartment

Fig. 9.1 Three kinds of immature afferent projections during development. A. The projection of one afferent is shown with its arborization centered at the topographically correct position in the target. However, the arbors initially extend too far, and two local branches are eliminated during development. B. A single neuron is shown to receive input from three afferents initially, and two of these inputs are eliminated during development. C. A projection is shown to innervate the soma and dendrite of a postsynaptic neuron initially, but the dendritic innervation is eliminated during development.

alterations support the storage of memories (see **Box 9.1**, "Remaining Flexible"). The first studies to draw a strong causal relationship between environmental stimulation and the development of connections were performed on the cat

visual system (Wiesel and Hubel, 1963a, 1965; Hubel and Wiesel, 1965). In control animals, extracellular recordings from the cortex show that most neurons fire action potentials in response to stimulation of both eyes. However, when visual stimulation to one eye is decreased during development, there is a dramatic reduction in the ability of that eye to activate cortical neurons. The result suggests that synapses driven by the closed eye are either eliminated or weakened. Thus, even though the initial connectivity of the visual pathway is quite accurate (Chapter 6), it is apparently not stable, and can be altered permanently by a developmental mechanism that makes use of neural activity.

FUNCTIONAL SYNAPSES ARE ELIMINATED

What is the evidence that synapses are eliminated in the developing nervous system? How widespread is this mechanism? Two experimental approaches have been taken to address these questions. First, intracellular recordings show changes in the number of functional afferents per postsynaptic neuron. Second, anatomical studies reveal that single axonal arbors become spatially restricted within the target population. The loss of synaptic contacts is observed throughout the developing nervous system, from the nerve–muscle junction of invertebrates to the cerebral cortex of primates.

How is it possible to count the number of afferents per postsynaptic neuron? One imaginative approach to this problem uses intracellular recordings and electrical stimulation of the afferent pathway. The basic assumptions are that each axon will evoke a postsynaptic potential (PSP) when stimulated, and that the PSPs will summate linearly, in discrete steps, as each additional fiber is recruited by increasing the electric stimulus current (**Figure 9.2**A). Therefore, the increments in PSP size provide an estimate of the number of axons making a functional contact on a single muscle fiber or neuron. When this experiment is performed at the mature neuromuscular junction, a single large PSP is recorded, indicating that the muscle fiber is innervated by a single motor nerve terminal. However, when the same experiment is performed in neonatal animals, the PSP size increases in quantal steps as the stimulus activates two, then three, motor axons (Figure 9.2B). The elimination of convergent motor axons at the rat soleus muscle results in a decrease from three axons per muscle fiber

Box 9.1   Remaining Flexible: Adult Mechanisms of Learning and Memory—Cont'd

in the abdominal ganglion, and stimuli were delivered to both afferent pathways simultaneously. Following paired stimulation, one of the synapses produces much larger EPSPs, and this effect lasts for up to 40 minutes (Kandel and Tauc, 1965b). The increase was termed *heterosynaptic facilitation* because synaptic transmission at one set of synapses modified the functional status of a second, independent set.

One of the most compelling examples of synaptic plasticity, called *long-term potentiation* (LTP), was first identified in the early 1970s. By recording extracellularly from the hippocampus of anesthetized rabbits, it was found that a brief, high-frequency stimulus to the afferent pathway results in an enhancement of the evoked potential that lasted for hours to days (Bliss and Lømo, 1973). Over the next few years, intracellular recordings from mammalian brain slice preparations demonstrated that the size of EPSPs also increased following tetanic afferent stimulation. LTP is now thought to be one mechanism by which synapses store information because manipulations that block LTP are able to impair spatial learning in rodents.

The discovery of a cellular analog of learning suggested that the underlying molecular pathways could be characterized. One likely scenario for LTP in the hippocampus has glutamatergic transmission and postsynaptic depolarization combining to activate NMDARs, allowing calcium to flood the postsynaptic cell and activate calcium-dependent kinases such as CaMKII. NMDA receptor-dependent learning has also been demonstrated both in *Aplysia* (Murphy and Glanzman, 1999) and *Drosophila* (Xia et al., 2005). Thus, this molecular mechanism may be an evolutionarily conserved form of synaptic plasticity.

The influx of calcium activates one or more kinases which, in turn, phosphorylates proteins at the synapse. Although the specific proteins that are modified will vary between different areas of CNS, there is strong evidence that functional glutamate receptors are trafficked into the postsynaptic membrane, thus enhancing the response to the transmitter. A very simple form of learning in *Aplysia*, long-term facilitation of transmitter release, illustrates another important molecular pathway. An increase in presynaptic cAMP leads to the activation of a cAMP-dependent protein kinase (PKA). Once activated, the PKA subunit travels to the nucleus where it phosphorylates a transcription factor. The facilitated transmitter release involves new gene expression and protein synthesis (Kaang et al., 1993).

The cAMP signaling pathway seems to be a primary bridge to the formation of long-term memories in both fruit flies and mice. There are two cAMP-dependent transcription factors (CREB), one that activates gene expression and a second that represses it. Thus, when transgenic flies are bred to express the activator, they remember an odor with much less training. However, flies that express the repressor are unable to store long-term olfactory memories (Yin et al., 1994, 1995). The many experimental studies on CREB (in *Aplysia*, fly, mouse, and rat) established that the nucleus was involved in long-term memory formation. This fact present an interesting question: How is this nuclear signaling—common to all synapses of a given neuron—restricted to only those synapses that undergo structural and functional modifications? One possibility is that synapses undergoing LTP are "tagged" by a unique molecular event that serves to identify these synapses as the proper target for diffusely transported proteins. In fact, it has been suggested that CaMKII, which is activated at synapses undergoing LTP, serves this role (Frey and Morris, 1997; Redondo et al., 2010). The genetic approach to learning and memory clearly holds the promise of uniting cellular and behavioral findings, and it is likely that studies of developmental plasticity will profit as well.

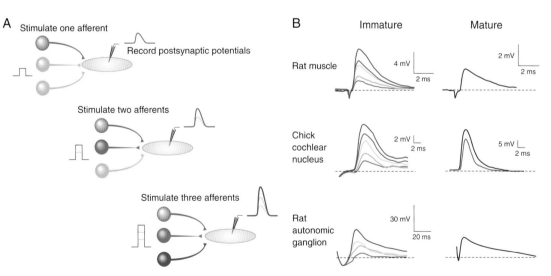

Fig. 9.2 Functional synapses are eliminated during development. A. An electrophysiological method for determining the number of inputs converging onto a neuron. A stimulating electrode is placed on the afferent population while an intracellular recording is obtained from the postsynaptic cell. As the stimulation current is increased, the afferent inputs are recruited to become active. When a single afferent is active (top), the postsynaptic potential (PSP) is small. When two (middle), and then three afferents are activated (bottom), the PSP become quantally larger. One can estimate the number of inputs by counting the number of quantal increases in PSP amplitude, in this case three. B. Three examples of decreased convergence as measured electrophysiologically. On the left are shown the increases in afferent-evoked PSP recorded in immature postsynaptic cells of the neuromuscular junction, the chick cochlear nucleus, and the rat autonomic ganglion. There are 3–5 quantal increases in PSP amplitude. On the right are shown the increases in afferent-evoked PSP amplitude in mature neurons. There are 1–2 quantal increases in PSP amplitude, indicating the functional elimination of inputs. *(Adapted from Jackson and Parks, 1982; Lichtman, 1977; O'Brien et al., 1978)*

to only one during the second postnatal week, and similar observations have been made in other species (Redfern, 1970; Bagust et al., 1973; Bennett and Pettigrew, 1974).

The precise time course over which synapse elimination occurs, and the number of connections that are lost, varies greatly between areas of the nervous system, even within a single species. In the rat cerebellum, the elimination of climbing fiber synapses onto Purkinje cells occurs during the second postnatal week (Mariani and Changeux, 1981). In contrast, the elimination of preganglionic synapses onto neurons of the rat submandibular ganglion occurs over at least five postnatal weeks (Lichtman, 1977), far longer than is required for elimination at the neuromuscular junction (Figure 9.2B). The number of cochlear nerve synapses on neurons of the chick cochlear nucleus declines rapidly, from about four to two afferents (Figure 9.2B), and reaches a mature state even before hatching (Jackson and Parks, 1982).

When a postsynaptic neuron receives a large number of inputs, it becomes difficult to resolve small differences in PSP size. However, functional estimates of synaptic convergence demonstrate that elimination occurs even in systems that remain multiply innervated in adulthood (Lichtman and Purves, 1980; Sanes, 1993; Kim and Kandler, 2003). Therefore, synapse elimination is a widespread phenomenon, although there are no general rules about the percentage of afferents that are lost or the duration of time required.

MANY AXONAL ARBORIZATIONS ARE ELIMINATED OR REFINED

When central neurons receive hundreds of contacts, as occurs in cortex, the PSPs are generally quite small. This makes it difficult to use steps in PSP amplitude as a measure of synapse number. An alternative approach is to count all of the synaptic contacts in a fixed volume of tissue, and determine whether a net reduction occurs with age (see Figure 8.6). Measures of synapse number taken from motor neurons suggest that 50%

of synaptic contacts are lost during development and similar estimates have been made for human cortex (Conradi and Ronnevi, 1975; Huttenlocher and de Courten, 1987).

Impressive as these changes are, there are two potential problems. First, they do not tell us whether the actual number of axons making synapses onto postsynaptic neurons change during development. For example, a single afferent could initially make 100 weak synapses on a postsynaptic neuron, and gradually transform to make 50 strong contacts. Second, *total* synapse number may increase in some systems during the time when a fraction of synapses are being eliminated, and this could go undetected.

If we assume that axons make synapses where they arborize, then it should be possible to study synapse elimination indirectly by measuring the territory occupied by individual axonal arbors. The most obvious case occurs when a projection is eliminated entirely due to the death of nerve cell bodies. For example, there is a projection from a chick brainstem nucleus, the isthmo-optic nucleus, to the retina in which nearly 60% of the projecting cells die, particularly those projecting to the wrong place in the retina (Casticas et al., 1987). The elimination of errant projections from the olfactory epithelium to the olfactory bulb may also result from the selective loss of sensory neurons (Zou et al., 2004). Furthermore, these errant projections remain into adulthood when one nostril is closed during development, suggesting that sensory experience is required for the formation of adult maps (Chapter 6).

In many cases, axonal arbors are eliminated in the absence of cell death. This was demonstrated for commissural axons that project from one side of the cerebral cortex to the other by labeling them twice during development with different retrograde tracers (Innocenti et al., 1977; O'Leary et al., 1981). Neurons that projected to the other hemisphere were first retrogradely labeled at birth by injecting a dye on one side of the brain and allowing commissural axons to transport it back to their cell bodies (**Figure 9.3**). Three weeks later, a second dye was injected in the same spot, and those commissural axons that remained retrogradely transported this second dye to their cell bodies.

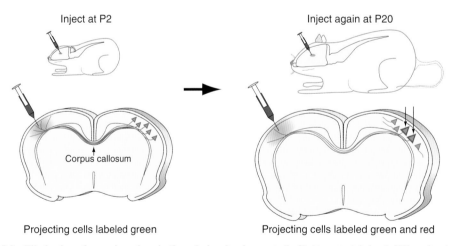

Inject at P2

Inject again at P20

Corpus callosum

Projecting cells labeled green

Projecting cells labeled green and red

Fig. 9.3 Elimination of commissural projections during development. (Left) At postnatal day 2 (P2), a dye (green) was injected into the cortex, and it was retrogradely transported through the corpus callosum to commissural neurons in the opposite cortex. (Right) A second (red) dye was injected at P20. When the tissue was processed, some commissural neurons were double labeled (green and red), whereas others contained only the green dye. The intepretation is that the green-labeled cells projected to the contralateral cortex at P2, but retracted their axons during postnatal development and maintained local connections. Cells labeled with both the green and red dyes retained their axonal projection to the opposite cortex. *(Adapted from O'Leary et al., 1981)*

When the tissue was examined, many cells were stained with the first dye only (Figure 9.3, green), but a fraction of these labeled cells also contained the second dye (Figure 9.3, red). The green-only cells must have sent axons through the commissure at birth, but apparently retracted their axons before the red dye was injected. Therefore, many cortical neurons generate transient projections through the cerebral commissure (corpus callosum), and some of these axons are eliminated even though the neurons survive.

A second system that displays massive axon withdrawal is the projection for motor cortex to spinal cord. For example, corticospinal axons initially project to the ipsilateral spinal cord, and these axons are largely eliminated during development (Stanfield, 1992). This period of refinement has been demonstrated in humans with a technique called transcranial magnetic stimulation (TMS), in which weak electric currents are induced in cortical neurons by a device that creates a magnetic field. When a specific region of motor cortex is activated by TMS, one can record the response that it produces in specific muscle groups. In newborn humans, a strong TMS-evoked response is observed in both ipsilateral and contralateral arm muscles. However, the ipsilateral response gradually wanes over the first 18 months, suggesting that the corticospinal projection to the ipsilateral spinal cord is largely eliminated. Furthermore, these ipsilateral responses persist in subjects with unilateral lesions of the corticospinal system acquired perinatally, suggesting that axon elimination is activity-dependent (Eyre, 2007).

The wholesale withdrawal of axons provides clear examples of developmental refinement, but this is not the norm. Changes in terminal arbor morphology are usually quite subtle. Axons tend to innervate the correct target region (Chapter 6), extend a bit beyond their correct topographic position, and then pull back slightly to the adult boundary. The best characterized examples of axon terminal elimination come from the developing visual pathway. In adult cats and primates, retinal ganglion cells from each eye project to separate layers in the lateral geniculate nucleus (LGN). The LGN neurons then project to Layer IV of the visual cortex, forming segregated eye-specific termination zones, called ocular dominance columns or "stripes" (**Figure 9.4**A).

It is possible to visualize the projection pattern of an entire eye by injecting ³H-proline into the eye cup. The amino acid is taken up by retinal ganglion cells and transported down their axons to the LGN where it crosses the synapse, enters the postsynaptic LGN neuron, and is carried by the LGN axons to the cortex. Autoradiographic images from the cortex show the termination pattern of LGN axons originating in one eye-specific layer (Figure 9.4B). In young animals, around the time of eye opening, LGN afferents arborize broadly in cortical Layer IV. Over the next several weeks, this diffuse label breaks up into discrete patches that represent eye-specific termination zones (LeVay et al., 1978; Crair et al., 2001). The light-evoked responses of Layer IV visual cortex neurons are consistent with the anatomy. By monitoring an intrinsic optical signal that is produced by electrically active brain tissue (see **Box 9.2**, "Watching Neurons Think"), it is possible to show that visual cortex responds uniformly to stimulation of one eye at postnatal day 8. However, the same stimulus to one eye begins to activate patches of cortex by postnatal day 14 (Crair et al., 2001). Individual LGN arbors from one eye retract a portion of their terminals during postnatal development,

consistent with the autoradiographic studies (Florence and Casagrande, 1990; Antonini and Stryker, 1993).

The development of retinal arbors in the LGN indicates that targeting errors are prevalent, yet subtle in appearance. In mammals, retinal axons grow to the correct area of the LGN from the outset, but they also make two kinds of inappropriate targeting projections. First, nearly all of the axons produce a few small collaterals, about 10 to 20 µm in length, in a part of the LGN that will eventually be innervated solely by axons from the other eye. Interestingly, the formation of an eye-specific innervation pattern in the LGN begins prior to visual experience. In the cat, these side branches are selectively eliminated in utero (**Figure 9.5**A and B).

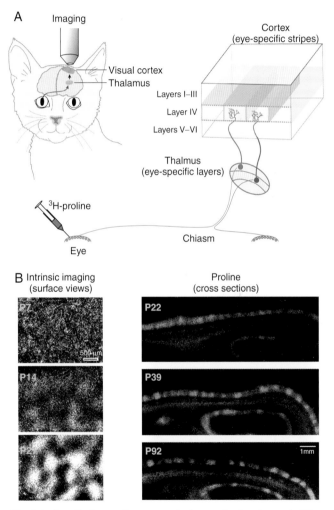

Fig. 9.4 Specific innervation patterns in the cat visual pathway. A. When ³H-proline is injected into one eye, it is transported to the visual thalamus (lateral geniculate nucleus, LGN), and reveals an eye-specific innervation pattern (contralateral eye, blue layers; ipsilateral eye, red layer). The proline moves transsynaptically into thalamic neurons and is transported to the cortex. LGN neurons project to eye-specific stripes in cortex Layer IV (contralateral eye, blue stripes; ipsilateral eye, red stripes). B. Images showing the emergence of eye-specific stripes in the visual cortex. The first set of images (left) show the activity pattern recorded at the surface of the cortex in response to stimulation of each eye (contralateral, black; ipsilateral, white). Segregation of afferent is first observed by about 14 days postnatal. The second set of images (right) show autoradiograms of ³H-proline in transverse section. The terminal fields continue to segregate for several weeks of development. *(Adapted from LeVay et al., 1978; Crair et al., 2001)*

Box 9.2   Watching Neurons Think: Functional Properties of Neuron Ensembles

A major goal of neurophysiology is to demonstrate a causal relationship between CNS function and animal behavior. While our success has been limited, there are a number of exciting strategies that should bring us closer to the goal. For almost a century, neurophysiologists have been recording electrical activity from the nervous system, at first from large populations of cells with scalp electrodes (cf. electroencephalograms) and eventually with small extracellular electrodes that monitor the action potentials from a single neuron. For those interested in sensory coding and perception, the extracellular electrodes are usually lowered into the brain of an anesthetized animal, and a neuron's activity is recorded while stimuli are delivered to the ears, eyes, or other receptor populations. In this way, we learn how environmental stimuli are converted into a neuronal code. For example, if an electrode is placed in the visual cortex and stimuli are delivered to each eye, we find that some neurons respond to bars of light that are vertically oriented, whereas others are driven best by horizontal bars. The neurophysiologist would call this "orientation selectivity," and such coding properties usually find their way into theories on the neural basis of visual perception. Therefore, single neuron recordings provide an extremely sensitive measure of whether synapse formation has occurred correctly during development.

Of course, it would be most compelling to record neural activity while the animal is actually perceiving a stimulus or moving a limb. In an early approach, animals were injected with a tritium-labeled sugar molecule, ^{3}H-2-deoxyglucose, that was taken up by nerve cells that were very active and required energy (Kennedy et al., 1975). It is now possible to measure neural activity and behavior simultaneously using several different techniques. Electrodes can be permanently mounted in the nervous system during an initial surgery, and these electrodes are then used to monitor neural activity when the animal recovers. Arrays of such electrodes have been used to record from the hippocampus of freely moving rats as they explore their environment and learn new tasks. It is also possible to stimulate or inactivate a region of the brain in awake-behaving animals, including humans during the course of neurosurgical treatment, and to monitor the effects on motor function

or sensory perception (Penfield and Rasmussen, 1950; Riquimaroux et al., 1991).

There are several ways to monitor brain activity in awake animals, including humans, that can be performed without exposing the brain. Although these techniques have not been applied widely to developing animals, they will probably play an important role in our future understanding of plasticity. One technique that offers <1 mm spatial resolution, called *functional magnetic resonance imaging* (fMRI), uses a very strong magnetic field (15,000 times the earth's magnetic field) to detect oxygen content. Since deoxyhemoglobin is paramagnetic relative to oxyhemoglobin and surrounding brain tissue, brain activity commonly produces a local increase in oxygen delivery. For example, it is possible to visualize activity in a single barrel field in rat somatosensory cortex (Yang et al., 1996). Another technique, magnetoencephalography (MEG), uses superconducting detectors to monitor the magnetic fields produced by a population of active neurons. This technique provides information about the timing, location, and magnitude of neural activity. For example, word-specific responses in the inferior temporo-occipital cortex are slow or absent in dyslexic individuals compared to control subjects, suggesting a specific neural impairment in this developmental disorder (Salmelin et al., 1996). Finally, there are changes in light absorbance that are well-correlated with neuronal activity, and it is possible to illuminate the surface of the brain and measure the reflected light while the system is processing information, referred to as differential optical imaging. By using these signals, many features of visual cortex development have been observed, including ocular dominance orientation selectivity (Blasdel et al., 1995).

Obviously, the challenge to find causality between brain function and behavior is magnified during development when nonspecific behavioral factors (cf. level of arousal) and the fragility of nerve cell function (cf. rapid fatigue) introduce great restraints. However, the study of immature brains, and the behaviors that they generate, should prove useful because it is the only means of correlating changes in the two without imposing surgery, drugs, or bad genes.

A second phase of refinement occurs postnatally when retinal terminals become more focused within a portion of the correct eye-specific layer (Figure 9.5C). Retinal ganglion cells that have small visual receptive fields (cf. X-cells) decrease the width of LGN tissue that they innervate by a small ($\approx$60 μm) but significant amount (Sur et al., 1984; Sretavan and Shatz, 1986). This would be the equivalent of pulling into your neighbor's driveway after completing a trip from a hundred miles away. Thus, immature projections are numerous but modest in size.

Are synapses actually being formed by these transient projections? The answer can be found by first filling entire axons with a tracer, such as horseradish peroxidase, and then examining the terminals with an electron microscope. In fact, labeled presynaptic terminals have been found in parts of the target where they never remain in the adult, indicating that these structurally mature synapses will eventually be removed (Reh and Constantine-Paton, 1984; Campbell and Shatz, 1992).

A particularly compelling example of synapse elimination occurs at the neuromuscular synapses, where the withdrawal of a motor axon can be observed in living animals (**Figure 9.6**).

Two strains of mice are genetically engineered to express unique fluorescent proteins in their motor neurons. When the two strains are mated, muscle cell fibers can be identified that are co-innervated by motor terminals that contain one or the other fluorescent proteins (Walsh and Lichtman, 2003). By imaging the muscle fiber in vivo over successive days, one can observe the withdrawal of one synapse and the enlargement of the other to occupy the entire motor endplate.

Axonal refinement or the loss of synaptic connections occurs in a great variety of species and neural structures. There is significant remodeling of retinal axons during topographic map formation in the visual midbrain (Reh and Constantine-Paton, 1984; Simon and O'Leary, 1992). In frogs, single retinal afferents innervate successively posterior locations within the tectum as development proceeds. In mammals, retinal axons innervate topographically incorrect positions at birth, and these terminals are gradually withdrawn over two weeks. Synapse elimination is also found in invertebrates. In the cricket, single sensory neurons are functionally connected with two different interneurons during an early stage of development. Gradually, the strength of one connection doubles while the other connection is completely eliminated

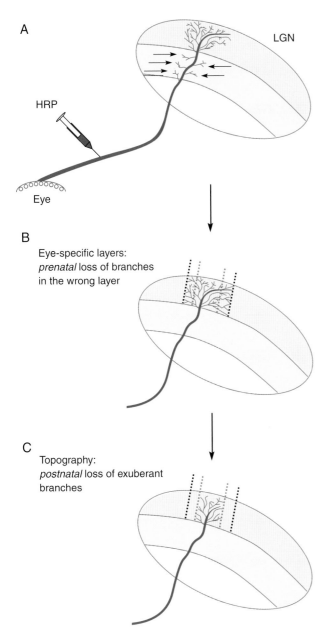

Fig. 9.5 Development of retinal ganglion cell terminals in the cat lateral geniculate nucleus (LGN). A. When individual retinal fibers are labeled with horseradish peroxidase (HRP) at embryonic days 43–55, they have many side branches in the inappropriate layer of LGN (arrows). B. By birth, most of the side branches have been eliminated, and terminal arborizations have been restricted to the correct layer. However, the terminal zone remained wider in the eye-specific lamina at 3–4 weeks postnatal (dotted lines). C. When fibers of retinal X-cells were filled in adult cats, they were found to have retracted (dotted lines). *(Adapted from Sur et al., 1984; Sretavan and Shatz, 1986)*

(Chiba et al., 1988). The refinement of neural connections is not confined to excitatory terminals; inhibitory afferents also undergo synapse elimination during development (see *Plasticity of Inhibitory Connections, page 277*).

There are afferent pathways where the refinement of axonal projections is either scarce or absent. For example, retinal ganglion cells with large visual receptive fields (Y-cells) have axonal arbors that expand within the LGN during development (Sur et al., 1984; Florence and Casagrande, 1990), in contrast

to the retinal arbors described above (Figure 9.5). Similarly, axons converging on a chick brainstem structure, the *nucleus laminaris*, spread out along the tonotopic axis during the embryonic period, suggesting that synapses are being added (Young and Rubel, 1986). The connections from sensory axons to motor neurons also appear to be accurate from the outset. Inappropriate monosynaptic connections from sensory axons to spinal motor neurons are not found with intracellular recordings, although there may be significant immaturities within the polysynaptic pathways (Seebach and Ziskind-Conhaim, 1994; Mears and Frank, 1997). While these examples suggest that axonal projections are not refined, it is possible that individual synaptic contacts are eliminated while new synapses are added, leaving no net change in the projection. This would be impossible to determine without following the life cycle of each synaptic connection.

THE SENSORY ENVIRONMENT INFLUENCES SYNAPTIC CONNECTIONS

We encounter use-dependent changes of our nervous system on a daily basis. Sensory information is processed by our nervous system, and stored as memories (see Box 9.1, "Remaining Flexible"). Our eye-hand coordination improves gradually as we practice with a newly purchased kitchen utensil. When the level of sound or light changes abruptly, our eyes, ears, and nervous system can adjust to maintain a certain fidelity of response. Similarly, our rearing environment can influence many aspects of our adult behavior (see Chapter 10). To take an obvious example, we speak and comprehend the language to which we were exposed as infants, whether it was Hindi, American Sign Language, or Spanish. But how much is the developing nervous system actually altered by the environment? Is synapse formation or elimination influenced directly by sensory experience?

An early approach to this problem, and one that is still regularly employed, involves the elimination of sensory structures. Denervation studies demonstrate how neuron growth and survival depend on intact connections during development (Chapter 7). The next experimental step was to change nervous system activity to find out whether improperly used synapses became weak or lost entirely. With this goal in mind, Wiesel and Hubel (1963a, 1965) began to explore the effects of monocular and binocular deprivation on the development of visual coding properties in the CNS. Their results demonstrated convincingly that visual experience during development influences the maintenance or elimination of neural connections.

Before we can understand the functional changes brought about by visual deprivation, it is worthwhile reviewing a few basic properties of the visual cortex. As described above, each neuron in Layer IV of the visual cortex receives projections that are largely driven by one eye or the other (Figure 9.4A). When visual stimuli are delivered to the appropriate eye, Layer IV neurons respond with a burst of action potentials. Cortical neurons that respond to only one eye are referred to as monocular. However, the majority of cortical neurons, particularly those located outside of Layer IV, are activated by both eyes, and are referred to as binocular. Thus, when an extracellular electrode is lowered through the visual cortex to record from many neurons in succession, most cells are found to be binocular (i.e., most

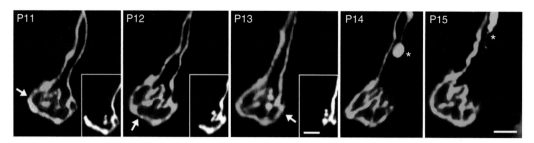

Fig. 9.6 In vivo imaging of the same multiple innervated neuromuscular junctions in a neonatal mouse. Two transgenic mouse lines, each expressing fluorescent proteins in their motor neurons, were mated, and the progeny examined. Images were obtained from the same sternomastoid muscle fiber over the course of several days. In this example, one of the motor terminals (blue and insets) occupies a larger percentage (70%) of the postsynaptic territory at P11. It gradually withdraws from the junction (arrows) over the next four days. The second motor terminal (green) gradually takes over the entire postsynaptic site. The withdrawing axon is marked by an asterisk at P14 and 15. Scale bars are 10 μm. *(From Walsh and Lichtman, 2003)*

neurons are recorded outside of Layer IV). Hubel and Wiesel (1962) divided the cortex neurons into seven groups, based on the relative ability of each eye to evoke a response. For example, if a neuron was driven solely by the contralateral eye, then it was assigned to group one. If it was driven equivalently by each eye, then it was assigned to group four, and so forth. These data can be conveniently represented as a histogram of ocular dominance (**Figure 9.7**). Judged by its continued use during the past 50 years, ocular dominance histograms provide a reliable measure of the innervation pattern in the visual cortex.

To examine the role of early visual experience, light-evoked activity was decreased by keeping the eyelid closed, referred to as visual deprivation (Wiesel and Hubel, 1963a, 1963b, 1965). This manipulation does not damage the retina, and LGN neurons remain responsive to visual stimulation after the eyelid is reopened. In an initial experiment, a single eyelid was kept closed for a few months, and recordings were made from the visual cortex after the eye was reopened. The effect was unmistakable: Most cortical neurons no longer respond to visual stimulation

of the eye that is deprived during development (**Figure 9.8**), although they continue to respond to the unmanipulated eye.

Reasoning that disuse must have weakened the synapses from the deprived eye, Wiesel and Hubel recorded from animals that were deprived of visual through both eyes during development, expecting to see a total absence of visually-evoked activity. It came as a great surprise, then, that most cortical neurons remain responsive to stimuli through both deprived eyes. That is, the ocular dominance histogram obtained from binocularly deprived animals resembles that of normal animals (**Figure 9.9**). Wiesel and Hubel wrote: "It was as if the expected ill effects from closing one eye had been averted by closing the other." However, a large fraction of neurons are completely unresponsive to light in binocularly deprived animals, somewhat consistent with the original prediction (Sherman and Spear, 1982).

The total amount of evoked activity does not necessarily determine whether a visual pathway will be strong or weak. Rather, differences in the amount of activity in the pathway from one eye, relative to the other, seems to determine the

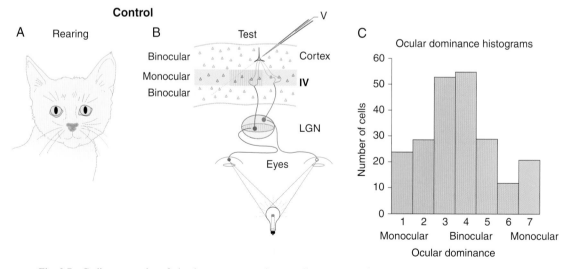

Fig. 9.7 Coding properties of visual cortex neurons in normally reared cats. A. The visual system received normal stimulation until the time of recording. B. Single neuron recordings were made with an extracellular electrode that passed tangentially through the cortex. Neurons respond to only one eye in Layer IV (monocular). In Layers I–III and V–VI, neurons respond to both eyes (binocular) due to convergent connections. In normal cats, the terminal stripes from each eye-specific layer of the LGN occupy a similar amount of space. C. Each neuron was characterized as responding to a single eye (monocular), or responding to both eyes (binocular). In normal adult cats most visual cortical neurons are binocular. *(Adapted from Hubel and Wiesel, 1962)*

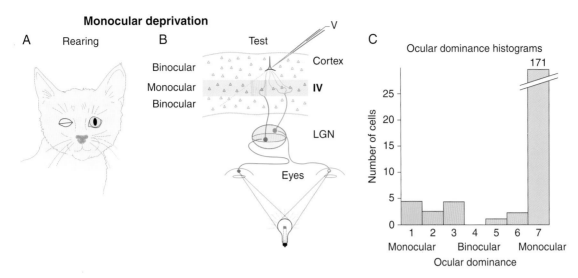

Fig. 9.8 Coding properties of visual cortex neurons in monocularly deprived cats. A. The visual system received normal stimulation through one eye, and the other eye was kept closed until the time of recording. B. The terminal stripes from the deprived eye become much narrower, and the visual response of neurons outside of Layer IV is more responsive to the open eye. C. In monocularly deprived cats, the vast majority of cortical neurons respond to the open eye only. *(Adapted from Wiesel and Hubel, 1963a)*

strength of the projection. This idea came to be known as the *competition hypothesis*. Under this proposal, retinal synapses in the LGN should not be affected by deprivation because LGN neurons are monocular and receive afferents that are either uniformly active or uniformly deprived of light. Similarly, binocular deprivation evens the playing field in the cortex because all afferents should have a similar low level of activity. Monocular deprivation creates a situation in which cortical neurons receive a set of active afferents from the open eye and a set of afferents with lowered activity from the closed eye, placing the latter at a disadvantage.

In a pivotal test of the competition hypothesis, kittens were raised with an artificial strabismus (cf. misalignment of the eyes), produced by surgically manipulating one of the extraocular muscles (Hubel and Wiesel, 1965). This manipulation mimics a clinical condition in humans, called *amblyopia*,

which commonly results in the suppression of vision through one of the eyes. In strabismic kittens, visual stimuli activate different positions on the two retinas, and cortical neurons are rarely activated by both eyes at the same time. Following several months of strabismus, recordings were once again made from the visual cortex. Following this procedure, both eyes effectively activate neurons in the cortex, but most cortical neurons respond to stimulation of one eye or the other (**Figure 9.10**). Few binocular neurons were observed. Therefore, an equivalent *amount* of activity in the two pathways is not sufficient to explain the results. Instead, it seems that the *timing* of synaptic activity must somehow be involved in allowing inputs to remain connected functionally to cortical neurons. Synapses from each eye must be active at nearly the same instant if both are to keep strong, functional contacts with the same postsynaptic neuron.

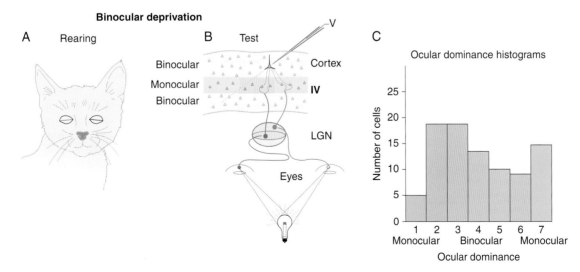

Fig. 9.9 Coding properties of visual cortex neurons in binocularly deprived cats. A. Both eyes were kept closed until time of recording. B. The terminal stripes from each eye occupy a similar amount of space. C. The majority of visually responsive neurons are driven by both eyes. *(Adapted from Wiesel and Hubel, 1965)*

257

Strabismus

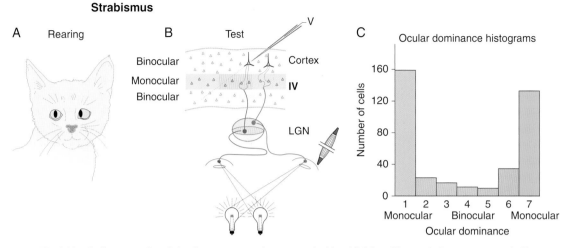

Fig. 9.10 Coding properties of visual cortex neurons in cats reared with artificial strabismus. A. One eye was surgically deflected, such that a visual stimulus activated different topographic positions on each eye. Thus, cortical neurons were not activated by both eyes at the same time. B. The terminal stripes from each eye occupy a similar amount of space. However, neurons outside of Layer IV do not receive convergent input from both eyes. C. In strabismic cats, the vast majority of cortical neurons respond to either one eye or the other. *(Adapted from Hubel and Wiesel, 1965)*

The ocular dominance columns (stripes) formed by the axon terminals of geniculate fibers have served as a valuable anatomical assay of synapse elimination. In normal animals, when ³H-proline is injected into one eye, there is a periodic variation in silver grain density in Layer IV of the cortex. Labeled and nonlabeled regions are about 500 μm wide (Figure 9.4). Following monocular deprivation, the LGN afferents from the nondeprived eye come to occupy the majority of Layer IV, while LGN afferents from the deprived eye occupy narrower regions (Figure 9.8B). This suggests that synapses from the nondeprived eye fail to undergo their normal process of elimination. In contrast, a greater than normal number of synapses from the deprived eye must be lost.

The emergence of stripes does occur in cat cortex during binocular deprivation over the first three postnatal weeks. However, when visual deprivation extends into the fourth postnatal week, the striping pattern begins to deteriorate (Crair et al., 1998). Pattern vision is unnecessary for the segregation of thalamic afferents into stripes, but maintenance of this striping pattern does require normal visual experience. Surprisingly, thalamic afferents can segregate into a striping pattern even when one or both eyes are removed (Crowley and Katz, 1999, 2000). One possible reason for this result is that spontaneous activity persists within corticothalamic circuit, independent of retinal input, and this has an impact on synapse development (discussed below).

The majority of binocular neurons in the cortex are created by local projections from one cortical neuron to its neighbors, and these projections also become refined during development. Since neurons in all layers of the cortex are monocular following strabismus, it would be interesting to know what happens to these intracortical projections. Do they now become more segregated than normal, extending the striped pattern throughout the entire cortical depth? This seems to be precisely what happens. Small injections of dye were made in the cortex, retrogradely labeling neurons that make local projections to this area (**Figure 9.11**). The animals were also injected with 2-deoxyglucose (see Box 9.2, "Watching Neurons Think") and stimulated through one eye to label the regions of cortex that were driven by that eye. With this double-labeling technique, one can learn whether

local projection neurons are found exclusively above one ocular dominance column or both. In normal animals, the local projections arise from both columns, and the cells that they innervate are binocularly driven. In strabismic cats, the local projections

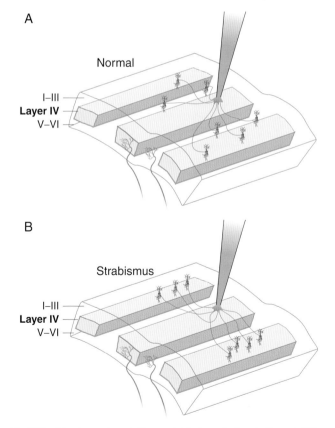

Fig. 9.11 Visual experience influences intrinsic cortical projections. A. When dye injections are made into superficial layers of the visual cortex of normal cats, the label is retrogradely transported by neurons in both ocular dominance columns. B. In cats reared with artificial strabismus, the label is retrogradely transported only by neurons that share the same ocular dominance. *(Adapted from Löwel and Singer, 1992)*

originate almost exclusively from above one column (Löwel and Singer, 1992). Furthermore, this alteration of horizontal projections occurs very rapidly; after only two days of strabismus, one can detect the loss of horizontal projections (Trachtenberg and Stryker, 2001). Thus, activity influences the development of synaptic connections not only in ascending sensory projections, but also in many of the intracortical projections.

A direct measure of synaptic pruning was achieved by imaging individual retinal ganglion cell (RGC) arbors as they compete for postsynaptic space within the frog optic tectum (Ruthazer et al., 2003). When RGC axons from both eyes are induced to innervate the same tectal lobe, the axons gradually segregate from one another (**Figure 9.12**A). To visualize this process, a single ipsilateral RGC was labeled with a fluorescent

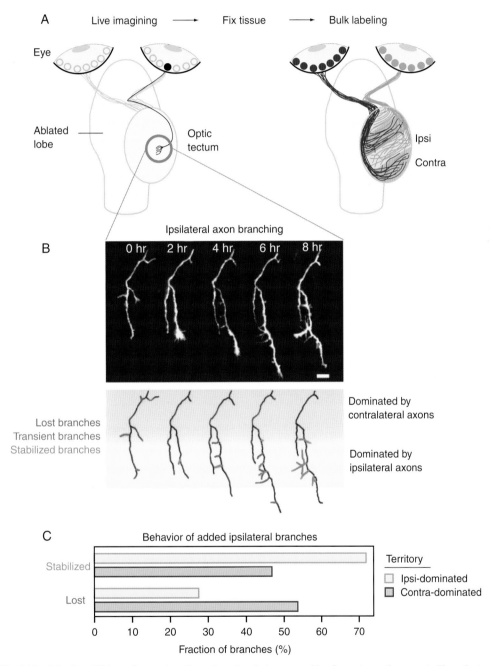

Fig. 9.12 Selective addition and retraction of axon branches during competition for postsynaptic space. A. The retinotectal projection in *Xenopus* tadpoles is normally to the contralateral tectum, but ablation of one tectal lobe causes RGC axons from the ipsilateral eye to innervate the spared tectum. A single RGC axon is labeled with a fluorescent dye and imaged over an eight-hour period. Following this, the brain is fixed, and the projection from each eye is completely labeled with two different dyes (purple and yellow) such that the innervation pattern in the tectum is evident. B. Time-lapse images (top) and drawings (bottom) of a single ipsilateral eye axon in a living tadpole are shown. Branches are color coded for their behavior during the recording period: lost branches (green), transient branches (red), and stabilized branches (blue). In this case, the largest number of stabilized branches form in the region dominated by ipsilateral axons. C. New ipsilateral axon branches are more likely to remain for the entire recording period (i.e., stabilized) if they are formed in territory dominated by ipsilateral axons. In contrast, they are preferentially eliminated from the contralateral territory. *(From Ruthazer et al., 2003)*

dye during the period of segregation, and the axon terminal was imaged over eight hours. The addition and retraction of each axonal branch was followed during this interval (Figure 9.12B). The brain was then fixed, and RGC projections from each eye were bulk-labeled with two different dyes. The procedure permitted one to characterize the innervation density of each eye, a measure that is analogous to ocular dominance columns. New branches that persist for the entire recording period, called *stabilized branches*, are more numerous when formed in a tectal region dominated by the same eye (Figure 9.12C). Conversely, RGC axons preferentially retract branches from territory that is already dominated by the other eye. This mechanism is prevented by the presence of an antagonist to a class of glutamate receptors called NMDA receptors (see below). Thus, branch addition and retraction (and presumably synapse formation and elimination) can occur simultaneously and depend on synaptic transmission.

The central concept to emerge from these studies is that coactive synapses are stabilized, while inactive synapses, particularly those that are inactive while others are firing, become weakened and in many cases are eliminated. As one might expect, the development of a complicated structure such as the cortex is unlikely to be explained by one tidy hypothesis. We have just learned that activity-dependent changes in connectivity are occurring simultaneously at several locations. Functional and structural changes are also found in the LGN following eyelid suture or strabismus, and the extent to which these changes influence cortical development is not clear. Proprioceptive feedback from the eye muscles also contributes, and its blockade somehow prevents monocular deprivation from altering synaptic connections in the cortex. This is not to lose sight of the forest, but rather to say that there are some large trees that remain to be carefully examined.

ACTIVITY INFLUENCES SYNAPSE ELIMINATION AT THE NMJ

Given the great complexity of cortex circuitry, it would be nice to have a simpler model system in which to explore synaptic plasticity. As we discovered in the last chapter, the neuromuscular junction (NMJ) is the most accessible and well-studied of all synapses, and it has served as an important model of synaptic plasticity for decades. In mammals, there is only a single type of synapse on fast muscle fibers, and only a single fiber ends up innervating each muscle cell in adults (Figure 9.2). There are also many technical advantages: it is easy to record intracellularly from muscle cells, and to manipulate neurons or muscle cells in a tissue culture system. Of course, these advantages also function as limitations. For example, there are no inhibitory neuromuscular synapses in vertebrates, and the postsynaptic cell does not have dendrites or spines, as found in many areas of the central nervous system.

Despite these differences, it is surprising how much NMJ developmental plasticity resembles that observed in the cortex. As we learned earlier, mammalian muscle cells are innervated by more than one axon at birth, but after synapse elimination only a single axon remains. When action potentials are blocked with TTX during the normal period of synapse elimination, muscle cells remain polyneuronally innervated (**Figure 9.13**) (Thompson et al., 1979). Similar results are obtained in the developing cat visual pathway, where TTX

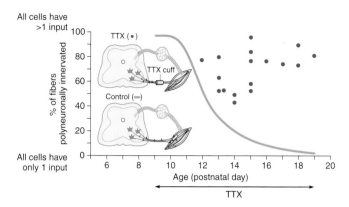

Fig. 9.13 Decreased activity prevents synapse elimination. At the rat nerve–muscle junction, polyneuronal innervation declines between 10 and 15 days postnatal (green line). When a TTX cuff is placed around the motor nerve root from postnatal day 9–19 and action potentials are eliminated, polyneuronal innervation remains high (red circles). *(Adapted from Thompson et al., 1979)*

blocks the segregation of retinal afferent in eye-specific layers of the LGN, and also the segregation of LGN afferents in the cortex. To test whether the temporal pattern of activity is important (as suggested by the strabismus results), two sets of motor axons innervating the same muscle were stimulated in synchrony (Busetto et al., 2000). This manipulation preserves polyneuronal innervation of muscle fibers.

Given that too many synapses remain when activity is blocked or synchronized, one might predict that unsynchronized postsynaptic activity could speed up the process of synapse elimination. In fact, direct electrical stimulation of the muscle induces the early loss of motor synapses (**Figure 9.14**). This result is particularly intriguing because it suggests that *postsynaptic* electrical activity can determine whether presynaptic terminals survive (O'Brien et al., 1978). Together, these experiments show that synaptic transmission can influence the process of synapse elimination at the NMJ. In fact, plasticity at the NMJ is a lifelong matter. Adult motor terminals will sprout to innervate adjacent muscle cells when neuromuscular transmission is blocked, and this polyneuronal innervation can be reduced when the muscle cells are stimulated directly (Jansen et al., 1973; Holland and Brown, 1980).

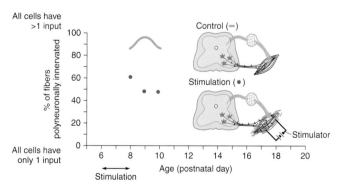

Fig. 9.14 Increased muscle activity accelerates synapse elimination. Most rat soleus muscles are polyneuronally innervated between postnatal days 8–10 (green line). When a stimulating electrode is implanted in the leg to activate the sciatic nerve and muscle from postnatal days 6–8, there is a severe decline in the number of polyneuronally innervated muscle cells (red circles). *(Adapted from O'Brien et al., 1978)*

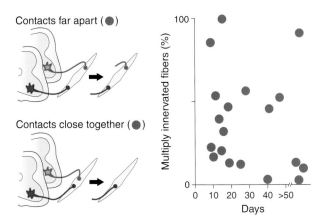

Fig. 9.15 Synapse elimination depends on distance between contacts. Two motor axons were positioned on the same muscle, either close to one another or at a distance. Intracellular recordings were made from muscle fibers to monitor polyneuronal innervation, as shown in Figure 9.2. When the synapses are close together (blue circles), synaptic elimination occurs within a few weeks. When the synapses are far apart (red circles), synaptic elimination fails to occur. *(Adapted from Kuffler et al., 1977)*

A synapse can only influence its neighbors within a short distance. This can be demonstrated by grafting two motor nerve terminals onto the same muscle in adult rats. Both nerves maintain a functional connection to the muscle fiber over several months, even though muscles are normally innervated by a single axon. The trick is to place the two motor terminals several millimeters from one another (Kuffler et al., 1977). If the terminals are placed within a millimeter or two, then one of the contacts is eliminated within about three weeks (**Figure 9.15**). In fact, some animals have muscle fibers that are normally innervated by more than one motor axon. In one polyneuronally innervated muscle in chicks, the distance between terminals can be reduced when synaptic transmission is blocked, presumably because competition between active terminals normally keeps them separated (Gordon et al., 1974). This may explain why terminal endings from a single axonal arbor often innervate a continuous region of the postsynaptic cell (Forehand and Purves, 1984; Glanzman et al., 1991). For example, when sensory neurons from the sea slug, *Aplysia*, are grown in culture along with a common target motor neuron, their terminals come to occupy separate regions of the postsynaptic cell. However, if the two sensory neurons are grown without a target, then they grow extensively along one another. Therefore, the antagonistic relationship between synapses must somehow be mediated by the postsynaptic neuron. Even if we know the amount and pattern of synaptic activity in a set of inputs, this may not be enough to predict whether one synapse will dislodge a second one. One must also know the spatial arrangement of these terminals.

SYNAPSE REFINEMENT IS REFLECTED IN SENSORY CODING PROPERTIES

How does the refinement of synaptic connections impact the computational properties of the nervous system? To answer this question, we will shift our attention to experiments that explore how individual neurons encode the sensory environment. The basic concept is that sensory neurons usually respond to a very specific range of stimuli. For example, a visual cortex neuron may fire the greatest number of action potentials to a vertically oriented bar moving to the right, and produce no responses if the bar moves to the left. An auditory midbrain neuron might display the largest discharge rate to a 4000 Hz tone located left of the midline, but fail to respond to a 1000 Hz tone. In each case, the neuron's response is limited to specific stimuli (called the neuron's *coding properties*), and this specificity reflects the number and strength of its many synaptic inputs. Thus, neuron coding properties serve as a sensitive assay of synaptic refinement.

To determine whether sensory experience influences the development of sensory coding properties, animals are reared in an abnormal sensory environment that alters the pattern of neural activity. This has been addressed in the central auditory system by experiments that used repetitive stimulation with sound. Central auditory neurons usually respond to a limited range of frequencies because the auditory nerve fibers from the cochlea project topographically in the central nervous system (cf. tonotopy). A plot of sound frequency versus the intensity at which each frequency evokes a threshold response from a neuron, called a *frequency tuning curve*, provides a good measure of afferent innervation. When many areas of the cochlea project to a central neuron, then its frequency tuning curve is broad. When only a small region of the cochlea projects to a neuron, then its frequency tuning curve is narrow.

To test whether the timing of neural activity influences the development of frequency coding, mice were reared in a sound environment consisting of repetitive clicks for a few weeks (Sanes and Constantine-Paton, 1985b). This type of sound evokes synchronous activity in a large population of cochlear nerve axons (**Figure 9.16**A and B). When frequency tuning curves were obtained from the inferior colliculus of normal mice and compared to those reared in repetitive clicks, the latter group had significantly broader tuning curves (Figure 9.16C). A similar experiment has been performed in rats using pulses of noise, and the normal sharpening of tuning curves fails to emerge (Figure 9.16D). Noise pulse-rearing has no effect on tuning curves from mice older than 30 days, indicating a specific developmental effect (Zhang et al., 2002a). Therefore, when afferents all have the identical pattern of activity, they are apparently unable to segregate properly along the frequency axis.

The acoustic rearing environment can also affect more complex auditory coding properties, including those that are relevant to animal communication, such as frequency modulation. Ordinarily, central auditory neurons respond to increasing or decreasing sound frequency. However, when juvenile rats were reared with frequency-modulated sounds that only swept from high to low frequency, auditory cortex neurons were subsequently found to be more selective for this direction of modulation (Insanally et al., 2009). Since bats emit frequency modulated sounds to navigate through space by echo-location, the role of early experience with these signals was tested by preventing juvenile bats from producing these sounds and rearing them in isolation. In the absence of this early experience, cortical neurons are less selective for either the direction or the rate of frequency modulation (Razak et al., 2008).

Neurons in the visual pathway display a similar malleability in response to the rearing environment. For example, many cortical neurons respond to stimuli of a specific orientation,

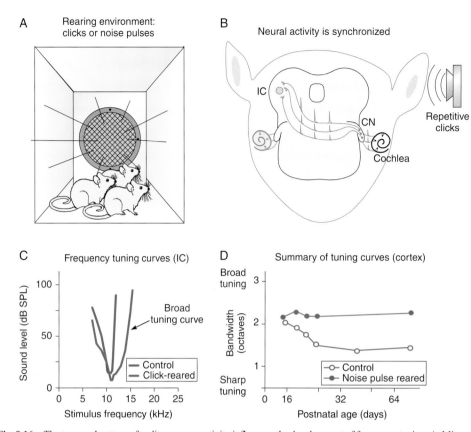

Fig. 9.16 The temporal pattern of auditory nerve activity influences the development of frequency tuning. A. Mice or rats were reared with repetitive broadband stimuli (clicks or noise pulses). B. Stimuli of this sort activate many hair cells in the cochlea and synchronize the discharge pattern of auditory nerve fibers. C. Single neurons were recorded in the mouse inferior colliculus (IC), and those from click-reared animals respond to a broader range of frequencies compared to those from controls. D. Recordings made in the rat auditory cortex demonstrate that broader frequency tuning properties are maintained during development in noise pulse-reared animals. A larger bandwidth, plotted in octaves, indicates that the neurons respond to a greater range of frequencies. *(Adapted from Sanes and Constantine-Paton, 1985; Zhang et al., 2002a)*

or to stimuli that move in a specific direction. However, when kittens are reared in a visual environment consisting entirely of vertical or horizontal stripes, and neurons are later recorded in the cortex, the majority of them respond selectively to the orientation that was present in the rearing environment (**Figure 9.17**A). To determine whether moving visual stimuli also influence the development of motion selectivity, kittens were raised in stroboscopic light. As nightclub enthusiasts know, this stimulus permits one to see a full range of shapes and colors, but it eliminates smooth motion. When cortical neurons were recorded after a period of strobe-rearing, the majority no longer respond selectively to the direction of movement (Figure 9.17B). As with ocular dominance columns, sensory experience appears to alter normal development, rather than shape the adult pattern. For example, orientation columns form normally in the kitten visual cortex, even in the absence of visual experience (Crair et al., 1998).

In contrast to ocular dominance and orientation coding properties, the selective response of visual cortex neurons to the direction of motion emerges after eye opening, and requires visual experience to mature properly (Li et al., 2006). This was demonstrated with a technique called *intrinsic signal imaging*, in which neural activity is monitored optically by detecting changes in light reflectance. At a light wavelength of 610 nm, the absorbance of oxyhemoglobin is very

low, while the absorbance of deoxyhemoglobin is relatively high. Therefore, images of cortex are acquired at 610 nm, and comparisons are made at rest and during stimulation. **Figure 9.18**A shows the results of one such experiment in which the activity of visual cortex was measured in response to stimuli that varied either in orientation or direction of motion. Regions of cortex that respond to a similar orientation or motion are represented as the same shade of gray. Stimulation with oriented bars evokes specific responses in ferret visual cortex by postnatal day 30, about the time of eye opening, and response selectivity increases over the next few days (images in blue boxes). Although visual deprivation (dark-rearing) results in diminished orientation selectivity, it is nonetheless present. In contrast, stimulation with moving bars does not elicit specific responses at the time of eye opening, and visual deprivation results in a complete absence of direction selectivity (images in red boxes). The developmental emergence of each coding property, and the effect of visual deprivation, are illustrated graphically (Figure 9.18A). The effect of experience can also be demonstrated by providing supplemental stimulation (Li et al., 2008). The emergence of direction selectivity is induced within several hours by presenting moving visual stimuli along one axis (Figure 9.18B). Together, these experiments demonstrate that environmentally driven activity is required for the full development of some coding properties.

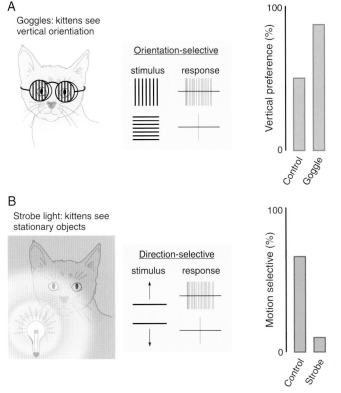

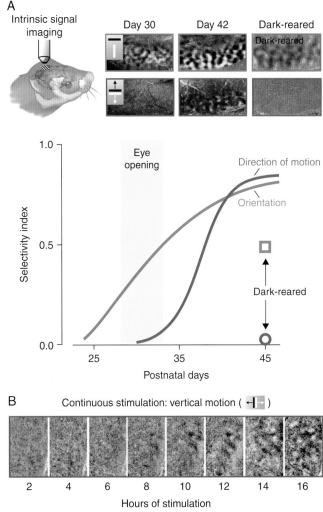

Fig. 9.17 The visual environment influences orientation and motion processing. A. Kittens were reared with goggles that permitted visual experience with only vertically oriented stimuli. The response to oriented stimuli was then obtained from single visual cortex neurons. The cell shown responds best to vertical stimuli. In goggle-reared cats, 87% of neurons are selective for vertical stimuli, compared to only 50% in controls. B. Kittens were reared in stroboscopic light that permitted visual experience with only stationary objects. The response to moving stimuli was then obtained from single visual cortex neurons. The cell shown responds best to stimuli moving down and to the right. In strobe-reared cats, only 10% of neurons are motion selective, compared to 66% in controls. *(Adapted from Stryker et al., 1978; Pasternak et al., 1985)*

These results, and many others like them, show that environmental stimuli influence a broad range of functional properties. Although this discussion has focused on the cortex, sensory experience may influence sensory coding properties at the most peripheral level of the nervous system. For example, mouse retinal ganglion cells (RGC) initially respond to both an increase and decrease in luminescence, called an *ON-OFF response*. During postnatal development, the RGC dendritic arbor is refined, and most neurons end up producing either ON or OFF responses to visual stimulation, but not both. When mice are reared in the dark, the RGC dendrites are not refined, and the cells continue to produce ON-OFF responses (Tian and Copenhagen, 2003). Therefore, environmental rearing studies may influence maturation at many levels, and the cumulative effect is assessed in the cortex.

ACTIVITY CONTRIBUTES TO TOPOGRAPHY AND THE ALIGNMENT OF MAPS

When peripheral sensory axons reach the central nervous system, they usually innervate the target in an orderly manner, forming topographic maps that are quite accurate due

Fig. 9.18 The selective response of visual cortex neurons to direction of motion emerges after eye opening, and requires visual experience. A. Stimulus-evoked cortical activity is obtained by monitoring changes in the reflectance of light by the cortex (left). The response of visual cortex to oriented stimuli (top row of images with blue borders) is shown at two ages, and in a visually deprived animal. There are already specific responses to vertical (white regions) and horizontal objects (black regions) at the time of eye opening, postnatal day 30. Specific response to orientation can emerge in the absence of visual experience (dark-rearing). The response of visual cortex to moving stimuli (bottom row of images with red borders) is also shown at two ages, and in a visually deprived animal. There is little response specificity to vertical motion at the time of eye opening, and it emerges over the next several days. Furthermore, visual deprivation prevents the emergence of direction-specific responses. The graph shows the average amount of response selectivity for orientation and direction of motion during development. Orientation selectivity begins to emerge prior to eye opening, and direction selectivity emerges afterwards. Orientation selectivity is intact, but somewhat reduced in dark-reared animals (blue square), while direction selectivity is abolished (red circle). B. Direction selectivity was monitored in the visual cortex of ferrets that had less than 1 day of visual experience, and moving visuals were presented over the course of several hours. As illustrated by this example, there is little evidence of direction selectivity in the first 8 hours, but strong selectivity emerges after 10–12 hours of stimulation. *(Adapted from Li et al., 2006; Li et al., 2008)*

to molecular gradients and activity-dependent mechanisms (Chapter 6). The influence of environmental activity is most evident when two maps must become aligned with one another in the same structure. For example, binocular neurons in the frog optic tectum respond to the same visual position in space when activated through each eye. Tectal neurons receive a direct projection from the contralateral retina and an indirect projection that is driven by the ipsilateral eye (**Figure 9.19**A, left panel). Tectal neurons project to a structure called *nucleus isthmus*, and isthmal neurons project to the contralateral tectal lobe. Therefore, there are two well-aligned maps of visual space in the tectum, one from the contralateral eye and one from the ipsilateral eye.

To test whether visual activity plays a role in this precise alignment, frogs were reared in the dark. Direct retinal projections from the contralateral eye continue to form a precise map, but the indirect projection via the nucleus isthmus is poorly organized (Keating and Feldman, 1975). That is, the formation of the contralateral retinotopic map depends largely on molecular cues, while the ipsilateral map depends on activity. The effect is so powerful that one can actually cause isthmal axons to innervate a new position in the tectum when

the contralateral map is disrupted by rotating the contralateral eye 180° (Figure 9.19A, center panel). Following rotation, a single point in space activates different positions in the tectum via each eye. When this manipulation is performed while the animals are still tadpoles, the isthmal projection to the tectum shifts such that the ipsilateral retinal map comes into register with the direct contralateral projection (Figure 9.19A, right panel). Thus, isthmotectal axons can be induced to innervate a part of the tectum that they would not ordinarily contact (Udin and Keating, 1981). When nucleus isthmus axons are labeled in normal animals they are oriented very accurately along the rostro-caudal axis of the tectum (Guo and Udin, 2000). In eye-rotated animals, these isthmal axons appear disordered, often terminating in two locations or meandering within the tectum to their new location (Figure 9.19B).

In some cases, maps from two different sensory systems are found within the same structure, and they must come into

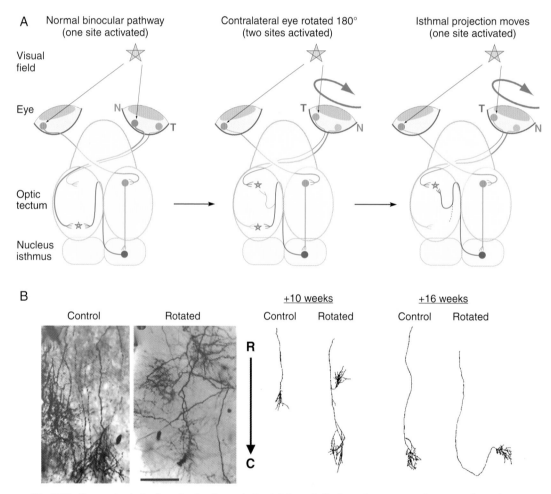

Fig. 9.19 Remapping in the frog visual pathway. A. In adult frogs (left), the optic tectum receives two retinotopic projections: a direct projection from the contralateral eye and an indirect projection from the ipsilateral eye (via the nucleus isthmus). These two projections are aligned such that a visual stimulus activates the same position in the tectum through either eye. In this example, the temporal eye activates the anterior tectum. When the contralateral eye is rotated 180°, a visual stimulus activates two different positions in the tectum, one via the eye and the other via the projection from nucleus isthmus (middle). This is because the retinal ganglion cells in nasal eye project to the posterior tectum. Over time, the projection from nucleus isthmus to the tectum adjusts its position so that a visual stimulus once again activates the same tectal position through either eye (right). B. Photographs of axons from nucleus isthmus as they course through the tectum of an adult control, and an eye-rotated animal (left). Note that control axons grow parallel to one another from rostral to caudal. Axons from eye-rotated animals display less ordered growth. Drawings of individual axons from nucleus isthmus in control and eye-rotated animals are shown at two postmetamorphic (PM) ages (right). Note that the axons of eye-rotated animals can make two terminal zones. *(Adapted from Udin and Keating, 1981; Guo and Udin, 2000)*

alignment during development. In the optic tectum of barn owls, there are maps of both the visual and auditory world. The maps are aligned such that neurons respond to acoustic and visual stimuli from the same position in space. For example, neurons that respond to visual stimuli directly in front of the animal (0°) will also respond to a sound stimulus that arrives at each ear simultaneously (i.e., 0 μs interaural time difference). Neurons that respond to visual stimuli at 20° to the right will also respond to a sound stimulus that arrives first at the right ear and then at the left (60 μs interaural time difference). To test whether the alignment of these maps depends on sensory experience, owls were reared with prismatic glasses that displace visual stimuli by 23° (**Figure 9.20**A). As in the retinotectal system, the direct connections between eye and tectum are unaltered in

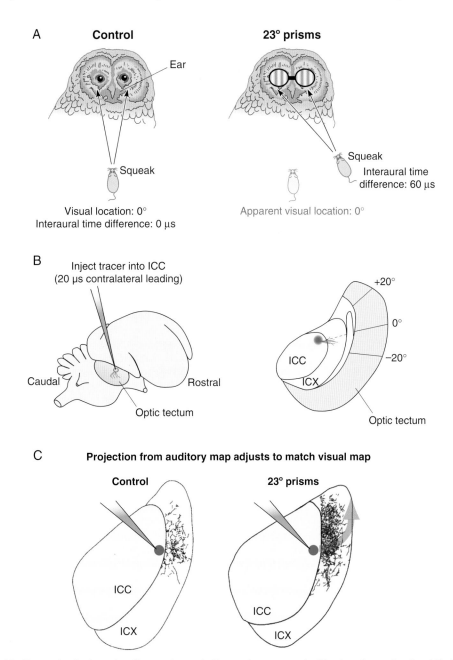

Fig. 9.20 Remapping in the owl auditory pathway. A. Barn owls were reared with prismatic goggles that shifted the visual field by 23°. In control animals, a squeaking mouse in front of the owl would appear at 0°, and the sound would reach both ears simultaneously (0 μs interaural time difference). With prisms in place, a squeaking mouse at 23° to the left would *appear* at 0°, but the squeak would reach the left ear first (60 μs interaural time difference). B. A tracer was injected into the auditory nucleus (ICC, central nucleus of the inferior colliculus) that encodes the ITD of the sound stimulus (left). The IC relays this information to the ICX where the map of auditory space is assembled. ICX then projects to the optic tectum where visual and auditory information is first integrated (right). C. In control animals, the projection from a region in ICC that responds to a 20 μs interaural time difference projects to a narrow region in the ICX (and ICX then projects to an optic tectal location that responds to visual cues about 8° from the midline). In prism-reared animals, ICC projects to a more rostral region of ICX. Thus, a new projection is formed within the auditory space map which compensates during prism rearing such that sound and sight can once again be integrated. *(Adapted from DeBello et al., 2001)*

prism-reared animals. If no compensation were to occur, then visual stimuli and auditory stimuli from the same position in space would activate different loci in the optic tectum. That is, maps of auditory space and visual space would be out of alignment. In fact, the auditory map adjusts to remain in register with the visual map. When the prisms are removed, tectal neurons that responded to visual stimuli directly in front of the animal (0°) are now found to respond to an auditory stimulus to one side. Therefore, auditory connections must have changed in response to visual activity (Brainard and Knudsen, 1993).

To determine the anatomical basis for this compensatory response, a tracer was injected in the central nucleus of the inferior colliculus, the structure that responds to interaural time differences (Figure 9.20B) (DeBello et al., 2001). In control animals, tracer injected at a specific position labels fibers that project to a unique location in a second auditory nucleus, called ICX (Figure 9.20C, left). It is the ICX neurons that will project to visually responsive neurons in the optic tectum. In prism-reared animals, the labeled ICC axons project to a different position within ICX, permitting optic tectum neurons to integrate a new auditory spatial position (Figure 9.20C, right). Thus, the elimination and addition of synapses depends on their activity pattern for some, but not all, afferent pathways. Even in the visual cortex, geniculate afferents from the contralateral eye appear to establish their innervation pattern first, whereas those from the ipsilateral eye may be more dependent on activity (Crair et al., 1998).

The pattern of evoked activity can also disrupt normal topography. This can be demonstrated in deafened cats that are stimulated electrically. This is a highly relevant experimental model because children with profound hearing loss are routinely implanted with cochlear prostheses, and they hear the acoustic world via electrical stimulation of their cochlea. In the experimental situation, auditory nerve fibers are stimulated with an electrode that is implanted within the cochlea. When animals are reared with repetitive electrical pulses to their cochlea, a synchronous activity pattern is evoked. To assess the effect, a single position along the cochlea was activated while an electrode was advanced through the inferior colliculus. In normal animals, a single point on the cochlea evokes a response within a limited region of tissue, referred to as a *spatial tuning curve*. In stimulated animals, the spatial tuning curves are much broader, as if a single position in the cochlea now projects to a much wider area of the tonotopic map in the inferior colliculus (Snyder et al., 1990). However, when animals are raised with alternating stimulation at two adjacent intracochlear electrodes, then the spatial tuning curves do not broaden (Leake et al., 2000). These experiments are consistent with findings from the visual pathway showing that coactive synapses establish strong connections during development.

SPONTANEOUS ACTIVITY AND AFFERENT REFINEMENT

We have seen that many sensory coding properties mature rapidly from the onset of sight or hearing, suggesting that normal neuronal activity simply maintains a precise but unstable set of connections (Blakemore and Van Sluyters, 1975). However, the developing nervous system is not silent before the eyelids and ear canals open; rather, spontaneous action potentials arise from many locations. Therefore, we next ask when the nervous system first displays electrical activity, and whether this activity influences synapse development.

Electrical activity has now been demonstrated in many, if not all, areas of the developing nervous system even in the absence of sensory stimulation (Blankenship and Feller, 2010). For example, retinal ganglion cells fire action potentials before the system is activated by light. In the embryonic rat, single retinal neurons discharge about once every second, but occasionally fire short bursts of almost 100 spikes per sec (Galli and Maffei, 1988). Since the competition hypothesis suggests that synapse modification depends on the pattern of activity, it was of interest to determine whether spontaneous activity has a temporal or spatial pattern. To address this question, the entire retina can be isolated from an embryo and placed in a recording chamber, where many retinal ganglion cells can be monitored (**Figure 9.21**A). Synchronous bursts of action potentials are recorded from many neurons about every minute or two, and regions of maximal activity sweep across the entire retina during the first few postnatal weeks (Meister et al., 1991; Wong et al., 1993; Demas et al., 2003). This patterned activity gradually breaks down at about the time of eye opening, but visual experience does not seem to terminate the waves. They are lost at about the same time even in dark-reared animals (Figure 9.21B). Similar patterns of spontaneous activity are found in the developing auditory system (Lippe, 1994; Kotak and Sanes, 1995; Jones et al., 2001; Jones et al., 2007; Kotak et al., 2007). In fact, the developing cochlea possesses an intrinsic mechanism that produces bursting activity prior to the onset of responses to sound (Tritsch et al., 2007). The activity of inner hair cells is, at first, driven by an ATP signal from adjacent supporting cells (Figure 9.21C and D). The ATP release leads to a local depolarization of the inner hair cell, which leads to glutamate release and evoked action potentials in the auditory nerve.

Spontaneous activity is also generated within the CNS. Internally generated waves of activity are found in the neonatal cerebellum, and this activity propagates along connected Purkinje cells in the sagittal plane (Watt et al., 2009). In brain slices through the cortex, large-scale waves of activity have been found to move slowly, from caudal to rostral, during the first postnatal week (**Figure 9.22**A). Spontaneous oscillatory activity can also be observed in vivo by imaging the calcium-dependent fluorescent signal. The bursts of activity appear about once every 10–20 sec, and are most apparent when the animal is not moving (Figure 9.22B). This activity pattern involves the vast majority of neurons and seems to end when inhibitory synapses become hyperpolarizing (Garaschuk et al., 2000; Adelsberger et al., 2005; Lischalk et al., 2009).

Does spontaneous activity influence synapse formation or elimination? Developing motor neurons initially display highly correlated firing patterns, due to coupling through gap junctions, and the loss of polyneuronal innervation at the muscle occurs when this temporally correlated activity ends (Personius et al., 2007; Personius et al., 2008).

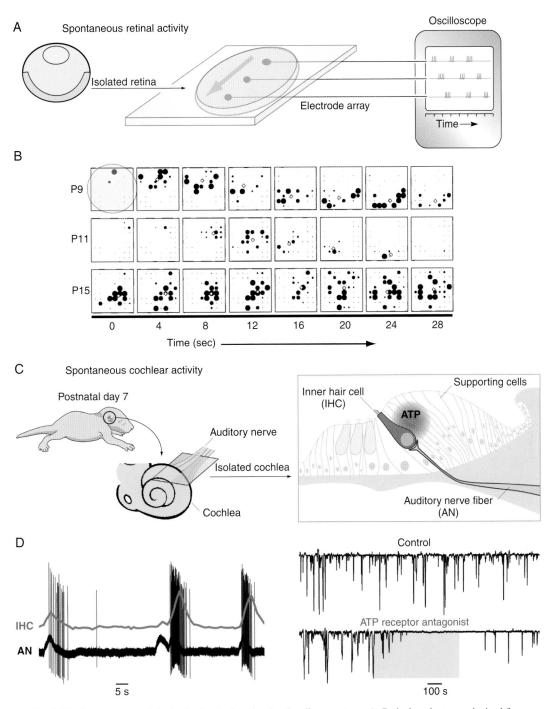

Fig. 9.21 Spontaneous activity in the developing visual and auditory systems. A. Retinal explants are obtained from neonatal ferrets and placed on an array of electrodes to record bursts of spontaneous activity from several retinal locations at the same time. (For clarity, only three locations are shown.) These bursts of activity move across the retina, as shown by the increasing response latency in each oscilloscope trace (right). B. Electrode array recordings are plotted for mouse retina at three postnatal ages (P9, P11, and P15). Each square shows the spatial pattern of activity across the two dimensions of a flattened retinal, with circles representing individual neurons. By P15, the spontaneous activity no longer travels across the retina. The spatial activity patterns were acquired every 4 s. C. The cochlea is removed prior to the onset of hearing and placed in vitro to visualize inner hair cells (IHC) and auditory neurons (AN). The schematic illustrates that supporting cells release ATP which activates adjacent IHCs via purinergic receptors. D. Recordings obtained simultaneously from an IHC (blue trace) and a nearby AN (black trace) display a correlated bursting response (left). When an antagonist of ATP receptors is added to the bath, spontaneous activity is eliminated (right). *(Adapted from Tritsch et al., 2007; Meister et al., 1991; Demas et al., 2003)*

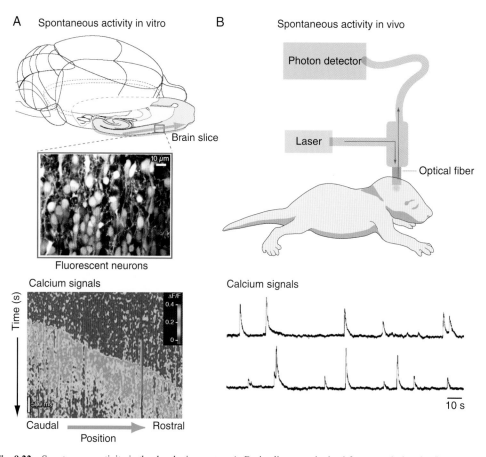

Fig. 9.22 Spontaneous activity in the developing cortex. A. Brain slices are obtained from rats during the first postnatal week, and intracellular calcium is monitored using a Ca-sensitive dye (top). Waves of spontaneous activity sweep across the cortex from caudal to rostral (bottom). B. A calcium indicator dye is injected into the cortex and an optical fiber is implanted under anesthesia to permit the monitoring of calcium signals in vivo (top). The dye is activated with a laser and the calcium-evoked light emission is collected by a photon detector. B. Calcium-evoked signals recorded from awake neonatal mice reveal oscillatory activity (bottom). *(Adapted from Garaschuk et al., 2000; Adelsberger et al., 2005)*

The segregation of thalamic afferents in visual cortex may also rely on spontaneous retinal activity (Stryker and Harris, 1986; Huberman et al., 2006). LGN afferents are known to segregate in visually deprived animals, suggesting that visually driven activity is not required (**Figure 9.23**A). However, when retinal action potentials are blocked with TTX, the LGN afferents fail to segregate into stripes (Figure 9.23B).

The temporal or spatial patterns of spontaneous activity may be important contributors to the formation of specific connections. This has been tested in the visual system, by disrupting spontaneous waves of retinal activity. The absolute levels of spontaneous activity from each eye are crucial to establishing the mature lamination pattern in the LGN. When the spontaneous activity of one eye is experimentally increased by elevating cAMP levels, its projection within the LGN is expanded (Stellwagen and Shatz, 2002). Furthermore, mice lacking the AChR β2 subunit display a disrupted spatiotemporal pattern of retinal activity, and the topographic projections from retina to superior colliculus do not undergo the normal period of refinement (McLaughlin et al., 2003).

CRITICAL PERIODS: ENHANCED PLASTICITY DURING DEVELOPMENT

Synaptic activity begins to exert an influence soon after synaptogenesis, but how long does this process continue? If we embrace learning and memory in our definition, then it lasts for our entire lifetime (see Box 9.1, "Remaining Flexible"). However, there are certain significant changes in nervous system structure and function that occur primarily during a limited period of development (Berardi et al., 2000; Hensch, 2004). A common example is language acquisition, which is accomplished most easily before age 10 and becomes a grueling task for most of us when attempted in adulthood (see Chapter 10). Thus some forms of neuronal plasticity are observed for only a limited period of development, often called a *sensitive* or *critical period*.

The first clear demonstration of a critical period considered the influence of visual experience on ocular dominance (Hubel and Wiesel, 1970). These studies illustrate that the neuronal property under discussion must be taken into account. In primates, monocular deprivation affects the segregation of LGN afferents in visual cortex for about two to three months, but

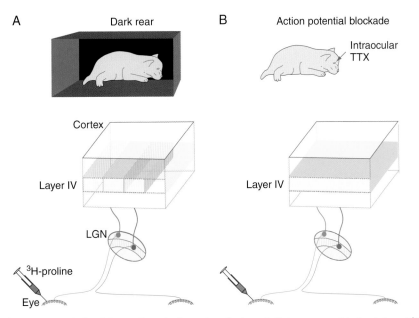

Fig. 9.23 Spontaneous retinal activity regulates the formation of stripes. A. Cats were reared in the dark, and ³H-proline was injected into one eye to visualize ocular dominance columns. Although the columns are slightly degraded, they do form. B. Bilateral intraocular TTX injections were performed beginning at postnatal day 14. When ³H-proline is injected into one eye to visualize ocular dominance columns, it is found that segregation of geniculate afferents into stripes fails to occur. Thus, spontaneous retinal activity is sufficient to influence stripe formation. *(Adapted from Stryker and Harris, 1986)*

continues to produce severe weakening of intrinsic cortical synapses for up to a year (**Figure 9.24**). Similarly, when retinal activity is blocked with TTX after ganglion cell arbors have segregated into eye-specific layers in the cat LGN, the coding properties of LGN neurons are nonetheless altered by the manipulation (Dubin et al., 1986).

Critical periods may occur during a very narrow time window in some brain areas. One region of rodent somatosensory cortex receives afferents from each of the facial whiskers and contains an array of barrel-shaped cell clusters that are activated selectively by each of the whiskers. If the whiskers are destroyed before postnatal day 5, their associated barrels do not form, but if the whiskers are destroyed after that time, then the manipulation has no effect (Van der Loos and Woolsey, 1973). There are also extensive intracortical connections between each of the whisker barrels, and this connectivity is dependent on continued use. When the sensory nerve to the whiskers is cut on postnatal day 7, after the barrel fields are formed, there is a dramatic reduction in the number of local projections (McCasland et al., 1992). In fact, whisker trimming significantly decreases the motility of dendritic spines and filopodia in rat barrel cortex only from postnatal days 11 to 13 (Lendvai et al., 2000). This manipulation also affects the development of whisker-evoked receptive fields in cortical Layer II–III until day 14, but has no effect on Layer IV receptive fields (Stern et al., 2001). Thus, the critical periods for barrel cortex differ by layer.

The influence of low levels of spontaneous activity on the refinement of synaptic connections may be facilitated by local inhibitory projections that further restrict the spatial pattern. In fact, local GABAergic circuits within the cortex regulate the binocular competition between thalamic afferents, and establish the precise dimensions of ocular dominance columns (Iwai et al., 2003; Hensch and Stryker, 2004). A series of

experiments in the mouse visual pathway have illustrated how inhibitory transmission regulates the ocular dominance critical period. As in other species, monocular deprivation in mice leads to a shift in ocular dominance toward the open eye, and this effect occurs during a critical period from about 19–32 days (**Figure 9.25**A). However, the effect of monocular deprivation can be initiated much earlier in development by simply increasing inhibitory transmission. In contrast, termination of the critical period can be delayed well after day 32 by decreasing inhibition (Figure 9.25B). The permissive influence of inhibition involves a particular type of inhibitory synapse that employs α1-containing $GABA_A$ receptors (Fagiolini and Hensch, 2000; Fagiolini et al., 2004). To determine whether developing inhibitory neurons could, themselves, sustain visual plasticity, embryonic inhibitory neurons from the medial ganglionic eminence of the forebrain were harvested and transplanted into postnatal mice (Southwell et al., 2010). The transplanted inhibitory neurons made numerous weak synapses, similar to that observed in younger animals, and permitted ocular dominance plasticity after the critical period (Figure 9.25C). Therefore, the maturation of cortical inhibitory neurons appears to delimit the critical period in visual cortex.

HETEROSYNAPTIC DEPRESSION AND SYNAPSE ELIMINATION

The activity of a synapse can be harmful to an inactive neighbor. This mechanism has been tested in vivo at an unusual muscle in the rat foot that receives its innervation via two separate peripheral nerves (**Figure 9.26**A). It is possible to electrically stimulate one group of motor axons during the normal period of synapse elimination, and then determine what happens to the

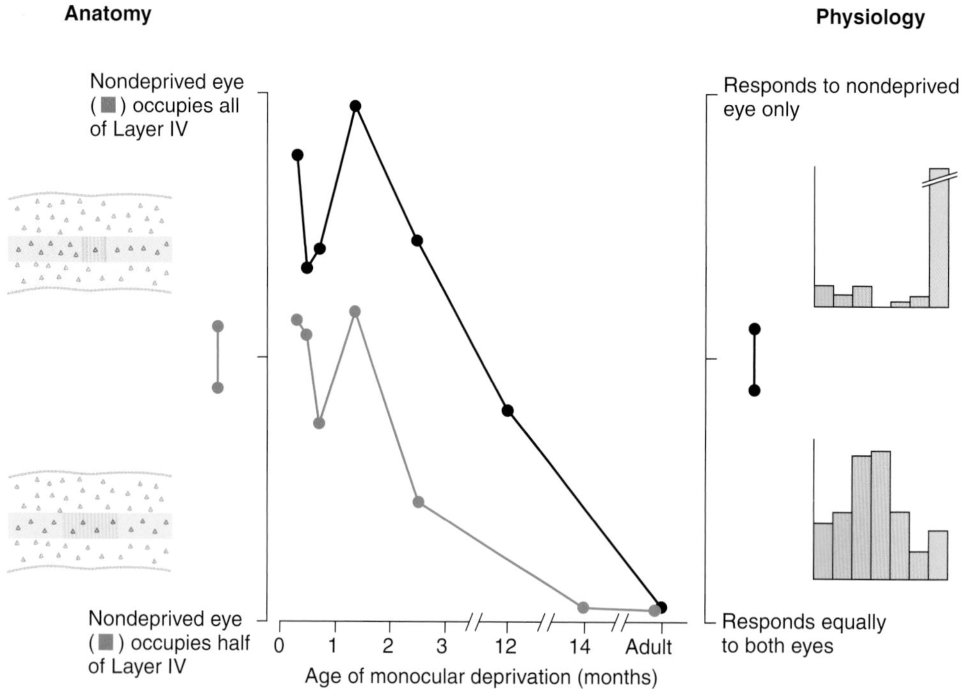

Anatomy

Physiology

Nondeprived eye
(■) occupies all
of Layer IV

Responds to nondeprived
eye only

Nondeprived eye
(■) occupies half
of Layer IV

Responds equally
to both eyes

Age of monocular deprivation (months)

Fig. 9.24 The effects of monocular deprivation are age-dependent. One eye was sutured shut in macaque monkeys at different postnatal ages. The sutures were removed after several months, and ocular dominance columns were examined with anatomical (left) and electrophysiological techniques (right). When deprivation begins between 0 and 2 months of age, most neurons subsequently respond to the open eye only, and the open eye occupies much more of Layer IV. The anatomical effect of deprivation on ocular dominance columns declines more rapidly than the physiological effect on ocular dominance histograms. *(Adapted from LeVay et al., 1980)*

synapses of unstimulated axons. The strength of both connections is determined indirectly by measuring the magnitude of a nerve-evoked muscle contraction, where a larger contraction signifies a stronger connection. When one nerve is stimulated for about six days, the *unstimulated* axons are subsequently

found to be much less effective at producing a muscle contraction (Ridge and Betz, 1984). That is, the unstimulated synapses have been weakened, and perhaps eliminated.

How long does it take for one synapse to snuff out its neighbor? The onset of synapse elimination can be studied in a

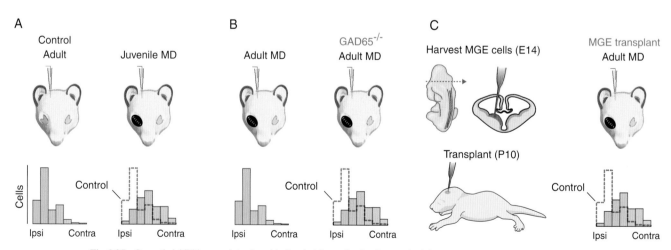

Fig. 9.25 Synaptic inhibition regulates the critical period for ocular dominance plasticity. A. In normal mice, recordings from the binocular region of visual cortex (top) yield an ocular dominance histogram that is biased toward the ipsilateral eye (bottom). See Figure 9.7 for a description of ocular dominance histograms. When juvenile mice are monocularly deprived (MD) between the ages of about 19–32 days, the ocular dominance histogram is shifted toward the nondeprived eye (blue-shaded bars). The distribution of cells in the control histogram is shown for comparison (gray dashed line). B. When adult animals are monocularly deprived, there is no change to the ocular dominance histogram. In contrast, mice lacking the GABA-synthesizing enzyme, GAD65, continue to display an effect of MD. C. Embryonic inhibitory neurons can be obtained from their zone of proliferation (MGE), and transplanted into the visual cortex at P10 (top). These animals also continue to display an effect of MD after the normal critical period has ended. *(Adapted from Fagiolini and Hensch, 2000; Southwell et al., 2010)*

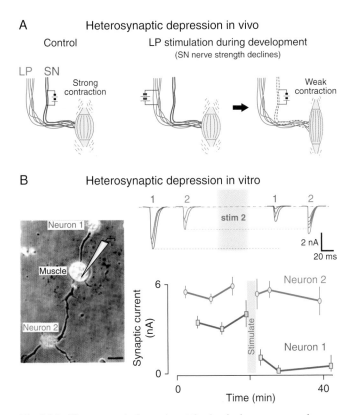

A   Heterosynaptic depression in vivo

Control   LP stimulation during development
(SN nerve strength declines)

B   Heterosynaptic depression in vitro

Fig. 9.26 Heterosynaptic depression at the developing neuromuscular synapse in vivo and vitro. A. The lumbrical muscle in the rat foot is innervated by two physically separate motor nerve roots, LP and SN. In control animals, stimulation of the SN motor nerve root (red) elicits a robust contraction of the lumbrical muscle (left). However, if the LP motor nerve root (blue) is repetitively stimulated during the developmental period when synapse elimination occurs, then SN nerve root stimulation elicits a much weaker contraction of the lumbrical muscle (right) when it is subsequently tested. Therefore, stimulation of one set of synapses leads to the a decrease in the strength of a second set of synapses, referred to as heterosynaptic depression. B. Whole-cell recordings were made from *Xenopus* muscle cells that were co-innervated by two neurons (left). Evoked synaptic currents are first measured in response to stimulation of each neuron (top left). A strong stimulus is then applied to neuron 2 (blue), and the strength of each neuron tested again. Stimulation of neuron 2 leads to smaller evoked synaptic currents from the *unstimulated* neuron 1 (red) but no change to the neuron 2-evoked response (top right). The graph of ESC amplitude before and after stimulation of neuron 2 shows that the depression is long-lasting (bottom right). *(Adapted from Ridge and Betz, 1984; Lo and Poo, 1991)*

simple culture system, containing two presynaptic neurons and one postsynaptic cell, a myocyte (Figure 9.26B). Whole cell recordings are made from the myocyte, while the activity of one or both presynaptic neurons is controlled with stimulating electrodes. In the first experiment, the synaptic currents elicited by each neuron are first measured, and then one of the neurons is stimulated for several seconds. Within moments of this brief procedure, the unstimulated neuron produces much smaller synaptic currents, and the effect lasts for the duration of the experiment (Lo and Poo, 1991). When one synapse is able to decrease the strength of a second one, the interaction is referred to as *heterosynaptic depression*.

Two additional observations seem to fit neatly into the puzzle. First, if the two synapses are separated from one another by >50 μm, then activation of one synapse is not able to

suppress its neighbor. Two synapses can apparently compete for the right to activate a postsynaptic cell, but the interaction occurs over a very short distance, consistent with results from the adult NMJ and *Aplysia* cultures (Kuffler et al., 1977; Glanzman et al., 1991). Second, when both neurons are stimulated at the same time, there is no change in the strength of either synapse. That is, synaptic activity is able to *safeguard* a synapse from the harmful effects of an active neighbor.

From this description, it is clear that synaptic terminals carry on their competition through an independent agent, the postsynaptic cell. It is a bit like choosing the winner of a boxing match by judging who punches the referee harder. The contribution of the postsynaptic cell to heterosynaptic depression can be demonstrated by activating it directly with depolarizing current injections or with transmitter application. In both cases, when a muscle cell is activated focally, then the size of synaptic responses declines dramatically (Dan and Poo, 1992; Lo et al., 1994).

Is a depressed synapse necessarily an eliminated synapse? Observations from the neuromuscular junction indicate that postsynaptic AChR clusters disappear before nerve terminals withdraw from the synapse (Role et al., 1987; Balice-Gordon and Lichtman, 1993). Furthermore, the reduction of functional AChRs is sufficient to produce synapse withdrawal from the inactivated region of the neuromuscular synapse (Balice-Gordon and Lichtman, 1994). To examine whether heterosynaptic depression and receptor turnover are correlated, a simple dual innervated nerve–muscle culture is again employed (Li et al., 2001). Electrical stimulation of one neuron leads to a depression of the excitatory response evoked by the unstimulated neuron, as discussed above. When the AChR density under each synapse is examined before and after stimulation, the unstimulated synapse is found to have lost more than 30% of its AChRs (**Figure 9.27**A). Since synaptic activity leads to the depression and receptor elimination, it follows that the active synapse must somehow be protected. There is evidence that two kinases, PKA and PKC, are involved in the distinct response of active and inactive terminals (Figure 9.27B). For example, PKC activators can produce synaptic depression in the absence of stimulation. In contrast, the PKC activator has no effect when the neuron is stimulated (Li et al., 2001). PKC may also regulate synapse elimination in the CNS. The normal loss of climbing fibers from cerebellar Purkinje cells is reduced in mice deficient for PKC (Kano et al., 1995).

A second mechanism whereby the postsynaptic cell could induce synapse stabilization or elimination involves retrograde signaling. In this scenario, activity may lead to the postsynaptic release of compounds that act directly on the presynaptic terminal. One such compound is the preprocessed neurotrophin, proBDNF (see Chapter 7). When proBDNF is applied directly to developing neuromuscular contacts in vitro, it elicits synaptic depression rapidly, and this is followed immediately by the withdrawal of presynaptic terminals (**Figure 9.28**). Blockade of the proBDNF receptor, p75NTR, on presynaptic terminals can prevent this activity-induced synaptic depression and withdrawal. Interestingly, a different scenario results if proBDNF is cleaved to produce the processed neurotrophin, BDNF. In this case, BDNF potentiates the synapse (Yang et al., 2009). This result suggests a model in which the postsynaptic cell releases a molecule that can either destabilize presynaptic terminals or stabilize them, depending on its proteolytic conversion.

A AChR loss at unstimulated synapse

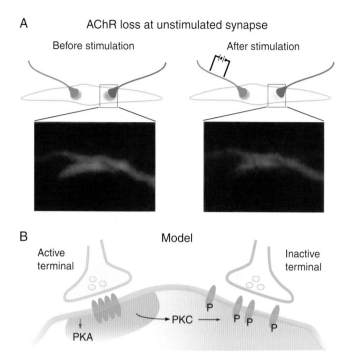

Fig. 9.27 Heterosynaptic depression is associated with a loss of postsynaptic AChRs. A. Muscle cells are co-innervated by two neurons in vitro. When one neuron is stimulated for 1–2 hours, the nonstimulated neuron displays heterosynaptic depression and a loss of AChRs beneath its synapse. The images show staining for nerve (green) and AChR clusters (red) beneath the unstimulated synapse before (left) and after (right) heterosynaptic depression. AChR aggregates are stained with rhodamine-bungarotoxin. B. Model of activity-dependent synapse elimination. The active terminal increases PKA and PKC locally. PKC phosphorylates AChRs beneath the inactive terminal, leading to their loss. However, PKA activity can protect the AChRs beneath the active terminal. PKC may become activated by calcium influx, whereas PKA may become activated by a ligand that is released along with ACh. *(Adapted from Li et al., 2001; Nelson et al., 2003)*

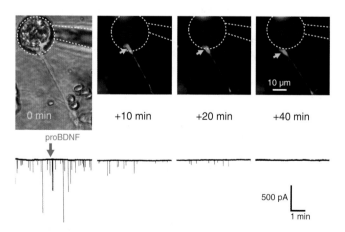

Fig. 9.28 Depression and presynaptic withdrawal induced by proBDNF. Recordings are made from an innervated muscle cell in vitro, and the preprocessed neurotrophin proBDNF is applied. Within 10 mins, the amplitude of spontaneous synaptic currents has declined. During the next 30 mins, synaptic currents are eliminated and the presynaptic terminal (yellow arrow) withdraws from the muscle cell body (white dashed circle). *(Adapted from Yang et al., 2009)*

INVOLVEMENT OF INTRACELLULAR CALCIUM

Clearly, neighboring presynaptic terminals carry out a rather harsh exchange via the postsynaptic cell, and there must be a molecular pathway that conveys the signal from one contact to another through the cytoplasm. One hypothesis is that depolarizing synaptic potentials open voltage-gated Ca^{2+} channels, and Ca^{2+}-dependent proteolytic enzymes are recruited to degrade the nonactive terminals, ultimately leading to their withdrawal. A variety of proteolytic enzymes have been discovered in the nerve terminals and somata of developing neurons. This hypothesis was first tested at the mammalian NMJ by decreasing extracellular Ca^{2+} or blocking specific Ca^{2+}-activated proteases, and both manipulations were able to slow down the process of synapse elimination (Connold et al., 1986).

Once again, there are important similarities between synapse elimination and heterosynaptic depression (above). First, it is possible to prevent heterosynaptic depression by injecting the muscle cell with a Ca^{2+} chelator that sops up free Ca^{2+}, suggesting that a rise in postsynaptic calcium is necessary for depression to occur. Second, it is possible to cause synaptic depression by momentarily raising postsynaptic calcium. This was accomplished by loading muscle cells with molecules of "caged" calcium, which can release the calcium into the cytoplasm when it is exposed to ultraviolet light (**Figure 9.29**). When the synaptic responses at one muscle cell are recorded while it is exposed to a brief pulse of UV light, synaptic transmission becomes depressed by 50% within seconds (Figure 9.29A) (Lo and Poo, 1994; Cash et al., 1996). Interestingly, when synapses are activated by stimulation of the neuron, simultaneously with the release of calcium, then synaptic depression is prevented (Figure 9.29B). This suggests that a rise in postsynaptic calcium recruits a retrograde signal that ultimately depresses presynaptic transmitter release; however, when presynaptic terminals are active, this depression can be prevented (Figure 9.29, right).

Calcium influx may occur directly through neurotransmitter receptor-coupled channels. As we learned in Chapter 8, NMDA-sensitive glutamate receptors (NMDAR) are highly expressed in the central nervous system during synaptogenesis. These receptors become active when glutamate and membrane depolarization are present at the same instant (**Figure 9.30**A). During depolarization, a magnesium ion is expelled, and the open channel permits Ca^{2+} to rush into the postsynaptic neuron. To determine whether NMDA receptors are involved in activity-dependent synapse plasticity, these receptors have been blocked during developmental manipulations. For example, we learned that monocular lid closure weakens synapses in the cortex that are driven by the deprived eye. However, when the same manipulation is performed during the chronic infusion of a NMDAR blocker, the strength of "deprived" synapses is preserved (Kleinschmidt et al., 1987). Thus, synapses from the open eye activate NMDARs in the cortex, allowing calcium to enter postsynaptic neurons. We will consider what calcium might be doing once it enters the postsynaptic neuron below.

NMDA receptors turn out to be broadly involved in the stability of developing excitatory synapses. In the cerebellum, where adult Purkinje cells are innervated by one climbing fiber, the chronic administration of an NMDAR blocker (AP5) results in 50% of Purkinje cells remaining multiply innervated

movement of bird embryos. Chicks, like many other shell-bound embryos, go through a very specific motor behavior pattern just at hatching (Bekoff and Kauer, 1984). These behaviors allow the embryo to break open the shell. If one places a newly post-hatch chick into the hatching position within a glass egg, the bird reinitiates its hatching behavior (**Figure 10.11**), but the capacity to do these hatching movements quickly fades as time goes on. What stimulates these hatching movements? If sensory input from the neck is eliminated with a local anesthetic, then hatching behavior is suppressed (Bekoff and Sabichi, 1987). Therefore, it appears that sensory receptors located in the neck provide a specific input signal for initiating hatching behavior. Another example of a transient embryonic behavior is the migration that marsupial embryos make from their womb to the mother's pouch. Born at an extremely early phase of development with their hind limbs little more than buds, these tiny embryos use their forelimbs to crawl tens of body lengths into the pouch where they attach onto a nipple and suckle, another transient but adaptive behavior of all mammals, for several months. During this time they complete their embryonic development. That suckling movements are not substrative is shown through human infants that do not suckle. They can be fed through a tube, and it is clear in these cases that the absence of these early suckling movements does not impair the development of adult eating.

Metamorphosis signals a dramatic change in the lifestyles and the nervous systems in insects and amphibians. Many very adaptive larval behaviors are lost at this transition while new adult behaviors are gained. For example, the larval behaviors of the swimming, filter-feeding tadpole are inappropriate for the hopping, bug-eating frog. The transition from the larval to the adult state is activated by specific hormones: ecdysone for insects and thyroxine for frogs. Each of these hormones has a widespread effect on gene expression and cellular function. As would be expected, the changes in nervous system structure that subserve these selective losses and gains of specific behaviors are regulated by ecdysone and thyroxine. In both insects and amphibia, larval neurons die upon exposure to metamorphic hormone, and some neuroblasts that have been quiescent throughout larval life begin to proliferate. Some larval neurons survive the transition to adult but are drastically reorganized. For example, the motor neurons that move the abdominal prolegs of the caterpillar do not die, even though these appendages are lost, but their axons and dendrites are remodeled, and old synapses are eliminated to support new behaviors (**Figure 10.12**).

In metamorphic insects, there are important transitional behaviors associated with building and emerging from the pupal state. In moths, the adult motor system is constructed primarily from remodeled larval components, whereas the adult sensory system is primarily composed of new neurons. Simple reflexes correlate these neuronal changes with the acquisition or loss of particular behaviors. The loss of the larval proleg retraction reflex is associated with the loss of the dendrites of the proleg motor neurons; the adult stretch receptor reflex begins when new adult-specific connections are added to new dendritic growth in an adult neuron (Levine and Weeks, 1990).

Drosophila larvae are eating machines. There are two styles of eating found in natural populations of *Drosophila* larvae, and this difference can be genetically mapped to differences in the activity of a single gene, which encodes cGMP-regulated protein kinase (PKG) (Osborne et al., 1997). Larvae with the more active *rover* allele of PKG have significantly longer foraging path lengths than do those homozygous for the less active *sitter* allele. The two *Drosophila* foraging variants do not differ in their general activity in the absence of food, but when food is available the two variants may fare differentially based on the density of other animals feeding in the vicinity. If it is crowded, larvae that forage further may do better. Natural selection experiments done in the laboratory showed that under high-density rearing conditions for several generations, the *rovers* did better, whereas the short path *sitter* phenotype was selected under low-density conditions (Sokolowski et al., 1997). Interestingly, the age-related transition of honeybees from hive work to foraging is also associated with an increase in the expression of the PKG gene and both experimentally added cGMP or elevated PKG activity cause premature foraging behavior in bees (Ben-Shahar et al., 2002). Thus, what appears to be a conserved genetic pathway plays on the foraging circuits of flies and bees, and in the latter case, it regulates a stage-specific transition in behavior.

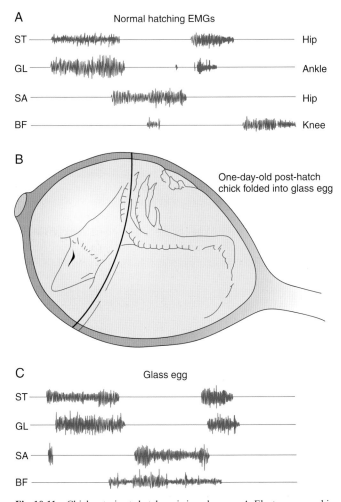

Fig. 10.11 Chicken trying to hatch again in a glass egg. A. Electromyographic (EMG) recordings of the normal hatching motor program in the chick embryo. (ST and SA are hip muscles, GL is an ankle muscle, and BF is a knee muscle.) Note the alternation between the activity of hip and lower leg. B. A one-day-old chick is crammed back inside a glass egg. C. EMG records from such a chick show that it reinitiates the hatching motor program. *(Adapted from Bekoff and Kauer, 1984)*

297

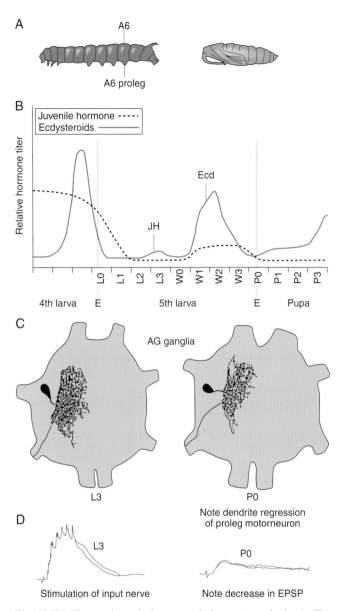

Fig. 10.12 Changes in a single neuron during metamorphosis. A. The caterpillar of the moth (*Maduca sexta*) showing abdominal segment 6 and the proleg associated with this segment. To the right is shown the pupal stage. B. Profile of hormonal changes through larval life and the transition between the larval and pupal development where ecdysone rises without an increase in juvenile hormone. C. Remodeling of the A6 proleg motor neuron du ing metamorphoses. There is a dramatic pruning of the dendritic tree. D. Correlated with this dendritic remodeling, there is a decrease in the activation of this neuron when the sensory input nerve was stimulated. These changes reflect the changing role of this neuron in postmetamorphic life where there are no prolegs. *(Adapted from Streichert and Weeks, 1995)*

Humans do not go through metamorphosis in the same way as flies and frogs, though like bees, we do go through distinct stages in our behavioral repertoires. The differing behaviors of babies, toddlers, teenagers, new parents, and aging adults must be largely due to changes in the nervous system that result not only from experience, but also from a variety of intrinsic influences, such as hormones and growth factors that are regulated throughout life. The drastic changes in behavior that accompany metamorphosis in insects and frogs, and the more progressive changes that happen as mammals go through the stages of their lives, lead us to wonder how the circuitry at any one stage can best be molded or added to, so that the appropriate behaviors can arise at the next stage.

GENETIC DETERMINANTS OF BEHAVIOR

We have discussed the development of motor behavior up till this point from the standpoint of how the circuits that underlie these behaviors develop, initiate, and become fine-tuned. The question we address in the next two sections is about nature and nurture, genes and environment, and their particular and interdependent roles in shaping early behavior. Donald Wilson, who worked on the neural basis of insect flight, was amazed that a locust, spreading its wings for the first time, could fly without practice and make appropriate adjustments to wind speed and visual signals. "How perfect is the motor score that is built into the thoracic ganglion?" he wondered (Wilson, 1968). Michael Bate, who works on the development of the *Drosophila* nervous system, expresses a similar surprise, but for a different reason (Bate, 1998). "How do we explain," Bate asks, "the remarkable fact that behavioral 'sense' of this kind is inherited and built into the nervous system as it develops?" It is difficult to comprehend the genetic basis of the neural circuits that underlie complicated behaviors, yet it is obvious that genetic factors are at the root of behavior. Flies do not behave as humans, no matter how similar their rearing environment. Yet the nervous systems of both insects and mammals are built of neurons, synapses, and neurotransmitters. They also obey the same developmental rules, often using homologous molecules.

In the 1960s, Seymour Benzer began to search for genes intimately associated with particular behaviors by screening for mutants with aberrant behaviors in fruit flies (Benzer, 1971). He and his colleagues discovered genes that were involved in reflexive visual behaviors such as phototaxis, motor behaviors such as the ability to jump and fly, rhythmic behaviors such as sleep-wake patterns, sexual behaviors such as courtship, and adaptive behaviors such as learning. Surprisingly, many of the genes he and his colleagues discovered by this process are conserved among different species. For instance, the genes found to govern the 24-hour circadian activity rhythm in flies were found to have homologs that are involved in the same process in mammals (Ishida et al., 1999). Do these genetic homologies extend to neural development in the sense that genetically conserved programs construct similar neural circuits that control homologous behaviors? One possible way to begin approaching this question is to consider how simple behaviors may differ between different regions of the same animal. A chicken flaps its wings synchronously yet moves its legs in an alternating gate. But if, during embryogenesis, the brachial cord that drives wing movements in chick embryo is exchanged with the lumbrosacral cord that drives leg movements, the result is a very mixed-up chicken in which the wings flap alternately and the legs hop synchronously (**Figure 10.13**) (Straznicky, 1967; Narayanan and Hamburger, 1971). The central pattern generators in the spinal cord that generate locomotion provide coordinated bilateral control over the normal limb alternation that underlies walking or synchronous wing flapping in the forelimbs of birds. The molecules that organize these central pattern generators are largely unknown, but spinal cords from mice lacking either the EphA4 receptor or its ligand EphrinB3

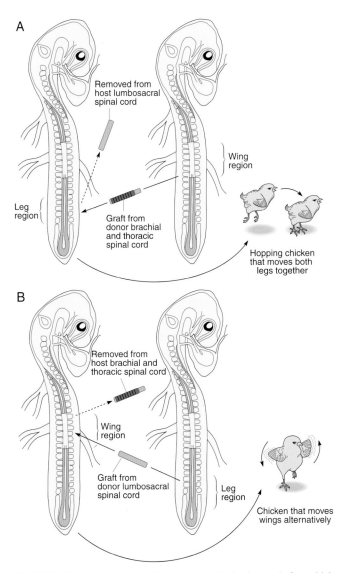

Fig. 10.13 Flapping legs and walking arms. A. The lumbar cord of one chick embryo is replaced by the brachial cord of a donor embryo. When the chick hatches, instead of walking, it jumps in the rhythm of flapping wings. B. The reciprocal experiment leads to a chick that has alternating wing movements instead of normal flapping ones. *(Adapted from Narayanan and Hamburger, 1971)*

and so are good candidates for being responsible for the motor circuit regionalization, especially as these genes are expressed in the cells of the nervous system as they begin to wire up. Indeed, in *Drosophila*, loss and gain of function experiments with the Hox genes, *Ubx* and *abdA*, which are expressed in thoracic and abdominal segments, show that these genes are necessary and sufficient to dictate the formation of the region-specific motor network involved in peristaltic locomotion (**Figure 10.14**) (Dixit et al., 2008).

V1 neurons are inhibitory interneurons that selectively express the transcription factor Engrailed1 (see **Figure 10.15**). Genetic techniques used to knock out the function of these Engrailed1 expressing neurons, show that V1 neurons are required for a different aspect of the motor circuit, i.e., for generating "fast" motor bursting. When these neurons are taken out of the circuit either developmentally, or by inactivation through expression of a conditional activity inhibitor, the rhythm cannot shift into a fast mode, and such mice are constrained to walking very slowly. Homologous neurons in the *Xenopus* are involved in generating the swimming pattern, suggesting that these neurons may play an evolutionarily conserved role in controlling the speed of vertebrate locomotor movements (Gosgnach et al., 2006). Excitatory V3-derived neurons specifically express the transcription factor Sim1, the promoter of which can then be used to express tetanus toxin and this takes V3 out of the locomotor circuit. In this case, something yet different happens: the left-right alternation regularity and the robustness of the locomotor rhythm is severely perturbed, suggesting that V3 may be responsible for distributing excitatory drive between both halves of the spinal cord (Zhang et al., 2008).

These studies show that it is now becoming possible to study how specific genes are used in developing the circuitry of locomotion. In worms, flies, fish, and mice, there has been a tremendous amount of work on genes that affect specific aspects of behavior, and on genes that regulate the development of the nervous system, many of which we have mentioned throughout the previous chapters. However, it is only more recently, through studies such as those discussed above, that we have made inroads into the genes used for regulating specific aspects of the circuits that underlie particular behaviors, such as how these circuits are genetically programmed, what components they are built from, and how they might evolve.

ENVIRONMENTAL DETERMINANTS OF BEHAVIORAL DEVELOPMENT

From the earliest embryonic stages of life, genetic influences operate in the context of specific environments. These environments place selective pressures on the embryo and influence development. The most obvious examples for humans come from studies of environmental hazards such as drug use or alcohol consumption during pregnancy. Embryonic exposure to high levels of these substances can lead to mental retardation (Johnson and Leff, 1999). When thinking of "determinants," the toxic effects of drugs and alcohol are not so much the point. The special and adaptively significant sensitivities of the nervous system to particular environmental inputs are much more interesting. As we will see later in this chapter, embryonic exposure to sex-specific steroids can alter

(see Chapter 5) lose left-right limb alternation and instead exhibit synchrony, so they hop rather than run (Kullander et al., 2003). In addition, certain transcription factors such as Vsx2 and Evx1, mentioned in Chapter 4 as factors that control the determination of specific spinal interneurons, are also critical for left-right limb movement coordination (Garcia-Campmany et al., 2010). Such work shows that dramatic locomotor changes can occur as a consequence of local genetic rewiring and is beginning to identify the genes required for the development of alternating versus synchronous gait. This leads us back to the question of whether there is a difference in the expression of these genes in the brachial versus the lumbrosacral region of a chick. And if so, what controls this differential expression? The answer to the first question is unknown at the time of writing, but the answer to the second question may well be related to Hox gene expression as Hox genes regulate the regional character of the nervous system (Chapter 2)

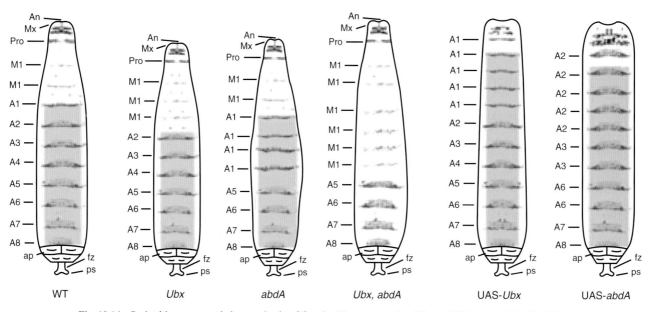

Fig. 10.14 Peristaltic movement in loss- and gain-of-function Hox mutants. In wild-type WT characteristically abdominal peristaltic movement begins posteriorly and extends to the first abdominal segment, A1. The domain where this movement is seen is shaded in this and other panels. The segmental transformations are shown in each case. An, antennal; Mx, maxillary; Pro, prothorax; Ms, mesothorax; Mt, metathorax. A1–A8 represent abdominal segments 1 to 8. ap, anal plates; fz, filzkörper; and ps, posterior spiracles. *(From Dixit et al., 2008)*

both physical and neural aspects of sexual maturation. In bird embryos, a brief exposure to a mother's call can imprint a preference for that call upon hatching. After hatching, songbirds learn to produce the father's song, and vocalization centers in its brain are strongly influenced by this process (see below).

Early deprivation of many kinds (visual, auditory, even emotional) is known to have permanent effects on behavioral development, just as early experiences can have profound effects on neural development. As we saw in Chapter 9, normal vision can never be restored to an adult mammal that has been blind from birth; the neural circuitry simply has not developed properly. Human infants who are blind also show delays in various motor skills, a fact that emphasizes the importance of vision as a sensory input for motor development (Levtzion-Korach et al., 2000). Lack of various aspects of education and emotional interactions early in life may also affect the ability to perform later in life because the neural substrates of these behaviors are wired in at particular stages. Thus, although we do not believe that human brains, or those of other animals, emerge as blank slates, it is nevertheless clear that experience shapes and adjusts the nervous system. An argument has been made that the brain is adaptive, in the Darwinian sense, simply because it is an organ that learns how to modify behavior in order to improve survival in a changing environment. In fact, learning is one of the main functions of the nervous system, and this process begins in the embryo. In dealing with the genetic and environmental influences on behavior, we must not only address how genes control cellular and molecular events to construct the neural substrate of behavior, but also how behavioral and sensory events feed back onto these molecular mechanisms.

A very interesting and instructive example of the relationship of nature and nurture is that of maternal care in the rat,

investigated by Michael Meaney and colleagues (Cameron et al., 2008). Maternal care involves feeding, licking, and grooming of the pups. Licking and grooming serve to enhance somatic growth and future behavior. These effects came to light when it was noticed that there was considerable variation in the amount of licking and grooming among rats. Some mothers are high groomers and others are low groomers. This difference seems to be maternally inherited as it is stably transmitted across many generations. Rats that are reared with low maternal care display an increased response to stress throughout their adult lives. However, the mode of inheritance is non-Mendelian, it is epigenetic. This is shown by cross-fostering, which reverses the pattern of inheritance. Females born to low-grooming mothers, but reared by high-grooming stepmothers, show increased pup-grooming when they become mothers and vice-versa. The effects of artificial rearing on maternal behavior are even worse than being reared by a low-grooming mother, but are greatly reduced by stroking the female pups during the first weeks of life, which mimics the tactile stimulation associated with maternal grooming. These findings suggest a direct relation between the quality of maternal care in early life and a variety of behaviors, including maternal care itself, that are expressed subsequently in adulthood.

What is the basis of this epigenetic inheritance of maternal behavior? Differences in grooming and sexual receptivity in females are related to the levels of estrogen receptor expression in the brain, whereas the differences in stress correlate with glucocorticoid receptor expression (**Figure 10.16**). The expression of both of these receptors is controlled by the methylation status of the promoters for these genes. DNA methylation represses gene expression as methylated DNA excludes transcription factors. When a pup is groomed, it stimulates the release of serotonin onto neurons involved in maternal behavior and anxiety

A

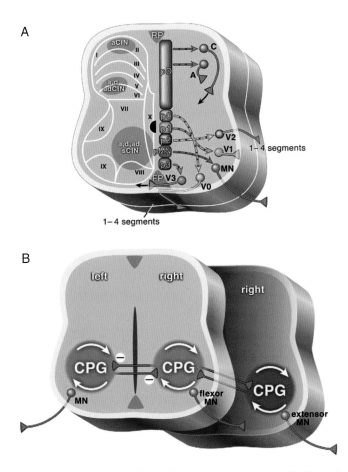

B

Grooming and licking

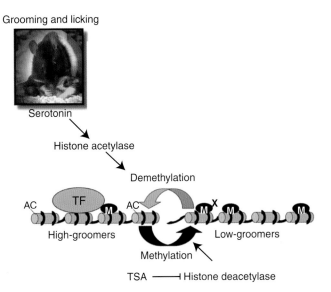

Fig. 10.16 Epigenetic reprogramming by maternal care; a model. Maternal licking and grooming in the rat triggers activation of serotonin receptors leading to increased recruitment of the histone acetylase to the glucocorticoid and estrogen receptor promoters. Acetylation of histone facilitates demethylation. In offspring of low-grooming mothers, this process is reduced in comparison with offspring of high-grooming mothers, leading to differential epigenetic programming of the receptor promoters. In the adult rat, the epigenetic state is reversible. TSA, a histone deacetylase inhibitor, increases histone acetylation and facilitates demethylation and epigenetic activation of these genes in the offspring of the low-grooming mothers. *(From Weaver et al., 2004)*

Fig. 10.15 Schematic of spinal cord development, cellular organization and CPG coordination. A. The right half of the spinal cord depicts the origin of the interneuron (IN) and motor neuron (MN) populations defined in developmental studies using molecular markers for cell identification. V0 INs are marked by Evx1, V1 INs by En1, V2 INs by Vsx2, V3 INs by Sim1 and MNs by Hb9. The left half of the spinal cord gives a summary of neuronal labeling studies defining the general location of commissural INs (CINs). Ascending (a), descending (d), ascending–descending (ad) and segmental (s) projection patterns are indicated for each group of CINs. B. A neuronal network located within each half of the ventral spinal cord controls rhythmic locomotor activity and is defined as a central pattern generator (CPG). The left and right CPGs are linked by cross-inhibitory connections that coordinate alternation. Likewise, intersegmental connections coordinate the activity of flexor–extensor motor control at different rostral–caudal levels of the spinal cord. The cellular basis for extensor–flexor coordination is less well understood. *(From Goulding and Pfaff, 2005)*

in the brain. Serotonin causes an increase in histone acetylase activity, which promotes the demethylated state. Indeed, there is increased methylation of the estrogen and glucocorticoid receptor promoters in the female offspring of low-grooming mothers, as compared with those of high-grooming mothers. The effects of DNA methylation on this key gene involved in manal behavior can be reversed by early postnatal cross-fostering. In fact, it can also be reversed pharmacologically by treating adults with Trichostatin A, a drug that demethylates DNA and turns behaviors associated with low-grooming into those more associated with high-grooming (Weaver et al., 2004).

The impact of epigenetic inactivation on neural development and behavior can be overwhelming. Methylation of the glucocorticoid receptor neuron-specific promotor is increased

in the hippocampus of suicide victims with a history of childhood abuse, leading to reduced expression of glucocorticoid receptors (McGowan et al., 2009). Thus, the epigenetic control of glucocorticoid signaling has a developmental origin in mice and humans, and this can exert a profound impact on adult behavior.

The nervous system is unique among organs in that it responds to a huge variety of environmental influences by changing itself structurally and functionally and even epigenetically, from an early age (Chapter 9). Neural activity modulates the expression of genes used in regulating behaviors. This interplay between the environment and the genome continues throughout life. And because of this, it is usually not reasonable to ascribe specific behaviors to purely genetic or environmental determinants. Genetic influences on behavior are as clear as experiential ones, and the two are interlinked. Identical twins who are separated at birth and reared in different families may show amazing similarities in attitude and taste, compared to nonidentical twins reared together. But these two human beings usually also show an enormous array of dissimilarities, reflecting lifelong interactions between the environment and the genome.

The normal behavior of an animal indicates a successfully assembled nervous system, and many human neurological, psychiatric, and behavioral problems have a mixture of genetic and environmental etiologies. By affecting the way that neurons develop, both environmental insults and genetic mutations have an enormous impact on the emergence of the functional circuitry underlying behavior, restricting an organism's ability to perceive the world and to respond to it appropriately.

BEGINNING TO MAKE SENSE OF THE WORLD

The nervous system becomes active well before animals experience the world around them or move about within it. In fact, this "spontaneous" activity initially promotes synapse maturation (Chapter 9). However, when animals first begin to hear, smell, see, taste, and feel, the sensory epithelia (e.g., hair cells, photoreceptors, etc.) and central nervous system connections are immature. When do visual and auditory perception reach maturity, and what is their relationship to neural development?

Studies of sensory perception are among the most difficult experiments in the field of development. Neonatal animals (especially human infants) tend to be slow, sleepy, cranky, inattentive, and forgetful. These are generally referred to as nonsensory factors. For example, adults pay better attention to novel stimuli than young animals. When adult primates are presented with two images, one of which they have never seen before, they spend about 70% of the time staring at this novel object. In contrast, infants spend an equal amount of time staring at the familiar and novel objects (Bachevalier, 1990). Even though adults are more attentive to novel stimuli, they can also focus narrowly on a stimulus of interest and ignore novel stimuli that may be distracting. For example, when taking an examination, we tend to "block out" extraneous noise. This was demonstrated by asking people to detect a tone when it was presented on 75% of trials. Several other tones were presented on the other 25% of trials. Adults come to "expect" the tone that is presented 75% of the time, and they detect it quite well, whereas they are very poor at detecting the other tones (i.e., those presented on 25% of trials). Infant perception differs rather dramatically: they detect all of the tones equally well (Bargones and Werner, 1994). Thus, infants and adults experience the world in quite different ways, and nonsensory factors play a prominent role for all developmental studies of sensory perception.

ASKING BABIES QUESTIONS (AND GETTING SOME ANSWERS!)

How can uncooperative little animals tell us about their sensory experiences? Various experimental tricks have been devised to determine how sensory information is processed in young animals. In one scenario, the behavioral scientist watches for a motor reflex while a sensory stimulus is presented. For example, we often respond to an unexpected noise with a startle, and this reflex provides a reliable measure that sound has been detected. Alternatively, we can take advantage of an animal's tendency to stop responding to a stimulus when it is presented many times, called habituation. Once an animal is habituated to a stimulus, we can determine whether or not a different stimulus is treated as novel; a response tells us that the animal can discriminate between two stimuli (e.g., yellow versus orange).

A more complicated procedure involves learning. Young animals can be trained to produce a stereotyped behavior, such as a head turn, when a stimulus is detected. Since most animals will work for a reward, even human infants, the head turn that an infant makes to a sound can be reinforced by showing her an interesting toy (**Figure 10.17**). The infant

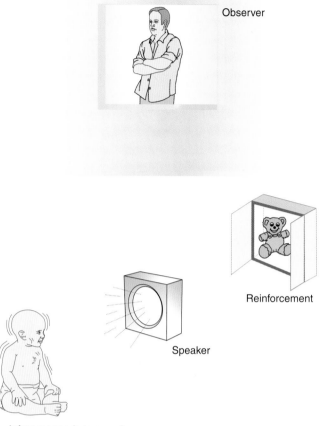

Fig. 10.17 Determining an infant's sensitivity to sound. Infants will make small movements in response to a sound they can hear, and this natural tendency has been exploited to measure the infant's ability to hear different sounds. To improve the sensitivity of this procedure, an adult observer watches the infant and judges whether the infant heard the stimulus based on any response that the infant makes. To increase the infant's responsiveness to sound, she can be reinforced for correct responses. In this case, the infant is rewarded with the appearance of a teddy bear.

learns to "work" for a visual reward (that is, turn her head when she hears a sound). This is somewhat below the minimum wage, but it serves the purpose. Infants as young as 3 months can learn this task, but scoring their performance is a bit tricky. Infants often rarely display crisp head-turns, making it difficult to decide whether they have performed correctly. Therefore, an adult observer, who cannot hear the test sounds, watches the baby and judges whether she makes a response to sound. The baby gets rewarded (with a viewing of the toy bear) whenever the observer determines that the baby has responded. In this manner, any possible response that a baby might make to sound (for example, an eye movement or a tongue wag) can be conditioned. If the adult observer can reliably determine whether a sound has been presented to the infant, when the only information available to the observer is the infant's behavior, then the infant must be responding to the sound. Of course, we believe that the infant is oblivious to this process, but she nonetheless ends up working for a reward and providing valuable information about sensory development along the way. Such training techniques require far more of an animal than sensory skills, and one may well end up studying the development of attention or memory rather than the development of sensory perception.

Behavioral scientists tend to concentrate on two criteria: absolute sensitivity and discrimination. Absolute sensitivity is a measure of the minimum stimulus amplitude that can be detected: the softest touch, the finest line, the quietest sound. Discrimination is a measure of our ability to perceive a difference between two similar stimuli: sky-blue versus turquoise, middle C versus C sharp, margarine versus butter. Below, we explore how developing animals initially process their sensory world.

ACUTE HEARING

As most animals mature, their auditory perceptual skills typically improve, both in sensitivity and resolution. Although kittens can hear at birth, they only respond to extremely loud sounds, well above the level of city traffic (>100 dB SPL, sound pressure level). Their auditory thresholds gradually decrease so that they can detect sounds at the level of a whisper (≈30 db SPL) by one month, and by adulthood they become even more sensitive than humans. A similar change is found in most developing animals. In humans, auditory thresholds drop rapidly during the first 6 months of life and are virtually adult-like by 2 years of age. Improved behavioral thresholds are well correlated to a decrease in sound level needed to evoke an electrical potential from the cochlea, suggesting that the major factor limiting detection in young animals is the ear (Werner and Gray, 1998).

As thresholds decrease, most animals also respond to higher sound frequencies. This is due, in part, to a physical change in the cochlea: a physical position along the basilar membrane responds to higher frequencies as the animal matures. Since the topographic projection from the cochlea to the central nervous system does not change significantly during this time, one might expect to find interesting changes in sound perception with development. In fact, 15-day-old rat pups can be trained to suppress activity when they hear an 8 kHz tone, but when tested three days later they suppress activity to a higher frequency. That is, the higher sound frequency activates a position on the cochlea that only days ago responded to 8 kHz (Hyson and Rudy, 1987; Rübsamen, 1992).

Although the basic sensitivity and frequency range mature rapidly, several features of sound remain difficult to detect. This is well-illustrated for tasks in which one must detect a very brief event, often referred to as temporal processing (**Figure 10.18**). For example, adults are able to detect a 5 ms gap of silence in an ongoing sound quite easily. In contrast, one-year-old infants, who have already begun to process and produce speech sounds, are only able to detect gaps that are 10 times longer (≈60 ms), and adult-like performance is not reached until about 5 years of age (Werner et al., 1992).

In humans, the maturation of adult-like performance on auditory perceptual tasks extends through the first decade of life (Stollman et al., 2004). The ability to detect small differences in the duration of a tone is more than an order of magnitude poorer in 4-year-olds, as compared to adults (Jensen and Neff, 1993). The ability to detect one sound in the presence of another may not mature until puberty. In one task, the listener is asked to recognize a long duration tone that is presented during an ongoing burst of noise. Even 10-year-old listeners

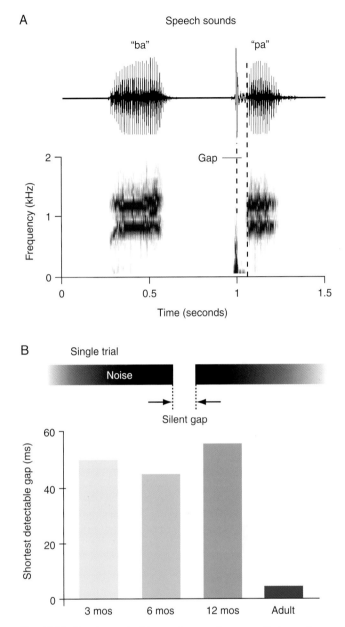

Fig. 10.18 Development of temporal processing may affect speech perception. A. An oscilloscope record of the human speech phonemes, /ba/ and /pa/. Below each record is a spectrogram of the phoneme showing the sound frequencies that compose the phoneme and their relative intensity (darker is louder). Note that the /ba/ is a continuous sound, whereas /pa/ consists of a nonvoiced component (in this case, the p sound), followed by a brief gap, and then a voiced sound (the "a"). Perception of this brief gap is critical to speech processing. B. The minimum gap that humans can perceive was assessed with a brief silent period embedded in a white noise stimulus. The bar graph shows that even at 12 months of age, human infants are almost 10 times less sensitive at detecting a gap than adults. *(Adapted from Werner et al., 1992)*

cannot perform as well as adults. Furthermore, children identified as learning disabled never reach the normal adult level of performance on this task, and it has been suggested that the onset of puberty may terminate the critical period during which neural maturation takes place (Wright and Zecker, 2004). Direct comparisons between juvenile and adult performance in nonhuman species are uncommon because a unique behavior is often used to assess the juveniles' abilities (e.g., approaching a maternal call). However, when identical procedures are used to test auditory perception, a prolonged maturation is observed, consistent with human studies (**Figure 10.19**). The threshold for detecting a sound's amplitude modulation, a major component of speech, continues to mature over 10 years in humans (Hall and Grose, 1994; Banai et al., 2007). When juvenile and adult gerbils are trained to detect amplitude modulations using an identical procedure, performance also improves in a gradual manner (Sarro and Sanes, 2010).

These behavioral measures of auditory processing are relevant to language development because human speech sounds are composed of rapid changes in frequency and intensity, and discrete periods of silence. In fact, children with learning disabilities that are due primarily to a difficulty with spoken language also perform poorly on simple auditory discrimination tasks that require temporal processing. For example, when normal children are exposed sequentially to two tones, they can report the correct sequence with delays as small as 8 ms. In contrast, the language-impaired group requires a silent interval of 300 ms in order to report the correct sequence. Performance can be improved when language-impaired children are trained to recognize speech sounds that are slowed down such that they can be discriminated. Apparently, once the nervous system learns to recognize this slower speech, it is better able to process the rapid temporal variations in normal speech (Tallal and Piercy, 1973; Tallal et al., 1996).

Visual and auditory perceptual skills that display a prolonged maturation must depend on central nervous system development. For example, we have seen that synaptic potentials are usually of much longer duration in young animals (see Chapter 8). Perhaps temporal processing is limited by long-lasting PSPs that effectively limit the "clock speed" of an organism. Processing may also be limited by low discharge rates. For example, adult LSO neurons can devote

twice as many action potentials to a given change of interaural level compared to juvenile animals (Sanes and Rubel, 1988). Striking developmental changes have now been documented for the human brain, many of which are expected to impact behavior. An anatomical study of the human auditory cortex shows that the pattern of neurofilament-positive axonal fibers continues to expand after 5 years of age (**Figure 10.20**A) (Moore and Guan, 2001). Similarly, a biochemical study of human visual cortex shows that the expression of two $GABA_A$ receptor α subunits continues to change over the first decade of life (Figure 10.20B) (Pinto et al., 2010).

Perhaps the most useful information that a developing animal gets from its ears is the location of significant objects, such as its mother or a predator. Although infants can tell whether a sound source is coming from the left or the right (sound lateralization), they are not able to make fine discriminations. Adult humans can detect a 1° change in the position of a speaker (at arms distance, your thumb occupies about one degree of visual space), but newborns can only detect a change of about 25°. In fact, it is fairly challenging for a newborn infant to determine whether a sound is coming from the left or right side (**Figure 10.21**). The sound stimulus must remain on for about one second if the infant is to make an appropriate head orientation response, whereas adults need only about a millisecond of sound, such as a finger snap (Clarkson et al., 1989). The ability of nonhuman mammals to lateralize sounds also improves with age. Rat pups can turn toward a noise a few days after the ear canal opens, but the percentage of correct turns toward the sound increases over the next week (Kelly et al., 1987).

Recordings from the developing central auditory system reveal functional properties that may help to explain immature sound localization. Maps of space are found in the superior colliculus (SC) of several mammals, and single SC neurons are selectively activated by sound from a specific

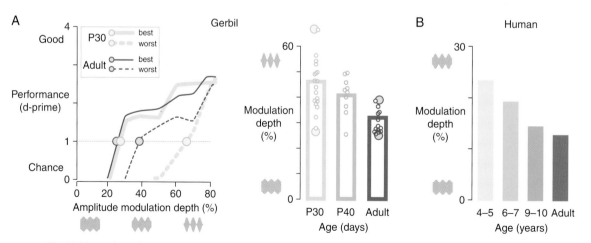

Fig. 10.19 Prolonged maturation of amplitude modulation detection in gerbils and humans. A. The graph contains psychometric functions from four individual juvenile and adult animals. Each curve shows the performance of an animal (an unbiased measure of sensitivity, called d-prime) plotted against the stimulus parameter being tested, amplitude modulation (AM) depth. Larger d-prime values indicate better performance, and larger stimulus values are easier to detect. The best juvenile and adult animal (solid lines) display the same sensitivity to AM depth. However, the worst juvenile animal performed much more poorly than the worst adult. The bar graph plots AM depth thresholds obtained at d-prime=1 for three age groups (each symbol is an individual animal, and the large symbols indicate values obtained from the four psychometric functions). There is a significant reduction in average AM depth threshold with age, and a decrease in intersubject variance. B. The bar graph shows a similar reduction in the average AM depth thresholds for humans as a function of age. *(Adapted from Hall and Grose, 1994; Sarro and Sanes, 2010)*

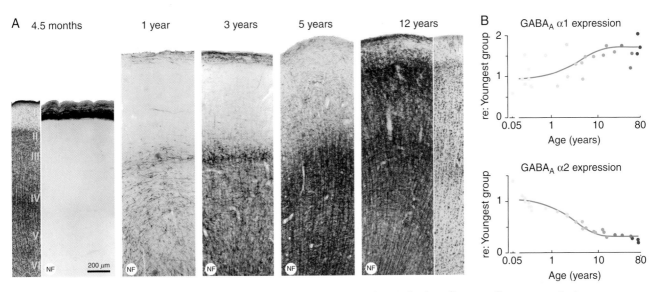

Fig. 10.20 Prolonged development of human auditory and visual cortices. A. Sections of human auditory cortex stained for neurofilament (NF) are shown for a range of postnatal ages. Nissl sections are shown for the youngest (left) and oldest (right) sample to illustrate the layers. A gradual increase in neurofilament-positive fibers is observed during the first ten years of life. B. The expression of two different GABA$_A$ receptor subunits, α_1 (top) and α_2 (bottom), for visual cortex as a function of postnatal age. There is a gradual increase in α_1 expression and a reduction in α_2 expression over the first ten years. *(Adapted from Moore et al., 2001; Pinto et al., 2010)*

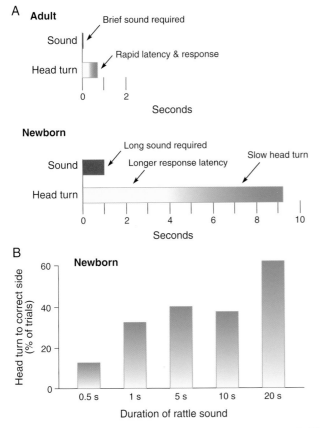

Fig. 10.21 Infants are poor at sound lateralization. A. When presented with a sound located to one side, adult humans turn their head toward the sound source within a fraction of a second. When human infants are presented with a sound to one side, they may take several seconds to respond, and the head movement can be quite slow. B. For infants, a sound stimulus must be presented for a long time period in order to elicit accurate lateralization. Whereas adults can localize sounds that last only a millisecond, newborn infants require at least 1 s of sound. *(Adapted from Clarkson et al., 1989)*

location. During the course of development, these SC neurons respond to a smaller part of the sensory world, and topography improves (Withington-Wray et al., 1990). Changes to both the external ears and the central nervous system contribute to this maturation. Thus, the shape and size of the pinna (i.e., the external ear) change dramatically during development, and this transforms sound arriving at the ear canal. Interestingly, infant neurons display adult-like spatial tuning when activated with sounds that are filtered through an adult pinna (Mrsic-Flogel et al., 2003). This suggests that the connections responsible for spatial tuning are mature at the outset of hearing. In contrast, the topography of the auditory map remains immature, even when sounds are filtered through an adult pinna (Campbell et al., 2008).

While it is easiest to consider the development of one system at a time, animals ordinarily integrate information from more than one system when making a decision. This integration can be crucial to perception and learning. Infants can learn to discriminate between different rhythms, but only if they receive simultaneous auditory and visual signals (Bahrick and Lickliter, 2000, 2002). Furthermore, this integration requires experience. Humans deprived of vision during early life due to cataracts display a reduced ability to integrate auditory speech cues with visual lip-reading cues (Putzar et al., 2007). Similarly, auditory-visual integration in speech perception is relatively poor when tested in congenitally deaf children who receive a cochlear implant (Schorr et al., 2005). The integration of auditory, visual, and somatosensory information has been studied in both the superior colliculus and association cortex (Wallace and Stein, 1997; Wallace et al., 2006). In cats, the average size of the sensory receptive fields decrease dramatically over many months, and the rate of development is much slower in cortex (**Figure 10.22**A). Receptive field size is mature by about 3 months in the colliculus, but continues to develop for about 5 months in cortex. Furthermore, neurons become responsive to more than one sensory input very gradually. In the colliculus, this process occurs between weeks 5 to 10, while in the cortex it occurs between week 10 and adulthood (Figure 10.22B). Multisensory integration by association cortex is perturbed when sensory experience is limited during development, consistent with the observations on humans who were deprived of vision or audition during development. Compared to control adults, animals reared in the dark have neurons that display much more sensory-evoked suppression of activity (Carriere et al., 2007).

SHARP EYESIGHT

Infants display clear behavioral evidence of being able to see at birth. For example, they stare for longer periods of time at a familiar face, such as their mother. However, their visual skills are quite poor as compared to adults (Teller, 1997). Visual acuity, or the ability to detect fine detail, is very poor at birth (**Figure 10.23**). One measure of visual acuity is the number of black and white lines that can be resolved per degree of visual space. (Recall that your thumb occupies about one degree of visual space at arm's length.) Adults can see about 30 black and white lines per degree, but babies can only see about one. In the more common language of an optometrist, the baby sees at 20 feet what a normal adult can see at 600 feet, and adult sensitivity is reached between 3 and 5 years of age (Birch et al., 1983). In principle, this level of acuity would permit an infant to distinguish the fingers of a hand, but their actual abilities remain somewhat of a mystery. Our best ideas come from "preferential looking" studies which tell us what babies prefer to look at, given a choice (e.g., a male face versus a female face). If infants do prefer to look at one of two visual objects, then they must be able to discriminate the difference between them. Therefore, looking preferences establish a lower bound on discrimination ability.

For primates, poor visual acuity at birth is partly due to an immature retina. Photoreceptors are relatively short and wide, meaning that less light is absorbed and a greater amount of visual space can activate each cell (Yuodelis and Hendrickson, 1986). Thalamic and cortical neurons may also impose limits on visual acuity in primates. If one compares

the theoretical acuity of the retina, based on the density of cone photoreceptors, with the acuity of single cortex neurons, then cortical performance is worse than expected (Jacobs and Blakemore, 1988; Kiorpes and Movshon, 2004). Furthermore, the animal's behavioral acuity is, at first, even worse than that of individual cortical neurons (Figure 10.23). Such results suggest that the development of accurate anatomical

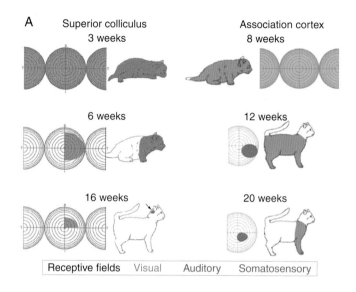

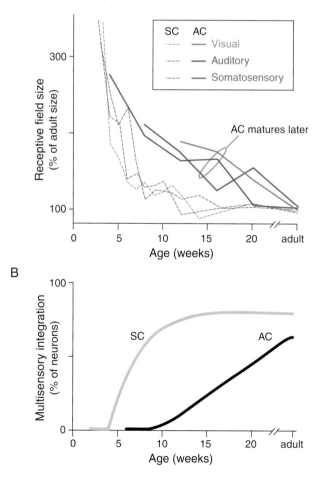

Fig. 10.22 Single neuron receptive field sizes can decrease dramatically during development. A. Recordings are made from single neurons in the superior colliculus (left) or association cortex (right) that respond to more than one sensory modality. The auditory (green), somatosensory (red), and visual (blue) receptive fields are shown for neurons from cats of increasing age. For a superior colliculus neuron at 3 weeks (top left), the neuron responds to auditory stimuli located anywhere in space, and to touch on any area of skin. By 6 weeks (middle left), the receptive fields are much smaller. In contrast, an association cortex neuron at 8 weeks (top right) remains unselective for somatosensory or auditory stimulus location. The relative size of visual, auditory, and somatosensory receptive fields are plotted for neurons recorded throughout development. For the superior colliculus (SC), there is a dramatic decrease during the first 8 weeks, and mature properties are attained by the 15th week. In contrast, the receptive field size of association cortex (AC) neurons display a more prolonged maturation maturity at 20 weeks or later. B. The age at which most neurons display multisensory integration (that is, respond to more than one sensory modality) is much earlier for SC, as compared to AC. *(Adapted from Wallace and Stein, 1997; Wallace et al., 2006)*

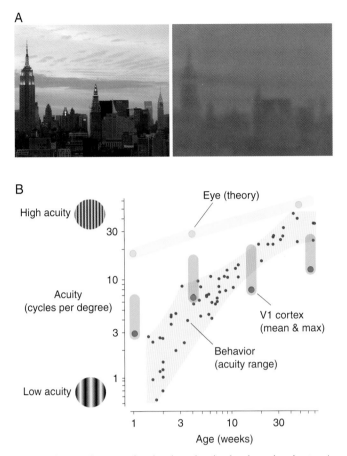

Fig. 10.23 Development of acuity along the visual pathway in primates. A. These two images show how a visual scene appear to an adult (left) and a neonatal primate (right). The "infant view" is spatially low pass filtered (blurred) and reduced in contrast to reflect the theoretical limitations of the retina. B. Acuity is measured by determining the number of visible bars per degree (left). The theoretical acuity of the eye is linked to the spacing of cones within the fovea (blue circles and shading). The behavioral acuity of individual animals is quite poor at birth and continues to mature until about 50 weeks postnatal (red circles and shading). These data suggest that acuity development is not limited by properties of the retina. Single neurons in the primary visual cortex respond to a broad range of spatial frequencies. (Green circle is the mean value recorded, and green bar extends to maximum value recorded at that age.) Behavioral acuity appears to track cortical neuron development until about 8 weeks, but there is a disparity in performance after this age. Therefore, it is possible that maturation of higher visual cortical areas is necessary for adult-like performance to emerge. *(From Kiorpes and Movshon, 2004)*

connections (Chapters 5 and 6) creates a system with only minimal capabilities, and optimal performance is acquired through widespread changes to synaptic architecture and function (Chapters 8 and 9).

Even though acuity improves dramatically in human infants during the first two years, they often fail to make proper use of visual information. When 18- to 30-month-old children were provided with a large object, followed by exposure to a miniature replica, they often failed to understand the concept of size. For example, children were first permitted to use a child-sized chair on which they could sit comfortably. They were then escorted from the room, and when they returned a miniature chair replica had replaced the original object. In many instances, children would attempt to sit on the miniature chair even though they could easily discriminate between objects

of different size, and would choose to sit in the large chair when given a choice. Such "scale errors" reach a maximum at about 2 years of age (DeLoache et al., 2004). The results suggest that visual information about object identity is not being integrated with information about its size.

Binocular vision involves the coordinated use of both eyes to judge the distance of an object, and this requires both sensory and motor development. To look at an object, the eyes must be aligned so that the images of the object of regard fall onto corresponding points on the two retinas. In addition, the brain must be able to decode the images on the two retinas into a representation of the 3D world. Cruel as it sounds, one of the simplest ways to determine whether depth perception is present in young animals is to ask whether they are willing to crawl off a cliff. To test this without the threat of litigation, an infant is placed on an elevated glass surface that is patterned on one half and clear on the other; beneath the clear section, several feet away, is a similarly patterned surface (**Figure 10.24**). If an infant is willing to crawl out over the clear surface—that is, off the "visual cliff"—then we conclude that depth perception is poor. By the time that they crawl, most infants do avoid the "visual cliff," indicating that depth perception is present (Walk and Gibson, 1961). To determine whether infants can perceive depth before they crawl, 1- to 4-month-old subjects were equipped with a heart rate monitor and suspended either above the shallow side or the deep side of a perceptual cliff. Interestingly, the heart rate was lower when the infants were suspended above the deep side, suggesting that they were interested but not fearful. In fact, an accelerated heart rate was measured at an age when infants begin to crawl (Campos et al., 1970). More precise measurements of depth perception, obtained in human and nonhuman species, suggest a rather sudden improvement in depth perception. For example, binocular perception in cats goes from being rather poor to almost adult-like between 4 and 6 weeks postnatal (Timney, 1981). In humans, the ability to perceive depth on the basis of binocular disparity emerges between 3 and 6 months (Birch et al., 1982).

Fig. 10.24 Investigating the process of depth perception in human infants. The testing device, called a *visual cliff*, consists of a sheet of plexiglas that covers a high-contrast checkerboard pattern. On one side of the device, the cloth is placed immediately beneath the plexiglas, and on the other side it is placed 4 feet below. The majority of infants would not crawl onto the seemingly unsupported surface, even when their mothers beckoned them from the other side. These results suggest that infants perceive depth by 6 months of age. (Walk and Gibson, 1961)

Why does binocular vision improve with age? Do neurons in the cortex suddenly become selective for binocular stimulation? Several neural mechanisms may contribute to the maturation of binocular vision (Daw, 1995). Since the visual system can discriminate smaller detail with age (discussed above), it is also likely to resolve smaller differences between the two eyes. Detecting differences between the two eyes is fundamental to depth perception (Barlow et al., 1967), and stereoacuity should improve as visual resolution improves. However, the gradually improved discrimination of fine detail does not explain the sudden onset of stereoscopic vision. During the period when binocular vision is improving, individual neurons respond more selectively to visual cues such as bar orientation (Chapter 9). Binocular neurons signal depth by becoming sensitive to the difference between the retinal positions of images in the two eyes, their "disparity" (Barlow et al., 1967). These neurons become more selective for disparity during development. In adult cats, a response to binocular stimulation is lost when the stimulus to one eye is offset by more than a few degrees. In contrast, few cortical neurons display this sensitivity within a few weeks of eye opening (Pettigrew, 1974; Bonds, 1979; Chapman and Stryker, 1993). It is plausible to suggest that the development of disparity sensitivity is the substrate for the contemporaneous improvement in depth discrimination, but we have yet to execute the experiments that establish the exact relationship to perception.

Color vision is a perceptual ability that improves significantly our ability to categorize objects. Interestingly, human infants seem to use color in a different way than adults. When 2- to 4-month-old infants are assessed with a special stimulus that requires subjects to use color information to detect motion, they are found to perform *better* than adults when both groups are compared to a reference stimulus in which only luminance information is needed to perceive motion (Dobkins and Anderson, 2002). These results suggest that the visual pathways that carry information about an object's color have a relatively strong input to motion processing areas of the visual cortex at first, and this is reduced during postnatal maturation.

The perceptual abilities discussed above are concerned primarily with basic sensory properties, yet there are many higher level skills that display a prolonged maturation. We are often called upon to integrate features of an object that are separated in space such that we perceive the global shape. Thus, when viewing a horse that is partially occluded by a tree, we nonetheless recognize that the front and back halves belong to a single animal. The relative alignment of separated features is an important cue for visual object recognition (for example, the curved horizontal contour of the horse's back). This ability can be tested in the laboratory by asking subjects to detect a contour that is composed of small, spatially separated, segments. Children demonstrate relatively weak spatial integration of these segments, and only reach adult levels of performance at about 14 years (Kovács et al., 1999; Hadad et al., 2010). Thus, many behavioral attributes of the visual system may only reach maturity well into adolescence.

SEX-SPECIFIC BEHAVIOR

The many differences between male and female behavior are a popular subject of conversation. They are also a source of considerable controversy, and the political stakes can be quite high. We primates tend to debate whether sex-specific behaviors are due to our "biology" or to the social environment in which we are raised. While the debate is seductive, the relationship between brain development and sexual behavior varies tremendously from species to species. Since mating and maternity have been most thoroughly explored at the neural level, we will mostly focus on these behaviors. However, it is worth mentioning some complex behaviors that differ between male and female animals. These differences in behavior are commonly referred to as *sexual dimorphisms*. Predatory behavior is sexually dimorphic in lions (females do more of it), urination posture is sexually dimorphic in dogs (males elevate one leg), and olfactory signaling is sexually dimorphic in moths (females produce a pheromone). In both rats and monkeys, young animals engage in play behavior that differs between the sexes, at least in its frequency of occurrence. Males tend to have more play fights than females. These "fights" will typically begin with one animal jumping onto the other, and they end with one animal on top of the other. When testosterone is given to a pregnant monkey, the play behavior of her female offspring becomes more male-like (Abbott and Hearn, 1978).

Behavioral evidence of a sexually distinct nervous system comes from the prevalence of certain neurological and psychiatric diseases in males versus females. For example, both dyslexia and schizophrenia are more prominent in males (about 75% of cases), while anorexia nervosa is exhibited primarily by females (over 90% of cases). Many studies have also focused on the cognitive abilities of normal adult humans (Kimura, 1996). When presented with two figures drawn at different orientations, males are better able to "mentally rotate" the objects to determine whether the two figures are the same. In contrast, when presented with a picture containing many objects, females are better able to say which objects have been moved in a second picture. While these results tend to fascinate us, the challenge will be to understand what exactly is being measured and what its relevance is to behavior.

The male-female differences in complex behavior patterns do raise a host of interesting questions: Are these differences due to biology or environment? If there is a biological signal, then is it genetic or hormonal? Are the differences irretrievably established at birth, or are they modifiable throughout life? Certain sexual characteristics emerge during embryonic development, such as differentiation of the genitals and the motor neurons that innervate them (see Chapter 7). However, this is only the first step of a lifelong process. The nervous system continues to respond to steroid hormones throughout life, making it important to ask whether a dimorphism is induced by early exposure to a hormone, or whether it can be elicited during adulthood, in either sex, merely by adjusting the amount of circulating hormone. For example, even before puberty, one region of the amygdala has a greater volume and 80% more excitatory synapses in male rats than in females, indicating that the dimorphism is induced developmentally. However, adult circulating hormones maintain this difference. Castration of adult males causes the volume to shrink to female values, and androgen treatment of adult females enlarges the structure to normal male size (Cooke et al., 1999; Cooke and Woolley, 2005). Therefore, we will begin by examining the early determinants of gender and then explore determinants of behavior and brain development.

GENETIC SEX

Animals seem to have two general ways of establishing gender. In fruit flies and other insects, the genetic sex of each cell is the key determinant. If a cell has a single X, then it is male. If it has two Xs, then the cell expresses a protein called Sex-lethal and becomes female. The nematode, *C. elegans*, also comes in two categories, but they are male and hermaphrodite (i.e., an animal with both types of gonad). As with fruit flies, sex is determined by the ratio of X chromosomes to auto-somes, and XO-lethal is the gene product that is activated in animals with a single X.

The genetic sex of each cell in a mammal is specified by the presence of either two X chromosomes (female) or one X and one Y (male). However, the genetic sex of most somatic cells is not thought to have an immediate influence on their development. It is the genetic sex of gonadal tissue that really matters. Primary sex determination refers to differentiation of the gonadal tissue, and this is determined by the *sry* gene on the Y chromosome, which encodes a transcription factor (Goodfellow and Lovell-Badge, 1993; Sekido and Lovell-Badge, 2009). If Sry is present, the gonads develop into testes and secrete testosterone; if Sry is absent, the gonads develop into ovaries. The Sry protein binds directly to a testis-specific enhancer of *sox9*, leading to its expression. Targeted deletion of *sox9* leads to ovary development in XY embryonic mice, while *sox9* duplication leads to male development in XX embryos. Sox9 controls the expression of downstream targets to induce Sertoli cells of the testes and prevent development along the female pathway. For example, a locus on the X chromosome, called Dax-1, is involved in ovary determination. Thus, it is thought that Sox9 represses Dax-1 in genetic males.

All sex differences, including those of the nervous system, were originally thought to originate from the gonads after primary sex determination was completed. However, the genetic sex of somatic cells may play a role in their differentiation, and genomic imprinting biases brain development toward one of the parents. Each of these issues is discussed below.

HORMONAL CONTROL OF BRAIN GENDER

In most vertebrates, as the gonads differentiate and begin to secrete hormones, tissues throughout the body respond by adopting a male or female phenotype. This is called *secondary sex determination*. The principal importance of gonadal hormones is powerfully demonstrated by removing the gonads before primary determination occurs (Jost, 1953). Without exception, animals develop as females (e.g., they have a vagina, a uterus, and oviducts). Furthermore, their sexual behavior is female-like, presumably because certain areas of the nervous system have developed female characteristics (Phoenix et al., 1959).

The testes masculinize the body by releasing the steroid hormone, testosterone. The level first rises during the perinatal period, goes down after birth, and rises again at puberty. When genetically female (XX) rats are treated with testosterone within a few days of birth, they will not display female sexual behaviors as adults. That is, they will not arch their back (lordose) when approached by a male, and they will mount a female rat if given a dose of testosterone. When genetic males (XY) are castrated soon after birth, they will not mount a female as an adult, even if supplied with testosterone.

To carry out some of its actions, testosterone must be converted to 5α-dihydrotestosterone (DHT) by the enzyme 5α-reductase. Genetic males who carry a disrupted form of the 5α-reductase gene, and cannot make DHT, develop female-like external genitals despite having functional testes and plenty of circulating testosterone (Imperato-McGinley et al., 1979; Thigpen et al., 1992). Interestingly, most of these individuals who were raised as girls nonetheless chose to adopt a male identity during or after puberty. The results suggest that DHT is involved in differentiation of external genitalia, but the development of gender identity is a more complex process that involves testosterone.

Testosterone and DHT can masculinize the mammalian brain directly, or following its conversion to estradiol (below). These ligands act through steroid hormone receptors, cytoplasmic proteins with a steroid-binding domain and a DNA binding domain, providing a very direct pathway to the genome (Beato and Klug, 2000). Estradiol binding to the estrogen receptor (ER) leads to dimerization, and migration into the nucleus where the dimer binds a hormone response element to regulate transcription. Similarly, the androgen receptor (AR) is activated by binding either testosterone or DHT in the cytoplasm and then translocating into the nucleus where it acts as a transcription factor. Estradiol receptors are found in neurons of the hypothalamus and amygdala, and they are expressed transiently in the cortex and hypothalamus. Androgen receptors are expressed at highest concentration in the hypothalamus and limbic structures.

The first masculinizing pathway involves the conversion of testosterone to the estrogen hormone, estradiol-17β, by an enzyme called aromatase; DHT is *not* converted. At first, this might seem puzzling because estradiol is secreted by the ovaries and promotes differentiation of the female reproductive organs. However, testosterone is also an intermediate metabolite of estradiol in the ovaries. Thus, we should probably not think of hormones as being "male" or "female." When an estrogen receptor gene (ER-α or ER-β) is deleted, sexual behavior is significantly attenuated. Males rarely achieve an intromission or an ejaculation when ER-α is deleted, and display less aggression than wild-type males, often failing to attack an "intruder" when it is placed in the male's home cage (Ogawa et al., 1997). Genetic male mice that lack both ER genes display no sexual behavior, including mounting and ultrasonic vocalizations (Ogawa et al., 2000).

There are at least two factors that allow estradiol to act selectively on the brains of genetic males. First, aromatase activity is higher in the brains of male mice, particularly during the prenatal and neonatal periods (Hutchison, 1997). Second, the blood of young animals contains an estradiol-binding protein, called α-fetoprotein, that may prevent estrogen secreted by the ovaries from reaching the brain (Uriel et al., 1976). This interaction is clearly revealed by a set of genetic experiments in mice. When the α-fetoprotein gene is deleted, genetic females display masculine sexual behavior and neurochemical characteristics. However, the female phenotype can be rescued by preventing the conversion of testosterone to estradiol (Bakker et al., 2006).

Testosterone can also have a direct masculinizing role, as revealed by androgen receptor (AR) null mice

(Sato et al., 2004). Genetic males with the AR mutation do not display male-typical sexual and aggressive behaviors. Treatment with DHT does not restore normal sexual behavior, but it does partially rescue male aggressive behavior. In primates, androgen appears to be the main source of brain masculinization (Swartz and Soloff, 1974; Wallen, 2005). Genetically male humans born with complete AR insensitivity display feminine behaviors (Imperato-McGinley et al., 1979). In contrast, genetic males with an aromatase dysfunction display normal male behavior (Grumbach and Auchus, 1999).

One might expect the hypothalamus to be a developmental target of gonadal hormones since these regions are involved directly in producing sex-specific behaviors. For example, medial preoptic neurons fire rapidly just prior to male copulation, and copulatory behavior is disrupted when this area is lesioned. Medial preoptic neurons are also known to take up more testosterone than any other brain region in adult animals. One of the first studies to show that male and female brains actually differ in a measurable way was an ultrastructural study in the preoptic area (Raisman and Field, 1973). One aptly named structure, the sexual dimorphic nucleus of the preoptic area (SDN-POA), is so much larger in male rats, than females, that one can actually see the difference in tissue sections without using a microscope (**Figure 10.25**A). A similar difference is found in the primate hypothalamus, including that of humans. Sexual dimorphisms have also been found in the nucleus accumbens (NAc), a structure that is involved in reward learning (Figure 10.25C). The NAc also plays a primary role in addictive behaviors, and females are more sensitive to psychostimulants, self-administering such drugs at lower doses than males and escalating to addiction more rapidly (Becker and Hu, 2008). In rats, females display a greater density of dendritic spines, particularly those with very large spine heads, suggesting that excitatory drive might be greater (Forlano and Woolley, 2010).

The sexual dimorphism of SDN-POA is an example of secondary sex determination in the nervous system. The hormonal environment of developing males yields a larger nucleus, and the dimorphism can be greatly reduced by castrating genetic males within a few days of birth (Figure 10.25B). Furthermore, this nucleus can be enlarged in genetic females when they are treated with testosterone as neonates (Gorski et al., 1978). Intracranial implants of estradiol turn out to be as effective as testosterone in masculinizing the SDN-POA, and such estradiol-treated females fail to lordose or ovulate. Presumably, testosterone is converted to estradiol in the male SDN-POA, whereas the circulating estradiol in females is bound by α-fetoprotein (Naftolin et al., 1975).

A second region of the hypothalamus, the ventromedial region (VMH), also participates in sexual behavior in rats. Damage to this region disrupts female copulatory behaviors in rats, such as lordosis, and stimulation of the region seems to facilitate such behaviors. Lesions also have a profound effect on food intake, particularly in females. Neurons of the VMH are selectively activated by the ovarian hormone, estrogen, and in female rats the cells respond by producing progesterone receptors. This does not occur in the male VMH. In primates, the hormonal signal may be somewhat different because loss of the adrenal glands, a source of androgen hormones, leads to reduction in copulatory behavior. Although there is little difference in the

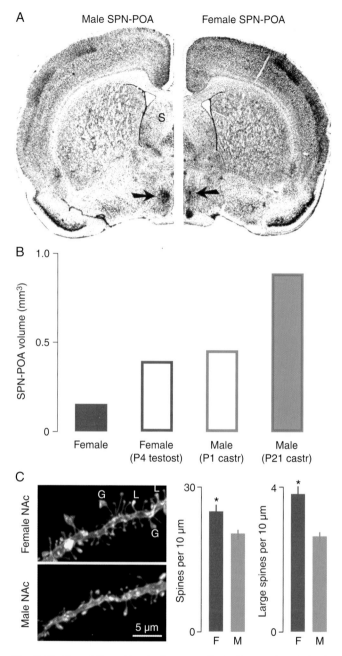

Fig. 10.25 Sexual dimorphism in the mammalian brain. A. A hypothalamic structure called the sexually dimorphic nucleus of the preoptic area (SPN-POA, arrows) is almost six times larger in male than in female rats. B. In genetic females, the size of SPN-POA can be increased by treating with testosterone at postnatal day 4. In genetic males, the SPN-POA can be decreased in size by castrating at postnatal day 1. C. Dendritic spines are more numerous and have larger spine heads in the nucleus accumbens of adult female rats, as compared to males. Photomicrographs show labeled spines, indicating large (L) and giant (G) spine heads. The bar graph quantifies the difference between males and females. *(Adapted from Gorski et al., 1978, 1980; Forlano and Woolley, 2010)*

absolute size of the male and female VMH, there is some reason to believe that it becomes sexually dimorphic during development (Sakuma, 1984). Estradiol and testosterone have a dramatic affect on both neurite outgrowth and dendritic branching in organotypic cultures of the mouse hypothalamus (Toran-Allerand, 1980; Toran-Allerand et al., 1983).

Gonadal hormones can induce sexual dimorphism in the brain by selectively preventing cell death, inducing growth, or inducing the production of new cells. The male SDN-POA nucleus is larger because estradiol decreases naturally-occurring neuron death (see Chapter 7). Selective cell death may also account for the sexual dimorphism in a human hypothalamic nucleus, called INAH 1. Until age 5, the number of INAH 1 neurons is about the same in males and females, but the number of neurons then declines more rapidly in females (Swaab and Hofman, 1988). Newly generated neurons are also produced in many areas of the nervous system during puberty. When the cell birth date marker, BrdU, is injected at around the time of puberty in rats, a significantly greater amount of label is found in nuclei that are soon to become sexually dimorphic, and many of the new cells are neurons. For example, a nucleus that is larger in females, the anteroventral periventricular nucleus of the hypothalamus (AVPV), displays about twice as much proliferation as the male counterpart. The selective addition of new neurons does not occur when gonadal hormones are eliminated before (Ahmed et al., 2008). The emergence of sexual dimorphism may even require social interaction. Naked mole rat colonies consist of one breeding female and one to three breeding males. Most mole rats do not display mating behavior, but can assume this role if removed from the colony and placed with an opposite sex partner. Central regions known to be sexually dimorphic in other mammals are larger only in breeding mole rats, be they female or male; however, males and females of a given breeding status do not differ from one another (Holmes et al., 2007). This suggests that social status can induce neural remodeling that facilitates breeding.

SINGING IN THE BRAIN

One of the most striking correlations between sexual behavior and brain anatomy is found in several species of songbirds. Male birds attract a mate of the same species with vocalizations, or songs, that are commonly learned during juvenile development (see below). Zebra finches learn one song during the first 80 days after hatching, while canaries add new phrases to their song each breeding season. When scientists first looked at the brains of these animals, they were startled to find brain regions of remarkably different size in males and females (Nottebohm and Arnold, 1976). The sexual dimorphism occurs in brain nuclei that are known to participate in song production (RA, HVc), and these structures are much larger in males (**Figure 10.26**). Furthermore, when hatchling females are treated with estradiol, they can grow up to sing almost as adeptly as male birds (Gurney and Konishi, 1980; Simpson and Vicario, 1991). In male canaries, the size of vocal control nuclei changes during the course of a single breeding season, getting larger as testosterone levels rise (Nottebohm, 1981). Hormone treatment can apparently enhance the size of brain nuclei both by increasing afferent innervation and promoting dendritic growth.

It seems odd that females do not vocalize, if only to facilitate the mating process. In fact, female tropical wrens do sing a "duet" with the males. Furthermore, when the song repertoire of a female wren becomes relatively large, then the size of its song-control nuclei is found to be similar to that of males (Brenowitz and Arnold, 1986). The vocal repertoire of *Xenopus* females is also important in guaranteeing

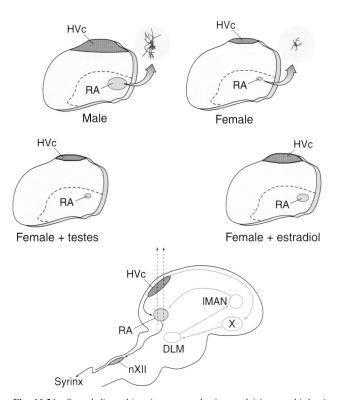

Fig. 10.26 Sexual dimorphism in song production nuclei in song birds. A sagittal view of the brain shows the nuclei involved in the learning and production of vocalizations. The pathway from HVc to RA to nXII is the primary output pathway to the song production apparatus (bottom). Both HVc (red) and RA (yellow) are much larger in adult male birds (top left) compared to adult females (top right). In addition, neurons in the male RA nucleus have a more elaborate dendritic architecture compared to those in females. Female zebra finches can be engineered to develop testes and little ovarian tissue, yet their HVc and RA nuclei do not become larger (middle left). In contrast, when female birds are treated with estradiol during development, the HVc and RA nuclei do become masculinized (middle right). *(Adapted from Schlinger, 1998)*

fertilized eggs. In this species, the male mating call has been well-characterized; like birds, there is a sexual dimorphism of both neural and muscular components related to song production (Kelley, 1997). However, a female vocal behavior, termed *rapping*, is thought to trigger the entire copulatory repertoire (Tobias et al., 1998). When the female frog is unreceptive, it produces a ticking sound, but when it is ready to lay eggs, it begins to rap. This call stimulates males to vocalize even more vigorously and to attempt copulation. Although it is not yet known how the female brain becomes specialized for this behavior during development, sex specific differentiation has been observed at the level of vocal motor neuron membrane properties, and this may lead to different output patterns for males and females (Yamaguchi et al., 2003).

GENETIC CONTROL OF BRAIN GENDER IN FLIES

Since the control of gender is cell autonomous in insects, sexual behavior can be explored from a genetic perspective. Male fruit flies recognize females based on an olfactory cue, called a *contact pheromone*, and males will perform stereotyped courtship behavior when they detect this signal. The male orients toward a female, taps her abdomen, flutters his wings in song, and places his proboscis (the mouthparts) on the female's genitals. If the female is receptive, the male will then mount her and copulate. How does the central nervous system create this complex set of sex-specific behaviors? One approach to the problem is to create unusual flies, called mosaics, that have some cells that are genetically female (XX) and some cells that are genetically male (XO). By studying many flies of this sort, each with a unique mosaic, it is possible to determine which brain cells must be male or female such that the proper behavior is displayed (Hall, 1977).

A genetic trick can be used to construct a line of animals in which a single part of the brain is female (**Figure 10.27**A). A piece of DNA, called PGAL4, is randomly inserted in the genome of many flies. By chance, it will occasionally insert

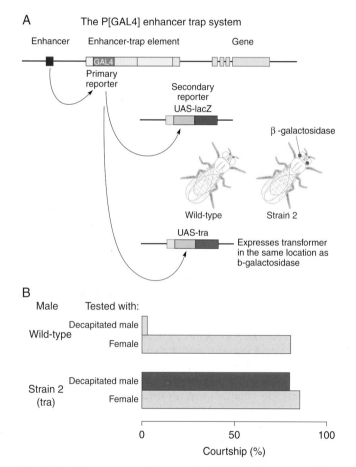

A

next to an enhancer, and this enhancer will then activate the GAL4 gene in the enhancer trap element. If the enhancer is only active in one part of the body, then GAL4 will be expressed only in that same part of the body. It turns out that GAL4 can activate another promotor, called *upstream activating sequence* (UAS). If an experimenter can hook up a gene of interest to the UAS promoter, then the gene of interest will be expressed wherever GAL4 is expressed. This enhancer trap system has been used to express a feminizing signal (*transformer*) in olfactory neurons that process a pheromonal signal (Ferveur et al., 1995).

Genetic males that express *transformer* are presented with flies of either sex to see whether they selectively court the female, as normal males do. Surprisingly, some strains of flies courted males with as much vigor as they did females (Figure 10.27B). The behavior of transformed animals may be due to their failure in discriminating the female pheromone. In fact, when the enhancer trap technique is used to make male flies that secrete only female pheromones, these flies are courted as if they are females (Ferveur et al., 1997). Thus, in flies, specific brain regions must have a gender if animals are to accurately interpret sensory information and produce sexually appropriate motor responses.

A separate tack has been used to explore what kinds of genes must be expressed in male or female nerve cells in order to produce correct sexual behaviors (Hall, 1994). For example, a gene product called *fruitless* is expressed in about 500 neurons of male flies only, and mutations of this gene also cause males to court one another. A mutation of the *dissatisfaction* gene leads virgin females to resist males during courtship, and they fail to lay mature eggs (Finley et al., 1997). Most mutations that affect sexual activity in flies are also found to affect other behaviors. Mutations of the *period* gene affects circadian rhythms, but they also change the temporal properties of the courtship song. Depending on the precise mutation, the interval between wingbeats can be shorter or longer than normal. That is, the song will have a lower or higher frequency, respectively.

FROM GENOME TO BRAIN GENDER IN VERTEBRATES?

For vertebrates, the simple hypothesis is that testosterone is secreted by the testes, and this leads to a masculinized nervous system in male animals. However, observations in birds and frogs suggest that other factors are involved (Wade and Arnold, 1996; Kelley, 1997; Arnold, 2009). They raise the possibility that female and male brains differ from one another even in the absence of gonadal signals, and suggest a primary role for genes encoded on the sex chromosomes. First, the level of estradiol required to masculinize the nervous system of female song birds is quite high, and even these high levels do not result in a fully masculinized phenotype. Furthermore, the level of circulating androgen is quite similar in male and female frogs during development. However, female frogs that receive transplanted testes develop a larger larynx and more laryngeal motor neurons than do females that are treated with a single androgen (Watson et al., 1993). Second, it has not been possible to block masculine development of the nervous system in male birds by manipulations designed to decrease estrogen. Third, when genetic female zebra finches are pharmacologically engineered to develop with testes, with little to

Fig. 10.27 Enhancer traps and the expression of the transformer gene. A. An enhancer trap element inserts into the fly genome between an enhancer region and the gene that it normally controls. Whenever the enhancer is activated by a transcription factor, a reporter gene within the enhancer trap is expressed. In this example, a yeast transcription factor called *GAL-4* gene is expressed. To visualize the anatomical location of *GAL-4* expression, the enhancer trap flies are crossed to flies that have a *UAS-lacZ* gene. Since *GAL-4* is a transcription factor that activates UAS (blue), the *lacZ* gene (red) is expressed, and it encodes a protein (β-galactosidase) that can be stained for (red). Thus, labeled cells are known to have *GAL-4* expression. The enhancer trap line can also be used to drive the expression of native genes, such as *transformer* (green). Expression occurs only in cells with an activated enhancer. B. When an enhancer trap line was used to express transformer in olfactory neurons, the male flies courted males and females equally. Normal males only court females. *(Adapted from Ferveur et al., 1995)*

no ovarian tissue, their vocal control nuclei continue to exhibit a female phenotype (Figure 10.26).

A particularly compelling example of genetic determination of brain sexual dimorphism was discovered in a rare zebra finch gynandromorph (i.e., an animal that is a mosaic of male and female structures because some cells are chromosomal females while others are chromosomal males). In this instance, the animal was genetically male on one side and genetically female on the other (**Figure 10.28**A). Since only females carry a W chromosome, it was possible to stain for mRNA encoding a W chromosome gene and to show that expression was limited to one side of the brain (Figure 10.28B). As expected, one side of the animal developed a male-like gonad, and the other side developed a female-like gonad (Figure 10.28C). Thus, the brain was exposed to an identical, if somewhat peculiar, gonade-produced hormone environment during development. However, the genetically male side of the brain had a much larger HVc nucleus, as compared to the genetically female side (Agate et al., 2003). Thus, the genetic sex of songbird brain cells appears to play a role in their differentiation.

These results suggest either that the gonads are a more complicated endocrine organ than we suspect, or that the nervous system contains intrinsic signals that bias its development towards a male or female phenotype. If so, then we might expect some neurons to exhibit a sexual dimorphism even before gonadal development. In fact, some diencephalic neurons express a sex-specific phenotype in vitro, even when explanted before a difference in plasma levels of testosterone emerge (**Figure 10.29**A). Tyrosine hydroxylase-expressing neurons are 30% larger in males, and the number of prolactin-expressing neurons is two to three times greater in female tissue, similar to adult animals (Kolbinger et al., 1991; Beyer et al., 1992). A similar experiment can be performed in mice carrying a Y chromosome that lacks the *sry* gene. These mice are genetically male (XY), but they do not develop testes Furthermore, mice can be genetically engineered to carry an active copy of the *sry* gene on an autosome. Thus, it is possible to create genetically female (XX) mice that expressed Sry and developed male gonads (Carruth et al., 2002). When mesencephalic neurons were explanted at embryonic day 14, the number of dopaminergic neurons was greater in *sry*[-/-] genetic males than genetic females, even though both groups had female gonads (Figure 10.29B). Further characterization of *sry*[-/-] genetic males indicated that genes on the sex chromosomes other than *sry* are responsible for several behavioral or neural traits (Arnold and Chen, 2009). Thus, aggression is higher in genetic males, even when they develop with ovaries, and genetic females display a more rapid response to painful stimuli, even when they develop with testes. Sry, itself, has a direct effect on neural phenotype. Down-regulation of Sry causes a specific reduction of tyrosine hydroxylase in the substantia nigra and an associated motor deficit (Dewing et al., 2006). Taken together, these experiments show that the vertebrate nervous system develops at least some sex-specific characteristics due to genes on the sex chromosomes, and independent of gonadal signaling.

GENOMIC IMPRINTING: THE ULTIMATE IN PARENTAL CONTROL

Parents generally contribute unevenly to the environmental influences on their developing offspring. In mammals, maternal care is usually greater, beginning with the in utero environment and continuing through postnatal lactation. This is also true at the genetic level where sex chromosomes provide the strongest parental influence, as discussed above. Moreover, mitochondrial DNA is maternally inherited, and has been implicated in regulating brain function and size (Roubertoux et al., 2003). Even at the level of autosomal genes there is a tremendous impact of the parent of origin. Through a process called *genomic imprinting*, hundreds of genes are inactivated in the germ cells produced by each parent (Wilkinson et al., 2007; Curley and Mashoodh, 2010). The process whereby alleles are in activated is called *epigenetic control*, and usually involves DNA methylation at specific control regions or histone modification. This process can result in the selective expression of one of the two alleles, depending on whether it pass through the egg or the sperm (**Figure 10.30**A).

What is the reason for genomic imprinting and how does it affect brain development? A general evolutionary theory holds that imprinting arose from a "conflict" that was due to differing interests of the maternal and paternal genomes. In this scenario, genomic imprinting by the parent is thought to increase the likelihood that his or her genes will be passed on. The clearest example of this idea is the allocation of resources

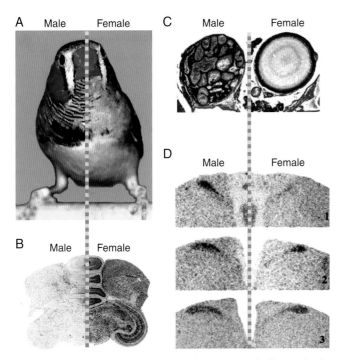

Fig. 10.28 Genetic sex of brain determines the pattern of differentiation in the zebra finch. A. A zebra finch gynandromorph with male plumage on one side (left) and female plumage on the other (right). B. The W chromosome is found normally only in females. The brain section shows *in situ* hybridization of mRNA encoding the W chromosome gene, *ASW*, to be ubiquitous on the female (right) side of the brain, but virtually absent on the male (left) side (dark areas show label). C. Histological sections of the gonads reveal dysmorphic testis on the genetically male side (left) and ovarian tissue containing a number of follicles on the female side (right). D. The HVc nucleus is normally larger in males than in females. The series of images shows *in situ* hybridization for androgen receptor mRNA (dark areas) to mark HVc at three caudal-to-rostral levels. The HVc is 82% larger on the male side of the brain (left) as compared to the female side (right). *(From Agate et al., 2003)*

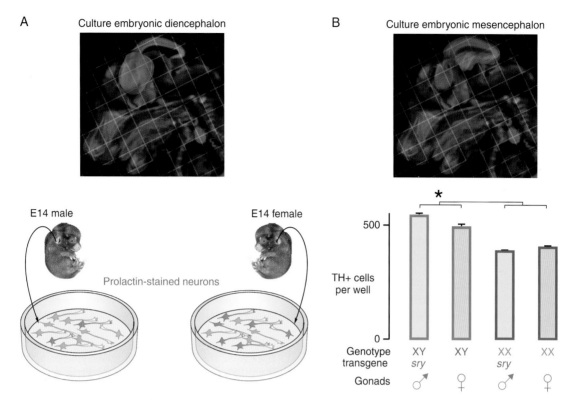

A Culture embryonic diencephalon

E14 male

Prolactin-stained neurons

E14 female

B Culture embryonic mesencephalon

TH+ cells
per well

500

0

Genotype	XY	XY	XX	XX
transgene	*sry*	♀	*sry*	♀
Gonads	♂	♀	♂	♀

Fig. 10.29 Development of sexual dimorphism independent of gonadal development. A. A horizontal section through the embryonic mouse brain shows the location of the diencephalons (green). When rat diencephalon is explanted at embryonic day 14, the tissue from female embryos produces two to three times more prolactin-expressing neurons (red) compared to cultures from E14 males. B. A horizontal section through the embryonic mouse brain shows the location of the mesencephalon (green). When rat mesencephalon is explanted at embryonic day 14.5, the cultures from genetic males (bars with blue borders) have significantly more TH-positive cells, as compared to cultures from genetic females (bars with red borders). Genetic males had the *sry* gene deleted on the Y chromosome (red shade), but had male gonads if a *sry* transgene was expressed by an autosome (blue shade). Genetic females (red borders) also produced male gonads if the *sry* transgene was expressed (blue shade). *(Adapted from Beyer et al., 1992; Caruth et al., 2002)*

between mother and offspring. The father's genes are more likely to be passed on if the mother's resources are maximized for the fetus in the current pregnancy (even if this harms the mother's long-term viability). In contrast, the mother's genes are more likely to be passed on if she distributes her resources to all offspring (which may be sired by a different father in the future). In mice, insulin-like growth factor 2 (IGF2) is paternally expressed (that is, the maternal copy is inactivated), and promotes growth in the offspring through binding to the IGF1 receptor (Wilkins and Haig, 2001). Greater growth would monopolize the mother's resources, but would yield an individual that is likely to survive and reproduce. IGF2 also binds to the IGF2 receptor which is maternally expressed (and inactivated by the father), and this receptor reduces the IGF2 effect on growth, potentially allowing the mother to allocate more resources to a future pregnancy.

The impact of genomic imprinting is also apparent in developmental disorders of the nervous system. In humans, two disorders that lead to mental retardation, Prader-Willi Syndrome and Angelman Syndrome, result from a deletion of the paternal or maternal copy of 15q11-13, respectively (Figure 10.30B). Both chromosomes contain genes for Necdin (*ndn*) and the ubiquitin protein ligase E3A (Ube3a). However, the maternal copy of *ndn* is inactivated through genomic imprinting, such that if the paternal region lacks

a functional copy of *ndn*, then there is a complete loss of Necdin function, and individuals display Prader-Willi Syndrome (mental retardation, hyperphagia, stubbornness, and compulsive traits). Conversely, the paternal copy of *ube3a* is inactivated through genomic imprinting, so that a deletion of a functional maternal copy of *ube3a* leads to Angelman Syndrome (mental retardation, absent speech, happy affect, and inappropriate laughter).

At least 1300 genes that are expressed in the mouse brain display some degree of bias toward the maternal or paternal allele (Gregg et al., 2010a, 2010b). Many of these genes are expressed in brain areas associated with feeding, mating, social interactions, and reward learning. In the cortex of adult females, almost 50% more neurons express genes on the maternal X chromosome, whereas expression of the paternal and maternal X did not differ in the hypothalamus (Figure 10.30C, left). When considering all imprinted genes in the fetal brain, 61% percent are maternal in origin, indicating that the mother exerts a primary influence over brain development. The opposite pattern is seen in adulthood, with about 70% of imprinted genes in cortex and hypothalamus coming from the father (Figure 10.30C, right). This indicates that fathers have a greater influence on adult brain function. Thus, maternal bias is greater for the X chromosome, and paternal bias may be greater for autosomes.

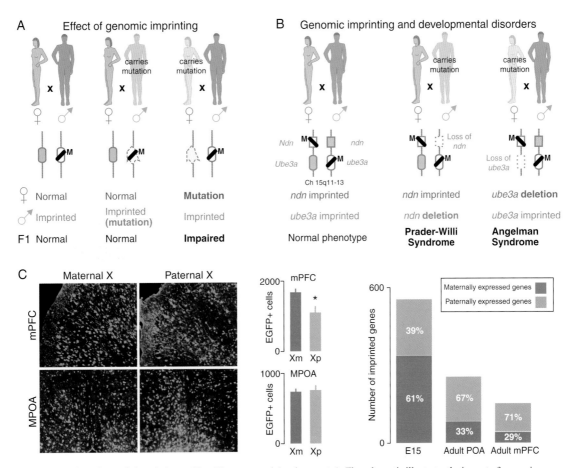

Fig. 10.30 Genomic imprinting and its effect on neural development. A. The schematic illustrates the impact of genomic imprinting. When both alleles are normal and the male allele is imprinted, then the offspring is normal. Similarly, if the male carries a mutant allele and it is imprinted, then the offspring is normal. However, if the female carries a mutation and the male allele is imprinted, then the offspring will express only the mutated protein and may be impaired. B. Genomic imprinting can lead to specific developmental disorders of the nervous system. Chromosome 15q 11-13 contains two genes that are either maternally inactivated (*ndn*) or paternally inactivated (*ube3a*). If the paternal *ndn* gene is deleted and the maternal allele is inactivated, then offspring will display Prader-Willi Syndrome. However, if the maternal *ube3a* gene is deleted and the paternal allele is inactivated, then offspring will display Angelman Syndrome. Both disorders include mental retardation as part of the phenotype. C. Genomic imprinting of the X chromosome in mice is studied with an X-linked transgene that expresses a fluorescent protein and is subject to inactivation. The medial prefrontal cortex (mPFC) displays a greater number of GFP+ neurons from the maternal X chromosome transgene, as compared to the male, indicating that there is more inactivation of the paternal X chromosome. There is no difference in the medial preoptic area (MPOA). The total number of maternal and paternally expressed genes is shown for the whole brain at embryonic day 15 (E15) and for the adult POA and mPFC. A greater percentage of maternal genes are expressed during development, while a greater percentage of paternal genes are expressed in adulthood. (Gregg et al., 2010a, 2010b)

HIT THE GROUND LEARNING

Many vertebrates are born with an ability to obtain food and warmth from their mother, when offered. Nestling herring gulls peck at the tip of their mother's beak for food, neonatal rodents assume a specific position in order to suckle at a nipple, and newly hatched jewel fish have a natural tendency to approach objects that are colored like the broody adult. Although these innate motor behaviors are very sophisticated in the apparent absence of any experience, many animals must learn to recognize and respond selectively to their mother. Konrad Lorenz, a corecipient of the 1973 Nobel Prize, made the rather dramatic observation that hatchling ducks and geese will follow the first moving object that they see, forming a very stable attachment (Lorenz, 1937). Ordinarily, the

mother goose fills this role, but hatchlings can also learn to follow inanimate objects, and even the experimenter himself (**Figure 10.31**A). This learned behavior is termed *filial imprinting* (there is no biological relationship to genomic imprinting, discussed above). Filial imprinting has the immediate advantage of keeping offspring with the provider, and it can also have implications much later in life. When mature, the male birds will court a member of the species on which they imprinted, whether it is a bird, dog, or human.

Filial imprinting is not unique to birds. Tree shrew pups will imprint on the nursing mother during the second postnatal week. If removed from the nest during this period, a pup will not learn to follow its real mother, and it can be induced to follow a cloth permeated with the odor of a foster mother (Zippelius, 1972). Similarly, rat pups come to prefer their

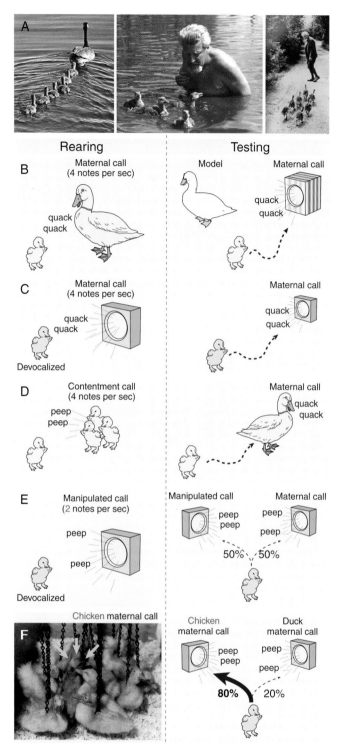

Fig. 10.31 Maternal imprinting. A. Goslings that have imprinted on their mother (left). Konrad Lorenz is shown with goslings (middle) and grown geese (right) that imprinted on him. B. When ducklings are exposed to maternal vocalizations (~4 notes per sec) in preference to a mallard model, they will subsequently approach an assembly call (~4 notes per sec) in preference to a mallard model. C. Exposure to the assembly call alone is sufficient to promote auditory imprinting. D. Exposure to the duckling's own contentment call, which is also ~4 notes per sec, is sufficient to promote auditory imprinting. E. When a duckling is exposed to an unnatural call (2 notes per sec) during development, it is not able to recognize and respond to the assembly call when tested subsequently. F. When a duckling (yellow arrows) is reared with inanimate stuffed ducklings and a chicken maternal call, the ducking subsequently prefers the chicken maternal call over that of its own species. *(Adapted from Gottlieb, 1980; Gottlieb, 1993)*

nest during the first few postnatal weeks based on the mother's odor, and this preference can be modified by providing a novel odorant (Brunjes and Alberts, 1979). Subantarctic fur seal pups must learn their mother's vocalization within five days after birth. The mother seals set out on two to three week foraging trips, and the pups locate their mother within minutes of her return using the sound of her vocalization (Charrier et al., 2001). Newborn humans also display a preference for their mother. When infants can trigger their mother's voice or the voice of another female by the rate at which they suck on a nipple, they preferentially activate their mother's voice (DeCasper and Fifer, 1980). What is the evidence that this preference is learned? When mothers read a story aloud during the last six weeks of pregnancy, their babies will subsequently prefer to hear that story over one that was not read aloud. Unexposed newborns display no preference between the two stories. Thus, even though an infant's hearing is quite limited in utero, she may already be forming certain auditory preferences (DeCasper and Spence, 1986).

What exactly is the nervous system learning during filial imprinting? Do infant animals simply learn their mother's smell or image? These questions have been explored in newly hatched ducklings, and the results suggest that several factors are necessary for filial imprinting to occur: visual cues, auditory cues, and social environment. When one-day-old mallard ducklings are allowed to follow a stuffed mallard hen for 30 minutes, they develop a preference for this replica, presumably based on its visual appearance (Johnston and Gottlieb, 1981). However, mother ducks also produce an "assembly" vocalization, and this auditory cue maintains filial imprinting as the ducklings begin to grow. The assembly call is such a powerful signal that ducklings will preferentially follow an unfamiliar red-and-white striped box that is producing this call rather than a familiar mallard hen model (Figure 10.31B). Devocalized and isolated ducklings do not develop proper sensitivity to the maternal calls, but these perceptual skills can be rescued by stimulating the animal with natural vocalizations (Figure 10.31C).

Why is the mother's assembly call such a powerful cue? One possibility is that the mother's call is necessary to maintain her duckling's attachment when the entire family leaves the nest and begins to move about the environment. Older ducklings become very attached to their siblings as they grow, and this "peer imprinting" can actually interfere with filial imprinting (Dyer et al., 1989). For example, socially reared ducklings will not preferentially follow a silent, familiar mallard model, although individually reared ducklings will do so. However, the mallard maternal call will induce socially reared ducklings to follow a familiar mallard or an unfamiliar pintail model (Dyer and Gottlieb, 1990).

This raises an important question: Do ducklings respond innately to their mother's call, or does it depend on sensory experience? Interestingly, ducklings have an innate preference for the mother's call rate, 4 notes per second (Gottlieb, 1980). However, to maintain this preference after hatching, the duckling must either hear its own "contentment" call or that of its siblings (Figure 10.31D). When ducklings are devocalized and reared in isolation with a "contentment" call that is slowed down to about 2 notes per second, they subsequently show no preference for the mother's "assembly" call (Figure 10.31E). Although ducks, geese, and chicks can visually imprint on an object after walking behind it, many other factors regulate this learning. For example, the early social environment exerts a powerful influence (Gottlieb, 1993). Tactile stimulation (with stuffed toy ducklings) is necessary and sufficient to cause mallard ducklings to imprint on a chicken maternal call, and to prefer this over its own species' maternal call (Figure 10.31F). By studying the developing animal in its natural setting, it becomes clear that it is prepared to learn certain cues that it is likely encounter, such as the sibling vocalizations.

Newly hatched domestic chicks also display filial imprinting, and the neural substrates have been studied. When chicks are presented with tones pulsed at about 3 Hz, they will develop a strong preference for this acoustic stimulus and selectively approach it (Wallhausser and Scheich, 1987). There is a dramatic reduction of spines in two different higher forebrain regions during the period of imprinting, and this is not observed in naive chicks. Furthermore, both imprinting and spine elimination depend on functional NMDARs within these forebrain regions (Bock et al., 1996; Bock and Braun, 1999a, 1999b). Chicks can also learn to imprint on the visual characteristics of an object and follow it around, just as ducklings do. Destruction of a specific forebrain area impairs imprinting, and this same region displays an increase in NMDARs following imprinting (Horn, 2004). Therefore, some of the cellular mechanisms that have been implicated in synaptic plasticity (Chapter 9) may have an important role in this early form of learning.

LEARNING PREFERENCES FROM AVERSIONS

Learning is often portrayed as an extension of neural development, and there are many similarities, particularly at the cellular level (Chapter 9). But this portrait does not capture an important fact: learning and memory, themselves, change dramatically during the course of development. Even extremely simple forms of learning, such as habituation and sensitization, can emerge at different times (Rankin and Carew, 1988). However, many developmental learning studies in nonhumans use paradigms that are not available in adults. Even though filial imprinting is a profound example of early learning, it cannot be used to compare learning in young and adult animals. Since our immediate goal is to relate nervous system development to behavior, the following discussion considers reasonably simple forms of associative learning through which animals can establish an aversion for dangerous situations.

At first, the nervous system promotes maternal attachment regardless of the quality of care. In fact, infant animals are far less likely to display aversion learning, possibly to safeguard against learning to avoid their only caregiver. This is demonstrated by a simple associative learning experiment in which an odor is paired with a painful stimulus. Adult animals learn rapidly to avoid the odor. In contrast, neonatal rats actually display a preference for the odor after it is paired with a shock or tail pinch (Sullivan et al., 1986; Camp and Rudy, 1988). After about 10 days postnatal, rat pups begin to display aversion to the odor when it is paired with a painful stimulus, and this is associated with endogenous glucocorticoid production and functional changes in the amygdala, a primary neural substrate for fear conditioning. Even at this later stage, pairing of odor and shock in the mother's presence can induce an odor preference. Apparently, the mother's presence can suppress glucocorticoid signaling in the pup that is induced by stress and amygdala activation (Moriceau and Sullivan, 2006; Barr et al., 2009).

Developing animals must acquire behavioral mechanisms to avoid danger, such as a poisonous plant or a predator. Animals are born with the innate ability to avoid certain things (even if not a negligent caregiver). For example, several species of birds will run from a black hawk-shaped silhouette that is moved over their heads. This occurs even when the birds are reared in isolation with no chance to learn that the hawk image represents danger (Tinbergen, 1948). However, recognition of most dangers is not instinctual. Animals must learn to avoid these situations through experience. A well-studied form of learning, called *fear conditioning*, is responsible for our skill at avoiding danger. During fear conditioning, an animal learns to associate an unconditioned stimulus and response (e.g., a snake bite and the painful withdrawal that it produces) with a neutral stimulus (e.g., the image of a snake). Obviously, an image of a snake can do no harm, but the animal learns that if it sees a snake, then it may be bitten. Thus, the sight of a snake becomes a conditioned stimulus, and it produces a conditioned response (e.g., running away or freezing in fear). Although unethical by modern standards, a nine-month-old baby with no fear of animals was apparently trained to fear rats by pairing the animal with a startling noise (a hammer striking metal just behind the baby's head). This conditioning eventually caused the baby to cry when he saw a rat (Watson and Raynor, 1920).

In a typical fear conditioning experiment, an animal is exposed to a painful stimulus such as a mild foot shock, and at the same time a pure tone is presented from a speaker. How do we know that the electric shock is frightening? Animals usually stop moving (i.e., they freeze) and their blood pressure goes up when they are in frightening situations, and this is precisely the response to mild foot shock. In contrast, the pure tone alone does not produce a change in movement or blood pressure. By presenting these stimuli together several times, the sound alone is gradually able to elicit a fear response.

Are developing animals able to form such associations? Actually, it seems to depend on the stimulus that the animal is asked to learn as well as the behavior that it is asked to perform. For example, rat pups at 15 days or older can learn to freeze in response to a tone that was paired previously with mild foot shock. However, there is a developmental improvement in the ability to associate events separated by longer periods of time (**Figure 10.32**A). This seems to be a general property of learning. Whether a sensory system becomes functional earlier (olfactory, gustatory) or later (auditory, visual), there is similar delay in the memory processes that permits associative learning to occur when events are separated in time (Moye and Rudy, 1987).

For many tasks, learning improves over a protracted period of time. This was explored in rats by first pairing a brief loud sound that elicited a startle response with a long-lasting pure tone at moderate intensity. Adult animals learn quickly that the pure tone predicts the arrival of the loud sound. During test trials, they produce a much larger startle response when the pure tone is present, and this is referred to as fear-potentiated startle. Thus, if the pure tone enhances the startle response, then one concludes that the animal learned the association. When 16-day rat pups are trained in the same paradigm, they do not show any sign of learning (Hunt et al., 1994). Their response to the tone plus noise is nearly identical to noise alone (Figure 10.32B). A similar delay in learning is demonstrated when a light stimulus is paired with foot shock (Hunt, 1999). For some learning tasks, neonates perform better than juvenile animals. Rat pups of

5–10 days can learn to avoid a sugar solution when it is paired with mild foot shock, yet 15-day pups fail to learn this task (Hoffman and Spear, 1988). Furthermore, when odors are paired with an interoceptive aversive stimulus (e.g., within the digestive tract), then associative learning can be demonstrated soon after birth (Haroutunian and Campbell, 1979). Thus, learning is not simply poor in young animals and robust in adults. Rather, it is a complex function of age, sensory modality, and the motor response that is being modified by training.

Studies that target simple forms of learning will be critical for linking behavior with underlying neural mechanisms. However, it is also interesting to ask how developing animals learn complex, multistep tasks, such as how to write a sentence or make a peanut butter sandwich. Of course, sophisticated learning tasks must also be studied with a rigorous paradigm. For example, in a delayed nonmatch to sample task, a primate is first shown an object that can be moved to reveal a reward, such as a food pellet (Bachevalier, 1990). After a delay, the animal is next presented with two objects, one of which it saw previously. In this case, the animal must learn to move the new object in order to obtain the reward (**Figure 10.33**). The task can be made more complicated by increasing the time between trials or by increasing the number of objects that must be memorized. Infant and adult primates were trained on a daily basis until they could perform the task correctly 90% of the time. Animals 2 to 3 years of age reached criterion after 8 days of training, but 3-month-old monkeys required 36 days of training (Figure 10.33). One possible limitation for younger animals may be the amount of sensory activity entering the central nervous system (Bachevalier et al., 1991). In 3-month-old animals, visually evoked activity is significantly lower in regions of the cortex thought to mediate this form of learning, as measured with the 2-deoxyglucose technique (see Box 9.2, "Watching Neurons Think," page 254).

When humans of different ages are challenged with a similar delayed nonmatch to sample task, they also display a gradual improvement with age (Overman, 1990). Children nearly 3 years of age take about 10 times longer to learn the task as compared to adults. Furthermore, children forget things more quickly. The duration of time that children can retain a simple associative learning task gradually increases between 2 and 18 months of age (Hartshorn et al., 1998). Another type of factual learning involves storing information about the spatial environment. Spatial memory was tested in 2- and 4-year-old children by asking them to retrieve candy from eight different locations in an unfamiliar room. It was found that 2-year-olds revisited locations where they had already procured the candy more often than did 4-year-olds. That is, the younger subjects did not remember where they had been. In a different type of factual learning, children were asked to recall details of a story that they had been read, and there was significant improvement between 5 and 10 years of age. Together, these

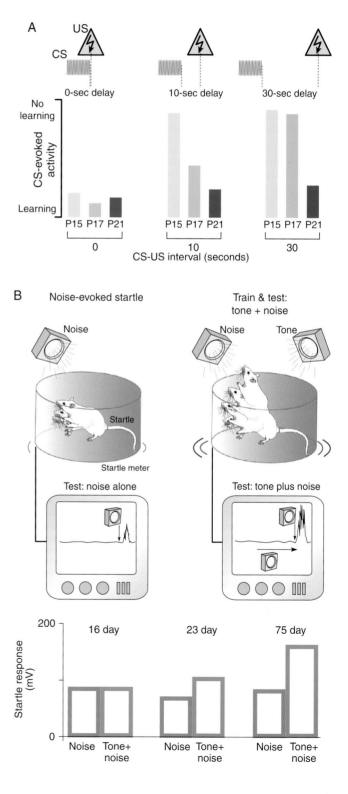

Fig. 10.32 Emergence of associative learning. A. The ability to associate a tone (CS, conditioned stimulus) with a shock (US, unconditioned stimulus) depends on their temporal proximity. When the interval between CS and US is 0 seconds, P15 rat pups learn the association, and display a suppression of ongoing activity when the CS is turned on. However, when the interval between CS and US increases to 10 or 30 seconds, P15 and P17 rat pups fail to learn the association, although P21 rats can learn it. B. When rats are exposed to a brief, loud noise (red speaker) they make a sudden movement, called a startle response (left). This can be recorded by a platform on which the animal stands and displayed on an oscilloscope. When a pure tone (green speaker) precedes the loud noise, the rats learn that the tone predicts the noise burst (right). In subsequent tests, the rats give a larger startle response to the paired tone plus noise, and this is called *fear-potentiated startle*. When trained in this paradigm, postnatal day 16 rat pups display no potentiation, indicating that they have not learned to associate the two signals. At 23 days, the animals do display a potentiation due to pairing, although the potentiation displayed by adults is greater still. *(Adapted from Moye and Rudy, 1987; Hunt et al., 1994)*

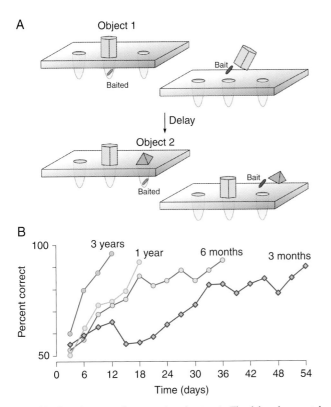

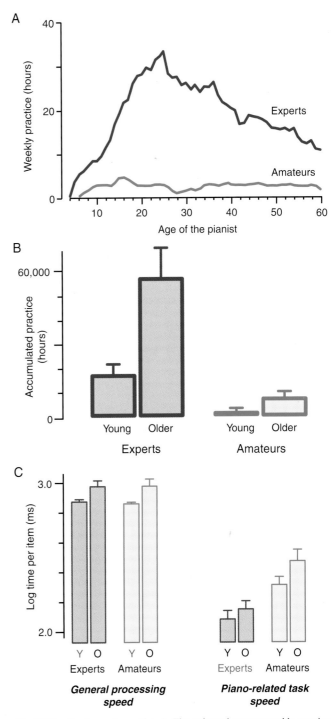

Fig. 10.33 Development of memory in primates. A. The delayed nonmatch to sample task involves remembering the object presented first and, after a delay, choosing the new object on a test trial. B. The time to learn this task is shown for primates of different ages. At 2–3 years of age, animals learn the task within eight days of training. However, 3-month-old monkeys do not reach criterion (90% correct responses) until they receive 36 days of training. *(Adapted from Bachevalier, 1990)*

Fig. 10.34 Practice and expertise. A. The estimated average weekly practice time for amateur and expert pianists as a function of age. B. The estimated accumulated hours of practice for amateur and expert pianists in two age groups (young = 20–31 years; older = 52–68 years). C. Measures of general processing speed (left) indicate that young participants (red Y) outperform older participants (O), independent of expertise on the piano. However, speed on tasks that are specifically related to proficiency on the piano (right) indicate that experts outperform amateurs, independent of age. *(Adapted from Krampe and Ericsson, 1996)*

studies point out the diversity of complex associative learning. Presumably, the improvements that are observed with age result from the maturation of sensory function, motor skills, and learning and memory systems themselves.

SKILL LEARNING: IT DON'T COME EASY

Children are generally considered to be outstanding learners. We think of certain skills (e.g., language, sports, games, music) as most easily learned prior to adulthood. However, this view does not acknowledge the overwhelming amount of time that young animals spend practicing. This is analogous to the immersion that turns out elite performers. For expertise to emerge, it has been estimated that about 1000 hours of practice per year must be sustained for about 10 years. Expert classical pianists practice almost an order of magnitude more than skilled amateurs, and this difference accelerates during the first decade of performance (**Figure 10.34**). Whereas age is a good predictor of general processing speed on manual dexterity tasks, expertise is a better predictor for tasks that are closely related to the skill (Krampe and Ericsson, 1996; Ericsson et al., 2009).

Elite performance is an extreme example of learning, yet it demonstrates that some forms of plasticity require an exceptionally long period of time. Learning to crawl and walk is a prime example. It has been estimated that human infants practice locomotion for more than 6 hours per day, traveling

up to 29 football fields in the process (Adolph et al., 2003). As infants learn to move about their environment, they encounter a range of surfaces, including variation in slope. To test how infants perceive and adjust to this challenge, they can be tested on a mechanized walkway (**Figure 10.35**). By adjusting the

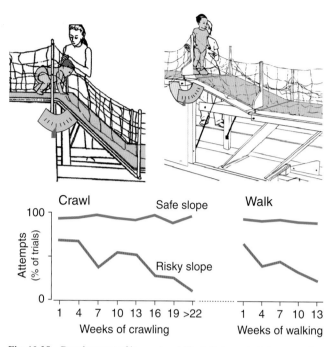

Fig. 10.35 Development of locomotor skills. Infants are tested on a mechanized walkway with adjustable slope (red and blue gauge), and their willingness to crawl (top left) or walk (top right) down the ramp is recorded over a series of visits to the laboratory. In all cases, an assistant stands close by for safety. At first, the infants attempt to crawl down a risky slope about 70% of the time. Over 10 or more weeks of practice, they gradually learned to judge the difficulty of the downhill ramp, and no longer plunged down slopes that they were unprepared for (red line, bottom left). Apparently, this learning is posture-specific. When the same infants begin to walk, and are again challenged with the downhill ramps, they initially attempt to walk down risky slopes. It is only after about 10 more weeks that they learn to gauge difficult terrain for their skill level. *(Adapted from Adolph, 1997)*

angle of a ramp, it is possible to measure skill level for crawling or walking downhill. The downhill slope that an infant is able and willing to descend increases from about 5 to 16 degrees over the first year of walking. During this period, infants slowly become better as gauging how steep of a slope they are capable of walking down safely. At first, crawling infants attempt to locomote down risky slopes on most trials, but they learn this skill within about 10 weeks of practice. When they begin to walk, this process begins anew, and they need another 10 weeks to learn to gauge difficult terrain (adapted from Adolph, 1997).

Many skills depend on fine sensory discrimination, yet we have seen that perception can be relatively slow to mature. Perhaps immature perception facilitates learning as suggested by computer simulations in which a low-resolution sensory system permits better learning than a mature system (Jacobs and Dominguez, 2003). In adults, it is possible to improve performance on a sensory task through training, a process called *perceptual learning*. When adults are tested for their ability to discriminate between the silent gap durations for two intervals, they gradually improve over several days of training. Surprisingly, adolescents are entirely unable to improve when challenged with an identical training regimen, even when their initial performance is adult-like (**Figure 10.36**). In fact, the performance of 11-year-old children actually becomes worse when the training progresses beyond 5 days (Huyck and Wright, 2010). While adolescents might learn if trained differently, this result demonstrates that they do not display adult learning capabilities.

One possibility is that animals derive some sort of an advantage from immature performance. Thus, pre-adolescents do not perform as well as 17-year-olds on a manual dexterity task, yet all ages display similar improvement with practice. Despite their poor performance, adolescents

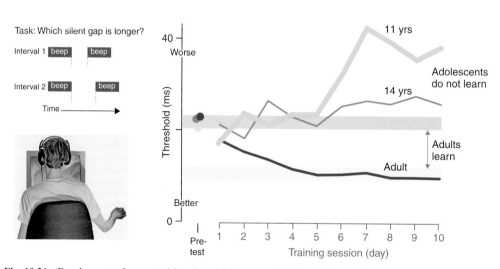

Fig. 10.36 Development of perceptual learning. Adolescents and adults perform an identical auditory task in which they are asked to judge which of two intervals contains the longer silent gap (left). The standard gap is 100 ms and can appear during interval 1 or 2; the test gap is varied in duration to determine the smallest detectable difference. This "duration threshold" is obtained for each individual during a pretest session. Each person then returns to the lab on 10 successive days to receive additional training on the task (right). Adults display a significant improvement in threshold over this period (dark blue line). In contrast, 11- and 14-year-old adolescents do not improve on the task, despite being given identical training. In fact, 11-year-old children become worse during the course of training (thick light blue line). *(Adapted from Huyck and Wright, 2010)*

can learn two motor tasks practiced sequentially, whereas 17-year-olds exhibit poorer learning on the first task when a second task is added (Dorfberger et al., 2007). A second possibility is that the effects of training during development may be advantageous to animals only after they have reached adulthood. When rats are trained on simple auditory tasks throughout juvenile development and into adulthood, they display better performance when retested as adults, as compared to a group trained only for the first time as adults (Threlkeld et al., 2009). This effect may be due to the influence that practice has on brain centers required for learning and memory. For example, many animals hide food, and require a superb spatial memory to recover the fruits of their labor. When marsh tits are reared in captivity, they will continue to hide the sunflower seeds that they are fed. However, if the birds are given powdered seeds that cannot be stored, they develop a smaller hippocampus (Clayton and Krebs, 1994). Thus, learning and memory skills require practice, and this process may influence nervous system development.

GETTING INFORMATION FROM ONE BRAIN TO ANOTHER

The development of animal communication is a fascinating mix of inherited traits and learning. For many of us animals, communication provides the foundation of existence. Some might say that it is a foundation of consciousness. Depending on our position in the food chain, it fetches us a mate, warns us of danger, informs of a food source, bonds us in society, and enriches us with artistry. Perhaps the best studied communication system is that of songbirds, where adult males produce courtship vocalizations to attract conspecific females. While many sex-specific behaviors are understood in terms of genetic and epigenetic factors (above), the individual songs require learning and practice.

When juvenile birds are reared in isolation such that they do not hear a normal adult male song, they develop abnormal vocalizations, and complete auditory deprivation results in severely degraded vocalizations (Marler and Sherman, 1983). Even the relatively tedious vocalizations of crows or roosters are affected by sensory experience. When male chicks are reared with a moderate hearing loss, they subsequently crow at a higher frequency as adults (Grassi et al., 1990). However, the vocalizations of untutored birds do retain a few species-specific characteristics, such as song duration. Moreover, bird colonies that contain only adult male isolates are able to regenerate the species-specific song over the course of generations. Juvenile males change a few characteristics of the degraded song that they learn from a male isolate, making it a bit more like the original wild-type song. When the juveniles mature, they tutor the next generation, and this process is repeated. Changes produced by each generation accumulate such that the song evolves back toward the original wild-type version after four generations (Fehér et al., 2009). Since all zebra finches don't sing an identical song, there must be some mechanism that leads to diversity. In the natural environment, where zebra finches live in social groups, the juvenile males are exposed to many songs, and may choose to copy from more than one tutor. It is also known that when juveniles are exposed to more songs, they subsequently sing a less accurate

copy as adults. Juveniles that hear 30 songs per day generate a song that is 70% like that of the tutor, whereas juveniles that hear >600 presentations per day produce a song that is only 33% similar (Tchernichovski et al., 1999). These studies illustrate a clear role for learning, but also suggest that there are intrinsic limitations on the song that any single species of bird is able to acquire.

Many neuroscientists have settled on the zebra finch as a model for experimental studies of behavior and nervous system development (**Figure 10.37**A). Juvenile birds leave the nest about 20 days after hatching, and they begin to sing a few days later. As with sparrows, male zebra finches must be exposed to the species-specific song, and they must be able to hear themselves sing if they are to produce an accurate rendition

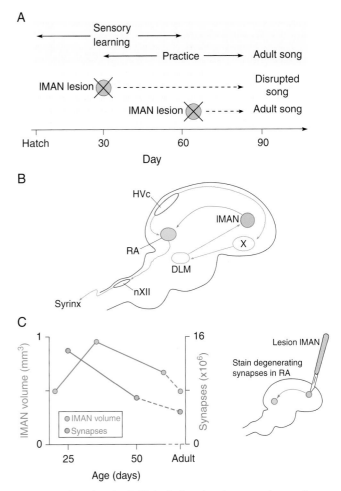

Fig. 10.37 Song learning in birds. A. Song learning occurs in two phases. During the first, juvenile males listen to the tutor song (sensory learning). During the second, juveniles practice singing. The telencephalic nucleus, lMAN, has been implicated in developmental song learning by lesioning it at different ages. If lesioned at 30 days, during sensory learning, the ability to produce song as an adult is disrupted. If lMAN is lesioned at around 60 days, then adult song is unaffected. B. A sagittal section through the songbird brain shows the major nuclei involved in song learning and production. C. The projection from lMAN to RA can be assessed by lesioning lMAN, waiting a day for the synapses to begin degenerating, and then performing a stain for degenerating terminals. Greater staining indicates a greater level of innervation. The number of synapses begins to decline after day 25. For comparison, the size of lMAN is plotted, and it also begins to decline after day 35. (*Adapted from Bottjer et al., 1984; Hermann and Arnold, 1991*)

as adults. When males reach about 90 days of age, they produce a stereotyped song that remains unchanged throughout life, providing they continue to hear themselves sing.

Lesions of the vocal control nuclei, HVc or RA, have a devastating effect on song production in adults. In contrast, lesions to a telencephalic nucleus, lMAN, has no effect on song production in adult birds, but when lesioned in animals before song learning has been completed, song learning and production is disrupted (Figure 10.37A). Together, these experiments suggest that certain nuclei participate in song learning while other areas support adult song production (Nottebohm et al., 1976; Bottjer et al., 1984). The anatomical development of lMAN is highly correlated with its role in song learning (Herrmann and Arnold, 1991; Johnson et al., 1995). lMAN size increases when birds first start to practice their tutor's song, and then decreases in adulthood. The synaptic projection from lMAN to the song motor output nucleus RA is also greatest during the early stages of learning (Figure 10.37C). The number of lMAN neurons that project to RA remains constant during this period, suggesting that terminals are being eliminated. Does lMAN really play such a limited role in zebra finch behavior? In fact, lMAN may participate in song recognition by adult females. Lesions of HVc are known to disrupt song recognition such that the females perform a precopulatory behavior in response to the song of another species (Brenowitz, 1991).

lMAN seems to be essential for song learning, and it would be interesting to know whether the synaptic mechanisms are similar to other forms of plasticity, including the involvement of NMDA receptors (see Chapter 9). Since the level of NMDA receptor expression is quite high in lMAN, their influence can be tested by infusing an NMDA receptor antagonist, AP5, during song learning (**Figure 10.38**). Juveniles that receive AP5 during tutoring sessions performe very poorly at day 90, producing only 20% of the tutor song. In contrast, infusion on alternate days allows learning of ≈50% of the tutor song (Basham et al., 1996). Direct measures of synaptic plasticity made in brain slices through the forebrain nuclei involved in song learning demonstrate NMDA receptor-dependent long-term potentiation (LTP). LTP of the lMAN synapses leads to the strong selectivity for a particular song structure, whereas LTP in area X may be responsible for the phase of learning during which birds practice (Boettiger and Doupe, 2001; Ding and Perkel, 2004). There are also signs of synaptic plasticity at the structural level at the level of individual dendritic spines within the HVc (Roberts et al., 2010). When HVc dendrites are imaged immediately before and after a juvenile songbird is first exposed to a live tutor, the spines are found to become more stable; that is, fewer spines are added and lost (**Figure 10.39**). Furthermore, most birds display a net addition of stable spines. Taken together, these results suggest that song learning in zebra finches shares synaptic mechanism with other forms of developmental plasticity (see Chapter 9).

LANGUAGE

Although human communication is far more complicated than bird song, there are some interesting similarities. Learning is certainly involved at every stage of development, from the production and perception of vowels to the syntax of a sentence. Humans generally speak their first words between 9 and 12

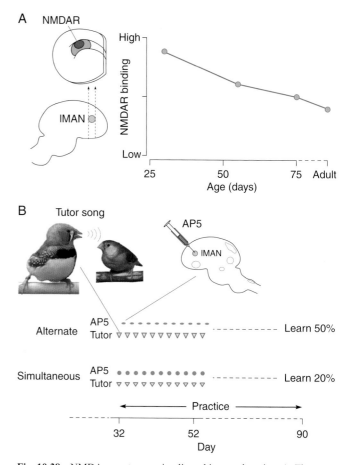

Fig. 10.38 NMDA receptors are implicated in song learning. A. The number of NMDA receptors was assessed by measuring the amount of a receptor antagonist bound to lMAN during development. NMDA receptors began to decrease after day 30. B. The influence of NMDA receptors on song learning was tested by injecting an antagonist (AP5) into lMAN while the bird was being exposed to a tutor song. In control experiments, AP5 was injected on days when the birds were not exposed to the tutor song. When AP5 and exposure to tutor song are delivered simultaneously, the learning is worse at day 90, as compared to the controls. *(Adapted from Aamodt et al., 1995; Basham et al., 1996)*

months and slowly acquire about 50 single words, mostly nouns, over the next eight months. As with birds, there is a period of development when communication skills are typically acquired. From 2 to 6 years of age, children learn about eight words per day. One indication of a sensitive period for language development comes from studies of humans who learn to produce and understand a second language. When English-language skills are analyzed in native Korean or Chinese speakers who arrived in the United States as children or adults, the youngest subjects perform best (Johnson and Newport, 1989). A second indication of a sensitive period comes from studies of deaf individuals who are exposed to sign language from birth to one year of age. Those individuals who are exposed to sign language from birth are more skilled than infants who are exposed even as early as 6 months of age (Newport, 1990).

In contrast to bird song, human communication is performed with equal precision in three sensory modalities. Those born with profound hearing loss can learn to communicate perfectly with their hands and visual system using sign language. Those born without sight can learn to read with

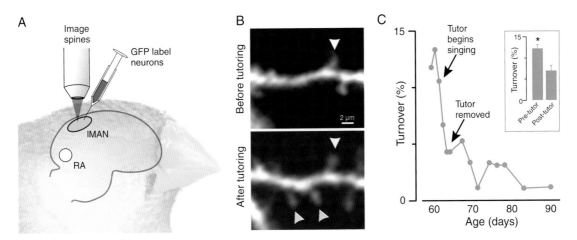

Fig. 10.39 Song learning is associated with stabilization of dendritic spines. A. HVc is transfected with a lentivirus-green fluorescent protein (GFP) construct and the neurons that subsequently express GFP can be imaged through a cranial window. Images were obtained at night when the birds were not active. B. A small section of dendrite is shown the night before a tutor is introduced for the first time (top) and the night after the first tutoring session (bottom). Two stable spines are added (yellow arrowheads) during the first day of tutoring. C. The graph plots spine turnover for one bird over the course of days, and shows a reduction in turnover after the tutor begins to sing. The inset shows average spine turnover the night before and the night after tutoring for five birds in which the turnover rates were initially high. *(Adapted from Roberts et al., 2010)*

their somatosensory system using Braille. Furthermore, the development of language seems to be quite natural in any of these modalities.

It has been known for some time that hearing infants begin to produce speech sounds well before they can understand words. These vocalizations, called vocal babbling, are commonly made up of repeated syllables (e.g., "dadadada"). Whereas deaf children usually do not produce perfect vocalizations as adults, similar to deafened songbirds, a remarkable thing happens: their language ability can be transferred to another sensory modality. Early stages of language acquisition can be studied in infants who are deaf from birth but continually exposed to American Sign Language (ASL) by their deaf parents (Petitto and Marentette, 1991). To determine whether the infants "babble" with their hands, the manual activity of each infant is codified in some detail, and their production of ASL hand shapes is analyzed (**Figure 10.40**). Deaf children devote about 50% of their manual activity to ASL hand shapes, while hearing children only produce about 10% of this activity, presumably by chance. Interestingly, the disparity between deaf and hearing children increases from 10 to 14 months of age, suggesting that deaf children learn language at the same stage of development as hearing children when given the opportunity to use their visual system. Finally, 98% of the manual babbling is performed in front of the body, presumably within the infant's visual field. Thus, imitation of a "tutor" and sensory feedback are important when learning to "speak" with one's hands, in general agreement with studies of bird song development.

Given the complexity of language, it is not surprising that we are only beginning to understand the neural mechanisms that support human communication and how it develops. The ability of infants from two countries to recognize their native vowel sounds was studied to find out whether early experience affects perception. Six-month-old infants from Sweden and the United States were asked to judge two vowel sounds, one from their own country and one from the other country (Kuhl et al., 1992). The English vowel was a /i/ sound, as in the word "fee". The Swedish vowel was a front rounded vowel /y/ sound, as in the Swedish word "fy." Vowel sounds are composed of a unique set of frequencies, called *formants*, and the /i/ sound has slightly higher formants than the /y/ sound (**Figure 10.41**A). Most adult English-speaking listeners can categorize a sound as being like a /i/ sound if the first and second formants are reasonably close to the ideal. This ability to generalize is thought

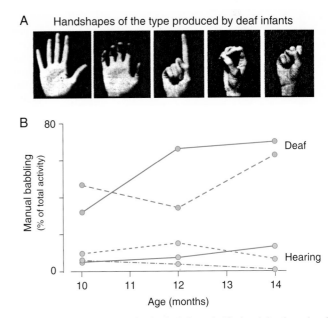

Fig. 10.40 Manual communication by infants. A. The hand signals produced by deaf infants were studied and codified in order to detect hand shapes that correspond to American Sign Language (ASL). B. Deaf infants produced hand shapes corresponding to ASL (manual babbling) more often than hearing infants of the same age. *(Adapted from Petitto and Marentette, 1991)*

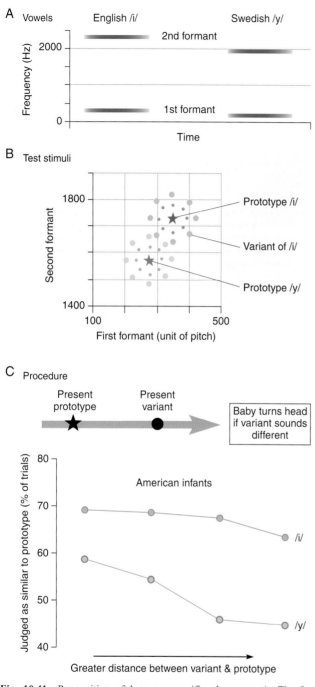

to prevent confusion since individual voice quality varies a good deal, particularly between children, adult females, and adult males. To see whether infants are able to categorize vowel sounds, ideal /i/ and /y/ vowels, called *prototypes*, were generated, and slight variations were made to formant frequencies in order to produce variants (Figure 10.41B). Do infants treat the variants as a member of the group, as adults do? Infants are first exposed to the prototype vowel and then presented with a variant. If she perceives the variant to be different from the prototype, the infant is trained to turn her head. The data show that American infants are more likely to treat variants of /i/ as a member of that group, but are less likely to treat variants of /y/ as members of that group. The opposite result is found for Swedish infants. These results suggest that experience with one's own language in the first 6 months of life allows for improved perception of unique speech sounds.

Is it possible that language-specific activity is present in the brain as the infant is becoming sensitive to the unique attributes of human speech? A functional magnetic resonance imaging study in 3-month-old infants suggests that it is (Dehaene-Lambertz et al., 2002). Human adults exhibit the greatest activation along the left superior temporal sulcus when exposed to their native language, but the response is much smaller when the speech is played in reverse (i.e., "my dog has fleas" versus "saelf sah god ym"). When 3-month-old infants are tested with their native language (French), they

Fig. 10.41 Recognition of language-specific phonemes. A. The frequency spectrum of an American vowel (/i/) and a Swedish vowel (/y/) are shown. Each vowel is composed of two major frequency bands, called *formants*. B. The ability of American and Swedish infants to recognize their native vowel sounds was examined with a range of computer-generated stimuli. An "ideal" version of each vowel, called a *prototype*, was produced, along with vowels with small changes to one of the formant frequencies (called *variants*). C. Infants were trained to turn their head if the second of two vowels sounded different than the first. The data for American infants show that their native /i/ sound can be recognized even when formant frequency changes a good deal. However, the same frequency changes for the Swedish /y/ led to decreased recognition. *(Adapted from Kuhl et al., 1992)*

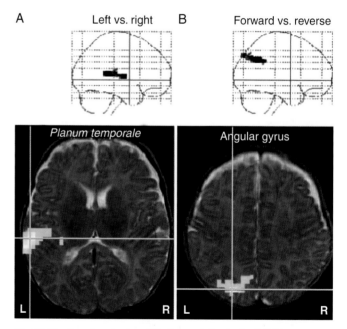

Fig. 10.42 Speech activation of the human infant brain. fMRI images were obtained from 2- to 3-month-old infants during presentation of native speech. A. A transparent brain view (top) and an axial section (bottom) map the relative sound-evoked activity for left versus right temporal cortex. The activation was significantly greater on the left side. B. An activation map showing the relative sound-evoked activity in response to forward speech versus reverse speech. While there was no difference in the temporal cortex, there was an asymmetry within the angular gyrus. *(From Dehaene-Lambertz et al., 2002)*

also display greater activation of the left superior temporal gyrus (**Figure 10.42**). Forward speech elicits greater activation for one brain region on the left side, the angular gyrus, as compared to reversed speech. However, forward and reverse speech are equally effective at activating the temporal lobe, which is quite different than adults. Thus, left-hemisphere dominance for speech processing appears to be already present by 3 months of age, yet the sensitivity to phonological cues (e.g., forward vs. reverse speech) are immature.

SUMMARY

The great strides we have made in molecular and cellular neurobiology have underscored the importance of revisiting the behavior of animals, particularly during development. The maturation of neural processing, or the ability of a neuron to respond accurately to its synaptic inputs, depends on all the building blocks being in place. What can be gained from studying the system as a whole? If we learn the alphabet, are we not able to read a novel? Of course, the blemish in this logic is simple to grasp: systems of molecules, or systems of nerve cells, take on new properties that are not expressed by the single molecule or nerve cell. While the genetic dissection of behavior is an important strategy, it should also be recognized that multiple gene products inevitably contribute to each phenotype, including behavior. Furthermore, the expression of many genes is influenced by the environment (i.e., neuronal activity). Therefore, a rich understanding of the relationship between brain and behavior is fundamental to interpreting many developmental results.

Studying animal behavior is one of the best ways to measure the properties of a system of nerve cells. It provides the most sensitive and universal indicator of a successfully assembled nervous system. All types of developmental errors (i.e., inappropriate fate, ion channel mutations, pathfinding errors, weak synapses) will affect the computational abilities of individual neurons. This is precisely why behavioral measures have long been used to tell clinicians when the nervous system is broken (e.g., schizophrenia, sleep apnea, delayed learning). It is, therefore, not too surprising that behavioral analyses are reemerging as one of the most powerful tools available to developmental neuroscientists.

REFERENCES

Aamodt, S. M., Nordeen, E. J., & Nordeen, K. W. (1995). Early isolation from conspecific song does not affect the normal developmental decline of N-methyl-D-aspartate receptor binding in an avian song nucleus. *J Neurobiol, 27*, 76–84.

Abbott, D. H., & Hearn, J. P. (1978). The effects of neonatal exposure to testosterone on the development of behaviour in female marmoset monkeys. *Ciba Found Symp, 62*, 299–327.

Adolph, K. E. (1997). Learning in the development of infant locomotion. *Monogr Soc Res Child Dev, 62*(3, Serial No. 251).

Adolph, K. E., Vereijken, B., & Shrout, P. E. (2003). What changes in infant walking and why. *Child Dev, 74*, 475–497.

Agate, R. J., Grisham, W., Wade, J., Mann, S., Wingfield, J., Schanen, C., et al. (2003). Neural not gonadal origin of brain sex differences in a gynandromorphic finch. *Proc Natl Acad Sci USA, 100*, 4873–4878.

Ahmed, E. I., Zehr, J. L., Schulz, K. M., Lorenz, B. H., DonCarlos, L. L., & Sisk, C. L. (2008). Pubertal hormones modulate the addition of new cells to sexually dimorphic brain regions. *Nat Neurosci, 11*, 995–997.

Armand, J., Edgley, S. A., Lemon, R. N., & Olivier, E. (1994). Protracted postnatal development of corticospinal projections from the primary motor cortex to hand motoneurones in the macaque monkey. *Exp Brain Res, 101*(1), 178–182.

Arnold, A. P. (2009). The organizational-activational hypothesis as the foundation for a unified theory of sexual differentiation of all mammalian tissues. *Horm Behav, 55*, 570–578.

Arnold, A. P., & Chen, X. (2009). What does the "four core genotypes" mouse model tell us about sex differences in the brain and other tissues? *Front Neuroendocrinol, 30*, 1–9.

Bachevalier, J. (1990). Ontogenetic development of habit and memory formation in primates. *Ann New York Acad Sci, 608*, 457–477.

Bachevalier, J., Hagger, C., & Mishkin, M. (1991). Functional maturation of the occipitotemporal pathway in infant rhesus monkeys. In N. A. Lassen, D. H. Ingvar, M. E. Raichle, & L. Friberg (Eds.), *Alfred Benzon symposium 31, Brain work and mental activity* (pp. 231–240). Copenhagen: Munksgaard.

Bahrick, L. E., & Lickliter, R. (2000). Intersensory redundancy guides attentional selectivity and perceptual learning in infancy. *Dev Psychol, 36*, 190–201.

Bahrick, L. E., & Lickliter, R. (2002). Intersensory redundancy guides attentional selectivity and perceptual learning in infancy. *Dev Psychol, 36*, 190–201.

Baines, R. A., & Bate, M. (1998). Electrophysiological development of central neurons in the Drosophila embryo. *The Journal of Neuroscience, 18*, 4673–4683.

Bakker, J., De Mees, C., Douhard, Q., Balthazart, J., Gabant, P., Szpirer, J., et al. (2006). Alpha-fetoprotein protects the developing female mouse brain from masculinization and defeminization by estrogens. *Nat Neurosci, 9*, 220–226.

Banai, K., Sabin, A. T., Kraus, N., & Wright, B. A. (2007). The development of sensitivity to amplitude and frequency modulation follow distinct time courses. *Assoc Res Otolaryngol, 30*, 283.

Bargones, J. Y., & Werner, L. A. (1994). Adults listen selectively: infants do not. *Psychol Sci, 5*, 170–174.

Barish, M. E., (1986). Differentiation of voltage-gated potassium current and modulation of excitability in cultured amphibian spinal neurones. *J Physiol, 375*, 229–250.

Barlow, H. B., Blakemore, C., & Pettigrew, J. D. (1967). The neural mechanism of binocular depth discrimination. *J Physiol, 193*, 327–342.

Barr, G. A., Moriceau, S., Shionoya, K., Muzny, K., Gao, P., Wang, S., et al. (2009). Transitions in infant learning are modulated by dopamine in the amygdala. *Nat Neurosci, 12*, 1367–1369.

Basham, M. E., Nordeen, E. J., & Nordeen, K. W. (1996). Blockade of NMDA receptors in the anterior forebrain impairs sensory acquisition in the zebra finch (Poephila guttata). *Neurobiol Learn Mem, 66*, 295–304.

Bate, M. (1998). Making sense of behavior. *Int J Dev Biol, 42*(3), 507–509.

Beato, M., & Klug, J. (2000). Steroid hormone receptors: an update. *Human Reproduction Update, 6*, 225–236.

Becker, J. B., & Hu, M. (2008). Sex differences in drug abuse. *Frontiers in Neuroendocrinology, 29*, 36–47.

Bekoff, A., & Kauer, J. A. (1984). Neural control of hatching: fate of the pattern generator for the leg movements of hatching in post-hatching chicks. *The Journal of Neuroscience, 4*(11), 2659–2666.

Bekoff, A., & Sabichi, A. L. (1987). Sensory control of the initiation of hatching in chicks: effects of a local anesthetic injected into the neck. *Developmental Psychobiology, 20*(5), 489–495.

Ben-Shahar, Y., Robichon, A., Sokolowski, M. B., & Robinson, G. E. (2002). Influence of gene action across different time scales on behavior. *Science, 296*, 741–744.

Bentley, D. R., & Hoy, R. R. (1972). Genetic control of the neuronal network generating cricket (Teleogryllus Gryllus) song patterns. *Animal Behaviour, 20*, 478–492.

Benzer, S. (1971). From the gene to behavior. *JAMA, 218*(7), 1015–1022.

Beyer, C., Kolbinger, W., Froehlich, U., Pilgrim, C., & Reisert, I. (1992). Sex differences of hypothalamic prolactin cells develop independently of the presence of sex steroids. *Brain Research, 593*, 253–256.

Birch, E., Gwiazda, J., & Held, R. (1982). Stereoacuity development for crossed and uncrossed disparities in human infants. *Vision Res, 22*, 507–513.

Birch, E. E., Gwiazda, J., Bauer, J. A., Naegele, J., & Held, R. (1983). Visual acuity and its meridional variations in children aged 7–60 months. *Vision Research, 23*, 1019–1024.

Bock, J., & Braun, K. (1999a). Blockade of N-methyl-D-aspartate receptor activation suppresses learning-induced synaptic elimination. *Proceedings of the National Academy of Sciences of the United States of America, 96*, 2485–2490.

Bock, J., & Braun, K. (1999b). Filial imprinting in domestic chicks is associated with spine pruning in the associative area, dorsocaudal neostriatum. *European Journal of Neuroscience, 11*, 2566–2570.

Bock, J., Wolf, A., & Braun, K. (1996). Influence of the N-methyl-D-aspartate receptor antagonist DL-2-amino-5-phosphonovaleric acid on auditory filial imprinting in the domestic chick. *Neurobiology of Learning and Memory, 65*, 177–188.

Boettiger, CA., Doupe, AJ (2001). Developmentally restricted synaptic plasticity in a songbird nucleus required for song learning. *Neuron, 31*, 809–818.

Bonds, A. B. (1979). Development of orientation tuning in the visual cortex of kittens. In R. D. Freeman (Ed.), *Developmental neurobiology of vision* (pp. 31–41). New York: Plenum Press.

Bottjer, S. W., Miesner, E. A., & Arnold, A. P. (1984). Forebrain lesions disrupt development but not maintenance of song in passerine birds. *Science, 224*, 901–903.

Brenowitz, E. A. (1991). Altered perception of species-specific song by female birds after lesions of a forebrain nucleus. *Science, 251*, 303–305.

Brenowitz, E. A., & Arnold, A. P. (1986). Interspecific comparisons of the size of neural song control regions and song complexity in duetting birds: evolutionary implications. *The Journal of Neuroscience, 6*, 2875–2879.

Brunjes, P. C., & Alberts, J. R. (1979). Olfactory stimulation induces filial preferences for huddling in rat pups. *Journal of Comparative and Physiological Psychology, 93*, 548–555.

Brustein, E., Saint-Amant, L., Buss, R. R., Chong, M., McDearmid, J. R., & Drapeau, P. (2003). Steps during the development of the zebrafish locomotor network. *Journal of Physiology, 97*, 77–86.

Cameron, N. M., Shahrokh, D., Del Corpo, A., Dhir, S. K., Szyf, M., Champagne, F. A., et al. (2008). Epigenetic programming of phenotypic variations in reproductive strategies in the rat through maternal care. *Journal of Neuroendocrinology, 20*, 795–801.

Camp, L. L., & Rudy, J. W. (1988). Changes in the categorization of appetitive and aversive events during postnatal development of the rat. *Developmental Psychobiology, 21*, 25–42.

Campbell, R. A., King, A. J., Nodal, F. R., Schnupp, J. W., Carlile, S., & Doubell, T. P. (2008). Virtual adult ears reveal the roles of acoustical factors and experience in auditory space map development. *The Journal of Neuroscience, 28*, 11557–11570.

Campos, J. J., Langer, A., & Krowitz, A. (1970). Cardiac responses on the visual cliff in prelocomotor human infants. *Science, 170*, 196–197.

Carmichael, L. (1954). The onset and early development of behavior. In L. Carmichael (Ed.), *Manual of child psychology* (pp. 60–214). New York: Wiley.

Carpenter, W. B. (1874). *Principles of mental physiology*. London: King.

Carriere, B. N., Royal, D. W., Perrault, T. J., Morrison, S. P., Vaughan, J. W., Stein, B. E., et al. (2007). Visual deprivation alters the development of cortical multisensory integration. *Journal of Neurophysiology, 98*, 2858–2867.

Carruth, L. L., Reisert, I., & Arnold, A. P. (2002). Sex chromosome genes directly affect brain sexual differentiation. *Nature Neuroscience, 5*, 933–934.

Chapman, B., & Stryker, M. P. (1993). Development of orientation selectivity in ferret visual cortex and effects of deprivation. *The Journal of Neuroscience, 13*, 5251–5262.

Charrier, I., Mathevon, N., & Jouventin, P. (2001). Mother's voice recognition by seal pups. *Nature, 412*, 873.

Chub, N., & O'Donovan, M. J. (1998). Blockade and recovery of spontaneous rhythmic activity after application of neurotransmitter antagonists to spinal networks of the chick embryo. *The Journal of Neuroscience, 18*(1), 294–306.

Clarkson, M. G., Clifton, R. K., Swain, I. U., & Perris, E. E. (1989). Stimulus duration and repetition rate influences newborns' head orientation towards sound. *Developmental Psychobiology, 22*, 683–705.

Clayton, N. C., & Krebs, J. R. (1994). Hippocampal growth and attrition in birds affected by experience. *Proceedings of the National Academy of Sciences of the United States of America, 91*, 7410–7414.

Coghill, G. E. (1929). *Anatomy and the problem of Behaviour*. London: Cambridge University Press.

Cooke, B. M., Tabibnia, G., & Breedlove, S. M. (1999). A brain sexual dimorphism controlled by adult circulating androgens. *Proceedings of the National Academy of Sciences of the United States of America, 96*, 7538–7540.

Cooke, B. M., & Woolley, C. S. (2005). Sexually dimorphic synaptic organization of the medial amygdala. *The Journal of Neuroscience, 25*, 10759–10767.

Crisp, S., Evers, J. F., Fiala, A., & Bate, M. (2008). The development of motor coordination in Drosophila embryos. *Development, 135*, 3707–3717.

Curley, J. P., & Mashoodh, R. (2010). Parent-of-origin and trans-generational germline influences on behavioral development: the interacting roles of mothers, fathers, and grandparents. *Developmental Psychobiology, 52*, 312–330.

Daw, N. W. (1995). *Visual development*. New York: Plenum Press.

DeCasper, A. J., & Fifer, W. P. (1980). Of human bonding: newborns prefer their mothers' voices. *Science, 208*, 1174–1176.

DeCasper, A. J., & Spence, M. J. (1986). Prenatal maternal speech influences newborns' perception of speech sounds. *Infant Behavior & Development, 9*, 133–150.

Dehaene-Lambertz, G., Dehaene, S., & Hertz-Pannier, L. (2002). Functional neuroimaging of speech perception in infants. *Science, 298*, 2013–2015.

DeLoache, J. S., Uttal, D. H., & Rosengren, K. S. (2004). Scale errors offer evidence for a perception-action dissociation early in life. *Science, 304*, 1027–1029.

Demarque, M., & Spitzer, N. C. (2010). Activity-dependent expression of lmx1b regulates specification of serotonergic neurons modulating swimming behavior. *Neuron, 67*, 321–334.

Dewing, P., Chiang, C. W., Sinchak, K., Sim, H., Fernagut, P. O., Kelly, S., et al. (2006). Direct regulation of adult brain function by the male-specific factor SRY. *Current Biology, 16*, 415–420.

Ding, L., & Perkel, D. J. (2004). Long-term potentiation in an avian basal ganglia nucleus essential for vocal learning. *J Neurosci, 24*, 488–494.

Dixit, R., Vijayraghavan, K., & Bate, M. (2008). Hox genes and the regulation of movement in Drosophila. *Developmental Neurobiology, 68*, 309–316.

Dobkins, K. R., & Anderson, C. M. (2002). Color-based motion processing is stronger in infants than in adults. *Psychological Science: A Journal of the American Psychological Society/APS, 13*, 76–80.

Dorfberger, S., Adi-Japha, E., & Karni, A. (2007). Reduced susceptibility to interference in the consolidation of motor memory before adolescence. *PLoS ONE, 2*(2), e240.

Dyer, A. B., & Gottlieb, G. (1990). Auditory basis of maternal attachment in ducklings (Anas platyrhynchos) under simulated naturalistic imprinting conditions. *Journal of Comparative Psychology, 104*, 190–194.

Dyer, A. B., Lickliter, R., & Gottlieb, G. (1989). Maternal and peer imprinting in mallard ducklings under experimentally simulated natural social conditions. *Developmental Psychobiology, 22*, 463–475.

Ericsson, K. A., Nandagopal, K., & Roring, R. W. (2009). Toward a science of exceptional achievement: attaining superior performance through deliberate practice. *Annals of the New York Academy of Sciences, 1172*, 199–217.

Fehér, O., Wang, H., Saar, S., Mitra, P. P., & Tchernichovski, O. (2009). De novo establishment of wild-type song culture in the zebra finch. *Nature, 459*, 564–568.

Ferveur, J. F., Savarit, F., O'Kane, C. J., Sureau, G., Greenspan, R. J., & Jallon, J. M. (1997). Genetic feminization of pheromones and its behavioral consequences in Drosophila males. *Science, 276*, 1555–1558.

Ferveur, J. F., Störtkuhl, K. F., Stocker, R. F., & Greenspan, R. J. (1995). Genetic feminization of brain structures and changed sexual orientation in male Drosophila. *Science, 267*, 902–905.

Fetcho, J. R., Higashijima, S., & McLean, D. L. (2008). Zebrafish and motor control over the last decade. *Brain Research Reviews, 57*, 86–93.

Finley, K. D., Taylor, B. J., Milstein, M., & McKeown, M. (1997). Dissatisfaction, a gene involved in sex-specific behavior and neural development of Drosophila melanogaster. *Proceedings of the National Academy of Sciences of the United States of America, 94*, 913–918.

Forlano, P. M., & Woolley, C. S. (2010). Quantitative analysis of pre- and postsynaptic sex differences in the nucleus accumbens. *The Journal of Comparative Neurology, 518*, 1330–1348.

Garcia-Campmany, L., Stam, F. J., & Goulding, M. (2010). From circuits to behaviour: motor networks in vertebrates. *Current Opinion in Neurobiology, 20*, 116–125.

Gonzalez-Islas, C., & Wenner, P. (2006). Spontaneous network activity in the embryonic spinal cord regulates AMPAergic and GABAergic synaptic strength. *Neuron, 49*, 563–575.

Goodfellow, P. N., & Lovell-Badge, R. (1993). SRY and sex determination in mammals. *Annual Review of Genetics, 27*, 71–92.

Gorski, R. A., Gordon, J. H., Shryne, J. E., & Southam, A. M. (1978). Evidence for a morphological sex difference within the medial preoptic area of the rat brain. *Brain Res, 148*, 333–346.

Gorski, R. A., Harlan, R. E., Jacobson, C. D., Shryne, J. E., & Southam, A. M. (1980). Evidence for the existence of a sexually dimorphic nucleus in the preoptic area of the rat. *The Journal of Comparative Neurology, 193*, 529–539.

Gosgnach, S., Lanuza, G. M., Butt, S. J., Saueressig, H., Zhang, Y., Velasquez, T., et al. (2006). V1 spinal neurons regulate the speed of vertebrate locomotor outputs. *Nature, 440*, 215–219.

Gottlieb, G. (1980). Development of species identification in ducklings: VI. Specific embryonic experience required to maintain species-typical perception in ducklings. *Journal of Comparative and Physiological Psychology, 94*, 579–587.

Gottlieb, G. (1993). Social induction of malleability in ducklings: sensory basis and psychological mechanism. *Animal Behaviour, 45*, 707–719.

Goulding, M., & Pfaff, S. L. (2005). Development of circuits that generate simple rhythmic behaviors in vertebrates. *Current Opinion in Neurobiology, 15,* 14–20.

Grassi, S., Ottaviani, F., & Bambagioni, D. (1990). Vocalization-related stapedius muscle activity in different age chickens (Gallus gallus), and its role in vocal development. *Brain Research, 529,* 158–164.

Gregg, C., Zhang, J., Butler, J. E., Haig, D., & Dulac, C. (2010a). Sex-specific parent-of-origin allelic expression in the mouse brain. *Science,* Jul 8 [Epub ahead of print].

Gregg, C., Zhang, J., Weissbourd, B., Luo, S., Schroth, G. P., Haig, D., et al. (2010b). High-resolution analysis of parent-of-origin allelic expression in the mouse brain. *Science,* Jul 8 [Epub ahead of print].

Grumbach, M. M., & Auchus, R. J. (1999). Estrogen: consequences and implications of human mutations in synthesis and action. *The Journal of Clinical Endocrinology and Metabolism, 84,* 4677–4694.

Gurney, M. E., & Konishi, M. (1980). Hormone-induced sexual differentiation of brain and behavior in zebra finches. *Science, 208,* 1380–1383.

Hadad, B., Maurer, D., & Lewis, T. L. (2010). The effects of spatial proximity and collinearity on contour integration in adults and children. *Vision Research, 50,* 772–778.

Hall, J. C. (1977). Portions of the central nervous system controlling reproductive behavior in Drosophila melanogaster. *Behavior Genetics, 7,* 291–312.

Hall, J. C. (1994). The mating of a fly. *Science, 264,* 1702–1714.

Hall, J. W., & Grose, J. H. (1994). Development of temporal-resolution in children as measured by the temporal modulation transfer function. *The Journal of the Acoustical Society of America, 96,* 150–154.

Hamburger, V. (1963). Some aspects of the embryology of behavior. *Q. Rev. Biol., 38,* 342–365.

Hamburger, V., Wenger, E., & Oppenhein, R. W. (1966). Motility in the chick and embryo in the absence of sensory input. *Journal of Experimental Zoology, 162,* 133–160.

Hanson, M. G., & Landmesser, L. T. (2003). Characterization of the circuits that generate spontaneous episodes of activity in the early embryonic mouse spinal cord. *The Journal of Neuroscience, 23,* 587–600.

Haroutunian, V., & Campbell, B. A. (1979). Emergence of interoceptive and exteroceptive control of behavior in rats. *Science, 205,* 927–929.

Harrison, R. G. (1904). An experimental study of the relation of the nervous system to the developing musculature in the embryo of the frog. *The American Journal of Anatomy, 3,* 197–220.

Hartshorn, K., Rovee-Collier, C., Gerhardstein, P., Bhatt, R. S., Wondoloski, T. L., Klein, P., et al. (1998). The ontogeny of long-term memory over the first year-and-a-half of life. *Developmental Psychobiology, 32,* 69–89.

Haverkamp, L. J. (1986). Anatomical and physiological development of the Xenopus embryonic motor system in the absence of neural activity. *The Journal of Neuroscience, 6(5),* 1338–1348.

Haverkamp, L. J., & Oppenheim, R. W. (1986). Behavioral development in the absence of neural activity: effects of chronic immobilization on amphibian embryos. *The Journal of Neuroscience, 6(5),* 1332–1337.

Herrmann, K., & Arnold, A. P. (1991). The development of afferent projections to the robust archistriatal nucleus in male zebra finches: a quantitative electron microscopic study. *The Journal of Neuroscience, 11,* 2063–2074.

Higashijima, S., Masino, M. A., Mandel, G., & Fetcho, J. R. (2004). Engrailed-1 expression marks a primitive class of inhibitory spinal interneuron. *The Journal of Neuroscience, 24,* 5827–5839.

Hoffmann, H., & Spear, N. E. (1988). Ontogenetic differences in conditioning of an aversion to a gustatory CS with a peripheral US. *Behavioral and Neural Biology, 50,* 16–23.

Holmes, M. M., Rosen, G. J., Jordan, C. L., de Vries, G. J., Goldman, B. D., & Forger, N. G. (2007). Social control of brain morphology in a eusocial mammal. *Proceedings of the National Academy of Sciences of the United States of America, 104,* 10548–10552.

Horn, G. (2004). Pathways of the past: the imprint of memory. *Nature Reviews. Neuroscience, 5,* 108–120.

Hunt, P. S. (1999). A further investigation of the developmental emergence of fear-potentiated startle in rats. *Dev Psychobiol, 34,* 281–291.

Hunt, P. S., Richardson, R., & Campbell, B. A. (1994). Delayed development of fear-potentiated startle in rats. *Behavioral Neuroscience, 108,* 69–80.

Hutchison, J. B. (1997). Gender-specific steroid metabolism in neural differentiation. *Cellular and Molecular Neurobiology, 17,* 603–626.

Huyck, J. J., & Wright, B. A. (2010). Late maturation of auditory perceptual learning. *Developmental Science,* published online: 11 Nov 2010.

Hyson, R. L., & Rudy, J. W. (1987). Ontogenetic change in the analysis of sound frequency in the infant rat. *Developmental Psychobiology, 20,* 189–207.

Imperato-McGinley, J., Peterson, R. E., Gautier, T., & Sturla, E. (1979). Male pseudohermaphroditism secondary to 5 alpha-reductase deficiency model for the role of androgens in both the development of the male phenotype and the evolution of a male gender identity. *Journal of Steroid Biochemistry, 11,* 637–645.

Ishida, N., Kaneko, M., & Allada, R. (1999). Biological clocks. *Proc Natl Acad Sci USA, 96(16),* 8819–8820.

Jacobs, D. S., & Blakemore, C. (1988). Factors limiting the postnatal development of visual acuity in the monkey. *Vision Research, 8,* 947–958.

Jacobs, R. A., & Dominguez, M. (2003). Visual development and the acquisition of motion velocity sensitivities. *Neural Computation, 15,* 761–781.

Jensen, J. K., & Neff, D. L. (1993). Development of basic auditory discrimination in preschool children. *Psychological Science: A Journal of the American Psychological Society/APS, 4,* 104–107.

Johnson, F., Sablan, M. M., & Bottjer, S. W. (1995). Topographic organization of a forebrain pathway involved with vocal learning in zebra finches. *The Journal of Comparative Neurology, 358,* 260–278.

Johnson, J. L., & Leff, M. (1999). Children of substance abusers: overview of research findings. *Pediatrics, 103(5 Pt 2),* 1085–1099.

Johnson, J. S., & Newport, E. L. (1989). Critical period effects in second language learning: the influence of maturational state on the acquisition of English as a second language. *Cognitive Psychology, 21,* 60–99.

Johnston, T. D., & Gottlieb, G. (1981). Visual preferences of imprinted ducklings are altered by the maternal call. *Journal of Comparative and Physiological Psychology, 95,* 663–675.

Jost, A. (1953). Problems of fetal endocrinology: The gonadal and hypophyseal hormones. *Recent Progress in Hormone Research, 8,* 379–418.

Kelley, D. B. (1997). Generating sexually differentiated songs. *Current Opinion in Neurobiology, 7,* 839–843.

Kelly, J. B., Judge, P. W., & Fraser, I. H. (1987). Development of the auditory orientation response in the albino rat (Rattus norvegicus). *Journal of Comparative Psychology, 101,* 60–66.

Kimura, D. (1996). Sex, sexual orientation and sex hormones influence human cognitive function. *Current Opinion in Neurobiology, 6,* 259–263.

Kiorpes, L., & Movshon, J. A. (2004). Neural limitations on visual development in primates. In L. M. Chalupa, & J. S. Werner (Eds.), *The visual neurosciences.* MIT Press.

Kolbinger, W., Trepel, M., Beyer, C., Pilgrim, C., & Reisert, I. (1991). The influence of genetic sex on sexual differentiation of diencephalic dopaminergic neurons in vitro and in vivo. *Brain Research, 544,* 349–352.

Kovács, I., Kozma, P., Fehér, A., & Benedek, G. (1999). Late maturation of visual spatial integration in humans. *Proceedings of the National Academy of Sciences of the United States of America, 96,* 12204–12209.

Krampe, R. T., & Ericsson, K. A. (1996). Maintaining excellence: deliberate practice and elite performance in young and older pianists. *Journal of Experimental Psychology. General, 125,* 331–359.

Kuhl, P. K., Williams, K. A., Lacerda, F., Stevens, K. N., & Lindblom, B. (1992). Linguistic experience alters phonetic perception in infants by 6 months of age. *Science, 255,* 606–608.

Kullander, K., Butt, S. J., Lebret, J. M., Lundfald, L., Restrepo, C. E., Rydstrom, A., et al. (2003). Role of EphA4 and EphrinB3 in local neuronal circuits that control walking. *Science, 299,* 1889–1892.

Levine, R. B., & Weeks, J. C. (1990). Hormonally mediated changes in simple reflex circuits during metamorphosis in Manduca. *Journal of Neurobiology, 21(7),* 1022–1036.

Levtzion-Korach, O., Tennenbaum, A., Schnitzer, R., & Ornoy, A. (2000). Early motor development of blind children. *Journal of Paediatrics and Child Health, 36,* 226–229.

Li, W. C., Cooke, T., Sautois, B., Soffe, S. R., Borisyuk, R., & Roberts, A. (2007). Axon and dendrite geography predict the specificity of synaptic connections in a functioning spinal cord network. *Neural Development, 2,* 17.

Lorenz, K. (1937). The companion in the bird's world. *Auk, 54,* 245–273.

Marin-Burgin, A., Eisenhart, F. J., Baca, S. M., Kristan, W. B., Jr., & French, K. A. (2005). Sequential development of electrical and chemical synaptic connections generates a specific behavioral circuit in the leech. *The Journal of Neuroscience, 25,* 2478–2489.

Marin-Burgin, A., Kristan, W. B., Jr., & French, K. A. (2008). From synapses to behavior: development of a sensory-motor circuit in the leech. *Developmental Neurobiology, 68,* 779–787.

Marler, P., & Sherman, V. (1983). Song structure without auditory feedback: emendations of the auditory template hypothesis. *The Journal of Neuroscience, 3,* 517–531.

Martin, J. H., Choy, M., Pullman, S., & Meng, Z. (2004). Corticospinal system development depends on motor experience. *The Journal of Neuroscience, 24,* 2122–2132.

McGowan, P. O., Sasaki, A., D'Alessio, A. C., Dymov, S., Labonté, B., Szyf, M., et al. (2009). Epigenetic regulation of the glucocorticoid receptor in human brain associates with childhood abuse. *Nature Neuroscience, 12,* 342–348.

McLean, D. L., & Fetcho, J. R. (2008). Using imaging and genetics in zebrafish to study developing spinal circuits in vivo. *Developmental Neurobiology, 68,* 817–834.

Moore, J. K., & Guan, Y. L. (2001). Cytoarchitectural and axonal maturation in human auditory cortex. *Journal of the Association for Research in Otolaryngology: JARO, 2,* 297–311.

Moriceau, S., & Sullivan, R. M. (2006). Maternal presence serves as a switch between learning fear and attraction in infancy. *Nature Neuroscience, 9,* 1004–1006.

Moye, T. B., & Rudy, J. W. (1987). Ontogenesis of trace conditioning in young rats: Dissociation of associative and memory processes. *Developmental Psychobiology, 20,* 405–414.

Mrsic-Flogel, T. D., Schnupp, J. W., & King, A. J. (2003). Acoustic factors govern developmental sharpening of spatial tuning in the auditory cortex. *Nature Neuroscience, 6,* 981–988.

Myers, C. P., Lewcock, J. W., Hanson, M. G., Gosgnach, S., Aimone, J. B., Gage, F. H., et al. (2005). Cholinergic input is required during embryonic development to mediate proper assembly of spinal locomotor circuits. *Neuron, 46,* 37–49.

Naftolin, F., Ryan, K. J., Davies, I. J., Reddy, V. V., Flores, F., Petro, Z., et al. (1975). The formation of estrogens by central neuroendocrine tissues. *Recent Prog Horm Res, 31,* 295–319.

Narayanan, C. H., & Hamburger, V. (1971). Motility in chick embryos with substitution of lumbosacral by brachial and brachial by lumbosacral spinal cord segments. *Journal of Experimental Zoology, 178*(4), 415–431.

Newport, E. (1990). Maturational constraints on language learning. *Cognitive Sci, 14,* 11–28.

Nottebohm, F. (1981). A brain for all seasons: cyclical anatomical changes in song control nuclei of the canary brain. *Science, 214,* 1368–1370.

Nottebohm, F., & Arnold, A. P. (1976). Sexual dimorphism in vocal control areas of the songbird brain. *Science, 194,* 211–213.

Nottebohm, F., Stokes, T. M., Leonard, C. M. (1976) Central control of song in the canary, *Serinus canarius. J Comp Neurol 165,* 457–486.

O'Donovan, M. J., Wenner, P., Chub, N., Tabak, J., & Rinzel, J. (1998). Mechanisms of spontaneous activity in the developing spinal cord and their relevance to locomotion. *Ann N Y Acad Sci, 860,* 130–141.

O'Malley, D. M., Kao, Y. H., & Fetcho, J. R. (1996). Imaging the functional organization of zebrafish hindbrain segments during escape behaviors. *Neuron, 17*(6), 1145–1155.

Ogawa, S., Chester, A. E., Hewitt, S. C., Walker, V. R., Gustafsson, J. A., Smithies, O., et al. (2000). Abolition of male sexual behaviors in mice lacking estrogen receptors alpha and beta (alpha beta ERKO). *Proceedings of the National Academy of Sciences of the United States of America, 97,* 14737–14741.

Ogawa, S., Lubahn, D. B., Korach, K. S., & Pfaff, D. W. (1997). Behavioral effects of estrogen receptor gene disruption in male mice. *Proceedings of the National Academy of Sciences of the United States of America, 94,* 1476–1481.

Oppenheim, R. W. (1981). Ontogenetic adaptation and regressive processes in the development of the nervous system and behavior: a neuro-embryological perspective. In K. Connolly, & H. Prechtl (Eds.), *Development and Maturation* (pp. 73–109). Philadelphia: J. Lippincott.

Oppenheim, R. W. (1982). The neuroembryological study of behavior: progress, problems, perspectives. *Current Topics in Devel. Biol, 17,* 257–309.

Osborne, K. A., Robichon, A., Burgess, E., Butland, S., Shaw, R. A., Coulthard, A., et al. (1997). Natural behavior polymorphism due to a cGMP-dependent protein kinase of Drosophila. *Science, 277,* 834–836.

Overman, W. H. (1990). Performance on traditional matching to sample, non-matching to sample, and object discrimination tasks by 12- and 32-month-old children. *New York Acad Sci, 608,* 365–385.

Perrins, R., Walford, A., & Roberts, A. (2002). Sensory activation and role of inhibitory reticulospinal neurons that stop swimming in hatchling frog tadpoles. *The Journal of Neuroscience, 22,* 4229–4240.

Petitto, L. A., & Marentette, P. F. (1991). Babbling in the manual mode: evidence for the ontogeny of language. *Science, 251,* 1493–1496.

Pettigrew, J. D. (1974). The effect of visual experience on the development of stimulus specificity by kitten cortical neurones. *The Journal of Physiology, 237,* 49–74.

Phoenix, C. H., Goy, R. W., Gerall, A. A., & Young, A. C. (1959). Organizing action of prenatally administered testosterone propionate on the tissues mediating mating behavior in the female guinea pig. *Endocrinology, 65,* 369–382.

Pinto, J. G., Hornby, K. R., Jones, D. G., & Murphy, K. M. (2010). Developmental changes in GABAergic mechanisms in human visual cortex across the lifespan. *Frontiers in Cellular Neuroscience, 4,* 16.

Preyer, W. (1885). *Specielle physiologie des Embryo.* Leipzig: Grieben.

Provine, R. R. (1972). Ontogeny of bioelectric activity in the spinal cord of the chick embryo and its behavioral implications. *Brain Research, 41,* 365–378.

Provine, R. R. (1982). Preflight development of bilateral wing coordination in the chick (Gallus domesticus): effects of induced bilateral wing asymmetry. *Developmental Psychobiology, 15,* 245–255.

Putzar, L., Goerendt, I., Lange, K., Rösler, F., & Röder, B. (2007). Early visual deprivation impairs multisensory interactions in humans. *Nature Neuroscience, 10,* 1243–1245.

Raisman, G., & Field, P. M. (1973). Sexual dimorphism in the neuropil of the preoptic area of the rat and its dependence on neonatal androgen. *Brain Research, 54,* 1–29.

Rankin, C. H., & Carew, T. J. (1988). Dishabituation and sensitization emerge as separate processes during development in Aplysia. *The Journal of Neuroscience, 8,* 197–211.

Reynolds, S. A., French, K. A., Baader, A., & Kristan, W. B., Jr. (1998). Development of spontaneous and evoked behaviors in the medicinal leech. *The Journal of Comparative Neurology, 402*(2), 168–180.

Roberts, A. (2000). Early functional organization of spinal neurons in developing lower vertebrates. *Brain Research Bulletin, 53,* 585–593.

Roberts, T. F., Tschida, K. A., Klein, M. E., & Mooney, R. (2010). Rapid spine stabilization and synaptic enhancement at the onset of behavioural learning. *Nature, 463,* 948–952.

Roubertoux, P. L., Sluyter, F., Carlier, M., Marcet, B., Maarouf-Veray, F., Chérif, C., et al. (2003). Mitochondrial DNA modifies cognition in interaction with the nuclear genome and age in mice. *Nature Genetics, 35,* 65–69.

Rübsamen, R. (1992). Postnatal development of central auditory frequency maps. *Journal of Comparative Physiology. A, Neuroethology, Sensory, Neural, and Behavioral Physiology, 170,* 129–143.

Sakuma, Y. (1984). Influences of neonatal gonadectomy or androgen exposure on the sexual differentiation of the rat ventromedial hypothalamus. *J Physiol (Lond), 349,* 273–286.

Sanes, D. H., & Rubel, E. W. (1988). The ontogeny of inhibition and excitation in the gerbil lateral superior olive. *The Journal of Neuroscience, 8,* 682–700.

Sarro, E. C., & Sanes, D. H. (2010). Prolonged maturation of auditory perception and learning in gerbils. *Dev Neurobiol, 70,* 636–648.

Sato, T., Matsumoto, T., Kawano, H., Watanabe, T., Uematsu, Y., Sekine, K., et al. (2004). Brain masculinization requires androgen receptor function. *Proceedings of the National Academy of Sciences of the United States of America, 101,* 1673–1678.

Schlinger, B. A. (1998). Sexual differentiation of avian brain and behavior: Current views on gonadal hormone-dependent and independent mechanisms. *Annual Review of Physiology, 60,* 407–429.

Schorr, E. A., Fox, N. A., van Wassenhove, V., & Knudsen, E. I. (2005). Auditory-visual fusion in speech perception in children with cochlear implants. *Proceedings of the National Academy of Sciences of the United States of America, 102,* 18748–18750.

Sekido, R., & Lovell-Badge, R. (2009). Sex determination and SRY: down to a wink and a nudge? *Trends in Genetics: TIG, 25,* 19–29.

Simpson, H. B., & Vicario, D. S. (1991). Early estrogen treatment alone causes female zebra finches to produce learned, male-like vocalizations. *Journal of Neurobiology, 22,* 755–776.

Sokolowski, M. B., Pereira, H. S., & Hughes, K. (1997). Evolution of foraging behavior in Drosophila by density-dependent selection. *Proceedings of the National Academy of Sciences of the United States of America, 94,* 7373–7377.

Stollman, M. H., van Velzen, E. C., Simkens, H. M., Snik, A. F., & van den Broek, P. (2004). Development of auditory processing in 6–12-year-old children: a longitudinal study. *International Journal of Audiology, 43,* 34–44.

Straznicky, K. (1967). The development of the innervation and the musculature of wings innervated by thoracic nerves. *Acta Biologica Academiae Scientiarum Hungaricae, 18*(4), 437–448.

Streichert, L. C., & Weeks, J. C. (1995). Decreased monosynaptic sensory input to an identified motoneuron is associated with steroid-mediated dendritic regression during metamorphosis in Manduca sexta. *The Journal of Neuroscience, 15,* 1484–1495.

Sullivan, R. M., Hofer, M. A., & Brake, S. C. (1986) Olfactory-guided orientation in neonatal rats is enhanced by a conditioned change in behavioral state. *Developmental Psychobiology, 19,* 615–623.

Suster, M. L., & Bate, M. (2002). Embryonic assembly of a central pattern generator without sensory input. *Nature, 416,* 174–178.

Swaab, D. F., & Hofman, M. A. (1988). Sexual differentiation of the human hypothalamus: ontogeny of the sexually dimorphic nucleus of the preoptic area. *Developmental Brain Research, 44,* 314–318.

Swartz, S. K., & Soloff, M. S. (1974). The lack of estrogen binding by human alpha fetoprotein. *The Journal of Clinical Endocrinology and Metabolism, 39,* 589–591.

Tallal, P., Miller, S. L., Bedi, G., Byma, G., Wang, X., Nagarajan, S. S., et al. (1996). Language comprehension in language-learning impaired children improved with acoustically modified speech. *Science, 271,* 81–84.

Tallal, P., & Piercy, M. (1973). Defects of non-verbal auditory perception in children with developmental aphasia. *Nature, 241,* 468–469.

Tchernichovski, O., Lints, T., Mitra, P. P., & Nottebohm, F. (1999). Vocal imitation in zebra finches is inversely related to model abundance. *Proceedings of the National Academy of Sciences of the United States of America, 96,* 12901–12904.

Teller, D. Y. (1997). First glances: the vision of infants. The Friedenwald lecture. *Investigative Ophthalmology & Visual Science, 38,* 2183–2203.

Thigpen, A. E., Davis, D. L., Gautier, T., Imperato-McGinley, J., & Russell, D. W. (1992). The molecular basis of steroid 5 alpha-reductase deficiency in a large Dominican kindred. *The New England Journal of Medicine, 327,* 1216–1219.

Threlkeld, S. W., Hill, C. A., Rosen, G. D., & Fitch, R. H. (2009). Early acoustic discrimination experience ameliorates auditory processing deficits in male rats with cortical developmental disruption. *International Journal of Developmental Neuroscience, 27,* 321–328.

Timney, B. (1981). Development of binocular depth perception in kittens. *Investigative Ophthalmology & Visual Science, 21,* 493–496.

Tinbergen, N. (1948). Social releasers and the experimental method required for their study. *The Wilson Bulletin, 60,* 6–51.

Tobias, M. L., Viswanathan, S. S., & Kelley, D. B. (1998). Rapping, a female receptive call, initiates male-female duets in the South African clawed frog. *Proceedings of the National Academy of Sciences of the United States of America, 95,* 1870–1875.

Toran-Allerand, C. D. (1980). Sex steroids and the development of the newborn mouse hypothalamus and preoptic area in vitro. II. Morphological correlates and hormonal specificity. *Brain Research, 189,* 413–427.

Toran-Allerand, C. D., Hashimoto, K., Greenough, W. T., & Saltarelli, M. (1983). Sex steroids and the development of the newborn mouse hypothalamus and preoptic area in vitro: III. Effects of estrogen on dendritic differentiation. *Brain Research, 283,* 97–101.

Uriel, J., Bouillon, D., Aussel, C., & Dupiers, M. (1976). Alpha-fetoprotein: the major high-affinity estrogen binder in rat uterine cytosols. *Proc Natl Acad Sci USA, 73,* 1452–1456.

Vince, M. A. (1979). Effects of accelerating stimulation on different indices of development in Japanese quail embryos. *Journal of Experimental Zoology, 208*(2), 201–212.

Vince, M. A., & Salter, S. H. (1967). Respiration and clicking in quail embryos. *Nature, 216*(115), 582–583.

Wade, J., & Arnold, A. P. (1996). Functional testicular tissue does not masculinize development of the zebra finch song system. *Proceedings of the National Academy of Sciences of the United States of America, 93,* 5264–5268.

Walk, R. D., Gibson, E. J. (1961). A comparative and analytical study of visual depth perception. *Psychol Monogr, 75,* 1–44.

Wallace, M. T., Carriere, B. N., Perrault, T. J., Jr., Vaughan, J. W., & Stein, B. E. (2006). The development of cortical multisensory integration. *The Journal of Neuroscience, 26,* 11844–11849.

Wallace, M. T., & Stein, B. E. (1997). Development of multisensory neurons and multisensory integration in cat superior colliculus. *The Journal of Neuroscience, 17,* 2429–2444.

Wallen, K. (2005). Hormonal influences on sexually differentiated behavior in nonhuman primates. *Frontiers in Neuroendocrinology, 26,* 7–26.

Wallhausser, E., & Scheich, H. (1987). Auditory imprinting leads to differential 2-deoxyglucose uptake and dendritic spine loss in the chick rostral forebrain. *Brain Research, 428,* 29–44.

Watson, J. B., & Raynor, R. (1920). Conditioned emotional reactions. *Journal of Experimental Psychology, 3,* 1–14.

Watson, J. T., Robertson, J., Sachdev, U., & Kelley, D. B. (1993). Laryngeal muscle and motor neuron plasticity in Xenopus laevis: testicular masculinization of a developing neuromuscular system. *Journal of Neurobiology, 24,* 1615–1625.

Weaver, I. C., Cervoni, N., Champagne, F. A., D'Alessio, A. C., Sharma, S., Seckl, J. R., et al. (2004). Epigenetic programming by maternal behavior. *Nature Neuroscience, 7,* 847–854.

Wenner, P., & O'Donovan, M. J. (2001). Mechanisms that initiate spontaneous network activity in the developing chick spinal cord. *Journal of Neurophysiology, 86,* 1481–1498.

Werner, L. A., & Gray, L. (1998). Behavioral studies of hearing development. In E. W. Rubel, A. N. Popper, & R. R. Fay (Eds.), *Development of the auditory system* (pp. 12–79). New York: Springer-Verlag.

Werner, L. A., Marean, G. C., Halpin, C. F., Spetner, N. B., & Gillenwater, J. M. (1992). Infant auditory temporal acuity: gap detection. *Child Development, 63,* 260–272.

Wilkins, J. F., Haig, D. (2001). Genomic imprinting of two antagonistic loci. *Proc Biol Sci, 268,* 1861–1867.

Wilkinson, L. S., Davies, W., & Isles, A. R. (2007). Genomic imprinting effects on brain development and function. *Nature Reviews. Neuroscience, 8,* 832–843.

Wilson, D. M. (1968). The flight-control system of the locust. *Scientific American, 218*(5), 83–90.

Withington-Wray, D. J., Binns, K. E., & Keating, M. J. (1990). The developmental emergence of a map of auditory space in the superior colliculus of the guinea pig. *Developmental Brain Research, 51,* 225–236.

Wright, B. A., & Zecker, S. G. (2004). Learning problems, delayed development, and puberty. *Proceedings of the National Academy of Sciences of the United States of America, 101,* 9942–9946.

Yamaguchi, A., Kaczmarek, L. K., & Kelley, D. B. (2003). Functional specialization of male and female vocal motoneurons. *The Journal of Neuroscience, 23,* 11568–11576.

Yuodelis, C., & Hendrickson, A. (1986). A qualitative and quantitative analysis of the human fovea during development. *Vision Research, 26,* 847–855.

Zhang, Y., Narayan, S., Geiman, E., Lanuza, G. M., Velasquez, T., Shanks, B., et al. (2008). V3 spinal neurons establish a robust and balanced locomotor rhythm during walking. *Neuron, 60,* 84–96.

Zippelius, H. M. (1972). Die Karawanenbildung bei Feld- und Hausspitzmaus. *Zeitschrift für Tierpsychologie, 30,* 305–320.

Molecules and Genes Index

Note: Page numbers followed by *b* indicate boxes and *f* indicate figures.

2-deoxyglucose (2-DG), 254, 258–259, 318
20-hydroxecdysone (20E), 187–188
5α-dihydrotestosterone (DHT), 309–310
5α-reductase, 309
5-HT (Serotonin), 132–133, 293

A

α 4-integrin, 70–71
AbdA, 298–299, 300*f*
Abl, 122–123
α-bungarotoxin (α-Btx), 222*f*, 226, 294–295
ACBD3, 57–58
Acetylcholine (ACh), 89, 234
  Agrin, 234
  sympathetic neurons, 89
Acetylcholine receptors (AChRs)
  accumulation in central region of uninnervated muscle, 222*f*
  clustering, inducing, 225*f*
  clustering and postsynaptic differentiation at NMJ, 226
  clustering on muscle fibers, 224*f*
  dispersing, 232*f*
  extrasynaptic, 231*f*
  heterosynaptic depression associated with loss of postsynaptic, 272*f*
  innervation, 229*f*, 232–233
  synapse transmission at target, 197–198
  treating chick embryo with, 197–198
Achaete scute
  Notch/Delta signaling pathway, 20–21
  role in neuroblast formation, 18–19
  role in neuroblast segregation, 17
Achaete-Scute complex (asc), 17*f*, 19*f*, 21*f*
Actin
  filaments, 110–111, 111*f*, 112*f*
  filopodia, growth cone, 110–111
  growth cone guidance, 112*f*, 113*f*
  growth cone structure, 111*f*
Activin, 12–13, 13*f*
Afadin, 219*f*, 223
Agrin
  acetylcholine (ACh), 234
  synapse formation and function, 226–228
AIF (Apoptosis-inducing factor), 193
AMPA receptor (AMPAR)
  AMPAR:NMDAR ratio, 275
  local synthesis of in dendrite, 230*f*
  Narp and, 229

Androgen receptor (AR), 309–310
Antennapedia, 25
APKC (Atypical protein kinase C), 83
Aplysia CAM (apCAM), 119–120
ApoER2, 65–66
Apolipoprotein E, receptor 2, 65–66
Apoptosis protease activating factor-1 (Apaf-1), 192, 193*f*, 194*f*
Apoptosis-inducing factor (AIF), 193
Arc, 233*f*, 235
Asc (Achaete-Scute complex), 17*f*, 19*f*, 21*f*
Asense, 17
Ath5, 93–94, 95–96
Atonal (ato), 17, 85–86, 86*f*, 96–97
ATP, 197, 266, 267*f*
Atypical protein kinase C (aPKC), 83
Aurora-A, 83–84
Axin, 30

B

β 1-integrin, 70–71
β2 laminin, 216
β-actin, 133
Bad, 194*f*, 195
Bar, 85, 86*f*
Bax, 193*f*, 194–195, 194*f*, 195*f*
Bazooka (Baz), 83
β-catenin, 16, 30, 88, 219*f*
Bcl-2 (B-cell lymphoma 2), 194–196
Bcl-x, 194*f*, 195, 195*f*
BDNF (Brain derived neurotophic factor), 145, 178
Beaten path (beat-1a), 143–144, 144*f*
Bicoid, 24
Bicuculline, 237*f*, 240, 290
Bid, 195
Bim, 190*f*, 191–192, 195
Bithorax, 25, 26, 26*f*
BMP2, 115–116
BMP4, 14, 14*f*
BMP6, 115–116
BMP7, 14
Bok, 120*f*
Bone Morphogenetic Factor (BMP)
  Dpp/BMP4 study, 14
  progenitor cells and, 55–56
  response elements, 16*b*
  Sonic hedgehog (Shh), 46*f*
  Wnt and, 16*b*
Brain derived neurotophic factor (BDNF), 145, 178
Bride of sevenless (Boss), 87*f*

C

CAD (Caspase-activated deoxyribonuclease), 194*f*
Cadherin, 119*f*, 155, 158, 222–223
Cadherin-8, 222–223
Calcitonin gene-related peptide (CGRP), 274
Calcium
  calcium-activated second messenger systems, 273–275
  growth cones and, 220–221
  neuron death, 199–201
  postsynaptic, synapse depression and, 273*f*
  synaptic connections, 272–273
Calcium calmodulin-dependent protein kinase II (CaMKII), 273–274
CAM kinase II (CaMKII), 273–274
CAMP, 134*f*
CAMP-dependent protein kinase (PKA), 136*f*, 221, 251, 271, 272*f*, 274
CAMs. *See* Cell adhesion molecules
Cardiotrophin-1 (CT-1), 184–185
Caspase-3 (Cysteine requiring aspartate protease-3), 192–193, 192*f*, 193*f*, 194*f*, 195, 197, 199, 201*f*
Caspase-9 (Cysteine requiring aspartate protease-9), 192, 193*f*, 194*f*, 195
Caspase-activated deoxyribonuclease (CAD), 194*f*
CAT (Choline acetyltransferase), 89, 295–296
Cdc42, 116, 129–130
Cdk (cyclin-dependent kinase), 53
Cdk5 (Cyclin-dependent kinase 5), 232*f*, 234
Cdk-inhibitor (Cdki), 53, 54–55, 54*f*
Ced-1 (cell death abnormal-1), 197
Ced-3 (cell death abnormal-3), 192
Ced-4 (cell death abnormal-4), 192
Ced-7 (cell death abnormal-7), 197
Ced-9 (cell death abnormal-9), 193*f*, 194
Cell adhesion molecules (CAMs)
  activity-dependent synapse formation and, 275*f*
  changing, 123–124, 124*f*
  homophilic, 124
  limbic-associated membrane protein (LAMP), 122
  neural cell adhesion molecule (NCAM), 122
  synapse formation and, 218*f*, 221
Cerberus, 30
CGMP, 132–133, 297

CGMP-regulated protein kinase (PKG), 297
CGRP (Calcitonin gene-related peptide), 274
Choline acetyltransferase (ChAT, CAT), 89, 295–296
Chordin
  BMP4 signaling and, 14, 14*f*
  compared to Noggin, 12
  eliminating, effect on neural development, 14–15, 15*f*
  frog embryo, 12*f*
Ciliary neuronotrophic factor, 55–56
Ciliary neurotrophic factor (CNTF), 55–56, 116, 184
Ciliary neurotrophic factor receptor α (CNTFRα), 56*f*, 184–185, 185*f*
C-Jun kinase (JNK), 183*f*, 184, 190*f*, 191–192, 191*f*
C-Jun N-terminal kinase (JNK), 184
CKI (Cyclin-dependent kinase inhibitors), 191
Cofilin, 114–115
Cog-1, 81–82, 81*f*
Collagen, 70–71
Collapsin, 124–125. *See also* Semaphorins
Collapsin response mediator protein (CRMP-2), 116
Commissureless (comm), 130–131, 130*f*, 131*f*
Concanavalin A, 118
Connectin, 158
Cre-recombinase, 70
C-Ret, 145
CSF (colony stimulating factor), 184
CT-1 (Cardiotrophin-1), 184–185
Ctip2, 94*f*
Cullin5 (Cul5), 66
Cut, 116–117
CXCL12, 120*f*
CXCR4, 120*f*
Cyclic AMP response element binding protein (CREB), 161, 190, 199–201
CyclinB, 53, 54*f*
CyclinD, 53, 54*f*, 55
Cyclin-dependent kinase 5 (cdk5), 232*f*, 234
Cyclin-dependent kinase (Cdk), 53
Cyclin-dependent kinase inhibitors (CKI), 191
CyclinE, 54*f*
Cysteine requiring aspartate protease-3 (caspase-3), 192–193, 192*f*, 193*f*, 194*f*, 195, 197, 199, 201*f*
Cysteine requiring aspartate protease-9 (caspase-9), 192, 193*f*, 194*f*, 195

Subject Index

Note: Page numbers followed by *b* indicate boxes and *f* indicate figures.